Methods in Enzymology

Volume 249

ENZYME KINETICS AND MECHANISM

Part D

Developments in Enzyme Dynamics

METHODS IN ENZYMOLOGY

Methods in Enzymology

Volume 249

Enzyme Kinetics and Mechanism

Part D
Developments in Enzyme Dynamics

EDITED BY

Daniel L. Purich

DEPARTMENT OF BIOCHEMISTRY
AND MOLECULAR BIOLOGY
UNIVERSITY OF FLORIDA
COLLEGE OF MEDICINE
GAINESVILLE, FLORIDA

ACADEMIC PRESS

San Diego New York Boston London Sydney Tokyo Toronto

This book is printed on acid-free paper. ♾

Academic Press, Inc.
A Division of Harcourt Brace & Company
525 B Street, Suite 1900, San Diego, California 92101-4495

United Kingdom Edition published by
Academic Press Limited
24-28 Oval Road, London NW1 7DX

International Standard Serial Numper: 0076-6879

International Standard Book Number: 0-12-182150-1

PRINTED IN THE UNITED STATES OF AMERICA
95 96 97 98 99 00 MM 9 8 7 6 5 4 3 2 1

Table of Contents

Section I. General Approaches to Biological Catalysis

Section II. Inhibitors as Mechanistic Probes

Section III. Isotopic Probes of Enzyme Action

Section IV. Kinetics of Specialized Systems

Contributors to Volume 249

Article numbers are in parentheses following the names of contributors. Affiliations listed are current.

BRIAN J. BAHNSON (14), *Department of Chemistry, University of California, Berkeley, Berkeley, California 94720*

STEPHEN J. BENKOVIC (19), *Davey Laboratory, Pennsylvania State University, University Park, Pennsylvania 16802*

OTTO G. BERG (21), *Department of Chemistry and Biochemistry, University of Delaware, Newark, Delaware 19716*

W. W. CLELAND (7, 13), *Institute of Enzyme Research, University of Wisconsin, Madison, Wisconsin 53706*

BRUCE F. COOPER (8), *Department of Biochemistry and Cell Biology, Rice University, Houston, Texas 77251*

TERRI DAHNKE (16), *Departments of Chemistry and Biochemistry, Ohio State University, Columbus, Ohio 43210*

RONALD G. DUGGLEBY (3, 6), *Department of Biochemistry, Centre for Structure, Function, and Engineering, University of Queensland, Brisbane, Queensland 4072 Australia*

CAROL A. FIERKE (1), *Department of Biochemistry, College of Medicine, Duke University Medical Center, Durham, North Carolina 27710*

HERBERT J. FROMM (5), *Department of Biochemistry and Biophysics, Iowa State University, Ames, Iowa 50011*

MICHAEL H. GELB (21), *Department of Chemistry and Biochemistry, University of Delaware, Newark, Delaware 19716*

GORDON G. HAMMES (1), *Department of Biochemistry, Duke University Medical Center, Durham, North Carolina 27710*

MAHENDRA KUMAR JAIN (21), *Department of Chemistry and Biochemistry, University of Delaware, Newark, Delaware 19716*

RU-TAI JIANG (16), *Departments of Chemistry and Biochemistry, Ohio State University, Columbus, Ohio 43210*

KENNETH A. JOHNSON (2), *Department of Molecular and Cell Biology, Pennsylvania State University, University Park, Pennsylvania 16802*

JUDITH P. KLINMAN (14), *Department of Chemistry, University of California, Berkeley, Berkeley, California 94720*

IRENE LEE (19), *Lavey Laboratory Pennsylvania State University, University Park, Pennsylvania 16802*

LEISHA S. MULLINS (15), *Department of Chemistry, Texas A&M University, College Station, Texas 77843*

KENNETH E. NEET (20), *Department of Biological Chemistry, Chicago Medical School, North Chicago, Illinois 60064*

DEXTER B. NORTHROP (9), *Division of Pharmaceutical Biochemistry, School of Pharmacy, University of Wisconsin, Madison, Wisconsin 53706*

BRYCE V. PLAPP (4), *Department of Biochemistry, University of Iowa, Iowa City, Iowa 52242*

BRUCE A. POSNER (19), *Davey Laboratory, Pennsylvania State University, University Park, Pennsylvania 16802*

ANNA RADZICKA (11), *Department of Biochemistry, University of North Carolina, Chapel Hill, North Carolina 27514*

FRANK M. RAUSHEL (15), *Department of Chemistry, Texas A&M University, College Station, Texas 77843*

KAREN L. REBHOLZ (9), *Division of Pharmaceutical Biochemistry, School of Pharmacy, University of Wisconsin, Madison, Wisconsin 53706*

JOSEPH ROGERS (21), *Department of Chemistry and Biochemistry, University of Delaware, Newark, Delaware 19716*

IRWIN A. ROSE (12), *Institute for Cancer Research, Fox Chase Cancer Center, Philadelphia, Pennsylvania 19111*

FREDERICK B. RUDOLPH (8), *Department of Biochemistry and Cell Biology, Rice University, Houston, Texas 77251*

ZHENGTAO SHI (16), *Departments of Chemistry and Biochemistry, Ohio State University, Columbus, Ohio 43210*

RICHARD B. SILVERMAN (10), *Department of Chemistry, Northwestern University, Evanston, Illinois 60208*

DAVID N. SILVERMAN (18), *Department of Pharmacology, Health Science Center, University of Florida, Gainesville, Florida 32610*

JON D. STEWART (19), *Davey Laboratory, Pennsylvania State University, University Park, Pennsylvania 16802*

STEFAN E. SZEDLACSEK (6), *Department of Enzymology, Institute of Biochemistry, 77748 Bucharest, Romania*

MING-DAW TSAI (16), *Departments of Chemistry and Biochemistry, Ohio State University, Columbus, Ohio 43210*

FREDERICK C. WEDLER (17), *Department of Molecular and Cell Biology, Althause Laboratory, Pennsylvania State University, University Park, Pennsylvania 16802*

RICHARD WOLFENDEN (11), *Department of Biochemistry, University of North Carolina, Chapel Hill, North Carolina 27514*

Preface

Since their appearance in the late 1970s and early 1980s, Parts A, B, and C of Enzyme Kinetics and Mechanism, Volumes 63, 64, and 87 in the *Methods in Enzymology* series, have achieved broad acceptance and use by practicing enzymologists and molecular and cell biologists. Nonetheless, during the decade or so following the publication of Part C, significant new developments have expanded the scope of enzyme chemistry. At the most fundamental level, Nature's use of nucleic acids as enzymes (ribozymes) became evident as did the ability of enzyme chemists to design catalytic antibodies (abzymes) as synthetic enzymes. Moreover, advances in structural biology as well as the application of such molecular biological techniques as site-directed mutagenesis have provided far reaching insights about enzyme structure and molecular design. Another force driving the development of enzyme chemistry has been the broad use of enzymes as both intermediate and final products in the agricultural and pharmaceutical industries under the rubric of "biotechnology." Obviously, one cannot escape mentioning that there have also been conceptual breakthroughs regarding the kinetic and mechanistic properties of biological catalysts. While Parts A–C remain useful guides for the characterization of enzyme action and regulation, the above-mentioned prominent developments have necessitated the publication of Part D (Volume 249) which serves to both update and expand the range of approaches available in the armamentarium of those practicing scientists in this field. Accordingly, an extraordinary group of experts were invited to share their knowledge on the methods that have proved most useful in their own work and that of other enzyme mechanics. While one cannot achieve in a single book the comprehensive coverage presented in the earlier volumes, the strategy has been to refer readers to Parts A–C whenever practical. Authors have also provided extensive reference to other key chapters and/or monographs, such as the most recent volumes of "The Enzymes."* Plans for additional volumes on Enzyme Kinetics and Mechanism are also underway. I would greatly value receiving comments about Parts A–D as well as suggestions for topics that might be worthy of further development.

DANIEL L. PURICH

* D. S. Sigman and P. D. Boyer, "The Enzymes, Volume XIX: Mechanisms of Catalysis." Academic Press, San Diego, 1990. D. S. Sigman, "The Enzymes, Volume XX: Mechanisms of Catalysis." Academic Press, San Diego, 1992.

METHODS IN ENZYMOLOGY

VOLUME I. Preparation and Assay of Enzymes
Edited by SIDNEY P. COLOWICK AND NATHAN O. KAPLAN

VOLUME II. Preparation and Assay of Enzymes
Edited by SIDNEY P. COLOWICK AND NATHAN O. KAPLAN

VOLUME III. Preparation and Assay of Substrates
Edited by SIDNEY P. COLOWICK AND NATHAN O. KAPLAN

VOLUME IV. Special Techniques for the Enzymologist
Edited by SIDNEY P. COLOWICK AND NATHAN O. KAPLAN

VOLUME V. Preparation and Assay of Enzymes
Edited by SIDNEY P. COLOWICK AND NATHAN O. KAPLAN

VOLUME VI. Preparation and Assay of Enzymes (*Continued*)
Preparation and Assay of Substrates
Special Techniques
Edited by SIDNEY P. COLOWICK AND NATHAN O. KAPLAN

VOLUME VII. Cumulative Subject Index
Edited by SIDNEY P. COLOWICK AND NATHAN O. KAPLAN

VOLUME VIII. Complex Carbohydrates
Edited by ELIZABETH F. NEUFELD AND VICTOR GINSBURG

VOLUME IX. Carbohydrate Metabolism
Edited by WILLIS A. WOOD

VOLUME X. Oxidation and Phosphorylation
Edited by RONALD W. ESTABROOK AND MAYNARD E. PULLMAN

VOLUME XI. Enzyme Structure
Edited by C. H. W. HIRS

VOLUME XII. Nucleic Acids (Parts A and B)
Edited by LAWRENCE GROSSMAN AND KIVIE MOLDAVE

VOLUME XIII. Citric Acid Cycle
Edited by J. M. LOWENSTEIN

VOLUME XIV. Lipids
Edited by J. M. LOWENSTEIN

VOLUME XV. Steroids and Terpenoids
Edited by RAYMOND B. CLAYTON

VOLUME XVI. Fast Reactions
Edited by KENNETH KUSTIN

VOLUME 73. Immunochemical Techniques (Part B)
Edited by JOHN J. LANGONE AND HELEN VAN VUNAKIS

VOLUME 74. Immunochemical Techniques (Part C)
Edited by JOHN J. LANGONE AND HELEN VAN VUNAKIS

VOLUME 75. Cumulative Subject Index Volumes XXXI, XXXII, XXXIV–LX
Edited by EDWARD A. DENNIS AND MARTHA G. DENNIS

VOLUME 76. Hemoglobins
Edited by ERALDO ANTONINI, LUIGI ROSSI-BERNARDI, AND EMILIA CHIANCONE

VOLUME 77. Detoxication and Drug Metabolism
Edited by WILLIAM B. JAKOBY

VOLUME 78. Interferons (Part A)
Edited by SIDNEY PESTKA

VOLUME 79. Interferons (Part B)
Edited by SIDNEY PESTKA

VOLUME 80. Proteolytic Enzymes (Part C)
Edited by LASZLO LORAND

VOLUME 81. Biomembranes (Part H: Visual Pigments and Purple Membranes, I)
Edited by LESTER PACKER

VOLUME 82. Structural and Contractile Proteins (Part A: Extracellular Matrix)
Edited by LEON W. CUNNINGHAM AND DIXIE W. FREDERIKSEN

VOLUME 83. Complex Carbohydrates (Part D)
Edited by VICTOR GINSBURG

VOLUME 84. Immunochemical Techniques (Part D: Selected Immunoassays)
Edited by JOHN J. LANGONE AND HELEN VAN VUNAKIS

VOLUME 85. Structural and Contractile Proteins (Part B: The Contractile Apparatus and the Cytoskeleton)
Edited by DIXIE W. FREDERIKSEN AND LEON W. CUNNINGHAM

VOLUME 86. Prostaglandins and Arachidonate Metabolites
Edited by WILLIAM E. M. LANDS AND WILLIAM L. SMITH

VOLUME 87. Enzyme Kinetics and Mechanism (Part C: Intermediates, Stereochemistry, and Rate Studies)
Edited by DANIEL L. PURICH

VOLUME 88. Biomembranes (Part I: Visual Pigments and Purple Membranes, II)
Edited by LESTER PACKER

VOLUME 89. Carbohydrate Metabolism (Part D)
Edited by WILLIS A. WOOD

VOLUME 107. Posttranslational Modifications (Part B)
Edited by FINN WOLD AND KIVIE MOLDAVE

VOLUME 108. Immunochemical Techniques (Part G: Separation and Characterization of Lymphoid Cells)
Edited by GIOVANNI DI SABATO, JOHN J. LANGONE, AND HELEN VAN VUNAKIS

VOLUME 109. Hormone Action (Part I: Peptide Hormones)
Edited by LUTZ BIRNBAUMER AND BERT W. O'MALLEY

VOLUME 110. Steroids and Isoprenoids (Part A)
Edited by JOHN H. LAW AND HANS C. RILLING

VOLUME 111. Steroids and Isoprenoids (Part B)
Edited by JOHN H. LAW AND HANS C. RILLING

VOLUME 112. Drug and Enzyme Targeting (Part A)
Edited by KENNETH J. WIDDER AND RALPH GREEN

VOLUME 113. Glutamate, Glutamine, Glutathione, and Related Compounds
Edited by ALTON MEISTER

VOLUME 114. Diffraction Methods for Biological Macromolecules (Part A)
Edited by HAROLD W. WYCKOFF, C. H. W. HIRS, AND SERGE N. TIMASHEFF

VOLUME 115. Diffraction Methods for Biological Macromolecules (Part B)
Edited by HAROLD W. WYCKOFF, C. H. W. HIRS, AND SERGE N. TIMASHEFF

VOLUME 116. Immunochemical Techniques (Part H: Effectors and Mediators of Lymphoid Cell Functions)
Edited by GIOVANNI DI SABATO, JOHN J. LANGONE, AND HELEN VAN VUNAKIS

VOLUME 117. Enzyme Structure (Part J)
Edited by C. H. W. HIRS AND SERGE N. TIMASHEFF

VOLUME 118. Plant Molecular Biology
Edited by ARTHUR WEISSBACH AND HERBERT WEISSBACH

VOLUME 119. Interferons (Part C)
Edited by SIDNEY PESTKA

VOLUME 120. Cumulative Subject Index Volumes 81–94, 96–101

VOLUME 121. Immunochemical Techniques (Part I: Hybridoma Technology and Monoclonal Antibodies)
Edited by JOHN J. LANGONE AND HELEN VAN VUNAKIS

VOLUME 122. Vitamins and Coenzymes (Part G)
Edited by FRANK CHYTIL AND DONALD B. MCCORMICK

VOLUME 123. Vitamins and Coenzymes (Part H)
Edited by FRANK CHYTIL AND DONALD B. MCCORMICK

VOLUME 124. Hormone Action (Part J: Neuroendocrine Peptides)
Edited by P. MICHAEL CONN

VOLUME 125. Biomembranes (Part M: Transport in Bacteria, Mitochondria, and Chloroplasts: General Approaches and Transport Systems)
Edited by SIDNEY FLEISCHER AND BECCA FLEISCHER

VOLUME 126. Biomembranes (Part N: Transport in Bacteria, Mitochondria, and Chloroplasts: Protonmotive Force)
Edited by SIDNEY FLEISCHER AND BECCA FLEISCHER

VOLUME 127. Biomembranes (Part O: Protons and Water: Structure and Translocation)
Edited by LESTER PACKER

VOLUME 128. Plasma Lipoproteins (Part A: Preparation, Structure, and Molecular Biology)
Edited by JERE P. SEGREST AND JOHN J. ALBERS

VOLUME 129. Plasma Lipoproteins (Part B: Characterization, Cell Biology, and Metabolism)
Edited by JOHN J. ALBERS AND JERE P. SEGREST

VOLUME 130. Enzyme Structure (Part K)
Edited by C. H. W. HIRS AND SERGE N. TIMASHEFF

VOLUME 131. Enzyme Structure (Part L)
Edited by C. H. W. HIRS AND SERGE N. TIMASHEFF

VOLUME 132. Immunochemical Techniques (Part J: Phagocytosis and Cell-Mediated Cytotoxicity)
Edited by GIOVANNI DI SABATO AND JOHANNES EVERSE

VOLUME 133. Bioluminescence and Chemiluminescence (Part B)
Edited by MARLENE DELUCA AND WILLIAM D. MCELROY

VOLUME 134. Structural and Contractile Proteins (Part C: The Contractile Apparatus and the Cytoskeleton)
Edited by RICHARD B. VALLEE

VOLUME 135. Immobilized Enzymes and Cells (Part B)
Edited by KLAUS MOSBACH

VOLUME 136. Immobilized Enzymes and Cells (Part C)
Edited by KLAUS MOSBACH

VOLUME 137. Immobilized Enzymes and Cells (Part D)
Edited by KLAUS MOSBACH

VOLUME 138. Complex Carbohydrates (Part E)
Edited by VICTOR GINSBURG

VOLUME 139. Cellular Regulators (Part A: Calcium- and Calmodulin-Binding Proteins)
Edited by ANTHONY R. MEANS AND P. MICHAEL CONN

VOLUME 140. Cumulative Subject Index Volumes 102–119, 121–134

VOLUME 141. Cellular Regulators (Part B: Calcium and Lipids)
Edited by P. MICHAEL CONN AND ANTHONY R. MEANS

VOLUME 142. Metabolism of Aromatic Amino Acids and Amines
Edited by SEYMOUR KAUFMAN

VOLUME 143. Sulfur and Sulfur Amino Acids
Edited by WILLIAM B. JAKOBY AND OWEN GRIFFITH

VOLUME 144. Structural and Contractile Proteins (Part D: Extracellular Matrix)
Edited by LEON W. CUNNINGHAM

VOLUME 145. Structural and Contractile Proteins (Part E: Extracellular Matrix)
Edited by LEON W. CUNNINGHAM

VOLUME 146. Peptide Growth Factors (Part A)
Edited by DAVID BARNES AND DAVID A. SIRBASKU

VOLUME 147. Peptide Growth Factors (Part B)
Edited by DAVID BARNES AND DAVID A. SIRBASKU

VOLUME 148. Plant Cell Membranes
Edited by LESTER PACKER AND ROLAND DOUCE

VOLUME 149. Drug and Enzyme Targeting (Part B)
Edited by RALPH GREEN AND KENNETH J. WIDDER

VOLUME 150. Immunochemical Techniques (Part K: *In Vitro* Models of B and T Cell Functions and Lymphoid Cell Receptors)
Edited by GIOVANNI DI SABATO

VOLUME 151. Molecular Genetics of Mammalian Cells
Edited by MICHAEL M. GOTTESMAN

VOLUME 152. Guide to Molecular Cloning Techniques
Edited by SHELBY L. BERGER AND ALAN R. KIMMEL

VOLUME 153. Recombinant DNA (Part D)
Edited by RAY WU AND LAWRENCE GROSSMAN

VOLUME 154. Recombinant DNA (Part E)
Edited by RAY WU AND LAWRENCE GROSSMAN

VOLUME 155. Recombinant DNA (Part F)
Edited by RAY WU

VOLUME 156. Biomembranes (Part P: ATP-Driven Pumps and Related Transport: The Na,K-Pump)
Edited by SIDNEY FLEISCHER AND BECCA FLEISCHER

VOLUME 157. Biomembranes (Part Q: ATP-Driven Pumps and Related Transport: Calcium, Proton, and Potassium Pumps)
Edited by SIDNEY FLEISCHER AND BECCA FLEISCHER

VOLUME 158. Metalloproteins (Part A)
Edited by JAMES F. RIORDAN AND BERT L. VALLEE

VOLUME 159. Initiation and Termination of Cyclic Nucleotide Action
Edited by JACKIE D. CORBIN AND ROGER A. JOHNSON

VOLUME 160. Biomass (Part A: Cellulose and Hemicellulose)
Edited by WILLIS A. WOOD AND SCOTT T. KELLOGG

VOLUME 161. Biomass (Part B: Lignin, Pectin, and Chitin)
Edited by WILLIS A. WOOD AND SCOTT T. KELLOGG

VOLUME 162. Immunochemical Techniques (Part L: Chemotaxis and Inflammation)
Edited by GIOVANNI DI SABATO

VOLUME 163. Immunochemical Techniques (Part M: Chemotaxis and Inflammation)
Edited by GIOVANNI DI SABATO

VOLUME 164. Ribosomes
Edited by HARRY F. NOLLER, JR., AND KIVIE MOLDAVE

VOLUME 165. Microbial Toxins: Tools for Enzymology
Edited by SIDNEY HARSHMAN

VOLUME 166. Branched-Chain Amino Acids
Edited by ROBERT HARRIS AND JOHN R. SOKATCH

VOLUME 167. Cyanobacteria
Edited by LESTER PACKER AND ALEXANDER N. GLAZER

VOLUME 168. Hormone Action (Part K: Neuroendocrine Peptides)
Edited by P. MICHAEL CONN

VOLUME 169. Platelets: Receptors, Adhesion, Secretion (Part A)
Edited by JACEK HAWIGER

VOLUME 170. Nucleosomes
Edited by PAUL M. WASSARMAN AND ROGER D. KORNBERG

VOLUME 171. Biomembranes (Part R: Transport Theory: Cells and Model Membranes)
Edited by SIDNEY FLEISCHER AND BECCA FLEISCHER

VOLUME 172. Biomembranes (Part S: Transport: Membrane Isolation and Characterization)
Edited by SIDNEY FLEISCHER AND BECCA FLEISCHER

VOLUME 173. Biomembranes [Part T: Cellular and Subcellular Transport: Eukaryotic (Nonepithelial) Cells]
Edited by SIDNEY FLEISCHER AND BECCA FLEISCHER

VOLUME 174. Biomembranes [Part U: Cellular and Subcellular Transport: Eukaryotic (Nonepithelial) Cells]
Edited by SIDNEY FLEISCHER AND BECCA FLEISCHER

VOLUME 175. Cumulative Subject Index Volumes 135–139, 141–167

VOLUME 176. Nuclear Magnetic Resonance (Part A: Spectral Techniques and Dynamics)
Edited by NORMAN J. OPPENHEIMER AND THOMAS L. JAMES

VOLUME 177. Nuclear Magnetic Resonance (Part B: Structure and Mechanism)
Edited by NORMAN J. OPPENHEIMER AND THOMAS L. JAMES

VOLUME 178. Antibodies, Antigens, and Molecular Mimicry
Edited by JOHN J. LANGONE

VOLUME 179. Complex Carbohydrates (Part F)
Edited by VICTOR GINSBURG

VOLUME 180. RNA Processing (Part A: General Methods)
Edited by JAMES E. DAHLBERG AND JOHN N. ABELSON

VOLUME 181. RNA Processing (Part B: Specific Methods)
Edited by JAMES E. DAHLBERG AND JOHN N. ABELSON

VOLUME 182. Guide to Protein Purification
Edited by MURRAY P. DEUTSCHER

VOLUME 183. Molecular Evolution: Computer Analysis of Protein and Nucleic Acid Sequences
Edited by RUSSELL F. DOOLITTLE

VOLUME 184. Avidin–Biotin Technology
Edited by MEIR WILCHEK AND EDWARD A. BAYER

VOLUME 185. Gene Expression Technology
Edited by DAVID V. GOEDDEL

VOLUME 186. Oxygen Radicals in Biological Systems (Part B: Oxygen Radicals and Antioxidants)
Edited by LESTER PACKER AND ALEXANDER N. GLAZER

VOLUME 187. Arachidonate Related Lipid Mediators
Edited by ROBERT C. MURPHY AND FRANK A. FITZPATRICK

VOLUME 188. Hydrocarbons and Methylotrophy
Edited by MARY E. LIDSTROM

VOLUME 189. Retinoids (Part A: Molecular and Metabolic Aspects)
Edited by LESTER PACKER

VOLUME 190. Retinoids (Part B: Cell Differentiation and Clinical Applications)
Edited by LESTER PACKER

VOLUME 191. Biomembranes (Part V: Cellular and Subcellular Transport: Epithelial Cells)
Edited by SIDNEY FLEISCHER AND BECCA FLEISCHER

VOLUME 192. Biomembranes (Part W: Cellular and Subcellular Transport: Epithelial Cells)
Edited by SIDNEY FLEISCHER AND BECCA FLEISCHER

VOLUME 193. Mass Spectrometry
Edited by JAMES A. MCCLOSKEY

VOLUME 194. Guide to Yeast Genetics and Molecular Biology
Edited by CHRISTINE GUTHRIE AND GERALD R. FINK

VOLUME 195. Adenylyl Cyclase, G Proteins, and Guanylyl Cyclase
Edited by ROGER A. JOHNSON AND JACKIE D. CORBIN

VOLUME 196. Molecular Motors and the Cytoskeleton
Edited by RICHARD B. VALLEE

VOLUME 197. Phospholipases
Edited by EDWARD A. DENNIS

VOLUME 198. Peptide Growth Factors (Part C)
Edited by DAVID BARNES, J. P. MATHER, AND GORDON H. SATO

VOLUME 199. Cumulative Subject Index Volumes 168–174, 176–194 (in preparation)

VOLUME 200. Protein Phosphorylation (Part A: Protein Kinases: Assays, Purification, Antibodies, Functional Analysis, Cloning, and Expression)
Edited by TONY HUNTER AND BARTHOLOMEW M. SEFTON

VOLUME 201. Protein Phosphorylation (Part B: Analysis of Protein Phosphorylation, Protein Kinase Inhibitors, and Protein Phosphatases)
Edited by TONY HUNTER AND BARTHOLOMEW M. SEFTON

VOLUME 202. Molecular Design and Modeling: Concepts and Applications (Part A: Proteins, Peptides, and Enzymes)
Edited by JOHN J. LANGONE

VOLUME 203. Molecular Design and Modeling: Concepts and Applications (Part B: Antibodies and Antigens, Nucleic Acids, Polysaccharides, and Drugs)
Edited by JOHN J. LANGONE

VOLUME 204. Bacterial Genetic Systems
Edited by JEFFREY H. MILLER

VOLUME 205. Metallobiochemistry (Part B: Metallothionein and Related Molecules)
Edited by JAMES F. RIORDAN AND BERT L. VALLEE

VOLUME 206. Cytochrome P450
Edited by MICHAEL R. WATERMAN AND ERIC F. JOHNSON

VOLUME 207. Ion Channels
Edited by BERNARDO RUDY AND LINDA E. IVERSON

VOLUME 208. Protein–DNA Interactions
Edited by ROBERT T. SAUER

VOLUME 209. Phospholipid Biosynthesis
Edited by EDWARD A. DENNIS AND DENNIS E. VANCE

VOLUME 210. Numerical Computer Methods
Edited by LUDWIG BRAND AND MICHAEL L. JOHNSON

VOLUME 211. DNA Structures (Part A: Synthesis and Physical Analysis of DNA)
Edited by DAVID M. J. LILLEY AND JAMES E. DAHLBERG

VOLUME 212. DNA Structures (Part B: Chemical and Electrophoretic Analysis of DNA)
Edited by DAVID M. J. LILLEY AND JAMES E. DAHLBERG

VOLUME 213. Carotenoids (Part A: Chemistry, Separation, Quantitation, and Antioxidation)
Edited by LESTER PACKER

VOLUME 214. Carotenoids (Part B: Metabolism, Genetics, and Biosynthesis)
Edited by LESTER PACKER

VOLUME 215. Platelets: Receptors, Adhesion, Secretion (Part B)
Edited by JACEK J. HAWIGER

VOLUME 216. Recombinant DNA (Part G)
Edited by RAY WU

VOLUME 217. Recombinant DNA (Part H)
Edited by RAY WU

VOLUME 218. Recombinant DNA (Part I)
Edited by RAY WU

VOLUME 219. Reconstitution of Intracellular Transport
Edited by JAMES E. ROTHMAN

VOLUME 220. Membrane Fusion Techniques (Part A)
Edited by NEJAT DÜZGÜNEŞ

VOLUME 221. Membrane Fusion Techniques (Part B)
Edited by NEJAT DÜZGÜNES

VOLUME 222. Proteolytic Enzymes in Coagulation, Fibrinolysis, and Complement Activation (Part A: Mammalian Blood Coagulation Factors and Inhibitors)
Edited by LASZLO LORAND AND KENNETH G. MANN

VOLUME 223. Proteolytic Enzymes in Coagulation, Fibrinolysis, and Complement Activation (Part B: Complement Activation, Fibrinolysis, and Nonmammalian Blood Coagulation Factors)
Edited by LASZLO LORAND AND KENNETH G. MANN

VOLUME 224. Molecular Evolution: Producing the Biochemical Data
Edited by ELIZABETH ANNE ZIMMER, THOMAS J. WHITE, REBECCA L. CANN, AND ALLAN C. WILSON

VOLUME 225. Guide to Techniques in Mouse Development
Edited by PAUL M. WASSARMAN AND MELVIN L. DEPAMPHILIS

VOLUME 226. Metallobiochemistry (Part C: Spectroscopic and Physical Methods for Probing Metal Ion Environments in Metalloenzymes and Metalloproteins)
Edited by JAMES F. RIORDAN AND BERT L. VALLEE

VOLUME 227. Metallobiochemistry (Part D: Physical and Spectroscopic Methods for Probing Metal Ion Environments in Metalloproteins)
Edited by JAMES F. RIORDAN AND BERT L. VALLEE

VOLUME 246. Biochemical Spectroscopy
Edited by KENNETH SAUER

VOLUME 247. Neoglycoconjugates (Part B: Biomedical Applications)
Edited by Y. C. LEE AND REIKO T. LEE

VOLUME 248. Proteolytic Enzymes: Aspartic and Metallo Peptidases (in preparation)
Edited by ALAN J. BARRETT

VOLUME 249. Enzyme Kinetics and Mechanism (Part D: Developments in Enzyme Dynamics)
Edited by DANIEL L. PURICH

VOLUME 250. Lipid Modifications of Proteins (in preparation)
Edited by PATRICK J. CASEY AND JANICE E. BUSS

VOLUME 251. Biothiols (Part A: Monothiols and Dithiols, Protein Thiols, and Thiyl Radicals) (in preparation)
Edited by LESTER PACKER

VOLUME 252. Biothiols (Part B: Glutathione and Thioredoxin; Thiols in Signal Transduction and Gene Regulation) (in preparation)
Edited by LESTER PACKER

VOLUME 253. Adhesion of Microbial Pathogens (in preparation)
Edited by RON J. DOYLE AND ITZHAK OFEK

VOLUME 254. Oncogene Techniques (in preparation)
Edited by PETER K. VOGT AND INDER M. VERMA

VOLUME 255. Small GTPases and Their Regulators (Part A: RAS Family) (in preparation)
Edited by W. E. BALCH, CHANNING J. DER, AND ALAN HALL

VOLUME 256. Small GTPases and Their Regulators (Part B: RHO Family) (in preparation)
Edited by W. E. BALCH, CHANNING J. DER, AND ALAN HALL

VOLUME 257. Small GTPases and Their Regulators (Part C: Proteins Involved in Transport) (in preparation)
Edited by W. E. BALCH, CHANNING J. DER, AND ALAN HALL

VOLUME 258. Redox-Active Amino Acids in Biology (in preparation)
Edited by JUDITH P. KLINMAN

VOLUME 259. Energetics of Biological Macromolecules (in preparation)
Edited by MICHAEL L. JOHNSON AND GARY K. ACKERS

VOLUME 260. Mitochondrial Biogenesis and Genetics, Part A (in preparation)
Edited by GIUSEPPE M. ALTARDI AND ANNE CHOMYN

Section I

General Approaches to Biological Catalysis

[1] Transient Kinetic Approaches to Enzyme Mechanisms

By CAROL A. FIERKE and GORDON G. HAMMES

Introduction

The elucidation of enzymatic mechanisms through kinetic investigations has been an important area of research for many years.[1,2] Enzymes have a special fascination for many reasons: their role in physiological processes; their often unstable nature; their incredible efficiency as catalysts; and their unusual kinetic patterns. Initially kinetic studies were carried out at very low enzyme concentrations relative to substrate concentrations. This was due to the difficulty in obtaining large quantities of high-purity enzymes and the relative simplicity of the kinetic equations for this situation. Under such conditions, all of the enzyme species can be assumed to be in a steady state so that any mechanism, regardless of complexity, can be described by a single rate equation. If initial rates are measured, with no products present, the rate measurements are usually easy to carry out and easy to interpret.

Steady-state kinetics can provide information about the overall reaction pathway for multiple substrate reactions, the specificity of the enzyme for specific substrate structures, and lower bounds for the specific rate constants in a mechanism.[1-4] However, the kinetic parameters measured are complex functions of the specific rate constants so that individual rate constants rarely can be determined, and little information is obtained about mechanistic intermediates. If the pH dependencies of the steady-state kinetic parameters are measured, the nature of the ionizable groups of the enzymes that are important for catalysis can be inferred; in addition, the use of isotopes sometimes can shed light on the reaction intermediates.

To elucidate the details of an enzymatic mechanism and to measure rate constants for elementary steps, kinetic experiments must be carried out at high enzyme concentrations where enzyme–substrate species are significantly populated and can be directly detected. Because enzymes are very efficient catalysts, fast reaction techniques are often needed to study the transient kinetics. Such studies are vital for the establishment

[1] G. G. Hammes, "Enzyme Catalysis and Regulation." Academic Press, New York, 1982.
[2] A. Fersht, "Enzyme Structure and Mechanism," 2nd Ed. Freeman, San Francisco, 1985.
[3] W. W. Cleland, *Adv. Enzymol.* **45,** 273 (1977).
[4] L. Peller and R. A. Alberty, *J. Am. Chem. Soc.* **81,** 5907 (1959).

of detailed reaction mechanisms. The primary difficulties in carrying out transient kinetics are the special technologies and the large amounts of enzyme required. However, commercial stopped-flow and rapid quench equipment are now available, and cloning methods can be used to produce large amounts of most proteins. Therefore, the role of transient kinetics in understanding enzyme mechanisms should markedly expand in the near future.

This chapter is concerned with introducing transient kinetic studies to biochemists. First, the theoretical bases for analyzing the rate equations associated with typical mechanisms will be considered. Second, a brief description of the various types of stopped-flow, rapid quench, and temperature jump equipment is given. Finally, some examples of transient kinetic studies are presented. Additional information can be found in other reviews.[1,5–7]

Theory of Transient Kinetics

A complete treatment of all the kinetic complexities that may be encountered in carrying out transient studies obviously is not possible. Instead, the principles associated with the kinetic analysis of complex mechanisms are presented, with specific examples of commonly encountered situations. In analyzing a complex mechanism, the first question that must be asked is how many rate equations are required to describe the mechanism. The answer is deceptively simple: the number of independent rate equations is equal to the number of independent concentration variables. The number of independent concentration variables in turn is equal to the total number of concentration variables minus the number of conservation relationships that exist between the concentration variables. The conservation relationships are usually simple mass conservation, but they also include more subtle relationships such as those determined by steady-state situations, that is, the rate of change of a concentration with time may equal zero.

The experimental design of transient kinetic experiments is important since significant simplification in the kinetic analysis often can be accomplished. For example, whenever possible, reactions should be carried out under conditions where the observed kinetics are first order. This is usually

[5] G. G. Hammes and P. R. Schimmel, *in* "Enzymes" (P. D. Boyer, ed.), 3rd Ed., Vol. 2, p. 67. Academic Press, New York, 1970.

[6] K. A. Johnson, *in* "Enzymes" (P. D. Boyer, ed.), 4th Ed. Vol. 20, p. 1. Academic Press, New York, 1992.

[7] C. F. Bernasconi (ed.), "Investigation of Rates and Mechanisms of Reactions," Part 2, 4th Ed. Wiley(Interscience), New York, 1986.

done by making all concentrations large relative to the concentration of the species whose rate of change with time is being measured. Other potential simplifications are discussed throughout the theoretical analysis.

The simplest type of reaction that might be anticipated is an irreversible first-order reaction. For example, a conformational change of an enzyme from state E to state E′, characterized by a rate constant k_1, might be triggered by a change in pH:

$$\mathrm{E} \xrightarrow{k_1} \mathrm{E}' \tag{1}$$

This mechanism is obviously characterized by a single rate equation since there are two species and the total enzyme concentration, $(\mathrm{E})_0$, is conserved:

$$(\mathrm{E}) + (\mathrm{E}') = (\mathrm{E})_0 \tag{2}$$

For this mechanism

$$-d(E)/dt = k_1(\mathrm{E}) \tag{3}$$

or after integration

$$(\mathrm{E}) = (\mathrm{E})_0\, e^{-k_1 t} \tag{4}$$

if $(\mathrm{E}) = (\mathrm{E})_0$ at $t = 0$, where t is the time. Data analysis is simple as a plot of ln(E) versus t is linear with a slope of $-k_1$, or k_1 can be obtained by fitting the time course of E to Eq. (4) with a nonlinear least squares analysis. If this reaction is reversible,

$$\mathrm{E} \underset{k_{-1}}{\overset{k_1}{\rightleftharpoons}} \mathrm{E}' \tag{5}$$

the rate equation becomes

$$-d(\mathrm{E})/dt = k_1(\mathrm{E}) - k_{-1}[(\mathrm{E})_0 - (\mathrm{E})] \tag{6}$$

which can be integrated with the same boundary conditions as above to give

$$(\mathrm{E}) = (\mathrm{E})_e + [(\mathrm{E})_0 - (\mathrm{E})_e]\, e^{-(k_1 + k_{-1})t} \tag{7a}$$

or

$$\ln\left(\frac{[(\mathrm{E}) - (\mathrm{E})_e]}{[(\mathrm{E})_0 - (\mathrm{E})_e]}\right) = -\,(k_1 + k_{-1})t \tag{7b}$$

where the subscript e denotes the equilibrium concentration. A plot of ln $[(\mathrm{E}) - (\mathrm{E})_e]$ versus time is a straight line with a slope of $-(k_1 + k_{-1})$.

If the equilibrium constant, $K_1 = (E')_e/(E)_e = k_1/k_{-1}$, is known, both rate constants can be calculated.

Concentrations are usually not measured directly. Instead some property that is related to the concentration such as absorbance or fluorescence is measured. If the property, P_E, is a linear function of the concentration,

$$P_E = a(E) + b \tag{8}$$

where a and b are constants, then P_E can be substituted for (E) in the integrated rate equations. For example, the absorbance and fluorescence are directly proportional to the concentration.

As a second example, consider the combination of enzyme, E, and substrate, S, to form a complex, X:

$$E + S \underset{k_{-2}}{\overset{k_2}{\rightleftharpoons}} X \tag{9}$$

A single independent rate equation is required to describe this mechanism and can be written as

$$-d(E)/dt = k_2(E)(S) - k_{-2}(X) \tag{10}$$

If the mass conservation relationships

$$(E)_0 = (E) + (X) \tag{11}$$

$$(S)_0 = (S) + (X) \tag{12}$$

are utilized, the concentrations (S) and (X) can be expressed in terms of (E), $(E)_0$, and $(S)_0$, and Eq. (10) can be integrated. The result is very complex and rarely used.[8]

This integrated rate equation can be considerably simplified by adjusting the concentrations so that either E or S remains essentially constant during the experiment. In practice, this means that one of the concentrations must be about an order of magnitude greater than the other. For example, if $(S)_0 \gg (E)_0$, the integration of Eq. (10) gives Eq. (7a), except that $k_1 = k_2(S)_0$ and $k_{-1} = k_{-2}$. This is frequently called a "pseudo"-first-order rate equation where the observed rate constant, $k_{obs} = k_2(S)_0 + k_{-2}$. In this case, if $k_2(S)_0 + k_{-2}$ is determined at several concentrations of S_0, a plot of k_{obs} versus $(S)_0$ is linear with a slope of k_2 and an intercept of k_{-2}. The ratio of rate constants is equal to the equilibrium constant. Reduction of higher order kinetic equations to first-order rate equations by experimental design should be done whenever possible.

[8] G. G. Hammes, "Principles of Chemical Kinetics," p. 72. Academic Press, New York, 1978.

An alternative method of reducing higher order rate equations to first order is to carry out kinetic measurements near equilibrium, that is, chemical relaxation experiments.[1,9,10] Consider the combination of enzyme and substrate [Eq. (9)] near equilibrium. New concentration variables can be defined as

$$\begin{aligned}(\mathrm{E}) &= (\mathrm{E})_e + \Delta\mathrm{E}\\ (\mathrm{S}) &= (\mathrm{S})_e + \Delta\mathrm{S}\\ (\mathrm{X}) &= (\mathrm{X})_e + \Delta\mathrm{X}\end{aligned} \tag{13}$$

where Δ designates the deviation from equilibrium and $\Delta\mathrm{E} = \Delta\mathrm{S} = -\Delta\mathrm{X}$ because of mass conservation. Equation (10) becomes

$$-\mathrm{d}\Delta\mathrm{E}/dt = \{k_2[(\mathrm{E})_e + (\mathrm{S})_e] + k_{-2}\}\Delta\mathrm{E} + k_2(\mathrm{E})_e(\mathrm{S})_e - k_{-2}(\mathrm{X})_e + k_2(\Delta\mathrm{E})^2 \tag{14}$$

Note that $k_2(\mathrm{E})_e(\mathrm{S})_e = k_{-2}(\mathrm{X})_e$ by definition, and near equilibrium $(\Delta\mathrm{E})^2$ can be neglected because it is small. Thus Eq. (14) becomes

$$-d\Delta\mathrm{E}/dt = \Delta\mathrm{E}/\tau \tag{15}$$

where $1/\tau = k_2[(\mathrm{E})_e + (\mathrm{S})_e] + k_{-2}$ and τ is the chemical relaxation time. This example illustrates that near equilibrium all rate equations become first order. Equation (15) can be integrated to give

$$\Delta\mathrm{E} = (\Delta\mathrm{E})_0\, e^{-t/\tau} \tag{16}$$

so that a plot of ln $\Delta\mathrm{E}$ versus t gives a straight line with a slope of $-1/\tau$. If τ is determined at a series of equilibrium concentrations, a plot of $1/\tau$ versus $[(\mathrm{E})_e + (\mathrm{S})_e]$ has a slope of k_2 and an intercept of k_{-2} [Eq. (15)]. If the equilibrium constant is not known, a value can be assumed to get preliminary values of k_2 and k_{-2}. Then the ratio of the rate constants obtained can be used to calculate a new equilibrium constant, and the process can be repeated until the ratio of rate constants equals the equilibrium constant used to calculate the equilibrium concentrations.

This linearization of the rate equations is typically valid up to 10% from equilibrium and can be used to analyze stopped-flow data, as well as those obtained from more conventional relaxation techniques. Typical

[9] M. Eigen and L. deMaeyer, *in* "Investigation of Rates and Mechanisms of Reactions," (S. L. Friess, E. S. Lewis and A. Weissberger, eds.) Part 2, 2nd Ed. p. 895. Wiley, New York, 1963.

[10] C. F. Bernasconi, "Relaxation Kinetics." Academic Press, New York, 1976.

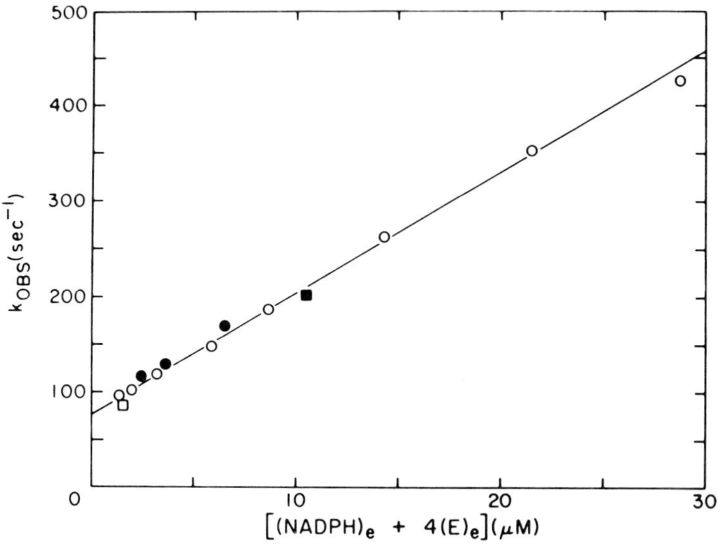

FIG. 1. Plot of the first-order rate constant for binding of NADPH to fatty-acid synthase, k_{obs}, versus the sum of the equilibrium concentrations of unbound NADPH and unoccupied NADPH binding sites on the enzyme, $[(NADPH)_e + 4(E)_e]$. [Reprinted with permission from J. A. H. Cognet, B. G. Cox, and G. G. Hammes, *Biochemistry* **22,** 6281 (1983). Copyright 1983 American Chemical Society.]

experimental data are shown in Fig. 1, where the reciprocal relaxation time ($1/\tau = k_{obs}$) is plotted versus the sum of the equilibrium concentrations for the reaction of NADPH with fatty-acid synthase.[11]

Thus far only single-step reaction mechanisms have been considered. However, reaction mechanisms of interest usually have multiple steps. A two-step mechanism is now considered in detail, with the methodology developed being applicable to more complex mechanisms.

Consider a mechanism whereby an enzyme and substrate combine to form a complex which undergoes a conformational change:

$$E + S \underset{k_{-1}}{\overset{k_1}{\rightleftharpoons}} X_1 \underset{k_{-2}}{\overset{k_2}{\rightleftharpoons}} X_2 \tag{17}$$

Two independent rate equations are needed to describe this mechanism as four concentration variables and two conservation relationships exist:

$$(E)_0 = (E) + (X_1) + (X_2) \tag{18}$$

[11] J. A. H. Cognet, B. G. Cox, and G. G. Hammes, *Biochemistry* **22,** 6281 (1983).

or

$$\Delta E + \Delta X_1 + \Delta X_2 = 0$$

and

$$(S)_0 = (S) + (X_1) + (X_2) \tag{19}$$

or

$$\Delta S + \Delta X_1 + \Delta X_2 = 0$$

The rate equations can be written as

$$-d(E)/dt = k_1(E)(S) - k_{-1}(X_1) \tag{20}$$

$$-d(X_2)/dt = -k_2(X_1) + k_{-2}(X_2) \tag{21}$$

The same procedures previously utilized can be used to linearize Eqs. (20) and (21) in the neighborhood of equilibrium, namely, $(E) = (E)_e + \Delta E$, etc. The result is

$$\begin{aligned} -d\Delta E/dt &= \{k_1[(E)_e + (S)_e] + k_{-1}\}\Delta E + k_{-1}\Delta X_2 \\ &= a_{11}\Delta E + a_{12}\Delta X_2 \end{aligned} \tag{22}$$

$$\begin{aligned} -d\Delta X_2/dt &= k_2\Delta E + (k_{-2} + k_2)\Delta X_2 \\ &= a_{21}\Delta E + a_{22}\Delta X_2 \end{aligned} \tag{23}$$

where the a_{ij} terms are defined by these equations. The integration of coupled first-order linear homogeneous differential equations is described in textbooks on differential equations, and only the results are presented here. On integration of Eqs. (22) and(23), the two concentration variables are given as a sum of two exponentials:

$$\Delta E = A_1 e^{-t/\tau_1} + A_2 e^{-t/\tau_2} \tag{24}$$

$$\Delta X_2 = A_3 e^{-t/\tau_1} + A_4 e^{-t/\tau_2} \tag{25}$$

where the A_i values are constants and the τ_i are obtained by solving the determinant

$$\begin{vmatrix} a_{11} - 1/\tau & a_{12} \\ a_{21} & a_{22} - 1/\tau \end{vmatrix} = 0 \tag{26}$$

The τ_i are equivalent to reciprocal first-order rate constants and are functions of equilibrium concentrations and rate constants. The values of A_i are determined by the boundary conditions and are not explored further here. The determinant can be expanded to give

$$(1/\tau)^2 - (a_{11} + a_{22})(1/\tau) + a_{11}a_{22} - a_{12}a_{21} = 0 \tag{27}$$

This quadratic equation can be solved for the relaxation times:

$$1/\tau_{1,2} = \frac{(a_{11} + a_{22})}{2}\left\{1 \pm \left[1 - \frac{4(a_{11}a_{22} - a_{12}a_{21})}{(a_{11} + a_{22})^2}\right]^{1/2}\right\} \quad (28)$$

where one relaxation time corresponds to the positive square root and the other to the negative square root.

These relationships can be directly related to experimental measurements: observing concentrations as a function of time should yield a plot of two exponentials characterized by the two relaxation times which can be obtained from the data, for example, by a nonlinear least squares analysis. The relaxation times are complex functions of equilibrium concentrations. However, the analysis of the relaxation times is usually not difficult. For example,

$$1/\tau_1 + 1/\tau_2 = a_{11} + a_{22} = k_1[(\mathrm{E})_e + (\mathrm{S})_e] + k_{-1} + k_2 + k_{-2} \quad (29)$$

$$(1/\tau_1)(1/\tau_2) = a_{11}a_{22} - a_{12}a_{21} = k_1(k_2 + k_{-2})\,[(\mathrm{E})_e + (\mathrm{S})_e] + k_{-1}k_{-2} \quad (30)$$

Linear plots of the sum and product of the reciprocal relaxation times versus $[(\mathrm{E})_e + (\mathrm{S})_e]$ can be used to obtain all four individual rate constants.

Often the bimolecular step is much more rapid than the conformational change. In fact, this can usually be made the case by working at sufficiently high concentrations of enzyme and/or substrate. In this case, $k_1[(\mathrm{E})_e + (\mathrm{S})_e] + k_{-1} \gg (k_2 + k_{-2})$ (i.e., $a_{11} \gg a_{22}, a_{21}, a_{12}$), and the square root in Eq. (28) can be expanded as $(1 - \mathrm{X})^{1/2} \approx 1 - \mathrm{X}/2$ for $\mathrm{X} \ll 1$. This expansion for $\mathrm{X} = 4(a_{11}a_{22} - a_{12}a_{21})/(a_{11} + a_{22})^2$ gives

$$1/\tau_1 = k_1[(\mathrm{E})_e + (\mathrm{S})_e] + k_{-1} \quad (31)$$

$$1/\tau_2 = k_{-2} + \frac{k_2}{1 + k_{-1}/\{k_1[(\mathrm{E})_e + (\mathrm{S}_e)]\}} \quad (32)$$

The expression for $1/\tau_1$ is the same as for an isolated bimolecular reaction because that step essentially equilibrates before the conformational change occurs. The second relaxation time primarily characterizes the rate of the conformational change because the bimolecular step is essentially at equilibrium while the conformational change occurs. When the sum of the equilibrium concentrations $[(\mathrm{S})_e + (\mathrm{E})_e]$ is large relative to the equilibrium dissociation constant for the first step (k_{-1}/k_1), then $1/\tau_2 = k_2 + k_{-2}$, the relaxation time for the isolated second step, because all of the enzyme is present as X_1 and X_2.

From the analysis above, it should be evident that in the neighborhood of equilibrium, all mechanisms can be described by a set of independent rate equations that are homogeneous first-order (linear) differential equations. They can be written symbolically as

$$-d\Delta c_i/dt = \sum_{j=1}^{n} a_{ij}\Delta c_j \tag{33}$$

where n is the number of independent concentration variables. The solution to this set of equations is a sum of n exponential terms:

$$\Delta c_i = \sum_{j=1}^{n} A_{ij}\, e^{-t/\tau_j} \tag{34}$$

where the τ_j are relaxation times and the A_{ij} are constants determined by the boundary conditions. The relaxation times are functions of equilibrium concentrations and rate constants and can be calculated from the following determinant:

$$\begin{vmatrix} a_{11} - 1/\tau & \cdots & a_{1n} \\ \vdots & & \vdots \\ a_{n1} & \cdots & a_{nn} - 1/\tau \end{vmatrix} = 0 \tag{35}$$

Application of this analysis to experiments is quite feasible since multiexperimental curves can be deconvoluted readily with standard statistical computer packages. In practice, the deconvolution often can be accomplished experimentally by adjusting the concentrations and conditions such that the rate of the different steps are quite different. In a study of the interaction of erythro-β-hydroxyaspartate with aspartate aminotransferase, nine relaxation times associated with transamination were resolved ranging from microseconds to seconds.[12] Although the amplitude constants, A_{ij}, contain useful information, they are difficult to calculate and are rarely utilized in kinetic analyses of experimental data.

An analysis of the mechanism in Eq. (17) far from equilibrium is now considered. The coupled rate equations [Eqs. (20) and (21)] cannot be solved analytically. Numerical integration can be carried out, but this is difficult to apply to experimental situations. Instead, the common practice is to make the coupled reactions "pseudo" first order by having the substrate concentration much greater than the enzyme concentration, $(S)_0 \gg (E)_0$. [Of course, $(E)_0 \gg (S)_0$ works equally well.] In this case, Eqs. (20) and (21) become

$$-d(E)/dt = [k_1(S)_0 + k_{-1}](E) + k_{-1}(X_2) - k_{-1}(E)_0 \tag{36}$$

[12] G. G. Hammes and J. L. Haslam, *Biochemistry* **8,** 1591 (1969).

$$-dX_2/dt = k_2(E) + (k_{-2} + k_2)(X_2) - k_2(E)_0 \tag{37}$$

These equations are identical in form to Eqs. (22) and (23), except that $k_1(S)_0$ appears rather than $k_1[(E)_e + (S)_e]$ and constant terms containing $(E)_0$ are present. The solution to Eqs. (36) and (37), which are nonhomogeneous analogs of Eqs. (22) and (23), is a sum of two time-dependent exponentials and a constant:

$$(E) = A_1 e^{-\lambda_1 t} + A_2 e^{-\lambda_2 t} + (E)_e \tag{38}$$
$$(X_2) = A_3 e^{-\lambda_1 t} + A_4 e^{-\lambda_2 t} + (X_2)_e \tag{39}$$

where the λ_i values are obtained by solving the same determinant as Eq. (26), except that $k_1(S)_0$ is substituted for $k_1[(E)_e + (S)_e]$ and λ for $1/\tau$. The analysis of experimental data proceeds exactly as for the near-equilibrium situation.

A special case worth further consideration is the slower time constant when the bimolecular reaction equilibrates rapidly relative to the conformational change. In this case

$$\lambda_2 = k_{-2} + \frac{k_2}{1 + K_1/(S)_0} \tag{40}$$

with $K_1 = k_{-1}/k_1$. This is a general result for a binding equilibrium preceding a first-order reaction, namely, the "apparent" first-order rate constant is dependent on the ligand concentration. Measurement of the dependence of λ_2 on the ligand concentration permits determination of the equilibrium dissociation constant and the two first-order rate constants characterizing the second step.

If all rate equations can be made "pseudo" first order, an analytical solution of the coupled differential equations can be obtained, analogous to the situation near equilibrium. The solutions are sums of time-dependent exponentials and constants. The number of exponential terms observed establishes the minimum member of independent rate equations characterizing the mechanism.

One additional mechanism is considered, namely, consecutive first-order reactions. This mechanism illustrates the commonly observed situation of an intermediate appearing and then disappearing. The simplest case is

$$A \xrightarrow{k_1} B \xrightarrow{k_2} C \tag{41}$$

If the experiment starts with an initial concentration of A equal to $(A)_0$, then

$$(A) = (A)_0\, e^{-k_1 t} \tag{42}$$

and

$$\begin{aligned} d(B)/dt &= k_1(A) - k_2(B) \\ &= k_1(A)_0\, e^{-k_1 t} - k_2(B) \end{aligned} \tag{43}$$

Integration of Eq. (43) for $k_1 \neq k_2$ gives

$$(B) = \frac{(A)_0 k_1}{k_2 - k_1}(e^{-k_1 t} - e^{-k_2 t}) \tag{44}$$

Equation (44) is not valid for $k_1 = k_2$ since B becomes indeterminant. If Eq. (43) is integrated for $k_1 = k_2$, then

$$(B) = k_1(A)_0 t\, e^{-k_1 t} \tag{45}$$

The time dependencies of A and B completely describe this system since only two independent concentration variables exist. If the time dependence of C is desired, it can be obtained from the mass conservation relationship $(A)_0 = (A) + (B) + (C)$:

$$(C) = (A_0)\left(1 - \frac{k_2}{k_2 - k_1}\, e^{-k_1 t} + \frac{k_1}{k_2 - k_1}\, e^{-k_2 t}\right) \tag{46}$$

If $k_2 \gg k_1$, B decays to C more rapidly than it is formed, and the concentration of B is very low at all times. Under these conditions, direct observation of the intermediate is virtually impossible, and the concentration of C essentially increases exponentially with k_1 being the characteristic rate constant (see Fig. 2). For the other extreme case, $k_1 \gg k_2$, the intermediate rapidly accumulates until essentially all of A has been converted to B which then decays to C as a first-order process with rate constant k_2. The buildup and decay occur in different time domains and can be treated essentially as independent processes. Likewise, the time dependence of the appearance of C shows two phases; a "lag" at short times while the intermediate (B) is accumulating followed by an exponential decay of (B) to (C). If k_1 and k_2 are comparable, the concentration of B goes through a well-defined maximum, and its time dependence requires consideration of both the rate of formation and the rate of decay. This behavior is shown graphically in Fig. 2.

The analysis of consecutive reactions can be applied directly to enzymatic processes as illustrated by consideration of a simple Michaelis–Menten mechanism:

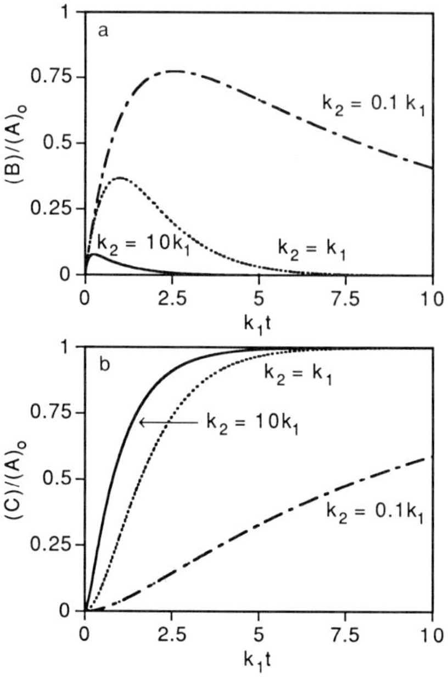

FIG. 2. Time dependence of the concentration ratio of either $(B)/(A)_0$ (a) or $(C)/(A)_0$ (b) for irreversible consecutive first-order reactions [Eqs. (41), (44), and (46)] where the ratio k_2/k_1 is 10 (——), 1 (· · ·), or 0.1 (— - —).

$$\mathrm{E} + \mathrm{S} \underset{k_{-1}}{\overset{k_1}{\rightleftharpoons}} \mathrm{X} \overset{k_2}{\rightarrow} \mathrm{E} + \mathrm{P} \tag{47}$$

For this mechanism

$$d(\mathrm{X})/dt = k_1(\mathrm{S})(\mathrm{E}) - (k_2 + k_{-1})(\mathrm{X}) \tag{48}$$

If the initial substrate concentration, $(\mathrm{S})_0$, is much greater than the total enzyme concentration, $(\mathrm{E})_0$, then (S) can be assumed to remain constant at early times so that

$$d(\mathrm{X})/dt = k'_1(\mathrm{E})_0 - (k'_1 + k_{-1} + k_2)(\mathrm{X}) \tag{49}$$

where $k'_1 = k_1(\mathrm{S})$ and the conservation relationship, $(\mathrm{E})_0 = (\mathrm{E}) + (\mathrm{X})$, has been utilized. Equation (49) can be integrated to give

$$(\mathrm{X}) = \frac{k'_1(\mathrm{E})_0}{k'_1 + k_{-1} + k_2}[1 - e^{-(k'_1+k_{-1}+k_2)t}] \tag{50}$$

Furthermore, the initial rate of product formation is given by

$$d(\mathrm{P})/dt = k_2(\mathrm{X}) \tag{51}$$

If Eq. (50) is inserted into Eq. (51) and the latter is integrated, the time dependence of product formation can be written as follows:

$$(\mathrm{P}) = \frac{k'_1 k_2(\mathrm{E})_0 t}{k'_1 + k_{-1} + k_2} + \frac{k'_1 k_2(\mathrm{E})_0}{(k'_1 + k_{-1} + k_2)^2}[e^{-(k'_1+k_{-1}+k_2)t} - 1] \tag{52}$$

Note that at long times the second term becomes much smaller than the first, and the time dependence of (P) is given by the usual steady-state Michaelis–Menten equation. At very short times, if k_1 (S) $\gg (k_{-1} + k_2)$, there is a "lag" in the formation of product as the concentration of the intermediate, (X), increases to a steady-state concentration (see Fig. 3), and then the concentration of product increases linearly with time. However, if the concentration of both X and P are assayed as product, as in the case where X is a species in which product is noncovalently bound to the enzyme, the observed product concentration, [(X) + (P)], is described by

$$[(\mathrm{X}) + (\mathrm{P})] = \frac{k'_1(\mathrm{E})_0}{k'_1 + k_{-1} + k_2}\left[k_2 t + \frac{k'_1 + k_{-1}}{k'_1 + k_{-1} + k_2}(1 - e^{-(k'_1+k_{-1}+k_2)t})\right] \tag{53}$$

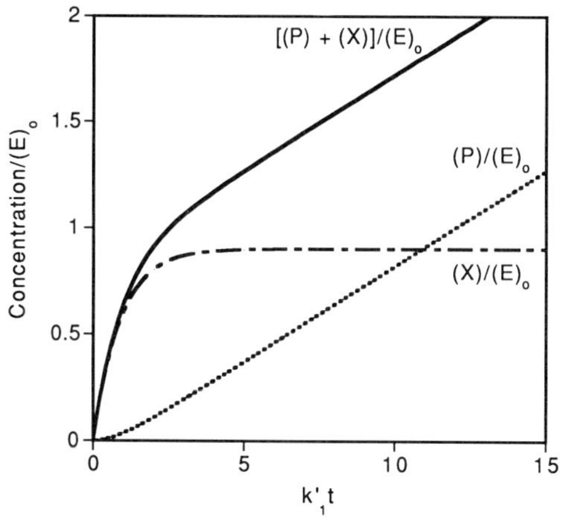

FIG. 3. Time dependence of the concentration ratios $(\mathrm{P})/(\mathrm{E})_0$ (· · ·), $(\mathrm{X})/(\mathrm{E})_0$ (— - —), and $[(\mathrm{P}) + (\mathrm{X})]/(\mathrm{E})_0$ (——) for a Michaelis–Menten mechanism [Eq. (47)] where the ratio $k_2/k'_1 = 0.1$ and $k_{-1}/k'_1 = 0.1$. These variations in concentration ratios are described by Eqs. (50), (52), and (53).

In this case when $k'_1 \gg (k_2 + k_{-1})$, the observed product concentration initially increases rapidly with time with a maximum amplitude of $[(X) + (P)]/(E)_0 = 1$ and then becomes a linear function of time (see Fig. 3). Similarly, because $\Delta(S) = -\Delta[(X) + (P)]$, the substrate concentration as a function of time displays a rapid, initial decrease followed by a linear decrease. The initial phase is called a "burst." Burst kinetics are observed for any mechanism in which product is formed in a rapid, first step followed by regeneration of active enzyme in a slower, second step as illustrated by chymotrypsin.[13]

A well-known paradigm in chemical kinetics is that the number of possible mechanisms is limitless. Therefore, only some general procedures for carrying out analyses have been presented, with specific solutions to some relatively simple cases. Application of the principles developed to other mechanisms, although not always obvious, is possible.

Computer Simulation of Transient Kinetics

As discussed, the time dependence of substrate disappearance for simple kinetic mechanisms can be determined analytically by integration of the differential equations describing this scheme. Furthermore, two methods for simplifying these equations, relaxation kinetics and pseudo first-order kinetics, allow explicit integration of the coupled differential equations for kinetic mechanisms with several steps. As the kinetic mechanisms become more complex, however, the differential equations describing the time dependence of the concentration terms become more difficult to integrate, and analytic solutions are impossible without simplification. For these reasons, it is often useful to integrate numerically the differential equations for complex reactions. Even though these numerical solutions are approximate, in practice they can be calculated as accurately as desired.

Advances in computing have allowed investigators to develop computer algorithms that will numerically integrate differential equations,[14-18] providing rapid solutions to even complex reactions. The computer programs are readily available, flexible, easy to use, and run on a variety of microcomputers. Originally the programs simply allowed simulation of

[13] H. Gutfreund and J. M. Sturtevant, *Biochem. J.* **63,** 656 (1956).

[14] B. A. Barshop, R. F. Wrenn, and C. Frieden, *Anal. Biochem.* **130,** 134 (1983).

[15] C. T. Zimmerle and C. Frieden, *Biochem. J.* **258,** 381 (1989).

[16] C. Frieden, *Trends Biochem. Sci.* **18,** 58 (1993).

[17] J. P. Hecht, J. M. Nikonov, and G. L. Alonso, *Comput. Methods Prog. Biomed.* **33,** 13 (1990).

[18] H. G. Holzhütter and A. Colosimo, *Comput. Appl. Biosci.* **6,** 23 (1990).

the time dependence of the concentration of substrate, intermediates, and products; however, some programs have been extended to allow the fitting of the reaction kinetics based on simulation.[15,18] This analysis provides an estimate of the confidence limits of each kinetic parameter.

Computer simulation of complex kinetic data has the following distinct advantages: (i) no simplifying assumptions are required to fit the data; (ii) the solutions must account for the concentration dependence of both the rate and the amplitude of a reaction; (iii) the simulation can explicitly fit data from a variety of experiments; and (iv) the simulations clearly demonstrate that a specific mechanism can account for the given data. However, a good fit between a simulated mechanism and the data does not prove that the mechanism is correct. Simulations are most useful for establishing a minimal mechanism required to fit the data and for quantitatively eliminating other possible mechanisms. To obtain the maximum information from numerical simulation, experiments should be designed to isolate specifically a few rate constants in order to provide narrow limits for their value in the simulation. Additional experiments can then build on this initial mechanism in an iterative fashion until all of the rate constants in the proposed mechanism are sufficiently constrained. This type of analysis has been essential for determination of the complete kinetic mechanism for a number of enzymes including dihydrofolate reductase,[19,20] alcohol dehydrogenase,[21] and 5-enolpyruvoylshikimate 3-phosphate synthase (EPSP synthase; 3-phosphoshikimate 1-carboxyvinyltransferase).[22]

Rapid Mixing Methods

Stopped-flow instrumentation is useful for measuring rapid transient kinetics whenever there is an observable signal change that is easily related to concentration changes, including absorbance, fluorescence, light scattering, circular dichroism, nuclear magnetic resonance (NMR), and conductivity. The kinetics of enzyme-catalyzed reactions are determined most frequently by monitoring changes in absorbance or fluorescence. Instruments are equipped with a monochromator for determining the excitation wavelength and interference filters for the emission wavelength. A variety of band-pass and cutoff filters optimized for different wavelengths and

[19] C. A. Fierke, K. A. Johnson, and S. J. Benkovic, *Biochemistry* **26,** 4085 (1987).
[20] M. H. Penner and C. Frieden, *J. Biol. Chem.* **262,** 15908 (1987).
[21] V. C. Sekhar and B. V. Plapp, *Biochemistry* **29,** 4289 (1990).
[22] K. S. Anderson, J. A. Sikorski, and K. A. Johnson, *Biochemistry* **27,** 7395 (1988).

purposes are available commercially.[23] There are several commercially available instruments designed for use with biological materials that combine high sensitivity with low sample volumes.

The design of a stopped-flow instrument is quite simple and has been discussed in detail in previous reviews.[6,23] The reaction is initiated by pushing two or more solutions (maintained in separate syringes) through a mixing chamber and into an observation cell. The flow of liquids is limited by a stop syringe (hence the name stopped-flow), and spectral data are collected by a computer after the flow has stopped. Stopped-flow enzyme kinetic data can be obtained under conditions of either excess substrate (pre-steady-state experiment) or excess enzyme (single turnover experiment).

The maximum rate measurable by the stopped-flow technique is limited by the amount of time it takes to mix the reactants and to transport the solutions from the mixing chamber, where the reaction is initiated, to the observation point. This time, also called the "dead time" of the instrument, is dependent on the efficiency of mixing, the flow rate, and the distance between the mixer and the observation port. The dead time can be varied by changing the flow velocity; however, at high pressures the shear force through the mixer may damage some biological samples. In modern commercial instruments the dead time is approximately 1–2 msec. This sets a practical upper limit of about 700 sec^{-1} for first-order rate constants observable by this technique, since at 1 msec half of the reactants have been converted to products for this rate constant [Eq. (4)]. For second-order (or higher) rate constants, the maximum rate constant observable is dependent on the concentration of reactants. For example, the observed pseudo-first-order rate constant for a simple association reaction [Eq. (9)] is dependent on the second-order association rate constant, the concentration of the reactant in excess, and the first-order dissociation rate constant [Eq. (7)].

Another limitation of stopped-flow spectroscopy is that the workable concentration of substrates is dictated by the extinction coefficients and quantum yields. The enzyme and substrate concentrations must be large enough to produce an observable signal, yet the total absorbance of the sample must be low enough so as to not interfere with the observed fluorescence or absorbance signal (via the inner filter effect or deviations from the Beer–Lambert law[24]). Variation in the path length of the observa-

[23] K. A. Johnson, this series, Vol. 134, p. 677.

[24] J. R. Lakowicz, "Principles of Fluorescence Spectroscopy," p. 44. Plenum, New York, 1983.

tion cell in modern instruments (0.2–2.5 cm) allows some flexibility in the design of these experiments. Additionally, averaging of several reaction traces significantly increases the ratio of signal to noise (*S/N*). Although in many cases a fluorescence signal is more sensitive, this measurement has the disadvantage that it is difficult to obtain stoichiometric amplitude data. The development of a rapid scanning diode array stopped-flow spectrophotometer allows observation of absorbance at multiple wavelengths at a rate of about 800 spectra/sec. This technology can determine high-resolution difference spectra of transient intermediates and has been utilized for detecting and identifying reaction intermediates with pyridoxal 5′-phosphate-dependent enzymes.[25,26]

A second technique for obtaining nonequilibrium transient kinetic data is the quench-flow approach. This method has several advantages: an optical signal is not necessary for following the reaction; the chemical nature of intermediates and products can be probed directly by a variety of methods, including NMR spectroscopy[27,28]; and the amplitude of the reaction is more straightforward to determine. However, it requires more time and material since each reaction measures only a single time point in a kinetic curve, rather than determining the entire kinetic transient, as in stopped-flow spectroscopy.

Several commercial quench-flow instruments are available for use with biological materials that combine gentle mixing methods with either a minimal sample volume (20–100 μl/reaction)[6] for measurement of enzyme kinetics or a larger volume used for determination of protein folding kinetics.[29] The design of a quench-flow instrument[6] is reminiscent of the stopped-flow instrument. The reaction is started by pushing two or more solutions (maintained in separate syringes) through a mixing chamber. At this point the solution flows through a tube of defined length where it is allowed to react for various amounts of time before being mixed with a second solution, or quench, that stops the reaction. The reaction time is controlled by varying both the length of the reaction tube (i.e., the distance between the first and second mixing chambers) as well as the velocity of the liquid traveling through this tube. In this case the dead time of the

[25] R. S. Phillips, *Biochemistry* **30,** 5927 (1991).

[26] P. S. Brzović, Y. Sawa, C. C. Hyde, E. W. Miles, and M. F. Dunn, *J. Biol. Chem.* **267,** 13028 (1992).

[27] P. N. Barlow, R. J. Appleyard, B. J. O. Wilson, and J. N. S. Evans, *Biochemistry* **28,** 7985 (1989).

[28] K. S. Anderson, R. D. Sammons, G. D. C. Leo, J. A. Sikorski, A. J. Benesi, and K. A. Johnson, *Biochemistry* **29,** 1460 (1990).

[29] H. Roder, this series, Vol. 176, p. 446.

instrument (~2 msec) is determined by the minimum length of tubing needed to connect the two mixing chambers and the maximum flow velocity.

In chemical quench-flow experiments the choice of quench reagents is critical. To observe the correct composition of intermediates, the reaction must be quenched quickly relative to the observed rate constants ($k_{quench} \gg 700\ sec^{-1}$), and the materials must be stable under the conditions of the quench. In practical terms this means that the quench usually involves denaturation of the enzyme by acid or base since proton transfer reactions are fast.[30] However, metal chelators (such as EDTA) have been used to quench magnesium-catalyzed reactions, such as those catalyzed by ribozymes,[31,32] where the substrates are unstable under both basic and acidic conditions. Verification that the same intermediates are obtained using different quench conditions is an essential control. For example, a tetrahedral intermediate formed in the EPSP synthase-catalyzed reaction of shikimic 3-phosphate with phosphoenol pyruvate is observed only when quenched by base (either triethylamine or 0.2 *M* NaOH) since it is unstable in acidic conditions (0.2 *M* HCl).[22]

A variant of the chemical-quench experiment, called a pulse–chase experiment, can be used to measure rate constants for formation of noncovalent enzyme intermediates.[33] In this experiment the enzyme is mixed with radioactive substrate under pseudo-first-order conditions, and the reaction is quenched at various times with a large excess (≥10-fold) of either unlabeled substrate, product, or inhibitor. The quench stops additional reaction of the enzyme with radiolabeled substrate by either diluting the label or binding to and inhibiting the enzyme. However, noncovalent enzyme complexes are able to partition between dissociation (k_{-1}) and reaction to form products (k_2) under these conditions:

$$E + S \underset{k_{-1}}{\overset{k_1}{\rightleftharpoons}} E \cdot S \xrightarrow{k_2} E + P \tag{54}$$

Finally, after allowing the reaction to occur for 5–10 half-times, the reaction is stopped with a chemical quench and the products analyzed. If substrate dissociation is faster than the chemical transformation, the kinetics of product formation should be identical to those observed in the chemical quench. On the other hand, if substrate dissociation is slow, additional product formation should be observed, reflecting the noncova-

[30] M. Eigen and G. G. Hammes, *Adv. Enzymol.* **25,** 1 (1963).
[31] D. Herschlag and T. R. Cech, *Biochemistry* **29,** 10159 (1990).
[32] J. A. Beebe and C. A. Fierke, *Biochemistry* **33,** 10294 (1994).
[33] I. A. Rose, this series, Vol. 64, p. 47.

lent enzyme complexes [such as E · S in Eq. (54)]. The percentage of radiolabeled substrate observed as product is indicative of the ratio $k_2/(k_{-1} + k_2)$. The rate constant for formation of E · S can be calculated from the slope of a plot of k_{obs} versus concentration of substrate (or enzyme, depending on which is in excess).

Relaxation Methods

Rapid mixing methods have been predominantly used for transient kinetic studies of enzymes. However, these methods are inherently limited by how rapidly solutions can be mixed. This limitation is due to the hydrodynamics of mixing, and mixing times less than about 1 msec are difficult to achieve. Many of the elementary steps in enzymatic reactions require a shorter resolution time, which can be obtained with relaxation techniques.[5,9,10] With relaxation methods, the reactants are mixed and allowed to come to equilibrium. If the equilibrium is shifted by rapidly changing an external parameter, such as the temperature, the rate of change of the concentrations of the reactants to new equilibrium values permits the kinetics of the reactions occurring to be studied.

The temperature jump method has proved to be the most useful relaxation technique for studying enzymatic reactions. (Rapid mixing methods also can be used as relaxation techniques through concentration jumps, pH jumps, etc.) The equilibrium constant, K, for a chemical reaction changes with temperature, T, at constant pressure, according to the relationship

$$d \ln K/dT = \Delta H°/RT^2 \tag{55}$$

where $\Delta H°$ is the standard enthalpy change for the chemical reaction and R is the gas constant. The simplest way to change the temperature of an electrolyte solution rapidly is to discharge a high voltage across the solution. The energy discharged is $\frac{1}{2}CV^2$ where C is the capacitance used to store the high voltage and V is the voltage. For example, if an 0.1 μF capacitor is used to store 10,000 V, a temperature jump of 7.5° can be obtained in 0.15 ml. The time constant for the exponential discharge is $rC/2$ where r is the resistance of the solution.[34] Typically r is 100–200 ohms, which gives a discharge time constant of 5–10 μsec. An inert electrolyte, such as 0.1 M KNO_3, is used to achieve such low resistances.

[34] T. C. French and G. G. Hammes, this series, Vol. 16, p. 3.

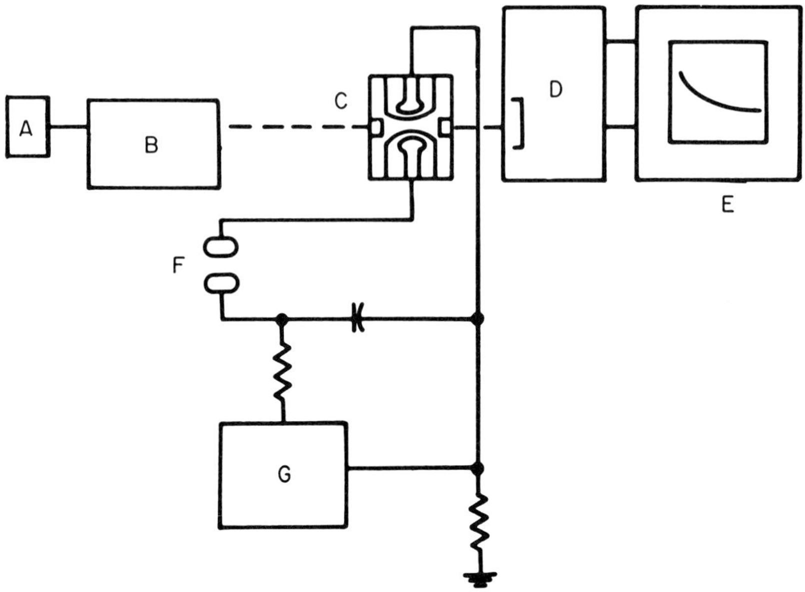

FIG. 4. Schematic diagram of a temperature jump apparatus utilizing absorption spectrophotometry for the detection of concentration changes. A, Light source; B, monochromator; C, observation cell; D, photomultiplier; E, oscilloscope and/or computer; F, spark gap; and G, high voltage. (From T. C. French and G. G. Hammes, this series, Vol. 16, p. 3, with permission.)

A schematic diagram of a temperature jump apparatus is shown in Fig. 4. The practical problems associated with building such an apparatus have been discussed elsewhere and are not presented here.[9,10,34] Although commercial equipment is not readily available, a temperature jump apparatus can be built relatively easily. The typical time resolution is a few microseconds although submicroseconds are feasible. Other methods have been used to obtain temperature jumps, such as dielectric heating and laser flashes, but have not proved as generally useful as a high-voltage discharge.[10,34]

External parameters other than temperature can be used to perturb chemical equilibria (e.g., pressure and electric field) and have been extensively utilized in nonenzymatic systems.[9,10,35] Chemical equilibria, however, are generally more sensitive to temperature changes than to pressure and electric field changes. Consequently, only temperature jumps have been extensively used to study enzyme mechanisms. In using this method,

[35] K. Kustin (ed.), this series, Vol. 16.

the enzymatic activity must be constantly monitored to be sure the enzyme is not denatured by the temperature/electric field jumps.

Relaxation methods are useful only if appreciable concentrations of intermediates, reactants, and/or products are present at equilibrium. Obviously, if a reaction goes essentially to completion (i.e., all products), no equilibria are present to perturb. However, relaxation methods also can be used to perturb steady states. Thus, for example, reactants might be rapidly mixed and a temperature jump applied to the reactants and intermediates during a steady state, even if the reaction ultimately goes to completion. This method is applicable for steady states with time constants longer than about 10 msec, with the temperature jump time resolution being a few microseconds. The methodology for carrying out such an experiment is somewhat difficult,[34,36] but a stopped-flow–temperature jump method has been used to study the ribonuclease reaction.[37]

Dihydrofolate Reductase

Dihydrofolate reductase (DHFR) catalyzes the reduction of 7,8-dihydrofolate (H_2F) by NADPH to form 5, 6, 7, 8-tetrahydrofolate (H_4F).[38–40] The enzyme maintains the intracellular pools of H_4F and derivatives, which are essential cofactors for the biosynthesis of thymidylate, purines, and several amino acids. Hence, DHFR is the target enzyme of both antitumor and antimicrobial drugs. Transient kinetics have been essential for investigation of the roles of amino acid residues in catalysis and inhibitor binding.[39,41] Complete kinetic schemes have been described for dihydrofolate reductases from a variety of sources, including *Escherichia coli*,[19,20] *Lactobacillus casei*,[42] mouse, [43] and human,[44,45] and they show remarkable similarities. Salient features of the kinetics of *E. coli* DHFR are described here.

[36] J. E. Erman and G. G. Hammes, *Rev. Sci. Instrum.* **37,** 746 (1966).

[37] J. E. Erman and G. G. Hammes, *J. Am. Chem. Soc.* **88,** 5607 (1966).

[38] R. L. Blakeley, *in* "Folates and Pterins" (R. L. Blakely and S. J. Benkovic, eds.), Vol. 2, p. 91. Wiley, New York, 1985.

[39] S. J. Benkovic, C. A. Fierke, and A. M. Naylor, *Science* **239,** 1105 (1988).

[40] R. A. Brown and J. Kraut, *Faraday Discuss. Chem. Soc.* **93,** 217 (1992).

[41] K. Taira, C. A. Fierke, J.-T. Chen, K. A. Johnson, and S. J. Benkovic, *Trends Biochem. Sci.* **12,** 275 (1987).

[42] J. Andrews, C. A. Fierke, B. Birdsall, G. Ostler, J. Feeney, G. C. K. Roberts, and S. J. Benkovic, *Biochemistry* **28,** 5743 (1989).

[43] J. Thillet, J. A. Adams, and S. J. Benkovic, *Biochemistry* **29,** 5195 (1990).

[44] J. R. Appleman, W. A. Beard, T. J. Delcamp, N. J. Prendergast, J. H. Freisheim, and R. L. Blakely, *J. Biol. Chem.* **265,** 2740 (1990).

[45] J. R. Appleman, W. A. Beard, T. J. Delcamp, N. J. Prendergast, J. H. Freisheim, and R. L. Blakely, *J. Biol. Chem.* **264,** 2625 (1989).

Determination of the association (k_1) and dissociation (k_{-1}) rate constants for substrate and product was particularly illuminating for the kinetic mechanism of dihydrofolate reductase. The rate of binding ligands to DHFR can be measured under conditions of excess ligand by following either the quenching of intrinsic enzyme fluorescence or fluorescence energy transfer between the enzyme and NADPH.[19,46,47] In the formation of binary complexes of DHFR two exponentials of equal amplitude are observed, consistent with a mechanism in which substrate (L_1, L_2) binds rapidly to only one of two enzyme conformers (E_1 and E_2) and interconversion between these conformers is slow:

$$\begin{array}{c} E_1 + L_1 \underset{k_{-1}}{\overset{k_1}{\rightleftharpoons}} E_1 \cdot L_1 + L_2 \underset{k'_{-1}}{\overset{k'_1}{\rightleftharpoons}} E_1 \cdot L_1 \cdot L_2 \\ k_{-2} \Big\downarrow\Big\uparrow k_2 \qquad\qquad\qquad\qquad\qquad\qquad \\ E_2 \qquad\qquad\qquad\qquad\qquad\qquad\qquad \end{array} \tag{56}$$

The fast exponential phase is primarily due to the rate of formation and dissociation of the binary complex, $E_1 \cdot L_1$, whereas the slow phase reflects the interconversion of E_1 and E_2. The association and dissociation rate constants can be determined from the slope and intercept, respectively, of a plot of the observed pseudo-first-order rate constants characterizing the ligand binding reaction versus ligand concentration [Eq. (7a)]. Typical association and dissociation rate constants determined using this technique are listed in Table I.[19]

A single, ligand-dependent exponential is observed in the formation of ternary complexes from binary complexes. The binding rate constants for formation of the reactive $E \cdot NADPH \cdot H_2F$ ternary complex are estimated from $E \cdot I \cdot S$ complexes where I is methotrexate for binding NADPH and either $NADP^+$ or thio-NADPH for binding H_2F (Table I). More direct measurements using a pulse–chase technique[33] would be useful since these inhibitors bind differently than substrate.[48,49]

Dissociation rate constants can also be measured from the fluorescence change observed when an enzyme–ligand complex ($E \cdot L_1$) is mixed with a large excess of a second ligand (L_3) that competes for the binding site:

$$E \cdot L_1 \underset{k_{-1}}{\overset{k_1}{\rightleftharpoons}} E + L_1 \underset{k''_{-1}}{\overset{k''_1[L_3]}{\rightleftharpoons}} E \cdot L_3 \tag{57}$$

[46] S. M. J. Dunn and R. W. King, *Biochemistry* **19,** 766 (1980).

[47] P. J. Cayley, S. M. J. Dunn, and R. W. King, *Biochemistry* **20,** 874 (1981).

[48] C. Bystroff, S. J. Oakley, and J. Kraut, *Biochemistry* **29,** 3263 (1990).

[49] M. A. McTigue, J. F. Davie II, B. T. Kaufman, and J. Kraut, *Biochemistry* **32,** 6855 (1993).

TABLE I
ASSOCIATION AND DISSOCIATION RATE CONSTANTS FOR LIGANDS BINDING TO *Escherichia coli* DIHYDROFOLATE REDUCTASE[a]

Ligand	Enzyme species	k_{on}[b,c] (μM^{-1} sec^{-1})	k_{off}[b,c] (sec^{-1})	k_{off}[b,d] (sec^{-1})
NADPH	E	20	3.5	3.6
	E · H_4F	8	85	85
	E · MTX[e]	20	≤2	—
H_2F	E	42	47	22
	E · TNADPH[e]	25	40	43
	E · $NADP^+$	26	10	7
$NADP^+$	E	13	300	290
	E · H_2F	5	60	50
	E · H_4F	—	—	200
H_4F	E	24	≤1	1.4
	E · $NADP^+$	28	5	2.4
	E · NADPH	2	10	12

[a] Measured at pH 6.0, 25°. Reprinted with permission from C. A. Fierke, K. A. Johnson, and S. J. Benkovic, *Biochemistry* **26,** 4085 (1987). Copyright 1987 American Chemical Society.

[b] Here, k_{on} is k_1 or k'_1 and k_{off} is k_{-1} or k'_{-1} [Eq. (56)], depending on whether a binary or ternary complex is formed.

[c] The values of k_{on} and k_{off} are determined from the slope and intercept, respectively, of a plot of k_{obs} for the formation of E · L versus [L].

[d] Determined from mixing E · L_1 with excess competitive ligand (L_3) to form E · L_3 [Eq. (57)].

[e] MTX, Methotrexate; TNADPH, thio-NADPH.

To measure k_{-1} accurately, the effective rate constant for binding the second ligand ($k''_1[L_3]$) must be significantly faster than $k_1[L_1]$, k_{-1}, and k''_{-1}. Experiments of this type (see Table I) demonstrate the following[19]: (1) association rate constants for the binding of all ligands are near the diffusion-controlled limit, about $2 \times 10^7\ M^{-1}\ \text{sec}^{-1}$; (2) dissociation of tetrahydrofolate is the rate-limiting step for steady-state turnover; and (3) NADPH binding increases the rate constant for dissociation of H_4F, leading to the following preferred pathway of product dissociation:

$$\text{E}\cdot\text{NADP}+\cdot\text{H}_4\text{F} \xrightarrow{\nearrow \text{NADP}^+} \text{E}\cdot\text{H}_4\text{F} \xrightarrow[\text{NADPH}\nearrow]{} \text{E}\cdot\text{NADPH}\cdot\text{H}_4\text{F} \xrightarrow{\nearrow \text{H}_4\text{F}} \text{E}\cdot\text{NADPH} \quad (58)$$

At saturating substrate concentrations, the kinetic mechanism of DHFR can be simplified to two steps, namely, reduction of H_2F by NADPH to form products followed by dissociation of H_4F:

$$\mathrm{E \cdot NADPH \cdot H_2F \xrightarrow{k_2} E \cdot NADP^+ \cdot H_4F \xrightarrow[\text{NADPH}]{\text{NADP}^+}}$$

$$\mathrm{E \cdot NADPH \cdot H_4F \xrightarrow{k_3} E \cdot NADPH + H_4F} \quad (59)$$

Equilibration between the $E \cdot NADP^+ \cdot H_4F$ and $E \cdot NADPH \cdot H_4F$ species is rapid at saturating NADPH. Because H_4F dissociation is slow ($k_2 \gg k_3$),[19] a burst of disappearance of NADPH may be observed during the first turnover by measuring either ultraviolet absorbance, fluorescence, or fluorescence energy transfer (Fig. 5). The fluorescence energy transfer experiments provide the highest signal-to-noise ratio, and therefore the kinetic measurements were done using this technique. However, the absolute amplitude of the burst was determined from absorbance data. At low pH the rate constant of the burst is about 950 sec^{-1}, with an amplitude approaching 1 mol NADPH/mol DHFR. The large fluorescence energy transfer signal allows measurement of this fast rate constant even though more than half of the signal disappears within the dead time of the instrument. This observed rate constant is independent of the substrate concentration, indicating that association is not the rate-limiting step. Furthermore, a kinetic deuterium isotope effect (k_{NADPH}/k_{NADPD}) of 3 on the rate constant of the burst indicates that this fluorescence signal directly reflects hydride transfer from NADPH to H_2F (Fig. 5). If the rate constant for hydride transfer is significantly faster than product dissociation, it can be

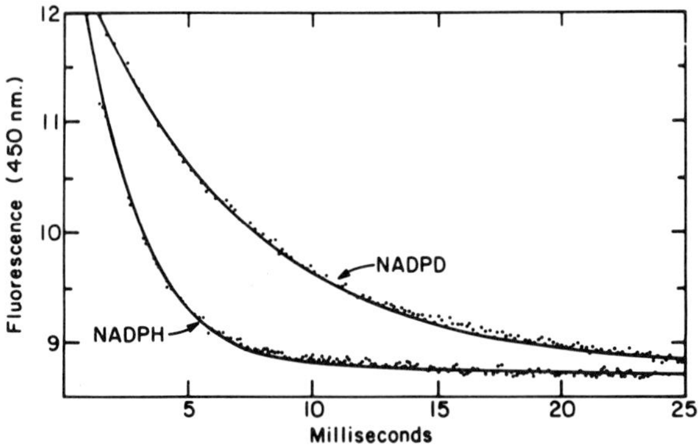

FIG. 5. Measurement of the pre-steady-state transient by stopped-flow fluorescence energy transfer for the reaction of DHFR with substrates. The enzyme was preincubated with either NADPH or [4′ (*R*) $-^2$H]NADPD, and the reaction was initiated by addition of H_2F. Final conditions were as follows: 15 μM DHFR, 125 μM NADPH(D), 100 μM H_2F, pH 6.5, 25°. The data were fit by a single exponential decay followed by a linear, steady-state rate, k_{cat} = 12 sec^{-1}. [Reprinted with permission from C. A. Fierke, K. A. Johnson, and S. J. Benkovic, *Biochemistry* **26**, 4085 (1987). Copyright 1987 American Chemical Society.]

determined directly by fitting the burst data to a single exponential followed by a linear rate. If not, simulation of the pre-steady-state data to more complicated models is required.[19]

Both the rate constant and the amplitude of the pre-steady-state burst show a marked pH dependence.[19] At low pH the observed rate constant decreases, with a pK of 6.5 characterizing ionization of the ternary E · NADPH · H_2F complex, while the burst amplitude is invariant. These data indicate that the proton transfer component of the reaction is fast relative to hydride transfer. The pK of approximately 8 observed in the burst amplitude reflects the pH at which the observed rate constant for hydride transfer equals the rate constant for H_4F dissociation.

The mechanism of DHFR was completed by determining the overall equilibrium constant for the reaction, and by measuring the rate constant for hydride transfer from H_4F to $NADP^+$. In the *E. coli* enzyme, hydride transfer is the rate-limiting step for steady-state turnover in the reverse direction ($H_4F + NADP^+ \rightarrow H_2F + NADPH$).[19] However, in mouse DHFR dissociation of NADPH and/or H_2F is the rate-limiting step in this direction so that the rate constant for hydride transfer must be determined from pre-steady-state transients.[43]

A necessary requirement for a kinetic mechanism derived using transient kinetic techniques is that it predict the observed steady-state behavior. The proposed kinetic mechanism for *E. coli* DHFR predicts the observed steady-state kinetic parameters (k_{cat}, k_{cat}/K_m, and K_m) in both directions as well as the pH dependence and isotope effects.[19] Furthermore, this model also predicts full time course kinetic curves.[19,20] Finally, this mechanism is essential for understanding the effects of single amino acid variants in DHFR and allows evaluation of the importance of a given amino acid for both binding ligands and catalyzing hydride transfer.

Ribonuclease P

The ribonucleoprotein complex ribonuclease P (RNase P) catalyzes an essential step in tRNA maturation, namely, the cleavage of precursor-tRNA (pre-tRNA) to produce a mature 5′ terminus.[50–52] In diverse organisms the enzyme is composed of two subunits; one is RNA (about 400 nucleotides) and the other is protein (about 120 amino acids). Although both are essential for *in vivo* activity, the RNA component from bacterial RNase P is sufficient to catalyze specific cleavage of pre-tRNA *in vitro*

[50] S. C. Darr, J. W. Brown, and N. R. Pace, *Trends Biochem. Sci.* **17,** 178 (1992).
[51] S. Altman, *J. Biol. Chem.* **265,** 20053 (1990).
[52] N. R. Pace and D. Smith, *J. Biol. Chem.* **265,** 3587 (1990).

in the presence of high salt. Transient kinetics approaches have been useful for investigating the catalytic mechanism of RNase P[32,53,54] as well as the *Tetrahymena*[31] and hammerhead ribozymes.[55] A minimal kinetic description of a single turnover of RNase P RNA-catalyzed tRNA maturation includes (i) binding of pre-tRNA to RNase P; (ii) cleavage of the phosphodiester bond generating a 5′ product with a 3′-hydroxyl terminus and a 3′-tRNA product with a 5′-phosphate terminus[50,51]; and (iii) independent dissociation of both products.

The time course for hydrolysis of pre-tRNA catalyzed by the RNA component of RNase P (RNase P RNA) is characterized by two phases: a burst of tRNA product at short times followed by a linear increase in the concentration of tRNA[32,53] (see Fig. 3). The data are consistent with a two-step mechanism in which hydrolysis occurs in the first step followed by regeneration of the active catalyst in the second step. The size of the pre-steady-state burst is dependent on the RNase P RNA concentration, with an amplitude of approximately 0.85 mol tRNA/mol RNase P RNA. This indicates that at least 85% of the enzyme molecules are catalytically competent and that the rate constant of the first step is at least 10-fold faster than that of the second step. Assuming two consecutive first-order (or pseudo-first-order) reactions, the amplitude and observed rate constant of the burst are dependent on both k_1 and k_2, as in Eq. (53).

To investigate whether product dissociation is the rate-limiting step for steady-state turnover catalyzed by RNase P RNA, the rate and equilibrium constants for product binding were measured. Bound and free product were resolved by gel shift or gel filtration centrifuge column chromatography.[32,56] These experiments demonstrate the following: (i) association rate constants are quite large, about 5×10^6 M^{-1} $\sec^{-1}$; (ii) dissociation of tRNA is the rate-limiting step for steady-state turnover with a rate constant of 0.013 $\sec^{-1}$; and (iii) dissociation of the 5′ fragment is rapid.

The hydrolytic cleavage step catalyzed by the RNA component of RNase P can be isolated by measuring a single turnover in the presence of excess enzyme ([E]/[S] $\geq$ 5). Under these conditions product dissociation is not observable since E · tRNA and tRNA are indistinguishable on the denaturing polyacrylamide gels used for assaying cleavage. The reaction is initiated by mixing RNase P RNA and pre-tRNA in a quench-flow instrument and is quenched by the addition of either EDTA or acid.[32] At low concentrations of RNase P RNA, the appearance of tRNA product

[53] C. Reich, G. J. Olsen, B. Pace, and N. R. Pace, *Science* **239,** 178 (1988).

[54] A. Tallsjo and L. A. Kirsebom, *Nucleic Acids Res.* **21,** 51 (1993).

[55] M. J. Fedor and O. C. Uhlenbeck, *Biochemistry* **31,** 12042 (1992).

[56] W.-D. Hardt, J. Schlegl, V. A. Erdmann, and R. K. Hartmann, *Nucleic Acids Res.* **21,** 3521 (1993).

can be fit by a single first-order exponential with the rate constant linearly dependent on the concentration of RNase P RNA (Fig. 6A). However, at higher concentrations there is a lag in product formation, and the observed rate constant is independent of the concentration of RNase P RNA (Fig. 6B). The data of Fig. 6B are best explained by two consecutive first-order irreversible reactions [Eq. (46)] where the first step is association of RNase P RNA and pre-tRNA to form a binary complex, $k_1 \approx 5 \times 10^6$ M^{-1} sec^{-1}, and the second step is the chemical cleavage step, $k_2 = 6$ sec^{-1}:

$$E + \text{pre-tRNA} \xrightarrow{k_1} E \cdot \text{pre-tRNA} \xrightarrow{k_2} E \cdot \text{tRNA} \cdot P \quad (60)$$

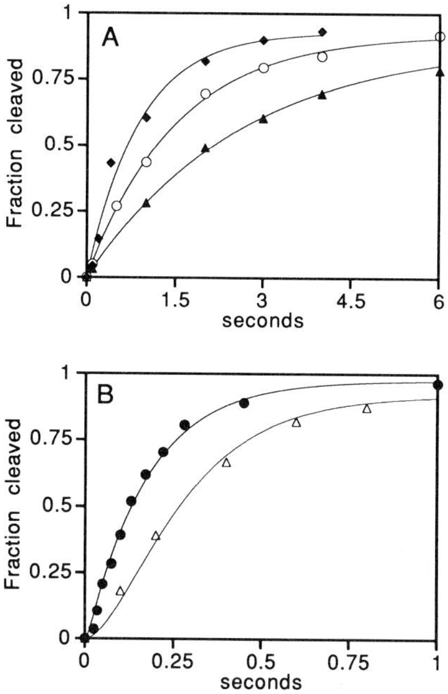

FIG. 6. Single turnover measurements of the hydrolysis of pre-tRNA^{Asp} catalyzed by the RNA component of *Bacillus subtilis* RNase P. Pre-tRNA^{Asp} (24 nM) was mixed with varying concentrations of excess RNAse P RNA in a quench-flow instrument at 37° in 100 mM $MgCl_2$, 800 mM NH_4Cl, 50 mM Tris, pH 8.0, 0.05% Nonidet P-40, and 0.1% sodium dodecyl sulfate. The reaction was quenched by a 3-fold dilution into 90 mM EDTA followed by addition of urea to 5 M. (A) At low concentrations of RNase P RNA [0.12 μM (▲), 0.24 μM (○), and 0.45 μM (◆)], the data are fit by a single exponential [Eq. (4)]. (B) At high concentration of RNase P RNA [1.4 μM (△) and 19 μM (●)], the data are fit to a mechanism of two consecutive first-order reactions [Eq. (46)] with $k_1 = 6 \times 10^6$ M^{-1} sec^{-1}[RNase P RNA]$_0$ and $k_2 = 6$ sec^{-1}. [Reprinted with permission from J. A. Beebe and C. A. Fierke, *Biochemistry,* **33,** 10294 (1994). Copyright 1994 American Chemical Society.]

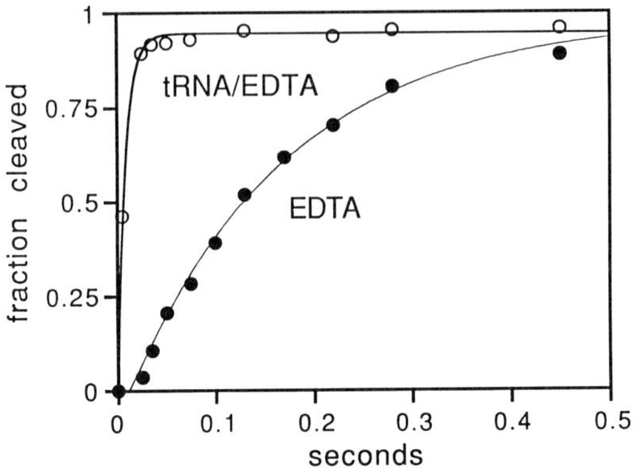

FIG. 7. Formation of an RNase P RNA · pre-tRNA^{Asp} binary complex assayed by a cold tRNA trap. Pre-tRNA^{Asp} (24 n*M*) was mixed with excess RNase P RNA (19 μM) in a quench-flow instrument at 37° in 100 m*M* $MgCl_2$, 800 m*M* NH_4Cl, 50 m*M* Tris, pH 8.0, 0.05% Nonidet P-40, and 0.1% sodium dodecyl sulfate. The reaction was quenched by a 3-fold dilution into either (●) 90 m*M* EDTA followed by 5 *M* urea or (○) 58 μM tRNA followed by addition of 90 m*M* EDTA and 5 *M* urea after 10 sec. The cold tRNA trap data are fit by a single exponential [Eq. (4)], whereas the EDTA data are fit to a mechanism of two consecutive first-order reactions [Eq. (46)], $k_1 = 6 \times 10^6\ M^{-1}\ \text{sec}^{-1}$[RNase P RNA$)_0$ and $k_2 = 6\ \text{sec}^{-1}$. [Reprinted with permission from J. A. Beebe and C. A. Fierke, *Biochemistry,* **33,** 10294 (1994). Copyright 1994 American Chemical Society.]

Measurement of the effect of thiophosphate substitution at the cleavage site on k_2 would verify that it directly reflects hydrolysis.[57] Furthermore, at high RNase P RNA concentration more than 95% of the pre-tRNA is hydrolyzed in the first turnover, indicating that either the equilibrium for hydrolysis is very favorable or the dissociation rate constant of the 5′ fragment is significantly faster than religation. This sets an upper limit for the rate constant for religation.

The model predicts that at high RNase P RNA concentrations a significant amount of the E · pre-tRNA binary complex forms and does not exchange readily with ligands in solution. This can be tested by stopping the reaction with the addition of a high concentration of tRNA before the EDTA quench.[32] In this case, any intermediate enzyme–substrate complexes which do not dissociate rapidly should react to form products before the reaction is quenched with EDTA; this is observed as additional product formation in the tRNA trap (Fig. 7). The tRNA trap data can be fit by a single exponential with a rate constant equal to k_1[RNase P RNA], where k_1 is similar to the enzyme-dependent rate constant observed in the EDTA quench (Fig. 6A). Furthermore, since at least 95% of the

[57] D. Herschlag, J. A. Piccirilli, and T. R. Cech, *Biochemistry* **30,** 4844 (1991).

radiolabeled pre-tRNA is observed as labeled tRNA product in the tRNA trap experiment, hydrolysis of pre-tRNA is at least 20-fold faster than dissociation.[33]

This set of experiments allowed development of a scheme which describes the kinetic behavior of a single turnover and simulates the steady-state turnover number, k_{cat}. However, it predicts that the steady-state kinetic parameter measuring the second-order rate constant for hydrolysis of pre-tRNA catalyzed by RNase P RNA at low substrate concentration, k_{cat}/K_m, should be about 30-fold faster than the observed rate constant.[53] Because the single turnover experiments measure the reaction up to product formation, the only model consistent with all the data is that an additional enzyme species (E′) is produced during the first turnover that binds substrate significantly more slowly and does not equilibrate rapidly with the original conformer (E). This different enzyme species could vary in RNA conformation, bound metals, or bound products. This example clearly illustrates the utility of testing whether a kinetic mechanism determined from transient kinetic methods is sufficient to explain the observed steady-state behavior.

These experiments have determined a minimal kinetic scheme for the hydrolysis of pre-$tRNA^{Asp}$ catalyzed by the RNA component of *Bacillus subtilis* RNase P under high salt conditions.[32] This scheme will allow pinpointing of the role of metals, pH, and the protein component on each step of the reaction. Furthermore, it is an essential background for comprehending the effect of structural changes in both the substrate and enzyme on catalytic activity. Additional studies should lead to a greater understanding of the chemistry and mechanisms of rate acceleration utilized by ribozymes.

Fatty-Acid Synthase

Chicken liver fatty-acid synthase is a multifunctional enzyme that catalyzes the synthesis of palmitic acid according to the following overall reaction:

$$\text{Acetyl-CoA} + 7\ \text{malonyl-CoA} + 14\ \text{NADPH} + 14\ H^+ \rightarrow \text{palmitic acid} + 8\ \text{CoA} + 14\ \text{NADP}^+ + 6\ H_2O + 7\ CO_2 \quad (61)$$

The enzyme has been extensively reviewed,[58-60] and only some aspects of transient kinetic studies are discussed here. The enzyme consists of

[58] S. J. Wakil, J. K. Stoops, and V. C. Joshi, *Annu. Rev. Biochem.* **52,** 537 (1983).
[59] S. J. Wakil, *Biochemistry* **28,** 4523 (1989).
[60] S.-I. Chang and G. G. Hammes, *Acct. Chem. Res.* **23,** 363 (1990).

two identical polypeptide chains and two independent catalytic sites composed of two different polypeptide chains. During the catalysis, reaction intermediates are covalently bound to three different sites, a serine hydroxyl, 4′-phosphopantetheine, and a cysteine sulfhydryl. The sequence of reactions in a catalytic cycle is shown in Fig. 8.

The reaction is initiated by nucleophilic attack of serine on acetyl-CoA and formation of an acetyl oxyester; the acetyl moiety is then passed sequentially to 4′-phosphopantetheine and cysteine, where thioesters are formed. Subsequently the enzyme binds malonyl-CoA, which reacts with serine, and a malonyl residue is passed from serine to 4′-phosphopantetheine. The acetyl then condenses onto the malonyl, with the release of carbon dioxide, to form a ketothioester. The ketothioester is reduced to an alcohol by NADPH, then dehydrated, and the double bond is reduced by NADPH to give a saturated four-carbon thioester (Fig. 9). The saturated acid is then transferred to the cysteine from 4′-phosphopantetheine, malonyl is loaded onto the enzyme, and the reaction sequence is repeated until

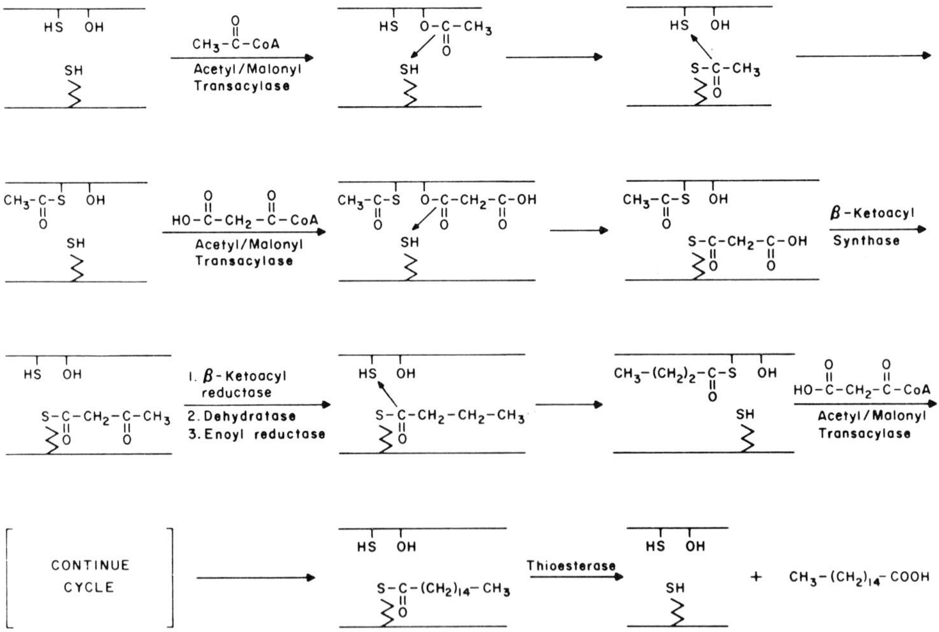

FIG. 8. Schematic representation of palmitic acid synthesis by chicken liver fatty-acid synthase. The symbol ⌇SH represents 4′-phosphopantetheine, —SH is a cysteine sulfhydryl, and —OH is a serine hydroxyl. The initial sequence of reactions, involving malonyl transacylase, β-ketoacyl synthase, β-ketoacyl reductase, dehydratase, and enoyl reductase, is repeated seven times to give enzyme-bound palmitic acid.

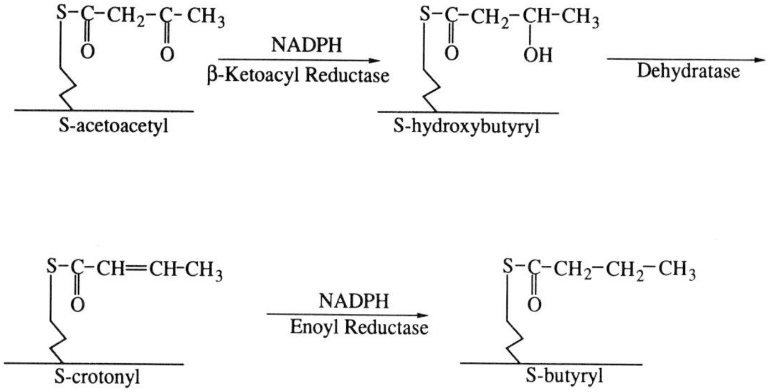

FIG. 9. Reduction of enzyme-bound acetoacetate to butyrate by fatty-acid synthase. The cycle is repeated for each addition of malonate to the growing fatty acid chain. Here S is part of 4′-phosphopantetheine that is covalently coupled to the enzyme.

the thioester of palmitic acid is present on the 4′-phosphopantetheine. The thioester is cleaved by a thioesterase activity to release palmitic acid. The separate activities are acetyl/malonyl transacylase, β-ketoacyl synthase, β-ketoacyl reductase, dehydratase, enoyl reductase, and palmitoyl thioesterase.

Transient kinetic methods have been used to investigate the individual steps in the first catalytic cycle. The reaction of acetyl-CoA with the enzyme was studied with a quench-flow apparatus utilizing radioactive acetyl, with perchloric acid as the quench.[61] The kinetics were first order because the acetyl-CoA concentration was much greater than that of the enzyme. The apparent first-order rate constant increased with increasing acetyl-CoA concentration, ultimately leveling off at high concentrations. This is indicative of a first-order reaction preceded by a binding equilibrium. Because this reaction is slightly reversible, the analysis of the kinetics also must take into account dissociation of the acetylated enzyme; however, this complication is not considered here. The deacetylation of the enzyme by reaction with CoA was studied independently with the quench-flow approach. The overall process was fit to the following mechanism:

$$\text{AcCoA} + \text{E} \overset{K_1}{\rightleftharpoons} \text{E} \cdot \text{AcCoA} \underset{k_{-2}}{\overset{k_2}{\rightleftharpoons}} \text{EAc} \cdot \text{CoA} \overset{K_3}{\rightleftharpoons} \text{EAc} + \text{CoA} \quad (62)$$

At pH 7.0, 0.1 M potassium phosphate, 23°, the equilibrium dissociation constants, K_1 and K_3, are estimated to be 85 and 70 μM, respectively,

[61] J. A. H. Cognet and G. G. Hammes, *Biochemistry* **22,** 3002 (1983).

and the rate constants, k_2 and k_{-2}, are 43 and 103 sec^{-1}. Note that the rate constant for deacetylation of the enzyme is larger than the acetylation rate constant. However, the deacetylation does not interfere with the formation of palmitic acid under physiological conditions owing to the low concentration of CoA.

Although the observed acetylation kinetics are deceptively simple, the chemistry is actually complex because the acetyl group has three different binding sites on the enzyme. Therefore, the apparent first-order rate constant characterizes the rate of acetyl transfer from enzyme-bound CoA to three different sites and/or the rate of internal transfer between the sites. The rate of acetylation of individual sites can be determined by specifically blocking either the cysteine or cysteine and 4′-phosphopantetheine through chemical modification of the protein.[62] The results obtained with both sulfhydryls blocked demonstrate that formation of the oxyester is relatively rapid, $k_2 = 150\ sec^{-1}$. If the transfer between sites is strictly ordered (Fig. 8), then results with other modified enzymes indicate that intramolecular transfer of acetyl from serine to 4′-phosphopantetheine has a rate constant of less than 110 sec^{-1} and that for transfer to cysteine is less than 43 sec^{-1}. (If binding to the three sites is random, these rate constants represent lower bounds for the transfer of acetyl from enzyme-bound CoA to each of the binding sites.) Similar studies of the binding of malonyl to the enzyme have not been carried out owing to the expense of radioactive substrate and the complications of side reactions.

Kinetic investigation of the next step in the reaction sequence, namely, the condensation of acetyl and malonyl, is problematic because spectral changes specific to this reaction do not occur and isolation of the reaction product is difficult. However, the binding of NADPH to the enzyme is accompanied by a large enhancement in the fluorescence of NADPH. The kinetics of binding is consistent with a bimolecular reaction characterized by a second-order rate constant of $1.27 \times 10^7\ M^{-1}\ sec^{-1}$ at pH 7.0, 25°.[11] If the substrate concentrations, acetyl-CoA, malonyl-CoA, and NADPH, are large relative to the enzyme concentration, the enzyme will cycle through the entire reaction, and determination of specific rate constants is not possible. If the concentration of NADPH is much less than that of the enzyme, however, the reaction will stop with formation of the 3-hydroxybutyryl/crotonyl– or butyryl–enzyme intermediate (Fig. 9). All that remains to be done is to adjust the substrate reactions so that the kinetics for the formation of the acetoacetyl and 3-hydroxybutyryl/crotonyl intermediates are pseudo first order.

[62] Z. Y. Yuan and G. G. Hammes, *J. Biol. Chem.* **260,** 13532 (1985).

If the concentrations of acetyl-CoA and malonyl-CoA are large relative to the enzyme concentration, the formation of acetoacetyl–enzyme, EAcAc, is first order with respect to the enzyme E:

$$d(\mathrm{EAcAc})/dt = k_1(\mathrm{E}) \tag{63}$$

where k_1 is a function of the concentrations of acetyl-CoA and malonyl-CoA owing to the binding equilibria of the substrates with the enzyme. Integration of Eq. (63) gives

$$(\mathrm{EAcAC}) = (\mathrm{E})_0\,(1 - e^{-k_1 t}) \tag{64}$$

where $(\mathrm{E})_0$ is the enzyme concentration at $t = 0$.

If the formation of the acetoacetyl–enzyme is coupled to the subsequent β-ketoacyl reductase reaction under conditions where the enzyme concentration is much greater than that of NADPH, the reduction is first order with respect to NADPH:

$$\begin{aligned} -d(\mathrm{NADPH})/dt &= k_2(\mathrm{EAcAc})(\mathrm{NADPH}) \\ &= k_2(\mathrm{E})_0(1 - e^{-k_1 t})(\mathrm{NADPH}) \end{aligned} \tag{65}$$

Integration of Eq. (65) gives

$$(\mathrm{NADPH}) = (\mathrm{NADPH})_0 \exp\{-k_2(\mathrm{E})_0[t + (1/k_1)(e^{-k_1 t} - 1)]\} \tag{66}$$

where $(\mathrm{NADPH})_0$ is the NADPH concentration at $t = 0$. Note that $k_2(\mathrm{E})_0$ is a pseudo-first-order rate constant that is a function of the total enzyme concentration because of the rapid binding equilibrium between enzyme and NADPH. This analysis depends on several assumptions: (i) binding of acetyl-CoA, malonyl-CoA, and NADPH is rapid relative to the formation of acetoacetyl– and 3-hydroxybutyryl–enzyme; and (ii) further reaction of 3-hydroxybutyryl does not interfere with the observed kinetics. All these assumptions have been shown to be valid.[11]

The time course for the reduction of NADPH should display a lag as acetoacetyl–enzyme is formed, followed by a first-order decay with rate constant $k_2(\mathrm{E})_0$. A typical time course is shown in Fig. 10. If the binding equilibria are taken into account by extrapolating the pseudo-first-order rate constants to the approximate limits at high substrate concentrations, the rate constants for acetoacetyl– and 3-hydroxybutyryl–enzyme formation are 30.9 and 17.5 $\sec^{-1}$, respectively, at pH 7.0, 0.1 *M* potassium phosphate, 25°.

The reduction of acetoacetyl–enzyme can be independently studied because this intermediate can be directly formed by reaction of acetoacetyl-CoA with the enzyme.[63] This intermediate can then be reacted directly

[63] J. A. H. Cognet and G. G Hammes, *Biochemistry* **24,** 290 (1985).

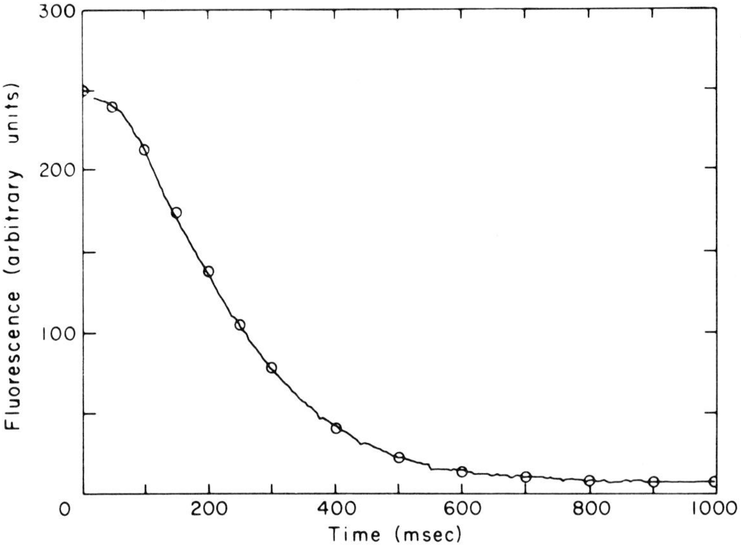

FIG. 10. Typical stopped-flow kinetic trace of the change in fluorescence following mixing of fatty-acid synthase (2.14 μM)–NADPH (0.75 μM) with acetyl-CoA (56 μM) and malonyl-CoA (98 μM) in 0.1 M potassium phosphate buffer–1 mM EDTA, pH 7.0, at 25°. The circles have been calculated with the best fit parameters $k_1 = 4.3\ \mathrm{sec}^{-1}$, $k_2 = 9.5\ \mathrm{sec}^{-1}$ [Eq. (66)]. [Reprinted with permission from J. A. H. Cognet, B. G. Cox, and G. G. Hammes, *Biochemistry* **22,** 6281 (1983). Copyright 1983 American Chemical Society.]

with NADPH, with (NADPH) $\gg$ (E) so the reaction is pseudo first order. The first-order rate constant obtained when the enzyme is saturated with NADPH is 20 sec^{-1}, in good agreement with the result from the coupled reaction. Similarly the reduction of the hydroxybutyryl/crotonyl–enzyme by NADPH can be studied directly. The rate constant obtained is 36.6 sec^{-1}.[63] In both cases, the dependence of the pseudo-first-order rate constant on the concentration of NADPH was determined in order to obtain the first-order rate constants cited. Unfortunately no method has been devised to study the kinetics of the interconversion of the hydroxybutyryl and crotonyl intermediates. In fact, these intermediates are in equilibrium on the enzyme so that the reaction of NADPH is with an equilibrium mixture.

The studies described above have permitted delineation of most of the rate constants for the first catalytic cycle. However, the rate constants may change as the fatty acid chain length of the covalently bound intermediate increases. An assessment of how the rate constants vary with chain length can be obtained by examining the distribution of chain lengths covalently bound during catalysis.[64] Because this chapter is primarily concerned with transient kinetics, these studies are not described here. The results suggest that the rate constants for the condensation and reduction steps become larger as the chain length increases, by approximately a factor of two per cycle.

Finally the turnover number for the thioesterase has been estimated as about 2 sec^{-1} with palmitoyl-CoA as the substrate.[65]

Although the mechanism of action of fatty-acid synthase is quite complex, transient kinetics have been used to elucidate many of the elementary steps in the mechanism. These studies illustrate the experimental design and interpretation that are often needed for transient kinetic investigations.

Conclusion

This chapter has presented a description of transient kinetic approaches to the study of enzyme mechanisms. In particular, transient kinetic analysis of three enzymes, DHFR, RNase P, and fatty-acid synthase, demonstrates that determination of the individual rate constants in an enzyme pathway provides novel insights into enzyme mechanisms, particularly if chemical transformation is not rate limiting for steady-state turnover. Transient kinetic analysis is an essential method for determination of enzyme mechanism because of the vast wealth of unique information it provides. Although this chapter could not cover all possible kinetic mechanisms, it provides a useful blueprint for the utilization of these methods to elucidate the molecular bases of enzyme catalysis.

Acknowledgments

We thank Teaster Baird, Michael Been, Shawn Zinnen, Jane Beebe, and Lora LeMosy for help with the preparation of this review and the National Institutes of Health (GM40602) and the Office of Naval Research (C.A.F.) for support of this work. C.A.F. received an American Heart Association Established Investigator Award and a David and Lucile Packard Foundation Fellowship in Science and Engineering.

[64] V. E. Anderson and G. G. Hammes, *Biochemistry* **24,** 2147 (1985).

[65] W. Liu and G. G Hammes, unpublished results (1988).

[2] Rapid Quench Kinetic Analysis of Polymerases, Adenosinetriphosphatases, and Enzyme Intermediates

By KENNETH A. JOHNSON

Introduction

Transient-state kinetic methods allow definition of the sequence of reactions occurring at the active site of an enzyme following substrate binding and leading up to product release. It is precisely these reaction steps that define the elementary steps accounting for the high fidelity of DNA polymerization or establishing the pathway for coupling of ATP hydrolysis to energy transduction by molecular motors. In each case, analysis of the reactions occurring at the active site of the enzyme by single turnover kinetic methods has allowed definition of the reaction sequence.[1-4] On the other hand, steady-state kinetic analysis only establishes the order of substrate binding and the order of product release; the steady-state kinetic parameters k_{cat} and k_{cat}/K_m only define the maximum rate of substrate to product conversion and a lower-limit estimate of the rate of substrate binding, respectively. Steady-state kinetic analysis cannot address questions regarding the pathway of events occurring at the active site following substrate binding and prior to product release.[1]

This chapter summarizes the methods of kinetic analysis used to define the sequence and rates of reactions catalyzed by DNA polymerases and by force-transducing adenosinetriphosphatases (ATPases) and gives a brief review of the detection of enzyme intermediates. The methods are of general utility in studying enzyme reaction mechanisms and have led to the discovery of several new enzyme intermediates.[5-9] For a more detailed

[1] K. A. Johnson, *in* "The Enzymes" (P. D. Boyer, ed.), 4th Ed., Vol. 20, p. 1. Academic Press, New York, 1992.

[2] K. A. Johnson, *Annu. Rev. Biochem.* **62,** 685 (1993).

[3] K. A. Johnson, this series, Vol. 134, p. 677.

[4] W. M. Kati, K. A. Johnson, L. F. Jerva, and K. S. Anderson, *J. Biol. Chem.* **267,** 25988 (1992).

[5] K. S. Anderson, J. A. Sikorski, and K. A. Johnson, *Biochemistry* **27,** 7395 (1988).

[6] K. S. Anderson and K. A. Johnson, *Chem. Rev.* **90,** 1131 (1990).

[7] K. S. Anderson, J. A. Sikorski, A. J. Benesi, and K. A. Johnson, *J. Am. Chem. Soc.* **110,** 6577 (1988).

[8] K. S. Anderson, E. W. Miles, and K. A. Johnson, *J. Biol. Chem.* **266,** 8020 (1991).

[9] K. A. Anderson, W. M. Kati, C. Z. Ye, J. Liu, C. T. Walsh, A. J. Benesi and K. A. Johnson, *J. Am. Chem. Soc.* **113,** 3198 (1990).

analysis of transient kinetic methods in general, the reader is referred to [1] in this volume and to Johnson.[1]

Principles of Transient-State Kinetic Analysis

The rapid quench transient kinetic methods described in this chapter involve the rapid mixing of enzyme with substrate to initiate the reaction. The reaction is then terminated after a short time interval by mixing with a quenching agent, such as a denaturant, and the amount of product formed is then quantified. By analyzing the time course of product formation the reaction can then be analyzed. If the experiment is done over a sufficiently short time interval and at a high enough enzyme concentration, then the kinetics of a single enzyme turnover can be measured.

Basically, two types of experiments can be performed. With substrate in excess over enzyme, one may observe a rapid "burst" of product formation, followed by steady-state turnover. Analysis of the rate and amplitude of the burst provides information to define the rate of the chemical reaction occurring at the active site, while the linear steady-state phase defines the rate-limiting steps leading to product release. However, a burst of product formation is only seen if the release of product, or some other step after chemistry, is slower than catalysis. If chemistry or a preceding step is rate limiting, then there is no burst and the initial rate would then define the rate of the chemical reaction. In this case, the absence of a burst could serve to establish that chemistry (or a step preceding chemistry) is rate limiting in the steady state.

A single enzyme turnover can be observed if the experiment is performed with enzyme in excess over substrate. If the enzyme is at sufficiently high concentration so that substrate binding is fast, then the kinetics measure the rate of the chemical conversion of substrate to product at the active site. Alternatively, at lower concentrations, analysis of the enzyme concentration dependence of the observed rate defines the kinetics of reactions involved in substrate binding and any steps leading to the observed chemical reaction. A single turnover experiment is most useful in attempts to define new enzyme intermediates because the 100% conversion of substrate to product within a single pass through the enzyme-bound species provides the most sensitivity to allow observation of any potential intermediates.[1,5,6,8–12]

[10] E. D. Brown, J. L. Marquardt, J. T. Lee, C. T. Walsh, and K. S. Anderson, *Biochemistry* in press (1994).

[11] E. D. Brown, J. L. Marquardt, C. T. Walsh, and K. S. Anderson, *J. Am. Chem. Soc.* **115,** 10398 (1993).

In the analysis of transient-state reaction kinetics, the enzyme is considered as a stoichiometric reactant rather than a trace catalyst. Therefore, the reactions occurring at the active site of the enzyme can be observed and quantified directly. Moreover, because the data define the kinetics of the reaction in terms of absolute concentration units that can be related to enzyme active site concentrations, the pathway of events occurring at the active site can be established without the indirect inference typical of steady-state kinetic methods. It is important to stress the value of information provided by the amplitude of a chemical reaction in absolute concentration units. Such information is often just as valuable as measuring the rate of a reaction in order to establish the pathway.[1,5,6]

In this chapter, I summarize the logic and the methods of analysis used to establish the reaction pathways of polymerases and force-transducing ATPases. Previous studies on these two systems provide a recipe for kinetic analysis in terms of the order and type of kinetic experiments to perform to examine other enzymes in these classes. In addition, I briefly summarize the key elements of kinetic analysis leading to the identification of new enzyme intermediates.

Chemical Quench-Flow Methods

Rapid chemical quench-flow methods are based on the rapid mixing of substrate with enzyme to initiate the reaction, followed at a specific time interval by the mixing with a quenching agent to terminate the reaction. The product formed in this time interval can then be quantified. In principle these experiments are done by driving syringes first to mix enzyme with substrate such that a reaction occurs while the mixture is flowing through a reaction loop of tubing (see Fig. 1). The reaction is then terminated when the enzyme–substrate mixture is combined with the quenching agent from a third syringe. The time of reaction is determined by the time it takes the enzyme–substrate mixture to flow from the point of mixing, through the reaction loop, and to the point of quench. The reaction time can be varied by changing the length of the reaction loop and the rate of flow. In practice there are limits on the rate of flow between the maximum rate that can be achieved at reasonable drive pressures and the minimum flow rate which is necessary to maintain efficient mixing. Within a factor of 2 or 3 in flow rate, a given length of tubing can be used to obtain a range of reaction times.

[12] H. Cho, R. Krishnaraj, E. Kitas, W. Bannwarth, C. T. Walsh, and K. S. Anderson, *J. Am. Chem. Soc.* **114,** 7296 (1992).

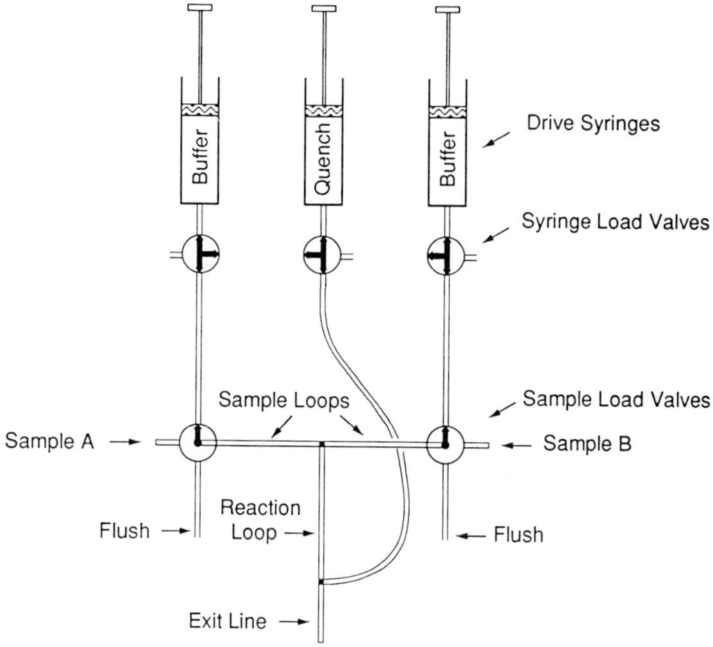

FIG. 1. Schematic of a pulsed-quench-flow apparatus. (Courtesy of KinTek Corporation, State College, PA.)

The slowest rate of flow is limited by the conditions to maintain turbulence. Turbulent flow through a tube can be predicted by calculation of a Reynolds number, which is a dimensionless parameter dependent on the flow rate, V, the viscosity, v, and the diameter of the tube, d: $R = Vd/v$. Turbulent flow occurs when the Reynolds number exceeds 2000.[13] Thus, for a 0.08 cm diameter tubing, and a viscosity of 0.01 poise (g/cm-sec), a linear flow rate of 2.5 m/sec (1.25 ml/sec) must be maintained. Although there has been some debate as to the validity of this calculation, in practice an instrument designed within these limits succeeds in measuring the rate of known chemical reactions,[3,14,15] and the failures of both private and commercial ventures in the manufacture of quench-flow instruments can be traced to a flow rate too slow to maintain turbulence by this criterion.

[13] H. K. Wiskind, *in* "Rapid Mixing and Sampling Techniques in Biochemistry" (B. Chance, R. H. Eisenhardt, Q. H. Gibson, and K. K. Lonberg-Holm, eds.), p. 355. Academic Press, New York, 1964.

[14] R. W. Lymn and E. W. Taylor, *Biochemistry* **9,** 2975 (1970).

[15] J. P. Froehlich, J. V. Sullivan, and R. L. Berger, *Anal. Biochem.* **73,** 331 (1976).

Quench-Flow Design

Most quench-flow instruments utilize a design based on loading enzyme and substrate into the drive syringes. This design is extremely wasteful of the precious reagents, however, because at the end of each drive the reaction loop still contains the two reactants. This solution, often 200 μl in volume, must be discarded in order to flush the system prior to collecting the next time point. An alternative design is shown in Fig. 1 (KinTek Corporation, State College, PA). In this instrument, the reactants are loaded into small sample loops, and then buffer in the drive syringes is used to force the reactants together, through the reaction loop to the point of mixing with the quenching agent, and then out into a collection tube. With this design, sample volumes as low as 15 μl can be obtained with nearly 100% sample recovery, making these experiments feasible for almost any enzyme system. An eight-way valve is used to select various reaction loop volumes, and a computer-controlled stepper motor is used to drive the reactants at a precise speed. By using eight different reaction loop volumes and with variation of the flow rate over a factor of about 2, reaction times ranging from 2 to 100 msec can be obtained. Longer reaction times are obtained in a push–pause–push mode, where the reactants are forced together, held in the reaction loop for a defined period of time, and then forced out to mix with the quench solution. Using this mode, reaction times from 100 msec to 100 sec (or longer) can be obtained.

Choice of Quenching Agent

Most commonly employed quenching agents are 1 *N* HCl and 1 *N* NaOH (final concentrations). In some experiments with 5-enolpyruvoylshikimate 3-phosphate synthase (EPSP synthase; 3-phosphoshikimate 1-carboxyvinyltransferase), 1 *N* trifluoroacetic acid gave better results than HCl. After stopping the reaction with acid, the reaction mixtures can be neutralized by addition of a small volume of a high concentration of buffer and base (i.e., 4 *M* Tris, 2 *M* NaOH). In some cases it is necessary to vortex the acid-quenched solutions with chloroform prior to neutralization to prevent renaturation of the enzyme;[5] that is, the enzyme must be killed twice to keep it from coming back to life. To first isolate the tetrahedral intermediate in EPSP synthase, neat triethylamine is used as a quenching agent, which when mixed 1 : 2 with sample gives an aqueous layer with approximately pH 12.[7] In experiments with DNA polymerases, 0.3 *M* EDTA has often been used to chelate the Mg^{2+} to prevent further reaction without the complications involved with the use of acids.

Regardless of the reagent used to quench the reaction, it is important to perform the appropriate control experiments involving mixing enzyme with quenching agent prior to adding substrate. This sample provides the most appropriate "blank" for measurement of background in the assay for product in addition to providing a necessary control for stopping the reaction.

Analysis of Reaction Products

Ultimately, the quality of the data is limited most by the methods used to separate substrate from product and quantify the amount of product formed. Methods have ranged from ion-exchange high-performance liquid chromatography (HPLC)[5] to the use of charcoal columns with batch elution[16] to isolate [^{32}P]phosphate from [γ-^{32}P]ATP. For studies on ATPase reactions, the best results are obtained using [α-^{32}P]ATP and resolving ADP from ATP by thin-layer chromatography (TLC) on poly(ethylenimine) (PEI) cellulose F plates (EM Separations, Gibbstown, NJ). Small aliquots (1–2 μl) are applied to the TLC plates. The plates can be washed with methanol to remove salts and then dried and developed with 0.6 *M* KH_2PO_4 at pH 3.4. ADP and ATP standards run in the outside lanes can be visualized with a hand-held fluorescent lamp so that appropriate areas of each lane of the plate can be cut and quantified by scintillation counting. Alternatively, the TLC plate can be quantified directly using a two-dimensional radioactivity detector, such as a Betascope (Betagen, Waltham, MA).

The kinetics of DNA polymerization can be followed on a sequencing gel by first labeling the DNA primer on the 5′ end with ^{32}P using T4 polynucleotide kinase. The products of reaction are then separated by electrophoresis of a 16% (w/v) polyacrylamide gel containing 8 *M* urea. The gel can then be quantified using a Betascope. Alternatively, by first exposing and developing a sheet of X-ray film, the gel can be placed on the X-ray film and the approriate bands cut with a razor blade and quantified by scintillation counting.

Regardless of the method, it is important to have an internal standard for the recovery of sample from the rapid quench-flow instrument. For HPLC, TLC, or gel analysis, the total radioactivity in a given sample or lane is used to normalize the data such that the analysis is a dependent on determining the fraction of product formed relative to starting reactant concentration. For example, in the case of gel analysis for DNA polymerization, the ratio of the elongated DNA (i.e., 26-mer) product relative to total labeled DNA (25-mer plus 26-mer) establishes the fraction of the reaction that has occurred. This fraction times the concentration of DNA in the original reaction mixture gives the concentration of product formed.

[16] K. A. Johnson, *J. Biol. Chem.* **258,** 13825 (1983).

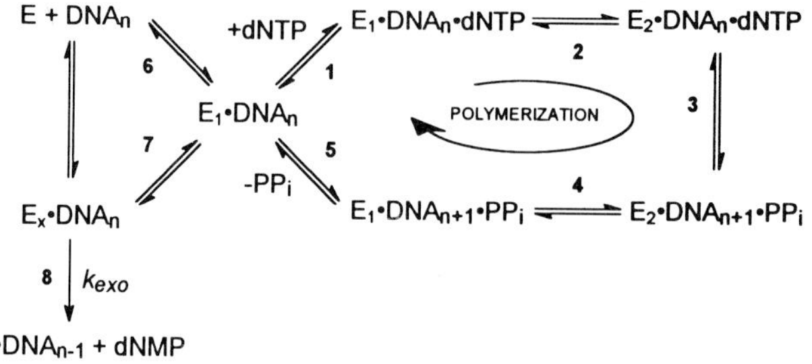

FIG. 2. Pathway of DNA polymerization. (Redrawn from Patel *et al.*[17])

Note on Utility of Concentrations Rather than Moles of Product

Reaction rates are governed by the concentrations of reactants present during the reaction. Accordingly, to interpret any experiment properly, one needs to know the concentrations of enzyme present and the concentrations of products formed as a function of time. In most laboratories performing enzyme kinetics, it appears to be the convention to graph moles (nanomoles or picomoles) of product formed as a function of time and to provide details as to the number of moles of enzyme present in the reaction mixture. However, to interpret these results, the reader must determine the total volume of the reaction (if it is even given) and then calculate the concentration of product formed. It is much more straightforward to always operate in concentration units, referring to the concentrations of reactants after mixing to initiate the reaction and the concentrations of product formed during the reaction. Quantitative analysis of the data is greatly facilitated by maintaining this convention.[1,5,6]

Kinetics of DNA Polymerization

The pathway of DNA polymerization is shown in Fig. 2. The pathway was derived in studies on T7 DNA polymerase[17–19] and defines the steps in the pathway catalyzed by reverse transcriptase[4] except for the absence of a proofreading exonuclease. Enzyme first binds DNA tightly, and then processive synthesis involves multiple rounds of polymerization without dissociation of the DNA. The binding of a correct deoxynucleoside triphosphate (dNTP) leads to a rate-limiting conformational change from an open to a closed state ($E_1 \cdot DNA_n \cdot dNTP \rightarrow E_2 \cdot DNA_n \cdot dNTP$) which is then

[17] S. S. Patel, I. Wong, and K. A. Johnson, *Biochemistry* **30,** 511 (1991).
[18] I. Wong, S. S. Patel, and K. A. Johnson, *Biochemistry* **30,** 526 (1991).
[19] M. J. Donlin, S. S. Patel, and K. A. Johnson, *Biochemistry* **30,** 538 (1991).

followed by a fast chemical reaction. Relaxation of the closed state is then followed by release of pyrophosphate and translocation, all of which are much faster than the rate-limiting conformational change.

The important questions to address in studying polymerases center on understanding the extraordinarily high fidelity of DNA replication in terms of the elementary steps of the reaction sequence. The selectivity of the correct dNTP is a function of the steady-state kinetic parameter k_{cat}/K_m. However, measurement of k_{cat} and K_m does not define the mechanistic basis for fidelity; it only quantifies the problem. In fact, most often steady-state kinetic analysis of DNA polymerases is complicated by the nature of the processive synthesis whereby a single encounter of the polymerase with the DNA template results in multiple rounds of synthesis. Once the polymerase has filled in the template to which it is bound, the dissociation of the enzyme from the DNA is several orders of magnitude slower than polymerization. Accordingly, steady-state measurements of incorporation are a poorly defined average of the rate of synthesis and the rate of dissociation of the DNA according to the length of the DNA template.

Steady-state reaction measurements do provide an important check on the validity of measurement of partial reactions along the pathway by transient kinetic methods. For example, "processivity" is defined as the number of bases incorporated per encounter of the enzyme with the DNA. Therefore, processivity is simply the ratio of the rate of polymerization divided by the rate of dissociation of the enzyme–DNA complex. For T7 DNA polymerase, a maximum rate of polymerization of 300–500 sec^{-1} and a dissociation rate of 0.2 sec^{-1} predicts a processivity of 1500–2500, a value observed by bulk solution measurements.

To define the reactions leading to the high fidelity of replication, one needs to identify the rate-limiting step for incorporation and determine the contributions to fidelity for each step in the reaction sequence. This can be accomplished by analyzing the kinetics of single nucleotide incorporation using synthetic oligonucleotides.

Kinetics of Single Nucleotide Incorporation

Experiments can be performed using synthetic oligonucleotides to examine the kinetics of addition of a single nucleotide after the addition of only one dNTP to the reaction mixture. For example, the following 25/45-mer has a 25 base primer complexed to a 45 base template such that the addition of dATP results in the elongation by one residue only:

```
5'-GCCTCGCAGCCGTCCAACCAACTCA
   |||||||||||||||||||||||||
3'-CGGAGCGTCGGCAGGTTGGTTGAGTTCCAGCTAGGTTACGGCAGG-5'
```

The kinetics of elongation can be examined by mixing 50–100 n*M* enzyme with 200–300 n*M* DNA. The reaction shows a stoichiometric burst of product formation (26-mer) followed by slow steady-state turnover (see Fig. 3), which is limited by the release of the DNA product from the enzyme.[4] The reaction time course can then be analyzed by the burst equation, $Y = A \exp(-k_2 t) + k_{ss}t$. The concentration of E–DNA active sites is obtained from the amplitude, A, and the rate of product formation from the exponential rate constant, k_2. The steady-state rate, k_{ss}, divided by the amplitude, A, defines the rate constant for DNA release, k_6 (Fig. 2).

The analysis of the amplitude in such simple terms is dependent on the assumption that the rate of the reaction, k_2, is at least 20-fold faster than the rate of product release, k_3, for a simple three-step pathway.[1] In general, the amplitude of the burst is reduced owing to more rapid product release according to $A = [k_2/(k_2 + k_3)]^2$. In either case, the data can be fit to the burst equation and then interpreted according to the mechanism. It is not generally useful to include the equation for the amplitude in the

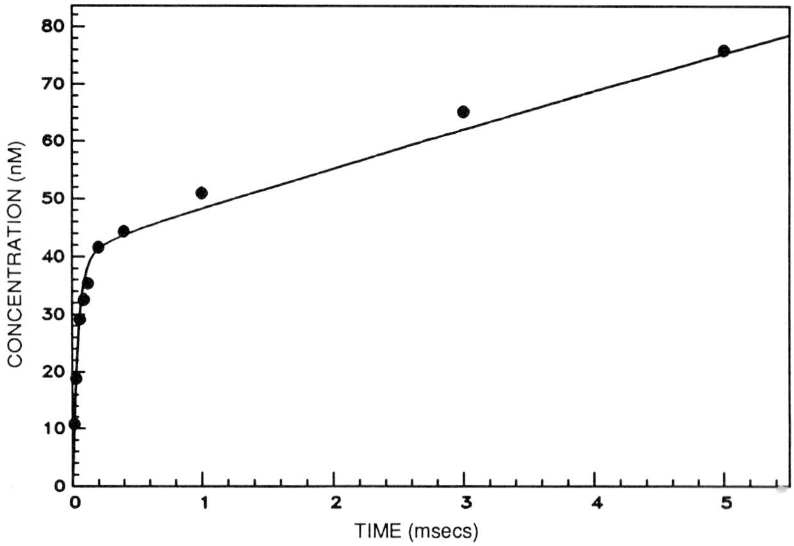

FIG. 3. Kinetics of a burst of DNA polymerization. The kinetics of single nucleotide incorporation were examined by reacting 40 n*M* reverse transcriptase (active site concentration) with 200 n*M* DNA (25/45-mer synthetic oligonucleotide) for the times indicated. The reaction was initiated by the addition of Mg^{2+} (10 m*M*) and dATP (25 μM) and terminated by the addition of 0.3 *M* EDTA. Products were analyzed on a sequencing gel to quantify the formation of the product 26/45-mer. The data were fitted to a burst equation with rate constants of 20 and 0.18 sec^{-1} for the rate of incorporation and release of DNA, respectively. (Reprinted with permission from Kati *et al.*[4])

fitting process because the amplitude may often be lower than that predicted because of dead enzyme or an internal equilibrium between substrate and product at the active site. Ultimately, data fitting based on computer simulation provides the most complete analysis to relate the reaction pathway directly to the observed kinetics (see Data Analysis).

Active Site Titration and Determination of K_d for DNA Binding of Polymerase Site

The rate of dissociation of the E · DNA complex is slower than polymerization; therefore, when polymerization is measured on a fast time scale, only those enzyme molecules with DNA bound at the polymerase site will react to form product. Accordingly, the amplitude of the fast polymerization reaction provides a direct measurement of the concentration of E · DNA complexes at the start of the reaction. By measuring the variation in the burst amplitude with increasing concentration of DNA, one can establish the concentration of active enzyme sites and measure the dissociation constant for the binding of DNA. It is important to stress that this experiment provides a K_d for DNA binding in a productive mode at the active site of the enzyme because it is based on the ability of the DNA to be elongated rapidly.

Typically, the K_d values for DNA are in the range of 5–20 n*M*.[2,4,17] The experiment can then be performed at 50 n*M* enzyme and concentrations of DNA ranging from 10 to 250 n*M*, if the data are analyzed by fitting to the quadratic equation (see Data Analysis). The hyperbolic equation is derived based on the assumption that substrate is always in excess of the enzyme concentration, and it is therefore not valid for this experiment.

Deoxynucleoside Triphosphate Concentration Dependence

Analysis of the dNTP concentration dependence of the observed rate of the burst of incorporation provides an estimate of the K_d for ground state nucleotide binding for formation of the productive complex poised for catalysis. The reaction kinetics follow a two-step sequence:

$$E \cdot DNA_n + dNTP \underset{}{\overset{K_d}{\rightleftharpoons}} E \cdot DNA_n \cdot dNTP \underset{}{\overset{k_{pol}}{\rightleftharpoons}} E \cdot DNA_{n+1} \cdot PP_i$$

Although the reaction pathway contains additional steps, the rate of incorporation is governed solely by the nucleotide binding followed by a rate-limiting step leading to elongation of the DNA.[4,17] As illustrated by the data in Fig. 4A, the rate increases with increasing concentration of dNTP, reaching a maximum.[4] As shown in Fig. 4B, the fit of the concentration dependence of the rate to a hyperbola yields the maximum rate k_{pol} and the dissociation constant, K_d.

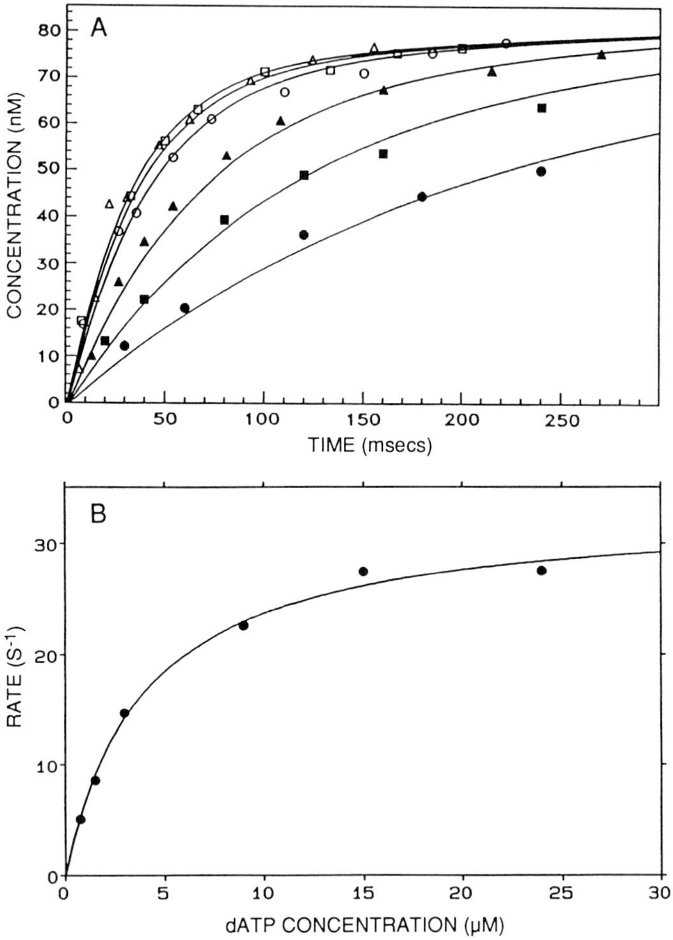

FIG. 4. Nucleotide concentration dependence of the rate of incorporation. (A) A solution of reverse transcriptase (80 n*M*) was incubated with 25/45-mer DNA (200 n*M*) and then mixed with Mg^{2+} (10 m*M*) and varying concentrations of dATP to initiate the reaction. The following dATP concentrations were used: 0.75 μM (●); 1.5 μM (■); 3.0 μM (▲); 9.0 μM (○); 15 μM (□); and 24 μM (△). The solid lines represent the best global fit to the data. (B) At each dATP concentration, the rate of polymerization was obtained by fit to a single exponential. The dATP concentration dependence was then fit to a hyperbola (solid line) to obtain a K_d of 4.0 ± 0.4 μM and a maximum rate of 33 ± 1.2 sec^{-1}. (Reprinted with permission from Kati *et al.*[4])

The fit to a hyperbola is dependent on two assumptions. First, the enzyme concentration must be much less than the K_d, so that the dNTP concentration added can be assumed to equal the free dNTP concentration. Second, the binding of dNTP must be a rapid equilibrium; that is, k_{-1} (the dNTP dissociation rate) must be much greater than k_{pol}. It is not always easy to prove this experimentally. The rapid equilibrium assumption is supported by the observation that the polymerization reaction is a single exponential at all dNTP concentrations. A two-step irreversible mechanism predicts a lag phase clearly observable at low dNTP concentrations, most prominent at the half-maximal rate. In addition, pulse–chase type experiments can be used to examine the kinetics of formation of a tight nucleotide complex.[17,20]

Kinetics of Processive Synthesis

It is important to establish that the rate of the first enzyme turnover is not different than subsequent turnovers. Most often this can be accomplished by careful comparison of the steady-state parameters with the rates and mechanism determined by transient kinetic methods. In the case of DNA polymerases, one can perform an experiment to measure the kinetics of processive synthesis involving the incorporation of a small number of nucleotides. For example, using the synthetic oligonucleotide described above, the addition of dATP, dCTP, and dTTP (leaving out dGTP) results in the elongation of the DNA by 5 base pairs (see Fig. 5). The kinetics of processive elongation can then be examined by following the polymerization reaction on a sequencing gel and quantifying the formation and decay of each intermediate from $DNA_{25} \rightarrow DNA_{26} \rightarrow DNA_{27} \rightarrow DNA_{28} \rightarrow DNA_{29} \rightarrow DNA_{30}$. The reaction kinetics are complex and require analysis by numerical integration.[4,17,21,22] Nonetheless, using this analysis, the rates of polymerization for each step of elongation can be obtained.[4]

Pyrophosphorolysis

The kinetics of the reverse of polymerization, pyrophosphorolysis, can be analyzed by following the shortening of DNA on a sequencing gel if there is no exonuclease in the reaction mixture. If an exonuclease is present, then the kinetics of hydrolysis to produce dNMP can be distinguished from the pyrophosphorolysis to produce dNTP by labeling the 3-terminal base. It is necessary to run the reaction in the presence of

[20] J.C. Hsieh, S. Zinnen, and P. Modrich, *J. Biol. Chem.* **268,** 24607 (1993).
[21] B. A. Barshop, R. F. Wrenn, and C. Frieden, *Anal. Biochem.* **130,** 134 (1983).
[22] C. T. Zimmerlie and C. Frieden, *Biochem. J.* **258,** 381 (1989).

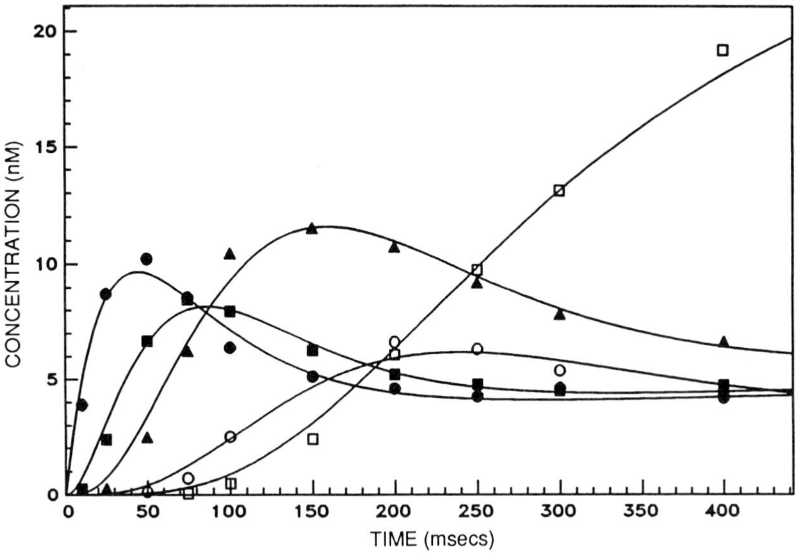

FIG. 5. Kinetics of processive polymerization. A solution of reverse transcriptase (40 n*M*) was preincubated with 25/45-mer DNA (100 n*M*) and then reacted with dATP, dCTP, and dTTP (20 μM each) with 10 m*M* Mg^{2+} for the times indicated before quenching with 0.3 *M* EDTA. The data show the formation of 26-mer (●), 27-mer (■), 28-mer (▲), 29-mer (○), and 30-mer (□). The solid lines represent the best fit to the data obtained by computer simulation using rates of 18, 31, 26, 9, and 13 sec^{-1}, respectively, for the five consecutive nucleotide incorporation reactions and DNA dissociation rates of 1.8, 3.4, 3.5, 1.2, and 1.2 sec^{-1}. (Reprinted with permission from Kati *et al.*[4])

excess unlabeled dNTP to trap any radiolabeled dNTP. Analysis of the kinetics of production of dNMP and dNTP can give the relative rates of pyrophosphorolysis and exonuclease hydrolysis. However, the time dependence of the reaction is a composite of both reactions such that the rate of pyrophosphorolysis, obtained as a fit to a single exponential, will be the sum of the rates of hydrolysis and pyrophosphorolysis even if care is taken to monitor the production of dNTP only.[17]

Another complication in measuring the kinetics of pyrophosphorolysis is due to the precipitation of magnesium pyrophosphate at concentrations required to saturate the rate of the reaction (above 2 m*M*). The problem is only partially overcome by initiating the reaction with Mg^{2+} in one syringe and pyrophosphate in the other. Precipitation begins almost immediately on mixing.

Exonuclease Kinetics

The kinetics of exonuclease hydrolysis can be easily measured by monitoring the shortening of DNA as observed on a sequencing gel with

5′-labeled DNA. Although multiple products will be observed owing to multiple rounds of hydrolysis, the quantification is based on the kinetics of disappearance of the starting material. In the case of T7 polymerase, different experimental designs have given different results that have been interpreted mechanistically. Single-stranded DNA is hydrolyzed quite rapidly (700–900 sec^{-1}).[19] In contrast, duplex DNA is hydrolyzed at rates of 0.2 and 2.3 sec^{-1} for correctly and incorrectly base-paired 3′ termini, respectively.[19] These results have been taken to imply that the movement of the duplex DNA from the polymerase site to the exonuclease site is rate limiting and that the hydrolysis of single-stranded DNA represents the rate of hydrolysis of the DNA once it occupies the exonuclease site. In each case, the DNA is first incubated with the polymerase in the absence of Mg^{2+}, and the reaction is initiated by the addition of Mg^{2+}; the data can be fit to a single exponential (see Data Analysis).

Biphasic kinetics have been observed when the reaction is initiated by mixing duplex DNA containing a mismatch with DNA.[19] There is a fast initial hydrolysis of 50% of the DNA (with enzyme in excess), whereas the remainder of the DNA is hydrolyzed at a rate of 2.3 sec^{-1}. These data have been interpreted in terms of a model whereby the DNA first binds to the exonuclease site and then partitions between hydrolysis and sliding into the polymerase site during the fast phase of the reaction. During the slow phase, the rate is limited by the sliding of the DNA from the polymerase site to the exonuclease site.

Complete Polymerase Pathway

The complete reaction pathway can be obtained by combining information from each experiment measuring the kinetics in the forward and reverse directions including estimates of the overall equilibrium constant for the reaction to constrain the fitting process. In the final fitting process, computer simulation can be used to model all of the data to a single model and a single set of rate constants.[1,4,17,21,22]

Adenosinetriphosphatase Mechanisms

In general, a mechanism for coupling ATP hydrolysis to do useful work can be analyzed in terms of the pathway shown in Fig. 6A. The enzyme exists in two states, E_1 and E_2, and the interconversion of the two forms is driven by the cycle of ATP hydrolysis coupled to do work. The pathway of coupling ATP hydrolysis to work is established by specifying the route through the maze shown in Fig. 6A by determining the kinetics of ATP binding, hydrolysis, and product release and the kinetics of the interconversion of the two enzyme forms. In the two examples

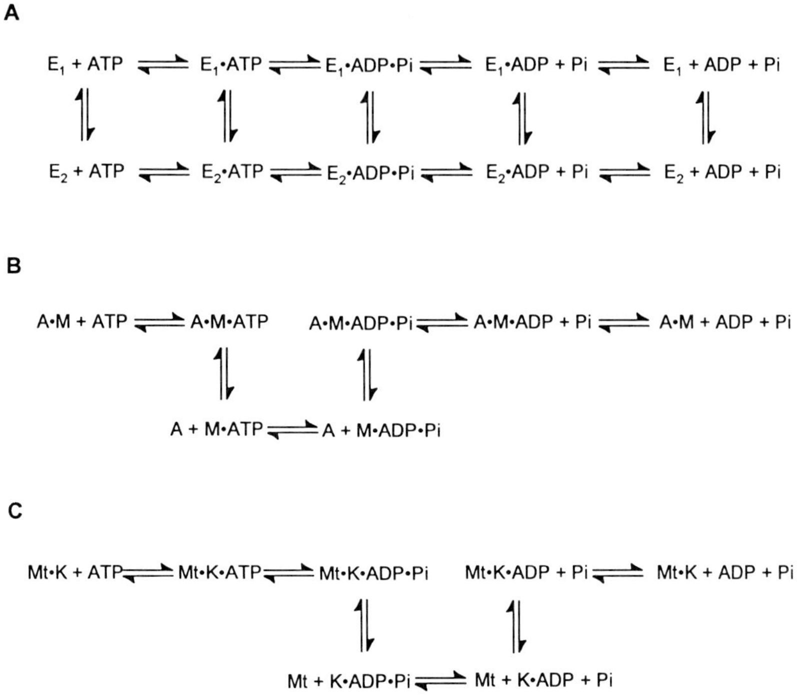

FIG. 6. Pathways of ATPase coupling. Pathways for ATPase coupling in general (A) and for actomyosin (B) and microtubule—kinesin (C) are shown. A represents actin; M, myosin; K, kinesin; Mt, microtubules.

given, actomyosin (Fig. 6B) and the microtubule–kinesin ATPase (Fig. 6C), the E_1 states are bound to the filament ($A \cdot M$ is actomyosin, $Mt \cdot K$ is microtubule–kinesin), and the E_2 states are dissociated from the filament (M is myosin, K is kinesin). Force is produced during conformational changes coupled to the changes in state while rebinding to the filament.

For actomyosin, ATP binding leads to rapid release of myosin from the actin (2000 sec^{-1}), which is followed by hydrolysis of the ATP on the free myosin molecule.[23,24] The myosin–products complex ($M \cdot ADP \cdot P_i$) rebinds to the actin, and the release of products (ADP and P_i) is coupled to force production. The dissociation of products from $A \cdot M \cdot ADP \cdot P_i$ is the rate limiting step ($\sim$20 sec^{-1}) for the maximum actin-activated ATPase.

[23] R. W. Lymn and E. W. Taylor, *Biochemistry* **10,** 4617 (1971).
[24] K. A. Johnson and E. W. Taylor, *Biochemistry* **17,** 3432 (1978).

A similar pathway has been observed for the dynein ATPase except phosphate release precedes rebinding to the microtubule.[16,25]

In contrast, in the kinesin pathway, ATP hydrolysis precedes the dissociation of the kinesin from the microtubule.[26] Phosphate release occurs from the free kinesin molecule, and the pathway is completed by the binding of the kinesin–ADP species (K · ADP) to the microtubule. Microtubules accelerate the release of ADP greater than 2000-fold. The release of kinesin from the microtubule following ATP hydrolysis is the rate-limiting step (20 sec^{-1}).

In these two examples, the pathways are similar and the maximum rates of the fully activated ATPase are similar. In each case, however, a different step is rate limiting. For actomyosin, the extremely fast dissociation of the motor from the filament leads to a pathway where the motor spends very little of the duty cycle attached to the filament. In contrast, kinesin spends most of the time attached to the microtubules. These observations are in keeping with the biological roles of these motors wherein myosin operates in ordered arrays, whereas kinesin works to transport vesicles along tracks of individual microtubules through the cytoplasm.

The experiments described below allow the pathway of coupling to be established by direct measurement of partial reactions.

Kinetics of ATP Hydrolysis

The kinetics of ATP hydrolysis can be measured in a rapid-quench experiment by first mixing enzyme with ATP and then quenching with 1 *N* HCl. Because the acid liberates any products bound at the enzyme active site, the amount of ADP observed is given by the sum $[ADP]_{obs} = [E \cdot ADP \cdot P_i] + [E \cdot ADP] + [ADP]$. Accordingly, the time dependence of the formation of ADP follows burst kinetics. The burst phase provides a measurement of the rate of formation and amount of $E \cdot ADP \cdot P_i + E \cdot ADP$ at the active site.

To achieve a maximal burst rate and amplitude so as to measure the rate of hydrolysis at the active site, a high concentration of ATP must be mixed with enzyme to initiate the reaction such that ATP binding is not rate limiting. The experiment is difficult because the K_m for ATP is typically much larger than the concentrations of enzyme that can be obtained, and the signal is limited by the background of hydrolyzed ATP (1%) in the most purified preparations. An ATP concentration equal to 5 times the

[25] K. A. Johnson, *Annu. Rev. Biophys. Biophys. Chem.* **14,** 161 (1985).

[26] S. P. Gilbert, M. R. Webb, M. Brune, and K. A. Johnson *Nature (London)* submitted (1994).

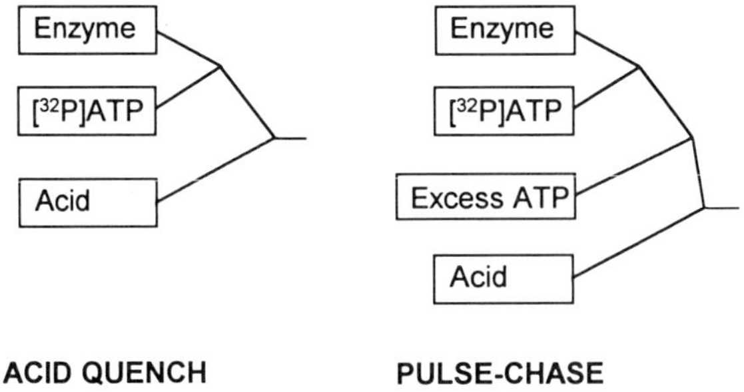

FIG. 7. Design a pulse–chase experiment.

K_m provides only a lower limit on the concentration of ATP that may be required to achieve a binding rate faster than catalysis. The steady-state K_m only defines the concentrations required for the binding rate to equal the steady-state turnover rate. Because hydrolysis may be an order of magnitude faster than net turnover, a correspondingly higher concentration of ATP is required.

To satisfy the most stringent requirements for optimal signal-to-noise ratios (S/N), we have found that the kinetics of ATP hydrolysis can be monitored most accurately using [α-^{32}P]ATP and quantifying the appearance of [α-^{32}P]ADP following separation by TLC on PEI cellulose F plates.[27] The concentration of ADP formed is then calculated from the counts per minute in the ADP spot (cpm_{ADP}) relative to counts in the ATP spot (cpm_{ATP}):

$$[ADP] = [ATP]_0 cpm_{ADP}/(cpm_{ADP} + cpm_{ATP})$$

The data can be fit to the burst equation (see Data Analysis) to obtain estimates of the rate and amplitude of the hydrolysis reaction. However, if the kinetics of ATP binding are known and the rate is less than 10-fold faster than hydrolysis, then the hydrolysis kinetics will exhibit a lag phase which is a function of ATP binding. Accordingly, the rates of hydrolysis will be underestimated unless the lag phase predicted by the binding kinetics is included in the fitting process.[16] This is most easily accomplished by computer simulation (see Data Analysis).

Kinetics of ATP Binding: Pulse–Chase Experiments

The kinetics ATP binding can be measured using a pulse–chase protocol outlined in Fig. 7. The reaction is initiated by mixing with [α-^{32}P]ATP

[27] S. P. Gilbert and K. A. Johnson, *Biochemistry* **33,** 1951 (1994).

and then, at various times, an excess of unlabeled ATP is added. The reaction is allowed to continue for 5–8 half-lives of the hydrolysis reaction, a time sufficient to convert all tightly bound ATP to ADP. The reaction is then terminated by the addition of acid. The amount of ADP formed is equal to the sum $[\mathrm{ADP}]_{\mathrm{obs}} = \theta[\mathrm{E}\cdot\mathrm{ATP}] + [\mathrm{E}\cdot\mathrm{ADP}\cdot\mathrm{P_i}] + [\mathrm{E}\cdot\mathrm{ADP}] + [\mathrm{ADP}]$, where θ is the fraction of bound ATP that is hydrolyzed before it is released from the enzyme. As described above, the data can be fit to a burst equation to a first approximation, and then fit more completely by computer simulation to the complete mechanism.

It is important that the chase period be as short as possible to prevent the accumulation of product owing to the binding and hydrolysis of the diluted radiolabeled ATP. A control experiment must be performed by mixing the enzyme with the diluted ATP for the time of the chase to ensure that the rate of dilution of the label is sufficient for the time required.

Kinetics of Phosphate Release

The kinetics of phosphate release from the enzyme can be measured by stopped-flow using a method newly developed by Webb using fluorescently modified phosphate-binding protein.[28] The details of this method, however, are beyond the scope of this review. Care must be taken to scrub the system of contaminating phosphate using a phosphate "mop." With appropriate attention to details given in the original manuscript, a burst of phosphate release of 50 n*M* can easily be observed at a concentration of 1 m*M* ATP.

Kinetics of ADP Release

The rate of ADP release can be measured by several methods. The easiest method relies on the use of fluorescently labeled ADP, mant–ADP.[29] On excitation at 350 nm, mant–ADP emits light at 440 nm. Mant–ADP fluorescence is quenched by solvent, and therefore there is a decrease in fluorescence on dissociation of the mant–ADP from the enzyme. When kinesin–ADP is mixed with microtubules, a single exponential decrease in fluorescence is observed, providing a direct measurement of the rate of ADP release from the microtubule–kinesin–ADP complex.

Kinetics of Motor Dissociation and Rebinding to Filament

The kinetics of the dissociation and rebinding of the motor to the filament can be measured by stopped-flow light-scattering measurements.

[28] M. Brune, J. L. Hunter, J. E. T. Corrie, and M. R. Webb, *Biochemistry* **33,** 8262 (1994).
[29] A. Sadhu and E. W. Taylor, *J. Biol. Chem.* **267,** 11352 (1994).

For example, the mixing of ATP with actomyosin or microtubule–kinesin leads to a reduction in light-scattering intensity that can be monitored in a stopped-flow apparatus. The method is relatively sensitive so that concentrations of 0.1 μM protein can be used and any concentration of ATP from 1 μM to 5000 mM can be employed. The data fit a single exponential. Plotting rate versus ATP concentration in the case of actomyosin yields a straight line, with the rate approaching 2000 sec^{-1}.[24] In contrast, the ATP concentration dependence observed for the microtubule–kinesin complex follows a hyperbola.[26]

Putting It All Together

The pathway of ATPase coupling is finally obtained by combining information from all of the partial reactions. For example, for kinesin, the rate of ATP hydrolysis is faster than the rate of dissociation of the kinesin from the microtubule, thereby establishing that ATP hydrolysis precedes kinesin–microtubule dissociation. Phosphate release occurs at a rate equal to the rate of kinesin–microtubule dissociation; therefore, one reaction must be much faster and follow the other. To complete the cycle without proposing additional intermediates, the most simple pathway involves the fast release of phosphate from kinesin after its dissociation (as $K \cdot ADP \cdot P_i$) from the microtubule. This conclusion is then supported by measurements of the rate of binding of kinesin–ADP ($K \cdot ADP$) to the microtubules. Finally, measurement of the rate of ADP release from the microtubule–kinesin–ADP complex ($Mt \cdot K \cdot ADP$) shows that ADP release is fast, supporting the identification of the single rate-limiting step involving the dissociation of the kinesin from the microtubule. The overall pathway can then be modeled by computer simulation to confirm that it accounts for all of the observed reactions.

Detection of Enzyme Intermediates

The search for enzyme intermediates is often a frustrating and difficult task because when one begins the investigation one may not know the properties of the intermediate or the time and substrate concentrations required for its formation, and it may be formed in low amounts. Experiments based on single turnover kinetics overcome many of the difficulties and have led to the identification and isolation of several new enzyme intermediates.[5–12]

We consider a single intermediate on the pathway from substrate to product:

$$E + S \rightleftharpoons E \cdot S \rightleftharpoons E \cdot I \rightleftharpoons E \cdot P \rightleftharpoons E + P$$

To prove the existence of the intermediate, I, it must be shown to be kinetically competent, and, if possible, it must be isolated to prove its structure. Kinetic competence is most precisely addressed by showing that the intermediate is formed and is broken down at rates sufficient to account for net conversion of substrate to product. A single turnover experiment with enzyme in excess over the substrate, S, provides the most sensitive means to look for intermediates and allows the kinetic competence of the intermediate to be established directly.[5,6,30]

The kinetics of a single turnover of EPSP synthase is shown in Fig. 8A.[5] The reaction involves the combination of shikimate 3-phosphate (S3P) with phosphoenolpyruvate (PEP) to form a tetrahedral intermediate, which decays by elimination of phosphate to yield the product enolpyruvoylshikimate 3-phosphate (EPSP). The reaction shown in Fig. 8A was initiated with 10 μM enzyme in excess over limiting [^{14}C]PEP (3.5 μM) in the presence of excess S3P (100 μM). The key to the success of the experiment lies in the use of enzyme excess over the limiting, radiolabeled substrate. This allowed the direct observation of the intermediate although it formed on only one-third of the enzyme sites. The kinetics of the reverse reaction are shown in Fig. 8B, again showing the formation of the intermediate from EPSP. The kinetics of the reactions could be fitted to the reaction sequence to obtain a unique set of rate constants describing the rates of formation and decay of the tetrahedral intermediate in each direction.[5] The kinetic analysis led directly to defining conditions under which the intermediate could be synthesized in quantities sufficient for structure determination.[7]

Data Analysis

Kinetic data can be fit to one of several simple equations involving either one or two exponential terms. The observed rates and amplitudes can then be interpreted in terms of the pathway of the reaction. As the pathway becomes more complex, then assumptions must be made to allow the use of the simplified equations. In general, there should be one exponential term for each step of the reaction,[1] except for steps in rapid equilibrium. Although fitting data to analytical solutions of the rate equations is useful and is often a necessary first step, it is usually desirable to fit the data directly to the complete mechanism by numerical integration. Computer simulation allows all of the kinetic data to be fit directly to the enzyme reaction pathway by global analysis without the need for simplifying assumptions. This section first describes the fitting of data to

[30] K. S. Anderson and K. A. Johnson, *J. Biol. Chem.* **265,** 5567 (1990).

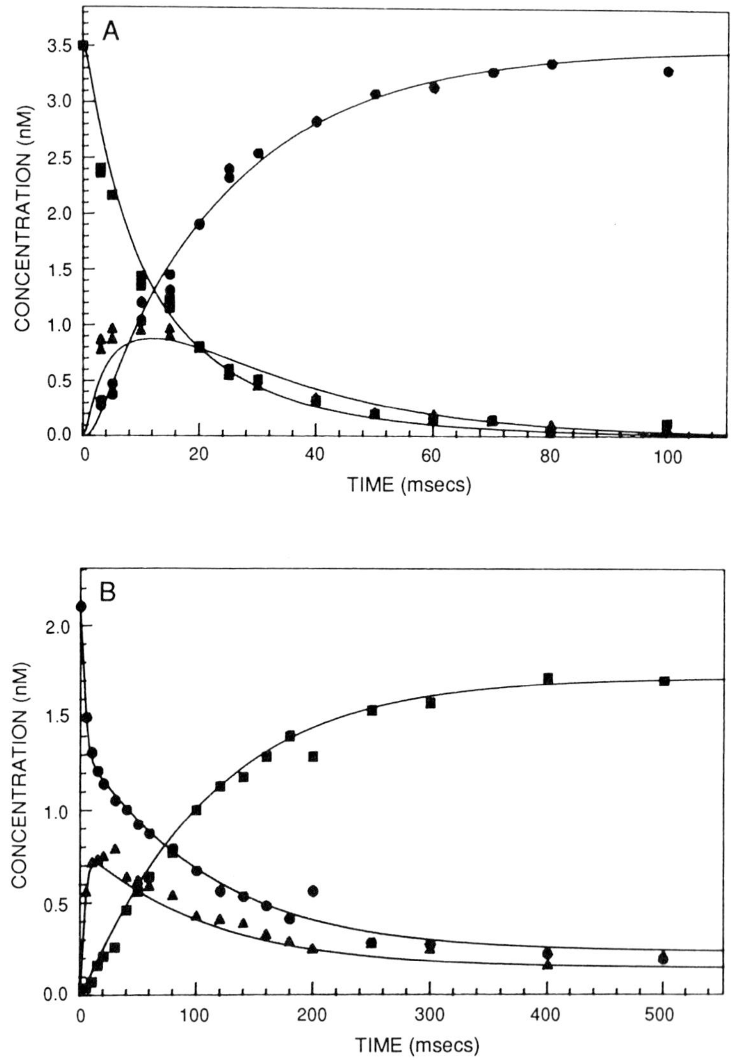

FIG. 8. Kinetics of a single turnover of EPSP synthase. (A) A single turnover in the forward reaction was observed by mixing a solution containing enzyme (10 μM) and S3P (100 μM) with [^{14}C]PEP (3.5 μM). The reaction was terminated by the addition of 1 *N* HCl, and then the solution was vortexed with chloroform and neutralized. Products were analyzed by ion-exchange HPLC. (B) Kinetics of the reverse reaction. The time course is shown for formation and disappearance of PEP (■), EPSP (●), and pyruvate (▲), a breakdown product of the intermediate. The curves were calculated by numerical integration according to the complete solution of the reaction pathway. (Reprinted with permission from Anderson *et al.*[5] Copyright 1988 American Chemical Society.)

standard solutions of simplified rate equations, then discusses the methods of global analysis.

Equilibrium Binding Measurements

It is often necessary and useful as part of the kinetic analysis to examine the equilibrium binding of substrates to the enzyme:

$$\mathrm{E} + \mathrm{S} \rightleftharpoons \mathrm{E} \cdot \mathrm{S}$$

If the binding is weak relative to the concentration of enzyme required to measure the binding, then the data can be fit to a hyperbola:

$$[\mathrm{E} \cdot \mathrm{S}] = [\mathrm{E}]_0[\mathrm{S}]_0/([\mathrm{S}]_0 + K_\mathrm{d})$$

where $[\mathrm{E}]_0$ and $[\mathrm{S}]_0$ are the total enzyme and substrate concentrations, respectively. Data should be fit by nonlinear regression, including $[\mathrm{E}]_0$, the starting enzyme concentration, as an unknown, if possible. If the data are in the form of an optical signal with an unknown extinction coefficient, a, then the signal must be fit including these additional unknowns, and the enzyme concentration must be assumed to be known:

$$Y = a[\mathrm{E} \cdot \mathrm{S}] + c = a[\mathrm{E}]_0[\mathrm{S}]_0/([\mathrm{S}]_0 + K_\mathrm{d}) + c$$

Several graphics programs containing nonlinear regression routines are available for personal computers (GraFit, Erithacus Software, Ltd., Staines, U.K.; KaleidaGraph, Synergy Software, Reading, PA, USA). After entering the approriate equation and initial estimates of the parameters, the program will find the best fit by minimizing the sum of squares error between the calculated and observed curves. A failure to get the program to converge to a solution is most often due to errors in the initial estimates. Take note of the form of the equation and the values of y at $x = \infty$ to estimate the parameters from the graph.

If the binding is tight relative to the enzyme concentration, then the data must be fit to a quadratic equation:

$$[\mathrm{E} \cdot \mathrm{S}] = \{(K_\mathrm{d} + [\mathrm{E}]_0 + [\mathrm{S}]_0) - \mathrm{SQRT}[(K_\mathrm{d} + [\mathrm{E}]_0 + [\mathrm{S}]_0)^2 - 4[\mathrm{E}]_0[\mathrm{S}]_0]\}/2$$

Again, care should be taken in considering the relationship between any optical signal and the concentration of E · S and in making initial estimates of the unknown parameters.

Single Exponential

Most kinetic experiments can be fit to a single exponential of the general form

$$Y = A \exp(-kt) + C$$

where A is the amplitude, k is the rate, and C is the end point. All three parameters should be taken as unknowns in fitting by nonlinear regression. In making initial estimates, note that a reaction with a signal increasing with time has a negative amplitude. At $t = 0$, $Y = A + C$, where as $t \rightarrow \infty$, $Y = C$.

Double Exponential

The equation for a double exponential follows the general form

$$Y = A_1 \exp(-k_1 t) + A_2 \exp(-k_2 t) + C$$

which is analogous to a single exponential except that there are two amplitude and rate terms. This equation is used whenever the reaction shows two phases, such as a fast and slow phase or a lag preceding a reaction.

Burst Kinetics

The kinetics of a pre-steady-state burst followed by linear steady-state turnover can be fit to the standard burst equation:

$$Y = A \exp(-k_1 t) + k_2 t + C$$

where A is the amplitude of the burst and k_1 is the rate of the burst (in units of $\sec^{-1}$), whereas k_2 is the steady-state rate (in units of concentration/sec). Divide k_2 by the enzyme concentration or amplitude of the burst to convert it to a first-order rate constant defining product release rates.

Global Analysis

Ultimately the goal of kinetic analysis is to derive a single mechanism to account completely for the results without introducing steps or intermediates not required by the data. It is now possible to fit all of the kinetic experiments directly to the mechanism by nonlinear regression based on numerical integration of the rate equations.[21,22] Computer simulation is also a useful learning tool that can aid in the design of experiments to optimize conditions to test opposing models.

In fitting data by numerical integration, no simplifying assumptions are needed, but it is necessary to be close to the best fit before beginning the nonlinear regression. In addition, it is necessary to allow only those constants to float which are sensitive parameters pertaining to each experiment. For example, if an experiment is done at high substrate concentration so the binding rate is fast and the observed reaction rate is a function of the chemistry step, then the binding rate must be fixed at a known value while floating the rate of the chemical reaction to obtain the best fit. Alternatively, for a series of experiments done as a function of substrate concentration, one may float the substrate binding rate while maintaining a fixed value for the rate of the chemical reaction. In the final analysis, the overall pathway can then be fit to all of the data and including in the fitting process estimates of rate constants even though they may pertain to different parts of the mechanism that are not tested by a given experiment. In this way, even subtle effects on the kinetics can be taken into consideration in the fitting process to achieve a global best fit.

Using these methods, the pathway of enzyme-catalyzed reactions can be established by direct measurement of individual steps of the reaction. By comprehensive analysis of the reaction kinetics, the partial reactions can be assembled to provide a simple, complete reaction pathway.

[3] Analysis of Enzyme Progress Curves by Nonlinear Regression

By RONALD G. DUGGLEBY

Introduction

The usual way in which an enzyme-catalyzed reaction is monitored is by measuring the *amount* of reactant remaining or product formed at one or more times. By contrast, most kinetic models are formulated in terms of *rates* of reaction. There is, therefore, a fundamental incompatibility between the data and the underlying kinetic model. This incompatibility can be resolved in two ways. Either the model can be integrated to give a description of the time course of the reaction, or the data can be differentiated to determine rates.

Traditionally, enzyme kinetic studies have employed the second approach and have focused on determining rates, especially initial rates, by measuring tangents to the reaction progress curves. There are several reasons for this preference for initial rates. (1) The substrate concentration is exactly equal to that which is added. (2) There is a vanishingly small concentration of potentially inhibitory reaction products present, unless these are added deliberately. (3) There is no opportunity for the enzyme to have undergone partial inactivation.

There are, however, several advantages in studying progress curves, and, ironically, the first two of these are restatements of what were mentioned above as reasons for preferring initial rates. (1) The substrate concentration is automatically varied as the reaction proceeds, providing information about the dependence on substrate concentration of the enzyme. (2) Products of the correct stereochemistry and of complete purity are formed automatically as the reaction proceeds, providing information about the dependence on product concentration of the enzyme. (3) As a consequence of the above, considerably more information is obtained from each assay so a complete description of the kinetic properties of the enzyme can be obtained from fewer experiments.

Given these advantages, one may inquire why all enzyme kinetic studies do not exploit progress curves. Again there are several reasons. (1) Frequently we may be interested only in a limited subset of the kinetic properties, the Michaelis constant for a particular substrate, for example. Progress curves contain information about the complete rate equation, and there is often no simple way to dissect out a subset of this information unless the enzyme has fortuitous properties, such as a lack of significant inhibition by the products. (2) The data analysis is considerably more complex than that for rate measurements, and it is only over the last few years that reliable methods have been developed that cover most of the common types of reactions. (3) The shape of the progress curve may not depend solely on the variations in reactant concentrations that are due to the catalyzed reaction. The enzyme may be subject to progressive inactivation, and substrates and products may undergo uncatalyzed side reactions. There is a reasonable prospect that it will be possible to handle such complications as methods of data analysis are developed further.

This chapter focuses mainly on the second point immediately above, the methods of data analysis. Because these are still being developed, the presentation is largely chronological and concentrates on the methods and experimental systems that have been developed and used in our laboratory. This personal perspective is in no way intended to diminish the contributions that others have made to the area of progress curve analysis. Later I shall indicate the type of developments that can deal with the complications mentioned in the third point. For a discussion of much of the earlier literature, see a previous article in this series by Orsi and Tipton.[1]

It is worth emphasizing at this point that analyses based on progress curves are not preferable to those based on rates in all circumstances. Indeed, there are many instances where rate measurements are the method of choice. Most commonly this will be where one is interested solely in characterizing the kinetics toward the substrate(s) for an enzyme that either (a) has cosubstrates whose kinetics are not of immediate interest,

[1] B. A. Orsi and K. F. Tipton, this series, Vol. 63, p. 159.

(b) catalyzes a readily reversible reaction, or (c) shows significant inhibition by its product(s). Applying progress curve analysis leads to unnecessary complications as an accurate description of the data requires that the integrated form of the full rate equation is used. For example, a simple ordered reversible reaction with two substrates and products has an integrated rate equation involving eight independent kinetic constants (maximum velocities for the forward and reverse reactions, Michaelis constants for all four reactants, and inhibition constants for the first substrate bound in each direction). The experiments needed to characterize the system completely will be excessive if all one is interested in is the Michaelis constant for one of the substrates.

Simple Michaelian Enzyme

Fitting Equations

In any situation where an equation is to be fitted to some experimental data there must be some sense in which the data lie close to the line or surface predicted by the equation. Fitting the equation will involve manipulating certain constants whose values are not fixed by the nature of the experiment; these constants are referred to as the parameters of the equation. If the data are perfect and the equation represents a theoretically correct description of the experimental system, it should be possible to find values of the parameters for which the line or surface agrees exactly with the data. In practice this does not occur because the data are inexact.

It is common practice to regard one of the coordinates of each datum as a measured or dependent variable (y) and all the others as controlling or independent variables (x_1, x_2, x_3, etc.). Under this convention, each of the independent variables is regarded as being known exactly, while any experimental uncertainties are associated with the dependent variable. Thus, to fit an equation to the data, the degree of closeness to the line or surface is assessed from the distances along the y coordinate only.

Unless one is simply fitting by eye, there has to be some objective measure of closeness, and this is usually taken as the sum of the squares of these distances. The justification for this is that if the experimental errors follow a Gaussian distribution, minimizing the sum of squares yields the most likely values of the fitted parameters. In practice, however, the distribution of errors is rarely determined, and the independent variables may also contain some uncertainties. The widespread use of the least squares criterion is more a reflection of its mathematical convenience than a firm belief in its assumptions.

Integrated Michaelis–Menten Equation

The basic principles, and the difficulties, of progress curve analysis can be illustrated by considering the simplest situation, namely, that of an enzyme catalyzing an irreversible reaction in which one substrate is converted to a noninhibitory product.

The rate equation for the reaction is given as Eq. (1), where V_m is the maximum velocity, K_a is the Michaelis constant, and $[A]_t$ is the substrate concentration at any time:

$$\frac{d[A]_t}{dt} = \frac{-V_m[A]_t}{K_a + [A]_t} \tag{1}$$

It is useful to rewrite Eq. (1) in terms of the initial substrate concentration ($[A]_0$) and a variable y, the amount of product formed by reaction:

$$[A]_0 = [A]_t + y \tag{2}$$

Combining Eqs. (1) and (2) yields Eq. (3), which then may be integrated to give Eq. (4):

$$\frac{dy}{dt} = \frac{V_m([A]_0 - y)}{K_a + [A]_0 - y} \tag{3}$$

$$V_m t = y - K_a \ln(1 - y/[A]_0) \tag{4}$$

In Eq. (4), the quantity that is normally measured experimentally is y itself, or some linear function of y. However, Eq. (4) is an implicit function of y, meaning that it cannot be rearranged to express y as an explicit function of the other variables. As a result, direct calculation of y is impossible, and fitting Eq. (4) to obtain values for V_m and K_a from an enzyme progress curve is not straightforward. There are, of course, simple rearrangements of Eq. (4) that put it into a form from which V_m and K_a can be obtained, such as Eq. (5):

$$y/t = V_m + K_a \ln(1 - y/[A]_0)/t \tag{5}$$

A plot of y/t versus $\ln(1 - y/[A]_0)/t$ should be a straight line from which V_m and K_a are easily obtained by linear regression. There are two objections to this approach. The first is a statistical one; y is the measured variable that is considered to contain any experimental error. However, y appears on both sides of Eq. (5), and estimates of V_m and K_a obtained by least squares fitting will be statistically biased. The second objection is more fundamental, although not evident for this simple single-substrate model. As we move to more complex enzymatic reactions that may involve multiple substrates or inhibitory products, or where the back reaction must be taken into account, no linear transformation may be possible.

To overcome these difficulties it is necessary to fit Eq. (4) to an enzyme progress curve by nonlinear regression. This will involve minimizing the sum of the squares of the difference between the measured value of y and that expected from Eq. (4); thus, it is necessary to solve Eq. (4) for y, given particular values for V_m, K_a, $[A]_0$, and t.

Solving Integrated Rate Equations

A practical method for solving integrated rate equations was first mentioned by Nimmo and Atkins[2] and described more fully by Duggleby and Morrison.[3] This method, the Newton–Raphson procedure, is a standard mathematical algorithm for solving implicit equations and is described briefly.

Equation (4) may be rearranged to give Eq. (6), in which the problem now becomes one of finding a value of y for which $F(y) = 0$:

$$F(y) = y - K_a \ln(1 - y/[A]_0) - V_m t \tag{6}$$

Equation (6) describes a curve with the general shape shown in Fig. 1A. That the curve will have this shape is readily seen when it is understood that Eq. (6) describes a progress curve that has had the axes interchanged, followed by a downward displacement of the curve by an amount $V_m t$.

The solution (i.e., the point where the curve crosses the abscissa) may be found using the differential of $F(y)$ with respect to y [Eq. (7)] and an initial estimate of y (y_i), as illustrated in Fig. 1B. The dashed lines are tangents to the curve obtained with different initial estimates. Each allows a "refined" estimate (y_{i+1}) to be calculated from the point where the line has $F(y) = 0$. This point may be obtained using Eq. (8):

$$F'(y) = \frac{dF(y)}{dt} = \frac{1}{1 + K_a/([A]_0 - y)} \tag{7}$$

$$y_{i+1} = y_i - F(y_i)/F'(y_i) \tag{8}$$

Several repetitions of this procedure using the successive refined estimates in place of the initial estimate will, in most cases where a reasonable starting value has been chosen, yield the required solution to any desired degree of numerical accuracy.

The problem with this procedure is that there is no guarantee that the refined estimate is closer to the solution than initial estimate, nor any automatic constraint to prevent it from exceeding $[A]_0$ (Fig. 1B, curve a). This is a computational disaster as the logarithmic term in Eq. (6) becomes

[2] I. A. Nimmo and G. L. Atkins, *Biochem. J.* **141,** 913 (1974).

[3] R. G. Duggleby and J. F. Morrison, *Biochim. Biophys. Acta* **481,** 297 (1977).

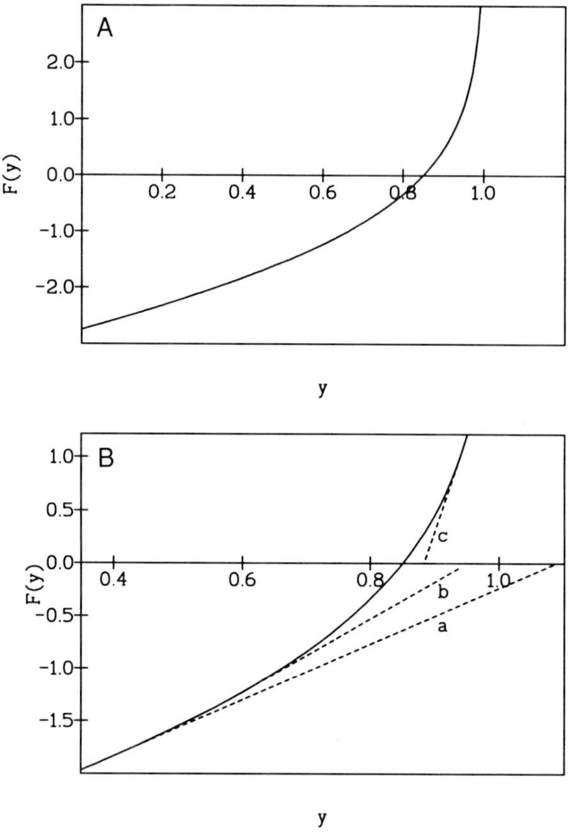

FIG. 1. Solving the integrated Michaelis–Menten equation by the Newton–Raphson method. Given a value of $K_a = [A]_0$, the time to reach 85% of $[A]_0$ was calculated from Eq. (4). Using this value ($t = 2.7471/V_m$), $F(y)$ was calculated using Eq. (6) for a range of values of y. (A) Graph over the entire range of y; (B) enlargement of the central region. Also shown in (B) are the results of the first cycle in the Newton–Raphson solution for initial estimates of 0.4 (line a), 0.6 (line b) and 0.94 (line c).

undefined. Although this can never occur if the initial estimate is greater than the solution[4] (Fig. 1B, curve c), there is no simple method of choosing an initial estimate in the appropriate range if the solution is unknown.

Duggleby[4] proposed an alternative definition of $F(y)$ that contains an exponential term rather than a logarithmic term [Eq. (9)] and a corresponding definition [Eq. (10)] of $F'(y)$:

$$F(y) = y/[A]_0 - 1 + \exp[(y - V_m t)/K_a] \tag{9}$$

[4] R. G. Duggleby, *Biochem. J.* **235,** 613 (1986).

TABLE I
NEWTON–RAPHSON SOLUTION OF INTEGRATED MICHAELIS–MENTEN EQUATION[a]

	Eqs. (6)–(8)			Eqs. (8)–(10)		
y_i	y_{i+1}	$F(y)$	Cycles	y_{i+1}	$F(y)$	Cycles
0.00	1.3736	−2.7471	Fail	0.8795	−0.9359	3
0.05	1.3390	−2.6458	Fail	0.8769	−0.8826	3
0.10	1.3040	−2.5418	Fail	0.8743	−0.8291	3
0.15	1.2686	−2.4346	Fail	0.8718	−0.7755	3
0.20	1.2329	−2.3240	Fail	0.8693	−0.7217	3
0.25	1.1970	−2.2094	Fail	0.8669	−0.6677	3
0.30	1.1608	−2.0904	Fail	0.8646	−0.6135	3
0.35	1.1246	−1.9663	Fail	0.8624	−0.5590	3
0.40	1.0886	−1.8363	Fail	0.8603	−0.5044	3
0.45	1.0530	−1.6993	Fail	0.8584	−0.4495	3
0.50	1.0180	−1.5540	Fail	0.8566	−0.3943	3
0.55	0.9841	−1.3986	6	0.8550	−0.3389	3
0.60	0.9517	−1.2308	6	0.8536	−0.2832	3
0.65	0.9215	−1.0473	5	0.8523	−0.2272	3
0.70	0.8946	−0.8431	5	0.8514	−0.1709	3
0.75	0.8722	−0.6108	5	0.8506	−0.1143	2
0.80	0.8563	−0.3377	5	0.8502	−0.0573	2
0.85	0.8500	0.0000	1	0.8500	0.0000	1
0.90	0.8586	0.4555	4	0.8502	0.0577	2
0.95	0.8929	1.1986	5	0.8507	0.1158	2
0.99	0.9618	2.8480	6	0.8514	0.1625	3
0.999	0.9938	5.1596	7	0.8516	0.1731	3
0.9999	0.9992	7.4630	8	0.8516	0.1742	3

[a] Given a value of $K_a = [A]_0$, the time to reach 85% of $[A]_0$ was calculated from Eq. (4) (the integrated Michaelis–Menten equation). Using this value ($t = 2.7471/V_m$) and initial estimates (y) shown in the first column, values of y_{i+1} and $F(y)$ were calculated using Eqs. (6)–(8). This process was repeated until y_{i+1} was equal to the correct solution (i.e., 85% of $[A]_0$), and the number of iterations required was counted. A similar procedure was applied using Eqs. (8)–(10).

$$F'(y) = 1/[A]_0 + \exp[(y - V_m t)/K_a]/K_a \qquad (10)$$

As shown in Table I, this variation provides an almost completely foolproof method for solving Eq. (4), and it is usually faster than the method using Eqs. (6)–(8). Moreover, it can be extended in a simple manner to more complex integrated rate equations. However, it is not completely general and cannot be applied to solving the integrated rate equations that describe many reactions with two substrates.

```
5100 Y = 0: GOSUB 5180: IF F = 0 THEN GOTO 5190
5110 Y = (Y + A0) / 2: IF Y > .99999 * A0 THEN GOTO 5190
5120 GOSUB 5180: IF F = 0 THEN GOTO 5190
5130 IF F < 0 THEN GOTO 5110
5140 F1 = 1 + KA / (A0 - Y)
5150 DY = F / F1: IF DY / A0 < .00001 THEN GOTO 5190
5160 Y = Y - DY: GOSUB 5180: IF F = 0 THEN GOTO 5190
5170 GOTO 5140
5180 F = Y - KA * LOG(1 - Y / A0) - VM * T: RETURN
5190 'end
```

FIG. 2. BASIC code to solve the integrated Michaelis–Menten equation by the modified Newton–Raphson method. It is assumed that on entry to this routine, values of V_m, K_a, $[A]_0$, and time have been defined. First (line 5100) the solution $y = 0$ is tested by calculating $F(y)$ in line 5180. If this is not a solution then successive values of $[A]_0/2$, $3[A]_0/4$, $7[A]_0/8$, etc., are tried (lines 5110 to 5130) until the solution is found or $F(y)$ is positive. In the latter case, the standard Newton–Raphson method is then used to refine the solution; $F'(y)$ is calculated (line 5140) then the correction $F(y)/F'(y)$ is compared to $[A]_0$ to ensure that the correction is not insignificant. The correction is made (line 5160) and the new value used in the next cycle unless it is already an exact solution.

A method that is virtually immune to failure was suggested by Boeker[5] and generalized further to more complex reactions by Duggleby and Wood.[6] The value of y that solves any integrated rate equation clearly lies between certain limits; it cannot be less than 0 nor greater than the equilibrium value (y_∞), which is easily calculated. If $F(y)$ is calculated at a value of y midway between these limits, the sign will indicate whether the estimate is above or below the solution. If it is below [$F(y)$ is negative], this value of y is taken as a new lower limit and the process repeated until $F(y)$ is positive. The value is then refined using the standard Newton–Raphson procedure. The computer code necessary to implement this method in BASIC is shown in Fig. 2 for the case $y_\infty = [A]_0$.

Nonlinear Regression

Fitting an integrated rate equation to a progress curve involves manipulating the parameters so as to minimize the sum of the squares of the difference between the measured value of y and that expected from the equation. In Eq. (4), the parameters that may be varied are V_m and K_a, although $[A]_0$ may be included as well if there is some uncertainty in its true value. Often it may be judicious to apply weighting factors to the

[5] E. A. Boeker, *Biochem. J.* **245,** 67 (1987).

[6] R. G. Duggleby and C. Wood, *Biochem. J.* **258,** 397 (1989); correction *Biochem. J.* **270,** 843 (1990).

individual squared differences if there is reason to believe that all measurements are not equally accurate.

Nonlinear regression methods fall into two classes: search methods and gradient methods.[7] In search methods, the parameters are varied systematically and at each set of values the sum of squares is calculated. This process continues until the minimum (or at least a local minimum) in the sum of squares surface is located. Gradient methods use information about the slope of the sum of squares surface at a particular point to predict the approximate location of the minimum. The gradient is then recalculated at this new location and used to improve the prediction of the location of the minimum.

Generally, gradient methods are faster than search methods. Moreover, they include information that will yield estimates of the standard errors of the fitted parameters. The principal disadvantage is that the gradients represent additional information that must be supplied. For most enzyme kinetic studies, including progress curve analysis, gradient methods have been used almost exclusively.

Perhaps the most well-known gradient procedure is the Gauss–Newton method that was first introduced to enzyme kineticists by Wilkinson,[8] and the reader wishing to understand nonlinear regression will find that article well worthwhile studying. A brief outline of this method is presented in Appendix 1 at the end of this chapter. The technique was applied to progress curve analysis by Nimmo and Atkins,[2] and much of this chapter is simply an extension of their work.

The gradient information is in the form of the differential of the fitted variable with respect to each of the fitted parameters. In the case of the integrated Michaelis–Menten equation, these partial differentials with respect to V_m and K_a are given by Eqs. (11) and (12), respectively, where $F'(y)$ is defined by Eq. (7):

$$\frac{\delta y}{\delta V_m} = \frac{t}{F'(y)} \tag{11}$$

$$\frac{\delta y}{\delta K_a} = \frac{\ln(1 - y/[A]_0)}{F'(y)} \tag{12}$$

Sometimes one may wish to treat $[A]_0$ as a third parameter to be estimated from the data, and to do so requires the partial differential with respect to $[A]_0$, which is given as Eq. (13):

[7] W. H. Swann, *FEBS Lett.* **2,** (Suppl.) S39 (1969).

[8] G. N. Wilkinson, *Biochem. J.* **80,** 324 (1961).

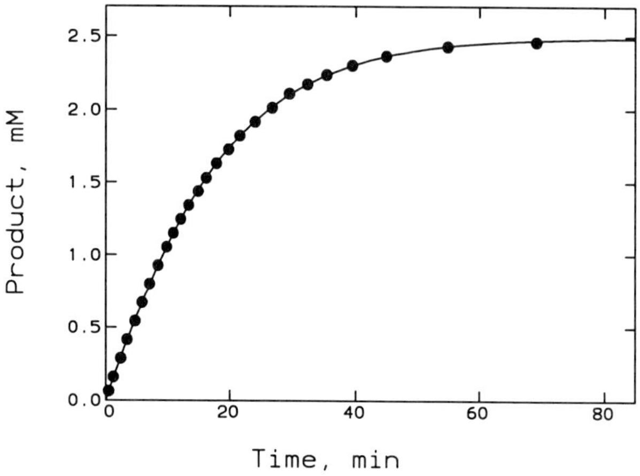

FIG. 3. Fit of the integrated Michaelis–Menten equation to a progress curve. The data represent the hydrolysis of 2.494 mM phenyl phosphate catalyzed by human prostate acid phosphatase, and were taken from Schønheyder.[9] The line is the fit by Eq. (4), with V_m = 0.4562 mM/min and K_a = 6.146 mM.

$$\frac{\delta y}{\delta[\mathrm{A}]_0} = \frac{K_a y}{([\mathrm{A}]_0 - y)F'(y)} \tag{13}$$

Although these partial differential equations are not hard to obtain for the integrated Michaelis–Menten equation, with more complex mechanisms their derivation becomes quite cumbersome. However, they can be approximated by numerical differentiation with little loss of accuracy in the final results. This process of numerical differentiation involves calculating y, changing the value of a parameter (say, K_a) by a small fraction of its value, then recalculating y. The differential $\delta y/\delta K_a$ is then approximated by $\Delta y/\Delta K_a$. It is more accurate if the parameter is perturbed both up and down by equal amounts and the partial derivative calculated from the total changes in y and K_a.

The effectiveness of numerical differentiation is illustrated by an analysis of data on the hydrolysis of phenyl phosphate by human prostate acid phosphatase (taken from Schønheyder[9]). Fitting the data using the algebraic expressions for the partial derivatives gave $V_m = 0.4562 \pm 0.0139$ mM/min and $K_a = 6.146 \pm 0.235$ mM. Using numerical differentiation, identical values and very similar standard errors were obtained: $V_m = 0.4562 \pm 0.0140$ mM/min and $K_a = 6.146 \pm 0.236$ mM. The fitted line gives an excellent description of the data (Fig. 3).

[9] F. Schønheyder, *Biochem. J.* **50,** 378 (1952).

TABLE II
EFFECT OF VARIATIONS IN $[A]_0$ ON THE FITTED VALUES OF V_m AND K_a

Variation	Concentration (mM)	V_m (mM/min)	K_a (mM)
−2.0%	2.4441	0.3410 ± 0.0172	4.024 ± 0.278
−1.0%	2.4691	0.3894 ± 0.0132	4.912 ± 0.218
−0.5%	2.4815	0.4199 ± 0.0121	5.475 ± 0.202
Fitted	2.4891 ± 0.0047	0.4412 ± 0.0191	5.868 ± 0.338
Nominal	2.4940	0.4562 ± 0.0139	6.146 ± 0.235
+0.5%	2.5065	0.5001 ± 0.0207	6.958 ± 0.352
+1.0%	2.5189	0.5541 ± 0.0332	7.960 ± 0.571
+2.0%	2.5439	0.7102 ± 0.0851	10.864 ± 1.493

[a] The expected value of $[A]_0$, taken from Schønheyder,[9] was 2.4940 mM. The data were fitted using Eq. (4) (the integrated Michaelis–Menten equation) using this nominal concentration and values that differed by the amounts shown in the first column. The data were also fitted using Eq. (4) treating $[A]_0$ as a third fitted parameter.

The results just described were obtained using DNRP53, a general nonlinear regression computer program written in our laboratory.[10] Most of the results to be described here are based on analyses with this program, copies of which may be obtained by writing to the author.

As noted earlier, it is sometimes desirable to treat $[A]_0$ as another fitted parameter to be estimated from the data rather than as a constant. The reason for this is that the fitted values of V_m and K_a are sensitive to quite small inaccuracies in the value of $[A]_0$, particularly when $[A]_0/K_a$ is not high.[11] This point is illustrated in Table II in which the data of Schønheyder[9] are further analyzed. Even small changes of $[A]_0$ lead to very large errors in the values of V_m and K_a. These uncertainties can be eliminated by including $[A]_0$ as a third fitted parameter; although the results obtained are not necessarily exactly correct, they will not be biased by an incorrect value of $[A]_0$.

Effects of Inhibitors

Dead-End Inhibition

A compound that is neither a substrate nor a product may combine with either the free enzyme or the enzyme–substrate complex, diverting a fraction of the enzyme into an inactive form. The effect of such dead-

[10] R. G. Duggleby, *Comput. Biol. Med.* **14,** 447 (1984).
[11] P. F. J. Newman, G. L. Atkins, and I. A. Nimmo, *Biochem. J.* **143,** 779 (1974).

end inhibitors will be to reduce V_m/K_a or V_m by a constant factor of $1 + [I]/K_i$, leading to competitive and uncompetitive inhibition, respectively. These effects can be analyzed using progress curve analysis to determine V_m and K_a in the presence of one or more concentrations of the inhibitor. However, it is the view of this author that these analyses do not properly fall within the ambit of progress curve analysis insofar as the inhibitor is not a compound whose concentration varies as the reaction proceeds. In much the same way, interactions of the enzyme with solvent ions and molecules (including H^+) may well affect the kinetics but will not be considered further in this chapter. The reader is referred to the review of Orsi and Tipton[1] for a discussion of the effects of dead-end inhibitors on progress curves.

Product Inhibition

So far it has been assumed that the accumulating product is noninhibitory, which in many cases is an unsafe assumption. The most common situation is where the product is a competitive inhibitor with respect to the single substrate. In this case the integrated rate equation is identical to Eq. (4), except that V_m and K_a are now apparent values (V_m' and K_a') that depend on a product inhibition constant (K_{ip}), $[A]_0$, and the concentration ($[P]_0$) of any product that is initially present:

$$V_m' = V_m K_{ip}/(K_{ip} - K_a) \quad (14)$$
$$K_a' = K_a(K_{ip} + [A]_0 + [P]_0)/(K_{ip} - K_a) \quad (15)$$

Equations (14) and (15) lead to a number of interesting points. First, because any given progress curve obtained with a particular combination of initial substrate and product concentrations can be defined by two parameters (V_m' and K_a'), it is clearly impossible to estimate all three of V_m, K_a, and K_{ip} from such a curve. Second, V_m' and K_a' must have the same sign, but both can be negative if the product is a strong inhibitor ($K_{ip} < K_a$). Third, K_a' is a linear function of the sum $[A]_0$ and $[P]_0$, whereas V_m' is independent of these concentrations. This then leads to the fourth point; it should be possible to estimate V_m, K_a, and K_{ip} using several progress curves, provided that they are obtained over a range of $[A]_0$ (or, less usually, of $[P]_0$).

The last point was illustrated by Duggleby and Morrison[3] in an analysis of five progress curves for the reaction catalyzed by prephenate dehydratase from which were obtained values for V_m (18.0 ± 0.8 U/mg), K_a (0.472 ± 0.008 m*M*), and K_{ip} (4.59 ± 0.62 m*M*). This reaction has a single product (apart from water), but reactions with two or more products can be analyzed in a similar manner when only one is inhibitory. Such is the case for the hydrolysis of *p*-nitrophenyl phosphate by potato acid phosphatase[3] which is not inhibited by *p*-nitrophenol (or by the *p*-nitrophenolate ion) but is strongly affected by phosphate ion. From eight progress curves, values for V_m (117 ± 8 U/mg), K_a (2.12 ± 0.20 m*M*), and K_{ip} (0.262 ± 0.011 m*M*) were obtained.

Note that because K_{ip} (0.262 m*M*) is less than K_a (2.12 m*M*), it would be predicted that both V_m' and K_a' from the analysis of a single curve should be negative. This prediction is confirmed by a reanalysis of the data obtained at 0.988 m*M* substrate, where the best fit progress curve (Fig. 4A) has values of $V_m' = -15.58 \pm 0.31$ U/mg and $K_a' = -1.443 \pm 0.014$ m*M*. The linear dependence of K_a' on ($[A]_0 + [P]_0$) that is predicted from Eq. (15) is illustrated in Fig. 4B.

In the case of potato acid phosphatase, the inhibitory product is competitive with the substrate. However, if the first product to leave was the stronger inhibitor, or if the enzyme has an iso mechanism,[12] more complex inhibition is observed. For these cases the integrated rate equation is a function of two inhibition constants (K_{is} and K_{ii}). In addition to the linear and logarithmic terms in y seen in Eq. (4), there is now a term in y^2 as well [Eq. (16)]:

$$\rho t = y + \gamma y^2/2 - \delta \ln(1 - y/[A]_0) \tag{16}$$

$$\rho = V_m K_{is} K_{ii}/(K_{is} K_{ii} - K_a K_{ii} + [P]_0 K_{is}) \tag{17}$$

$$\gamma = K_{is}/(K_{is} K_{ii} - K_a K_{ii} + [P]_0 K_{is}) \tag{18}$$

$$\delta = K_a K_{ii}(K_{is} + [A]_0 + [P]_0)/(K_{is} K_{ii} - K_a K_{ii} + [P]_0 K_{is}) \tag{19}$$

Reactions with Two Inhibitory Products

The next level of complexity beyond those exemplified by prephenate dehydratase and acid phosphatase is where there is one substrate but two inhibitory products. Despite this added complexity, the integrated rate equation has the same form as Eq. (16) although the definitions of ρ, γ, and δ are altered.[13]

Cox and Boeker[14] studied the enzyme-catalyzed decarboxylation of arginine to agmatine as an example of such a system. Their initial analysis was based on Eq. (20) which is identical in form to Eq. (16):

$$[E]_0 t = C_1 y + C_2 y^2/2 - C_f \ln(1 - y/[A]_0) \tag{20}$$

By separately fitting each progress curve to Eq. (20), they concluded that the C_2 term did not contribute significantly to the shape of the progress

[12] K. L. Rebholz and D. B. Northrop, this volume [9].
[13] R. G. Duggleby and J. F. Morrison, *Biochim. Biophys. Acta.* **568,** 357 (1979).
[14] T. T. Cox and E. A. Boeker, *Biochem. J.* **245,** 59 (1987).

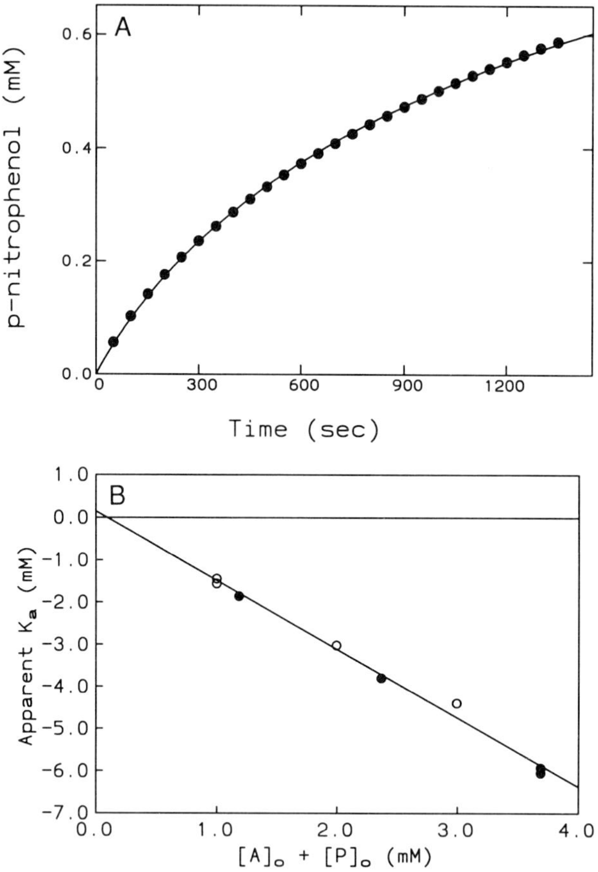

FIG. 4. Progress curve analysis of the reaction catalyzed by potato acid phosphatase. (A) Hydrolysis of 0.988 m*M* *p*-nitrophenyl phosphate and the fit of Eq. (4) to the data with the apparent values $V_m' = -15.58$ U/mg and $K_a' = -1.443$ m*M*. (B) Linear dependence of the apparent values on the total initial concentration of substrate plus product (phosphate), where the filled symbols represent reactions that contained added phosphate.

curve. By analyzing the relationship between C_1, C_f, and the initial concentrations of arginine and agmatine, they were able to determine several of the kinetic constants. Although the approach seems to have been successful, there is some doubt as to whether such a procedure would be useful in general. This is because fits to individual curves can give parameter values that vary wildly as a result of random fluctuations in the data.

A simple example illustrates this point. An exact progress curve was simulated for a single-substrate, single-product, irreversible reaction showing uncompetitive production inhibition, using $V_m = 1$, $K_a = 1$, $K_{ii} = 0.3$, and $[A]_0 = 3$. Nineteen points, equally spaced on the concentration axis, were selected from the curve, and these were perturbed by a random error with a standard deviation of 1% $[A]_0$. The appropriate progress curve equation was then fitted to the simulated curve; the entire process was repeated 100 times. Despite the fact that the simulated noise was quite small, 15 of the curves could not be fitted at all. For the remaining 85, V_m ranged from 0.68 to 42, K_a from −0.16 to 206.2, and K_{ii} from 0.12 to 0.42. In only 19 cases were all three parameters within 50% of the correct values, and in only four cases were they within 10%.

Thus, individual fits are not expected to give very reliable estimates, and we would recommend combining the entire set of progress curves and analyzing them using the complete integrated rate equation for the system. An example of a reaction that behaves kinetically as a one-substrate and two-product reaction and that was analyzed by overall fitting of the complete set of progress curves is described in the next section.

Complex Reactions

Reactions with Two Substrates

The oxidation of NADH catalyzed by lactate dehydrogenase has been studied[3] as an example of a reaction with two substrates. To simplify the analysis, the enzyme can be made to behave as a single-substrate system by raising the concentration of pyruvate to a constant value that is well in excess of that of NADH. The integrated rate equation is identical to Eq. (16) but with new definitions of ρ, γ, and δ [Eqs. (21)–(25)] where A, B, P, and Q refer to NADH, pyruvate, lactate, and NAD^+, respectively:

$$\rho = V_m/\Omega \tag{21}$$

$$\gamma = (1/K_{ip} - K_a/K_p{}^*K_{iq})/\Omega \tag{22}$$

$$\delta = K_a\{1 + ([A]_0 + [Q]_0)/K_{iq} + ([A]_0 + [P]_0)([A]_0 + [Q]_0)/K_p{}^*K_{iq}\}/\Omega \tag{23}$$

$$K_p{}^* = K_aK_{ib}K_p/K_{ia}K_b \tag{24}$$

$$\Omega = 1 + [P]_0/K_{ip} - (K_a/K_{iq})\{1 + ([A]_0 + [P]_0 + [Q]_0)/K_p{}^*K_{iq}\} \tag{25}$$

As it later turned out, the kinetic constants for lactate ($K_p{}^*$ and K_{ip}) are quite high relative to those for NAD^+ and NADH, which are in the micromolar range. Consequently, progress curves obtained in the absence of lactate will not accumulate enough of this product for it to have a significant influence. In effect, the reaction behaves exactly like acid phosphatase with one inhibitory product. If some reactions are conducted in the presence of added lactate, however, then the inhibition will be expressed; using this approach, all five of the kinetic constants for this reaction could be estimated.

For two-substrate reactions, using one substrate at a much higher concentration than the other provides a general means of simplifying the analysis, by eliminating some terms from the integrated rate equation. However, the choice of the substrate that is used in excess is constrained by the kinetic properties of the enzyme and the assay system. In the case of lactate dehydrogenase, pyruvate was used at a constant concentration of 4 m*M* while NADH never exceeded 0.15 m*M*. However, if one wanted to use NADH as the excess substrate some difficulties would arise. Because the Michaelis constant for pyruvate is approximately 0.2 m*M* under the reaction conditions employed, it would be necessary to run at least some of the progress curves with an initial pyruvate concentration well above this value, perhaps as high as 1 m*M*. Consequently, the initial NADH concentration, if it was to be considered constant as the reaction proceeded, would need to be 10 m*M* or higher. Although the resulting high absorbancies could be overcome by measurements away from the λ_{max}, one would still have to measure small absorbance changes against a large background, which would certainly introduce some noise into the data. Moreover, these high NADH concentrations might well have adverse effects on the enzyme.

Another two-substrate reaction that has been manipulated experimentally to simplify the progress curve analysis is the reaction catalyzed by aspartate aminotransferase.[15] The reaction is readily reversible, and the complete integrated rate equation contains kinetic constants for all four reactants. The system was simplified by using glutamate and oxaloacetate as substrates and by coupling the reaction to the oxidation of NADH by the addition of glutamate dehydrogenase and NH_4Cl. This coupling reaction has three functions: (a) it provides a means of monitoring the reaction; (b) it recycles all formed 2-ketoglutarate back to glutamate, causing the concentration of this substrate to remain constant at its initial value; and (c) it prevents the accumulation of 2-ketoglutarate, making the reaction irreversible and eliminating any kinetic constants associated with this reactant and with the back reaction from the integrated rate equation. The third point has one consequence that may not always be desirable;

[15] R. G. Duggleby and J. F. Morrison, *Biochim. Biophys. Acta* **526,** 398 (1978).

there is no way that the kinetic properties with respect to 2-ketoglutarate or the back reaction can be determined from the data.

The integrated rate equation for this reaction is identical to Eq. (16) where ρ is V_m/Ω [Eq. (21)], γ, δ, and Ω are defined by Eqs. (26)–(28), and A, B, and P refer to oxaloacetate, glutamate, and aspartate, respectively:

$$\gamma = K_b/[B]_0 K_{ip}\Omega \quad (26)$$
$$\delta = \{K_a + K_{ia}K_b([A]_0 + [P]_0)/[B]_0 K_{ip}\}/\Omega \quad (27)$$
$$\Omega = 1 + K_b\{1 + ([P]_0 - K_{ia})/K_{ip}\}/[B]_0 \quad (28)$$

All five kinetic parameters were estimated from the data obtained from 20 progress curves containing varying amounts of oxaloacetate, glutamate, and aspartate; two further repetitions of the entire experiment showed that each parameter was estimated reliably.

This recycling assay offers a general means to simplify progress curve analysis of any aminotransferase that uses glutamate as an amino donor, which includes most of the enzymes of this type. Similar types of recycling assays could be devised for kinases to study them in the direction of ATP utilization (using pyruvate kinase) or ATP synthesis (using hexokinase). Even when recycling is not feasible, it still may be possible to eliminate terms from the integrated rate equation by removing one or more products using a coupling reaction. The potential of these methods for product removal do not appear to have been exploited in practice.

Reversible Reactions

Very few reversible reactions have been studied by progress curve analysis; the principal exception is that catalyzed by fumarase (fumarate hydratase), which has been investigated by several groups.[16-18] The integrated rate equation for this one-substrate and one-product reaction contains four kinetic parameters: a maximum velocity for each direction (V_f and V_r) and a Michaelis constant for each reactant (K_f and K_r). The equation is identical in form to Eq. (4) except that V_m and K_a are replaced with apparent values (V_m' and K_a') while the logarithmic term is $\ln(1 - y/y_\infty)$; that is, instead of the reaction proceeding until y equals $[A]_0$, it stops when y reaches the equilibrium value. The definitions of V_m' and K_a' are given as Eqs. (29) and (30);

$$V_m' = (V_f K_r + V_r K_f)/(K_r - K_f) \quad (29)$$

[16] M. Taraszka and R. A. Alberty, *J. Phys. Chem.* **68,** 3368 (1964).
[17] I. G. Darvey, R. Shrager, and L. D. Kohn. *J. Biol. Chem.* **250,** 4696 (1975).
[18] R. G. Duggleby and J. C. Nash, *Biochem. J.* **257,** 57 (1989).

$$K_a' = K_f\{K_r + ([A]_0 + [P]_0)(1 + V_r/V_m')\}(K_r - K_f) \qquad (30)$$

The value of y_∞ depends on the equilibrium constant which, in turn, can be expressed in terms of the four kinetic parameters through the Haldane relationship. One such expression for y_∞ is given as Eq. (31):

$$y_\infty = ([A]_0 V_f K_r - [P]_0 V_r K_f)/(V_f K_r + V_r K_f) \qquad (31)$$

Using ten progress curves containing a range of fumarate concentrations but no added malate, it was shown that all four kinetic parameters could be estimated, although those associated with the substrate (fumarate) were determined with more accuracy than those associated with the product (malate). A similar result was obtained using progress curves measured with a range of malate concentrations but no added fumarate; the kinetic parameters associated with the substrate had smaller standard errors than those associated with the product. Overall, somewhat better results were obtained by using fumarate as the starting reactant, and this is probably because the equilibrium constant is approximately 4.0 in this direction. The reaction proceeds further so more information is obtained from each progress curve. Similar conclusions have been drawn from a study of the reaction catalyzed by enolase.

General Approach to Deriving Integrated Rate Equations

The complexity of integrated rate equations escalates rapidly as one moves from one to two substrates, from one to two products, and from irreversible to reversible reactions. Sometimes, owing to a fortuitous combination of kinetic parameters, the reaction can be treated as a simpler system as for the acid phosphatase example mentioned above. In other cases, the experimental conditions can be manipulated to achieve a similar result by maintaining one substrate at a constant concentration, as for lactate dehydrogenase[3] and aspartate aminotransferase.[15]

In general one would like to be able to deal with any reaction, and Boeker[19–21] has provided a suitable theoretical framework for reactions with up to two substrates and products. This framework was further developed by Duggleby and Wood[6] as follows.

Unbranched kinetic mechanisms with two or fewer substrates and products have a kinetic mechanism that is no more complex than Eq. (32):

[19] E. A. Boeker, *Biochem. J.* **223,** 15 (1984).

[20] E. A. Boeker, *Experientia* **40,** 453 (1984).

[21] E. A. Boeker, *Biochem. J.* **226,** 29 (1985).

$$\frac{dy}{dt} = \frac{V_f J_{ab}([A]_t[B]_t - [P]_t[Q]_t/K_{eq})}{\begin{aligned}(J_0 &+ J_a[A]_t + J_b[B]_t + J_p[P]_t + J_q[Q]_t + J_{ab}[A]_t[B]_t + J_{ap}[A]_t[P]_t \\ &+ J_{aq}[A]_t[Q]_t + J_{bp}[B]_t[P]_t + J_{bq}[B]_t[Q]_t + J_{pq}[P]_t[Q]_t \\ &+ J_{abp}[A]_t[B]_t[P]_t + J_{abq}[A]_t[B]_t[Q]_t + J_{apq}[A]_t[P]_t[Q]_t \\ &+ J_{bpq}[B]_t[P]_t[Q]_t + J_{abpq}[A]_t[B]_t[P]_t[Q]_t)\end{aligned}} \quad (32)$$

In Eq. (32), the various J terms represent combinations of the usual kinetic constants (Michaelis and inhibition constants), and for most mechanisms several of these terms will be absent from the rate equation. Except for the unusual case where $[A]_0$ and $[B]_0$ are exactly equal, the integrated rate equation is given by Eq. (33):

$$\begin{aligned}Ct = C_1Cy + C_2Cy^2/2 + C_3Cy^3/3 - C_fC\ln(1 - y/y_\infty) \\ + C_sC\ln[1 - y/(y_\infty + D)]\end{aligned} \quad (33)$$

Boeker[19-21] has given expressions that define all the variables of Eq. (33) in terms of combinations of initial concentrations and the various J constants [Eq. (32)], although it should be noted that there are some typographical errors that have been corrected.[6] There is a common factor (C) that could be canceled throughout, but this term is included because it simplifies the definitions (e.g., C_1C is a less complicated expression than C_1).

Thus, to obtain the integrated rate equation for a specific kinetic mechanism, the J terms and the Haldane relationship are defined, then substituted into the appropriate expressions for C, C_1, C_2, C_3, C_f, y_∞, and D. This is a purely mechanical process, and manipulation of the symbols is readily automated. Duggleby and Wood[6] have written a computer program (AGIRE) to achieve this automation. Given definitions of each of the J constants, the program generates computer code that defines both $F(y)$ and $F'(y)$. This code is directly compatible with the DNRP53[10] nonlinear regression program. Thus, fitting of progress curves requires only a knowledge of the appropriate differential rate equations, and these have been catalogued extensively (see, e.g., Segel[22]).

As noted earlier, Eqs. (32) and (33) may not be applicable to random mechanisms as the branches can introduce squared terms in substrate and product. This will make the general rate equation somewhat more complex, and the definitions of Boeker[19-21] may not apply. The principles re-

[22] I. H. Segel, "Enzyme Kinetics." Springer-Verlag, Berlin, 1975.

```
          P R O G R A M   A G I R E

Automatic Generation of Integrated Rate Equations

Please specify the type of reaction catalysed.

Enter the number of substrates (1 or 2): 1                       ←
Enter the number of  products  (1 or 2): 2                       ←
Reversible/Irreversible reaction  (R/I): I                       ←

The reaction has 1 substrate, 2 products and is irreversible

Is this all correct (Y/N): Y                                     ←

The differential form of the rate equation is taken to be:

                              VfJaA
v = ──────────────────────────────────────────────────────────
    Jo + JaA + JpP + JqQ + JapAP + JaqAQ + JpqPQ + JapqAPQ

You must supply the combinations of kinetic constants which
go to make up these J terms and the Haldane relationship.

 Term        Combination
───────      ───────────

( 1) Jo:     Ka                                                  ←
( 2) Ja:     1                                                   ←
( 3) Jp:                                                         ←
( 4) Jq:     Ka/Kiq                                              ←
( 5) Jap:    1/Kip                                               ←
( 6) Jaq:                                                        ←
( 7) Jpq:    Ka/KpKiq                                            ←
( 8) Japq:                                                       ←

Terms for correction (0 if all OK): 0                            ←

The following  5  kinetic constants have been used
Vf  Ka  Kiq  Kip  Kp
Are you happy with the analysis so far (Y/N): Y                  ←

In the integrated rate equation the various kinetic
constants will be referred to as follows.

B( 1) represents Vf
B( 2) represents Ka
B( 3) represents Kiq
B( 4) represents Kip
B( 5) represents Kp

BASIC code generation in progress
Completed and saved in 12I.MDL
```

main the same, however, and it should not be overwhelmingly difficult to extend this method to random mechanisms.

The use of the AGIRE program is illustrated with the lactate dehydrogenase reaction considered earlier. Under conditions where the pyruvate concentration is high, the rate equation simplifies to the particular form [Eq. (34), where A, P, and Q refer to NADH, lactate, and NAD^+, respectively] of the general equation for an irreversible, one-substrate, two-product reaction that is given as Eq. (35):

$$v = \frac{V_m[A]_t}{K_a + [A]_t + [Q]_t K_a/K_{iq} + [A]_t[P]_t/K_{ip} + [P]_t[Q]_t K_a/K_p{}^*K_{iq}} \tag{34}$$

$$\frac{dy}{dt} = \frac{V_f J_a[A]_t}{(J_0 + J_a[A]_t + J_p[P]_t + J_q[Q]_t + J_{ap}[A]_t[P]_t + J_{aq}[A]_t[Q]_t + J_{pq}[P]_t[Q]_t + J_{apq}[A]_t[P]_t[Q]_t} \tag{35}$$

This leads to the following definitions: $J_0 = K_a$; $J_a = 1$; $J_p = 0$; $J_q = K_a/K_{iq}$; $J_{ap} = 1/K_{ip}$; $J_{aq} = 0$; $J_{pq} = K_a/K_p{}^*K_{iq}$; and $J_{apq} = 0$. As illustrated in Fig. 5, it is now a simple matter to create a nonlinear regression program to fit the integrated form of Eq. (34) to the progress curve data. All that needs to be done to create the fitting program is to merge the code generated by AGIRE (which is stored in a file called 12I.MDL in this example) with the DNRP53 program. AGIRE is available from the author.

Practical Considerations

Selwyn's Test

As noted earlier, progress curve analysis may be applied only when the shape of the progress curve depends solely on the variations in reactant concentrations that are due to the catalyzed reaction. It is important to show that the enzyme does not undergo a progressive inactivation in the

FIG. 5. Use of the AGIRE computer program. The lactate dehydrogenase reaction behaves as a one-substrate, two-product, irreversible system when pyruvate is added at a high concentration. Comparison of the general rate equation [Eq. (35)] with the specific rate equation [Eq. (34)] allows the J terms to be defined. Using this information, the AGIRE program generates appropriate computer code for solving and fitting the integrated rate equation. User responses follow the colon on the lines indicated arrows; where there is nothing following the colon (e.g., for the Jp term), the "enter" key was pressed without any other response.

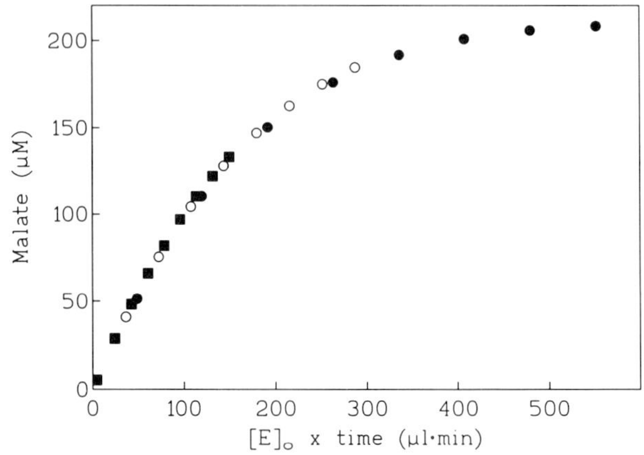

FIG. 6. Selwyn's test on the reaction catalyzed by fumarase. Assays contained 253 μM fumarate and 6 (■), 12 (○), or 24 μl (●) of fumarase, where 1 μl corresponds to a final assay concentration of 18 pM.

assay and that the reactants do not participate in any uncatalyzed side reactions. There is a simple method to show whether the system is suitable for progress curve analysis. This method, known as Selwyn's test,[23] relies on the fact that integrated rate equations have the general form given by Eq. (36):

$$[\mathrm{E}]_0 t = f(y, k, [\mathrm{X}]_0) \tag{36}$$

where k represents the various rate constants, $[\mathrm{X}]_0$ is the initial reactant concentrations, and $[\mathrm{E}]_0$ is the total enzyme concentration. Plots of y versus $[\mathrm{E}]_0 t$ obtained at several enzyme concentrations will be superimposable unless there is enzyme, substrate, or product instability. The results of Selwyn's test on the reaction catalyzed by fumarase[18] are illustrated in Fig. 6.

Preliminary Fitting

Integrated rate equations are formulated in terms of product formed or substrate utilized. However, what is measured experimentally is usually a linearly related quantity such as an absorbance, and conversion to concentration requires the measurements to be multiplied by a constant factor, frequently after subtraction of a zero-time value. The zero-time values have to be extremely reliable: a small inaccuracy can introduce a systematic error that can have quite substantial effects on subsequent analysis.[11]

[23] M. J. Selwyn, *Biochim. Biophys. Acta* **105,** 103 (1965).

One approach to this problem is to fit a generalized progress curve to the individual curves, and a suitable form might be a modified form of Eq. (33) in which y and y_∞ include an offset term (y_o) that is estimated from the data. In practice, Eq. (33) contains far too many parameters to be estimated from a single progress curve, and we have found empirically that Eq. (37) describes almost any progress curve very well, despite the fact that it lacks terms in y^2 and y^3 and has only a single logarithmic term:

$$V_m' t = \{y - y_o\} - K_a' \ln(1 - \{y - y_o\}/\{y_\infty - y_o\}) \qquad (37)$$

If we are to use Eq. (37) to estimate y_o we must fit this to the data, which implies that we will also be estimating V_m' and K_a', and perhaps y_∞ as well, from the same progress curve. The difficulty here is that nonlinear regression requires initial estimates of the fitted parameters and, although it should not be difficult to make reasonable guesses about the values of y_o and y_∞, initial values for V_m' and K_a' may not be so easy to obtain. For example, it was shown earlier that the progress curve illustrated in Fig. 4A has values of V_m' and K_a' of -15.58 U/mg and -1.443 m*M*, respectively, but it would be almost impossible to guess at these values by visual inspection of the curve.

Duggleby[24] has reparameterized Eq. (37) in terms of quantities that have a more straightforward relationship to the shape of a progress curve. These are the initial velocity (v_0) and the time for the reaction to reach half-completion ($t_{1/2}$). The resulting expression is shown as Eq. (38):

$$t = \frac{\{y - y_o\}}{v_0} + \frac{2v_0 t_{1/2} - \{y_\infty - y_o\}}{v_0(1 - \ln 4)}\left[\frac{\{y - y_o\}}{\{y_\infty - y_o\}} + \ln\left(\frac{\{y_\infty - y\}}{\{y_\infty - y_o\}}\right)\right] \qquad (38)$$

The value of v_0 so calculated can be used in conventional initial velocity analysis, and it was for this purpose that Eq. (38) was first developed and shown to perform well even when the integrated rate equation is not strictly valid owing to enzyme inactivation. It was later used to estimate y_∞ in an analysis of progress curves of an unstable enzyme.[25] If required, V_m' and K_a' can be calculated using Eqs. (39) and (40):

$$V_m' = v_0(1 - \ln 4)/(2v_0 t_{1/2}/y_\infty - \ln 4) \qquad (39)$$
$$K_a' = y_\infty(1 - 2v_0 t_{1/2}/y_\infty)/(2v_0 t_{1/2}/y_\infty - \ln 4) \qquad (40)$$

The reason for performing a fit to Eq. (38) is to estimate, and correct for, y_o; however, the value of y_∞ is often useful as well. It has been mentioned previously that systematic errors in the substrate concentration

[24] R. G. Duggleby, *Biochem. J.* **228,** 55 (1985).
[25] S. J. Pike and R. G. Duggleby, *Biochem. J.* **244,** 781 (1987).

can cause substantial errors in the subsequent analysis; estimating y_∞ from the data offers a solution to this problem.

Experimental Design

It is relatively easy to collect and analyze individual progress curves. If the exercise is to be at all useful, however, then the initial reactant concentrations for each progress curve must be considered carefully. Except in the simplest of cases, this question of experimental design has relied more on intuition than on detailed analysis.

When we are dealing with a system that obeys Eq. (4) then there is just one reactant, and the question to be addressed is what initial substrate concentration should be used. Defining $R = y/[A]_0$ (i.e., the fractional reaction, which will range from zero at $t = 0$ to unity as $t \to \infty$), Eq. (4) is rewritten as Eq. (41):

$$t = R[A]_0/V_m - (K_a/V_m)\ln(1 - R) \tag{41}$$

If $[A]_0$ is high, the first term in Eq. (41) will dominate, and the curve will contain information about V_m only. For example, if $[A]_0$ is five times K_a then the first term will exceed the second for more than 99% of the progress curve. In contrast, if $[A]_0$ is low, the second term will dominate and the curve will contain information about K_a/V_m only. Clearly, $[A]_0$ must take an intermediate value, and R must subtend a sufficient range that both terms have a sizable influence. Suppose that at the end of the reaction we want the first and second terms to be equal; it is easy to show that Eq. (42) is true:

$$[A]_0/K_a = -(1/R)\ln(1 - R) \tag{42}$$

Values calculated from this relationship are shown in Table III, from which it is seen that if the reaction is stopped when it is 80% complete, the starting substrate concentration should be twice K_a. If the reaction is allowed to proceed to 95% completion, $[A]_0$ should be three times K_a. Very similar results were obtained by Duggleby and Clarke[26] (Table III), based on calculations of the expected standard error of K_a. Moreover, the correctness of this design was verified experimentally using pyruvate kinase under conditions where it behaves as a single-substrate (phosphoenolpyruvate) reaction that is not inhibited by products.

A quantitative analysis of the experimental designs appropriate for more complex reactions has not been published. However, some general remarks can be made from an examination of the relevant integrated rate

[26] R. G. Duggleby and R. B. Clarke, *Biochim. Biophys. Acta* **1080,** 231 (1991).

TABLE III
EFFECT OF EXTENT OF REACTION ON OPTIMAL INITIAL SUBSTRATE CONCENTRATION

	$[A]_0/K_a$	
R	Eq. (42)	Optimal
0.50	1.39	1.38
0.55	1.45	1.44
0.60	1.53	1.52
0.65	1.62	1.61
0.70	1.72	1.70
0.75	1.85	1.82
0.80	2.01	1.96
0.85	2.23	2.14
0.90	2.56	2.38
0.95	3.15	2.71

[a] The extent of reaction, R, represents the total fraction of substrate utilized at the end of the progress curve. Using the values of R shown in the first column, $[A]_0/K_a$ was obtained using Eq. (42), or the optimal value was calculated as described by Duggleby and Clarke.[26]

equations. For an irreversible single-substrate reaction showing competitive product inhibition, it has already been pointed out that it is necessary to measure the progress curves obtained over a range of concentrations of $[A]_0 + [P]_0$. This is evident from Eqs. (14) and (15); the simultaneous estimation of V_m, K_a, and K_{ip} requires that K_a' is not a constant, and this will occur only when $[A] + [P]_0$ is varied. It will also be evident that these concentrations must be varied over a range that encompasses the value of K_{ip}. Normally this would be done by varying $[A]_0$ only, but if K_{ip} is much greater than K_a, addition of product may be advisable. For a reversible reaction involving one substrate and one product [Eqs. (29) and (30)], a similar design is applicable although the range of $[A]_0 + [P]_0$ is required is such that $([A]_0 + [P]_0)(1 + V_r/V_m')$ spans the value of K_r. Clearly, more work is needed to quantify the experimental designs for progress curve analysis.

Unstable Enzymes

When an enzyme fails Selwyn's test because of instability, the advantages of progress curve analysis are lost. This is unfortunate, and we have tried previously to overcome this limitation.[25,27] It has been shown that it is possible, from the residual substrate concentration after prolonged incubation, to determine the inactivation rate constants of various complexes along the catalytic pathway. However, the normal kinetic constants (maximum velocity, Michaelis constants, and inhibition constants) are not obtained readily from the data. Using an analysis that includes provision for enzyme inactivation, it has now been shown[28] that progress curves can be used for kinetic characterization of an unstable enzyme.

Consider an enzyme that obeys Eq. (1) but where the free enzyme and the enzyme–substrate complex are each unstable, being converted to inactive forms with rate constants of j_1 and j_2, respectively. The kinetics of this system are described by two differential equations defining the rate of loss of active enzyme $[\mathrm{E}']_t$ [Eq. (43)] and the rate of utilization of substrate $[\mathrm{A}]_t$ [Eq. (44), where $k_3[\mathrm{E}']_t$ equals V_m at $t = 0$]:

$$\frac{d[\mathrm{E}']_t}{dt} = \frac{-[\mathrm{E}']_t(j_1 + j_2[\mathrm{A}]_t/K_\mathrm{a})}{1 + [\mathrm{A}]_t/K_\mathrm{a}} \tag{43}$$

$$\frac{d[\mathrm{A}]_t}{dt} = \frac{-k_3[\mathrm{E}']_t[\mathrm{A}]_t/K_\mathrm{a}}{1 + [\mathrm{A}]_t/K_\mathrm{a}} \tag{44}$$

Applying progress curve analysis to this system requires calculation of $[\mathrm{A}]_t$ at any time, but these differential equations are coupled; that is, to obtain an expression for $[\mathrm{E}']_t$ from Eq. (43) requires a knowledge of $[\mathrm{A}]_t$ while to obtain an expression for $[\mathrm{A}]_t$ from Eq. (44) requires a knowledge of $[\mathrm{E}']_t$. Except in some special cases (e.g., when j_1 is equal to zero or to j_2), Eqs. (43) and (44) cannot be solved algebraically. They can, however, be solved numerically; given particular values of the parameters, as well as defined starting conditions ($[\mathrm{E}']_0$ and $[\mathrm{A}]_0$), Eqs. (43) and (44) can be evaluated. These rates are applied over a small time interval to calculate $[\mathrm{E}']_t$ and $[\mathrm{A}]_t$ after this time. The process is then repeated as often as required to obtain the entire progress curve for substrate utilization.

This is a relatively crude method, and more sophisticated algorithms have been described.[29] The third/fourth-order Runge–Kutta–Fehlberg numerical integration algorithm with automatic step size control has been incorporated into the DNRP53 computer program to analyze reaction systems described by coupled differential equations. The experimental system that has been used to evaluate the method is bovine intestinal

[27] R. G. Duggleby, *J. Theor. Biol.* **123,** 67 (1986).

[28] R. G. Duggleby, *Biochim. Biophys. Acta* **1205,** 268 (1994).

[29] J. Stoer and R. Bulirsch, "Introduction to Numerical Analysis." Springer-Verlag, New York, 1980.

mucosal alkaline phosphatase. This is a slightly more complex system than that described by Eq. (43) and (44) because the enzyme is inhibited by one of its products, phosphate ion. Although the enzyme is normally quite stable, the essential zinc ion can be removed by EGTA, leading to inactivation. This instability is illustrated by applying Selwyn's test (Fig. 7A) where the curves are seen clearly not to be superimposable. It was shown previously[25] that only the free enzyme is susceptible to inactivation; in the enzyme–substrate complex the zinc is locked in, as it is in the

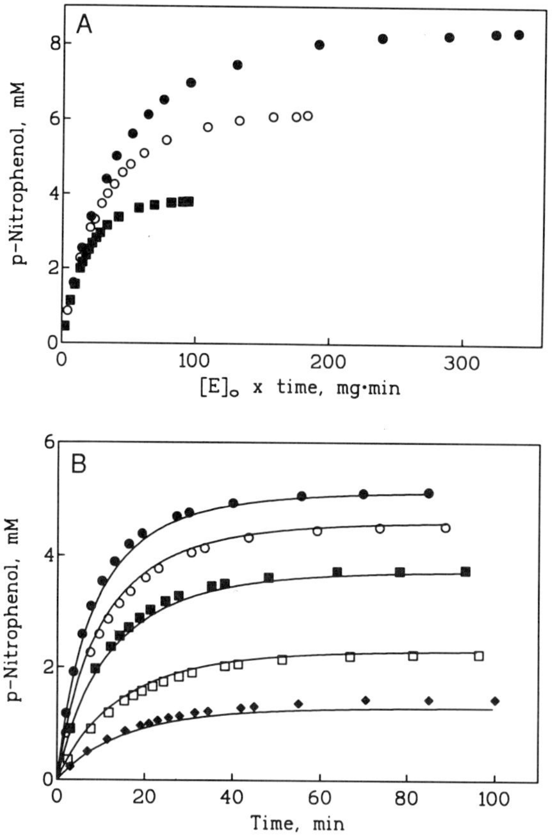

FIG. 7. Progress curve analysis of an unstable enzyme, alkaline phosphatase. (A) Results of Selwyn's test. Each reaction contained 10 m*M* *p*-nitrophenyl phosphate and 0.846 (■), 1.692 (○), or 3.384 mg (●) of enzyme. (B) Data obtained with 6 m*M* substrate and 0.423 (◆), 0.846 (□), 1.692 (■), 2.538 (○), or 3.384 mg (●) of enzyme. The curves are the results of fitting Eqs. (45) and (46) to the data shown plus similar data obtained at 10, 8, 4, and 2 m*M* substrate.

enzyme–phosphate complex. The system is described by Eqs. (45) and (46):

$$\frac{d[E']_t}{dt} = \frac{-j_1[E']_t}{1 + [A]_t/K_a + [P]_t/K_p} \tag{45}$$

$$\frac{d[P]_t}{dt} = \frac{k_3[E']_t[A]_t/K_a}{1 + [A]_t/K_a + [P]_t/K_p} \tag{46}$$

Some representative progress curves for the hydrolysis of *p*-nitrophenyl phosphate are shown in Fig. 7B. Fitting Eqs. (45) and(46) to the data gave kinetic parameters that are in excellent agreement with those obtained previously, and, as shown in the curves in Fig. 7B, describe the data quite well.

More complex schemes than that described by Eqs. (45) and (46) can be analyzed in much the same way. For example, the added possibility of instability of both the enzyme–substrate and enzyme–product complexes could be included. Further, there are a number of other possible situations where progress curve analysis should be useful but has not be applied since the underlying model leads to coupled differential equations. For example, it is known that a number of compounds act as slow-binding enzyme inhibitors.[30] The slow onset of inhibition is best studied by following progress curves, but the analyses currently available require that this is be done under conditions where substrate depletion is insignificant. This may necessitate high concentrations of substrate that tend to obscure the very inhibition one wishes to observe. The present approach offers a facile solution to this problem that will allow the progress curves for slow-binding enzyme inhibitors to be analyzed even when there is considerable substrate depletion.

Conclusions

The progress curves of enzyme-catalyzed reactions are a rich source of kinetic information. Provided suitable methods are used to analyze the data, relatively few such curves can allow a full quantitative description of the dependence of activity on the concentrations of substrates and products. Although difficulties in data analysis have impeded the development of this technique, there is now little reason why it could not be applied to a wide variety of enzyme kinetic studies. Problems of enzyme instability that distort progress curves from the shape dictated purely from the effects of substrate depletion and inhibitory product accumulation can now be overcome by including enzyme decay in the kinetic scheme.

[30] S. E. Szedlacsek and R.G. Duggleby, this volume [6].

Straightforward methods are available for analyzing all unbranched reaction mechanisms with two or fewer substrates and products, and there is no reason to suppose that more complex situations, such as random mechanisms, will not become amenable to progress curve analysis in the near future.

Appendix 1: Regression Analysis

The fundamentals of regression analysis are not difficult to understand and are best considered in relation to linear regression. The basic problem is to fit an equation to experimental data by minimizing the sum of the squares (*SSQ*) of the differences between a series of experimental values (y_i) and those predicted (μ_i) by the fitted line:

$$SSQ = \Sigma(y_i - \mu_i)^2 \tag{A1}$$

Suppose that the equation for a fitted line is a function of a single parameter (β) and some combination (X) of independent variables:

$$\mu_i = f(\beta, X) \tag{A2}$$

Equation (A2) is said to be *linear* if β appears as a constant multiplier; if this is so, we may partition Eq. (A2) as shown by Eq. (A3):

$$\mu_i = \beta f_1(X_1) + f_2(X_2) \tag{A3}$$

Substituting Eq. (A3) into Eq. (A1), introducing $y_i' = y_i - f_2(X_2)$, and expanding leads to Eq. (A4):

$$SSQ = \Sigma(y_i')^2 - 2\beta\,\Sigma[y_i'f_1(X_1)] + \beta^2\,\Sigma[f_1(X_1)]^2 \tag{A4}$$

From simple calculus we can find the value of β that yields the minimum *SSQ* by differentiating with respect to β and setting the resulting expression equal to zero:

$$0 = -2\,\Sigma[y_i'f_1(X_1)] + 2\beta\,\Sigma[f_1(X_1)]^2 \tag{A5}$$

$$\beta = \Sigma[y_i'f_1(X_1)]/\Sigma[f_1(X_1)]^2 \tag{A6}$$

This procedure is easily generalized if Eq. (A2) has several linear parameters to be estimated; instead of a single linear equation [Eq. (A5)] we obtain a set of linear simultaneous equations that may be solved readily.

In the Gauss–Newton method of nonlinear regression we start with an expression of the same form as Eq. (A2), but, because it is nonlinear, it cannot be rewritten in the same form as Eq. (A3). However, suppose we have an approximate value (B) of the nonlinear parameter β that differs from the best fit value by an unknown amount (b). We may rewrite Eq. (A2) as Eq. (A7):

$$\mu_i = f(B + b, X) \tag{A7}$$

From Taylor's theorem, the right-hand side of Eq. (A7) may be expanded as an infinite series [Eq. (A8)] involving successive derivatives (f', f'', f''', etc.) of the function evaluated at the current value B:

$$\mu_i = f(B, X) + bf'(B, X) + b^2 f''(B, X)/2! + b^3 f'''(B, X)/3! + \ldots \tag{A8}$$

Provided b is small (i.e., B is a reasonably good estimate of β), the terms in b^2 and higher powers can be ignored, giving Eq. (A9):

$$\mu_i \approx f(B, X) + bf'(B, X) \tag{A9}$$

Subtracting the experimental values (y_i) from both sides and defining the residual r_i as the different between y_i and the value calculated from the equation at the current value of the parameter [$f(B, X)$], we get Eq. (A10):

$$y_i - \mu_i = r_i - bf'(B, X) \tag{A10}$$

An expression for SSQ can now be written as before, then differentiated and solved for b [Eqs. (A11)–(A13)]:

$$SSQ = \Sigma(r_i)^2 - 2b\,\Sigma[r_i f'(B, X)] + b^2\,\Sigma[f'(B, X)]^2 \tag{A11}$$
$$0 = -2\,\Sigma[r_i f'(B, X)] + 2b\,\Sigma[f'(B, X)]^2 \tag{A12}$$
$$b = \Sigma[r_i f'(B, X)]/\Sigma[f'(B, X)]^2 \tag{A13}$$

From the relationship $\beta = B + b$, we would be able to use Eq. (A13) to calculate the best fit value, except for the fact that an approximation was made in going from Eq. (A8) to Eq. (A9). Provided the approximation is good, $B + b$ will be a better estimate of β than was B, and we can repeat the whole process using the refined estimate. The process of refinement is continued until no further improvement is obtained.

The procedure can be generalized to several nonlinear parameters by observing that Eq.(A13) has the same form as Eq. (A6) with b substituted for β, r_i replacing y_i', and the derivative $f'(B, X)$ used instead of $f_1(X_1)$. Thus, if the original function has several parameters we will obtain a system of linear simultaneous equations instead of the single equation [Eq. (A12)]. These may be solved to yield corrections for each of the parameters, and, as before, repeated application of the process will yield the best fit values.

Acknowledgments

This work supported by the Australian Research Council. The author is indebted to Prof. Dexter Northrop for useful comments on this chapter.

[4] Site-Directed Mutagenesis: A Tool for Studying Enzyme Catalysis

By BRYCE V. PLAPP

Introduction

Enzymes catalyze reactions by binding and orienting substrates in an active site formed with many amino acid residues that participate in the chemical transformation. Three-dimensional structures determined by X-ray crystallography show that enzymes bind substrates with a combination of ionic and hydrogen bonds and van der Waals interactions. The specificity of enzymes is due to multiple interactions that define the size and shape of the active site. The enzyme catalyzes the reaction by affecting solvation and the electrostatic environment and by facilitating proton transfers and covalent chemistry at the reaction center. Understanding the principles of enzyme action should allow us to design specific catalysts.

This chapter considers the purposes for using site-directed mutagenesis, some of the precautions for such studies, and an overall approach that is designed to provide maximal information. Some studies will be selected to illustrate what can be learned about the marvels of enzymes and to show why some conclusions are uncertain. Other articles have described the strategies and details for the techniques of site-directed mutagenesis and recombinant DNA technology.[1] Results with some enzymes have been reviewed previously.[2–7]

Purposes

The ability to substitute selected amino acid residues by site-directed mutagenesis and to express large amounts of enzyme with recombinant DNA technology enables a powerful approach for testing hypotheses for catalysis derived from knowledge of enzyme structures, kinetics, chemical modifications, and bioorganic chemistry. What can we learn from site-directed mutagenesis? A primary goal has been to identify which amino

[1] This series, Vols. 68, 100, 101, 153, 154, 155, 216, and 218.
[2] J. A. Gerlt, *Chem. Rev.* **87,** 1079 (1987).
[3] A. R. Fersht, *Biochemistry* **26,** 8031 (1987).
[4] M.-D. Tsai and H. Yan, *Biochemistry* **30,** 6806 (1990).
[5] E. W. Miles, *Adv. Enzymol.* **64,** 93 (1991).
[6] F. C. Hartman and M. R. Harpel, *Adv. Enzymol.* **67,** 1 (1993).
[7] D. N. Silverman, this volume [18].

METHODS IN ENZYMOLOGY, VOL. 249

acid residues in the active site are "essential," or critical, for catalysis. These usually include residues that appear to participate in acid–base catalysis or interact directly with the substrates. The residues may be conserved in homologous enzymes from different species or be identified by chemical modification or by inspection of the three-dimensional structure. An implicit goal of such studies is an estimation of the relative importance of a residue for catalysis. However, the change in activity cannot be the only criterion for assessing involvement of a residue, and circular arguments arise. If we are attempting to determine the magnitude of the contribution to catalysis, we cannot use the change in activity to judge whether the residue contributes. Some investigators seem to expect that substitution of an essential residue should totally inactivate the enzyme, but a 10-fold change in activity could represent the actual contribution. What we have here is not a failure to communicate, but rather a crisis in comprehension of catalysis.

A second purpose is to understand the evolution of catalytic efficiency, which requires chemical catalysis and specificity of binding. Bringing the substrates together and utilizing the binding energy provide the major contribution to catalysis.[8] Thus, residues that make van der Waals contacts or hydrogen bonds can be just as important for catalysis as reactive residues that participate in the chemistry.[3] Substituting amino acid residues permits the creative scientist to attempt to reproduce the evolution of catalytic function.[9] For a modest and practical goal, the origins of the different substrate specificities can be determined. Today, remodeling of substrate specificity is still difficult and unpredictable.[10,11] Additional purposes have included studies on structural features that control conformational changes, allosteric effects, and protein dynamics.

Precautions

Site-directed mutagenesis studies raise as many questions as they answer. There are some "successes," that is, results consistent with the predictions, and many surprises. Interpretation of results is difficult because many amino acid side chains contact the substrates, and it is not readily possible to assign a specific role to a particular group. A side-chain functional group involved in acid–base catalysis also participates in binding substrates and in the topography of the active site.

[8] W. P. Jencks, *Adv. Enzymol.* **43,** 219 (1975).

[9] "Evolution of Catalytic Function," *Cold Spring Harbor Symp. Quant. Biol.* **52,** 1987).

[10] R. Bone and D. A. Agard, this series, Vol. 202, p. 643.

[11] L. Hedstrom, L. Graf, C.-B. Stewart, W. J. Rutter, and M. A. Phillips, this series, Vol. 202, p. 671.

Some amino acid residues identified as being essential because chemical modification "inactivated" the enzymes have now been substituted with other amino acid residues without loss of activity.[12,13] This is not a real surprise, as it was recognized that attaching a large substituent to a reactive residue in the active site could simply block substrate binding. Furthermore, substitutions distant from the active site may decrease activity indirectly. H. J. Fromm pointed out one example. Modification of Lys-140 with pyridoxal 5′-phosphate or substitution with Ile inactivated *Escherichia coli* adenylosuccinate synthase, as did modification of Arg-147 with phenylglyoxal or substitution with Leu.[14,15] Substrates protected against the chemical modifications, and it was concluded that these residues were at the active site. Nevertheless, a three-dimensional structure suggests that these residues are at the interface between monomers and far from the active site, which was identified by homology to $p21^{ras}$.[16] At this time, the modification results are not explained. Other mutagenesis studies on the consensus phosphate-binding site gave results consistent with expectations.[17]

As compared to chemical modifications, where residual activity below a few percent could be attributed to unmodified enzyme, site-directed mutations have lowered the limits so that activities reduced by six or more orders of magnitude can be meaningful in careful studies. If enzymes accelerate reactions by 9 to 12 orders of magnitude, the residual activities should be determined if possible. The criterion that an amino acid residue is essential if its substitution totally inactivates is obsolete. Furthermore, the magnitude of the change in activity should not be the only criterion used to judge the significance of the residue for catalysis. There are many amino acid residues at the active site (see, e.g., Fig. 1 below), and it is reasonable to suppose that each residue participates in catalysis. If each of ten residues contributed a factor of 10 and the effects were additive, the enzymatic rate enhancement could be explained. Thus, site-directed substitutions could allow estimations of each contribution, but this analysis can be confounded by the cooperative nature of catalysis.

Removal of a critical functional group typically produces an impaired enzyme, whose activity reflects the new constellation of residues. For instance, if a group involved in acid–base catalysis is substituted, the

[12] A. T. Profy and P. Schimmel, *J. Biol. Chem.* **261,** 15474 (1986).

[13] H. Teng, E. Segura, and C. Grubmeyer, *J. Biol. Chem.* **268,** 14182 (1993).

[14] Q. Dong and H. J. Fromm, *J. Biol. Chem.* **265,** 6235 (1990).

[15] Q. Dong, F. Liu, A. M. Myers, and H. J. Fromm, *J. Biol. Chem.* **266,** 12228 (1990).

[16] B. W. Poland, M. M. Silva, M. A. Serra, Y. Cho, K. H. Kim, E. M. S. Harris, and R. B. Honzatko, *J. Biol. Chem.* **268,** 25334 (1993).

[17] F. Liu, Q. Dong, and H. J. Fromm, *J. Biol. Chem.* **267,** 2388 (1992).

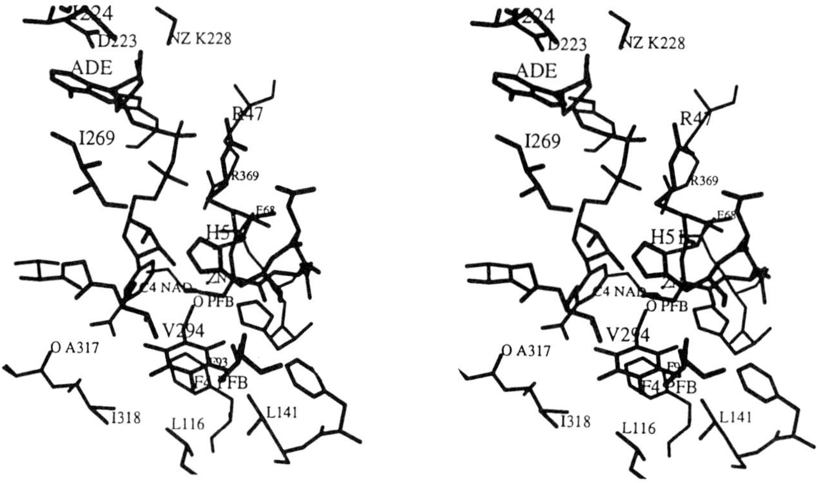

FIG. 1. Three-dimensional structure of horse liver alcohol dehydrogenase complexed with NAD^+ and 2,3,4,5,6-pentafluorobenzyl alcohol. The structure has been determined at 2.1 Å resolution with an *R* value of 18.3%. Some of the many interactions with the coenzyme and substrate analog are illustrated. The adenine ring is sandwiched between the side chains of Ile-224 and Ile-269; the adenosine ribose hydroxyl groups form hydrogen bonds with the carboxyl group of Asp-223 and the ε-amino group of Lys-228. The pyrophosphate moiety is neutralized by the guanidino groups of Arg-47 and Arg-369. The nicotinamide ribose sits on the side chain of Val-294, and its hydroxyl groups form hydrogen bonds with Ser-48, His-51, and Ile-269. The nicotinamide ring is hydrogen bonded to main-chain atoms of residues 317, 319, and 292. The catalytic zinc is tightly bound to the enzyme by Cys-46, Cys-174, and His-67. Near the zinc is the carboxyl group of Glu-68, which interacts with Arg-369. The hydroxyl group of the alcohol binds the zinc and is hydrogen bonded to the hydroxyl group of Ser-48, which is linked through the 2′-hydroxyl of the nicotinamide ribose to the imidazole of His-51. This system could shuttle a proton from the buried alcohol to His-51. The benzene ring of the alcohol is in a hydrophobic barrel that includes the side chains of Leu-57, Phe-93, Leu-116, Leu-141, Val-294, and Ile-318. Although the pentafluorobenzyl alcohol is not oxidized by the enzyme because of electron withdrawal by the fluorine atoms, the *pro-R* hydrogen is in a position that appears poised for direct transfer to C-4 of the nicotinamide ring. [From S. Ramaswamy, H. Eklund, and B. V. Plapp, *Biochemistry* **33,** 5230–5237 (1994).]

pH dependence of the modified enzyme will reflect the ionization of the remaining residues. Because active sites are different, the substitution of the same residue, for example, histidine, in different enzymes will naturally lead to effects of various magnitudes. The "context" complicates the estimation of catalytic contributions for particular interactions.[18] For instance, changing His-47 to Arg in three different alcohol dehydrogenases

[18] B. W. Matthews, *Annu. Rev. Biochem.* **62,** 139 (1993).

increased affinity for the coenzymes from 2- to 20-fold.[19] Nevertheless, the results from sufficient studies should lead to a consensus. Eventually, computational biochemistry should be able to account for the interaction energies. Site-directed mutagenesis is one way to perturb the energies.

The definition of the active site is also being reconsidered. Once it included only the amino acid residues in contact with the substrates, but now we appreciate that interactions from nearby residues modulate activity. Indeed, the whole protein structure is involved in catalysis, and distant changes can indirectly affect activity by long-range electrostatic effects or repositioning of residues near the reaction center. Substitutions of residues in mobile loops that form the active site during substrate-induced conformational changes also can significantly affect activity.[20,21] Proteins are dynamic molecules that are sensitive to global effects.

Distinguishing between direct effects of a substitution on catalysis and indirect effects due to structural changes cannot be resolved easily. Every substitution affects at least the local structure of the protein. It has been suggested that substitution of a residue at the active site decreases activity because of large changes in structure.[2,22] However, substitutions of surface residues distant from the active site probably have relatively small effects on catalysis, as there are many substitutions (e.g., in homologous enzymes) that do not affect activity. X-Ray crystallography of mutant forms of various proteins shows that the effects of local structural changes on nearby residues diminish with distance.[10,18,23,24] Proteins are structurally flexible ("plastic") and can accommodate changes readily, even with substitutions in the interior core.[25-28] How small changes at the active site affect catalysis is the critical question.

The kinetic mechanisms of enzymes have several steps that are energetically balanced for optimal efficiency. Each amino acid residue con-

[19] D. R. Light, M. S. Dennis, I. J. Forsythe, C.-C. Liu, D. W. Green, D. A. Kratzer, and B. V. Plapp, *J. Biol. Chem.* **267,** 12592 (1992).

[20] L. Li, C. J. Falzone, P. E. Wright, and S. J. Benkovic, *Biochemistry* **31,** 7826 (1992).

[21] Q.-W. Pan, S. Tanase, Y. Fukumoto, F. Nagashima, S. Rhee, P. H. Rogers, A. Arnone, and Y. Morino, *J. Biol. Chem.* **268,** 24758 (1993).

[22] D. W. Hibler, N. J. Stolowich, M. A. Reynolds, J. A. Gerlt, J. A. Wilde, and P. H. Bolton, *Biochemistry* **26,** 6278 (1987).

[23] J. S. Kavanaugh, P. H. Rogers, D. A. Case, and A. Arnone, *Biochemistry* **31,** 4111 (1992).

[24] Y. Xue, A. Liljas, B.-H. Jonsson, and S. Lindskog, *Proteins: Struct. Funct. Genet.* **17,** 93 (1993).

[25] R. Bone, J. L. Silen, and D. A. Agard, *Nature (London)* **339,** 191 (1989).

[26] K. P. Wilson, B. A. Malcolm, and B. W. Matthews, *J. Biol. Chem.* **267,** 10842 (1992).

[27] D. I. Roper, K. M. Moreton, D. B. Wigley, and J. J. Holbrook, *Protein Eng.* **5,** 611 (1992).

[28] D. E. Anderson, J. H. Hurley, H. Nicholson, W. A. Baase, and B. W. Matthews, *Protein Sci.* **2,** 1285 (1993).

tributes to some extent to the rate of each step. Thus, a site-directed substitution can change the rates differentially, and the rate of one step can be affected by various substitutions. In some cases, the effects of substitution of important residues may not be apparent from a standard assay with a fixed concentration of substrate.[29] This means that modified enzymes must be thoroughly studied to determine the rates of each step. Each mutant enzyme should become the subject of the various studies done on the wild-type enzyme.

Strategy

Design

A three-dimensional structure is invaluable for selecting amino acid residues to replace. At least a structure could be modeled on the basis of a homolog. Protein modeling programs such as O are useful for obtaining typical side-chain rotamers and interaction distances.[30,31] Comparison of amino acid sequences should also provide targets of conserved residues that presumably have similar functions or varied residues that could explain differences in specificity.[32] If a three-dimensional structure is not known, sequence alignments and chemical modification studies provide a secondary rationale for selection, and the interpretation of results is more speculative. "Alanine-scanning" mutagenesis, for instance, where all charged residues are systematically substituted with Ala, can identify residues that appear to be important and allow more detailed studies.[33,34]

The purpose of the experiment influences the choice of amino acid residues to use for substitution. There is considerable flexibility in the choice, as surface residues that make up the active site can often be changed without affecting protein stability significantly.[18] Consideration of the substitutions that have been accepted in the evolution of homologous proteins suggests that some interchanges are relatively conservative (Table I). However, any change has the potential to affect structure and catalysis. Each site-directed substitution provides a new enzyme that will

[29] Z. Shi, I.-J. L. Byeon, R.-T. Jiang, and M.-D. Tsai, *Biochemistry* **32,** 6450 (1993).

[30] T. A. Jones, J.-Y. Zou, S. W. Cowan, and M. Kjeldgaard, *Acta Crystallogr. Sect. A: Struct. Sci.* **A47,** 110 (1991).

[31] J. J. Langone (ed.), this series, Vol. 202.

[32] A. J. Ganzhorn, D. W. Green, A. D. Hershey, R. M. Gould, and B. V. Plapp, *J. Biol. Chem.* **262,** 3754 (1987).

[33] B. C. Cunningham and J. A. Wells, *Science* **244,** 1081 (1989).

[34] C. S. Gibbs and M. J. Zoller, *J. Biol. Chem.* **266,** 8923 (1991).

TABLE I
RELATIVELY FREQUENT ACCEPTED POINT MUTATIONS[a]

Cys → Ser	Ser → Ala, Thr, Asn, Gly, Pro
Thr → Ser, Ala, Val	Pro → Ala, Ser
Ala → Ser, Gly, Thr, Pro	Gly → Ala, Ser
Asn → Asp, Ser, His, Lys, Gln, Glu	Asp → Glu, Asn, Gln, His, Gly
Glu → Asp, Gln, Asn, His	Gln → Glu, His, Asp, Asn, Lys, Arg
His → Asn, Gln, Asp, Glu, Arg	Arg → Lys, His, Trp, Gln
Lys → Arg, Gln, Asn	Met → Leu, Ile, Val
Ile → Val, Leu, Met	Leu → Met, Ile, Val, Phe
Val → Ile, Leu, Met	Phe → Tyr, Leu, Ile
Tyr → Phe, His, Trp	Trp → Phe, Tyr

[a] The relative frequency is adapted from the mutation data matrix derived from comparison of homologous sequences given by M. O. Dayhoff, W. C. Barker, and L. T. Hunt, this series, Vol. 91, p. 524.

be compared to the wild-type enzyme, and it is the differences between the contributions of two residues that will be evaluated.

For studies on the roles of functional side chains in the mechanism, nearly isosteric substitutions will minimize structural changes. Thus, His can be changed to Asn or Gln, Cys to Ser, Thr to Val, Asp to Asn, Glu to Gln, and vice versa. However, these changes can alter the pattern of hydrogen-bonding interactions, as was found for rat trypsin when Asp-102 was changed to Asn.[35] The charge can be maintained with interchanges of Arg, Lys, or His and Glu with Asp. However, the accompanying change of size means that the contacts will be altered, which becomes the experimental variable. Removing the side chain by substituting with Ala is a large change in size and a more drastic test of the role of a residue since structural effects may be larger. Increasing the size, for instance, Asp to Glu, or Lys to Arg, risks the generation of steric interference, which is energetically more serious than creating a void. For tests of flexibility of the peptide backbone, which can be important when conformational changes are critical for activity, substitution of Pro or Gly with Ala, or another residue with Gly or Pro, will loosen or stiffen, respectively, the structure.[36] Substitutions of residues in conformational hinges or loops may give measurable changes in activity.[37-39] Another approach is to use

[35] S. Sprang, T. Standing, R. J. Fletterick, R. M. Stroud, J. Finer-Moore, N.-H. Xuong, R. Hamlin, W. J. Rutter, and C. S. Craik, *Science* **237,** 905 (1987).
[36] B. W. Matthews, *Biochemistry* **26,** 6885 (1987).
[37] M. T. Mas, J. M. Bailey, and Z. E. Resplandor, *Biochemistry* **27,** 1168 (1988).
[38] P. M. Ahrweiler and C. Frieden, *Biochemistry* **30,** 7801 (1991).
[39] A. M. E. Bullerjahn and J. H. Freisheim, *J. Biol. Chem.* **267,** 864 (1992).

random or saturation mutagenesis, which eliminates bias in the choice of substitutions and produces a large number of enzymes that may have diverse activities.[40–44]

Integrity

Once the mutations are made, the integrity of the product should be established. The complete sequence of the coding region of the modified DNA should be verified in order to minimize errors from adventitious mutations arising from mutagenesis, subcloning, or biological artifacts. Because sequencing results and experimental missteps and failures are not usually reported, the origins of published discrepancies are difficult to trace. After the proteins are expressed and purified,[45] some criteria of homogeneity should be applied, such as chromatography and electrophoresis in two different systems. Confirmation of the molecular weight by sodium dodecyl sulfate–polyacrylamide gel electrophoresis or gel filtration, with comparison to the wild-type protein, is often sufficient. When charged residues are changed, evidence of the altered properties should be seen from ion-exchange chromatography or electrophoresis. Although rarely reported, peptide mapping and amino acid sequencing can add considerable confidence to the identification of the mutated enzyme and to the elimination of artifacts. When the proteins are expressed in heterologous systems, the processing may be different, which could also alter enzymatic characteristics. Special care is required to ensure that wild-type, another mutant, or host enzymes do not contaminate the desired enzyme. This can happen in laboratories where different enzymes are expressed in microbes and sterile practices are insufficient. Because the mutations are unique biological materials, investigators should store stocks of the mutated plasmids or transformed microbes to be able to provide them to others who wish to replicate the published work.

The structural integrity of the protein should be studied. Simple procedures, such as denaturation by heat, pH, or storage, can be useful. Spectral measurements of UV absorption, fluorescence, or circular dichroism (CD) may reveal some relevant differences.[46] In most cases, if the enzyme can be isolated by the usual methods, the structure is similar to that of the wild

[40] D. A. Estell, T. P. Graycar, and J. A. Wells, *J. Biol. Chem.* **260,** 6518 (1985).

[41] D. M. Hampsey, G. Das, and F. Sherman, *J. Biol. Chem.* **261,** 3259 (1986).

[42] S. Climie, L. Ruiz-Perez, D. Gonzalez-Pacanowska, P. Prapunwattana, S.-W. Cho, R. Stroud, and D. V. Santi, *J. Biol. Chem.* **265,** 18776 (1990).

[43] S. R. Wente and H. K. Schachman, *J. Biol. Chem.* **266,** 20833 (1991).

[44] J. F. Krebs and C. A. Fierke, *J. Biol. Chem.* **268,** 948 (1993).

[45] This series, Vols. 22, 34, and 104.

[46] A. J. Ganzhorn and B. V. Plapp, *J. Biol. Chem.* **263,** 5446 (1988).

type. Instability is an indication of deleterious changes. Some practitioners insist that the three-dimensional structure must be determined by X-ray crystallography or nuclear magnetic resonance (NMR), but this is often not possible and does not resolve the fundamental issue of distinguishing between direct and indirect effects. As noted above, mutations cause local structural changes, and these can be observed. However, after a structure of the enzyme complexed with substrates, inhibitors, or transition state analogs is determined, one must still ask if the structure is relevant to one of the complexes in the catalytic pathway. This is not a question of whether the structure of enzyme in crystalline form is the same as that in solution, since there is little evidence to suggest that they are different. Rather, it is an issue because an enzymatic reaction has many steps, with side chains changing position and states of protonation or covalent chemistry, and substrates may bind differently at the multiple steps of the reaction.

For multisubunit enzymes, where the active site is formed from two subunits, structural and cooperative effects can be assessed by complementation studies in which two forms of inactive enzyme are rehybridized.[47,48] If activity is regained to the statistically expected level, it is reasonable to conclude that the structures of the subunits in the inactive enzymes are not grossly distorted. However, the regain of activity in heterodimers of the inactive carboxymethylated derivatives of the monomeric pancreatic ribonuclease indicates that structural rearrangements can be productive.[49] Furthermore, an altered subunit can inactivate a wild-type subunit in an oligomeric enzyme, as in a "dominant negative" mutant.[50]

Kinetics

Enzyme kinetics and mechanistic studies, using the whole arsenal of techniques,[51] are required to determine the changes in the altered enzymes. Initially, the activity of the enzyme in a standard assay and Michaelis–Menten parameters are determined. The K_m and V_{max} values for one of the substrates may be estimated with other substrates at a fixed "saturating" level. However, more complete studies are necessary if the kinetics studies are to be meaningful. The V_{max} values are underestimated if the

[47] S. R. Wente and H. K. Schachman, *Proc. Natl. Acad. Sci. U.S.A.* **84,** 31 (1987).

[48] F. W. Larimer, E. H. Lee, R. J. Mural, T. S. Soper, and F. C. Hartman, *J. Biol. Chem.* **262,** 15327 (1987).

[49] A. M. Crestfield, W. H. Stein, and S. Moore, *J. Biol. Chem.* **238,** 2421 (1963).

[50] S. Tsirka and P. Coffino, *J. Biol. Chem.* **267,** 23057 (1992).

[51] This series, Vols. 63, 64, and 87.

concentrations of all substrates are not saturating or if inhibitory levels of one of the substrates are used. The K_m values determined at a fixed level of other substrates can be larger or smaller than the true values, and they should not be interpreted as measures of affinity in the absence of mechanistic studies. A complete kinetic characterization can define the mechanism, true turnover numbers, K_m values, catalytic efficiencies (V/K_m), and dissociation constants for some substrates and inhibitors. For multisubstrate enzymes, this means that initial velocity studies in both the forward and reverse reactions should be performed by varying in a systematic way the concentrations of all substrates. For a bisubstrate mechanism, a 5 by 5 array of values, determined in duplicate, can provide good estimates of the kinetic constants. Fitting the appropriate equations to the data provides confidence in the precision of the results and permits comparisons with the wild-type enzyme.[52] The necessary computer programs are readily available.

For the interpretation of the kinetic data, the enzyme concentration must be determined. The protein concentration can be estimated by using amino acid analysis, dry weight, or UV absorption with a calculated extinction coefficient. Colorimetric procedures must be standardized with the wild-type enzyme, even if serum albumin is used as a working standard. Without proper standardization, maximum velocities and the stoichiometry of ligand binding can be seriously in error. As a control for enzyme purity and functionality, the concentration of active sites should be titrated with a ligand that binds tightly and gives a measurable signal. Although a protein may appear to be homogeneous by electrophoresis or chromatography, active sites may be incapable of binding an exogenous ligand because of chemically altered side chains, metal loss or poisoning, or tightly bound impurities. Such sites may be inactive or have altered kinetic properties. The concentration of enzyme is used to calculate turnover numbers for the forward and reverse reactions (V/E_t or k_{cat}), which have units of reciprocal seconds (sec^{-1}). For some mechanisms, combinations of turnover numbers and kinetic constants can provide estimates of the rates of binding and dissociation of substrates for some steps in the reactions.

Establishing the kinetic mechanism for a mutated enzyme is accomplished by using the techniques described in this series. Ideally, each mutated enzyme should be studied in as much detail as the wild-type enzyme. Product and dead-end inhibition studies are useful, as various mechanisms can be distinguished, and the inhibition constants give estimates of the affinities of ligands for the enzyme. Inhibitors that resemble the substrates are useful for examining changes in different parts of the

[52] W. W. Cleland, this series, Vol. 63, p. 103.

active site. Substrate specificities may have changed. Kinetic isotope effects and isotope exchange experiments can help establish which steps are rate limiting for the new enzyme. Studies on the pH dependence of kinetic parameters can identify changes in groups involved in acid–base catalysis. Indeed, without pH studies and identification of rate-limiting steps, there is insufficient evidence to support conclusions of the involvement of ionizable groups in the mechanism.

Enzymologists describe a mechanism by identifying all of the intermediates and assigning rate constants for each of the steps. As steady-state kinetics analysis does not provide all of the relevant information, transient (pre-steady-state) kinetics is an important complementary tool. Isomerizations of transitory and central complexes can be detected and characterized when high concentrations (e.g., 1–10 μM) of enzymes and complexes are used. Kinetic simulation, in which the differential equations for the enzymatic reaction are numerically integrated and progress curves for various reactions are fitted, has become an important and accessible tool.[53–55] The analysis of site-directed mutants reaches new heights when the effect on each step of the reaction can be determined.

Illustrations

Interpretation of Steady-State Kinetic Parameters

Horse liver and yeast alcohol dehydrogenases are used as primary examples since a variety of mutations have been made, and the kinetics have been extensively studied. The three-dimensional structure of the horse liver EE isoenzyme complexed with NAD^+ and a substrate analog has been determined to high resolution (Fig. 1). Inspection of the structure and studies of the mechanism have suggested that His-51 participates in acid–base catalysis, but it is apparent that many other amino acid residues participate in binding the substrates. Amino acid sequences of more than 47 members of the family have been determined,[56] and the yeast enzyme is similar enough to the liver enzyme to justify building models that represent the active site. Thus, the enzymes offer many opportunities to explore structure–function relationships and to test the roles of different residues in catalysis.[57]

[53] C. Frieden, *Trends Biochem. Sci.* **18,** 58 (1993).

[54] C. T. Zimmerle and C. Frieden, *Biochem. J.* **258,** 381 (1989).

[55] C. Frieden, this series, Vol. 240.

[56] H.-W. Sun and B. V. Plapp, *J. Mol. Evol.* **34,** 522 (1992).

[57] B. V. Plapp, A. J. Ganzhorn, R. M. Gould, D. W. Green, T. Jacobi, E. Warth, and D. A. Kratzer, *in* "Enzymology and Molecular Biology of Carbonyl Metabolism 3" (H. Weiner, B. Wermuth, and D. W. Crabb, eds.), p. 241. Plenum, New York, 1990.

The kinetic mechanisms for the enzymes were established as Ordered Bi Bi,[58] which can be written as follows:

$$\mathrm{E} \underset{k_{-1}}{\overset{k_1\mathrm{NAD}^+}{\rightleftharpoons}} \mathrm{E}\cdot\mathrm{NAD}^+ \underset{k_{-2}}{\overset{k_2\mathrm{Alc}}{\rightleftharpoons}} \mathrm{E}\cdot\mathrm{NAD}^+\cdot\mathrm{Alc}$$

$$\underset{k_{-3}}{\overset{k_3}{\rightleftharpoons}} \mathrm{E}\cdot\mathrm{NADH}\cdot\mathrm{Ald} \underset{k_{-4}\mathrm{Ald}}{\overset{k_4}{\rightleftharpoons}} \mathrm{E}\cdot\mathrm{NADH} \underset{k_{-5}\mathrm{NADH}}{\overset{k_5}{\rightleftharpoons}} \mathrm{E} \qquad (1)$$

The complete steady-state rate equation for the mechanism has ten kinetic constants, which can be estimated from initial velocity and product inhibition studies with both substrates in each direction. The reaction in the forward direction in the absence of products is described by the Sequential Bi rate equation: $v = V_1\mathrm{AB}/(K_{ia}K_b + K_aB + K_bA + AB)$, where A and B represent the concentrations of the substrates NAD^+ and alcohol, K_a and K_b are the Michaelis constants for these substrates, K_{ia} is the inhibition or dissociation constant for NAD^+, and V_1 is the maximum velocity (proportional to turnover number). A similar equation applies for the reverse reaction, where P and Q represent aldehyde and NADH. These kinetic constants can be estimated by varying the concentrations of both substrates in a systematic way or from product inhibition studies.[58,59]

A variety of site-directed substitutions have been made in these enzymes, and Table II summarizes kinetic data for some of them.[60–65] The substitutions were made in the coenzyme- and substrate-binding sites, in the environment of the zinc, and in the acid–base system, as identified by reference to Fig. 1.

The kinetic constants provide fundamental information about the mechanism and will be used to illustrate how the data may be interpreted for the various mutants. It is helpful to know the definitions of the kinetic constants in terms of the rate constants given in Eq. (1). Analogous expressions apply to the reverse reactions, and Haldane relationships relate these kinetic constants to the equilibrium constant for the overall reaction.

[58] C. C. Wratten and W. W. Cleland, *Biochemistry* **2,** 935 (1963).
[59] K. Dalziel, *J. Biol. Chem.* **238,** 2850 (1963).
[60] Mutations are conveniently designated with the single-letter abbreviations for the amino acids, wild-type first and mutant last, separated by the residue number.
[61] D.-H. Park and B. V. Plapp, *J. Biol. Chem.* **267,** 5527 (1992).
[62] K. Kim, Ph.D. Thesis, The University of Iowa, Iowa City (1994).
[63] F. Fan, Ph.D. Thesis, The University of Iowa, Iowa City (1994).
[64] R. M. Gould and B. V. Plapp, *Biochemistry* **29,** 5463 (1990).
[65] R. M. Gould, Ph.D. Thesis, The University of Iowa, Iowa City (1988).

TABLE II
KINETIC CONSTANTS FOR LIVER AND YEAST ALCOHOL DEHYDROGENASES[a]

Constant	*Eq*ADH-E Wild type	*Eq*ADH-E D115Δ	*Eq*ADH-E F93A	*Eq*ADH-E I269S	*Sc*ADH1 Wild type	*Sc*ADH1 H51Q	*Sc*ADH1 E68Q
K_a (μM)	3.9	11	3.3	1000	160	96	410
K_b (mM)	0.35	36	0.21	11	21	18	41
K_p (mM)	0.40	87	41	11	0.74	15	56
K_q (μM)	5.8	24	1.9	570	95	150	160
K_{ia} (μM)	27	29	44	9500	950	460	3500
K_{iq} (μM)	0.50	0.91	0.036	180	31	6	29
V_1 (sec^{-1})	3.5	7.5	0.35	90	360	27	9.9
V_2 (sec^{-1})	47	240	20	1500	1800	2800	730
V_1/K_b (mM^{-1} sec^{-1})	10	0.21	1.7	8.4	21	1.5	0.24
$V_1/K_{ia}K_b$ (mM^{-2} sec^{-1})	370	7.2	38	0.88	22	3.3	0.068
V_2/K_p (mM^{-1} sec^{-1})	120	2.8	0.49	140	2400	190	13
$V_2/K_{iq}K_p$ (μM^{-2} sec^{-1})	0.24	0.0030	0.013	0.00078	0.080	0.032	0.00045
K_i, CF_3CH_2OH (μM)	8.4	630	53	7.8	2500	33,000	25,000

[a] Kinetic constants were determined in 33 mM sodium phosphate buffer, pH 8.0, at 25° for the horse liver enzymes, and in 83 mM sodium phosphate, 40 mM KCl buffer, pH 7.3, 30° for the yeast enzymes. NAD^+ and ethanol or NADH and acetaldehyde were the substrates. *Eq*ADH-E is the horse (*Equus*) liver EE isoenzyme, mutant forms of which include D115Δ with Asp-115 deleted (Ref. 61), F93A (Ref. 62), and I269S (Ref. 63); *Sc*ADH1 is the yeast (*Saccharomyces cerevisiae*) cytoplasmic isoenzyme I (Ref. 64), mutants of which include H51Q (Ref. 65) and E68Q (Ref. 46).

$$
\begin{aligned}
K_a &= k_3k_4k_5/k_1(k_4k_5 + k_3k_4 + k_{-3}k_5) \\
K_b &= k_5(k_{-2}k_4 + k_{-2}k_{-3} + k_3k_4)/k_2(k_4k_5 + k_3k_4 + k_3k_5 + k_{-3}k_5) \\
V_1 &= k_3k_4k_5/(k_4k_5 + k_3k_4 + k_3k_5 + k_{-3}k_5) \\
K_{ia} &= k_{-1}/k_1 \\
V_1/K_a &= k_1 \\
V_1K_{ia}/K_a &= k_{-1} \\
V_1/K_b &= k_2k_3k_4/(k_{-2}k_4 + k_{-2}k_{-3} + k_3k_4) \\
V_1/K_{ia}K_b &= k_1k_2k_3k_4/k_{-1}(k_{-2}k_4 + k_{-2}k_{-3} + k_3k_4) \\
K_i, CF_3CH_2OH &= k_{-2}/k_2 \\
K_{eq} &= V_1K_pK_{iq}/V_2K_bK_{ia} = k_1k_2k_3k_4k_5/k_{-1}k_{-2}k_{-3}k_{-4}k_{-5}
\end{aligned}
$$

Inspection of the data in Table II shows that the K_m values for coenzyme are not necessarily the same as the K_d values ($K_a \neq K_{ia}$ and $K_q \neq K_{iq}$ in most cases). Furthermore, changes in a K_m value due to mutation are not in general directly proportional to changes in a K_d value. The K_m for alcohol (K_b) would approximate the K_d (k_{-2}/k_2) only with certain values of rate constants, so that a better measure of affinity of the enzymes for alcohol is given by the K_i value for the competitive inhibitor trifluoroethanol. Comparison of the changes in K_b and K_i show that there can be

10-fold changes in one of these values and not in the other (F93A, I269S, H51Q enzymes). It is apparent that changes in K_m values should not be described as changes in affinity.

For the Ordered Bi Bi mechanism, V_1, or k_{cat} for the forward reaction, is controlled by several rate constants for the conversion of the central E·NAD$^+$·alcohol complex to products. For the F93A enzyme, V_1 is decreased 10-fold relative to the wild-type enzyme. This change could conceivably result from an effect on the rate of hydrogen transfer, but in this case it is due to a slower release of the product NADH ($k_5 = V_2K_{iq}/K_q = 0.37\ sec^{-1}$). The V_1 for the I269S enzyme is increased by 25-fold, owing to an increased rate of release of NADH ($k_5 = 470\ sec^{-1}$), whereas the rate of binding of NADH ($k_{-5} = V_2/K_q = 2.6 \times 10^6\ M^{-1}\ sec^{-1}$) is decreased 3-fold. The V_2 for the I269S enzyme is increased 32-fold, because NAD$^+$ dissociates faster. The rate constants for binding were confirmed by transient kinetic studies using stopped-flow techniques.

Catalytic Efficiency

The increases in turnover numbers observed with some mutations raise the issues of catalytic efficiency and evolution of catalytic perfection. Can site-directed mutations or biological selection procedures produce "more active" enzymes? For analyzing the results of such studies, the ratio of constants, V/K_m or k_{cat}/K_m, is the important measure. When the concentration of substrate is low relative to the K_m value, the Michaelis–Menten equation reduces to $v = V[A]/K_m$ and describes the rate when the reaction is first order in substrate. The logarithm of this parameter is proportional to the energy change in attaining the transition state, and the logarithmic scale in figures is most informative for comparing a variety of substrates or mutant enzymes.

For alcohol dehydrogenase, V_1/K_b is the bimolecular rate constant describing the rate of reaction of the alcohol with the enzyme–NAD$^+$ complex to produce aldehyde and enzyme–NADH complex, including steps 2, 3, and 4 in Eq. (1). It is interesting that the liver and yeast enzymes have about the same catalytic efficiencies for ethanol even though these enzymes have very different Michaelis constants and turnover numbers. All of the mutant enzymes in Table II have decreased efficiencies, except for the I269S enzyme. The deletion of Asp-115 in the substrate binding pocket of the E isoenzyme (D115Δ) decreases activity on ethanol but confers activity on steroids.[61] The F93A substitution removes a phenyl group that interacts with ethanol. The H51Q change removes the catalytic base that facilitates loss of the proton from the hydroxyl group of the alcohol. In contrast, the I269S substitution affects the binding of the adenine ring of coenzyme, but not the catalytic efficiency with ethanol. Thus, many of the kinetic results are consistent with the proposed roles of the amino acid residues in reactions with ethanol. The decreased efficiency of the E68Q enzyme is not explained in the same way, however, since this residue is near the catalytic zinc and on the opposite side of the binding site for substrate. The E68Q substitution seems to affect the electrostatic environment of the zinc and indirectly affect catalysis. Although the changes in kinetic parameters may be consistent with simple interpretations, the ultimate explanation may be more complicated.

There are a few examples where catalytic efficiency is increased by mutation. Changing Thr-51 to Cys or Pro in tyrosyl-tRNA synthetase (tyrosyl-tRNA ligase) increased the efficiency of formation of the tyrosyl–adenylate complex 15- to 36-fold but decreased the overall k_{cat} and catalytic efficiency.[66,67] Substitution of Asp-153 with Ala in an alkaline phosphatase led to a 6-fold increase in k_{cat} and a 14-fold increase in k_{cat}/K_m for *p*-nitrophenyl phosphate.[68] The increased turnover is due to faster release of the product inorganic phosphate, but the increased efficiency is more difficult to explain as Asp-153 does not seem to have a direct role in catalysis. Another mutation, K328A, also activates the enzyme in the presence of Tris buffer, which can stimulate transphosphorylation.[69] The D101S substitution increased the efficiency 5-fold of an alkaline phosphatase on *p*-nitrophenyl phosphate by altering the interactions with an Arg and decreasing affinity for phosphate.[70] Insertion of three amino acid residues into maize aldolase increased efficiency with fructose 1,6-bisphosphate by about 3-fold.[71]

The magnitude of *V*/*K* depends on the substrate used and may be largest with the physiological reactant. It is a challenge to improve activity with the best substrate of the enzyme or to use protein engineering to make an enzyme more active on some substrate than the native enzyme is with its best substrate. There are some examples of mutations that increase activity with poor substrates, as would be expected if the specificity is re-engineered. For instance, the Q102R substitution converted the

[66] C. K. Ho and A. R. Fersht, *Biochemistry* **25,** 1891 (1986).
[67] J. M. Avis and A. R. Fersht, *Biochemistry* **32,** 5321 (1993).
[68] A. R. Matlin, D. A. Kendall, K. S. Carano, J. A. Banzon, S. B. Klecka, and N. M. Solomon, *Biochemistry* **31,** 8196 (1992).
[69] X. Xu and E. R. Kantrowitz, *Biochemistry* **30,** 7789 (1991).
[70] L. Chen, D. Neidhart, W. M. Kohlbrenner, W. Mandecki, S. Bell, J. Sowadski, and C. Abad-Zapatero, *Protein Eng.* **5,** 605 (1992).
[71] L. Berthiaume, D. R. Tolan, and J. Sygusch, *J. Biol. Chem.* **268,** 10826 (1993).

Bacillus stearothermophilus lactate dehydrogenase to malate dehydrogenase.[72] Changing Asp-179 to Asn in deoxycytidylate hydroxymethylase decreased k_{cat}/K_m for dCMP by 10^4-fold and increased activity on dUMP 60-fold. However, dCMP reacts 1.8×10^5 times faster than dUMP with native enzyme, and the efficiency of the D179N enzyme for dUMP is still 3000-fold less than the efficiency of native enzyme with dCMP.[73] The T48S and W93A substitutions in the alcohol-binding site of yeast alcohol dehydrogenase increased activity on octanol 16-fold, yielding 2.7-fold more activity than the native enzyme has with its best substrate, ethanol.[74] Deletion of Asp-115 and substitutions of three nearby residues in the substrate binding pocket of horse liver alcohol dehydrogenase E isoenzyme decreased V/K_b for ethanol 25-fold, but increased activity on hexanol by 2.7-fold. The changes generated activity on steroids that is about 6-fold higher than the activity of natural horse S isoenzyme on steroids.[61] Thus, some modest improvements in catalytic efficiency are possible.

The V/K_m parameter can be applied for reactions with one substrate, or for the second substrate as in the Ordered Bi reaction of alcohol dehydrogenase, but the evaluation of catalytic efficiency should be extended when there is more than one substrate. Using the kinetic principle that efficiency is the activity at low, limiting substrate concentrations, the catalytic efficiency for both substrates in the Sequential Bi reaction is expressed by the term $V_1/K_{ia}K_b$. This expression describes the termolecular reaction of enzyme with both substrates and is proportional to the free energy of activation of the overall reaction.[67]

A few studies have determined overall catalytic efficiencies. The results in Table II show that all of the mutations of alcohol dehydrogenase decreased catalytic efficiencies for the reactions in both directions. Studies on changing the specificity for nicotinamide coenzymes provide further examples. The specificity of NAD-dependent dehydrogenases for coenzyme appears to be controlled by an aspartic acid that interacts with the adenosine ribose hydroxyl groups (Fig. 1). Changing Asp-223 to Gly in the NAD-dependent yeast alcohol dehydrogenase made the enzyme almost equally reactive with NAD^+ or $NADP^+$, but reduced overall catalytic efficiency ($V_1/K_{ia}K_b$) by about 1000-fold.[75] When the corresponding Asp (residue 37) was changed to Ile in rat NADH-preferring dihydropteridine reductase, a complete kinetic study showed that selectivity for NADPH was improved by 170-fold, but the wild-type enzyme was still 1300-fold

[72] H. M. Wilks, K. W. Hart, R. Feeney, C. R. Dunn, H. Muirhead, W. N. Chia, D. A. Barstow, T. Atkinson, A. R. Clarke, and J. J. Holbrook, *Science* **242,** 1541 (1988).

[73] K. L. Graves, M. M. Butler, and L. W. Hardy, *Biochemistry* **31,** 10315 (1992).

[74] D. W. Green, H.-W. Sun, and B. V. Plapp, *J. Biol. Chem.* **268,** 7792 (1993).

[75] F. Fan, J. A. Lorenzen, and B. V. Plapp, *Biochemistry* **30,** 6397 (1991).

more reactive with NADH than the D37I enzyme was with NADPH.[76] These results show that the conserved Asp is not the only residue that determines specificity for coenzymes. Two studies show that multiple changes designed on the basis of homologous enzymes can reverse coenzyme specificity. Changing seven amino acid residues in the NADPH-preferring *E. coli* glutathione reductase inverted the preference for coenzyme (k_{cat}/K_a) by 18,000-fold, but the mutant was 30-fold less reactive with NADH than wild-type enzyme was with NADPH.[77] Conversely, changing seven amino acid residues in the NAD-dependent *E. coli* dihydrolipoamide dehydrogenase produced an enzyme that was 1.3-fold more reactive with $NADP^+$ than the wild-type enzyme was with NAD^+.[78] Overall catalytic efficiencies and dissociation constants for the coenzymes should be determined for these multiply substituted enzymes.

Another complete kinetic study, on chloramphenicol acetyltransferase, showed that changing Thr-174 to Ala decreased k_{cat} by 2-fold, k_{cat}/K_m by 4-fold, K_d by 12-fold, and $k_{cat}/K_{ia}K_b$ (which is ϕ_{AB}) by 42-fold.[79] Similar numbers were obtained for the Val-174 and Ile-174 enzymes. From these studies and inspection of the three-dimensional structure, it was concluded that a water molecule that formed a hydrogen-bonded bridge between Thr-174 and the tetrahedral intermediate stabilized the transition state of the reaction.

Transient Kinetics and Estimation of Individual Rate Constants

Steady-state kinetic analysis is useful to establish a mechanism and to indicate which steps are affected by mutations. It is conveniently applied when amounts of enzyme are limiting. Nevertheless, as shown above, V/K_m or catalytic efficiency parameters are collections of rate constants. When large amounts of enzyme are available, transient kinetics can be used to estimate magnitudes of individual rate constants.[80] Moreover, advances in computer simulation make it feasible to estimate the rate constants from progress curves.[55] With good data from a few experiments, a mechanism can be defined and rate constants estimated with small errors.

[76] C. E. Grimshaw, D. A. Matthews, K. I. Varughese, M. Skinner, N. H. Xuong, T. Bray, J. Hoch, and J. M. Whiteley, *J. Biol. Chem.* **267,** 15334 (1992).
[77] N. S. Scrutton, A. Berry, and R. N. Perham, *Nature (London)* **343,** 38 (1990).
[78] J. A. Bocanegra, N. S. Scrutton, and R. N. Perham, *Biochemistry* **32,** 2737 (1993).
[79] A. Lewendon and W. V. Shaw, *J. Biol. Chem.* **268,** 20997 (1993).
[80] K. A. Johnson, *in* "The Enzymes" (P. D. Boyer, ed.), 3rd Ed., Vol. 20, p. 1. Academic Press, New York, 1992.

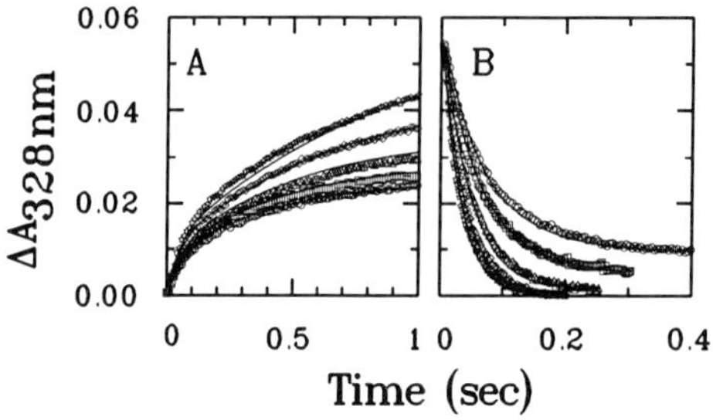

FIG. 2. Transient reactions of horse liver alcohol dehydrogenase, L57F enzyme. (A) The oxidation of benzyl alcohol (8 to 25 μM from lowest to highest curve) with 2 mM NAD^+ and 9 μM enzyme was determined with a BioLogic SFM3 stopped-flow instrument with a 1 cm light path. (B) The single turnover reduction of benzaldehyde (9.6 to 50 μM for highest to lowest curve) with 10 μM NADH and 7 μM enzyme was followed. The data points are shown, and the lines are simulated fits to all of the data together using KINSIM and FITSIM with the mechanism given in Eq. (1) and the rate constants given in Table III (Ref. 62).

Kinetic simulation has been used to analyze a complete mechanism for liver alcohol dehydrogenase acting on benzyl alcohol,[81] and some results with a mutant enzyme are given in Fig. 2 and Table III. The L57F substitution is in the substrate-binding pocket and has small effects on coenzyme binding and on the steady-state kinetic constants with ethanol and acetaldehyde. In contrast, rate constants for binding and reaction of benzyl alcohol and benzaldehyde are significantly affected. It is interesting that the rate constants for hydrogen transfer in the ternary complex (step 3) are increased 2- to 5-fold. This enzyme was found to have the greatest evidence for hydrogen tunneling, which is masked in the native enzyme because of kinetic complexity.[82,83] Decreasing the size of the substrate-binding pocket causes substrates to dissociate more rapidly (k_{-2} and k_4) and exposes the tunneling to observation.

Rate constants for a complete kinetic mechanism of *E. coli* dihydrofolate reductase were also analyzed with KINSIM.[84,85] This was the beginning of detailed studies on catalytic energetics and the effects of mutations.[86] Substitutions of the conserved Leu-54 in the dihydrofolate binding

[81] G. L. Shearer, K. Kim, K. M. Lee, C. K. Wang, and B. V. Plapp, *Biochemistry* **32,** 11186 (1993).

[82] B. J. Bahnson, D.-H. Park, K. Kim, B. V. Plapp, and J. P. Klinman, *Biochemistry* **32,** 5503 (1993).

[83] B. J. Bahnson and J. P. Klinman, this volume [14].

[84] M. H. Penner and C. Frieden, *J. Biol. Chem.* **262,** 15908 (1987).

[85] C. A. Fierke, K. A. Johnson, and S. J. Benkovic, *Biochemistry* **26,** 4085 (1987).

[86] S. J. Benkovic, C. A. Fierke, and A. M. Naylor, *Science* **239,** 1105 (1988).

TABLE III
RATE CONSTANTS FOR REACTIONS OF LIVER ALCOHOL DEHYDROGENASES WITH NAD$^+$ AND BENZYL ALCOHOL OR NADH AND BENZALDEHYDE[a]

Constant	Native[b]	L57F[c]
k_1 (μM^{-1} sec^{-1})	1.2	6.4
k_{-1} (sec^{-1})	90	270
k_2 (μM^{-1} sec^{-1})	3.7	35
k_{-2} (sec^{-1})	58	110
k_3 (sec^{-1})	38	60
k_{-3} (sec^{-1})	310	1700
k_4 (sec^{-1})	66	160
k_{-4} (μM^{-1} sec^{-1})	0.83	4.2
k_5 (sec^{-1})	5.5	6.4
k_{-5} (μM^{-1} sec^{-1})	11	9.9

[a] The rate constants are defined in Eq. (1). The reactions were studied in 33 m*M* sodium phosphate buffer, pH 8.0, at 25°. Coenzyme binding to free enzyme (steps 1 and 5) was determined in the absence of substrates. Rate constants for steps 2, 3, and 4 were estimated using KINSIM and FITSIM with the family of progress curves for the transient reactions, such as shown in Fig. 2. Errors were usually less than 10% of the values.

[b] V. C. Sekhar and B. V. Plapp, *Biochemistry* **29,** 4289 (1990).

[c] Ref. 62.

site of *E. coli* dihydrofolate reductase with Ile, Gly, and Asn decreased the hydride transfer rate by a constant factor of about 30-fold and increased dissociation constants for dihydrofolate by different factors. Thus, the contributions of residue 54 to binding and catalysis are separable.[87] Further study of a variety of mutated enzymes would be useful for developing our understanding of hydrogen transfer and substrate specificity.

Kinetic simulation has also been applied to mutants of bacterial luciferase, which has a complex mechanism described by 16 rate constants. Substitutions of Cys-106 were shown to decrease the stability of the hydroperoxy flavin intermediate most significantly.[88]

[87] D. J. Murphy and S. J. Benkovic, *Biochemistry* **28,** 3025 (1989).

[88] H. M. Abu-Soud, A. C. Clark, W. A. Francisco, T. O. Baldwin, and F. M. Raushel, *J. Biol. Chem.* **268,** 7699 (1993).

Acid–Base Catalysis

A central focus in studying enzymatic reactions is the evaluation of the contribution of residues to proton transfer steps. Imidazole, carboxyl, sulfhydryl, phenol, and amino groups are likely candidates. After substitution with a group that cannot be protonated in the physiological pH range, activity may decrease and the pH profiles should change. The magnitude of activity change is not a sufficient criterion to judge the participation of the residue, since (1) it is the magnitude we wish to determine, and (2) activity can decrease because of local structural effects. If the mutated enzyme has activity, it is important to study the mechanism of the enzyme and to determine at least the pH dependence for the steps using acid–base catalysis. Because enzymatic reactions are usually not limited by a single step in the mechanism and the pH-dependent step may be masked by kinetic complexity,[89] the analysis may require additional studies, for instance, with isotopes or NMR spectroscopy. Furthermore, removal of one ionizing group may expose the effects of other ionizable groups that can less directly affect catalysis and change the pH dependencies.

Altered pH Dependencies. The involvement of an amino acid residue in proton transfer steps is often inferred from a pH dependence study that shows evidence of pK values for free or complexed enzyme. Substitution of an involved residue should change the pH profile of the logarithm of V or V/K against the pH to a slope of 1.0 if hydroxide or hydronium were the catalyst or to a slope of zero if the reaction became pH independent. However, these expectations are not usually realized.

For alcohol dehydrogenases, substitution of His-51 with Gln illustrates some different results (Fig. 3). His-51 is part of a hydrogen-bonded system that includes the water or alcohol ligated to the catalytic zinc and the hydroxyl groups of Ser-48 (or Thr in yeast) and the nicotinamide ribose. The system could shuttle a proton from the buried alcohol to solvent. The S48A or T48A substitutions block the system and greatly reduce activity.[57,62,65] The pH dependencies observed with these enzymes could originate from His-51, the water that is ligated to the catalytic zinc, or other groups. The pH dependence of V/K_b provides pK values for the E·NAD$^+$ complex. Significant substrate deuterium isotope effects indicated that the chemical reactions were predominantly rate determining. For the horse enzyme, the native enzyme exhibits a bell-shaped pH dependence with pK values of 6.7 and 9.0, whereas the H51Q enzyme has only a single pK of 8.6. The wild-type yeast enzyme has a wavelike pH dependence with finite activity at low and high pH with a pK value of

[89] W. W. Cleland, *Adv. Enzymol.* **45,** 273 (1977).

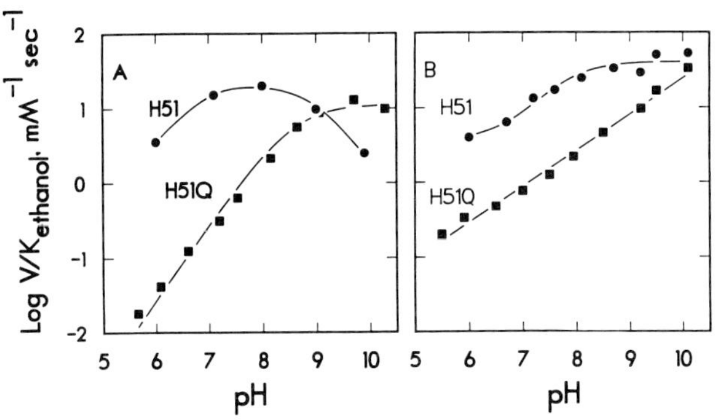

FIG. 3. Dependencies on pH for horse liver (A) and yeast (B) alcohol dehydrogenases and the H51Q mutants. Catalytic efficiencies for ethanol were determined with saturating concentrations of NAD^+. The data were fitted to the appropriate equations (Ref. 52), where the K terms refer to acid dissociation constants and k_x is a pH-independent rate. (A) Natural horse liver enzyme data (●) were determined by Dalziel at 23.5° (Ref. 59) and fitted to the BELL equation, $k = k_b/(1 + K_1/[H^+] + [H^+]/K_2)$. The data for the H51Q enzyme (■) were determined by D.-H. Park [Ph.D. Thesis, The University of Iowa, Iowa City (1991)] at 25° using 10 mM $Na_4P_2O_7$ buffers adjusted to the desired pH and an ionic strength of 0.1 with H_3PO_4 or sodium phosphate for the pH range from 5.5 to 9.5 and 10 mM $Na_4P_2O_7$ and 5 mM sodium carbonate at the higher pH values. Data were fitted to the equation $k = k_b/(1 + K_1/[H^+])$. (B) Natural yeast enzyme data (●) were determined in the pyrophosphate buffer system at 30° (Ref. 64) and fitted to the equation $k = (k_a + k_b K_1/[H^+])/(1 + K_1/[H^+])$. Data for the H51Q enzyme (■) were determined by Gould (Ref. 65) using the pyrophosphate buffer system and fitted to the equation for a straight line, although the data could also be described by equations with five parameters.

7.7.[64] In contrast, the H51Q enzyme has a linear pH dependence with a slope of 0.45, and no certain pK values. The straight line can be fitted to an expression for a mechanism with pK values of 6.8 and 8.7, but the errors of ±0.2 are not acceptable. For both enzymes, catalytic efficiency at pH 7.3 is reduced by a factor of 10, which could be significant physiologically. (Because His-51 is replaced by Thr or Tyr in alcohol dehydrogenase from some species, the residue is not required for activity.) This study does not directly measure the contribution of His-51 to catalysis, however, since the Gln isosterically replaces His and can form a hydrogen bond to the 2′-hydroxyl group of the nicotinamide ribose and block proton release. Substitution with another amino acid could change activity by a different magnitude. We conclude that His-51 contributes to catalysis and affects the pH dependence. Nevertheless, the assignment of a pK value to His-51 is problematic because it is part of the interacting system (with the zinc–water) that must be described by microscopic pK values.

Acid catalysis in *E. coli* dihydrofolate reductase was studied by substitution of Asp-27 with Asn or Ser.[90] Crystallography of the mutant enzymes complexed with methotrexate showed only minor changes in structure except for the positions of some water molecules at the site of substitution. Activities (V/K_m for dihydrofolate) of the two mutant enzymes were reduced about 4 orders of magnitude. The bell-shaped pH dependence of wild-type enzyme, with pK values of 5 and 8, was converted to amazingly linear dependencies with slopes between 0.55 and 0.75 describing increasing activity with pH decreasing from 8.7 to 4.3. The pH dependence was attributed to protonation of the substrate, with a pK of 3.5, but the linearity was not interpreted. Significant substrate deuterium isotope effects showed that hydrogen transfer steps were at least partially rate limiting in catalysis. It was concluded that Asp-27 participates in protonation of the substrate, but it is interesting that the proton appears to be shuttled to its destination on N-5 of the substrate through a hydrogen-bonded system that includes two water molecules and O-4 of the substrate.[91] Interpretation of the effects of the D27N substitution may be complicated by the kinetic evidence that the interconversion between two conformational states of the enzyme is also affected.[92] Furthermore, other changes of charge at the active site of the mouse enzyme with the L22R, W24R, and F31R substitutions also alter the pH profiles.[93]

Substitution of Asp-102 with Asn in rat trypsin also altered the pH dependencies.[94] The pH profiles for k_{cat} and k_{cat}/K_m with wild-type enzyme show that activity is maximal above a pK of 6.8, but the D102N enzyme shows a complex dependence for k_{cat}/K_m with pK values of 5.4 and 9.9 and evidence for hydroxide catalysis on k_{cat}. Catalytic efficiency at pH 7.1 was decreased 11,000-fold for hydrolysis of a peptide analog and for reaction with diisopropyl fluorophosphate, which reacts with Ser-195. However, reactivity of His-57 with the active-site directed reagent α-*N*-tosyl-L-lysylchloromethane was hardly affected. His-57 may be in the wrong tautomeric state to activate Ser-195, as suggested by X-ray crystallography of the D102N enzyme.[35]

The effects of changing the charge balance in the active site were studied in muscle lactate dehydrogenase.[95] Asn-140, which is a hydrogen

[90] E. E. Howell, J. E. Villafranca, M. S. Warren, S. J. Oatley, and J. Kraut, *Science* **231,** 1123 (1986).

[91] M. A. McTigue, J. F. Davies, II, B. T. Kaufman, and J. Kraut, *Biochemistry* **31,** 7264 (1992).

[92] J. R. Appleman, E. E. Howell, J. Kraut, and R. L. Blakley, *J. Biol. Chem.* **265,** 5579 (1990).

[93] J. Thillet, J. Absil, S. R. Stone, and R. Pictet, *J. Biol. Chem.* **263,** 12500 (1988).

[94] C. S. Craik, S. Roczniak, C. Largman, and W. J. Rutter, *Science* **237,** 909 (1987).

[95] A. Cortes, D. C. Emery, D. J. Halsall, R. M. Jackson, A. R. Clarke, and J. J. Holbrook, *Protein Sci.* **1,** 892 (1992).

bond donor to the pyruvate carbonyl oxygen, was changed to Asp. His-195 acts as an acid by donating a proton to the same keto group. Catalytic efficiency (V/K_m for pyruvate) was decreased at least two orders of magnitude at pH 7, and the pK value of 7, attributed to His-195, was abolished. Modeling of the resulting pH dependence, with an apparent pK of about 4.5, suggested that the pK of His-195 was shifted to about 10 and that only the form of the enzyme with both His and Asp protonated was active.

Elusive Bases. The combination of kinetics, specific substitutions, X-ray crystallography, and NMR spectroscopy provides different lines of evidence in support of catalytic roles for particular amino acid residues. However, such studies may lead to differing conclusions and may fail to identify "the base."

Ribulose-bisphosphate carboxylase/oxygenase from *Rhodospirillum rubrum* has been extensively studied with site-directed mutagenesis, and the results have been related to three-dimensional structures and chemical modification studies.[6] The enzyme is attractive for study because the base-catalyzed enolization and carboxylation reactions can be studied as partial reactions. Each of seven substitutions decreased carboxylase activity by more than 10-fold, and at least four of these had much higher rates for enolization as compared to carboxylase activities, providing some evidence for separate catalytic roles. Hartman and Harpel suggest that substitution of a base involved in deprotonation of C-3 of ribulose bisphosphate should reduce activity by at least six orders of magnitude, as was found for proton transfers in glyceraldehyde-3-phosphate dehydrogenase, aspartate aminotransferase, Δ^5-3-ketosteroid isomerase, and mandelate racemase. By this criterion, only Lys-166 could be considered to be the base in ribulose-bisphosphate carboxylase, but according to a three-dimensional structure, the amino group is 6.5 Å away from the C-3 hydrogen of ribulose bisphosphate and poorly oriented. Substitutions of closer residues, His-321 and Ser-368, decreased k_{cat}/K_m for ribulose bisphosphate to about 1%, but enolization rates were about 10% of those of the wild type.[96] Thus, identification of the base remains elusive. Analysis of a structure for the activated *Synechococcus* enzyme complexed with CO_2, Mg^{2+}, and 2-carboxyarabinitol 1,5-bisphosphate (analog of an intermediate) has suggested that the carbamate oxygen on Lys-191 (201 in *Synechococcus*) acts as the base and that His-287 (294) facilitates hydration of the carbonyl group formed at C-3.[97] Alternative interpretations of the three-dimensional structures are possible, and it is difficult to predict the effects of substitutions of the potential bases.

In a similar enolization, catalyzed by pig heart aconitase, Ser-642 is

[96] M. R. Harpel, F. W. Larimer, and F. C. Hartman, *J. Biol. Chem.* **266,** 24734 (1991).
[97] J. Newman and S. Gutteridge, *J. Biol. Chem.* **268,** 25876 (1993).

proposed to act as a base.[98] The S642A substitution decreased V_m by almost 5 orders of magnitude. X-Ray crystallography of the enzyme complexed with the substrate isocitrate shows that Arg-452 can stabilize the serine alkoxide, which apparently forms on transfer of the proton to the Fe-OH of the [4Fe–4S] cluster.[99] The mechanism for this proton transfer is not clear, in contrast to the situation in chymotrypsin, where Ser-195 is activated by His-57 in the catalytic triad. In one of the first "site-directed mutations," conversion of Ser-195 to dehydroalanine in chymotrypsin decreased activity by at least 4 orders of magnitude, and the S221A mutation in the catalytic triad of subtilisin reduced catalytic efficiency by almost 6 orders of magnitude.[100,101]

The role of tyrosine residues 248 and 198 in pancreatic carboxypeptidase was tested by replacing them with Phe and determining the pH dependencies of k_{cat} and k_{cat}/K_m for a variety of substrates.[102,103] Because both the native and mutant enzymes showed similar bell-shaped pH dependencies with pK values of about 6 and 10, it appears that the tyrosine residues do not mediate acid catalysis. Certainly the pK values near 10 for the mutant enzymes are not due to these tyrosine residues. The catalytic efficiencies decreased with the substitutions, attributed to local structural effects. Although the work is exemplary, the results do not prove that a tyrosine does not participate in acid–base catalysis. If the proton transfer steps were not partially rate limiting and the pH dependencies were due to binding or conformational isomerizations, the role of tyrosine could be masked. The origins of the observed pH dependencies still need to be explained.

Substitution of Tyr with Phe could permit a water to compensate for the phenolic hydroxyl group. This possibility was tested by X-ray crystallography of Y52F and Y73F mutants of phospholipase A_2.[104] The tyrosines are involved in a hydrogen-bonded system at the reaction center. The substitutions did not significantly affect activity, and a void, rather than a water site, was created.

Inspection of the structure of *Lactobacillus casei* thymidylate synthase

[98] L. Zheng, M. C. Kennedy, H. Beinert, and H. Zalkin, *J. Biol. Chem.* **267,** 7895 (1992).

[99] H. Lauble, M. C. Kennedy, H. Beinert, and C. D. Stout, *Biochemistry* **31,** 2735 (1992).

[100] H. Weiner, W. N. White, D. G. Hoare, and D. E. Koshland, Jr., *J. Am. Chem. Soc.* **88,** 3851 (1966).

[101] P. Carter and J. A. Wells, *Nature (London)* **332,** 564 (1988).

[102] D. Hilvert, S. J. Gardell, W. J. Rutter, and E. T. Kaiser, *J. Am. Chem. Soc.* **108,** 5298 (1986).

[103] S. J. Gardell, D. Hilvert, J. Barnett, E. T. Kaiser, and W. J. Rutter, *J. Biol. Chem.* **262,** 576 (1987).

[104] M. M. G. M. Thunnissen, P. A. Franken, G. H. de Haas, J. Drenth, K. H. Kalk, H. M. Verheij, and B. W. Dijkstra, *Protein Eng.* **5,** 597 (1992).

suggested that His-199 was the residue most likely to serve as a base in removing the hydrogen from C-5 of 2′-deoxyuridylate. The His is highly conserved in various species. Substitution of the corresponding His-147 in the *E. coli* enzyme to make the Gly, Asn, Gln, Val, or Leu enzymes decreased k_{cat} and k_{cat}/K_m by 10-fold.[105] The wild-type and the Gly, Asn, and Gln enzymes gave pH dependencies with a pK of about 5.7. Substitutions with Asp or Glu decreased k_{cat} by 10^3-fold, and with Arg or Lys by 10^4. Similar changes in activity were observed for the *L. casei* enzyme, and it was concluded that the results rule out a role for the His as an "essential" base, "even though there are no current alternative candidate residues to serve this function."[42] An alternative conclusion would be that the contribution of His-147 to catalysis is about 10-fold.

Aspartate transcarbamoylase (aspartate carbamoyltransferase) from *E. coli* has been studied by a variety of techniques. X-Ray crystallography of the complex with the bisubstrate analog *N*-(phosphonacetyl)-L-aspartate shows that His-134 and Arg-105 form hydrogen bonds to the carbonyl oxygen of the ligand, that Arg-54 binds OP-2 of the phosphate, and that Lys-84 is near the NH.[106] In the catalytic subunit, the H134A, K84R, R105A, and R54A substitutions decreased V/K_m for aspartate by 59, 580, 7900, and 60,000-fold, respectively.[107] The pH dependence of wild-type enzyme shows that V and V/K_m for aspartate is maximal above a pK of about 7.2, and this pK could be due to His-134.[108] However, the H134N enzyme, which has V/K_m reduced by 75-fold, retains a pK of 6.3.[109] Furthermore, NMR studies with ^{13}C-enriched enzyme suggest that His-134 does not titrate in the pH range from 6.2 to 8.5 and that the imidazole ring is neutral, from which it was concluded that His-134 does not participate in acid or base catalysis.[110] Instead, as supported by ^{31}P NMR studies on the native and R54A enzymes and by the failure of site-directed mutagenesis studies "to identify a residue in the active site that could act as a general base to accept the proton from aspartate," it is suggested that the phosphate of carbamoyl phosphate accepts the proton from the α-amino group of aspartate.[111] However, modeling of the reactive complex based on

[105] I. K. Dev, B. B. Yates, J. Atashi, and W. S. Dallas, *J. Biol. Chem.* **264,** 19132 (1969).

[106] H. Ke, W. N. Lipscomb, Y. Cho, and R. B. Honzatko, *J. Mol. Biol.* **204,** 725 (1988).

[107] R. C. Stevens, Y. M. Chook, C. Y. Cho, W. N. Lipscomb, and E. R. Kantrowitz, *Protein Eng.* **4,** 391 (1991).

[108] D. Léger and G. Hervé, *Biochemistry* **27,** 4293 (1988).

[109] X. G. Xi, F. Van Vliet, M. M. Ladjimi, R. Cunin, and G. Hervé, *Biochemistry* **29,** 8491 (1990).

[110] C. Kleanthous, D. E. Wemmer, and H. K. Schachman, *J. Biol. Chem.* **263,** 13062 (1988).

[111] J. W. Stebbins, D. E. Robertson, M. F. Roberts, R. C. Stevens, W. N. Lipscomb, and E. R. Kantrowitz, *Protein Sci.* **1,** 1435 (1992).

the complexes with inhibitors is uncertain, and the mechanism for the deprotonation of the aspartate ammonium group is not explained. Perhaps all four basic residues at the reaction center are involved in the dynamics of proton transfer.

Novel Catalysis. It is usually assumed that unprotonated imidazole is a base. Studies on triose-phosphate isomerase lead to the possibility that neutral imidazole acts as an acid. A three-dimensional structure of the yeast enzyme complexed with phosphoglycolohydroxamate, an analog of the enediolate intermediate, provides a detailed picture of the active site, consistent with Glu-165 acting as a base and His-95 as an acid.[112] The H95Q and H95N enzymes are about 200- and 3500-fold less reactive in both directions of reaction (k_{cat}/K_m), the E165D enzyme is about 300-fold less reactive, and the E165G or E165A enzymes are 10^6-fold less active than native enzyme.[113–115] The H95Q enzyme should most closely mimic the hydrogen-bonding interactions of native enzyme and introduce the least structural perturbation. The NMR results show that His-95 is uncharged between pH 5 and 9.9 and suggest that a neutral imidazole participates in the transfer of the proton between O-1 and O-2 of the substrate.[116] The structure and computer simulations suggest that other residues also participate in catalysis.[112,117]

The catalytic function of a base removed from an enzyme can be restored by added amines. Transaminases have a multifunctional lysine residue that forms an internal aldimine with pyridoxal 5′-phosphate and acts as a base in catalyzing a 1,3-prototropic shift in the conversion of the external aldimine to the ketimine. Substitution of Lys-258 with Ala in *E. coli* aspartate aminotransferase reduced activity by 6 to 8 orders of magnitude, but the loss of the ε-amino group permitted a Brønsted analysis of the proton transfer with exogenous amines. Ammonia could restore activity equivalent to 2% of the rate for the wild-type enzyme acting on cysteine sulfinate.[118] In the K258A enzyme, the rate-determining step is abstraction by the exogenous amine of the Cα proton from the amino acid. The enzyme-bound aldimine is 2×10^5-fold more reactive with

[112] R. C. Davenport, P. A. Bash, B. A. Seaton, M. Karplus, G. A. Petsko, and D. Ringe, *Biochemistry* **30,** 5821 (1991).

[113] E. B. Nickbarg, R. C. Davenport, G. A. Petsko, and J. R. Knowles, *Biochemistry* **27,** 5948 (1988).

[114] S. C. Blacklow and J. R. Knowles, *Biochemistry* **29,** 4099 (1990).

[115] J. R. Knowles, *Nature (London)* **350,** 121 (1991).

[116] P. J. Lodi and J. R. Knowles, *Biochemistry* **30,** 6948 (1991).

[117] P. A. Bash, M. J. Field, R. C. Davenport, G. A. Petsko, D. Ringe, and M. Karplus, *Biochemistry* **30,** 5826 (1991).

[118] M. D. Toney and J. F. Kirsch, *Protein Sci.* **1,** 107 (1992).

general base catalysts than a model aldimine in the imidazole-catalyzed transamination.

In contrast, substitution of the comparable Lys-245 with Gln in *Bacillus* D-amino-acid transaminase, which reduced k_{cat} to 1.5% and k_{cat}/K_m for amino acids or keto acid by 10^4- or 10^5-fold, was not rescued by exogenous amines, although ethanolamine stimulated the transaldimination.[119,120] The residual activity in the "attenuated" enzyme was suggested to result from a compensating role by Lys-267.[121] In this enzyme, the spectral changes on complexing substrates are so slow that the transients could be studied with ordinary spectrophotometers. Lys-229 binds the pyridoxal phosphate in *E. coli* serine hydroxymethyltransferase and was substituted with Gln.[122] The observation that rates of partial reactions were similar to those for the wild-type enzyme led to the conclusion that the ε-amino group is not the base that removes the α-hydrogen of the amino acid substrate. Given the uncertainties in identifying the roles of particular residues in the native or mutant enzymes, the results from "rescue" experiments should be interpreted cautiously.

Activity has been restored to other enzymes that had functional groups removed. Guanidine derivatives increased by about 100-fold the activity (k_{cat}/K_m) of R127A carboxypeptidase on some substrates to produce about 1% of the activity of native enzyme.[123] Because the reaction becomes bireactant, the concentrations of substrate and activator were varied so that correct K_m and k_{cat} values and dissociation constants for the guanidino compounds were determined. Such experiments test the importance of precise positioning of a group that stabilizes the transition state. The observation in this study that the R127A enzyme had 1% of the normal activity on one substrate and that guanidine derivatives did not restore activity limits the quantitative and structural interpretations.

The H51Q form of human liver alcohol dehydrogenase β_1 has about 6-fold lower V/K_b values for ethanol at pH 7 than the natural enzyme, and activity is restored by 200 m*M* glycylglycine.[124] A Brønsted plot showed that buffering components affect activity, a cautionary note for

[119] M. B. Bhatia, S. Futaki, H. Ueno, J. M. Manning, D. Ringe, T. Yoshimura, and K. Soda, *J. Biol. Chem.* **268,** 6932 (1993).

[120] S. Futaki, H. Ueno, A. Martinez del Pozo, M. A. Pospischil, J. M. Manning, D. Ringe, B. Stoddard, K. Tanizawa, T. Yoshimura, and K. Soda, *J. Biol. Chem.* **265,** 22306 (1990).

[121] T. Yoshimura, M. B. Bhatia, J. M. Manning, D. Ringe, and K. Soda, *Biochemistry* **31,** 11748 (1992).

[122] D. Schirch, S. D. Fratte, S. Iurescia, S. Angelaccio, R. Contestabile, F. Bossa, and V. Schirch, *J. Biol. Chem.* **268,** 23132 (1993).

[123] M. A. Phillips, L. Hedstrom, and W. J. Rutter, *Protein Sci.* **1,** 517 (1992).

[124] T. Ehrig, T. D. Hurley, H. J. Edenberg, and W. F. Bosron, *Biochemistry* **30,** 1062 (1991).

interpretation of results of replacing groups that participate in acid–base catalysis.

Additivity

A review of the results of combining multiple mutations in a protein suggests that the effects on free energy of reactions are often additive.[125] This can be expected when the mutations are distant from one another and do not grossly affect the protein structure.[126] When the groups interact with one another or when the mutations change the reaction mechanism or rate-limiting steps, effects may not be additive.

The large effects (4 or 5 orders of magnitude) of the Y14F and D38N mutations in a Δ^5-3-ketosteroid isomerase appeared to be additive for k_{cat} or k_{cat}/K_m, and the two residues appear to account for the total enzymatic rate acceleration, acting as acid–base catalysts for the rate-limiting enolization of the substrate.[127] The possible effects of mutations on individual steps in the mechanism were analyzed to show how different kinetic results can be interpreted.

In contrast, the studies on the Ser, His, and Asp residues in the catalytic triad of subtilisin are of interest because the k_{cat}/K_m values for the single, double, and triple substitutions with Ala all decreased activity by about 5 or 6 orders of magnitude.[101] The enzyme hydrolyzes a peptide substrate at a rate about 9 orders of magnitude faster than the uncatalyzed rate. The lack of additivity apparently arises because the residues interact to facilitate nucleophilic attack on the carbonyl carbon of the peptide substrate. The results emphasize the difficulty of determining contributions to catalysis of a particular residue. Moreover, it was found that substitution of Asn-155, which is thought to stabilize the oxyanion tetrahedral intermediate, with Gly increased the activity of the S221A enzyme by a factor of 10.[128] This result led to the conclusion that mutagenesis results may exaggerate the importance of catalytic groups. A similar lack of additivity (named "antagonistic") was observed for the effects on V_m of the single and double D21E and R87G mutations in staphylococcal nuclease.[129] However, additive or synergistic effects were seen for binding of a substrate or substrate analog.

An extreme example of nonadditivity is found in the restoration of

[125] J. A. Wells, *Biochemistry* **29,** 8509 (1990).

[126] P. J. Carter, G. Winter, A. J. Wilkinson, and A. R. Fersht, *Cell (Cambridge, Mass.)* **38,** 835 (1984).

[127] A. Kuliopulos, P. Talalay, and A. S. Mildvan, *Biochemistry* **29,** 10271 (1990).

[128] P. Carter and J. A. Wells, *Proteins: Struct. Funct. Genet.* **7,** 335 (1990).

[129] D. J. Weber, E. H. Serpersu, D. Shortle, and A. S. Mildvan, *Biochemistry* **29,** 8632 (1990).

activity of the inactive W191F mutant of cytochrome-*c* peroxidase by a second mutation, H175Q. Residue 175 is the proximal ligand to the heme and interacts with residue 191.[130] Further studies on multiple mutations in other enzymes are of interest for distinguishing between additivity and cooperativity in enzymatic catalysis.

Concluding Remarks

In combination with other methods, site-directed mutagenesis is a useful tool for enzymologists. As with other experimental approaches, the procedures must be applied carefully and the results interpreted cautiously. The examples discussed here outline some uncertainties in such investigations. The original literature should be read critically for details. Further studies should continue to illuminate different catalytic mechanisms and resolve ambiguities in current understanding. Each enzyme offers a different challenge, and generalizations about the contributions of various amino acid residues to catalysis must await more comprehensive results. Enzymes are cooperative macromolecules, and the functions of individual residues must be integrated with the structural and dynamic aspects of catalysis.

Acknowledgments

I thank co-workers Robert M. Gould, David W. Green, Andrew D. Hershey, Axel J. Ganzhorn, Darla Ann Kratzer, Doo-Hong Park, Hong-Wei Sun, James A. Lorenzen, Tobias Jacobi, Edda Warth, Keehyuk Kim, Fan Fan, Susan K. Souhrada, and Suresh Pal, who developed and applied site-directed mutagenesis for the study of alcohol dehydrogenases, and the National Institute on Alcohol Abuse and Alcoholism, the National Science Foundation, and The University of Iowa Center for Biocatalysis and Bioprocessing for support of our work.

[130] K. Choudhury, M. Sundaramoorthy, J. M. Mauro, and T. L. Poulos, *J. Biol. Chem.* **267,** 25656 (1992).

Section II

Inhibitors as Mechanistic Probes

[5] Reversible Enzyme Inhibitors as Mechanistic Probes

by HERBERT J. FROMM

Introduction

A number of experimental protocols have been employed by kineticists over the years in order to choose between bireactant ordered and random mechanisms. Unfortunately, most suffer from technical limitations. Among these, the most popular methods are the use of the Haldane relationship, product inhibition, and isotope exchange. The first method, which compares kinetically determined parameters with the apparent equilibrium constant for a reaction, requires precise initial rate data, which are frequently difficult to obtain. As this writer pointed out in 1962,[1] interpretation of results from product inhibition studies is complicated by the formation of abortive complexes of enzyme, substrate, and product. Finally, isotope exchange cannot be used with systems, such as fructose-1,6-bisphosphatase, in which the equilibrium constant is extremely large.

Although competitive inhibitors have been used extensively for a variety of reasons in enzyme experiments, their value as tools for making a choice of kinetic mechanism from among possible alternatives was not realized until 1962, when Fromm and Zewe[2] suggested that competitive inhibitors of substrates could be used to differentiate between random and ordered mechanisms. Furthermore, in the latter case, a determination of the substrate binding order could be made from such experiments. This protocol is quite likely the simplest approach for differentiating between Ordered and Random Bi Bi possibilities.* In addition, it has the advantage of permitting the kineticist to come to definitive conclusions from studies of reactions in a single direction only. Its obvious limitation involves the requirement that a competitive inhibitor be available for each substrate. However, when other initial rate data are available, a good deal of information concerning the kinetic mechanism may be provided from experiments with only one substrate analog even for Bi and Ter reactant systems.

In this discussion it is assumed that the competitive inhibitor, which is usually a substrate analog, when bound to the enzyme, will not permit

[1] H. J. Fromm and D. Nelson, *J. Biol. Chem.* **237,** 215 (1962).

[2] H. J. Fromm and V. Zewe, *J. Biol. Chem.* **237,** 3027 (1962).

* The nomenclature of Cleland will be used throughout this chapter. See W. W. Cleland, *Biochim. Biophys. Acta* **67,** 104 (1963).

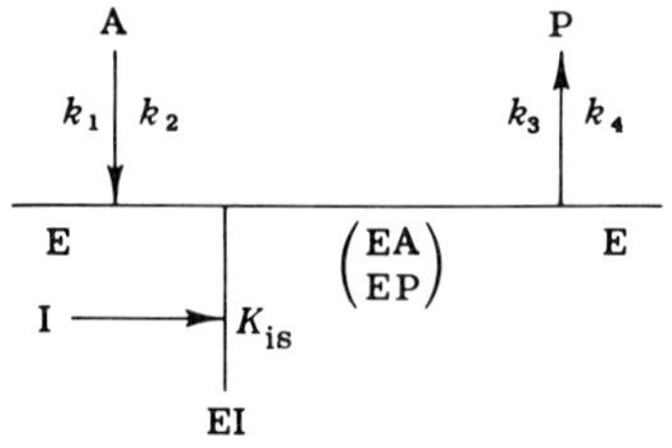

SCHEME 1

product formation to occur. These inhibitors are then dead-end inhibitors as contrasted with "partial" competitive inhibitors, which when associated with the enzyme allow formation of product either at a reduced or accelerated rate.[3] In addition, it will be assumed that enzyme–inhibitor complex formation occurs rapidly relative to other steps in the reaction pathway.

Theory

One-Substrate Systems

Before describing how competitive inhibitors are used to make a choice of mechanism in the case of multisubstrate systems, some discussion is warranted concerning reversible dead-end inhibition for one-substrate reactions. Scheme 1 illustrates a typical Uni Uni mechanism and how a linear competitive inhibitor enters into the reaction mechanism.

By definition, a competitive inhibitor competes with the substrate for the same site on the enzyme. Identical kinetic results are obtained, however, if the inhibitor and substrate compete for different sites, but where binding is mutually exclusive. Although there is usually a structural similarity between the substrate and the competitive inhibitor, this is not always the case. Finally, competitive inhibition is reversed when the enzyme is saturated with substrate.

The derivation of the rate equation for a competitive inhibitor can be done algebraically or by using any of the more sophisticated procedures described elsewhere.[4,5] Derivation by the former method is as follows.

[3] H. J. Fromm, "Initial Rate Enzyme Kinetics," p. 86. Springer-Verlag, Berlin and New York, 1975.

[4] C. Y. Huang, this series, Vol. 63, p. 54.

[5] H. J. Fromm, this series, Vol. 63, p. 84.

The velocity expression is $v = k_3 \begin{pmatrix} \text{EA} \\ \text{EP} \end{pmatrix}$, and the conservation of enzyme equation is $\text{E}_0 = \text{E} + \begin{pmatrix} \text{EA} \\ \text{EP} \end{pmatrix} + \text{EI}$, when E_0 is total enzyme. The dissociation constant for the enzyme–inhibitor complex is taken as $K_{is} = (\text{E})(\text{I})/(\text{EI})$. Thus,

$$\text{E}_0 = \text{E} + \begin{pmatrix} \text{EA} \\ \text{EP} \end{pmatrix} + \frac{(\text{E})(\text{I})}{K_{is}} = (\text{E})\left(1 + \frac{\text{I}}{K_{is}}\right) + \begin{pmatrix} \text{EA} \\ \text{EP} \end{pmatrix} \tag{1}$$

From the expression $\text{E} = (K_a/\text{A}) \begin{pmatrix} \text{EA} \\ \text{EP} \end{pmatrix}$, and the equation for initial velocity,

$$v = \frac{V_1}{1 + (K_a/\text{A})(1 + \text{I}/K_{is})} \tag{2}$$

where K_a is the Michaelis constant, $(k_2 + k_3)/k_1$, and V_1 is the maximal velocity, $k_3\text{E}_0$.

It can be seen from Eq. (1) that the free enzyme component in the conservation of enzyme equation is multiplied by the factor $(1 + \text{I}/K_{is})$. A shorthand method for including the effect of a reversible dead-end inhibitor in the rate equation is simply to first identify the enzyme species in the noninhibited rate equation that reacts with the inhibitor, then multiply that enzyme form by the proper factor. This will perhaps be somewhat clearer when the rate equation for noncompetitive inhibition is discussed.

Figure 1 illustrates a double-reciprocal plot for an enzyme system in

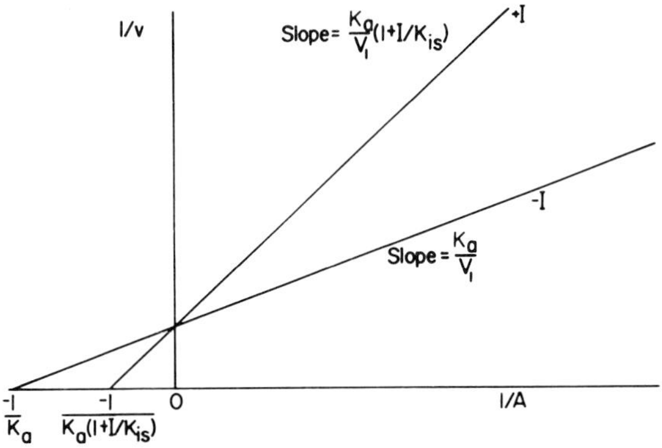

FIG. 1. Plot of 1/initial velocity (v) versus 1/substrate concentration (A) in the presence and absence of a linear competitive inhibitor (I).

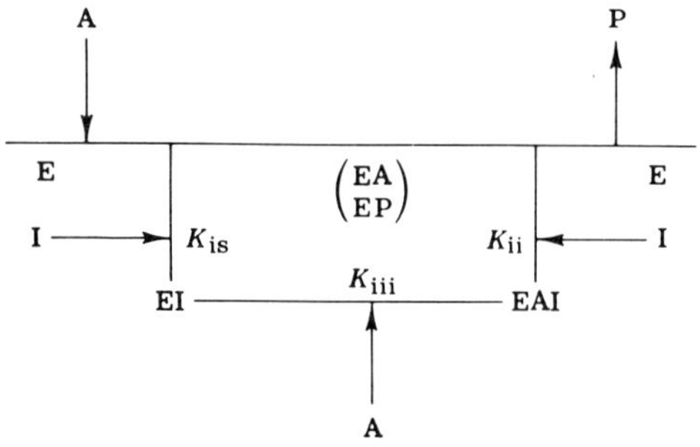

SCHEME 2

the presence and in the absence of a competitive inhibitor. The rationale for this particular plot comes by taking the reciprocal of both sides of Eq. (2):

$$\frac{1}{v} = \frac{1}{V_1}\left[1 + \frac{K_a}{A}\left(1 + \frac{I}{K_{is}}\right)\right] \tag{3}$$

The inhibition described in Eq. (3) is referred to as linear competitive, because a replot of slopes versus I gives a straight line. It will be seen later that competitive inhibition need not be linear.

A number of procedures are currently in vogue for linearizing initial rate plots for inhibitors. The method used most is that illustrated in Fig. 1. Another graphing method, the Dixon plot, is frequently used; however, it suffers from very serious inherent limitations,[6,7] and its use is not recommended.

Scheme 2 describes linear noncompetitive enzyme inhibition. In this model the substrate and the inhibitor are not mutually exclusive binding ligands, and complexes of enzyme, inhibitor, and substrate are thus possible. The rate equation for linear noncompetitive inhibition, in double-reciprocal form, is

$$\frac{1}{v} = \frac{1}{V_1}\left[\left(1 + \frac{I}{K_{ii}}\right) + \frac{K_{ia}}{A}\left(1 + \frac{I}{K_{is}}\right)\right] \tag{4}$$

where K_{ia} is the dissociation constant of the EA complex. In the derivation of Eq. (4), it is assumed that the inhibitory complexes are dead-end and

[6] D. L. Purich and H. J. Fromm, *Biochim. Biophys. Acta* **268,** 1 (1972).

[7] H. J. Fromm, "Initial Rate Enzyme Kinetics," p. 99. Springer-Verlag, Berlin and New York, 1975.

that all steps shown in Scheme 2 are in rapid equilibrium relative to the breakdown of the central complex to form product.

In the derivation of Eq. (4) the enzyme forms containing inhibitor must be added to the conservation of enzyme equation. Thus, $E_0 = E + EA + EI + EAI$. From the interactions shown in Scheme 2, it can be seen that EI arises from reaction of the inhibitor and E, whereas EAI *may* arise from interaction of the inhibitor and EA. In deriving Eq. (4), it is then necessary merely to multiply the factor $(1 + I/K_{ii})$ by the EA term of the uninhibited rate equation and the factor $(1 + I/K_{is})$ by the E term of the uninhibited rate expression. The very same result will be obtained if the K_{iii} step, rather than the K_{ii} pathway, is used. This is true because the four dissociation constants are not independent, but are related by the following expression: $K_{is}K_{iii} = K_{ia}K_{ii}$.

A typical plot of $1/v$ versus $1/A$ in the presence and in the absence of the noncompetitive inhibitor is shown in Fig. 2. It can be seen from the graph that the lines converge in the second quadrant. Convergence of the curves may also occur in the third quadrant or on the abscissa, depending on the relationship between the dissociation constants for the enzyme–inhibitor complexes. The *x, y* coordinates of the intersection point for noncompetitive inhibition are $-K_{is}/K_{ia}K_{ii}$, $(1/V_1)(1 - K_{is}/K_{ii})$. If $K_{is} = K_{ii}$, that is, the binding of the inhibitor to the enzyme is not affected by the presence of the substrate on the enzyme, the curves will intersect on the 1/A axis. When $K_{is} < K_{ii}$, intersection will be above the abscissa, whereas when $K_{is} > K_{ii}$, intersection of the curves will be in the third quadrant.

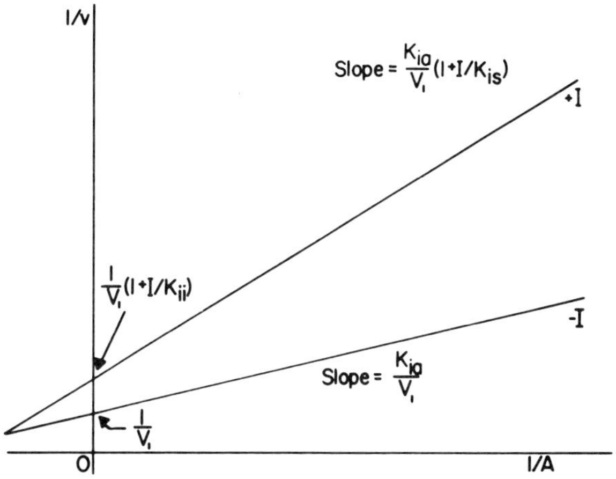

FIG. 2. Plot of 1/initial velocity (v) versus 1/substrate concentration (A) in the presence and in the absence of a linear noncompetitive inhibitor (I).

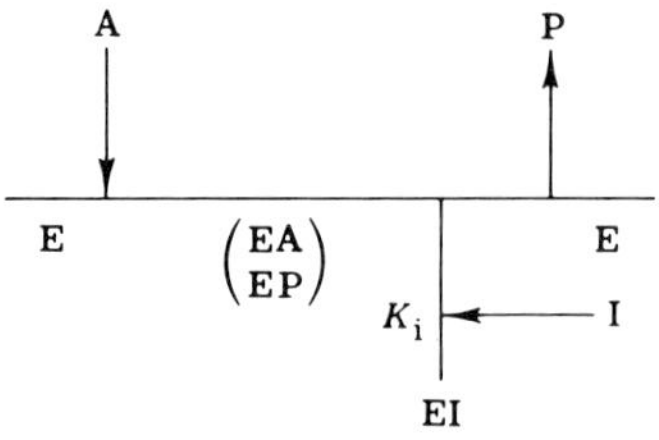

SCHEME 3

It can be seen from Eq. (4) that replots of slopes and intercepts against inhibitor concentration will give linear lines. This type of noncompetitive inhibition is more formally referred to as S-linear (slope) and I-linear (intercept) noncompetitive.

Another type of reversible dead-end enzyme inhibition that will be useful for later discussion is linear uncompetitive inhibition. Scheme 3 depicts the pathway that illustrates this phenomenon. The rate expression for linear uncompetitive inhibition is shown in Eq. (5), where K_i is the dissociation constant for the EAI complex:

$$\frac{1}{v} = \frac{1}{V_1}\left[\left(1 + \frac{I}{K_i}\right) + \frac{K_a}{A}\right] \tag{5}$$

Figure 3 describes the results of a plot of $1/v$ versus $1/A$ in the presence and absence of an uncompetitive inhibitor. It can be seen that linear uncompetitive inhibition gives rise to a family of parallel lines. It is often

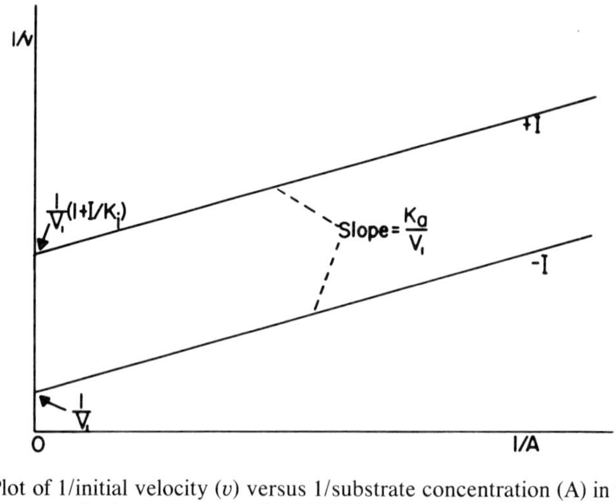

FIG. 3. Plot of 1/initial velocity (v) versus 1/substrate concentration (A) in the presence and in the absence of a linear uncompetitive inhibitor (I).

difficult to determine whether the lines in inhibition experiments are really parallel. Rather sophisticated computer programs are available that both test and fit *weighted* kinetic data in an attempt to address this problem.[8] A discussion of this point is not presented here; however, it may be helpful to point out that uncompetitive inhibition may be graphed as a Hanes plot (i.e., A/*v* versus A). This plot is illustrated in Fig. 4 and has the advantage, relative to the double-reciprocal plot, that the inhibited and uninhibited lines must converge at a common point on the A/*v* axis. This approach does not eliminate the necessity of model fitting and testing, but it does permit the investigator to make a preliminary judgment more easily on the nature of the inhibition.

The last type of reversible dead-end inhibition to be considered in conjunction with unireactant systems is nonlinear inhibition. Nonlinear enzyme inhibition may be obtained from replots of primary double-reciprocal plots as a result of multiple dead-end inhibition, substrate and product inhibition, partial inhibition, and allostery. This discussion is limited to multiple dead-end inhibition.

When an inhibitor adds reversibly to different enzyme forms (e.g., Scheme 2), inhibition is linear; however, when there is multiple inhibitor binding to a single enzyme form, or to enzyme forms that are sequentially connected, replots of either slopes or intercepts against inhibitor may be nonlinear. If the following equilibrium is added to the simple Uni Uni mechanism of Scheme 1, $EI + I = EI_2$, K_{ii}, the rate equation

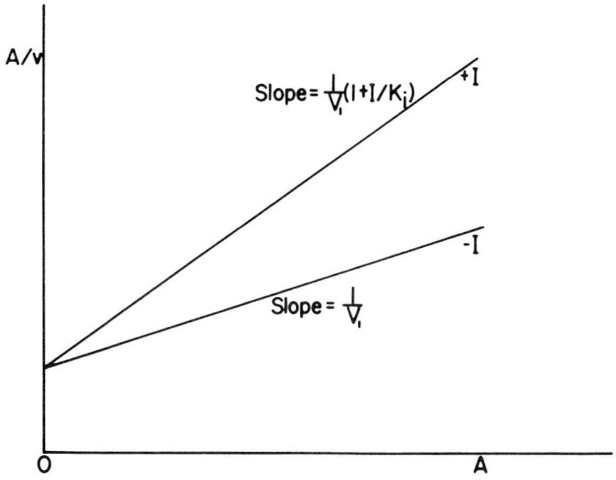

FIG. 4. Plot of substrate concentration (A) divided by initial velocity (*v*) versus substrate concentration (A) in the presence and in the absence of a linear uncompetitive inhibitor (I).

[8] H. J. Fromm, "Initial Rate Enzyme Kinetics," p. 63. Springer-Verlag, Berlin and New York, 1975.

obtained is

$$\frac{1}{v} = \frac{1}{V_1}\left[1 + \frac{K_{ia}}{A}\left(1 + \frac{I}{K_{is}} + \frac{I^2}{K_{is}K_{ii}}\right)\right] \tag{6}$$

Equation (6) describes parabolic competitive inhibition. The slope of Eq. (6) is

$$\text{Slope} = \frac{K_{ia}}{V_1}\left(1 + \frac{I}{K_{is}} + \frac{I^2}{K_{is}K_{ii}}\right) \tag{7}$$

and a plot of slope as a function of I will give rise to a parabola. It is also possible to obtain parabolic uncompetitive inhibition. In this case only the intercept will be affected by inhibitor. In the event that parabolic noncompetitive inhibition is encountered, either the slope or the intercept, or both, may contain inhibitor terms of greater than first degree.

Nonlinear noncompetitive inhibition may affect slopes and intercepts; in this case it is called S-parabolic I-parabolic noncompetitive inhibition. If the replots of intercept against I are linear, whereas the slope replot is parabolic, the inhibition is called S-parabolic I-linear noncompetitive inhibition.

Parabolic inhibition will also result from interactions of the following type:

$$E + I = EI,\ K_{is}; \qquad EI + A = EIA,\ K_{ii}; \qquad EIA + I = EI_2A,\ K_{iii} \tag{8}$$

Slope and intercept replots versus inhibitor may be of a more complicated nature. Cleland has referred to some of these as 2/1, 3/2, etc., functions. Equation (9) illustrates an example of a $S - 2/1$ function, in which a second-order polynomial is divided by a first-order polynomial:

$$\text{Slope} = \frac{K_{ia}(1 + aI + bI^2)}{V_1(1 + cI)} \tag{9}$$

It may be difficult to distinguish a plot of slope versus inhibitor concentration for Eq. (9) from linear replots.

Two-Substrate Systems

Let us now consider the case of a random mechanism to determine how the dead-end competitive inhibitor affects the kinetics of the system. The rapid-equilibrium random pathway of enzyme and substrate interaction (Random Bi Bi) is illustrated in Ref. 9. In the case of a competitive inhibitor for substrate A, the inhibitor, I, would participate at every step in the kinetic mechanism in which the substrate A normally reacts. Thus, the following interactions of enzyme with inhibitor might be expected:

$$\mathrm{E} + \mathrm{I} = \mathrm{EI},\ K_{\mathrm{i}}; \qquad \mathrm{EB} + \mathrm{I} = \mathrm{EIB},\ K_{\mathrm{ii}}; \qquad \mathrm{EI} + \mathrm{B} = \mathrm{EIB},\ K_{\mathrm{iii}} \tag{10}$$

When the expressions EI and EIB are added to the conservation of enzyme expression, and the rate equation derived for the effect of the competitive inhibitor of substrate A, the following relationship is obtained:

$$\frac{1}{v} = \frac{1}{V_1} + \frac{K_{\mathrm{a}}}{V_1(\mathrm{A})}\left(1 + \frac{\mathrm{I}}{K_{\mathrm{ii}}}\right) + \frac{K_{\mathrm{b}}}{V_1(\mathrm{B})} + \frac{K_{\mathrm{ia}}K_{\mathrm{b}}}{V_1(\mathrm{A})(\mathrm{B})}\left(1 + \frac{1}{K_{\mathrm{i}}}\right) \tag{11}$$

where K_{b} is the Michaelis constant for substrate B.

When double-reciprocal plots of $1/v$ as a function of $1/\mathrm{A}$ are made at different fixed concentrations of inhibitor, only the slope term of the rate expression is altered; that is,

$$\text{Slope} = \frac{K_{\mathrm{a}}}{V_1}\left(1 + \frac{\mathrm{I}}{K_{\mathrm{ii}}}\right) + \frac{K_{\mathrm{ia}}K_{\mathrm{b}}}{V_1(\mathrm{B})}\left(1 + \frac{\mathrm{I}}{K_{\mathrm{i}}}\right) \tag{12}$$

On the other hand, when B is the variable substrate, double-reciprocal plots at different fixed levels of inhibitor will exhibit increases in both slopes and intercepts:

$$\text{Intercept} = \frac{1}{V_1}\left[1 + \frac{K_{\mathrm{a}}}{\mathrm{A}}\left(1 + \frac{\mathrm{I}}{K_{\mathrm{ii}}}\right)\right] \tag{13}$$

$$\text{Slope} = \frac{1}{V_1}\left[K_{\mathrm{b}} + \frac{K_{\mathrm{ia}}K_{\mathrm{b}}}{\mathrm{A}}\left(1 + \frac{\mathrm{I}}{K_1}\right)\right] \tag{14}$$

Equation (11) predicts then that a dead-end competitive inhibitor for substrate A of the Random Bi Bi mechanism is a noncompetitive inhibitor of substrate B.

If now a dead-end competitive inhibitor for substrate B is used, the following interactions are to be expected;

$$\mathrm{E} + \mathrm{I} = \mathrm{EI},\ K_{\mathrm{i}}; \qquad \mathrm{EA} + \mathrm{I} = \mathrm{EAI},\ K_{\mathrm{ii}}; \qquad \mathrm{EI} + \mathrm{A} = \mathrm{EAI},\ K_{\mathrm{iii}} \tag{15}$$

The rate equation for the effect of a dead-end competitive inhibitor of substrate B is described by Eq. (16):

$$\frac{1}{v} = \frac{1}{V_1} + \frac{K_{\mathrm{a}}}{V_1(\mathrm{A})} + \frac{K_{\mathrm{b}}}{V_1(\mathrm{B})}\left(1 + \frac{\mathrm{I}}{K_{\mathrm{ii}}}\right) + \frac{K_{\mathrm{ia}}K_{\mathrm{b}}}{V_1(\mathrm{A})(\mathrm{B})}\left(1 + \frac{\mathrm{I}}{K_{\mathrm{i}}}\right) \tag{16}$$

It can be seen from Eq. (16) that a dead-end competitive inhibitor of substrate B will show noncompetitive inhibition relative to substrate A.

In summary, then, for the rapid-equilibrium Random Bi Bi mechanism, a competitive inhibitor for either substrate will act as a noncompetitive inhibitor for the other substrate. These observations are consistent with the symmetry inherent in the random mechanism. Similar inhibition patterns are to be expected for the rapid-equilibrium Random Bi Uni mechanism.

Very few Random BI BI mechanisms are truly rapid-equilibrium random in both directions; however, this condition will be approximated in the "slow direction." When steady-state conditions prevail, that is, when the interconversion of the ternary complexes is not slow relative to other steps of the kinetic mechanism, it may be supposed that the initial rate plots in double-reciprocal form would not be linear. This is to be expected because of the second-degree substrate terms generated under steady-state conditions; however, Schwert[10] has suggested that the deviation from linearity might be too subtle to discern. A similar point was also made by Cleland and Wratten,[11] and Rudolph and Fromm[12] concluded from computer simulations of the steady-state Random Bi Bi mechanism proposed for yeast hexokinase that the kinetics approximate the limiting equilibrium assumption. These workers also found that the competitive inhibition patterns proposed for the rapid-equilibrium case would be indistinguishable from the situation in which steady-state conditions prevail.

In the case of the Ordered Bi Bi mechanism, competitive dead-end inhibitors of the first substrate to add to the enzyme give inhibition patterns relative to the other substrate that are distinctively different from the pattern obtained when a competitive dead-end inhibitor of the second substrate is employed. It is this very point that permits the kineticist to make a choice between Random and Ordered Bi Bi mechanisms and allows one to identify the first and second substrates to add in the ordered mechanism.

In the case of the Ordered Bi Bi mechanism, substrate A adds only to free enzyme.[9] By analogy, the competitive dead-end inhibitor should add only to this enzyme form. In addition, it is assumed that the conformation of the enzyme has been distorted enough by the inhibitor so as to preclude addition of substrate B to the enzyme–inhibitor complex.

The competitive dead-end inhibitor for substrate A may react as follows with the enzyme:

[9] H. J. Fromm, this series, Vol. 63, p. 42.

[10] G. W. Schwert, *Fed. Proc., Fed. Am. Soc. Exp. Biol.* **13,** Abstr. 971 (1954).

[11] W. W. Cleland and C. C. Wratten, "The Mechanism of Action of Dehydrogenases," p. 103. Univ. Press of Kentucky, Lexington, 1969.

[12] F. B. Rudolph and H. J. Fromm, *J. Biol. Chem.* **246,** 6611 (1971).

$$\mathrm{E} + \mathrm{I} = \mathrm{EI}, \qquad K_{\mathrm{i}} = (\mathrm{E})(\mathrm{I})/\mathrm{EI} \tag{17}$$

If the conservation of enzyme equation for the Ordered Bi Bi mechanism is modified to account for the additional complex, EI, the initial rate expression is

$$\frac{1}{v} = \frac{1}{V_1} + \frac{K_{\mathrm{a}}}{V_1(\mathrm{A})}\left(1 + \frac{\mathrm{I}}{K_{\mathrm{i}}}\right) + \frac{K_{\mathrm{b}}}{V_1(\mathrm{B})} + \frac{K_{\mathrm{ia}}K_{\mathrm{b}}}{V_1(\mathrm{A})(\mathrm{B})}\left(1 + \frac{\mathrm{I}}{K_{\mathrm{i}}}\right) \tag{18}$$

Equation (18) predicts that the competitive inhibitor for substrate A, the first substrate to add in the ordered mechanism, will be noncompetitive relative to substrate B. On the other hand, for this mechanism, a dead-end competitive inhibitor for substrate B would be expected not to react with free enzyme, but rather with the EA binary complex. This interaction may be described by the following relationship:

$$\mathrm{EA} + \mathrm{I} = \mathrm{EAI}, \qquad K_{\mathrm{i}} = (\mathrm{EA})(\mathrm{I})/\mathrm{EAI} \tag{19}$$

The kinetic expression obtained when this effect is included in the Ordered Bi Bi mechanism is

$$\frac{1}{v} = \frac{1}{V_1} + \frac{K_{\mathrm{a}}}{V_1(\mathrm{A})} + \frac{K_{\mathrm{b}}}{V_1(\mathrm{B})}\left(1 + \frac{\mathrm{I}}{K_{\mathrm{i}}}\right) + \frac{K_{\mathrm{ia}}K_{\mathrm{b}}}{V_1(\mathrm{A})(\mathrm{B})} \tag{20}$$

It is quite clear that a dead-end competitive inhibitor for the second substrate will yield uncompetitive inhibition relative to substrate A. This unique inhibition pattern allows a distinction to be made between ordered and random bireactant kinetic mechanisms, and it permits determination of the substrate binding order in the former case. These points are summarized in Table I.

Three-Substrate Systems

The kinetic mechanisms for enzymes that utilize three substrates may be divided into Ping Pong and Sequential categories. It is then possible to make a choice of mechanism from among these terreactant systems using dead-end competitive inhibitors for the substrates. The kinetic mechanisms and expected inhibition patterns are illustrated in Table II.

Practical Aspects

Experimentally, it is very important that the fixed, or nonvaried, substrate be held at a subsaturating level, preferably in the region of the Michaelis constant. If, for example, when considering Eq. (11), the concentration of substrate A is held very high when B is the variable substrate,

TABLE I
USE OF DEAD-END COMPETITIVE INHIBITORS FOR DETERMINING BIREACTANT KINETIC MECHANISMS

Mechanism	Competitive inhibitor for substrate	1/A plot	1/B plot
Random Bi Bi, Ordered Bi Bi-Subsite mechanism[a]	A	C[b]	N[c]
and Random Bi Uni	B	N	C
Ordered Bi Bi	A	C	N[d]
and Theorell–Chance	B	U[e]	C
Ping Pong Bi Bi	A	C	U
	B	U	C

[a] See H. J. Fromm, this series, Vol. 63, p. 42.
[b] C refers to a double-reciprocal plot that shows competitive inhibition.
[c] N refers to a double-reciprocal plot that shows noncompetitive inhibition.
[d] In the ordered mechanism convergence may be on, above, or below the abscissa; however, the point of intersection with the inhibitor must have the same ordinate as a family of curves in which the other substrate is substituted for the inhibitor.
[e] U refers to a double-reciprocal plot that shows uncompetitive inhibition.

it is possible that the intercept increases to be expected in the presence of inhibitor may not be discernible, and the inhibition may appear to be competitive with respect to either substrate. It is important to note that, when replots of slopes and intercepts are made as a function of inhibitor concentration for the type of inhibition illustrated, the plots will be linear.

In studies in which dead-end competitive inhibitors are employed, it is often useful to evaluate the various inhibition constants. This can be done in a number of ways, and a few of the methods that may be used will be illustrated.

It is possible to evaluate either K_i or K_{ii} in Eqs. (11) and (16) from secondary plots of slopes and intercepts versus inhibitor concentration. It can be seen from Eq. (14) that a plot of slope versus I will give a replot in which the slope of the secondary plot is

$$\text{Slope} = \frac{K_{ia}K_b}{V_1K_1(\text{A})} \tag{21}$$

The value of K_i may also be evaluated by determining the intersection point of the secondary plot on the abscissa, that is, where slope = 0. In this case,

$$\text{I} = -K_i\left(1 + \frac{\text{A}}{K_{ia}}\right) \tag{22}$$

The advantage of using Eq. (22) rather than Eq. (21) is that it is not necessary to evaluate V_1. Presumably, data for A and K_{ia} will be in hand.

The value for K_{ii} can be determined with a knowledge of A and K_a by evaluating the point of intersection on the abscissa of data from Eq. (13). In this case, where the intercept equals 0, in the replot,

$$I = -K_{ii}\left(1 + \frac{A}{K_a}\right) \tag{23}$$

Methods similar to those described for the Random Bi Bi mechanism can be used to determine the dissociation constants in the case of the Ordered Bi Bi mechanism. For example, K_i can be evaluated from either a slope or intercept versus inhibitor replot using Eq. (18), in which B is the variable substrate. It is of interest to note that the inhibition constant must be the same for this mechanism regardless of whether the determination is made from the slope or intercept. This may or may not be true for the Random pathway, depending on whether $K_i = K_{ii}$. It will be possible to determine K_i in Eq. (20) readily, using the methods already described.

Examples

The use of dead-end competitive inhibitors for choosing between the Random and Ordered Bi Bi mechanisms has been employed with many enzyme systems. The basic protocol involves segregation of mechanisms into either the Ping Pong or Sequential class from initial rate experiments. After the Sequential nature of the system has been established, the dead-end competitive inhibitors may be used to establish whether the kinetic mechanism is Random or Ordered.

Two examples can be used to illustrate this point. Fromm and Zewe,[2] reported that yeast hexokinase is Sequential when they demonstrated that double-reciprocal plots of $1/v$ versus $1/[\mathrm{MgATP}^{2-}]$ at different fixed concentrations of glucose converged to the left of the axis of ordinates. From the same data, they observed that, when $1/v$ was plotted as a function of 1/[glucose] at different fixed concentrations of MgATP^{2-}, the resulting family of curves also converged to the left of the $1/v$ axis. In addition, both sets of primary plots intersected on the abscissa. These investigators also demonstrated that AMP, a competitive dead-end inhibitor for MgATP^{2-}, was a noncompetitive inhibitor with respect to glucose. From these experiments, it was concluded that the kinetic mechanism for yeast hexokinase was either Random Bi Bi or Ordered Bi Bi with MgATP^{2-} as the initial substrate to add to the enzyme. If glucose were to add to hexokinase before MgATP^{2-} in an Ordered Bi Bi mechanism, AMP inhibition would have been uncompetitive with respect to glucose.

TABLE II
COMPETITIVE INHIBITION PATTERNS FOR VARIOUS THREE-SUBSTRATE MECHANISMS[a]

Mechanism[b]	Competitive inhibitor for substrate	1/A plot	1/B plot	1/C plot
B 2a and 2b:	A	C[c]	N[d,e]	N[f]
Ordered Ter	B	U[g]	C	N[h]
Ter and Ordered Ter Bi	C	U	U	C
B 3: Random Ter	A	C	N	N
Ter and Ter Bi*	B	N	C	N
	C	N	N	C
B 4: Random AB*	A	C	N	C[i]
	B	N	C	C[j]
	C	U	U	C
B5: Random BC*	A	C	N	N
	B	U	C	N
	C	U	N	C
B 6: Random AC*	A	C	N	N
	B	N	C	N
	C	N	N	C
D 1: Hexa Uni Ping Pong	A	C	U	U
	B	U	C	U
	C	U	U	C
D 2: Ordered Bi Uni	A	C	N[k]	U
Uni Bi Ping Pong	B	U	C	U
	C	U	U	C
D 3: Ordered Uni	A	C	U	U
Uni Bi Bi Ping Pong	B	U	C	N[l]
	C	U	U	C
D 4: Random Bi Uni Uni	A	C	N	U
Bi Ping Pong[+]	B	N	C	U
	C	U	U	C
D 5 and D 6: Random	A	C	U	U
Uni Uni Bi Bi	B	U	C	N
Ping Pong[+]	C	U	N	C

[a] The various interactions of the competitive inhibitors are presented elsewhere along with the inhibited rate equations (H. J. Fromm, "Initial Rate Enzyme Kinetics," p. 102. Springer-Verlag, Berlin and New York, 1975).

[b] The numbers refer to the mechanisms listed in H. J. Fromm, this series, Vol. 63, p. 42. *It is assumed that all steps equilibrate rapidly relative to the breakdown of the productive quarternary complex to form products. [+]It is assumed that all steps equilibrate rapidly relative to the breakdown of the productive ternary complex to form products.

[c] C refers to a double-reciprocal plot that shows competitive inhibition.

[d] N refers to a double-reciprocal plot that shows noncompetitive inhibition.

[e] If EI reacts with B to form EIB, the plots would be nonlinear.

[f] If EIB reacts with C to form EIBC, the plots would be nonlinear.

[g] U refers to a double-reciprocal plot that shows uncompetitive inhibition.

The same investigators employed oxalate, a dead-end competitive inhibitor of L-lactate, to help establish the kinetic mechanism of the muscle lactate dehydrogenase reaction.[13] They observed that oxalate was uncompetitive with respect to NAD^+ and concluded from these findings, and other studies, that the kinetic mechanism was Iso Ordered Bi Bi with the nucleotide substrates adding to the enzyme first.

Lindstad *et al.*[14] studied the kinetic mechanism of sheep liver sorbitol dehydrogenase using a variety of kinetic techniques including the Haldane relationship,[15] the Dalziel relationship,[16] isotope effects, and a competitive inhibitor for sorbitol, dithiothreitol. Both the Haldane relationship and the Dalziel relationship require highly accurate data in order to obtain precise kinetic parameters. In addition, as has been pointed out, no Dalziel relation exists in the case of the rapid equilibrium Random Bi Bi mechanism, and, therefore, this mechanism cannot be excluded using Dalziel relationships.[16]

The initial rate data obtained by Lindstad *et al.*[14] apparently were not fit to an entire kinetic model; for example, in the case of Eq. (11), the data must be fit to the entire family of lines described by Eq. (11) rather than to a single line. This is obviously the case in Ref. 14 and often leads to the calculation of erroneous kinetic parameters. It is precisely these parameters that are used in the Haldane and Dalziel relationships to make a choice of mechanism. Fortuitously, the data obtained by Lindstad *et al.*[14] gave excellent agreement for the limiting Sequential Ordered Bi Bi case, the Theorell–Chance mechanism. In addition, the double-reciprocal plots in the forward direction intersect above the 1/substrate axis (Fig. 1) and below the 1/substrate axis in the reverse direction (Fig. 2). As pointed out by Lueck *et al.*,[17] these observations are also consistent with the Theorell–Chance mechanism.

A strong argument for the Ordered Bi Bi mechanism was obtained by Lindstad *et al.*[14] using dithiothreitol as a competitive inhibitor for sorbitol.

[13] V. Zewe and H. J. Fromm, *Biochemistry* **4,** 782 (1965).

[14] R. I. Lindstad, L. F. Hermansen, and J. S. KcKinley-McKee, *Eur. J. Biochem.* **210,** 641 (1992).

[15] J. B. S. Haldane, "Enzymes." Longmans, Green, London, 1930.

[16] K. Dalziel, *Acta Chem. Scand.* **11,** 1706 (1957).

[17] J. D. Lueck, W. R. Ellison, and H. J. Fromm, *FEBS Lett.* **30,** 321 (1973).

[h] If EAI reacts with C to form EAIC, the plot would be nonlinear.

[i] If EIB reacts with C to form EIBC, the plot would be noncompetitive.

[j] If EIA reacts with C to form EIAC, the plot would be noncompetitive.

[k] If EI reacts with B to form EIB, the plot would be nonlinear.

[l] If FI reacts with C to form FIC, the plot would be nonlinear.

Inhibition relative to NAD^+ was found to be uncompetitive. This unique inhibition pattern, namely, competitive–uncompetitive, shows that the kinetic mechanism is ordered and that sorbitol adds after NAD^+ to sorbitol dehydrogenase.

Dead-end competitive inhibitors have been used to study the kinetic mechanism of bireactant systems in which the inhibitor exhibits multiple binding to the enzyme.[18] Tabatabai and Graves[18] established that phosphorylase kinase has a sequential mechanism. They observed that adenyl-5′-yl(β,γ-methylene) diphosphate was a linear competitive inhibitor for ATP and a linear noncompetitive inhibitor relative to a tetradecapeptide, the other substrate. On the other hand, a heptapeptide was found to be nonlinear competitive (S-parabolic) with respect to the tetradecapeptide and nonlinear noncompetitive (I-parabolic, S-parabolic) relative to ATP. From these findings, it was concluded that the kinetic mechanism for phosphorylase kinase was Random Bi Bi. The rationalization of the nonlinear inhibition data was as follows. From the linear inhibition data provided by the ATP analog, the mechanism was either Random Bi Bi or Ordered Bi Bi with ATP as the first substrate to add to the enzyme. In the case of the ordered mechanism with multiple binding of the heptapeptide, if A and B are ATP and tetradecapeptide, respectively, then

$$\mathrm{EA} + \mathrm{I} = \mathrm{EAI}, K_\mathrm{i}; \qquad \mathrm{EAI} + \mathrm{I} = \mathrm{EAI_2}, K_\mathrm{ii} \tag{24}$$

and thus,

$$\frac{1}{v} = \frac{1}{V_1}\left[1 + \frac{K_\mathrm{a}}{\mathrm{A}} + \frac{K_\mathrm{b}}{\mathrm{B}}\left(1 + \frac{\mathrm{I}}{K_1} + \frac{\mathrm{I}^2}{K_\mathrm{i}K_\mathrm{ii}}\right) + \frac{K_\mathrm{ia}K_\mathrm{b}}{(\mathrm{A})(\mathrm{B})}\right] \tag{25}$$

It was possible to exclude the ordered mechanism for phosphorylase kinase from consideration, because the inhibition pattern relative to ATP was not nonlinear uncomeptitive (I-parabolic).

In the case of the Random Bi Bi mechanism, multiple binding by the inhibitor would be expected to occur as follows:

$$\begin{aligned} &\mathrm{E} + \mathrm{I} = \mathrm{EI}, K_\mathrm{i}; \quad \mathrm{EI} + \mathrm{I} = \mathrm{EI_2}, \mathrm{K_{ii}}; \quad \mathrm{EA} + \mathrm{I} = \mathrm{EAI}, K_\mathrm{iii}; \\ &\mathrm{EAI} + \mathrm{I} = \mathrm{EAI_2}, K_\mathrm{iv}; \quad \mathrm{EI} + \mathrm{A} = \mathrm{EAI}, K_\mathrm{v}; \quad \mathrm{EI_2} + \mathrm{A} = \mathrm{EAI_2}, K_\mathrm{vi} \end{aligned} \tag{26}$$

and thus,

$$\begin{aligned} \frac{1}{v} = \frac{1}{V_1}\Bigg[1 + \frac{K_\mathrm{a}}{\mathrm{A}} + \frac{K_\mathrm{b}}{\mathrm{B}}\left(1 + \frac{\mathrm{I}}{K_\mathrm{iii}} + \frac{\mathrm{I}^2}{K_\mathrm{iii}K_\mathrm{iv}}\right) \\ + \frac{K_\mathrm{ia}K_\mathrm{b}}{(\mathrm{A})(\mathrm{B})}\left(1 + \frac{\mathrm{I}}{K_\mathrm{i}} + \frac{\mathrm{I}^2}{K_\mathrm{iii}K_\mathrm{iv}}\right)\Bigg] \end{aligned} \tag{27}$$

[18] L. Tabatabai and D. J. Graves, *J. Biol. Chem.* **253,** 2196 (1978).

The initial rate results were consistent with Eq. (27), and Tabatabai and Graves[18] concluded that the kinetic mechanism for phosphorylase kinase is rapid-equilibrium Random Bi Bi.

Wolfenden[19] and Lienhard[20] have outlined how transition-state or multisubstrate (sometimes referred to as geometric) analogs may be used to provide information on the chemical events that occur during enzymatic catalysis. If it were possible to design an inactive compound that resembles the transition state, this analog would be expected to bind very tightly to the enzyme. In theory, a good deal of binding energy when enzyme and substrate interact is utilized to alter the conformation of both the enzyme and substrate so that proper geometric orientation for catalysis is provided between enzyme and substrate. Therefore, some of this binding energy is conserved because the multisubstrate analog more closely resembles the transition state than does the substrate. It certainly does not follow that, if the inhibition constant is lower than the dissociation constant for enzyme and substrate, the inhibitor is a transition-state analog. There are many examples in the literature where competitive inhibitors bind more strongly to enzymes than do substrates and yet are clearly not transition-state analogs.

These suggestions may be formalized by considering the following two reactions:

$$\begin{aligned} \mathrm{E + A = EA'}, &\qquad K_1 = 10^{-7}\,M \\ \mathrm{EA' = EA}, &\qquad K_2 = 10^{4} \end{aligned}$$

for the overall reaction

$$\mathrm{E + A = EA}, \qquad K_{\mathrm{ia}} = 10^{-3}\,M$$

The first reaction represents the thermodynamically favorable process of enzyme–substrate binding. The second reaction may be taken to be the enzyme-induced distortion of both the substrate and enzyme leading to the transition state.

From the perspective of kinetics, the use of geometric analogs may not be clearcut in the case of multisubstrate systems. The analog should bind free enzyme, and, in theory, for a two-substrate system, the analog and substrate should not be able to bind to the enzyme simultaneously. This situation is difficult to check experimentally because it is not easy to determine whether, for example, 50% of the enzyme has substrate bound and the other 50% of the enzyme is associated with both substrate

[19] R. Wolfenden, *Acc. Chem. Res.* **5,** 10 (1972).

[20] G. E. Lienhard, *Science* **180,** 149 (1973).

and analog or analog alone. For example, the kinetic pattern for a random mechanism where the analog can bind free enzyme and the EB complex will be the same as that for an ordered mechanism in which the analog binds exclusively to free enzyme.

It becomes fairly clear when considering the effect of multisubstrate or geometric analogs on the kinetics of bireactant enzyme systems that only in the case of the rapid-equilibrium Random Bi Bi mechanism may one obtain unequivocal results, and then only under certain circumstances. Consider, for example, the interaction of the analog and enzyme in an Ordered Bi Bi mechanism. The inhibitor will bind enzyme and will not permit the addition of the second substrate. Thus, the analog will act like any other competitive inhibitor of substrate A for this mechanism (see Table I); that is, it will be a noncompetitive inhibitor of substrate B. Therefore, the inhibition patterns provided by geometric analogs are identical to those to be expected for dead-end competitive inhibitors of the first substrate of the Ordered Bi Bi mechanism.

In the case of the Random Bi Bi mechanism, multisubstrate analogs may indeed give unique inhibition patterns, and this observation has been used to provide support for the Random Bi Bi mechanism for muscle adenylate kinase.[21,22] The multisubstrate analog used to test this theory with adenylate kinase was P^1,P^4-di(adenosine-5′) tetraphosphate (AP_4A).[21] It was subsequently shown that AP_5A binds even more strongly to the enzyme than does AP_4A.[22]

When considering the Random Bi Bi mechanism, the geometric analog should bind exclusively to free enzyme. This binding should effectively preclude binding of substrates A and B, and thus only the E term of the rate equation will be affected by the analog I. The rate expression is therefore

$$\frac{1}{v} = \frac{1}{V_1} + \frac{K_a}{V_1(\mathrm{A})} + \frac{K_b}{V_1(\mathrm{B})} + \frac{K_{ia}K_b}{V_1(\mathrm{A})(\mathrm{B})}\left(1 + \frac{\mathrm{I}}{K_i}\right) \tag{28}$$

where K_1 is the dissociation constant of the enzyme–multisubstrate inhibitor complex.

Equation (28) indicates that the multisubstrate analog will function as a competitive inhibitor for both substrates. This effect is unique to the Random mechanism and suggests that the inhibitor bridges both substrate-binding pockets.

[21] D. L. Purich and H. J. Fromm, *Biochim. Biophys. Acta* **276,** 563 (1972).

[22] G. E. Lienhard and I. I. Secemski, *J. Biol. Chem.* **248,** 1121 (1973).

The physiologically important regulator fructose 2,6-bisphosphate has been used to establish the kinetic mechanism of the mammalian fructose-1,6-bisphosphatase reaction.[23] It was well established that the inhibitor is competitive relative to the substrate, fructose 1,6-bisphosphate.[24] When fructose 2,6-bisphosphate was used as an inhibitor for the reverse reaction, it was found to be competitive relative to both of the reverse reaction substrates, orthophosphate and fructose 6-phosphate.[23] These observations are fully consistent with Eq. (28) and the Random Bi Bi mechanism previously proposed for the phosphatase.[25]

If the inhibitor binds only at one substrate site in either the Random or the Ordered Bi Bi cases, or if, for the latter mechanism, the compound does resemble the transition state and substrate B does not add, inhibition patterns will be competitive and noncompetitive relative to the two substrates. Thus, it will not be possible to differentiate between the two mechanisms based on these inhibition patterns, nor will it be possible to determine whether the inhibitor is really a transition-state analog in the Ordered mechanism. In the case of the Random pathway, the enzyme–inhibitor complex will permit binding of one substrate and the enzyme–substrate complex will allow analog to bind. In summary, then, only the unique inhibition pattern illustrated by Eq. (28) allows one to use multisubstrate analogs to differentiate unambiguously between kinetic mechanisms.

Frieden[26] has presented a rapid-equilibrium Ordered Bi Bi subsite mechanism[9] that cannot be differentiated from the rapid-equilibrium Random Bi Bi mechanism from studies of product inhibition, substrate analog inhibition, isotope exchange at equilibrium, and the Haldane relationship. Multisubstrate analogs can be used to make this differentiation, along with other kinetic procedures.[27]

In the case of Frieden's mechanism,[26] the transition-state analog should bind at both the active site and the subsite for B, that is, the EB site. This will result in modification of a number of terms in the rate equation. The dead-end complexes to be expected are EI_1, EI_2, EI_1I_2, and EAI_2, where I_1 and I_2 represent binding of the transition-state analog at the active and subsites, respectively. The resulting rate expression is described by Eq. (29), in which K_i, K_{ii}, K_{iii}, and K_{iv} represent dissociation constants for EI_1, EI_2, EI_1I_2, and EAI_2 complexes, respectively:

[23] N. J. Ganson and H. J. Fromm, *Biochem. Biophys. Res. Commun.* **108,** 233 (1982).

[24] E. Van Schaftigen and H. G. Hers, *Proc. Natl. Acad. Sci. U.S.A.* **78,** 2861 (1981).

[25] S. R. Stone and H. J. Fromm, *J. Biol. Chem.* **255,** 3454 (1980).

[26] C. Frieden, *Biochem. Biophys. Res. Commun.* **68,** 914 (1976).

[27] H. J. Fromm, *Biochem. Biophys. Res. Commun.* **72,** 55 (1976).

$$\frac{1}{v} = \frac{1}{V_1}\left[1 + \frac{K_{ia}K_b}{K_{ib}(A)} + \frac{K_b}{B}\left(1 + \frac{I}{K_{iv}}\right) + \frac{K_{ia}K_b}{(A)(B)}\left(1 + \frac{I}{K_i} + \frac{I}{K_{ii}} + \frac{I^2}{K_iK_{iii}}\right)\right] \quad (29)$$

Equation (29) indicates that the multisubstrate analog acts like a competitive inhibitor relative to substrate B and as a noncompetitive inhibitor with respect to substrate A. In addition, replots of the slopes from the primary plots versus inhibitor concentration will yield a concave-up parabola. This analysis indicates that a differentiation can be made between the Ordered Bi Bi subsite and the Random Bi Bi mechanisms.

Limitations

When considering competitive substrate inhibitors, the possibility is automatically excluded that the inhibitor may bind to an enzyme–product complex. In the case of the rapid-equilibrium Random Bi Bi mechanism, a competitive inhibitor for substrate B could in theory bind the EQ complex; however, this complex occurs after the rate-limiting step and is not part of the kinetic equation. Similarly, although this binary complex is kinetically important in the Ordered Bi Bi case, if an EQI complex did form, inhibition would be noncompetitive rather than competitive relative to substrate B. Under these conditions, the approach would not be a viable technique, and another inhibitor should be sought.

Huang[28] has suggested that it is not possible to differentiate between Ordered and Random Bi Bi mechanisms if in the Ordered case an inhibitor for B binds free enzyme and the EA complex to form complexes EI and EAI. This will be true only when all steps in the kinetic mechanism equilibrate rapidly relative to the breakdown of the productive ternary complex to form products. This mechanism is easily separated from those under discussion here by its unique double-reciprocal plot patterns in the absence of inhibitors.[29]

It should be pointed out that the dead-end competitive inhibitors cannot be used to differentiate between normal and Iso mechanisms. Nor can they be used to make a choice as to whether ternary complexes are kinetically important in Ordered mechanisms.

The competitive substrate inhibitors cited above have been referred to as "dead-end" inhibitors.[2] The question arises as to what happens if

[28] C. Y. Huang, *Arch. Biochem. Biophys.* **184,** 488 (1977).

[29] H. J. Fromm, "Initial Rate Enzyme Kinetics," p. 37. Springer-Verlag, Berlin and New York, 1975.

the inhibitors are not of the dead-end type, that is, if the enzyme–inhibitor complexes of the ordered mechanism act in a manner similar to those analogous complexes in the random mechanism. This possibility was considered by Hanson and Fromm.[30] If in the Ordered mechanism the EI complex permits substrate B to add, the additional reaction would be

$$\mathrm{EI} + \mathrm{B} = \mathrm{EIB},\ K_{\mathrm{ii}} \tag{30}$$

and Eq. (18) would be modified as shown in Eq. (31):

$$\frac{1}{v} = \frac{1}{V_1} + \frac{K_{\mathrm{a}}}{V_1(\mathrm{A})}\left[1 + \frac{\mathrm{I}}{K_{\mathrm{i}}} + \frac{(\mathrm{I})(\mathrm{B})}{K_{\mathrm{i}}K_{\mathrm{ii}}} + \frac{K_{\mathrm{ia}}K_{\mathrm{b}}(\mathrm{I})}{K_{\mathrm{a}}K_{\mathrm{i}}K_{\mathrm{ii}}}\right] + \frac{K_{\mathrm{b}}}{V_1(\mathrm{B})} + \frac{K_{\mathrm{ia}}K_{\mathrm{b}}}{V_1(\mathrm{A})(\mathrm{B})}\left(1 + \frac{\mathrm{I}}{K_{\mathrm{i}}}\right) \tag{31}$$

Inhibition relative to substrate A would of course be competitive; however, a $1/v$ versus 1/B plot would show concave-up hyperbolic inhibition. This effect is obviously readily distinguishable from the case in which a dead-end binary complex is formed.

Concluding Remarks

The use of competitive substrate inhibitors for studying kinetic mechanisms has received wide attention. If one is interested in determining the kinetic mechanism for an enzyme, this protocol should be used immediately after a determination is made as to whether the initial rate kinetics are Sequential or Ping Pong. Theoretically, using competitive substrate inhibitors to study the sequence of enzyme and substrate binding is far less ambiguous than either substrate or product inhibition or isotope exchange experiments. The protocol is no more complicated than initial velocity experiments in which inhibitors are not used.

The procedure does suffer from certain limitations, as do most kinetic methods. For example, it will not permit one to differentiate between Iso and conventional mechanisms, nor can it be used to choose between Theorell–Chance mechanisms and analogous mechanisms involving long-lived central complexes. However, on balance, using dead-end competitive inhibitors is a powerful tool for studying enzyme kinetics.

Acknowledgments

This work was supported by research grants from the National Institutes of Health (NS 10546) and the National Science Foundation (MCB-9218763).

[30] T. L. Hanson and H. J. Fromm, *J. Biol. Chem.* **240,** 4133 (1965).

[6] Kinetics of Slow and Tight-Binding Inhibitors

By STEFAN E. SZEDLACSEK and RONALD G. DUGGLEBY

Introduction

It is over 50 years since Easson and Stedman[1] first mentioned the existence of an unusual type of enzyme inhibitor that does not obey the well-established kinetics of classical reversible inhibitors. Initially, the so-called tight-binding inhibitors were described[1-3] that, owing to the very low dissociation constants, are used at concentrations comparable to that of the enzyme. As a result, the usual approximation that the free and total inhibitor concentrations are equal cannot be made. Morrison[4] derived a general rate equation for tight-binding inhibitors, covering a broad range of inhibitory mechanisms. Later, the kinetics of slow binding inhibitors were considered by Cha[5] who noted that sometimes the equilibrium between the enzyme and the enzyme–inhibitor complexes may be established slowly. Again, the classical theoretical treatment of this equilibrium as rapid is not valid. Cha[6] has also derived the kinetic equations for inhibitors that are both tightly bound and slow in establishing an equilibrium with the enzyme. These types of inhibitors, called slow, tight-binding inhibitors, have been analyzed kinetically and the corresponding rate and progress curve equations derived for frequently encountered mechanisms (see, e.g., Williams *et al.*[7]).

The interest in both slow and tight-binding inhibitors has been increasing, mainly owing to their importance as chemotherapeutic agents, herbicides, and transition state analogs (reaction intermediate analogs).[8-12] Al-

[1] L. H. Easson and E. Stedman, *Proc. R. Soc. London B* **127,** 142 (1936).
[2] O. H. Straus and A. Goldstein, *J. Gen. Physiol.* **26,** 559 (1943).
[3] W. W. Ackerman and V. R. Potter, *Proc. R. Soc. Exp. Biol. Med.* **72,** 1 (1949).
[4] J. F. Morrison, *Biochim. Biophys. Acta* **185,** 269 (1969).
[5] S. Cha, *Biochem. Pharmacol.* **24,** 2177 (1975); correction, *Biochem. Pharmacol.* **25,** 1561 (1976).
[6] S. Cha, *Biochem. Pharmacol.* **25,** 2695 (1976).
[7] J. W. Williams, J. F. Morrison, and R. G. Duggleby, *Biochemistry* **18,** 2567 (1979).
[8] J. F. Morrison and C. T. Walsh, *Adv. Enzymol. Relat. Areas Mol. Biol.* **59,** 201 (1988).
[9] K.-H. Röhm, *FEBS Lett.* **250,** 191 (1989).
[10] C.-Y. J. Hsu, P. E. Persons, A. P. Spada, R. A. Bednar, A. Levitzki, and A. Zilberstein, *J. Biol. Chem.* **258,** 4137 (1983).
[11] S. P. Jordan, L. Waxman, D. E. Smith, and G. P. Vlasuk, *Biochemistry* **29,** 11095 (1990).
[12] J. Urban, J. Kovalinka, J. Stehlíková, E. Gregorová, P. Majer, M. Souček, M. Andreánsky, M. Fábry, and P. Štrop, *FEBS Lett.* **298,** 9 (1992).

though the quantitative description of the effects of these inhibitors poses special problems, it is worth emphasizing that there is nothing intrinsically different between classical inhibitors and tight-binding inhibitors. Rather, tight binding is simply an experimental condition that sometimes occurs. Even a very weak inhibitor might be tight binding if used with a very high enzyme concentration, as might occur when using a very poor substrate. Similarly, slow binding inhibition merely means that the delay in attainment of the steady state is observable on a time scale of seconds to minutes; it is transient-phase kinetics observable on an ordinary spectrophotometer.

One of the most important purposes of the kinetic analyses of these inhibitors is the determination of the inhibition constants. In addition to the necessity of determining the dissociation constants of the enzyme–inhibitor complexes for both slow and tight-binding inhibitors, in the case of slow binding inhibitors it is also important to estimate the rate constants for formation and dissociation of the inhibitor complexes.

Usually, the derivation of the kinetic equations for these nonclassical inhibitors is carried out in the following way: first, an inhibition mechanism is proposed; second, some simplifying assumptions are set up; and, third, the corresponding kinetic equations are deduced. On the other hand, when the kinetic constants of a new (slow and/or tight-binding) inhibitor are to be determined, the following procedure is used: one or two alternative inhibition mechanisms are considered; the corresponding kinetic equations are fitted to the experimental data; and, finally, the model giving the best fit is accepted as correct and the corresponding kinetic constants taken as the real values.

Unfortunately, there is a limited range of kinetic models that have been considered for this type of inhibitor, especially in the case of slow binding inhibitors, where most frequently only the competitive slow binding model is used. Also, for both slow and tight-binding inhibition, very rarely does the kinetic mechanism take into account the "hyperbolic" component, that is, the possibility that the enzyme–substrate–inhibitor complex is still able to form the product. In addition, when a kinetic equation is applied to a given set of experimental data, relatively little attention is paid to the validity of the simplifying assumptions used to derive the kinetic equation.

There are two extensive reviews about this type of inhibition, both of which include the kinetic aspects: the review by Williams and Morrison of tight-binding inhibition[13] and the review by Morrison and Walsh about slow binding inhibitors.[8] Thus, the present review focuses on aspects

[13] J. W. Williams and J. F. Morrison, this series, Vol. 63, p. 437.

that have received less attention, as well as some new developments concerning the kinetics of slow and tight-binding inhibitors. Thus, we analyze a more general kinetic mechanism, discuss the validity of commonly used simplifying assumptions, suggest a procedure to determine the inhibition constants, and consider some more recent models for the analysis of slow and tight-binding inhibition.

Terminology, Definitions, and Current Models

Terminology

Slow and tight-binding inhibitors are involved mainly in the following types of equilibria:

$$E + I \underset{k_{-3}}{\overset{k_3}{\rightleftharpoons}} EI \tag{1}$$

$$EA + I \underset{k_{-4}}{\overset{k_4}{\rightleftharpoons}} EAI \tag{2}$$

$$E + I \underset{k_{-3}}{\overset{k_3}{\rightleftharpoons}} EI \underset{k_{-7}}{\overset{k_7}{\rightleftharpoons}} EI^* \tag{3}$$

The reason for the apparently arbitrary designation of these steps as reactions 3, 4, and 7 will become clear when they are put in the context of a more general mechanism (see Scheme 1).

Dissociation Constants. The ratios $K_3 = k_{-3}/k_3$ and $K_4 = k_{-4}/k_4$ define the competitive and the uncompetitive aspects of the effects of an inhibitor I, respectively. An inhibitor exhibiting both the competitive and uncompetitive components will be termed a noncompetitive inhibitor. In the particular case where an inhibitor has $K_3 = K_4$, it will be referred to as a "pure noncompetitive" one. The series of reactions defined by Eq. (3) is frequently encountered for slow binding inhibitors. One of the most important characteristics is that an inhibitor having a moderately low value of the dissociation constant K_3 may become a tight-binding inhibitor as a result of a slow isomerization of the initial enzyme–inhibitor complex, EI. This isomerization is quantitatively characterized by the isomerization constant $K_7 = k_{-7}/k_7$. A global equilibrium constant can be also defined if the following formal equilibrium is considered:

$$E + I \overset{K^*_i}{\rightleftharpoons} (EI + EI^*) \tag{4}$$

An expression for K^*_i, termed the overall dissociation constant, can be derived simply: using the relationships $K^*_i = E \cdot I/(EI + EI^*)$, $K_3 =$

$E \cdot I/EI$ and $K_7 = EI/EI^*$, where italics indicates concentrations, one can immediately deduce that

$$K^*_i = K_3/(1 + 1/K_7) \tag{5}$$

If K_7 is large, the overall dissociation constant K^*_i is approximately equal to K_3. However, if $K_7 \ll 1$ then K^*_i is considerably smaller than K_3, and there are two situations under which this occurs. One is when $k_7 \gg k_{-7}$ but k_7 is of comparable magnitude to k_3I. In the other situation, $k_{-7} \ll k_7 \ll k_3I$, which is equivalent to the slow development of the inhibition. Provided K^*_i is sufficiently small, the first situation corresponds to a two-step tight-binding inhibition and the second to a slow, tight-binding inhibition.

Another measure of the potency of inhibition is the I_{50}, which represents the total inhibitor concentration giving a decrease in the rate by 50% compared to the uninhibited enzyme reaction. Sometimes the I_{50} is used to characterize the tightness of the enzyme–inhibitor interaction; although there is always a relationship between I_{50} and the corresponding dissociation constant(s) (see below), the use of the latter is generally more informative. This is because the I_{50} can depend on both the total enzyme concentration E_t and the substrate concentration A. Accurate values of E_t may not be available, especially when the purity of the enzyme preparation used is not very high. Moreover, measurement of the rates needed to determine I_{50} may be difficult when the enzyme–inhibitor equilibrium is established slowly.

Rate Constants. The velocity of an elementary kinetic step is characterized by the corresponding first- or second-order rate constant. As an alternative for a first-order rate constant k, the half-time $t_{1/2}$ may also be used. For instance, if k is the rate constant for the kinetic step EI → E + I, then the exponential decay of EI produces the decrease of the concentration of EI to half, after a time period of $(\ln 2)/k$.

In the case of slow binding inhibitors that act according to the two-step process given by Eq. (3), the rate of EI* formation is frequently characterized by the observed rate constant, k_{obs}. If the isomerization step is much slower than the step of inhibitor binding to the enzyme, that is, $k_7 \ll k_3I$ and $k_{-7} \ll k_{-3}$, the expression of k_{obs} (see Appendix 1 for derivation) is given by

$$k_{obs} = k_{-7} + k_7I/(I + K_3) \tag{6}$$

Similarly, when the equilibrium given by Eq. (1) is slowly established (see Appendix 1), the formation of the enzyme–inhibitor complex is characterized by

$$k'_{obs} = k_3I + k_{-3} \tag{7}$$

Definitions

In many situations it is difficult to draw a clear demarcation between classical inhibition on one hand and slow or tight-binding inhibition on the other. However in the following, some definitions are given for slow and tight-binding inhibition, based on the most relevant characteristics.

Tight-Binding Inhibition. There are different ways to identify tight-binding inhibition. One possibility[2] is to evaluate the ratio between the total enzyme concentration E_t and the inhibition (dissociation) constants K_i. Another[8] is to compare the concentrations of inhibitor and enzyme that are used in the experiment. Two aspects need to be considered in deciding whether we are dealing with tight-binding inhibition.

The first concerns a consequence of the tight-binding inhibition: the proportion of inhibitor involved in the enzyme–inhibitor complexes is not negligible. For this reason, the usual approximation made for the classical inhibitors (that the total concentration of inhibitor I_t is equal to the free inhibitor concentration I_f) is no longer valid. Let us illustrate this with the simple case of a competitive inhibitor. As K_i is equal to $E_f I_f / EI$, we have $I_t = I_f + EI = I_f(1 + E_f/K_i)$. Supposing that the concentration of the free enzyme E_f is of a similar order of magnitude to E_t, we can conclude that if $E_t/K_i \ll 1$, the proportion of inhibitor involved in the EI complex is negligible and we have classical inhibition. On the other hand, if $E_t/K_i \gg 1$, virtually all the inhibitor is complexed with the enzyme. Straus and Goldstein[2] suggested a value of 0.1 as a numerical limit of this ratio; in other words, they consider the inequality $E_t/K_i > 0.1$ as a condition for tight-binding inhibition.

The second aspect concerns the reaction conditions that are needed for tight-binding inhibition to be demonstrated experimentally. Specifically, if the proportion of the enzyme complexed by the inhibitor is low, then the inhibitory effect is too small to be detected. On the other hand, if a too large proportion of the enzyme is involved in the enzyme–inhibitor complex, then the basic enzyme-catalyzed conversion of the substrate will not be detected. Considering the same example of a competitive inhibitor, we can write $E_t = E_f + EA + EI = E_f(1 + A/K_a + I_f/K_i)$, where K_a is the Michaelis constant of the substrate A. Appropriate experimental conditions are that EA and EI have comparable concentrations. In other words, the ratios A/K_a and I_t/K_i should be similar in magnitude.

Similar arguments apply to more complicated situations such as non-competitive inhibition, and we conclude that the following two conditions hold for tight-binding inhibition:

$$\left.\begin{array}{l}(1)\ E_t \text{ and } K_i \\ (2)\ I_t/K_i \text{ and } A/K_a\end{array}\right\} \text{should be of a similar order of magnitude} \qquad (8)$$

If the above conditions are met, clearly E_t and I_t will also be of the same order of magnitude; this situation (i.e., $E_t \approx I_t$) has been suggested by Morrison and Walsh[8] as diagnostic for tight-binding inhibition.

Slow-Binding Inhibition. Several types of kinetic models have been suggested for slow binding inhibitors.[5,8] However, a unique definition of slow binding inhibition can be stated as being a process of reversible enzyme inhibition that is slow when compared to the enzyme-catalyzed substrate conversion reaction. It is clear that the substrate conversion should take place within the time scale of the experimental measurements. That is why slow binding inhibition is time dependent. If slow binding inhibition is a multistep process, then at least one of the component steps should be slow in comparison with the substrate conversion reaction. For instance, in the case of a slow binding inhibitor that obeys the process described by Eq. (3), the slow step can be the formation of EI or its isomerization to EI*. However, if the inhibitor binding step is slow and the isomerization fast, then formally both EI and EI* can be considered as a single enzyme–inhibitor complex and the inhibition can be treated as a single-step process. If one of the component steps of the inhibition process is much slower than the other steps, then this step becomes rate-determining for the whole inhibition. In principle, it is possible that both steps are somewhat slower than the substrate conversion, but their combination leads to an inhibition that is clearly slow binding. This last possibility has not been subjected to close theoretical scrutiny.

Slow, Tight-Binding Inhibition. From the above definitions of slow and tight-binding inhibition, respectively, it can be inferred that the two types of inhibition may exist independently. It is especially important to stress that slow binding inhibition can occur in the absence of tight-binding inhibition. Nevertheless, the combination of slow and tight-binding inhibition is encountered frequently. It can be defined as being an inhibition process that satisfies simultaneously the definitions given above for both slow and tight-binding inhibition. In other words, slow, tight-binding inhibition is a time-dependent enzyme inhibition that meets the conditions of Eq. (8).

Current Models

Some general kinetic models and results concerning slow and tight-binding inhibition are discussed in the section below. For the most part, the results are given without corresponding derivations.

Tight-Binding Inhibition. Morrison[4] derived a steady-state rate equation covering a broad range of tight-binding inhibitions. He started by considering an enzyme-catalyzed reaction that is in a steady state for the enzyme–substrate complex, having the initial rate v_0 in the absence of inhibitor of the following form:

$$v_0 = NE_t/D \tag{9}$$

where N contains the rate constants and substrate concentrations that determine the maximum velocity and D is a sum of several terms related to the distribution of the enzyme in its complexed forms. Morrison used the following assumptions: (i) steady-state conditions are established between the free enzyme and enzyme–substrate complex; (ii) only "dead-end" enzyme–inhibitor complexes are formed, that is, complexes that cannot participate in the reaction; and (iii) rapid equilibrium is reached between the inhibitor, the enzyme form E_i, and the complex E_iI. Under these conditions, the rate equation can be written as

$$v_i = \frac{NE_t}{D + I_f \Sigma N_i/K_i} \tag{10}$$

where K_i is the dissociation constant for the equilibrium $E_i + I \rightleftharpoons E_iI$ and N_i is part of D representing the distribution of the enzyme form E_i. Combining Eq. (10) with the conservation equation for the inhibitor, Morrison demonstrated that v_i is the solution of the following equation:

$$v_i^2 + N\left[\frac{1}{\Sigma(N_i/K_i)} + \frac{I_t - E_t}{D}\right]v_i - \frac{N^2E_t}{D\Sigma(N_i/K_i)} = 0 \tag{11}$$

Later, Henderson[14] proposed a linearized form of Eq. (11):

$$\frac{I_t}{1 - v_i/v_0} = \frac{D}{\Sigma(N_i/K_i)}\frac{v_0}{v_i} + E_t \tag{12}$$

Equation (12) forms the basis of the graphical analysis of steady-state velocity data: plots of $I_t/(1 - v_i/v_0)$ as a function of v_0/v_i should be linear having the intercept equal to E_t and the slope equal to a so-called apparent inhibition constant $K_{i,app}$. Thus,

$$K_{i,app} = \frac{D}{\Sigma(N_i/K_i)} \tag{13}$$

In the case of a tight-binding noncompetitive inhibitor that complexes both E and EA according to Eqs. (1) and (2), the expression of $K_{i,app}$ is

$$K_{i,app} = K_3(1 + A/K_a)/(1 + K_3A/K_aK_4) \tag{14}$$

A plot of $K_{i,app}$ against substrate concentration can give information about K_3 and K_4, although estimation of their values from a purely graphi-

[14] P. J. F. Henderson, *Biochem. J.* **127,** 321 (1972).

cal analysis is unlikely to be very accurate. The graph is interpreted as follows: when $A \to \infty$, then $K_{i,app} \to K_4$; when $A \to 0$, then $K_{i,app} \to K_3$; if the plot is concave-up, then $K_3 > K_4$; and if $K_{i,app}$ does not depend on A, then $K_3 = K_4 = K_{i,app}$ (pure, noncompetitive inhibition). For a competitive inhibitor ($K_4 = \infty$), the expression for $K_{i,app}$ simplifies to

$$K_{i,app} = K_3(1 + A/K_a) \tag{15}$$

and a plot of $K_{i,app}$ against A is linear, allowing the estimation of the inhibition constant. If the inhibition is uncompetitive ($K_3 = \infty$), we obtain for $K_{i,app}$ the expression

$$K_{i,app} = K_4(1 + K_a/A) \tag{16}$$

and a plot of $K_{i,app}$ against $1/A$ is linear, allowing the estimation of K_4.

Henderson's equation [Eq. (12)] can also be used to obtain the expression of I_{50}, if we set v_i equal to $v_0/2$:

$$I_{50} = K_{i,app} + E_t/2 \tag{17}$$

The above expression for I_{50} [Eq. (17)] is useful in estimating the inhibition constants only if accurate values of E_t are available. In these circumstances, plots of I_{50} against E_t allow the estimation of $K_{i,app}$ which in turn can be used further to get information about K_3 and K_4, as described above.

Equation (11) can also be directly used to study the dependence of the reaction rate on E_t and I_t. Morrison[4] made a detailed analysis of the shape of the plots of v_i against E_t or against I_t as well as the plots of $1/v_i$ against I_t. The plots of v_i against E_t at different values of I_t are, according to Eq. (11), concave-up curves[4,13] in contrast to those for classical inhibitors, which are linear. This characteristic of tight-binding inhibitors has an important diagnostic role.

Plots of v_i as a function of substrate concentration are particularly complicated, more so because in many situations the inhibition studies are concerned with enzymatic systems of two substrates. The corresponding equations can be derived by using Morrison's general equation [Eq. (11)], by replacing N, D, and N_i/K_i with the appropriate expression that corresponds to the kinetic mechanism involved. This yields an implicit equation that relates v_i to both substrates, which can be rearranged into one of the common graphical forms (e.g., double-reciprocal plots of velocity as a function of the concentration of one reactant). Table I is useful in this respect, by giving the expressions of N, D, and N_i for four typical enzyme mechanisms involving reactions of two substrates.

As an illustration, we can consider the situation analyzed by Williams and Morrison[13] in which two substrates, A and B, react according to a

TABLE I
DISTRIBUTION OF ENZYME FORMS FOR DIFFERENT MECHANISMS

Mechanism	Rate equation	N_i			
		E	EA	EB	EQ
Random sequential (rapid equilibrium) 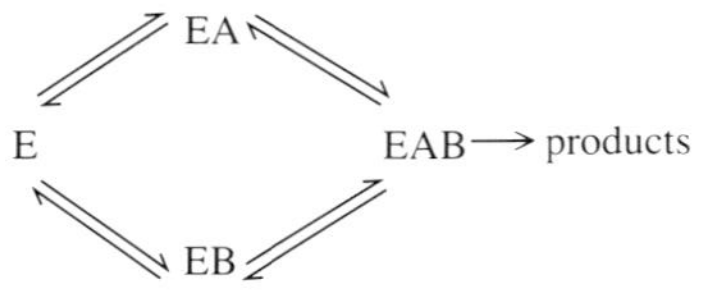	$v = k_{cat}E_tAB/(K_{ia}K_b + K_aB + K_bA + AB)$ K_a, K_{ia}, and K_b are dissociation constants of EAB to EB, EA to E, and EAB to EA	$K_{ia}K_b$	K_bA	K_aB	0
Ordered $E \rightleftharpoons EA \rightleftharpoons EAB \rightarrow EQ \xrightarrow{k_7} E$	$v = k_{cat}E_tAB/(K_{ia}K_b + K_aB + K_bA + AB)$ K_{ia} and K_{iq} are dissociation constants of EA and EQ; K_a, K_b, and K_q are Michaelis constants for A, B, and Q; V_1 and V_2 are forward and reverse maximum velocities	$K_{ia}K_b + K_aB$	K_bA	0	$V_1K_qAB/V_2K_{iq} = k_{cat}AB/k_7$
Ordered Theorell–Chance $E \rightleftharpoons EA \rightarrow EQ \rightarrow E$	$v = k_{cat}E_tAB/(K_{ia}K_b + K_aB + K_bA + AB)$ K_{ia} is the dissociation constants of EA; K_a and K_b are Michaelis constants for A and B	$K_{ia}K_b + K_aB$	K_bA	0	AB
Ping Pong $E \rightleftharpoons EA \rightarrow F \rightleftharpoons FB \xrightarrow{k_4} E$	$v = k_{cat}E_tAB/(K_aB + K_bA + AB)$ K_a and K_b are Michaelis constants for A and B	K_aB	$(1 - k_{cat}/k_4)AB$	0	K_bA (form F)

random order-type mechanism, under rapid equilibrium conditions. If a tight-binding analog of B combines with the free enzyme E and the EA complex:

$$\mathrm{E} + \mathrm{I} \underset{}{\overset{K_\mathrm{i}}{\rightleftharpoons}} \mathrm{EI} \tag{18}$$

$$\mathrm{EA} + \mathrm{I} \underset{}{\overset{K_\mathrm{I}}{\rightleftharpoons}} \mathrm{EAI} \tag{19}$$

then, using Table I, we obtain:

$$N = k_{\mathrm{cat}}AB; \qquad D = K_{\mathrm{ia}}K_\mathrm{b} + K_\mathrm{a}B + K_\mathrm{b}A + AB \tag{20}$$

Because the proportions of the total enzyme as E and EA are $K_{\mathrm{ia}}K_\mathrm{b}$ and $K_\mathrm{b}A$, respectively (Table I), we get

$$(N_\mathrm{i}/K_\mathrm{i}) = K_{\mathrm{ia}}K_\mathrm{b}/K_\mathrm{i} + K_\mathrm{b}A/K_\mathrm{I} \tag{21}$$

Substituting Eqs. (20) and (21) into Eq. (11), the following equation results:

$$v_\mathrm{i}^2\, K_\mathrm{b}(K_{\mathrm{ia}}K_\mathrm{b} + K_\mathrm{a}B + K_\mathrm{b}A + AB)(K_{\mathrm{ia}}/K_\mathrm{i} + A/K_\mathrm{I}) + k_{\mathrm{cat}}AB[K_{\mathrm{ia}}K_\mathrm{b} + K_\mathrm{a}B + K_\mathrm{b}A + AB + K_\mathrm{b}(I_\mathrm{t} - E_\mathrm{t})(K_{\mathrm{ia}}/K_\mathrm{i} + A/K_\mathrm{I})]v_\mathrm{i} - k_{\mathrm{cat}}{}^2E_\mathrm{t}A^2B^2 = 0 \tag{22}$$

Equation (22) may be used as such in numerical analyses of data or can be rearranged according to the type of plot that is to be analyzed.

Mixture of Tight-Binding Inhibitors. Kuzmič *et al.*[15] showed that a mixture of n tight-binding inhibitors acting simultaneously on the same enzyme yields an implicit polynomial equation of degree $n + 1$ with respect to the velocity of the inhibited reaction v_i. They assume, as is the case for the derivation of the Morrison equation [see Eqs. (9)–(11)], that the enzyme-catalyzed reaction is at steady state, having the rate equation of the form $v_0 = NE_\mathrm{t}/D$. It was also supposed that each of the n inhibitors (I_j) is in rapid equilibrium with several enzyme forms yielding the complexes $E_{1\mathrm{j}}I_\mathrm{j}, E_{2\mathrm{j}}I_\mathrm{j}, \ldots, E_{\mathrm{mj}}I_\mathrm{j}$, and that the dissociation constants of these complexes are $K_{1\mathrm{j}}, K_{2\mathrm{j}}, \ldots, K_{\mathrm{mj}}$. The distribution equations of the enzyme forms $\mathrm{E_{ij}}$ that are subject to inhibitor complexation can be written:

$$E_{\mathrm{ij}}/E_\mathrm{t} = N_{\mathrm{ij}}/D \tag{23}$$

Denoting the free and the total concentration of the inhibitor by I_j and I_j^t respectively, the following conservation equations were introduced: $I_\mathrm{j} = I_\mathrm{j}^\mathrm{t} - \Sigma_\mathrm{i}\, E_{\mathrm{ij}}I_\mathrm{j}$. Under these circumstances, an implicit expression of the rate equation results [Eq. (24)] and this may be rearranged to give Eq. (25).

[15] P. Kuzmič, K.-Y. Ng, and T. D. Heath, *Anal. Biochem.* **200,** 68 (1992).

$$v_i = E_t \frac{N}{D + \Sigma_j I_j^t \dfrac{\Sigma_i N_{ij}/K_{ij}}{1 + (v_i/N)\,\Sigma_i N_{ij}/K_{ij}}} \tag{24}$$

$$\frac{1}{v_i} = \frac{1}{v_0} + \frac{I_1^t/E_t}{v_i + \dfrac{N}{\Sigma_i N_{i1}/K_{i1}}} + \frac{I_2^t/E_t}{v_i + \dfrac{N}{\Sigma_i N_{i2}/N_{i2}}} + \ldots + \frac{I_n^t/E_t}{v_i + \dfrac{N}{\Sigma_i N_{in}/K_{in}}} \tag{25}$$

Equation (25) shows that the reciprocal of the inhibited reaction velocity is a formal sum of reciprocal of v_0 and of increments of apparent reciprocal velocities corresponding to each component inhibitor. Equation (25) can be further rearranged in the following form:

$$(1 - v_i/v_0) = \frac{I_1^t}{E_t + \dfrac{v_0}{v_i}\dfrac{D}{\Sigma_i N_{i1}/K_{i1}}} + \frac{I_2^t}{E_t + \dfrac{v_0}{v_i}\dfrac{D}{\Sigma_i N_{i2}/K_{i2}}} + \ldots + \frac{I_n^t}{E_t + \dfrac{v_0}{v_i}\dfrac{D}{\Sigma_i N_{in}/K_{in}}} \tag{26}$$

As shown in the previous section, according to the Henderson representation for a single tight-binding inhibitor [Eq. (12)], the plot of $I_t/(1 - v_i/v_0)$ against v_0/v_i should be linear. It is clear from Eq. (26) that this type of representation, frequently used in tight-binding inhibition studies, is not linear when there are two or more tight-binding inhibitors. However, there are two particular situations in which the Henderson representation is still linear. The first corresponds to the fortuitous case in which the ratio $\Sigma_i N_{ij}/K_{ij}$ is equal for all the inhibitors present in the mixture. The second possibility is when the term in the right-hand side of Eq. (26) for one inhibitor from the mixture is much greater than all the other terms. In this case, the sum $\Sigma_i N_{ij}/K_{ij}$ for this inhibitor is much greater than the sums corresponding to the other inhibitors. In other words, if a mixture of competitive tight-binding inhibitors has one with a much lower dissociation constant, then it is possible for the Henderson plot for the mixture to be linear. Thus, it is possible for one of the components in the mixture to mask the inhibitory effect of the remaining components, and linearity of the Henderson plot does not prove that there is only one tight-binding inhibitor in the mixture.

In the general case of a mixture of tight-binding inhibitors, the analysis of the experimental data requires the use of an appropriate computational technique based on Eq. (24) or Eq. (25). Kuzmič *et al.*[15] used an iterative numerical procedure based on Eq. (24) to analyze the experimental data for inhibition of rat liver dihydrofolate reductase by mixtures of the antitumor drug methotrexate and its metabolic precursor form, methotrexate α-aspartate.

Slow Binding Inhibition. The simplest competitive slow-binding inhibition model is the following:

$$\begin{array}{ccc} E & \underset{k_{-1}}{\overset{k_1 A}{\rightleftharpoons}} & EA \overset{k_2}{\rightarrow} E + P \\ {\scriptstyle k_3 I}\Big\downarrow\Big\uparrow{\scriptstyle k_{-3}} & & \\ EI & & \end{array} \tag{27}$$

where the inhibitor binding step is taken to be slow when compared to the enzymatic conversion of A to P. It is implicit in this model that there is no inhibition by the product, P. Cha[5] has analyzed this system under the following assumptions: (i) the complex EA reaches the steady state instantaneously; (ii) Eq. (28) is true;

$$K_a = (k_{-1} + k_2)/k_1 \gg K_3 = k_{-3}/k_3 \tag{28}$$

(iii) $k_{-1} \gg k_{-3}$ and $k_1A \gg k_3I$; and (iv) the depletion of free substrate by binding to the enzyme or by conversion to the product is negligible. It was shown that the corresponding rate equation for this system is

$$v_i = v_s + (v_0 - v_s)e^{-kt} \tag{29}$$

where v_0 is the initial velocity of the reaction, v_s is the steady-state velocity of the reaction, and k is an exponential decay constant. Equation (29) can be integrated, giving the following expression:

$$P = v_s t - (v_s - v_0)(1 - e^{-kt})/k \tag{30}$$

The progress curve defined by Eq. (30) has an asymptote; its equation may be obtained if we observe that the exponential term vanishes at high values of t:

$$P = v_s t - (v_s - v_0)/k \tag{31}$$

One notable feature of slow binding inhibitors is that the kinetics depend on whether the reaction is started by addition of enzyme or of substrate (i.e., preincubation of the enzyme with the inhibitor). When the reaction is started with enzyme, the expressions of v_0, v_s, and k are

$$v_0 = \frac{VA}{A + K_a} \tag{32}$$

$$v_s = \frac{VA}{A + K_a(1 + I/K_3)} \tag{33}$$

$$k = k_{-3}\frac{1 + A/K_a + I/K_3}{1 + A/K_a} = k_{-3} + \frac{k_3 I}{1 + A/K_a} \tag{34}$$

where $V = k_2E_t$ is the maximum velocity. From Eqs. (32) and (33) it is clear that $v_0 > v_s$ and thus, according to Eq. (29), v_i will decrease continuously with time. That is why the plot of P against t will be concave-down. Alternatively, when the enzyme is preincubated with the inhibitor, Eq. (29) still holds, but v_0 will be smaller than v_s. Consequently, the progress curve $P = f(t)$ will be concave-up. Another important characteristic of slow binding inhibition can be inferred if we notice that the exponential decay constant k increases when I increases [Eq. (34)]. This means that at higher inhibitor concentrations the product progress curve should approach the asymptote more quickly.

The procedure to estimate the constants k_3 and k_{-3} from the progress curves may involve the following steps (see also Williams and Morrison[13]). First, estimate the initial and steady-state velocities, v_0 and v_s, respectively, from the slopes of the tangent to the progress curves at $t = 0$ (for v_0) and of the asymptote at high values of t (for v_s). Second, extrapolate the asymptote back to the ordinate and, using Eq. (31), calculate the value for k from the intercept. Alternatively, the constant k can be calculated by estimating the velocity at several points along the progress curve, which are then plotted using a rearranged form of Eq. (29):

$$\ln[(v_0 - v_s)/(v_i - v_s)] = kt \tag{35}$$

Equation (35) shows that the plot of $\ln[(v_0 - v_s)/(v_i - v_s)]$ against t is linear, having the slope equal to k. Third, repeat the first and second steps for several inhibitor concentrations I. Fourth, plot k as a function of I; from the slope and intercept, the values for k_3 and k_{-3} are obtained using a value for A/K_a and Eq. (34).

The subjectivity that is inherent in these procedures is clearly evident. The values obtained may not be accurate, and, moreover, no estimate of their reliability is obtained. These criticisms may be overcome by fitting Eqs. (30) and (32)–(34) to the data by nonlinear regression, although it may be useful to perform a preliminary graphical analysis to obtain initial estimates of the various constants.

Another frequently used model for competitive tight-binding inhibitors is

$$\begin{array}{ccl} \mathrm{E} & \underset{k_{-1}}{\overset{k_1A}{\rightleftharpoons}} & \mathrm{EA} \overset{k_2}{\rightarrow} \mathrm{E} + \mathrm{P} \\ {\scriptstyle k_3I}\downarrow\uparrow{\scriptstyle k_{-3}} & & \\ \mathrm{EI} & \underset{k_{-7}}{\overset{k_7}{\rightleftharpoons}} & \mathrm{EI}^* \end{array} \tag{36}$$

where the slow step is the isomerization between EI and EI*. As with mechanism (27), this model does not include provision for possible product inhibition by P. The rate and progress curve equations for the model have been presented by Morrison and Walsh,[8] derived from the following conditions: (i) the enzyme–substrate complex EA is at steady state; (ii) the magnitude of the values for k_1A, k_{-1}, k_3I, and k_{-3} is very much greater than that of k_7 and k_{-7}; (iii) $k_{-7} \ll k_7$; and (iv) the depletion of the free substrate owing to binding to the enzyme or conversion to the product can be neglected. Under the above assumptions, the resulting rate equation is identical to Eq. (29), but the definitions of v_i, v_s, and k are changed:

$$v_0 = \frac{k_2E_tA}{A + K_a(1 + I/K_3)} = \frac{VA}{A + K_a(1 + I/K_3)} \tag{37}$$

$$v_s = \frac{k_2k_{-7}E_tA}{kK_a} = \frac{VA}{K_a(1 + I/K_i^*) + A} \tag{38}$$

$$k = k_{-7}\frac{1 + A/K_a + I/K_i^*}{1 + A/K_a + I/K_3} = k_{-7} + k_7\frac{I/K_3}{1 + A/K_a + I/K_3} \tag{39}$$

Morrison and Walsh[8] suggested ways to analyze the experimental data obtained for a slow binding inhibitor that obeys mechanism (36). The procedure to estimate the constants K_3, k_7, and k_{-7} from progress curves is as follows. First, estimate v_0 and v_s from the progress curve. Second, estimate the value of k using one of the following methods: (a) from the intercepts of the progress curve asymptote—the abscissa intercept is $(v_s - v_0)/kv_s$ whereas that on the ordinate is $(v_0 - v_s)/k$; or (b) from the intersection of the tangent to the progress curve at $t = 0$ (having the equation $P = v_0t$) and the asymptote of the curve—the time coordinate of the intersection point [according to Eq. (31)] is equal to $1/k$. Third, repeat the first and second steps for several inhibitor concentrations I. Fourth, calculate K_3 from the slope and intercepts of the plot of $1/v_0$ against I. Fifth, calculate K_i^* from the slope and intercepts of the plot of $1/v_s$ against I. Sixth and finally, calculate k_{-7} from the relationship $k_{-7} = kv_s/v_0$ [obtained from Eqs. (37)–(39)] and k_7 from $k_7/k_{-7} = (K_3/K_i^*) - 1$ [see Eq. (5)]. Alternatively, k_7 and k_{-7} can be estimated from plots of k against I. According to Eq. (39), at $I = 0$ and at $I = \infty$, k has the limiting values of k_{-7} and $k_7 + k_{-7}$, respectively. As before, the values so obtained are best used as initial estimates in nonlinear regression analysis.

Comparison of Eqs. (32) and (34) with Eqs. (37) and (39) reveals two important features that may be used to distinguish mechanism (27) from mechanism (36). First, v_0 is independent of the inhibitor concentration in mechanism (27) but not in mechanism (36). Although this offers an apparently simple method for distinguishing the two mechanisms, the effect on

v_0 will be discernible only to the extent that I/K_3 is greater than $1 + A/K_a$. Because I may be comparable in magnitude to $K_i{}^*$, which itself may be much smaller than K_3 [Eq. (5)], the expected variation of v_0 over the range of inhibitor concentrations employed may be slight.

Mechanisms (27) and (36) may also be distinguished from the shape of the plot of k against I, which is linear for the former and hyperbolic for the latter. Again, the hyperbolic nature of the plot may not always be apparent as it depends on the relative magnitudes of I/K_3 and $1 + A/K_a$.

In this context, it is worth mentioning a third possible slow binding mechanism in which isomerization precedes inhibitor binding.

$$\begin{array}{ccc} \mathrm{E} & \underset{k_{-1}}{\overset{k_1 A}{\rightleftharpoons}} & \mathrm{EA} \overset{k_2}{\rightarrow} \mathrm{E} + \mathrm{P} \\ k_3{}' \downarrow\uparrow k'_{-3} & & \\ \mathrm{E}^* & \underset{k'_{-7}}{\overset{k'_7}{\rightleftharpoons}} & \mathrm{EI}^* \end{array} \tag{40}$$

If the attainment of equilibrium between E* and EI* is slow while the enzyme isomerization is fast, mechanism (40) is indistinguishable from mechanism (27). However, if the equilibrium between E and E* is the slow step, then mechanisms (27) and (40) lead to different predictions about the dependence of k on I[16]:

$$k = \frac{k'_3}{1 + A/K_a} + \frac{k'_{-3}}{1 + I/K_i} \tag{41}$$

$$K_i = k'_{-7}/k'_7 \tag{42}$$

As for mechanism (36), there is a hyperbolic relationship between k and I; the difference is that k increases with I in mechanism (36) whereas it decreases in mechanism (40).

Slow, Tight-Binding Inhibition. Let us consider the simplest case of competitive slow binding inhibition, as illustrated in mechanism (27), when tight binding of the inhibitor must also be taken into account. Cha[6] analyzed this system, deriving the expression of the rate equation. Using the same assumptions as for mechanism (27), it was shown that the rate equation is given by

$$v_i = \frac{v_s + [v_0(1 - \gamma) - v_s]e^{-kt}}{1 - \gamma e^{-kt}} \tag{43}$$

[16] R. G. Duggleby, P. V. Attwood, J. C. Wallace, and D. B. Keech, *Biochemistry* **21**, 3364 (1982).

where

$$\gamma = \frac{K_i' + E_t + I_t - Q}{K_i' + E_t + I_t + Q} \tag{44}$$

$$K_i' = K_3(1 + A/K_a) \tag{45}$$

$$v_o = \frac{VA}{K_a + A} \tag{46}$$

$$v_s = v_0 \frac{-(K_i' - E_t + I_t) + Q}{2E_t} \tag{47}$$

$$k = \frac{k_3 Q}{1 + A/K_a} \tag{48}$$

$$Q = [(K_i' - E_t + I_t)^2 + 4K_i'E_t]^{1/2} \tag{49}$$

Despite the apparent complexity of the above expressions, it is possible to determine the unknown constants by using graphical methods only, provided that some additional relationships are utilized. Williams *et al.*[7] pointed out that γ is a function of v_0 and v_s:

$$\gamma = \frac{E_t}{I_t}\left(1 - \frac{v_s}{v_0}\right)^2 \tag{50}$$

Further, by using Eq. (50) as well as Eqs. (44), (47), and (48), we can express Q and k as functions of v_s/v_0. As a consequence, provided that v_0 and v_s can be determined from the progress curves, then the two unknowns, namely, k_3 and k_{-3} (or, equivalently, K_i' and k_3) can be also determined. To accomplish this, the following equations, derived from Eqs. (43), (44), (45), (47), and (48), can be used:

$$\frac{I_t}{1 - v_s/v_0} = E_t + K_i'\frac{v_0}{v_s} \tag{51}$$

$$\ln\left\{\left(\frac{v_0 - v_s}{v_i - v_s}\right)\left[1 - \frac{E_t}{I_t}\left(1 - \frac{v_s}{v_0}\right)\left(1 - \frac{v_i}{v_0}\right)\right]\right\} = kt \tag{52}$$

$$\frac{I_t}{1 - v_s/v_0} - E_t(1 - v_s/v_0) = \frac{1 + A/K_a}{k_3}k \tag{53}$$

The following procedure may be used in order to determine the inhibition parameters of slow, tight-binding inhibition that obeys the mechanism (27). First, estimate v_0 and v_s from the progress curve at each of several different inhibitor concentrations. Second, plot $[I_t/(1 - v_s/v_0)]$ against (v_0/v_s) [see Eq. (51)]; the slope of the straight line obtained will give the value of K_i'. Third, measure the rate at several points along each of the progress curves and plot

$$\ln\left\{\left(\frac{v_0 - v_s}{v_i - v_s}\right)\left[1 - \frac{E_t}{I_t}\left(1 - \frac{v_s}{v_0}\right)\left(1 - \frac{v_i}{v_0}\right)\right]\right\}$$

as a function of time [see Eq. (52)]; the slopes equal the experimental decay constants k. Fourth, plot $[I_t/(1 - v_s/v_0) - E_t(1 - v_s/v_0)]$ against k [see Eq. (53)]; from the slope, the rate constant k_3 can be obtained. Fifth and finally, calculate K_3 using Eq. (45) and k_{-3} from $K_3 = k_{-3}/k_3$. The values of the parameters obtained as above may then be refined by a nonlinear regression method, using Eq. (43).

A more direct method to determine the inhibition parameters for this type of inhibition would be to process the progress curve data by a nonlinear regression method based on the integrated form of Eq. (43) (Williams *et al.*[7]):

$$P = v_s t + \frac{(1 - \gamma)(v_0 - v_s)}{k\gamma} \ln\left(\frac{1 - \gamma e^{-kt}}{1 - \gamma}\right) \tag{54}$$

In applying nonlinear regression to Eq. (54), it is important to treat γ as a function of v_0 and v_s according to Eq. (50), rather than as an independent quantity. In this way, there will be only three parameters to be evaluated, namely, v_0, v_s, and k.

For slow, tight-binding competitive inhibitors with a slow isomerization of the initially formed enzyme–inhibitor complex [mechanism (36)], a detailed procedure is described by Sculley and Morrison,[17] based on a combination of rate and progress curve measurements. The time course equation for a slow, tight-binding noncompetitive inhibitor obeying the kinetic mechanism of Scheme 1 (see below) for $k_6 = k_{-6} = k_7 = k_{-7} = 0$ has also been derived (Takai *et al.*[18]).

Determination of Kinetic Model and Inhibition Parameters

Generally, the principal aim of a kinetic study concerning slow and tight-binding inhibition is the determination of the inhibition parameters. The accuracy of the values obtained depends primarily on whether the chosen kinetic model is correct. In the following, we suggest steps that might be followed when dealing with an unknown inhibitor in order to set up the corresponding kinetic model and to determine the inhibition parameters.

[17] M. J. Sculley and J. F. Morrison, *Biochim. Biophys. Acta* **874**, 44 (1986).
[18] A. Takai, Y. Ohno, T. Yasumoto, and G. Mieskes, *Biochem. J.* **287**, 101 (1992).

Characterization of Uninhibited Reaction

It is important to make a thorough analysis of the enzyme reaction in the absence of any inhibitor, because the expressions of the inhibition parameters are functions of parameters of the uninhibited reaction. The most important are the maximal velocities (in the forward and backward direction) and the Michaelis constants (for both substrates, in the case of a two-substrate enzyme reaction). They can be determined by using initial rate data or, preferably, progress curve data (see, e.g., Nimmo and Atkins,[19] Duggleby and Wood,[20] or Duggleby[21]). At this step it is also important to establish, for two-substrate reactions, what type of kinetic mechanism is obeyed by the enzyme reaction. Finally, if the inhibitors to be studied are of the tight-binding type, it is necessary to establish a method for the accurate determination of the active enzyme concentration.

Experiments to Gain Preliminary Information about Inhibitor

Five essential questions should be answered at this point, in order to identify the class to which the inhibitor belongs.

Is Inhibition Reversible? Reversibility can be established by simple experiments in which the rate is observed after lowering the concentration of the inhibitor by dilution or dialysis. In the case of slow, tight-binding inhibitors, it is also possible to separate the enzyme–inhibitor by size-exclusion chromatography. Subsequent prolonged dialysis of the enzyme–inhibitor complex against a competitive ligand with a higher dissociation constant and then against buffer should result in the recovery of the entire initial enzyme activity, provided that the inhibitor is reversible. In this chapter we restrict our attention to reversible inhibitors.

Is Inhibition Slow Binding? The characteristic feature of slow binding inhibition is that the progress curves at a constant inhibitor concentration are not linear (as in the case for classical inhibitors) but concave-down when the reaction is started by addition of enzyme. It is essential that, in the time interval in which this nonlinearity of the progress curves is observable, the corresponding progress curve for the noninhibited reaction be perfectly linear. Otherwise, the nonlinear progress curve could arise from causes unrelated to the inhibitor, such as depletion of the substrate or enzyme inactivation. It may be useful to lower the enzyme concentration in order to provide a sufficiently long time for establishment of the equilibrium between enzyme and inhibitor (before the substrate depletion be-

[19] I. A. Nimmo and G. L. Atkins, *Biochem. J.* **141,** 913 (1974).
[20] R. G. Duggleby and C. Wood, *Biochem. J.* **258,** 397 (1989).
[21] R. G. Duggleby, this volume, [3].

comes significant). Additional evidence for slow-binding inhibition can be obtained by recording the progress curves at different inhibitor concentrations: at higher concentrations, it might be more evident that the curves exhibit asymptotes [see, e.g., Eq. (34)]. Usually, at low inhibitor concentrations, the establishment of enzyme–inhibitor equilibrium is so slow that it may not be complete before substrate depletion is significant.

Is Inhibition Linear? The inhibition is called linear when the reciprocal of the steady-state velocity is a linear function of the inhibitor concentra tion. Thus, we should first be sure that accurate steady-state velocity data are available, for different inhibitor concentrations. Plots of the reciprocal of the steady-state velocity against the total inhibitor concentration should indicate whether the inhibitor is linear. These plots should remain linear at different total enzyme concentrations.

Nonlinear inhibition may arise mainly for the following reasons: (1) hyperbolic (partial) inhibition when the plots of $1/v_i$ against I_t are hyperbolas, where inhibitor complexes still are able to participate in the catalytic process but at a lower efficiency; (2) tight-binding inhibition, where, as shown by Morrison[4] [see also Eq. (11)], plots of $1/v_i$ against I_t in the case of tight-binding inhibition are nonrectangular hyperbolas, that is, concave-up curves exhibiting asymptotes; and (3) parabolic inhibition when the plots of $1/v_i$ against I_t curve upward parabolically (no asymptotes can be detected); this form of inhibition, which is quite rare but not unknown,[22] occurs when two inhibitor molecules bind to the same form of the enzyme. The first two of the above-mentioned types of nonlinear inhibition, by far the most frequently encountered situations, are considered further below.

Is Inhibition Hyperbolic? When the enzyme–inhibitor complex is catalytically active, plots of $1/v_i$ against I_t are concave-down, having a horizontal asymptote. Generally, it is important to increase experimentally the concentration of the inhibitor as much as possible, in order to determine the value of the residual enzyme activity at saturating concentrations of the inhibitor. If the residual activity is not zero, then the inhibitor is a hyperbolic one.

Is Inhibition Tight-Binding? A characteristic feature of tight-binding inhibitors is that plots of $1/v_i$ as a function of I_t have a concave-up form and asymptotic behavior at high inhibitor concentration. However, there is another test that can confirm the tight-binding inhibition: plots of the steady-state velocity against the total enzyme concentration at different I_t concentrations are asymptotic, concave-up curves, whereas the plot of v_i against E_t for a non-tight-binding inhibitor is linear.

[22] J. W. Williams, R. G. Duggleby, R. Cutler, and J. F. Morrison, *Biochem. Pharmacol.* **29,** 589 (1980).

Introduction of Appropriate Kinetic Mechanism of Inhibition

Once the previous steps have been completed there should be enough information to design a kinetic scheme that accounts for all the findings obtained. It is advisable to choose the least complicated mechanism that fits all the requirements of the preliminary experimental findings.

As an example, let us consider that preliminary experiments suggest that the given inhibitor is a reversible one, giving rise to a hyperbolic (partial) inhibition with tight binding between enzyme and inhibitor. No information is available about the enzyme forms that are complexed by the inhibitor. The kinetic scheme that might be appropriate for this situation is that depicted in Scheme 1, corresponding to a hyperbolic, slow (tight) binding inhibition. It is worth mentioning that the most frequently used models of slow and tight-binding inhibition can be obtained as particular cases of Scheme 1. It should also be noted that Scheme 1 is not completely general and could be expanded to include a complex EAI* that could be derived from EAI and from EI*.

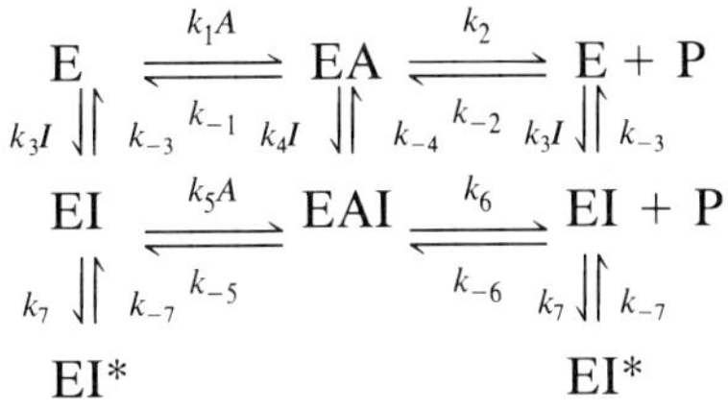

SCHEME 1

Introduction of Simplifying Assumptions

The analytical solution of the system of differential equations associated with the kinetic model cannot be obtained, or is very complicated, unless some simplifying assumptions concerning the model are introduced. The central problem of these simplifying assumptions is the following: when the equations obtained for a kinetic model are applied to an experimental system, the inhibition constants thus obtained will be correct only if the simplifying assumptions are also rigorously valid for the experimental system. Though this problem is a general one, it has a particular importance in the case of slow and tight-binding inhibitors. Indeed, in this case some of the usually accepted simplifying assumptions can be violated, and consequently the classical kinetic equations are frequently not valid for these categories of inhibitors. Generally, three types of simplifying assumptions have been applied.

Steady-State Assumptions. Assumptions of steady state usually concern the intermediates on the normal catalytic pathway, for which it is considered that, after a short initial period, the rates of formation and breakdown are equal. It was shown many years ago by Morales[23] that for pure noncompetitive inhibitors the steady-state assumption cannot be compatible with linearity of double-reciprocal plots of v against the substrate concentration unless, coincidentally, the Michaelis constant equals the enzyme–substrate dissociation constant. A slow binding step may complicate the situation further. As an illustration, let us consider the simple case of competitive slow binding inhibition as depicted in mechanism (27). Usually, the experimental measurements are carried out under pre-steady-state conditions, in which the enzyme–inhibitor complex is formed slowly with a rate constant k_3. The inhibition can be detected provided that the rate of EI formation is greater than the rate of EI decomposition. Thus, the free enzyme molecules will be slowly removed and converted to the EI complex. Consequently, the concentration of the EA complex will also decrease; this fact may invalidate the steady-state assumption for this enzyme–substrate intermediate.

Rapid Equilibrium in Some Steps. It is assumed that any steps which have much higher rate constants than the other steps may be considered as being at equilibrium. A particular situation encountered in slow binding inhibition studies is that in which one step is much slower than all of the remaining steps. The difficulty lies in deciding which of the component steps is fast or slow when limited or no information is available about the rate constants. In some situations, the analytical expression of the rate and/or the progress curve equation for the given kinetic mechanism can be derived. In many situations these expressions involve exponential terms; the expression of their relaxation times can be analyzed, however, and some conditions may be found in which one or more exponential terms can be neglected. For instance, in Appendix 1, the expression of k_{obs} is derived for the sequence

$$\mathrm{E} \underset{k_{-3}}{\overset{k_3 I}{\rightleftharpoons}} \mathrm{EI} \underset{k_{-7}}{\overset{k_7}{\rightleftharpoons}} \mathrm{EI}^*$$

In the general case, EI^* is a linear combination of two exponentials, $\exp(-r_1 t)$ and $\exp(-r_2 t)$ [see Eq. (A3) in Appendix 1]. However, if $k_7 \ll k_3 I$ and $k_{-7} \ll k_{-3}$ (as shown in Appendix 1) one of the relaxation times $(1/r_1)$ is much smaller than the other $(1/r_2)$. Thus, on the time scale of $1/r_2$, the contribution of $\exp(-r_1 t)$ is negligible and EI^* will evolve as if only the second exponential term, $\exp(-r_2 t)$, exists. In this situation, the expression of k_{obs} may be derived as being $k_{-7} + k_7 I/(I + K_3)$. It should

[23] M. F. Morales, *J. Am. Chem. Soc.* **77,** 4169 (1955).

be stressed that this frequently used formula of k_{obs} is valid only if $k_7 \ll k_3 I$ and $k_{-7} \ll k_{-3}$. These restrictions on k_7, $k_3 I$, k_{-7}, and k_{-3} will occur when the enzyme binding step is in rapid equilibrium as compared to the second, slow isomerization step of the EI complex.

Analyzing a general kinetic mechanism for an enzyme–substrate–modifier system similar to that given in Scheme 1, but with $k_7 = k_{-7} = 0$, Frieden[24] described several situations when the steady-state and the rapid equilibrium treatment lead to similar expressions of the rate equation. The same author pointed out that the experimental conditions can be so designed that certain steps are forced to be in equilibrium. For instance, at high inhibitor concentration and low substrate concentration, the steps of EI formation from E and I and of EAI formation from EA and I may be much faster than the formation of EA from E and A and the formation of EAI from EI and A.

Constant Concentration of Some Components. A common simplifying assumption is that the product concentration is zero. In this way the complications related to the reverse reaction are avoided. Sometimes, however, the evolution of the reaction should be observed over an extended period, and the concentration of the product formed can no longer be neglected.

Related in part to the above assumption is one concerning the constancy of the substrate concentration. In the case of initial rate experiments that would be used for studying a fast, tight-binding inhibitor, a high substrate concentration (as compared to the enzyme concentration) may be used, and this assumption would be clearly valid. In the case of slow binding inhibition, the constancy of substrate is usually assumed, though it is possible that substrate depletion becomes significant before the steady state is reached. If this situation cannot be avoided experimentally, then a model should be used that accounts for substrate depletion.

An assumption regarding the constancy of the inhibitor concentration is also used whenever a tight-binding component to the considered inhibition is not suspected. In these "non-tight-binding" cases, it is still important to use an excess of inhibitor as compared to the enzyme concentration. In this respect it is essential to check if the experimental conditions used, and the inhibition constants finally determined, are consistent with the conditions for a tight-binding inhibition as given earlier in the section on definitions. In relation to these conditions, we may also note that, at least theoretically, by raising both the substrate and inhibitor concentrations tight binding can be eliminated.

[24] C. Frieden, *J. Biol. Chem.* **239,** 3522 (1964).

We should also mention at this point the utility of using the relationships between the parameters of the model in order to define the number of independent parameters. Thus, for reversible reactions, the Haldane relationships are very useful. Also, in the case of kinetic schemes which contain a closed cycle, the relationship between the rate constants (or the equilibrium constants) of the steps forming the cycle should be considered. For instance, in the kinetic mechanism illustrated in Scheme 1, the following relationship holds:

$$K_1K_4 = K_3K_5 \tag{55}$$

It should be noted that the above relationship involves the equilibrium constants K_1 and K_5 and not the corresponding Michaelis constants K_a and K_a'.

In addition, for the sake of generality, we suppose that in the case of this example (Scheme 1), the only simplifying assumption to be considered is that the contribution of the backward enzyme-catalyzed reaction is negligible, that is, $k_{-2} = k_{-6} \approx 0$. In other words, if a slow, tight-binding inhibition is suspected, the substrate depletion cannot be neglected, and no prior assumptions can be made about which steps are slow and which are not.

In general, we would recommend that the use of simplifying assumptions which have no real support for a given enzyme–inhibitor system be avoided. It is advisable to use a less restrictive model, based on the particular data available for the system studied. A procedure is being elaborated[25] for deciding in a more rigorous way which simplifying assumptions are valid, having at hand some information about the inhibition being considered.

Writing Sets of Differential Equations

First, we must identify all of the molecular species involved in the kinetic scheme we have chosen, that is, all the forms of the enzyme, substrate, product, and inhibitor. For each molecular species, a differential equation can be written describing the rate of change of its concentration. In addition, three conservation equations can be written: one for the enzyme forms, one for the inhibitor species, and one for the substrate (and product) forms. If the constancy of inhibitor concentration and/or substrate concentration can be assumed, the corresponding conservation

[25] S. E. Szedlacsek and R.G. Duggleby, unpublished work (1994).

relationship need not be considered. In addition, the differential equation corresponding to the molecular species which are assumed to have constant concentrations should be eliminated. Both differential and conservation equations can be used, their total number equaling the number of molecular species considered to have variable concentrations. In selecting the equations to use, it is vital to check the linear independence of the chosen equations; that is, none of them can be obtained by simply rearranging or combining several of the others. It is advantageous to use the maximum number of conservation equations, in order to minimize number of differential equations.

Each differential equation selected corresponds to a variable for which it was initially written. These variables will be considered as being the independent variables of the system. All the other variables will be substituted in the differential equations by using the conservation equations. Alternatively, it may be useful at this step to introduce nondimensional variables. For instance, by dividing each independent variable by its maximum concentration, a system would be obtained in which all variables would vary between 0 and 1, and this fact may be helpful if a numerical solution of the equations is necessary. If a steady-state or a rapid equilibrium situation is also considered, the equation expressing this assumption will replace the appropriate differential equation in the system we write.

In the case of the example we have considered (Scheme 1), and applying the simplifying assumption that $k_{-2} = k_{-6} \approx 0$, the following set of differential and conservation equations can be written:

$$\begin{aligned}
dEA/dt &= k_1A \cdot E + k_{-4}EAI - (k_{-1} + k_2 + k_4I)EA \\
dEAI/dt &= k_4I \cdot EA + k_5A \cdot EI - (k_{-4} + k_{-5} + k_6)EAI \\
dEI^*/dt &= k_7EI - k_{-7}EI^* \\
dI/dt &= k_{-3}EI + k_{-4}EAI - (k_3E + k_4EA)I \\
dA/dt &= k_{-1}EA + k_{-5}EAI - (k_1E + k_5EI)A \\
E &= E_0 - I_0 - EA + I \\
EI &= I_0 - I - EAI - EI^* \\
P &= A_0 - EA - EAI - A
\end{aligned} \tag{56}$$

Substituting the conservation equation for the enzyme and the inhibitor into the differential equations, as well as introducing the nondimensional variables,

$$x_1 = EA/E_0; \quad x_2 = EAI/E_0; \quad x_3 = EI^*/E_0; \quad x_4 = I/I_0; \quad x_5 = A/A_0; \quad x_6 = P/A_0 \tag{57}$$

we get the following set of equations:

$$
\begin{aligned}
dx_1/dt &= -(k_{-1} + k_2)x_1 + k_{-4}x_2 + k_1(1 - I_0/E_0)A_0x_5 - k_4I_0x_1x_4 \\
&\quad - k_1A_0x_1x_5 + k_1(A_0I_0/E_0)x_4x_5 \\
dx_2/dt &= -(k_{-4} + k_{-5} + k_6)x_2 + k_5(I_0A_0/E_0)x_5 + k_4I_0x_1x_4 \\
&\quad - k_5A_0x_5(x_2 + x_3) - k_5(A_0I_0/E_0)x_4x_5 \\
dx_3/dt &= k_7I_0/E_0 - k_7x_2 - (k_7 + k_{-7})x_3 - k_7(I_0/E_0)x_4 \\
dx_4/dt &= k_{-3} + (k_{-4} - k_{-3})(E_0/I_0)x_2 - k_{-3}(E_0/I_0)x_3 \\
&\quad - [k_{-3} + k_3(E_0 - I_0)]x_4 + (k_3 - k_4)E_0x_1x_4 - k_3I_0x_4^2 \\
dx_5/dt &= k_{-1}(E_0/A_0)x_1 + k_{-5}(E_0/A_0)x_2 - [k_1(E_0 - I_0) + k_5I_0]x_5 \\
&\quad + k_1E_0x_1x_5 + k_5E_0(x_2 + x_3)x_5 + (k_5 - k_1)I_0x_4x_5 \\
x_6 &= 1 - (E_0/A_0)x_1 - (E_0/A_0)x_2 - x_5
\end{aligned}
\tag{58}
$$

The above expressions [Eq. (58)] can be rewritten in a more convenient form by substituting the usual kinetic parameters such as dissociation constants (e.g., K_3, K_4, K_7) or Michaelis constants [$K_a = (k_{-1} + k_2)/k_1$ and $K_a' = (k_{-5} + k_6)/k_5$]. It is also worth noting that the first five differential equations may be solved as an independent system of differential equations.

Derivation of Rate and Progress Curve Equations

Sometimes it is possible to derive an analytical solution for the set of differential equations obtained in the previous step. For instance, in the case of slow binding inhibition obeying the general Scheme 1, an analytical solution can always be derived, provided that both the free substrate and the free inhibitor can be considered as having constant concentrations. Indeed, it is noteworthy that in Eq. (58) all nonlinear terms contain x_4 and/or x_5. Thus, if $x_4 = x_5 = 1$, one obtains a system of linear differential equations for which an analytical solution can be derived. This case is detailed below.

There are also some other particular situations when, at least for the reaction rate, an analytical expression can be derived. Some of them were mentioned earlier in this chapter, and two others are treated later. However, in many other cases, as, for instance, in the general case of Eq. (58), the derivation of an analytical expression for the reaction rate is not possible. In this case, the system of differential equations should be used as such in the analysis of the data.

Processing Experimental Data

If the system of differential equations obtained as described above can be solved analytically, then the data analysis should involve the fitting of the equation derived to the experimental data, by using an appropriate regression method. The best situation is when an analytical expression of the progress curve equation is available. In this case, the progress curve data can be used directly in the regression analysis. If a linear representation is also possible, the data could be processed graphically first. The parameters thus obtained should be used as starting estimates only in the regression analysis, which will improve considerably the accuracy of the parameters to be determined. If only the rate equation can be expressed analytically, then a similar analysis can be applied using velocity data instead of progress curve data. As an alternative, the rate equation can be numerically integrated and then fitted to the progress curve data (see e.g., Chandler *et al.*[26]), thereby exploiting the advantages of progress curve analysis, as discussed by Duggleby and Morrison.[27]

Instead of using progress curve or velocity data directly, it is sometimes possible to use other experimental values derived from the primary progress curve data. For instance, the equations for slow binding inhibition usually contain exponential terms, and their relaxation times can be easily estimated from the progress curves. As the expressions of the transient times may be simpler to handle, the data analysis may be simplified.

If an analytical solution for the system of differential equations cannot be derived, then numerical integration of the system is necessary; when used in combination with nonlinear regression, it may be possible to determine values for all of the inhibition parameters.

Checking Reliability of Inhibition Parameters

To check the reliability of the fitting procedure used and of the experimental data, it would be advisable to repeat several times the procedure of parameter determination using several sets of experimental data and, where possible, alternative methods of data analysis. For instance, if an initial procedure used the analytical expression of the relaxation time, then in a second procedure the whole expression of the progress curve equation can be used. The use of simulated data may be valuable at this stage. It goes without saying that all these combinations of sets of data and ways of processing should give similar values of the parameters.

Once satisfactory consistency has been achieved, it is very important to check if the parameter values obtained are in agreement with the simplifying assumptions introduced in deriving the equations that were used in the fitting procedure. If there is no agreement between the assumptions and the parameter values, the equations should be changed accordingly

[26] J. P. Chandler, D. E. Hill, and H. O. Spivey, *Comput. Biomed. Res.* **5**, 515 (1972).

[27] R. G. Duggleby and J. F. Morrison, *Biochim. Biophys. Acta* **481**, 297 (1977).

and the data refitted. This sequence of steps should be continued until there is no conflict between the assumptions and the parameter values determined.

Where possible, it is advisable to compare the inhibition parameters obtained in this way with those obtained by using totally different methods, such as direct enzyme–inhibitor binding studies. A good example in this respect is the work of Blakley and Cocco,[28] where stopped-flow measurements of protein fluorescence quenching were used in order to characterize the binding between dihydrofolate reductase and some of its inhibitors.

Applications and Models

In this section we deal with an application of procedure outlined in the previous section to slow binding inhibition, and with some more recent developments in the field of tight-binding inhibition.

Slow Binding Inhibition

Let us consider again the general slow binding inhibition model depicted in Scheme 1. Let us also assume that the uninhibited reaction has been fully characterized, that is, k_1, k_{-1}, and k_2 have been determined, and that preliminary inhibition experiments have demonstrated that the inhibition is reversible and hyperbolic (partial).

In addition, we suppose that the concentration of the substrate can be made sufficiently high that the tight-binding contribution to the inhibition need not be considered. Thus, we can make the simplifying assumption that the free and total concentrations of the inhibitor and the substrate are equal (i.e., $I_f = I_t$ and $A_f = A_t$). Given that no information is available about the relative rates of the component steps, no steady-state and rapid-equilibrium assumptions are used.

The total number of unknown parameters is reduced by determining the parameters of the uninhibited enzyme-catalyzed reaction, but the model still has a great number of parameters. Consequently, accurate parameter estimation could be seriously embarrassed if five or more parameters are to be estimated. It is thus advisable to adopt a "step-by-step" procedure to determine all the unknown inhibition parameters. One such procedure could be to determine initially the parameters related to the sequence $E \rightleftharpoons EI \rightleftharpoons EI^*$, that is, enzyme binding and isomerization.

[28] R. L. Blakley and L. Cocco, *Biochemistry* **24,** 4772 (1985).

That can be done by direct-binding measurements, using one of the standard methods: nonequilibrium dialysis, equilibrium gel filtration, ultracentrifugation, or a spectroscopic technique (for a general description, see Fersht[29]). Alternatively, a rapid kinetic technique may be used to determine the rate constants involved in the enzyme–inhibitor interaction. In this way, the total number of unknown inhibition parameters can be reduced by four; a subsequent kinetic study of the whole enzyme–inhibitor system should be feasible, as only four independent parameters remain to be estimated. Although there are a total of five rate constants to be evaluated (k_4, k_{-4}, k_5, k_{-5}, and k_6), one can be estimated according to Eq. (55).

Another possibility for simplifying the general Scheme 1 in order to facilitate the evaluation of the inhibition parameters is to use a saturating concentration of the inhibitor. From a plot of steady-state velocity against inhibitor concentration, it may be possible to find an inhibitor concentration at which the velocity levels off at its lowest value. Provided this inhibitor concentration is experimentally accessible (i.e., no solubility or other problems interfere), the enzyme can be considered to be saturated with inhibitor. Under these conditions, the kinetic mechanism of Scheme 1 is reduced to the following, much simpler, mechanism:

$$\begin{array}{ccccc} \mathrm{EI} & \underset{k_{-5}}{\overset{k_5 A}{\rightleftharpoons}} & \mathrm{EAI} & \xrightarrow{k_6} & \mathrm{EI} + \mathrm{P} \\ k_7 \downarrow\uparrow k_{-7} & & & & k_7 \downarrow\uparrow k_{-7} \\ \mathrm{EI}^* & & & & \mathrm{EI}^* \end{array} \qquad (59)$$

Mechanism (59) is equivalent to a Michaelis–Menten type reaction, where the active enzyme form is subject to a reversible isomerization into an inactive form. Suppose that the enzyme is preincubated with an excess of inhibitor for an extended time, and the reaction is started with the substrate. Then, at $t = 0$, it can be taken that an equilibrium mixture of EI and EI* exists. However, we cannot assume that the system is in a steady state as we do not know whether the isomerization is slow or fast. Thus the transient-phase approach should be used to estimate the unknown kinetic constants.

Once we have determined the values of as many parameters as possible by using one of the approaches described above, the whole inhibition system (Scheme 1) can be evaluated. Taking into account the aforementioned simplifying assumptions (i.e., $I_f = I_t$ and $A_f = A_t$), the following system of equations describes the kinetic mechanism of Scheme 1:

[29] A. Fersht, "Enzyme Structure and Mechanism." Freeman, New York, 1985.

$$\begin{aligned} dEA/dt &= k_1A_tE + k_{-4}EAI - (k_{-1} + k_2 + k_4I_t)EA \\ dEAI/dt &= k_4I_tEA + k_5A_tEI - (k_{-4} + k_{-5} + k_6)EAI \\ dEI^*/dt &= k_7EI - k_{-7}EI^* \\ dE/dt &= k_{-1}EA + k_{-3}EI + k_2EA - (k_1A_t + k_3I_t)E \\ dP/dt &= k_2EA + k_6EAI \\ E_t &= E + EA + EI + EI^* + EAI \end{aligned} \tag{60}$$

As the concentration of the product is not involved in the first four differential equations, they can be solved independently by substituting an expression for EI, obtained from the conservation equation for the enzyme forms. Introducing also the nondimensional variables

$$y_1 = EA/E_t; \qquad y_2 = EAI/E_t; \qquad y_3 = EI^*/E_t; \qquad y_4 = E/E_t \tag{61}$$

we obtain the system:

$$\begin{aligned} dy_1/dt &= a_{11}y_1 + a_{12}y_2 + a_{14}y_4 \\ dy_2/dt &= a_{20} + a_{21}y_1 + a_{22}y_2 + a_{23}y_3 + a_{24}y_4 \\ dy_3/dt &= a_{30} + a_{31}y_1 + a_{32}y_2 + a_{33}y_3 + a_{34}y_4 \\ dy_4/dt &= a_{40} + a_{41}y_1 + a_{42}y_2 + a_{43}y_3 + a_{44}y_4 \end{aligned} \tag{62}$$

where

$$\begin{aligned} a_{11} &= -(k_{-1} + k_2 + k_4I_t); \qquad a_{12} = k_{-4}; \qquad a_{14} = k_1A_t \\ a_{20} &= -a_{23} = -a_{24} = k_5A_0; \qquad a_{21} = k_4I_t - k_5A_t; \\ & a_{22} = -(k_{-4} + k_{-5} + k_6 + k_5A_t) \\ a_{30} &= -a_{31} = -a_{32} = -a_{34} = k_7; \qquad a_{33} = -(k_7 + k_{-7}) \\ a_{40} &= -a_{42} = -a_{43} = k_{-3}; \qquad a_{41} = k_{-1} + k_2 - k_{-3}; \\ & a_{44} = -(k_1A_t + k_3I_t + k_{-3}) \end{aligned} \tag{63}$$

System (62) of differential equations has constant coefficients, and an analytical solution can be derived by using a standard mathematical procedure. Then, the expression of the reaction velocity can be obtained according to the following relationship:

$$v_i = dP/dt = E_t(k_2y_1 + k_6y_2) \tag{64}$$

If velocity data are available, then Eq. (64) can be fitted directly to the experimental data. As we have stressed previously, the use of progress curve data is to be preferred. In this case, Eq. (64) should be integrated numerically and then fitted to the experimental progress curve data. Despite the complexity of the analytical expressions used, nonlinear regression analysis of the data should be possible owing to the low number of inhibition parameters which remain to be evaluated.

Tight-Binding Inhibition

Hyperbolic Tight-Binding Inhibition. As mentioned earlier, the equation of Morrison [Eq. (11)] for tight-binding inhibition is valid only in the case of "dead-end" inhibition. Many inhibitors do not obey this condition, however, exhibiting the characteristics of a hyperbolic inhibition. For this category of partial inhibitors, Szedlacsek *et al.*[30] derived the rate equation using the general modifier mechanism and assuming hyperbolic noncompetitive tight-binding inhibition. This mechanism is the same as that depicted in Scheme 1 for the particular case where $k_{-2} = k_{-6} = k_7 = k_{-7} = 0$. According to Eq. (55), we can define $K_4 = \alpha K_3$ and $K_5 = \alpha K_1$. Similarly, k_6 can be written as βk_2, where β is a positive number less than unity. As simplifying assumptions, it was supposed that EA was at steady state and the inhibitor binding steps were at equilibrium (i.e., no slow binding was evident). Also, it was assumed that the concentration of the substrate is sufficiently high to be considered as a constant ($A = A_t$).

Initially, it was shown that the steady-state velocity v_i, with saturating concentrations of the inhibitor, tends to a limiting value v_∞:

$$v_\infty = \frac{(1+\sigma)(\beta/\alpha K_3)}{\sigma/\alpha K_3 + 1/K_3} v_0 = \beta k_2 \frac{\sigma}{\alpha K_3} \frac{E_t}{\rho} \tag{65}$$

where $\sigma = A/K_a$ and $\rho = \sigma/\alpha K_3 + 1/K_3$. Using the equations corresponding to the steady-state condition for EA, the rapid equilibration of the inhibitor binding steps, as well as the conservation equation for the inhibitor, it can be shown that the reaction velocity obeys

$$I_t \frac{v_0 - v_\infty}{v_0 - v_i} = E_t + \frac{v_0 - v_\infty}{v_i - v_\infty} \frac{1+\sigma}{\rho} \tag{66}$$

Equation (66) is useful in the graphical analysis of the experimental data. We may notice that by increasing the concentration of inhibitor, the reaction rate will decrease and the extent of inhibition can be expressed in terms of ζ, defined as

$$\zeta = \frac{v_0 - v_i}{v_0 - v_\infty} \tag{67}$$

Clearly, when I_t increases from 0 to the saturating values, ζ will increase from 0 to 1. Thus, Eq. (66) can be rewritten in terms of ζ as

$$\frac{I_t}{\zeta} = E_t + \frac{1}{1-\zeta} \frac{1+\sigma}{1+\sigma/\alpha} K_3 \tag{68}$$

[30] S. E. Szedlacsek, V. Ostafe, M. Serban, and M. O. Vlad, *Biochem. J.* **254,** 311 (1988).

The relationship in Eq. (68) shows that a plot of I_t/ζ against $1/(1 - \zeta)$ is linear. Let us suppose that the kinetic parameters of the uninhibited reaction have been previously determined. In addition, for given values of $\sigma = A/K_a$, and E_t, the values of v_0 and v_∞ have been evaluated by measuring the steady-state rate at inhibitor concentrations ranging from 0 to a saturating value. If we were to plot I_t/ζ as a function of $1/(1 - \zeta)$ at each of several substrate concentrations, we can obtain the slope as a function of σ. As the slope is given by $K_3[(1 + \sigma)/(1 + \sigma/\alpha)]$ [Eq. (68)], the values of α and K_3 can be estimated.

If approximate values of the inhibition constants K_3 and αK_3 are known, it is possible to estimate how high the inhibitor concentration should be in order to get an acceptable estimate of v_∞. For example, suppose that $I_t = E_t$; from Eq. (68) we have

$$I_t = \frac{\zeta}{(1 - \zeta)^2} \frac{1 + \sigma}{1 + \sigma/\alpha} K_3 \tag{69}$$

If $A_t \approx K_a$ and if the two inhibition constants are expected to have comparable values ($\alpha \approx 1$), then from Eq. (69) we deduce that to reach $\zeta = 0.99$ requires an inhibitor concentration about 10^4 times greater than K_3. If the uncompetitive component of the inhibition is much stronger than the competitive one (i.e., $\alpha \ll 1$), the saturation of the enzyme should take place at much lower inhibitor concentration, given by $10^4 \alpha K_3$.

Accurate evaluation of the parameters needs further analysis of the data by nonlinear regression. For this purpose, the following expression of v_i derived from Eq. (66) is useful:

$$v_i = \frac{v_0 - v_\infty}{2} \left\{ \left[\left(\frac{1 + \sigma}{\alpha + \sigma} \frac{\alpha K_3}{E_t} + \frac{I_t}{E_t} - 1 \right)^2 + 4 \frac{1 + \sigma}{\alpha + \sigma} \frac{\alpha K_3}{E_t} \right]^{1/2} + \frac{v_0 + v_\infty}{v_0 - v_\infty} - \frac{1 + \sigma}{\alpha + \sigma} \frac{\alpha K_3}{E_t} - \frac{I_t}{E_t} \right\} \tag{70}$$

Thus, in the above-mentioned paper, the BASIC nonlinear regression program DNRP53 (Duggleby[31]) was used to analyze the experimental data of Baici[32] on the tight-binding pure noncompetitive inhibition of human leukocyte elastase by a polysulfated glycosaminoglycan. Initally, provisional parameter values were obtained by a linear least squares fit to Eq. (66). Using the values thus obtained as initial estimates, regression analysis yielded best fit values of all four parameters, namely, E_t, K_3, v_0 and v_∞.

Progress Curve Equations for Tight-Binding Inhibition. The analysis of the steady-state velocity data in the case of tight-binding inhibition is

[31] R. G. Duggleby, *Comput. Biol. Med.* **14,** 447 (1984).

[32] A. Baici, *Biochem. J.* **244,** 793 (1987).

complex because the typical representations for the classical inhibition are no longer valid. The equation of Henderson [Eq. (12)]—a linear form of the Morrison equation—although allowing the graphical determination of the type of inhibition, cannot be used for an accurate estimation of the inhibition constants for a noncompetitive inhibitor. As repeatedly emphasized in this chapter, the analysis of progress curve data has many advantages over using the velocity data since more data are available from fewer experiments, the data are more reliable, the possibility exists to study the effect of the product, and so on.

The progress curve equations for a tight-binding inhibitor were derived by Szedlacsek *et al.*[33] for a one-substrate–one-product reversible enzyme-catalyzed reaction:

$$\mathrm{E + A} \underset{k_{-1}}{\overset{k_1}{\rightleftharpoons}} \mathrm{EA} \underset{k_{-2}}{\overset{k_2}{\rightleftharpoons}} \mathrm{E + P} \tag{71}$$

For mechanism (71), the expression of the rate equation is

$$v = \frac{V_f A/K_A - V_r P/K_P}{1 + A/K_A + P/K_P} \tag{72}$$

where K_A and K_P are Michaelis constants for A and P and V_f and V_r are the maximum velocities for the forward and the reverse reaction, respectively. As simplifying assumptions, a steady state for the concentration of EA and the rapid equilibration of inhibitor-binding steps described by Eqs. (1) and (2) were included. The conservation equation for the substrate and product is:

$$A_0 + P_0 = A + P = A_e + P_e \tag{73}$$

where the subscripts "0" and "e" refer to the initial and the equilibrium values, respectively.

Under these conditions, the Morrison equation [Eq. (11)] is valid. To integrate it, first a new variable z is introduced:

$$z = P - P_0 \tag{74}$$

At equilibrium, z has the value z_∞, that is,

$$z_\infty = P_e - P_0 = (V_f K_P A_0 - V_r K_A P_0)/(V_f K_P + V_r K_A) \tag{75}$$

For this particular inhibition, the forms of N, D, and $\Sigma(N_i/K_i)$ from the Morrison equation [Eq. (11)] were derived and integrated, resulting in different progress curve equations depending on whether the inhibition

[33] S. E. Szedlacsek, V. Ostafe, R. G. Duggleby, M. Serban, and M. O. Vlad, *Biochem. J.* **265,** 647 (1990).

is of the pure noncompetitive type. Thus, in cases of competitive, uncompetitive, and noncompetitive inhibition, the following expression was derived:

$$t = \frac{1}{2\rho}\left[\sum_{i=1}^{4} A_i \ln\left(\frac{x - x_i}{x_0 - x_i}\right) + \left(\frac{B_1}{x_0 - 1}\right)\left(\frac{x - x_0}{x - 1}\right) + \left(\frac{B_2}{x_0 + 1}\right)\left(\frac{x - x_0}{x + 1}\right)\right] \tag{76}$$

where

$$x = \left(\frac{z_\infty - s_2 - z}{z_\infty - s_1 - z}\right)^{1/2}; \quad x_0 = \left(\frac{z_\infty - s_2}{z_\infty - s_1}\right)^{1/2}; \quad x_1 = -x_2 = 1; \quad x_3 = -x_4 = (s_2/s_1)^{1/2} \tag{77}$$

The expressions of ρ, s_1, s_2, A_1 to A_4, B_1, and B_2 are functions of I_t, E_t, V_r, V_f, K_A, K_P, S_0, P_0, and of the inhibition constants K_3 and K_4. The definitions of each of these symbols is given in Appendix 2. For a pure noncompetitive inhibition, a simpler equation results:

$$t = \frac{1}{\rho^*}[z - \delta \ln(1 - z/z_\infty)] \tag{78}$$

where

$$\rho^* = \frac{\rho}{2E_t}\{[(K_4 + I_t - E_t)^2 + 4K_4E_t]^{1/2} - (K_4 + I_t - E_t)\} \tag{79}$$

It can be shown that ρ^* represents the initial velocity of the reaction at saturating concentrations of the substrate in the presence of the pure noncompetitive inhibitor. A linearized form of Eq. (79), useful in data analysis, is

$$\frac{I_t}{1 - (\rho^*/\rho)} = E_t + K_4\frac{\rho}{\rho^*} \tag{80}$$

We may note that Eq. (78) is formally similar to the progress curve equation in the case of a classical inhibitor:

$$t = \frac{1}{\rho^{\text{app}}}[z - \delta^{\text{app}} \ln(1 - z/z_\infty)] \tag{81}$$

However, the significance of ρ^{app} and δ^{app} differ from that of ρ^* and δ in Eq. (78):

$$\begin{aligned} \rho^{\text{app}} &= \frac{\rho}{1 + bI_t} \\ \delta^{\text{app}} &= \frac{\delta + aI_t}{1 + bI_t} \end{aligned} \tag{82}$$

This finding suggests a procedure to determine the type of inhibition and to estimate the inhibition parameters by using progress curve data only. First, the parameters of the uninhibited reaction would be determined, and values of ρ, δ, z_∞, and V_f will therefore be known. Now the inhibited reaction may be studied. Initially, progress curves for the inhibited reaction are recorded and the data obtained are plotted as t/z versus $(1/z)\ln(1 - z/z_\infty)$. There are two possibilities:

1. The plot of t–z versus $(1/z)\ln(1 - z/z_\infty)$ is linear. This means that we are dealing with a pure noncompetitive tight-binding inhibitor [see Eq. (78)] or a classical type of inhibition [see Eq. (81)]. To discriminate between these two possibilities, ρ^{app} and δ^{app} are estimated by linear regression and the dependence of $1/\rho^{app}$ and of δ^{app}/ρ^{app} on I_t are examined. Two further possibilities may arise. (a) Linear dependence of $1/\rho^{app}$ and of δ^{app}/ρ^{app} on I_t is consistent with the existence of a classical-type of inhibition, and evaluation of the constants can be done by further processing of the progress curve data. A description of the methodology for this situation is given by Orsi and Tipton.[34] (b) Nonlinearity of plots of $1/\rho^{app}$ and of δ^{app}/ρ^{app} against I_t means that the inhibitor is a pure noncompetitive tight-binding one, and both plots should be concave-up. It is possible then to determine the inhibition constant K_4 as follows: first, for several different inhibitor concentrations, the values of ρ^* are determined graphically from the t/z versus $(1/z)\ln(1 - z/z_\infty)$ plots [see Eq. (81)]; second, plotting $I_t/[1 - (\rho^*/\rho)]$ versus (ρ/ρ^*) allows an estimate of K_4 to be obtained from the slope according to Eq. (80), and the intercept is equal to E_t; third, the value K_4 should be refined by fitting Eqs. (78) and (79) to the progress curve data using nonlinear regression.

2. The plot of t/z versus $(1/z)\ln(1 - z/z_\infty)$ is curved. This means that the inhibited reaction obeys the general Eq. (76) and any of the three types of inhibition may occur: competitive, uncompetitive, or noncompetitive. Equation (76) should be fitted by nonlinear regression to the progress curve data, assuming noncompetitive inhibition. According to the relative magnitudes of K_3 and K_4 so obtained, the following situations may arise: (a) The absolute value of K_4 is very high as compared with K_3 and the largest I_t. This suggests a competitive inhibition. Equation (76), in which a and b are defined by the expressions corresponding to this type of inhibition (Appendix 2), is refitted to obtain a more accurate value of K_3. (b) The absolute value of K_3 is very high as compared to K_4 and to the largest I_t. This suggests that the inhibition is uncompetitive. Here, a and b should be redefined assuming an uncompetitive inhibition (Appendix 2)

[34] B. A. Orsi and K. F. Tipton, this series, Vol. 63, p. 159.

and then introduced in Eq. (76). Refitting Eq. (76) to the progress curve data should lead to an accurate estimate of K_4. (c) The magnitudes of K_3 and K_4 are comparable. In this case, the inhibitor is a noncompetitive one and no refitting is necessary.

Using simulated progress curve data for the inhibition of bovine chymotrypsin A by aprotinin as well as by an imaginary noncompetitive tight-binding inhibitor, the processing of data according to Eq. (76) has been successful.[33]

Conclusions

There is a considerable literature concerning slow and tight-binding inhibition. Kinetic models for these nonclassical inhibitors have also been elaborated. Although the determination of the inhibition constants can be done by fitting model equations to either velocity and progress curve data, the latter alternative is preferable in many situations. Provided that some limitations of using progress curve data are avoided (see Orsi and Tipton[34]) and an appropriate nonlinear regression procedure is used, accurate estimates of the inhibition parameters can be obtained.

We consider that it is particularly important to check the validity of the simplifying assumptions used in deriving the model equations. The parameters estimated by using a kinetic model are reliable only if an agreement exists between the assumptions used in the model and the experimental inhibition system under study.

Appendix 1: Derivation of Expression of k_{obs}

Let us consider the series of two consecutive reactions:

$$\mathrm{E} \underset{k_{-3}}{\overset{k_3 I}{\rightleftharpoons}} \mathrm{EI} \underset{k_{-7}}{\overset{k_7}{\rightleftharpoons}} \mathrm{EI}^* \tag{A1}$$

Applying the enzyme conservation relationship $E + EI + EI^* = E_t$, we can write the following set of differential equations:

$$\begin{aligned} dEI/dt &= k_3 I E_t - (k_3 I + k_{-3} + k_7)EI + (k_{-7} - k_3 I)EI^* \\ dEI^*/dt &= k_7 EI - k_{-7} EI^* \end{aligned} \tag{A2}$$

It should be noted here that we are ignoring any possible tight-binding effects. Applying the Laplace transform to solve Eq. (A2) and considering that at $t = 0$, $EI = EI^* = 0$, we obtain

$$EI^* = EI^*_\infty \left[1 + \frac{r_1}{r_1 - r_2} \exp(-r_1 t) + \frac{r_2}{r_2 - r_1} \exp(-r_2 t) \right] \tag{A3}$$

where

$$EI^*_\infty = \frac{k_3k_7I}{k_{-7}(k_3I + k_{-3}) + k_3k_7I} = \frac{1}{1 + K_7(1 + K_3/I)} \quad \text{(A4)}$$

and r_1 and r_2 are the roots of the following quadratic equation:

$$r^2 + br + c = 0 \quad \text{(A5)}$$

in which

$$b = -(k_3I + k_{-3} + k_7 + k_{-7}); \qquad c = k_{-7}(k_3I + k_{-3}) + k_3k_7I \quad \text{(A6)}$$

Thus,

$$r_1 = \frac{-b + (b^2 - 4c)^{1/2}}{2}; \qquad r_2 = \frac{-b - (b^2 - 4c)^{1/2}}{2} \quad \text{(A7)}$$

As $b < 0$ and $c > 0$, both r_1 and r_2 are positive. Let us suppose that the rate constants satisfy the following inequalities:

$$k_7 \ll k_3I; \qquad k_{-7} \ll k_{-3} \quad \text{(A8)}$$

Then, we have

$$\begin{aligned} b^2 - 4c = {} & (k_3I - k_7)^2 + (k_{-3} - k_{-7})(2k_3I + k_{-3} - k_{-7}) \\ & + 2k_7(k_{-3} + k_{-7}) \gg 0 \end{aligned} \quad \text{(A9)}$$

Consequently, from Eq. (A7), we get

$$r_1 \gg r_2 \quad \text{(A10)}$$

and, because $r_1 + r_2 = -b$ and $r_1r_2 = c$, we have

$$-b = r_1 + r_2 \approx r_1; \qquad r_2 = c/r_1 = -c/b \quad \text{(A11)}$$

Again, using Eq. (A8), we finally obtain

$$k_{\text{obs}} = r_2 \approx \frac{k_{-7}(k_3I + k_{-3}) + k_3k_7I}{k_3I + k_{-3}} = k_{-7} + k_7\frac{I}{I + K_3} \quad \text{(A12)}$$

In the case of the simpler mechanism

$$\text{E} \underset{k_{-3}}{\overset{k_3I}{\rightleftharpoons}} \text{EI} \quad \text{(A13)}$$

using the same procedure as described above we get

$$EI = \frac{E_t}{1 + K_i/I}\{1 - \exp[-(k_3I + k_{-3})t]\} \tag{A14}$$

the expression of k'_{obs} being clearly

$$k'_{obs} = k_3I + k_{-3} \tag{A15}$$

Appendix 2: Significance of Notations Used in Equations (76)–(82)

$$\rho = \frac{V_rK_A + V_fK_P}{K_P - K_A}; \qquad \delta = \frac{K_AK_P}{K_p - K_A}\left[1 + \frac{(A_0 + P_0)(V_r + V_f)}{V_fK_P + V_rK_A}\right]$$

$$s_{1,2} = -p_3 \pm (D/p_2)^{1/2}; \qquad D = 4I_tE_t(a - b\delta)^2$$

$$A_1 = \frac{r^2(2q_1 + q_2) + q_2 + 2q_3}{2(r^2 - 1)}; \qquad A_4 = \frac{q_1r^2 - q_2r + q_3}{r^2 - 1}$$

$$A_2 = \frac{r^2(2q_1 - q_2) - q_2 + 2q_3}{2(r^2 - 1)}; \qquad B_1 = \frac{q_1 + q_2 + q_3}{2}$$

$$A_3 = \frac{q_1r^2 + q_2r + q_3}{r^2 - 1}; \qquad B_2 = \frac{q_1 - q_2 + q_3}{2}$$

$$q_1 = p_5s_1 + p_6; \qquad p_2 = [1 + b(I_t - E_t)]^2 + 4b_tE_t$$
$$q_2 = (p_2)^{1/2}(s_1 - s_2); \qquad p_3 = ab(I_t - E_t)^2 + (a + b\delta)(I_t + E_t) + \delta$$
$$q_3 = -(p_5s_2 + p_6); \qquad p_4 = [\delta + a(I_t - E_t)]^2 + 4a\delta E_t$$
$$r = (s_2/s_1)^{1/2}; \qquad p_5 = 1 + b(I_t - E_t)$$
$$p_1 = 1/(2\rho) \qquad p_6 = \delta + a(I_t - E_t)$$

The expressions of a and b depend on the type of inhibition. For a noncompetitive inhibitor:

$$a = \frac{K_AK_P}{K_P - K_A}\left[\frac{1}{K_3} + \frac{(A_0 + P_0)(V_r + V_f)}{V_fK_P + V_rK_A}\frac{1}{K_4}\right); \qquad b = \frac{1}{K_4}$$

The expressions for a and b for a competitive, a pure noncompetitive, or an uncompetitive inhibitor may be derived from the above expressions by setting, respectively, $K_4 = \infty$, $K_3 = K_4$, and $K_3 = \infty$.

Acknowledgments

S.E.S. thanks Professor B. H. Havsteen and the team at the Institute of Biochemistry, University of Kiel (Germany), for help and encouragement during preparation of the manuscript.

[7] Kinetic Method for Determination of Dissociation Constants of Metal Ion–Nucleotide Complexes

by W. W. CLELAND

Introduction

The dissociation constants of metal ion–nucleotide complexes are important parameters for biochemical studies, and a number of methods have been used for determining them (see [12] in Volume 63 of this series for a summary[1]). For nonparamagnetic ions, ^{31}P nuclear magnetic resonance (NMR) spectroscopy is an excellent method and has been used for Mg^{2+} and Cd^{2+} complexes of ATP, ADP, and sulfur-substituted analogs of ATP and ADP,[2] whereas for paramagnetic species, such as Mn^{2+}, electron paramagnetic resonance (EPR) spectroscopy is useful.[3] However, none of these methods work well if the dissociation constant is less than 1 μM.

Morrison and Cleland have introduced a kinetic method based on competition for a nucleotide between Mg^{2+} and a more tightly bound metal ion, where the Mg^{2+}–nucleotide complex is a substrate for an enzyme and the other metal ion–nucleotide complex is not.[4] The method assumes the following equation:

$$Mg^{2+} + MeATP \rightleftharpoons Me + MgATP^{2-} \tag{1}$$

where Me is a metal ion of +2 or +3 charge and the charge on the MeATP complex is −2 or −1, respectively. In the presence of a fairly high Mg^{2+} level (such as 5 mM) the equilibrium shifts to the right as the level of MgATP is decreased. Thus a reciprocal plot against MgATP at a fixed level of MeATP is concave downward, and the curvature is more pronounced at high MeATP levels. When reciprocal plots are determined as a function of MgATP at different MeATP levels, a fit of the velocities to the rate equation for this situation yields the ratio of K_d values for MgATP and MeATP. This method is capable of determining K_d values for MeATP complexes as low as 25 nM.

[1] W. J. O'Sullivan and G. W. Smithers, This series, Vol. 63, p. 294.
[2] V. L. Pecoraro, J. D. Hermes and W. W. Cleland, *Biochemistry* **23,** 5262 (1984).
[3] M. Cohn and J. Townsend, *Nature* (*London*) **173,** 1090 (1954).
[4] J. F. Morrison and W. W. Cleland, *Biochemistry* **19,** 3127 (1980).

Theory

The rate equation for the experiment (assuming competitive inhibition by MeATP versus MgATP) is[4]

$$v = \frac{V[b + 2c/K + (b^2 - 4ac)^{1/2}]}{2(aK + b + c/K)} \tag{2}$$

where

$$a = (1 + I/K_i)[1 + (S + I)/K_i] + K_d R/K_i \tag{3}$$

$$b = S[1 + (S + I)/K_i] - K_d R[1 - (S + I)/K_i] \tag{4}$$

$$c = -K_d R(S + I) \tag{5}$$

In Eqs. (2)–(5), v is initial velocity, V is maximum velocity, S and I are the concentrations of added MgATP and MeATP, respectively, R is the ratio of the free Mg^{2+} concentration and the dissociation constant of MgATP under the conditions of the experiment, K is the Michaelis constant of MgATP for the enzyme used, K_i is the competitive inhibition constant of MeATP, and K_d is the dissociation constant of MeATP.

A computer program to permit fitting data to Eqs. (2)–(5) is given in the appendix at the end of this chapter. It requires preliminary estimates of K, K_i, and K_d, the first two of which can be obtained from a competitive inhibition pattern at relatively high MgATP levels and low (~0.15 mM) Mg^{2+}. A preliminary value of K_d is obtained by calculating the actual concentration of MgATP in an experiment at low enough MgATP and high enough MeATP and free Mg^{2+} for the curvature in the reciprocal plot to be pronounced.[4] At this point,

$$[\text{MgATP}] = \frac{vK[1 + (I + S)/K_i]}{V - v(1 - K/K_i)} \tag{6}$$

$$K_d = \frac{[\text{MgATP}]([\text{MgATP}] - S)}{R(S + I - [\text{MgATP}])} \tag{7}$$

The computer program calculates the apparent dissociation constant of MgATP (or MgADP if one is dealing with ADP complexes) from a knowledge of pH, ionic strength, and K^+ and Na^+ concentrations,[5] and it then determines the apparent dissociation constant of MeATP from the ratio of dissociation constants of MeATP and MgATP given by the fit to Eq. (2).

Experimental Methods

So far hexokinase has been the enzyme used with MeATP complexes, and the values of dissociation constants determined by the kinetic method

[5] R. Adolfsen and E. N. Moudrianakis, *J. Biol. Chem.* **253,** 4378 (1978).

TABLE I
DISSOCIATION CONSTANTS FOR METAL ION–NUCLEOTIDE COMPLEXES DETERMINED BY KINETIC METHOD

	ATP complexes			ADP complexes		
Metal ion[a]	pH	Dissociation constant (μM)[b]	Ref.	pH	Dissociation constant (μM)[b]	Ref.
Al	7	0.67	*c*			
Sc	8	0.45	*d*			
La	8	0.33	*d*	7	7	*f*
Ce	8	0.35	*d*	7	4	*f*
Pr	8	0.31	*d*	7	6	*f*
Nd	8	0.29	*d*	7	12	*f*
Sm	8	0.22	*d*	7	6	*f*
Eu	8	0.16	*d*	7	8	*f*
Gd	6	0.91	*e*			
	8	0.087	*d*			
	8.65	1.0	*e*			
Tb	8	0.094	*d*	7	7	*f*
Dy	8	0.049	*d*	7	12	*f*
Ho	8	0.099	*d*	7	5.5	*f*
Er	8	0.086	*d*	7	5	*f*
Tm	8	0.038	*d*	7	6	*f*
Yb	8	0.024	*d*	7	5	*f*
Lu	8	0.044	*d*	7	13	*f*

[a] All ions have +3 charge.
[b] Standard errors of the measurements were generally 10–25% of the value for ATP complexes, and up to 50% for ADP ones.
[c] R. E. Viola, J. F. Morrison, and W. W. Cleland, *Biochemistry* **19,** 3131 (1980).
[d] J. F. Morrison and W. W. Cleland, *Biochemistry* **22,** 5507 (1983).
[e] J. F. Morrison and W. W. Cleland, *Biochemistry* **19,** 3127 (1980).
[f] D. E. Vanderwall and W. W. Cleland, unpublished experiments (1984).

are given in Table I. Typical reaction mixtures (3 ml) contain 2.5 mM glucose, 0.2 mM $NADP^+$, 4 units of glucose-6-phosphate dehydrogenase, 50 mM buffer at the appropriate pH, 5 mM Mg^{2+}, and variable MgATP and MeATP.[6] Sufficient yeast hexokinase is used to give easily measured rates.

For determination of the K_d values of MeADP complexes, creatine kinase has been used.[7] Reaction mixtures contain 10 mM phosphocreatine and creatine kinase, as well as hexokinase, glucose-6-phosphate dehydrogenase, and the other components of the hexokinase assay given above. MgADP and MeADP concentrations are varied.

[6] J. F. Morrison and W. W. Cleland, *Biochemistry* **22,** 5507 (1983).
[7] D. E. Vanderwall and W. W. Cleland, unpublished experiments (1984).

Results

Table I shows the values determined by this method to date. The lanthanide ions form tight complexes with ATP, with the higher lanthanides with the smaller ionic radii forming the tightest complexes (25–50 n*M* for Tm^{3+}, Yb^{3+}, and Lu^{3+}).[6] The inhibition constants versus MgATP with hexokinase show a similar but more pronounced trend, going from 400 μM for LaATP to 14 n*M* for LuATP (the latter the final steady-state value, since LuATP is a slow binding inhibitor with an initial K_i of 0.9 μM).[6]

Lanthanide–ADP complexes are very strong inhibitors of creatine kinase (5–20 n*M*), but the complexes have higher dissociation constants of about 6 μM, which do not vary appreciably from La^{3+} to Lu^{3+}.[7] Presumably the ionic radius is not critical for binding a bidentate ligand such as ADP, whereas ATP is probably bound to lanthanide ions as a tridentate ligand in a facial manner, where the radius of curvature will have an effect.

Conclusions

This kinetic approach should prove useful for determining the dissociation constants of metal ion–nucleotide complexes as long as: (1) the metal ion complex is not a substrate for the enzyme being inhibited (hexokinase or creatine kinase, although other enzymes could certainly be used) and (2) the metal ion does not form an insoluble hydroxide at the pH of interest and free concentration present in the experiment. The real advantage of the method is its ability to measure dissociation constants that are 1 μM or less. At the dilutions used in these experiments, one is also very unlikely to get complexes containing more than one nucleotide or metal ion, as is frequently the case at higher concentrations.

Computer Program

The FORTRAN program which follows is designed to determine the dissociation constant of a metal ion–nucleotide complex by the kinetic method of Morrison and Cleland,[4] where the metal ion is one not giving substrate activity but giving an inhibitory complex with ATP or ADP. The metal ion is referred to in the program as Ln and the complex as LnATP. The changes needed to analyze data for ADP complexes are given below.

The data input consists of several lines of information as follows:

1. Title line. In columns 1–3 (I3 format), number of data points. In columns 4–20, blank if velocities are used as input, or 1 in column 20 if reciprocal velocities are used as input. In columns 21–68, a title which will be printed out to identify the output.

2. Preliminary estimates (F format). Columns 1–10, K (Michaelis constant of MgATP, in μM). Columns 11–20, K_i (competitive inhibition constant of LnATP, in μM). Columns 21–30, K_d [dissociation constant of LnATP, estimated as in the text above with Eqs. (6) and (7)]. Columns 31–40, free [Mg^{2+}] (in mM). Columns 41–50, pH. Columns 51–60, sum of Na^+ and K^+ concentrations in experiments (in M). Columns 61–70, ionic strength (in M).

3. Data points (one line for each initial velocity, and the number must match that on the title line; F formats). Columns 1–10, initial velocity in any convenient units (V will have the same units), or $1/v$ if 1 is placed in column 20 of the title line. Columns 11–20, MgATP concentration in μM. This is the concentration predicted by the levels of ATP added and free Mg^{2+} present, and not the actual level after reaction of Mg^{2+} with LnATP. Columns 21–30, LnATP concentration in μM. This is the concentration of Ln and ATP added in equimolar amounts, and not the final level remaining after reaction with Mg^{2+}.

The program will process as many sets of data similar to that outlined in 1 to 3 above, and a blank field in columns 1–3 will then stop the program.

Data output consists of the following: (1) A listing of the equations fitted. (2) Title for the first data set, preceded by a counting number. (3) Listing of preliminary estimates read in. (4) Values of K, K_i, K_d, and V for each of the five cycles of iteration. These should show convergence. (5) A table of experimental data, calculated velocities from the least squares fit, and residuals, as well as the logarithms of calculated and experimental velocities, and residuals. (6) Fitted values of V (in the units used for velocity), K for MgATP, K_i for LnATP, K_d for LnATP, K_d for MgATP (all in μM), and the ratio of dissociation constants of LnATP and MgATP. (7) Variance and its square root, sigma (σ). (8) Items (2)–(7) will be repeated for each subsequent data set. (9) The statement PROGRAM COMPLETED.

When this program is used to determine LnADP dissociation constants, all references to MgATP and LnATP should be changed to MgADP and LnADP and the statement just after statement 105 should be changed to:

```
CKDMG = (1.+10.**(6.7-PH)+4.*SODIUM)*(10.**(1.5*CION))/12.
```

Appendix: Computer Code for Determining Apparent Dissociation Constants

```
c      Program LNATP
       DIMENSION V(100),A(100),CI(100),S(5,6),Q(5),SM(5),SS(5)
       write(6,100)
   100 FORMAT(70H Fit to LOG Y = LOG(V(B+2C/K+SQRT(B**2-4A*C))/(A*K+B+C/K
      1)/2    where     )
       write(6,101)
   101 FORMAT(40H A = (1+I/Ki)(1+(S+I)/Ki)+Kd*R/Ki          )
       write(6,102)
   102 FORMAT(40H B = S(1+(S+I)/Ki) - Kd*R(1-(S+I)/Ki)      )
       write(6,103)
   103 FORMAT(40H C = -Kd*R(S+I)        K is Km for MgATP. )
       write(6,104)
   104 FORMAT(84H S = MgATP, I = LnATP, Ki is for LnATP, Kd is for LnATP,
      1 R = free Mg/(Kd FOR MgATP)   /)
    11 FORMAT(I3,I17,48H  ANYTHING HERE WILL BE PRINTED DURING OUTPUT    )
     1 FORMAT(F10.3,11F10.5)
       JJ = 0
    14 READ(5,11) NP, NO
       IF(NP) 99, 99,12
    12 JJ = JJ + 1
       write(6,11) JJ, NO
       READ(5,1) CK, CKI, CKD, FMG, PH, SODIUM, CION
       write(6,1) CK, CKI, CKD, FMG, PH, SODIUM, CION
       write(6,105)
   105 FORMAT(40H      K        Ki         Kd          V   )
       CKDMG=  (1.+10.**(7.-PH)+17.*SODIUM)*(10.**(3.1*CION))/140.
       RM = FMG/CKDMG
       M = 1
       N = 4
       N1 = N + 1
       N2 = N + 2
       NT = 0
       GO TO 2
    15 READ(5,1) V(I), A(I), CI(I)
       IF (NO) 17,17,32
    32 V(I)= 1./V(I)
    17 SI = A(I) + CI(I)
       SIP = 1. + SI/CKI
       SIM = 1. - SI/CKI
       CA = (1.+CI(I)/CKI)*SIP + CKD*RM/CKI
       CB = A(I)*SIP - CKD*RM*SIM
       CC = -CKD*RM*SI
       SQ = SQRT(CB**2 - 4.*CA*CC)
       D = CA*CK + CB + CC/CK
       DN = CB + 2.*CC/CK + SQ
       Q(1) = 1.
       Q(2) = -2.*CC/CK**2/DN - (CA-CC/CK**2)/D
       DB = -SI*(A(I)+CKD*RM)/CKI**2
       DA = -(CI(I)+SI+CKD*RM)/CKI**2 - 2.*CI(I)*SI/CKI**3
       Q(3) = (DB + (CB*DB - 2.*CC*DA)/SQ)/DN - (CK*DA+DB)/D
       Q(4) = RM*(-SIM-2.*SI/CK+(-CB*SIM+2.*(1.+CI(I)/CKI)*SIP*SI+4.*CKD
      1 *RM*SI/CKI)/SQ)/DN - RM*(CK/CKI-SIM-SI/CK)/D
       Q(5) = LOG(V(I)*2.*D/DN)
       GO TO 13
    16 M = 2
    18 CV = EXP(S(1,1))
       CK = CK + S(2,1)
       CKI = CKI + S(3,1)
       CKD = CKD + S(4,1)
       write(6,1) CK, CKI, CKD, CV
       NT = NT + 1
       IF (NT - 5) 2, 87, 87
    87 S2 = 0
       write(6,81)
    81 FORMAT('    A conc    I conc    exptl v    calc v    Diff    exp
      1Log v calc Log v   Diff   1/exptl v  1/calc v     MgATP
```

```
  2 free Ln'/)
    DO 82 I = 1,NP
    SI = A(I) + CI(I)
    SIP = 1. + SI/CKI
    SIM = 1. - SI/CKI
    CA = (1.+CI(I)/CKI)*SIP + CKD*RM/CKI
    CB = A(I)*SIP - CKD*RM*SIM
    CC = -CKD*RM*SI
    D = CA*CK + CB + CC/CK
    DN = CB+  2.*CC/CK + SQRT(CB**2 - 4.*CA*CC)
    X1 = CV*DN/D/2.
    DX1 = V(I) - X1
    X2 = LOG(V(I))
    X3 = LOG(X1)
    DX2 = X2 - X3
    X4 = 1./V(I)
    X5 = 1./X1
    X=(A(I)-CKD*RM+SQRT((A(I)-CKD*RM)**2+4.*CKD*RM*SI))/2.
    FLN = X-A(I)
    write(6,1) A(I), CI(I), V(I), X1, DX1, X2, X3, DX2,X4,X5,X,FLN
 82 S2 = S2 + DX2**2
    P = NP - N
    S2 = S2/P
    S1 = SQRT (S2)
    DO 10 J = 2,N1
    DO 10 K = 1,N
 10 S(K,J) = S(K,J)*SM(K)*SM(J-1)
    SEV =S1*SQRT (S(1,2))*CV
    SEK =S1*SQRT (S(2,3))
    SEKI=S1*SQRT (S(3,4))
    SEKD=S1*SQRT (S(4,5))
    write(6,39) CV, SEV
    write(6,43) CK, SEK
    write(6,37) CKI, SEKI
    write(6,38) CKD, SEKD
    CKDMG = CKDMG*1000.
    RKD = CKD/CKDMG
    write(6,40) CKDMG, RKD
 39 FORMAT(6H  V = F12.6,15H   S.E.(V)   = F11.6)
 43 FORMAT(6H  K = F12.6,15H UM   S.E.(K) = F11.6)
 37 FORMAT(6H Ki = F12.6,16H UM   S.E.(Ki) = F11.6)
 38 FORMAT(6H Kd = F12.6,16H UM   S.E.(Kd) = F11.6)
 40 FORMAT(12H MgATP Kd = F8.3,30H micromolar.  Kd/(Kd MgATP) = F10.7)
    write(6,44) S2,S1
 44 FORMAT(12H VARIANCE = E14.5,13H    SIGMA =  F10.7///)
    GO TO 14
  2 DO 3 J = 1,N2
    DO 3 K = 1,N1
  3 S(K,J) = 0
    DO 4  I = 1,NP
    GO TO  (15,17), M
 13 DO 4 J = 1,N1
    DO 4 K = 1,N
  4 S(K,J) = S(K,J) + Q(K)  *Q(J)
    DO 5 K = 1,N
  5 SM(K) = 1./SQRT (S(K,K))
    SM(N1) = 1.
    DO 6 J = 1,N1
    DO 6 K = 1,N
  6 S(K,J) = S(K,J)*SM(K)*SM(J)
    SS(N1) = -1.
    S(1,N2) = 1.
    DO 8 L = 1,N
    DO 7 K = 1,N
  7 SS(K) = S(K,1)
    DO 8 J = 1,N1
```

```
      DO 8 K = 1,N
    8 S(K,J) = S(K+1,J+1) - SS(K+1)*S(1,J+1)/SS(1)
      DO 9 K = 1,N
    9 S(K,1) = S(K,1)*SM(K)
      GO TO (16,18), M
   36 FORMAT(18H PROGRAM COMPLETED//)
   99 write(6,36)
      STOP
      END
```

[8] Product Inhibition Applications

By BRUCE F. COOPER and FREDERICK B. RUDOLPH

Introduction

The study of inhibition of an enzyme using the products of the reaction is a natural extension of initial rate kinetic investigations and can provide valuable information about the kinetic mechanism of an enzyme. Generally, the orders of binding and release of reactants for sequential or Ping Pong mechanisms can be determined by product inhibition studies except when the product release is rate limiting (e.g., Iso and Theorell–Chance type mechanisms). With random binding order mechanisms, the technique may provide a distinction between steady-state and rapid equilibrium binding of substrates and products. Isomerization of stable enzyme forms that occur between the release of products and the binding of substrates can also be detected even in the case of single substrate and product reactions.[1] In addition, abortive complex formation is a common finding of these studies. An abortive complex refers to the reversible formation of an enzyme–substrate–product complex in which the presence of the product prevents binding of a substrate.[2] Because elucidation of the mechanisms of reactant binding provides much insight into the catalytic and physiological function of an enzyme, product inhibition studies should be included in any thorough kinetic investigation of an enzyme.

The information gained from product inhibition studies is varied and often unique. For aspartate kinase, studies revealed that the reaction had random addition of substrates in the forward direction but product release was ordered.[3] In the case of phenylalanine dehydrogenase, initial rate

[1] I. G. Darvey, *Biochem. J.* **128,** 383 (1972).
[2] A. D. Winer and G. W. Schwert, *Biochim. Biophys. Acta* **29,** 424 (1958).
[3] T. S. Angeles and R. E. Viola, *Arch. Biochem. Biophys.* **283,** 96 (1990).

analysis was consistent with random binding of substrates, but product inhibition studies of the reverse reaction revealed ordered binding of all reactants and identified the presence of abortive complexes.[4] The influence of activators or modifiers on the kinetic mechanism may also be revealed. When Mg^{2+} is present, the pyruvate kinase reaction has a random order of substrate addition as shown by product inhibition studies, but it shows ordered addition when Mn^{2+} is present.[5] Predictions for rational drug design resulted from kinetic investigations of human immunodeficiency virus (HIV) reverse transcriptase[6] when product inhibition data revealed two isomeric forms of the enzyme with different binding properties for nucleotides.

Information from product inhibition studies is valuable when the products of the reaction are present at levels *in vivo* similar to their inhibition constants (K_i values). This is particularly true for reactants that vary because of changes in metabolic or physiological state. The inhibition constants are obtained for the products in the same manner as for dead-end inhibitors, but the constants are usually a summation of several rate constants so they may not represent simple interaction of the product with the enzyme as is common for the dead-end inhibitors. Still, the K_i values determined are useful in comparing Haldane relationships for numerical solutions of the rate equation.[7]

Product inhibition analysis and theory have been reviewed extensively in a number of previous publications.[7-10] In this chapter we present more recent examples of its use and demonstrate a simple, generalized approach to the interpretation of results. Limitations of the technique, pitfalls to avoid, and suggestions for companion studies are made when necessary.

Approach and Overview

The use of products of the reaction to determine the binding order of reactants was first proposed by Alberty for bisubstrate systems,[11] and the

[4] H. Misono, J. Yonezawa, S. Nagata, and S. Nagasaki, *J. Bacteriol.* **171,** 30 (1989).
[5] N. Carvajal, R. Gonzalez, A. Moran, and A. M. Oyarce, *Comp. Biochem. Physiol. B: Comp. Biochem.* **82B,** 63 (1985).
[6] C. Majumdar, J. Abbotts, S. Broder, and S. H. Wilson, *J. Biol. Chem.* **263,** 15657 (1988).
[7] W. W. Cleland, *in* "The Enzymes" (P. D. Boyer, ed.), 3rd Ed., Vol. 2, p. 18. Academic Press, New York, 1970.
[8] H. J. Fromm, "Initial Rate Enzyme Kinetics." Springer-Verlag, Berlin and New York, 1975.
[9] S. A. Kuby, "A Study of Enzymes." CRC Press, Boca Raton and Ann Arbor, 1991.
[10] F. B. Rudolph, this series, Vol. 63, p. 411.
[11] R. A. Alberty, *J. Am. Chem. Soc.* **80,** 1777 (1958).

interpretations were quickly modified to include abortive complexes when the technique was first experimentally used with two dehydrogenases.[12,13] Product inhibition studies have historically relied on the interpretation of patterns generated from initial rate data in Lineweaver–Burk plots. Other plotting methods may be used as discussed below, but presentation with Lineweaver–Burk plots is readily understood by most bioscientists.

Specifically, the effects on the slopes ($K_{m_{app}}/V_{m_{app}}$) and intercepts ($1/V_{m_{app}}$) of the plots are key to the analysis. The reaction rate is determined as a substrate is varied about its K_m (typically 0.5 to 5 times K_m) at various fixed concentrations of a single product (including zero). Other substrates, if any, are held constant at nonsaturating (1–5 times K_m) or saturating levels (up to 100 times K_m). Substrate inhibition and other effects often can prevent achieving true saturation. The reaction can be investigated in either catalytic direction, and the ideal study involves varying each substrate at different levels of each product at both saturating and nonsaturating levels of any other substrate. The data are typically presented in the Lineweaver–Burk format for pattern analysis, and statistical fitting of the data is best achieved by software capable of properly weighting the reciprocal of the data.[8] The patterns of the lines, in particular the existence or the location of the intersections of the lines, are described in conventional kinetic terms: competitive, noncompetitive (often referred to as mixed, but no distinction should be made between the two terms), and uncompetitive. The type of inhibition caused by a product is defined toward the specific substrate being varied.

Competitive inhibition (*C*) is described when extrapolation of the varied substrate to infinite levels eliminates the inhibition by the product (Fig. 1a). In this case, any terms of the rate equation that contain the product also contain the varied substrate, resulting in only slope effects when the product is present. A common point of intersection of all lines on the *y* axis signifies competitive inhibition. These patterns occur when the product and the varied substrate bind to the same enzyme form(s).

Noncompetitive inhibition (*N*) results when the product binds to a different enzyme form than the varied substrate and the two enzyme forms are reversibly connected. Both the slope and the intercept of the plots are influenced by the presence of the product, and saturation with the varied substrate fails to relieve the inhibition by the product in this case. For sequential mechanisms, the lines generated from various fixed levels of product intersect at a common point to the left of the *y* axis (Fig. 1b). The location of the intersection on, above, or below the *x* axis may have

[12] H. J. Fromm and D. R. Nelson, *J. Biol. Chem.* **237,** 215 (1962).
[13] V. Zewe, H. J. Fromm, *J. Biol. Chem.* **237,** 1668 (1962).

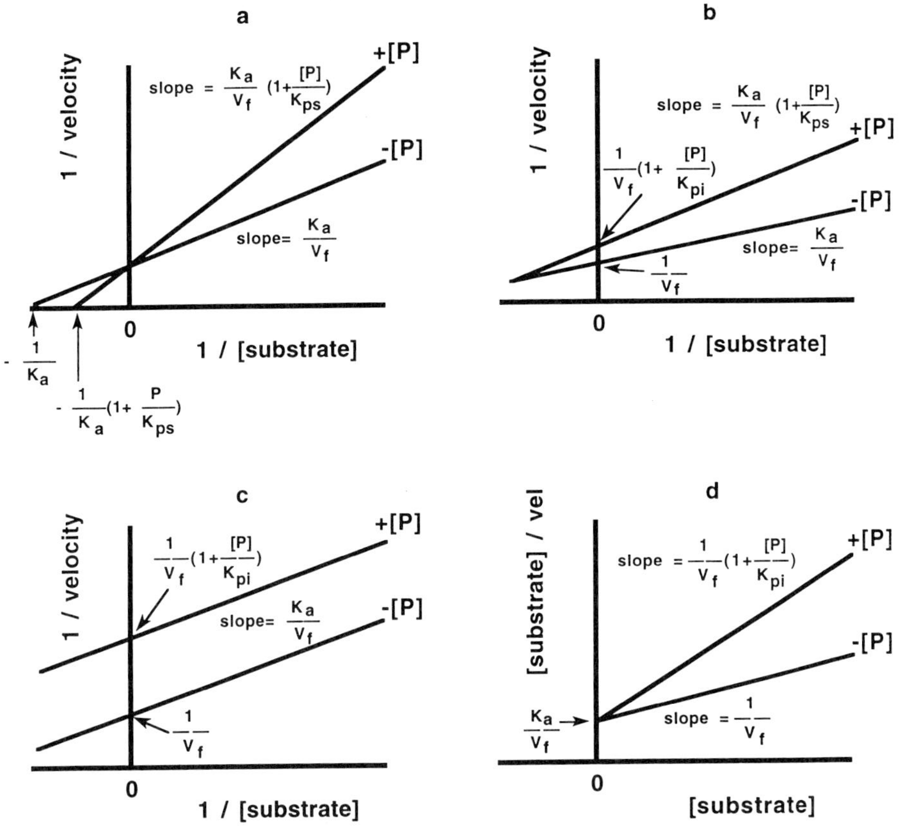

FIG. 1. Examples of plots generated when a product is present (+P) or absent (−P) from the reaction mixture. Patterns generated by competitive inhibition (a), noncompetitive inhibition (b), and uncompetitive inhibition (c) are described on a plot of the reciprocal of the initial velocity versus the reciprocal of the substrate concentration. K_a is the Michaelis constant for the substrate, and K_{ps} and K_{pi} are inhibition constants of the product from the slope and intercept terms, respectively. V_f is the maximum reaction velocity. (d) Patterns generated by uncompetitive inhibition when substrate concentration divided by the initial velocity ($[S]/v$) is plotted against the substrate concentration ([S]). Respective constants are labeled as in (c).

significance with respect to relative values of different rate constants but is not relevant to the interpretation of product inhibition data.

Uncompetitive interactions (*U*) yield patterns of parallel lines on Lineweaver–Burk plots (Fig. 1c). Terms of the rate equation that contain the product are independent of terms containing the varied substrate, and only the *y* intercept is affected. This pattern occurs when the product binds only after the varied substrate or when irreversible steps separate the binding of the product and the substrate. Irreversible steps can be the release of a product not present in the reaction solution (so its concentration is effectively zero), the binding of a substrate present at saturating levels (100 times K_m), or the isomerization of stable enzyme forms between the release of a product and the binding of the substrate. Parallel patterns are possibly the most definitive for diagnosing a mechanism, but, unfortunately, the determination of "parallel" lines also may be the most difficult pattern to prove. It can be argued that an observed pattern is actually noncompetitive with the lines intersecting far to the left of the axis. The term "nearly parallel" appears in the literature but has no diagnostic value (it is equivalent to the term "nearly pregnant"). A decision must be made to interpret the pattern as either parallel or intersecting.

Although the current interpretation is described for Lineweaver–Burk graphs, alternative plotting methods such as Hanes–Woolf ([S]/v versus [S]) or Eadie–Hofstee (v versus v/[S]) may be used to determine the effects on $K_{m_{app}}/V_{m_{app}}$ and $V_{m_{app}}$. The y intercept of the Hanes–Woolf method is described by $K_{m_{app}}/V_{m_{app}}$, which describes the slope of a Lineweaver–Burk plot. The Hanes–Woolf plot may be useful to determine if this term is constant because a common intercept is more easily observed than parallel lines (Fig. 1d). Of course, any plotting method requires accurate data points to be statistically valid. The problem of growing uncertainty at low concentrations and low velocity of points furthest from the axis in the double-reciprocal plot is also seen in the Hanes–Woolf plots except that this uncertainty now is near the axis where one may be trying to show a common intercept.[14] The Eadie–Hofstee plot gives V_m as the y intercept, the negative of K_m as the slope, and V_m/K_m as the x intercept. This plot magnifies nonlinearity that may not be seen in a double-reciprocal plot.

The graphical analysis of the data can yield further information about the binding and the number of enzyme species involved. Replots of the slope and intercepts determined from the initial inhibition plots should always be done. The curvatures of replots of the slope and/or intercept versus product concentration provide insight about the mechanism. Abortive complexes can generate hyperbolic replots by contributing alternative pathways in the scheme producing [P] terms in both the numerator and the denominator of the rate equation describing the enzyme behavior.[7] Parabolic replots can result if the product binds two or more reversibly connected enzyme forms in the steady state, or if more than one molecule of the product can bind to the enzyme.[7] The parabolic inhibition can be

[14] A. Cornish-Bowden, "Principles of Enzyme Kinetics." Butterworth, London and Boston, 1976.

readily shown using a Dixon plot of the reciprocal velocity versus the product concentration, especially if a wide range of product concentration is possible. A limitation is that linear plots may result if the binding constants for each enzyme form are similar even if more than one enzyme form is capable of binding the product. Although the analysis of replots is discussed often in the theoretical treatment of product inhibition, few examples are found in the literature where replots are analyzed for this information, even in cases where the replots are obviously nonlinear. The analysis of replots should be applied more often to decipher the mechanism. It really does not require new data, just better use of available information.

It is common to compile the qualitative findings from a study of product inhibition in a table presenting the type of interaction between each product and each substrate. Several sources present tables of predicted patterns for many different types of reactions.[7-10] Discrepancies among the tables result from different assumptions about the occurrence of abortive complexes and must be considered when using the tables. The apparent completeness of the tables covering a large number of possible reaction schemes can lead to the naive assumption that one merely completes a set of product inhibition experiments and reads the interpretation directly from the tables. Unfortunately, seldom do the data perfectly match predicted patterns. Many times abortive complexes must be considered, and attention must be given to the structure of related reactants[15] for the particular enzyme to properly interpret the results. However, the tables do demonstrate that the majority of possible mechanisms give unique patterns, and most remain unique even when abortive complexes are present. In practice, it is preferable to decipher the mechanism by analysis of the data to define the enzyme species that are likely present in the reaction pathway. A number of examples are given below that provide insight into the process of interpretation of product inhibition studies.

Application and Interpretation

The initial purpose of all kinetic studies, including product inhibition, is to identify enzyme species occurring during catalysis that contribute terms to the rate equation describing the particular reaction. When the results of product inhibition studies are combined with the information

[15] The term "reactants" refers to all of the compounds involved in the reaction including the products and substrates. Substrates are the reactants necessary for catalysis in the direction being studied, and the resulting compounds of the reaction are the products. These definitions are relative to the direction of the assay used for the reaction.

from the initial rate studies of the substrates, many terms of the rate equation for the enzyme are revealed. The terms are then correlated to enzyme forms that can be arranged into a reaction scheme. For some mechanisms, the assembly of the scheme is straightforward because all the species are defined by the kinetic studies. A number of specific examples for different types of enzymes follow.

One Substrate : Two Products Reactions

Elastase catalyzes a relatively simple model reaction involving the cleavage of a pseudosubstrate, 3-(2-furyl)acryloylglycyl-L-phenylalanyl-L-phenylalanine (faGly-Phe-Phe), yielding, the products faGly and Phe-Phe. A kinetic study was done with the enzyme involving variation of the substrate faGly-Phe-Phe about its K_m (0.5 to 5 times K_m) at various fixed concentrations of each of the products individually (including zero).[16] The product, faGly, was found to be a competitive inhibitor relative to faGly-Phe-Phe, whereas Phe-Phe was a noncompetitive inhibitor. The competitive interaction between substrate and faGly suggests that both ligands bind to the same enzyme species, indicating that E_{free}, E–faGly, and E–substrate forms are present. The noncompetitive effect of Phe-Phe indicates that this product binds a different enzyme form than the substrate. The scheme is completed with the addition of the central complex of (E–substrate—E–faGly–Phe-Phe), which is considered to be a single enzyme species for any initial rate kinetic study. The only mechanistic scheme that can be constructed with the indicated enzyme species is shown in Eq. (1):

$$\begin{array}{ccccccc} & \mathrm{A} & & \mathrm{P} & & \mathrm{Q} & \\ & \downarrow & & \uparrow & & \downarrow & \\ \hline \mathrm{E} & & \mathrm{(EA)} & & \mathrm{EQ} & & \mathrm{E} \\ & & \mathrm{(EPQ)} & & & & \end{array} \qquad (1)$$

A steady-state Ordered Uni Bi binding mechanism (Cleland notation) with Phe-Phe being the first product released is consistent with the data. The rate equation including the product terms is[8]

$$v = \frac{V_1V_2([\mathrm{A}] - [\mathrm{P}][\mathrm{Q}])}{V_2K_a + V_2[\mathrm{A}] + \dfrac{V_1K_q[\mathrm{P}]}{K_{eq}} + \dfrac{V_1K_p[\mathrm{Q}]}{K_{eq}} + \dfrac{V_2V_1[\mathrm{P}][\mathrm{Q}]}{K_{eq}} + \dfrac{V_2[\mathrm{A}][\mathrm{P}]}{K_{ip}}} \qquad (2)$$

For analysis any term containing the absent product is set equal to zero. Converting the equation to the reciprocal in the absence of P yields:

[16] L. Poncz, *Arch. Biochem. Biophys.* **266,** 508 (1988).

$$\frac{1}{v} = \frac{1}{V_1} + \frac{K_a}{V_1[A]} + \frac{K_p[Q]}{V_2K_{eq}[A]} \tag{3}$$

As [A] is extrapolated to infinity, the velocity dependence of [Q] is eliminated, yielding a common intercept on the y axis. In this case, and in general for sequential ordered mechanisms, the competitive interaction identifies the last product released and the first substrate bound as these compete for the same enzyme species, E_{free}.

The reciprocal rate equation when [Q] is zero is

$$\frac{1}{v} = \frac{1}{V_1} + \frac{K_a}{V_1[A]} + \frac{K_q[P]}{V_2K_{eq}[A]} + \frac{[P]}{V_1K_{ip}} \tag{4}$$

Here, even as [A] is taken to infinity, a term containing [P] remains to influence the V_m, resulting in both slope and intercept effects as [P] is increased, producing a noncompetitive interaction. The mechanism of elastase is assigned by substituting Phe-Phe for P and faGly for Q in the rate equation. Statistical analysis of the data fit to three possible mechanisms was used by the authors to assign the mechanism.

In this mechanism the product, P, decreases the amount of enzyme in the central complex by causing an increase in the steady-state levels of E–P–Q. By definition, steady-state binding results in concentration changes in the finite levels of each enzyme species as the availability of reactants changes. An increase in any reactant concentration can increase the concentration of a particular intermediate, thereby reducing the available productive enzyme in the central complex. This contrasts with the rapid equilibrium assumption which is satisfied if the reactants are released and bound many times before chemical conversion takes place. In steady state assumptions some enzyme species can be essentially zero depending on concentrations of available reactants. In rapid equilibrium the first product released in an ordered mechanism, such as with elastase, cannot cause inhibition because the release of the second product makes that step irreversible since no E–Q is now available.

Two other patterns of product inhibition are expected for Uni Bi mechanisms. Rapid equilibrium random binding of the products yields competitive interactions by both products, and steady-state random binding is indicated by two noncompetitive patterns.[10]

One should always look for alternative mechanisms that may be consistent with the data. The presence of abortive complexes can convert predicted competitive interactions to noncompetitive ones. Considering abortive complexes, the elastase data can be consistent with a rapid equilibrium random mechanism by the addition of one abortive complex. Reconsideration of the elastase example using rapid equilibrium assumptions allows another equation and scheme to be quickly constructed. For competitive interactions between faGly and substrate, all terms that contain [faGly] must also contain [substrate]. The noncompetitive interaction dictates that at least two terms contain [Phe-Phe], one with substrate and one without. The simplest rate equation in general terms is

$$\frac{1}{v} = \phi_1 + \frac{\phi_A}{[A]} + \frac{[P]}{\phi_p} + \frac{\phi_{qa}[Q]}{[A]} + \frac{\phi_{pa}[P]}{[A]} \tag{5}$$

A scheme consistent with Eq. (5) contains an abortive complex E–A–P:

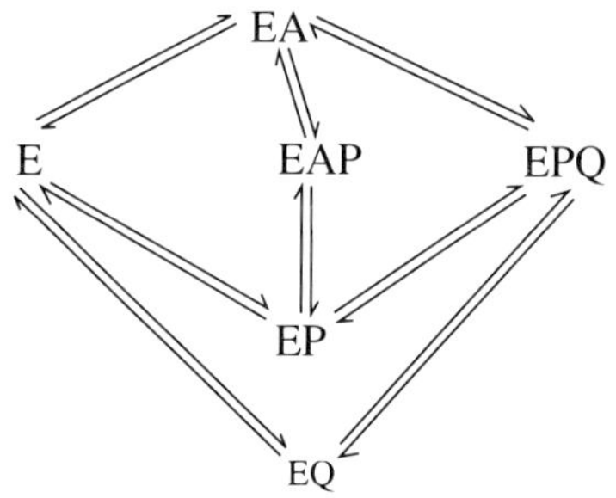

This abortive complex requires that Phe-Phe binds to the enzyme after the substrate is bound or that the substrate binds to the E–P complex and P must release before catalysis. Considering the similar structures of the two reactants, the abortive complex at first appears impossible. However, nonproductive substrate inhibition has been reported for elastase and was predicted to result from the binding of two molecules of substrate per elastase. This suggests the possibility of an E–substrate–Phe-Phe abortive complex that should be detected by nonlinearity of replots of the slope (K_m/V_m) versus the inhibitor concentration. In this case the alternative mechanism proves not to be applicable to elastase because linear replots were found consistent with the originally suggested steady-state mechanism. This type of analysis is necessary, however, whenever studying binding mechanisms since kinetic studies usually rule out possibilities but do not exclude the existence of other untested possibilities.

The above discussions are valid for Bi Uni reactions that can be analyzed in the reverse direction. Investigations of irreversible Bi Uni reactions are possible if the added product does not interfere with the detection method of the assay.

Two Substrates: Two Products Reactions

Two substrate: two product mechanisms present another layer of complexity owing to the presence of additional terms in the rate equation.

TABLE I
PATTERNS OF PRODUCT INHIBITION FOR BILIVERDIN REDUCTASE

Type of inhibition relative to	Product present	
	NADPH (P)	biliverdin (Q)
$NADP^+$ (A)	*C*[a]	*C*
Bilirubin (B)	*C*	*N*

[a] Competitive (*C*) and noncompetitive (*N*) interaction.

During the velocity measurements in the presence of one product, one substrate is held fixed either at nonsaturating or saturating levels while the other is varied in the region of its K_m. The experiment is then repeated with the substrate roles being reversed. These combinations can effectively quadruple the number of assays needed to investigate the inhibition by two products compared to the Uni Bi example. A complete set of data may be required to decipher some mechanisms, but, fortunately, a single set at either saturating or nonsaturating conditions is usually sufficient. The nonsaturating approach presents fewer obstacles to obtaining the data.

Product inhibition studies at nonsaturating levels of fixed substrate were used to elucidate the binding mechanism of biliverdin reductase.[17] In this reaction, bilirubin is reduced by $NAD(P)^+$ to form biliverdin and NAD(P)H. Initial rate studies with varied substrates indicated a sequential, non-Ping Pong mechanism. Product inhibition studies yielded the following results when the fixed substrate was held near its K_m value (Table I).

To construct a binding scheme, rapid equilibrium assumptions are initially applied to identify the minimum number of terms in the rate equation. The terms are then translated into the corresponding enzyme species that would exist in a given scheme. For many enzymes, such as in this example, this approach identifies the simplest mechanism consistent with the data. The nomenclature of Dalziel utilizing Ø values to represent combinations of rate constants associated with each term in a rate equation is used here for the sake of simplicity. One can determine $K_{m_{app}}$ values and V_{max} quite simply using Ø values without knowing the precise rate equation for a given binding scheme. Derivation of the rate equation from the derived binding scheme can be used to define the actual constants but qualitative interpretation is the initial concern.

[17] J. E. Bell and M. D. Maines, *Arch. Biochem. Biophys.* **263,** 1 (1988).

The initial rate studies of varying the two substrates relative to one another indicated that the following terms contribute to the rate equation and may result from the indicated enzyme species in Table II:

$$\frac{1}{v} = \frac{1}{V_m}\left(\phi_1 + \frac{\phi_a}{[A]} + \frac{\phi_b}{[B]} + \frac{\phi_{ab}}{[A][B]}\right) \quad (7)$$

Next, terms consistent with the product inhibition patterns are added. The competitive inhibition by bilirubin at varied levels of either substrate indicate that every term containing bilirubin also has both substrates ($\emptyset_{qab}$[Q]/[A][B]) because terms with only one of the substrates would yield noncompetitive behavior. The competitive interaction of $NADP^+$ versus NADPH suggests all [$NADP^+$]-containing terms also contain [NADPH]. The noncompetitive pattern for $NADP^+$ when biliverdin is the varied substrate indicates that terms with [$NADP^+$] are present with and without biliverdin. Together these patterns define the terms of [$NADP^+$]/[NADPH] ($\emptyset_{pa}$[P]/[A]) and [$NADP^+$]/[biliverdin] [NADPH] ($\emptyset_{pab}$[P]/[A][B]). A scheme containing all the indicated enzyme species is consistent with the random addition of substrates and products with formation of the abortive complex E–$NADP^+$–biliverdin:

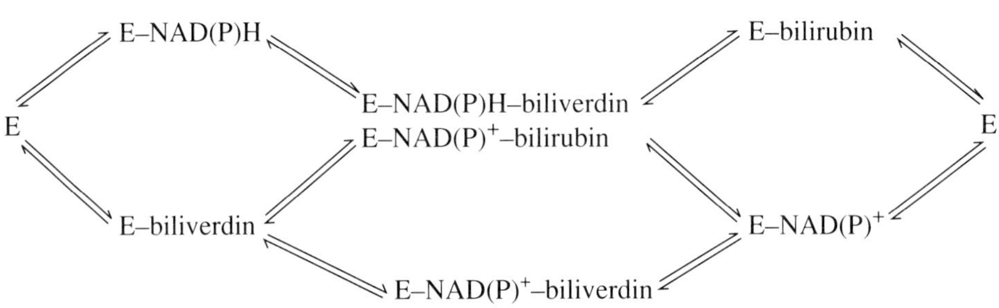

The presence of the proposed abortive complex E–$NADP^+$–biliverdin was supported by fluorescence studies that demonstrated the quenching of the enzyme-bound coenzyme analog etheno-$NADP^+$ by the addition of biliverdin. Hyperbolic replots of the slope of the Lineweaver–Burk plots for NADP inhibition versus 1/[$NADP^+$] also indicate the presence of the abortive complex.

Biliverdin reductase catalyzes an irreversible reaction, but this fact does not influence the interpretation of product inhibition studies. For example, orotate phosphoryltransferase yields identical product inhibition patterns of three competitive and one noncompetitive and catalyzes a

TABLE II
ENZYME SPECIES FOR BILIVERDIN REDUCTASE BY APPLICATION OF RAPID EQUILIBRIUM ASSUMPTIONS

Term	Enzyme species
From initial rate studies	
$\frac{\phi_{ab}}{[A][B]}$	E
$\frac{\phi_{a}}{[A]}$	E–B
$\frac{\phi_{b}}{[B]}$	E–A
ϕ_1	E–A–B
From product inhibition studies	
$\frac{\phi_{qab}[Q]}{[A][B]}$	E–Q
$\frac{\phi_{pab}[P]}{[A][B]}$	E–P
$\frac{\phi_{pa}[P]}{[A]}$	E–P–B

reversible reaction.[18] The mechanism of the phosphoryltransferase was investigated in both catalytic directions, and the rapid equilibrium random mechanism with one abortive complex, E–orotate–PP_i, was supported by the results of isotope exchange studies.

The abortive complexes formed in both these examples represent the classic abortive complex involving the substrate and product, each lacking the component transferred during the reaction. For the phosphoryltransferase reaction, a relatively bulky 5′-ribosyl group is absent in the abortive complex. For the biliverbin reductase only an absent hydride distinguishes the productive complex from the abortive complex. This type of complex is always predicted to appear in random binding mechanisms.[7] An abortive complex is not expected for the other reactant pair owing to overlap in the active site.

When two abortive complexes are present, additional studies may be required to characterize the binding mechanism fully. The assignment of a rapid equilibrium Random Bi Bi mechanism with two abortive complexes for glycerol-3-phosphate dehydrogenase required the use of competitive inhibitors.[19] Initial rate studies were consistent with a sequential mechanism, and the product inhibition results were consistent with a

[18] M. B. Bhatia, A. Vinitsky, and C. Grubmeyer, *Biochemistry* **29,** 10480 (1990).
[19] G. Klock and K. Kreuzberg, *Biochim. Biophys. Acta* **991,** 347 (1989).

TABLE III
PATTERNS OF PRODUCT INHIBITION FOR GLYCEROL-3-PHOSPHATE DEHYDROGENASE

Type of inhibition relative to	Product present	
	Glycerol phosphate (P)	NAD^+ (Q)
Dihydroxyacetone phosphate (A)	*C*[a]	*N*
NADH (B)	*N*	*C*

[a] Competitive (*C*) and noncompetitive (*N*) interaction.

Theorell–Chance mechanism or a rapid equilibrium Random Bi Bi mechanism with two abortive complexes (Table III). To distinguish between the two choices, competitive inhibitors to the two substrates were investigated at low and high levels of fixed substrate. If the product and the competitive inhibitor bind before the varied substrate and are connected by a reversible pathway, both will produce slope effects. However, if the product or the inhibitor binds after the varied substrate only the product will affect the slope.[7] In the case of the dehydrogenase, high levels of the fixed substrate eliminated the noncompetitive inhibition by the dead-end inhibitors, indicating that these inhibitors bind after the varied substrate. Only the rapid equilibrium random mechanism is consistent with the data, and the assignment of two abortive complexes completes the scheme. The dead-end inhibitor studies were necessary for the assignment because product inhibition studies at saturating levels of fixed substrate do not resolve a Theorell–Chance mechanism from the rapid equilibrium mechanism with two abortive complexes.[10]

The Ordered Bi Bi reaction mechanism for inosine 5′-monophosphate dehydrogenase was assigned from product inhibition studies and sup-

TABLE IV
PATTERNS OF PRODUCT INHIBITION FOR INOSINE 5′-MONOPHOSPHATE DEHYDROGENASE

Type of inhibition relative to	Product present	
	NADH (P)	XMP (Q)
IMP (A)	*N*[a]	*C*
NAD^+ (B)	*N*	*N*

[a] Competitive (*C*) and noncompetitive (*N*) interaction.

ported by dead-end inhibitor studies.[20] Product inhibition data were obtained at nonsaturating levels of the fixed substrate (Table IV). Rapid equilibrium assumptions applied as in the previous examples generate an unlikely abortive complex, E–NAD^+–NADH. Although both sites on the enzyme bind nucleotide substrates, no substrate inhibition is observed, indicating that this species likely does not form. The application of steady-state ordered binding assumptions predicts more reasonable enzyme species. The steady-state equation for an ordered Bi Bi reaction is

$$v = \frac{V_1V_2\left([\mathrm{A}][\mathrm{B}] - \dfrac{[\mathrm{P}][\mathrm{Q}]}{K_{\mathrm{eq}}}\right)}{\begin{array}{l} V_2K_{\mathrm{ia}}K_{\mathrm{b}} + V_2K_{\mathrm{b}}[\mathrm{A}] + V_2K_{\mathrm{a}}[\mathrm{B}] + \dfrac{V_1K_{\mathrm{q}}[\mathrm{P}]}{K_{\mathrm{eq}}} + \dfrac{V_1K_{\mathrm{p}}[\mathrm{Q}]}{K_{\mathrm{eq}}} + V_2[\mathrm{A}][\mathrm{B}] \\ \quad + \dfrac{V_1K_{\mathrm{q}}[\mathrm{A}][\mathrm{P}]}{K_{\mathrm{eq}}} + \dfrac{V_1[\mathrm{P}][\mathrm{Q}]}{K_{\mathrm{eq}}} + \dfrac{V_2K_{\mathrm{a}}[\mathrm{B}][\mathrm{Q}]}{K_{\mathrm{iq}}} + \dfrac{V_2[\mathrm{A}][\mathrm{B}][\mathrm{P}]}{K_{\mathrm{ip}}} + \dfrac{V_1[\mathrm{B}][\mathrm{P}][\mathrm{Q}]}{K_{\mathrm{ib}}K_{\mathrm{eq}}} \end{array}} \tag{9}$$

Equation (10) is produced by expressing Eq. (9) for the condition when only P is present and Q is absent. From this, [P] is predicted to always influence the intercept and to influence the slope at nonsaturating levels of B, leading to noncompetitive interactions of P with respect to both substrates:

$$\frac{1}{v} = \frac{1}{V_1}\left(1 + \frac{[\mathrm{P}]}{K_{\mathrm{ip}}}\right) + \frac{K_{\mathrm{a}}}{V_1[\mathrm{A}]} + \frac{K_{\mathrm{b}}}{V_1[\mathrm{B}]}\left(1 + \frac{K_{\mathrm{q}}[\mathrm{P}]}{K_{\mathrm{iq}}K_{\mathrm{p}}}\right) + \frac{K_{\mathrm{ia}}K_{\mathrm{b}}}{V_1[\mathrm{A}][\mathrm{B}]}\left(1 + \frac{K_{\mathrm{q}}[\mathrm{P}]}{K_{\mathrm{iq}}K_{\mathrm{p}}}\right) \tag{10}$$

When Q is present without P, all the terms containing [Q] also have [A], indicating that competitive inhibition by Q with respect to A should result. The inhibition will be noncompetitive with respect to B at nonsaturating levels of A:

$$\frac{1}{v} = \frac{1}{V_1} + \frac{K_{\mathrm{a}}}{V_1[\mathrm{A}]}\left(1 + \frac{[\mathrm{Q}]}{K_{\mathrm{iq}}}\right) + \frac{K_{\mathrm{b}}}{V_1[\mathrm{B}]} + \frac{K_{\mathrm{ia}}K_{\mathrm{b}}}{V_1[\mathrm{A}][\mathrm{B}]}\left(1 + \frac{[\mathrm{Q}]}{K_{\mathrm{iq}}}\right) \tag{11}$$

Steady-state ordered binding of IMP (A) before NAD^+ (B) followed by the release of NADH (P) prior to XMP (Q) is the simplest model consistent with the data. Steady-state assumptions allow a product to increase the concentration of the bound enzyme species and thereby reduce the available enzyme for conversion through the terniary complex even in the

[20] R. Verham, T. D. Meek, L. Hedstrom, and C. C. Wang, *Mol. Biochem. Parasitol.* **24,** 1 (1987).

absence of the other products. An effect on $V_{m_{app}}$ is produced, and noncompetitive patterns are generated.

The ordered mechanism for the dehydrogenase was supported by dead-end inhibitor studies at high and low levels of substrates. This example is a rare instance when comparison of the experimental results to a table of theoretical patterns yields a perfect match,[10] probably because the pattern of one competitive and three noncompetitive interactions remains unchanged even if abortive complexes are present. Abortive complexes may be detected by nonlinear replots if the binding constants to the two different enzyme species are significantly different.

The application of saturating and nonsaturating fixed substrate levels in product inhibition studies is also useful in diagnosing the mechanism. An inspection of the terms in Eq. (11) reveals that some of the product terms decrease to zero at saturating levels of the other substrate. This strategy depends on the feasibility of increasing the fixed substrate level to nearly 100 times the K_m. The traditional 10 times K_m for "saturation" in inital rate studies is certainly not always sufficient to mask the presence of terms in product inhibition studies.

An example of confusion caused by interpretations that exceed the limitations of the findings is found in a study of xylose reductase.[21] The initial rate studies for the enzyme that catalyzes the following reaction:

$$\text{NADH} + \text{xylose} \rightleftharpoons \text{NAD}^+ + \text{xylitol}$$

were consistent with a sequential binding mechanism, either random or ordered. A complete set of product inhibition results was obtained at nonsaturating fixed substrate levels, and a partial study was done at high fixed substrate concentrations (Table V). On the basis of these results, an Iso Bi Bi Ordered binding mechanism was reported for xylose reductase. The isomerization of a stable enzyme form between the release of the products and the substrate binding in the proposed mechanism created an irreversible step necessary to yield terms dependent only on the individual products. However, several limitations of product inhibition studies were not considered in the assignment of the mechanism of the reductase. A binding order for all the reactants was assigned based on the above data. Unfortunately, the Iso Bi Bi Ordered (and the Theorell–Chance) mechanisms only identify reactant pairs; the binding order cannot be defined by these studies because the rate equation is symmetrical for both products in these models.

[21] M. Rizzi, P. Erlemann, N.-A. Bui-Thanh, and H. Dellweg, *Appl. Microbiol. Biotechnol.* **29,** 148 (1988).

TABLE V
PATTERNS OF PRODUCT INHIBITION FOR XYLITOL REDUCTASE

Type of inhibition relative to	Product present	
	NAD$^+$ (P)	Xylitol
NADH (A)	C/N[a]	N $(U)_{\text{xylose}}$[b]
Xylose (B)	N $(U)_{\text{NADH}}$	N

[a] A distinction between noncompetitive and competitive was not made in the original paper for this data set. The intersection of three lines is on the y axis, and a fourth line intersects slightly to the left of the axis. Competitive inhibition interpretation is well within the experimental error of the data.

[b] Competitive (C), noncompetitive (N), and uncompetitive (U) interaction. Patterns in parentheses occur at saturating levels of the subscripted substrate.

Second, the fixed substrate at "saturating levels" was held at only 10 times the K_m level, and only a small change in the y intercept of the parallel lines was observed. The trend suggests that higher levels of fixed substrate may actually eliminate the apparent inhibition. Also, the apparent competitive interaction between NAD$^+$ and NADH at nonsaturating levels of xylose is inconsistent with the mechanism proposed. Furthermore, the initial rate studies of varying NADH at fixed xylose levels produced nonlinear Lineweaver–Burk plots. These questions make it clear that further analysis is required to define convincingly the binding mechanism for this enzyme.

Two Substrates: Three Products Reaction

Thus far, examples of two product reactions have been analyzed, and the problems associated with the analysis would suggest that more complex mechanisms involving three products would be difficult to evaluate with product inhibition. The following examples demonstrate the power and usefulness of the technique in solving even these complex mechanisms including the analysis of Ping Pong mechanisms. DNA ligase catalyzes the sealing of nicked DNA (nDNA) through an Uni Uni Bi Ordered Ping Pong mechanism, and product inhibition data were instrumental in assign-

TABLE VI
PATTERNS OF PRODUCT INHIBITION FOR DNA LIGASE

Type of inhibition relative to	Product present		
	PP_i (P)	AMP (Q)	sDNA (R)
nDNA (A)	*C*[a]	*U*	None
ATP (B)	*N*	*C*	None

[a] Competitive (*C*), noncompetitive (*N*), and uncompetitive (*U*) interaction. "None" indicates no inhibition was measured when the product was added to the reaction.

ing the reaction mechanism.[22] A Ping Pong mechanisms was suggested by the results of the initial rate studies involving the substrates, and the product inhibition results were interpreted for this type of mechanism. The three products PP_i, AMP, and sealed DNA (sDNA) were each investigated at nonsaturating levels of the fixed substrates (Table VI). As indicated, additional sealed DNA in the assay solution caused no inhibition versus either ATP or nicked DNA. The reaction is irreversible or very slow in the reverse direction.

A mechanism consistent with each part of the study can be assembled from the available patterns. The parallel Lineweaver–Burk plots observed in the initial rate studies suggests a Ping Pong mechanism and that the binding of ATP and nicked DNA are separated by an irreversible step that is likely the release of a product. Product inhibition results of a competitive interaction between ATP and AMP indicate that both bind to the same enzyme form. Pyrophosphate (PP_i) binding is reversibly connected to ATP binding in the absence of other products (noncompetitive PP_i versus ATP). Nicked DNA is separated from AMP binding by an irreversible step (uncompetitive AMP versus nDNA) and binds to the same enzyme form as PP_i (competitive PP_i versus nDNA); therefore, sealed DNA must be released after nDNA binds and before AMP is released:

$$\begin{array}{ccccccccc} \text{ATP} & & PP_i & & \text{nicked DNA} & & \text{sealed DNA} & & \text{AMP} \\ \downarrow & & \uparrow & & \downarrow & & \uparrow & & \uparrow \\ \hline \text{E} & \text{E–ATP} & & \text{E–AMP} & & \text{E–AMP–nDNA} & & \text{E*–AMP} & \text{E} \end{array} \quad (12)$$

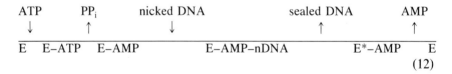

[22] H. Teraoka, M. Sawai, and K. Tsukada, *Biochim. Biophys. Acta* **747,** 117 (1983).

The scheme derived for mechanism (12) is consistent with both the kinetic and isotope exchange studies and is reasonable for the reaction that occurs. Two covalent enzyme–AMP complexes were isolated in concordance with the proposed mechanism.

Through the use of product inhibition, the true beauty of the mechanism is revealed. Because the product, sealed DNA, is naturally present in large abundance over nicked DNA, the enzyme must be active in the presence of a large excess of product (sDNA) over the substrate (nDNA). The Ping Pong mechanism creates a priming step in the formation of the E–AMP intermediate that binds only nicked DNA. Once the DNA is sealed and released, the release of AMP prevents the binding of sealed DNA that would inhibit the reaction. The competitive interaction of PP_i with nicked DNA suggests that the two E–AMP complexes are not equivalent. Therefore, in the absence (or in the presence of low levels) of AMP, no inhibition by sealed DNA can occur under what is likely the common intracellular condition.

In the case of an Ordered Bi Ter reaction, product inhibition can define the mechanism but not the order of the first two products released. A sequential Ordered Bi Ter mechanism with an abortive complex has been proposed for NAD^+–glutamate dehydrogenase[23] which catalyzes the following reaction:

$$NAD^+ + \text{glutamate} \rightleftharpoons NADH + NH_4^+ + \text{oxoglutarate}$$

The initial rate data when only the substrates were varied produced converging lines that intersected to the left of the y axis on a Lineweaver–Burk plot. Product inhibition studies were performed in both catalytic directions for all reactant combinations, and the inhibition between NAD^+ and NADH was found to be competitive. All other interactions were noncompetitive. An ordered mechanism with an abortive complex is consistent with these results. NAD^+ is the first substrate to bind and NADH the last product released, but the binding order for ammonia and oxoglutarate cannot be determined from these studies.

The use of competitive inhibitors provided the necessary insight to assign the order of substrate binding in the amination reaction.[23] Hydroxylamine is a competitive inhibitor relative to NH_4^+ but is uncompetitive relative to both NADH and oxoglutarate. The NH_4^+ must add last to be consistent with these findings, and this order is substantiated by a competitive inhibitor to oxoglutarate. An abortive complex of E–NAD^+–oxoglutarate coordinates the product inhibition data to the model.

[23] M. J. Bonete, M. L. Camacho, and E. Cadenas, *Biochim. Biophys. Acta* **990,** 150 (1989).

Three Substrates : Three Products Reactions

For a three substrates : three products system, the initial rate studies with the substrates usually provide evidence of sequential or Ping Pong mechanisms, and the interpretation of product inhibition studies is directed by this information. For indicated Ping Pong mechanisms, product inhibition studies can theoretically distinguish most conceivable Ping Pong binding mechanisms just using nonsaturating levels of fixed substrates.[10] The construction of a binding scheme from product inhibition patterns for a steady-state mechanism with random sequences in rapid equilibrium is achieved by applying the following rules[7]: (i) Competitive patterns occur when the varied substrate and the product bind to the same enzyme form(s). (ii) Noncompetitive interactions indicate that the product and the varied substrate bind different enzyme forms that are connected by a reversible pathway. (iii) Uncompetitive interactions arise when an irreversible step exists between the binding of the varied substrate and the product. Irreversible steps are created by the release of a substrate not present in the assay mixture, by the binding of a substrate present in saturating levels, or by a stable isomerization of the enzyme (e.g., Iso mechanisms). These rules are directly applicable to solve Ping Pong mechanisms as indicated in the previous DNA ligase example. When these results are combined with the initial rate data using the substrates, the binding mechanism is usually defined. Of course, exceptions may arise through the presence of abortive complexes that complicate the interpretation. If necessary, isotope exchange studies using partial reaction mixtures can provide strong evidence for the sequence of reactant binding and release in Ping Pong mechanisms.

Distinguishing among the sequential mechanisms requires information from product inhibition studies at saturating levels of fixed substrates to create unique patterns for each possible mechanism. The ability to attain information at saturating levels of substrates is severely limited not only by solubility and other effects but also by an increased possibility of multiple abortive complexes that may occur at high levels of substrates. Fortunately, competitive inhibition studies yield the same predicted results as for product inhibition in rapid equilibrium mechanisms without having the complication of abortive complex formation at saturating levels.[10] For sequential mechanisms with three substrates involving partially random mechanisms with ordered binding of one or two of the substrates, some of the decreased utility of product inhibition studies is balanced by the information available from the initial rate studies with the substrates. When some insight about the mechanism is available, product inhibition analysis is still useful to determine the presence of abortive complexes.

Thirty-six data sets would be required to investigate fully three substrate : three product reactions by product inhibition using the protocol as described for the two substrate reactions, but a strategic approach can significantly reduce the number of measurements needed to obtain the available information. Initially, nine sets of data are collected at nonsaturating levels of fixed substrates while each product is investigated versus each substrate. Competitive and uncompetitive interactions need not be investigated further, as these remain the same or the inhibition disappears as substrates are made saturating. Noncompetitive interactions may be investigated at saturating concentrations of both fixed substrates. Substrate–product interactions that convert to competitive or uncompetitive interactions can be analyzed at saturating levels of one of the fixed substrates and nonsaturating levels of the other to determine which substrate excluded or enhanced the binding of the product. Determining "no" inhibition is difficult, however, because to abolish the inhibition completely each fixed substrate may need to be raised to at least 100 times the K_m. These reaction conditions are typically impossible to attain, and several examples can be found in the literature where the scale of the plot was expanded to exaggerate a small degree of inhibition at 10 times the K_m. Interactions that tend toward reduced inhibition as the fixed substrate levels are increased should be investigated using competitive inhibitors to the substrate.

Theoretical manipulations consider the products R, Q, and P to be derived, respectively, from substrates A, B, and C and to bind at independent or separate sites. A Ter Ter Random rapid equilibrium binding mechanism with no abortive complexes yields competitive interaction between any substrate and any product.[7,8] The same binding mechanism considering abortive complexes results in competitive patterns for the companion parts and noncompetitive interactions for any other combination (competitive interactions; P versus A, Q versus B, and R versus C).[10]

A study in our laboratory on adenylosuccinate synthase illustrates the identification of reactant pairs from the interpretation of product inhibition patterns. Initial rate and isotope exchange studies are consistent with a random rapid equilibrium binding mechanism for the enzyme.[24] The synthase catalyzes the following reaction:

$$\text{IMP} + \text{L-aspartate} + \text{GTP} \rightleftharpoons \text{adenylosuccinate} + \text{GDP} + \text{P}_i$$

Note that substrates A (IMP) and B (aspartate) give rise to the same product, P (adenylosuccinate), whereas C (GTP) generates both products

[24] B. F. Cooper, H. J. Fromm, and F. B. Rudolph, *Biochemistry* **25,** 7323 (1986).

Q (GDP) and R (inorganic phosphate, P_i). Logically, the products Q and R are related reactants to the C site, and the presence of either or both may exclude C from binding. The product P may bind at one or both substrate sites for A and B and may be excluded by one or both substrates. If adenylosuccinate excludes both aspartate and IMP from the active site but has no effect on GTP binding and if GDP or P_i can prevent GTP from binding but has no effect on aspartate and IMP binding, the predicted patterns in Table VII would be generated. The noncompetitive interactions would result from the abortive complexes of E–IMP–Asp–GDP, E–IMP–Asp–P_i, and E–adenylosuccinate–GTP. The results of the experiment indicate that the additional abortive complexes of E–adenylosuccinate–Asp–GTP and E–IMP–Asp–GTP–P_i likely exist. This provides some information about the arrangement of the substrates in the active site. In other words, the binding of the succinyl portion of adenylosuccinate binds weakly in the aspartate site and does not exclude aspartate. Conversely, aspartate does not exclude the binding of adenylosuccinate. P_i appears not to exclude the binding of GTP, suggesting that the γ-phosphate of the nucleotide supplies little energy in the binding of GTP, although the relatively weak inhibition by P_i (K_i 8 mM) may arise from nonspecific effects or interference at the IMP binding site.

An alternative method for investigating reactions with three or more products involves providing two products during the reaction. This approach can be used whenever the addition of two products does not supply all the reactants for the reverse reaction. The presence of the second product may establish a reversible connection that can be diagnostic for ordered mechanisms. Abortive complexes involving more than one product also may be identified in this manner.

TABLE VII
PREDICTED AND DETERMINED PATTERNS OF PRODUCT INHIBITION FOR ADENYLOSUCCINATE

Type of inhibition relative to	Product present					
	Predicted			Determined[a]		
	AMPS[b] (P)	GDP (Q)	P_i (R)	AMPS (P)	GDP (Q)	P_i (R)
IMP (A)	*C*[b]	*N*	*N*	*C*	*N*	*N*
Aspartate (B)	*C*	*N*	*N*	*N*	*N*	*N*
GTP (C)	*N*	*C*	*C*	*N*	*C*	*N*

[a] B. F. Cooper and F. B. Rudolph unpublished results (1993).

[b] Adenylosuccinate (AMPS) and interactions characterized as competitive (*C*) and noncompetitive (*N*).

The presence of two products indicated the formation of an abortive complex in the binding mechanism of octopine dehydrogenase.[25] This enzyme catalyzes the following reaction:

$$\text{D}(+)\text{-Octopine} + \text{NAD}^+ + \text{H}_2\text{O} \rightleftharpoons \text{L-arginine} + \text{pyruvic acid} + \text{NADH} + \text{H}^+$$

Product inhibition data were collected in the usual manner except that a fixed low level of a second product (<1 times K_i) was added to the assay solution. When the products were added individually, linear replots of slope and intercept were produced. When arginine and pyruvate were both present the replot of the slopes were parabolic, whereas the replots of the intercepts remain linear. This effect on the slopes indicated that an inhibitory complex formed with the simultaneous binding of L-arginine and pyruvate to the enzyme forms that bind octopine, that is, the enzyme–NAD^+ complex, free enzyme, or both. Similar findings have been observed with other keto acid, amine dehydrogenases.[26–28]

To determine if the binding of one or more products is mutually exclusive, two products P and Q can be varied as the substrates are held constant.[29] A plot of $1/v$ versus [P] at various levels of Q (including zero) is constructed; if the binding is mutually exclusive, then parallel lines will result. The intersection of the lines above or below the axis indicates that Q raises or lowers the dissociation constant of P, respectively. Intersection on the y axis indicates that one product has no effect on the binding of the other.

Overview of Hurdles

The preceding examples are representative investigations of enzymatic reactions that employed product inhibition to analyze the binding mechanism. For some enzymes a complete product inhibition study is impossible owing to insoluble or unstable reactants or other assay difficulties, but even partial studies can provide information about the mechanism.[3,30,31] When substrate or product levels must be raised to millimolar ranges to achieve inhibition or saturation, dead-end inhibition, nonspecific binding, and ionic strength effects may further complicate the interpretation of the

[25] J. L. Schrimsher and K. B. Taylor, *Biochemistry* **23,** 1348 (1984).

[26] M. Fujioka, *J. Biol. Chem.* **250,** 8986 (1975).

[27] C. E. Grimshaw and W. W. Cleland, *Biochemistry* **20,** 5650 (1981).

[28] J. E. Rife and W. W. Cleland, *Biochemistry* **19,** 2321 (1980).

[29] T. Yonetoni and H. Theorell, *Arch. Biochem. Biophys.* **106,** 243 (1964).

[30] K. M. Bohren, J.-P. Von Wartburg, and B. Wermuth, *Biochem. J.* **244,** 165 (1987).

[31] J. S. Campbell and H. J. Karavolas, *J. Steroid Biochem.* **32,** 283 (1989).

studies. In practice, saturation is difficult to achieve for concentrations up to 100 times the K_m that may be necessary to achieve full saturation in product inhibition studies. Applications in the literature demonstrate that the traditional 10 times K_m definition of saturation for initial rate studies is not always sufficient to "saturate" product inhibition terms.

Reactions monitored with coupled assays usually preclude the addition of at least one product for inhibition studies. However, investigating the remaining products may clarify or define a possible mechanism. For putidaredoxin reductase the initial rate studies were ambiguous and reported to "Ping Pong like."[32] The noncompetitive results of product inhibition tipped the scale toward a sequential mechanism even though only one product could be investigated. Other alternatives for investigations employing coupled assays include designing alternative coupled systems to allow for the analysis of each product as was accomplished for hexokinases.[33]

The next hurdle to interpreting the studies is the analysis of the data. More than one example of questionably interpreted patterns can be found in the literature. When the primary data contain excessive scatter, the assignment of parallel or competitive interactions is difficult. Other problems include nonlinear effects from contamination of the reagents, substrate inhibition, and/or steady-state influences. A substrate or product also can act as an activator for the enzyme and complicate the interpretation.[34] Any of these problems can result in the wrong assignment of patterns in product inhibition studies and mislead the characterization of the mechanism(s).

Summary

A product inhibition study provides important insight into the binding mechanism of an enzyme, especially in the identification of abortive complexes, but seldom is it the only tool required to solve the mechanism completely. Always keep in mind that more than one mechanism may be consistent with the patterns, and several alternative schemes should be analyzed by including abortive complexes and, as a last resort, isomerization steps or slow product release steps in the interpretation. Support for the proposed mechanism should be garnered from other data, such as kinetic studies with alternative substrates and competitive inhibitors, positional and molecular isotope exchange studies, binding studies, and iso-

[32] P. W. Roome and J. A. Peterson, *Arch. Biochem. Biophys.* **266,** 41 (1988).
[33] W. T. Jenkins and C. C. Thompson, *Anal. Biochem.* **177,** 396 (1989).
[34] C. N. Cronin and K. F. Tipton, *Biochem. J.* **245,** 13 (1987).

tope effects. A well-characterized binding mechanism complete with the identification of abortive complexes is even more important as rational drug design becomes more prevalent. Product inhibition studies represent an important tool that is relatively easy to apply to gain significant information about the binding mechanism of most enzymes.

Acknowledgments

This work was supported by grants from the National Institutes of Health (GM42436) and from the Robert A. Welch Foundation (C1041).

[9] Kinetics of Iso Mechanisms

By KAREN L. REBHOLZ and DEXTER B. NORTHROP

Introduction

Many enzymes begin the chemistry of catalysis in one form and finish in a different form. To complete the catalytic cycle, they must return to the beginning form. Among them are enzymes engaging in general acid or general base catalysis that require reprotonations, enzymes catalyzing hydrolytic reactions that require rehydration, and isomerases that must assume different conformations to bind different isomers. Some enzymes designated as hysteretic[1] or mnemonic[2] also undergo a change in form. The process of converting the product form of an enzyme back to the substrate form falls under the broad classification of "isomerization of free enzyme," and enzymes which isomerize slowly enough to influence the rate of catalytic turnovers are said to have "Iso mechanisms."

The theoretical possibility of Iso mechanisms has been discussed for many years,[3-10] but there exists relatively little experimental evidence for

[1] N. Citri, *Adv. Enzymol.* **37,** 398 (1973).
[2] T. Keleti, J. Ovádi, and J. Batke, *J. Mol. Catal.* **1,** 173 (1975/1976).
[3] G. Medwedew, *Enzymologia* **2,** 53 (1937).
[4] E. L. King, *J. Phys. Chem.* **60,** 1378 (1956).
[5] J. Botts, *Trans. Faraday Soc.* **54,** 593 (1958).
[6] W. W. Cleland, *Biochim. Biophys. Acta* **67,** 104 (1963).
[7] H. G. Britton, *Arch. Biochem. Biophys.* **117,** 167 (1966).
[8] H. J. Fromm, "Initial Rate Enzyme Kinetics," Chap. 7. Springer-Verlag, New York, 1975.
[9] J. Albery and J. R. Knowles, *J. Theor. Biol.* **124,** 137 (1987).
[10] J. Albery and J. R. Knowles, *J. Theor. Biol.* **124,** 173 (1987).

this type of mechanism.[11] Furthermore, the possibility of a significant Iso step is considered only rarely in kinetic characterizations of specific enzyme mechanisms. The evidence is scant, not necessarily because this type of mechanism is uncommon, but because there were no simple methods to detect isomerizations of free enzyme. Britton [7] devised a brilliant kinetic experiment which he termed "induced transport" that can provide positive proof* for an Iso mechanism, but was unlucky in its application for he never found a good example.[12] The failure to consider an Iso step in the kinetic characterizations of particular enzymes is a fundamental oversight of our time.

The experimental methods which allow the detection and characterization of Iso mechanisms are developed most thoroughly for Uni Uni kinetic mechanisms (i.e., single-substrate and single-product reactions) but presently are being extended to more complex reactions, such as Uni Bi. The kinetic methods that are the focus of this chapter are product inhibition, dead-end inhibition, induced transport, and isotope effects. However, many other kinetic designs may provide useful clues for the detection of Iso mechanisms and may develop into specific methods for characterizing the nature of isomerizations. As examples, the many puzzling kinetic findings about fumarase (fumarate hydratase) which were not understood at the time can now be reconciled with a new Iso mechanism: biphasic Arrhenius plots,[13] anomalous progress curves,[14] shifts in pK values of maximal velocities,[15] unusual effects of viscosity on maximal velocities,[16] and biphasic effects of pressure.[17] As alternative experimental designs, Darvey derived equations for reaction equilibria measured at high concen-

[11] R. T. Raines and J. R. Knowles, *Biochemistry* **26,** 7014 (1987).

* Britton did postulate that interaction between two active sites could also result in countertransport. Nevertheless, the interacting sites still require that substrate and product bind to different forms of enzyme, even if not "free" enzyme. Hence, induced transport finally puts to rest the axiom that "you can't prove anything by kinetics, you can only disprove" or, as recently stated (D. Voet and J. G. Voet, "Biochemistry," p. 340, Wiley, New York, 1990.), "The steady-state kinetic analysis of a reaction cannot unambiguously establish its mechanism ... because ... there are an infinite number of alternate mechanisms ... that can account for these kinetic data equally well." This is not so. A positive result in the induced transport experiment *requires* a kinetically significant isomerization between substrate and product forms of enzyme; a negative result does not eliminate the possibility of isomerizations but *requires* that they not be kinetically significant.

[12] L. M. Fisher, J. W. Albery, and J. R. Knowles, *Biochemistry* **25,** 2529 (1986).

[13] V. Massey, *Biochem. J.* **53,** 72 (1953).

[14] M. Taraszka and R. A. Alberty, *J. Phys. Chem.* **68,** 3368 (1964).

[15] J. S. Blanchard and W. W. Cleland, *Biochemistry* **19,** 4506 (1980).

[16] W. L. Sweet and J. S. Blanchard, *Arch. Biochem. Biophys.* **277,** 196 (1990).

[17] P. Butz, K. O. Greulich, and H. Ludwig, *Biochemistry* **27,** 1556 (1988).

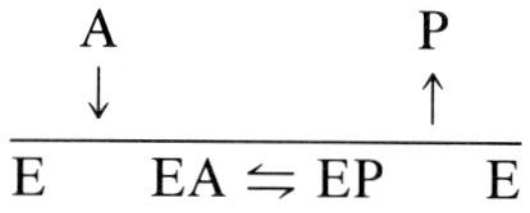

SCHEME I

trations of enzyme,[18] isotopic exchange at equilibrium,[19] and transient-state kinetics[20] and contrasted differences between the presence and absence of isomerizations of free enzyme. However, all of the Darvey methods require that rates of isomerizations of free enzyme be similar to or less than rates of conversions of substrate to product, and to our knowledge his designs have not been applied successfully.

Kinetic Description of Reversible Iso Uni Uni Reactions

A Uni Uni mechanism is diagrammed in the Cleland shorthand notation in Scheme I. It is governed by a steady-state rate equation of the form:

$$v_0 = \frac{\frac{V_f}{K_a}[A] - \frac{V_r}{K_p}[P]}{1 + \frac{[A]}{K_a} + \frac{[P]}{K_p}} \tag{1}$$

An Iso Uni Uni mechanism contains an Iso segment (F → E) in addition to the chemical segment (EA → P + F, Scheme II). It is governed by a steady-state rate equation of the form

$$v = \frac{V_f V_r\left([A] - \frac{[P]}{K_{eq}}\right)}{K_a V_r + V_r[A] + \frac{V_f[P]}{K_{eq}} + \frac{V_r[A][P]}{K_{iip}}} \tag{2}$$

The $[A][P]/K_{iip}$ term is the only term in Eq. (2) that is unique to the Iso Uni Uni mechanism, but when the velocity is represented as the flux in the forward direction minus the flux in the reverse direction,[7,21] an additional term, α_{iso}, is present:

[18] I. G. Darvey, *J. Theor. Biol.* **41,** 441 (1973).
[19] I. G. Darvey, *J. Theor. Biol.* **49,** 201 (1975).
[20] I. G. Darvey, *J. Theor. Biol.* **65,** 465 (1977).
[21] H. G. Britton, *Biochem. J.* **133,** 255 (1973).

$$\begin{array}{ccc} & \text{A} & & \text{P} & \\ & \downarrow & & \uparrow & \\ \hline \text{E} & & \text{EA} \leftrightharpoons \text{EP} & & \text{F} \leftrightharpoons \text{E} \end{array}$$

SCHEME II

$$v = \left[\frac{\frac{V_f}{K_a}\left(1 + \frac{\alpha_{iso}[P]}{K_{eq}}\right)[A]}{1 + \frac{[A]}{K_a} + \frac{[P]}{K_p} + \frac{[A][P]}{K_{iip}K_a}} \right] - \left[\frac{\frac{V_r}{K_p}(1 + \alpha_{iso}[A])\,[P]}{1 + \frac{[A]}{K_a} + \frac{[P]}{K_p} + \frac{[A][P]}{K_{iip}K_a}} \right] \quad (3)$$

Britton's α_{iso} was expressed directly in the magnitude of equilibrium perturbations of isotopic equilibria, but comparisons between Eqs. (2) and (3) require that α_{iso} be redundant, that it be a collection of kinetic parameters [see Eq. (20) below]. King[4] realized that with an Iso mechanism at high substrate concentration, F $\rightarrow$ E could be rate determining, which would lead to the peculiar situation that at saturation there would be no substrate on the enzyme. At the other extreme, at low substrate concentration, F $\rightleftharpoons$ E approaches equilibrium; the Iso step ceases to have any kinetic significance, and the chemical segment becomes all-important. Regardless of how rate determining the isomerizations are, both Eqs. (1) and (2) reduce to Eq. (4) under initial velocity conditions in the absence of added product:

$$v_0 = \frac{V_f[A_o]}{K_a + [A_0]} \quad (4)$$

Thus, the two mechanisms cannot be distinguished by measuring initial velocities as a function of substrate concentration, despite the fact that the presence of an Iso step does have specific effects on reaction rates and these effects change as a function of substrate concentration.

Definition of Kinetic Constants

Multistep Reversible Iso Uni Uni Reactions

Realistically, an isomerization segment should be considered to have more than the single step (and a chemical segment more than the two steps) shown in Scheme II, in order to ensure that rate equations assume

$$\mathrm{E} + \mathrm{A} \underset{k_2}{\overset{k_1}{\rightleftharpoons}} \mathrm{EA} \underset{k_4}{\overset{k_3}{\rightleftharpoons}} \mathrm{FP} \underset{k_6}{\overset{k_5}{\rightleftharpoons}} \mathrm{P} + \mathrm{F} \underset{k_8}{\overset{k_7}{\rightleftharpoons}} \mathrm{G} \underset{k_{10}}{\overset{k_9}{\rightleftharpoons}} \mathrm{E}$$

SCHEME III

a general and complete form.* Consequently, the derivations which follow consider a minimal Iso mechanism to have two steps in the isomerization segment, and the results are generalized to represent isomerization segments of any number of steps (Scheme III). The derivations assume that the steps represented in the Iso segment occur after release of the last product. The nomenclature used to designate the different forms of enzyme in a multistep Iso segment is derived from the standard suggested by Cleland[6]: the form of enzyme that binds substrate is labeled E, the form of enzyme that initially is present after product release is labeled F, and the steps from F to E are labeled consecutively with letters starting with G (excluding I). Catalytic turnovers are represented by a maximal velocity (e.g., V_f), in part for algebraic simplicity, but are equivalent to $V/[E_t]$ (or k_{cat}) in this discussion. Reference to the concentration of enzyme active sites was omitted in part to emphasize that the measures of rate limitation by the isomerization segment, represented by Eqs. (42)–(44) below, are independent of the amount of enzyme present in an experiment.

The concept of a rate-limiting step has been much abused and lacks an accepted definition.[22,23] Following Johnston[24] and Boyd,[25] the concept as used here is restricted to reaction processes which are irreversible under initial velocity conditions at saturating substrate and can be described by a single, apparent first-order rate constant. Because these processes may consist of multiple steps, they are referred to as reaction segments. For example, Scheme III has five steps, but only two segments: the chemical segment and the isomerization segment, rendered irreversible by zero product and infinite substrate, respectively. How rate limiting each segment is will be proportional to the amount of enzyme participating in it during steady-state turnovers at saturating concentrations of substrates, and equally proportional to the reciprocal of the net rate constant govern-

* For example, see Northrop [D. B. Northrop, *Biochemistry* **20,** 4056 (1981)] for a discussion of the importance considering multiple steps of chemical segments in the rate equations for isotope effects.

[22] D. B. Northrop, *Biochemistry* **20,** 4056 (1981).

[23] W. J. Ray, Jr. *Biochemistry* **22,** 4625 (1983).

[24] H. S. Johnston, "Gas Phase Reaction Theory," Chap. 16. Roland, New York, 1966.

[25] R. K. Boyd, *J. Chem. Educ.* **55,** 84 (1978).

ing the segment, which provides a definition and means of identification and quantitation.

The forward chemical segment (i.e., EA → F + P) of the reaction can be described by an apparent rate constant, k_{fchem}, composed of the reciprocal of the sum of reciprocals of net rate constants[26] (indicated by primes) for that segment, $k_3' = k_3k_5/(k_4 + k_5)$ and $k_5' = k_5$:

$$k_{\text{fchem}} = \frac{1}{1/k_3' + 1/k_5'} = \frac{k_3k_5}{k_3 + k_4 + k_5} \tag{5}$$

The forward isomerization segment can be described similarly by an apparent rate constant, k_{fiso}, also composed of the reciprocal of the sum of reciprocals of net rate constants for that segment;

$$k_{\text{fiso}} = \frac{1}{1/k_7' + 1/k_9'} = \frac{k_7k_9}{k_7 + k_8 + k_9} \tag{6}$$

Maximal velocities are determined by reciprocals of sums reciprocals of the apparent rate constants of the segments:

$$V_{\text{f}} = \frac{1}{\dfrac{1}{k_{\text{fchem}}} + \dfrac{1}{k_{\text{fiso}}}} \tag{7}$$

The apparent rate constants may be replaced by the component net rate constants[26] (indicated by primes):

$$V_{\text{f}} = \frac{1}{\dfrac{1}{k_3'} + \dfrac{1}{k_5'} + \dfrac{1}{k_7'} + \dfrac{1}{k_9'}} \tag{8}$$

or by the individual rate constants of Scheme III:

$$V_{\text{f}} = \frac{k_3k_5k_7k_9}{k_3k_5(k_7 + k_8 + k_9) + k_7k_9(k_3 + k_4 + k_5)} \tag{9}$$

Similarly, the maximum velocity in the reverse direction is labeled V_{r} and is comprised of k_{rchem} and k_{riso}:

$$V_{\text{r}} = \frac{1}{\dfrac{1}{k_{\text{rchem}}} + \dfrac{1}{k_{\text{riso}}}} = \frac{1}{\dfrac{1}{k_2'} + \dfrac{1}{k_4'} + \dfrac{1}{k_8'} + \dfrac{1}{k_{10}'}} = \frac{1}{\dfrac{k_2 + k_3 + k_4}{k_2k_4} + \dfrac{k_8 + k_9 + k_{10}}{k_8k_{10}}}$$
$$= \frac{k_2k_4k_8k_{10}}{k_2k_4(k_8 + k_9 + k_{10}) + k_8k_{10}(k_2 + k_3 + k_4)} \tag{10}$$

[26] W. W. Cleland, *Biochemistry* **14,** 3220 (1975).

The effect of the isomerization segment on the definitions of V_f and V_r is a *kinetic* function. In contrast, the effect of the isomerization segment on the definitions of V_f/K_a and V_r/K_p is a *thermodynamic* function, because V/K expresses catalysis at extremely low concentrations of substrate and isomerizations of free enzyme are at equilibrium. The fraction of free enzyme present as form E, termed f_E, which can react with substrate is described by

$$f_E = \frac{k_7k_9}{k_7k_9 + k_7k_{10} + k_8k_{10}} \tag{11}$$

Similarly, the fraction present as form F, termed f_F, which can react with product is

$$f_F = \frac{k_8k_{10}}{k_7k_9 + k_7k_{10} + k_8k_{10}} \tag{12}$$

V_f/K_a is defined as the product of k_1' and f_E:

$$\frac{V_f}{K_a} = k_1'f_E = \frac{k_1k_3k_5k_7k_9}{(k_2k_4 + k_2k_5 + k_3k_5)(k_7k_9 + k_7k_{10} + k_8k_{10})} \tag{13}$$

Similarly, V_r/K_p is defined as the product of k_6' and f_F:

$$\frac{V_r}{K_p} = k_6'f_F = \frac{k_2k_4k_6k_8k_{10}}{(k_2k_4 + k_2k_5 + k_3k_5)(k_7k_9 + k_7k_{10} + k_8k_{10})} \tag{14}$$

A distinction must be made between the equilibrium constant for the entire reaction, K_{eq}, and the equilibrium constant for the isomerization segment, $K_{eq(E/F)}$. K_{eq} contains the rate constants from both the chemical and the isomerization segments of the reaction as expressed in the Haldane relationship:

$$K_{eq} = \frac{V_fK_p}{K_aV_r} = \frac{k_1k_3k_5k_7k_9}{k_2k_4k_6k_8k_{10}} \tag{15}$$

$K_{eq(E/F)}$ contains only the rate constants in the isomerization segment:

$$K_{eq(E/F)} = \frac{f_E}{f_F} = \frac{k_7k_9}{k_8k_{10}} \tag{16}$$

K_{iip}, the noncompetitive inhibition term for the product, is defined as follows;

$$K_{iip} = \frac{k_3k_5(k_7 + k_8 + k_9) + k_7k_9(k_3 + k_4 + k_5)}{k_6(k_3 + k_4)(k_8 + k_9)} \tag{17}$$

A similar noncompetitive inhibition term, K_{iia}, is found in the equations of the reverse reaction and is related to K_{iip} by a second Haldane relationship, unique to Iso mechanisms:

$$K_{eq} = \frac{V_f K_{iip}}{V_r K_{iia}} \tag{18}$$

Britton's α is related to K_{iip} as shown in Eq. (19)[21]:

$$K_{iip} = \frac{V_f K_p}{\alpha K_a (V_f + V_r - \alpha K_a V_r)} \tag{19}$$

Equation (19) is quadratic in α as shown on rearrangement:

$$\alpha^2 \left(\frac{K_a}{V_f}\right) - \alpha \left(\frac{1}{V_f} + \frac{1}{V_r}\right) + \frac{K_{eq}}{K_{iip} V_f} = 0 \tag{20}$$

When Eq. (20) is solved for α using V_f, V_r, K_a, K_{eq}, and K_{iip}, two roots are obtained, and both roots can be used to estimate the kinetic importance of the isomerization segment of an enzymatic reaction.

The first root was recognized by Britton[21] and is here termed α_{iso} because of its relation to the isomerization segment as shown in Eq. (21):

$$\begin{aligned} \alpha_{iso} &= \frac{V_f}{K_a}\left(\frac{1}{k_{fiso}} + \frac{1}{k_{riso}}\right) \\ &= \frac{k_1'}{k_{10}'} = \frac{k_1 k_3 k_5 (k_8 + k_9)}{k_8 k_{10}(k_2 k_4 + k_2 k_5 + k_3 k_5)} = \frac{k_2 k_4 k_6 K_{eq}(k_8 + k_9)}{k_7 k_9 (k_2 k_4 + k_2 k_5 + k_3 k_5)} = \frac{k_6' K_{eq}}{k_7'} \end{aligned} \tag{21}$$

The α_{iso} term is derived from the net rate constant for the first step in the reaction divided by the net rate constant for the last reverse step in the isomerization segment. As the first net rate constant of isomerization, k_7', increases relative to the first net rate constant for the back reaction, k_6', thereby making the mechanism less Iso, α_{iso} becomes smaller. Conversely, as the rate of conversion of F to E decreases relative to the back reaction, making the mechanism more Iso, α_{iso} increases. As alternatives to α_{iso}, Albery and Knowles[9] use c_p and Cleland[27] uses x. The relation between α_{iso}, and x is

$$\alpha_{iso} = \frac{1 + K_{eq}}{c_p} = \frac{x(1 + K_{eq})}{A_0 + P_0} \tag{22}$$

[27] W. W. Cleland, *in* "The Enzymes" (D. S. Sigman and P. D. Boyer, eds.), 3rd Ed., Vol. 19, p. 99. Academic Press, San Diego, 1990.

The second root of α from Eq. (20) is termed α_{chem} because of its relation to the chemical segment:

$$\alpha_{\text{chem}} = \frac{V_f}{K_a}\left(\frac{1}{k_{\text{fchem}}} + \frac{1}{k_{\text{rchem}}}\right) = \frac{k_1k_7k_9[k_2k_4(k_3 + k_4 + k_5) + k_3k_5(k_2 + k_3 + k_4)]}{k_2k_4(k_2k_4 + k_2k_5 + k_3k_5)(k_7k_9 + k_7k_{10} + k_8k_{10})} \tag{23}$$

α_{iso} and α_{chem} are related:

$$\alpha_{\text{iso}} + \alpha_{\text{chem}} = \frac{1}{K_a} + \frac{K_{\text{eq}}}{K_p} \tag{24}$$

Comparing Eqs. (20), (21), and (23), it becomes obvious that if the chemical and isomerization segments are equally rate limiting, then α_{iso} will equal α_{chem}. In most cases to date, however, Iso mechanisms barely have been detectable with isomerizations that are only partially rate limiting. This condition requires that values of α_{iso} to be much larger than α_{chem}; hence, the larger root of Eq. (20) is usually associated with the isomerization segment.

Irreversible Iso Uni Uni Reactions

Previous discussions of Iso mechanisms have addressed reversible reactions only, perhaps because isomerizations were sought only in mutases and racemases.[28] However, Iso mechanisms have been extended to hydrolytic reactions,[29] which are effectively irreversible because of 55 *M* water. Rehydration of such enzymes during turnover may itself constitute an isomerization step in the kinetic mechanisms. Irreversibility may result from either an irreversible isomerization segment [e.g., either $k_8 = 0$ or $k_{10} = 0$ in Eqs. (9)–(17) and Scheme III] or an irreversible chemical segment (e.g., $k_4 = 0$), and both conditions reduce to the same general form of the rate equation for initial velocities:

$$v = \frac{V_f[\text{A}]}{K_a + [\text{A}] + \dfrac{K_a[\text{P}]}{K_p} + \dfrac{[\text{A}][\text{P}]}{K_{\text{iip}}}} \tag{25}$$

However, Eq. (19) no longer will be quadratic in α; instead, a single root exists:

[28] L. M. Fisher, J. W. Albery, and J. R. Knowles, *Biochemistry* **25,** 2538 (1986).
[29] K. L. Rebholz and D. B. Northrop, *Biochem. Biophys. Res. Commun.* **176,** 65 (1991).

$$\begin{array}{l} E + A \underset{k_2}{\overset{k_1}{\rightleftharpoons}} EA \underset{k_4}{\overset{k_3}{\rightleftharpoons}} FP \underset{k_6}{\overset{k_5}{\rightleftharpoons}} P + F \underset{k_8}{\overset{k_7}{\rightleftharpoons}} G \underset{k_{10}}{\overset{k_9}{\rightleftharpoons}} E \\ k_{11}[I] \downarrow\uparrow k_{12} \\ EI \end{array}$$

SCHEME IV

$$\alpha = \frac{K_p}{K_a K_{iip}} \tag{26}$$

For an irreversible chemical segment, the root is α_{iso} which combined with Eq. (21) yields

$$\frac{K_p}{K_{iip} V_f} = \frac{1}{k_{fiso}} + \frac{1}{k_{riso}} \tag{27}$$

For an irreversible isomerization segment, the single root is α_{chem}, which combined with Eq. (23) yields

$$\frac{K_p}{K_{iip} V_f} = \frac{1}{k_{fchem}} + \frac{1}{k_{rchem}} \tag{28}$$

Dead-End Inhibition

The kinetic constants for an inhibitor that resembles the substrate but not the product may be sensitive to the presence of an Iso step[30] (Scheme IV). Inhibition is controlled by the dissociation constant of I:

$$K_i = k_{12}/k_{11} \tag{29}$$

Initial velocities obey the equation for linear competitive inhibition in the forward reaction:

$$v_f = \frac{V_f[A]}{K_a(1 + [I]/K_{is}) + [A]} \tag{30}$$

and linear noncompetitive inhibition in the reverse reaction:

$$v_r = \frac{V_r[P]}{K_p(1 + [I]/K_{is}) + [P]\,(1 + [I]/K_{ii})} \tag{31}$$

where the slope and intercept inhibition constants are, respectively,

$$K_{is} = K_i/f_E \tag{32}$$
$$K_{ii} = K_i k'_{10}/V_r \tag{33}$$

[30] K. L. Rebholz and D. B. Northrop, *Arch. Biochem. Biophys.* **312,** 227 (1994).

Equations (30) and (31) show that identical values for K_{is} should be obtained in forward and reverse reaction kinetics. Substituting rate constants into Eqs. (32) and (33) yields the following relationship between inhibition constants:

$$\frac{K_{is}}{K_{ii}V_r} = \frac{K_i/f_E}{K_i k'_{10}} = \left(\frac{k_8 + k_9}{k_8 k_{10}}\right)\left(\frac{k_7 k_9 + k_7 k_{10} + k_8 k_{10}}{k_7 k_9}\right) = \frac{1}{k_{fiso}} + \frac{1}{k_{riso}} \tag{34}$$

Determining Rate Constants for Isomerization

Rearrangement of Eq. (21) isolates the apparent rate constants for the isomerization segment in the form of reciprocals[21]:

$$\frac{\alpha_{iso}}{V_f/K_a} = \frac{1}{k_{fiso}} + \frac{1}{k_{riso}} \tag{35}$$

Similarly, the kinetics for the chemical segment can be expressed as

$$\frac{\alpha_{chem}}{V_f/K_a} = \frac{1}{k_{fchem}} + \frac{1}{k_{rchem}} \tag{36}$$

If the isomerization is represented by a single step, or if intermediates such as form G of Scheme III are present in minor amounts, then $K_{eq(E/F)} = k_{fiso}/k_{riso}$. When $K_{eq(E/F)} = 1$, then

$$k_{fiso} = k_{riso} = \frac{V_f/K_a}{2\alpha_{iso}} \tag{37}$$

If $K_{eq(E/F)} \neq 1$, but is nevertheless known, it is possible to calculate the values of k_{fiso} and k_{riso} from the following relationships:

$$k_{fiso} = \frac{K_{eq(E/F)} + 1}{1/k_{fiso} + 1/k_{riso}} \tag{38}$$

$$k_{riso} = \frac{1/K_{eq(E/F)} + 1}{1/k_{fiso} + 1/k_{riso}} \tag{39}$$

If $K_{eq(E/F)} \ll 1$, then Eq. (37) reduces to

$$k_{fiso} = \frac{V_f/K_a}{\alpha_{iso}} \tag{40}$$

or if $K_{eq(E/F)} \gg 1$, then Eq. (37) reduces to

$$k_{riso} = \frac{V_f/K_a}{\alpha_{iso}} \tag{41}$$

Similarly, combining Eqs. (34) and (38) and again assuming that $K_{eq(E/F)} = k_{fiso}/k_{riso}$, it is possible to calculate k_{fiso} from dead-end inhibition kinetics:

$$k_{fiso} = \frac{K_{ii}V_r(K_{eq(E/F)} + 1)}{K_{is}} \tag{42}$$

Similarly, combining Eqs. (34) and (39),

$$k_{riso} = \frac{K_{ii}V_r(1/K_{eq(E/F)} + 1)}{K_{is}} \tag{43}$$

When $K_{eq(E/F)} \neq k_{fiso}/k_{riso}$, which occurs when the steady-state concentration of G is significant relative to [F] or [E], then Eqs. (37)–(43) do not hold, which underscores the need to treat isomerizations as segments with multiple steps and not as a single step as has been done previously. Nevertheless, the equations can be used to calculate useful limits for the apparent rate constants of isomerization. When $K_{eq(E/F)} = 1$, these equations hold despite the presence of significant [G]. If $K_{eq(E/F)} > 1$, then k_{fiso} will be overestimated and k_{riso} will be underestimated, and if $K_{eq(E-F)} <$ 1, then k_{fiso} will be underestimated and k_{riso} will be overestimated. In other words, if an isomerization is detected, the isomerization may be more rate limiting in the favorable direction than estimated if intermediate forms of free enzyme, such as G in Scheme III, are significant.

In the absence of knowledge about $K_{eq(E/F)}$, it still is possible to calculate a quantitative value of how rate limiting the isomerization segment is by comparing Eq. (35) to maximal velocities:

$$f_{iso} = \frac{\dfrac{\alpha_{iso}}{V_f/K_a}}{\dfrac{1}{V_f} + \dfrac{1}{V_r}} = \frac{\dfrac{1}{k_{fiso}} + \dfrac{1}{k_{riso}}}{\dfrac{1}{V_f} + \dfrac{1}{V_r}} \tag{44}$$

or by comparing dead-end inhibition constants to maximal velocities:

$$f_{iso} = \frac{K_{is}/K_{ii}}{V_r/V_f + 1} \tag{45}$$

or by comparing product inhibition constants if the chemical segment is irreversible:

$$f_{iso} = \frac{K_p}{K_{iip}} = \frac{\dfrac{1}{k_{fiso}} + \dfrac{1}{k_{riso}}}{\dfrac{1}{V_f}} \tag{46}$$

The fractional reduction of the maximal velocity, represented by f_{iso}, was introduced originally as f_v to quantify partially rate-limiting steps from isotope effects.[31] Here the symbol and concept indicate the average impact of the isomerization on the maximal velocities of the reaction in the forward and reverse directions together. The value of f_{iso} will range from zero, when an Iso mechanism in insignificant, to one, when the isomerization is rate limiting.

Britton[21] was frustrated by the impossibility of calculating exact values for individual rate constants—and thus an exact description of how rate limiting isomerizations are—from measurements of α_{iso}, V_f, V_r, K_a, and K_p, because of the quadratic nature of Eq. (19). As elaborated below in terms of interpreting K_{iip}, Britton's problem can also be described in terms of uncertainty in the value of $K_{eq(E/F)}$. For a given value of α_{iso}, two possibilities exist: one in which isomerizations are slow and $K_{eq(E/F)}$ is large, and the other in which isomerizations are fast and $K_{eq(E/F)}$ is small. However, combining data from induced transport, initial velocities, and product inhibition can yield a unique solution.

Expression of Isomerizations as Noncompetitive Product Inhibition

At saturating substrate Eq. (1) approaches the limit $v_0 = V_f$ while Eq. (2) approaches $v_0 = V_f(1 + [P]/K_{iip})$. Thus, in a Uni Uni mechanism such as Scheme I the product will be a competitive inhibitor of the substrate, whereas in an Iso Uni Uni mechanism such as Scheme II the product *may* appear to be a noncompetitive inhibitor, but only if k_{fiso} and k_{fchem} are of similar magnitude. The uncertainty of noncompetitive product inhibition was addressed first by Britton,[21] but the analysis was based on a confusing graph[32] and an unclear definition of K_{iip}.[33,34] This is a tricky problem, as illustrated in Fig. 1. Whether inhibition looks competitive or noncompetitive depends on the ratio of slope and intercept effects in classical inhibition patterns, expressed as K_{iip}/K_p in the present discussion. When this ratio is large (i.e., K_{iip} is indeterminate) the pattern appears to be competitive. As illustrated in the simulations of Fig. 1, the ratio K_{iip}/K_p becomes large when the ratio of k_{fchem}/k_{fiso} is either very large or very small. For a given set of values for K_{iip} and K_p, two solutions exist for the ratio of k_{fchem}/k_{fiso} such as indicated by the points represented by squares in Fig. 1, one for a relatively fast and the other for a relatively slow isomerization. Nevertheless, this uncertainty can be overcome by compar-

[31] D. B. Northrop, *in* "Isotope Effects on Enzyme-Catalyzed Reactions" (W. W. Cleland, M. H. O'Leary, and D. B. Northrop, eds.), p. 122. University Park Press, Baltimore, Maryland, 1977.

[32] K. L. Rebholz and D. B. Northrop, *Biochem. J.* **296,** 355 (1993).

[33] A. Cornish-Bowden, *Biochem. J.* **301,** 621 (1994).

[34] D. B. Northrop, *Biochem. J.* **301,** 623 (1994).

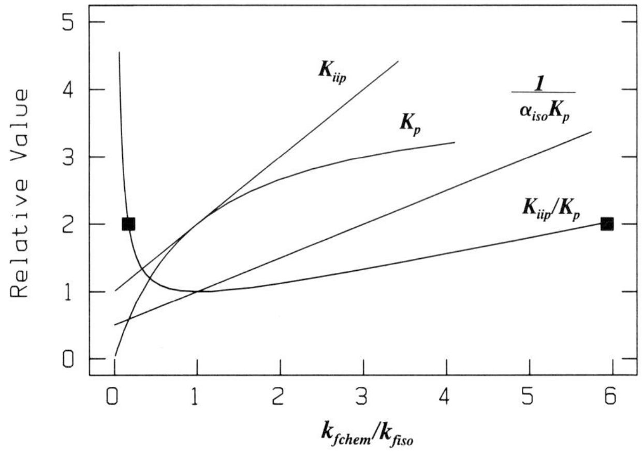

FIG. 1. Relationship between kinetic constants and the apparent rate constant for isomerization of free enzyme. Individual rate constants for the reaction mechanism in Scheme II were all set equal to a value of one, and then rate constants for the isomerization step were varied in a constant ratio. The results are plotted in terms of individual kinetic parameters or pairs of parameters as a function of the ratio k_{fchem}/k_{fiso}, which is a measure of how rate limiting the isomerization is. A classical noncompetitive product inhibition pattern has both slope ($1/K_p$) and intercept ($1/K_{iip}$) terms; hence, the detection of an intercept effect and therefore of noncompetitive inhibition depends on the ratio K_{iip}/K_p, which becomes large and indeterminant at both small and large values of k_{fchem}/k_{fiso}. The data points (■) illustrate Britton's ambiguity, that K_{iip}/K_p has the same value, in this case two, when k_{fchem}/k_{fiso} is either large and small.

ing kinetic constants with α_{iso} from the Britton experiment: the value of the dimensionless quantity $1/\alpha_{iso}K_a$ is also a linear function that falls below K_{iip}/K_p when isomerizations are slow and above when they are fast, as shown in the special case illustrated in Fig. 1 where $K_{eq(F \rightarrow E)} = 1$, but can be demonstrated for other cases as well. Hence, comparing values of $1/\alpha K_a$ against K_{iip}/K_p can identify which of the two ambiguous roots is the correct one.

Detection of Noncompetitive Product Inhibition in Uni Uni Mechanisms

The standard method to measure product inhibition is to generate a set of initial velocities at different cocnentrations of substrate in the presence and absence of different concentrations of product. Plotted as reciprocals of initial velocities versus reciprocals of substrate concentrations, the data generate lines that intersect on the *Y* axis if the product inhibition is competitive, but intersect left of the *Y* axis if inhibition is noncompetitive. For a reversible enzymatic reaction, however, it is often difficult to measure initial velocities in the presence of product. Furthermore, Darvey[35] realized that traditional product inhibition studies could not be used successfully with reversible Uni Uni reactions because double-reciprocal plots will not be linear and cannot be extrapolated to determine the presence or absence of intercept effects. Consequently, alternative methods have been developed to detect noncompetitive product inhibition associated with Iso Uni Uni kinetic mechanisms, and these are presented below. The most useful methods are presented first, followed by minor descriptions of older methods now mainly of historical interest. In relative terms for detecting Iso mechanisms, the method of nonlinear regression is 10 times as sensitive as are Foster–Niemann plots, which are in turn 10 times more sensitive than are oversaturation plots, and the latter supersede the remaining methods.[36]

Nonlinear Regression

The preferred method to detect noncompetitive inhibition and to determine K_{iip} is to fit a set of progress curves to the integrated forms of rate equations by nonlinear regression. A single progress curve cannot distinguish the mode or the presence of product inhibition, but a set of curves obtained at varied starting concentrations of substrate sometimes can. Greater precision and discrimination is possible if progress curves are collected in both the forward and reverse directions and fitted simultaneously, as a single data set, designated as positive and negative product formations, respectively. Figure 2 illustrates what a complete data set of forward and reverse progress curves should look like.

The integrated rate equation for the mechanism in Scheme I is

$$Vt = \left(1 - \frac{K_a}{K_p}\right)[P] + K_a\left(1 + \frac{[A_0]}{K_p}\right)\ln\frac{[A_0]}{[A_0] - [P]} \tag{47}$$

and that for Schemes II and III is

$$vt = \left(1 - \frac{K_a}{K_p}\right)[P] + \frac{[P]^2}{2K_{iip}} + K_a\left(1 + \frac{[A_0]}{K_p}\right)\ln\frac{[A_0]}{[A_0] - [P]} \tag{48}$$

where t represents time of reaction and $[A_0]$ represents the initial substrate concentration at time zero. Integrating rate equations for many, more complex, reaction mechanisms is not an easy matter. Fortunately, the computer program AGIRE (Automatic Generation of Integrated Rate Equations)[37] is available and can be used to write the lines of code neces-

[35] I. G. Darvey, *Biochem. J.* **128,** 383 (1972).

[36] D. B. Northrop and K. L. Rebholz *Anal. Biochem.* **216,** 285 (1994).

[37] R. G. Duggleby and C. Wood, *Biochem. J.* **258,** 397 (1989); erratum, *Biochem. J.* **270,** 843 (1990).

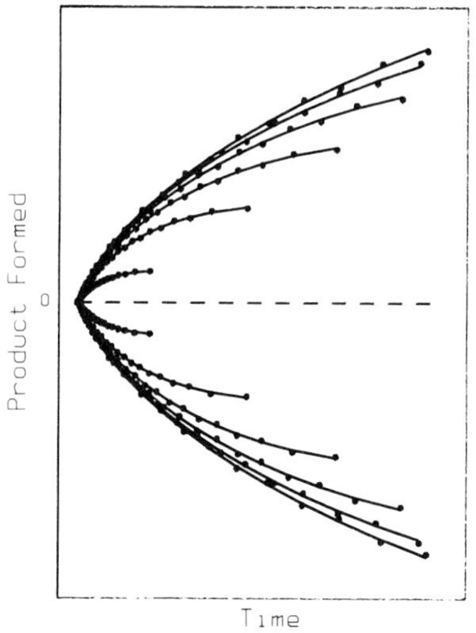

FIG. 2. Simulated progress curves with noncompetitive product inhibition. Data points in the top half of the plot were calculated using a pseudo-random number generator to simulate experimental error in the progress curves of an irreversible iso mechanism in which $K_{iip} = K_p$. These points therefore reflect prominent noncompetitive product inhibition. However, the lines were drawn from a fit to competitive product inhibition. Initial relative substrate concentrations were, top to bottom, 10, 7.5, 5, 2.5, 1.5, and 1. Data in the lower half of the plot were obtained by inverting the upper plots, to illustrate the relationship and appearance of a complete set of forward and reverse progress curves, originating from a common zero.

sary to perform numerical integrations on these and other rate equations. AGIRE is based on a common form of a rate equation and corresponding integrated rate equation, for unbranched enzyme kinetic mechanisms with two or fewer substrates and products.[38,39]

Foster–Nieman Plots

Plots of enzymatic progress curves presented as $[P]/t$ versus $(1/t)\ln(1 - [P]/[P_\infty])$ are linear when product inhibition is competitive[14,40–44] according to the following rearrangement of the integrated form of Eq. (1):

[38] E. A. Boeker, *Biochem. J.* **223,** 15 (1984).
[39] E. A. Boeker, *Biochem. J.* **226,** 29 (1985).
[40] A. C. Walker and C. L. A. Schmidt, *Arch. Biochem. Biophys.* **5,** 445 (1944).
[41] R. J. Foster and C. Niemann, *Proc. Natl. Acad. Sci. U.S.A.* **39,** 999 (1953).
[42] T. H. Applewhite and C. Niemann, *J. Am. Chem. Soc.* **77,** 4923 (1955).
[43] R. R. Jennings and C. Niemann, *J. Am. Chem. Soc.* **77,** 5432 (1955).
[44] B. A. Orsi and K. A. Tipton, this series, Vol. 63, p. 159.

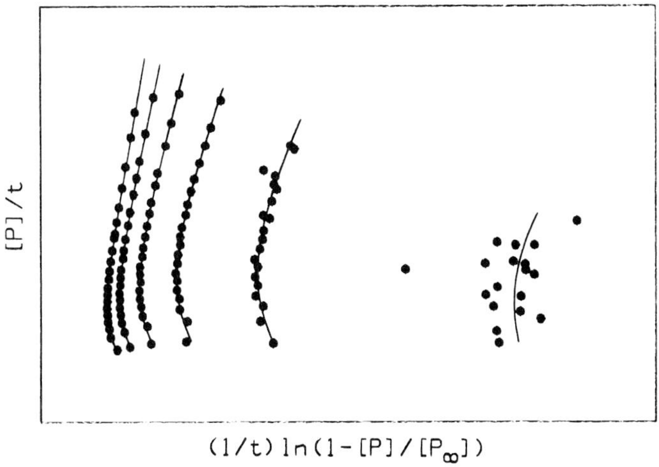

FIG. 3. Foster–Niemann plots of simulated progress curves with noncompetitive product inhibition. The data are derived from the points of Fig. 2 but are plotted in a form that normally generates linear relationships when competitive product inhibition is expressed (lines not shown). The lines represent a fit of the data to noncompetitive product inhibition.

$$\frac{[P]}{t} = \frac{V_f/K_a + V_r/K_p}{1/K_a - 1/K_p} + \left(\frac{1 + [A_0]/K_p}{1/K_a - 1/K_p} + \frac{[A_0]}{1 + K_{eq}}\right)\frac{1}{t}\ln\left(\frac{[P_\infty] - [P]}{[P_\infty]}\right) \quad (49)$$

where $[P_\infty]$ represents the final concentration of product. Similar plots are nonlinear* when product inhibition is noncompetitive,[14,44] as illustrated in the simulation in Fig. 3. Other plots based on transformations of progress curves are possible, but the Foster–Niemann plot is important because (a) it allows progress curves obtained at high and low concentrations of substrate to be presented on a normalized scale, (b) the presentation is in a visual format which easily distinguishes between mechanisms (e.g., curves versus straight lines), and (c) the goodness of fit between point and line can be seen and evaluated after fitting to alternative integrated rate equations. To illustrate these points, note that the noncompetitive progress curves in the top half of Fig. 2 are the same data used to calculate the plots of Fig. 3 and that the lines of Fig. 2 were calculated from a fit to competitive product inhibition. The deviations between point and line in Fig. 2 are significant, but barely discernible; in contrast, the nonlinearity

* Plots of $[P]/t$ versus $(1/t)\ln(1 - [P]/[P_\infty])$ may appear curved in the absence of a K_{iip} term if there is an error in zero time [M. Taraszka, Ph.D. Thesis, University of Wisconsin, Madison (1962)], so special care must be taken when recording the time of the start of the reaction.

of the individual Foster–Niemann plots of Fig. 3 is obvious and easily discernible from any sort of imaginary set of straight lines which would be consistent with competitive inhibition.

Oversaturation Plots

A normalized plotting method has been proposed by Fisher *et al.*[12] in which ([A] − [P])/[A_0] is plotted against t/[A_0]. The results are bowl-shaped curves that intersect the Y axis at a common value of 1.0, as illustrated in Fig. 4. The slopes of tangents to the curves at the Y axis are negative and become more negative in the "unsaturated" region as concentrations of substrate are increased. In the "saturated" region the curves superimpose. Then, as concentrations of substrate are increased further, the slopes of tangents become less negative—a process termed "oversaturation"—only if product inhibition is noncompetitive. The use of this method is limited to kinetic situations in which the concentration of substrate can be raised significantly above the value of K_{iip} to achieve oversaturation. For example, in the proline racemase study of Fisher *et al.*,[12] L-proline was elevated to 189 mM, more than 3 times the K_{iip} and more than 60 times K_a. Such an excess of substrate may not always be experimentally accessible.

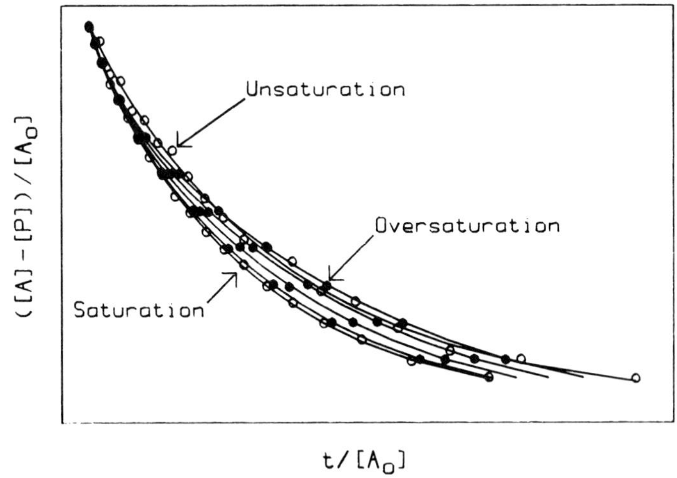

FIG. 4. Normalized progress curves of simulated progress curves with noncompetitive product inhibition. The data are derived from the points of Fig. 2 but are plotted in a form that illustrates oversaturation kinetics. Some data (○) represent unsaturation at [A] = 1, 1.5, and 2.5; other data (●) show oversaturation at [A] = 5, 7.5, and 10 (see text).

Darvey Plots

Darvey[35] proposed a graphical procedure based on initial velocity studies in the presence and absence of added product inhibitor. From Eq. (1) and competitive product inhibition, the following equation was derived:

$$(v_0)_{[P_0]=0} - (v_0)_{[P_0]>0} = \frac{K_a\{(V_f + V_r)[A_0] + V_rK_a\}[P_0]}{K_p(K_a + [A_0])^2 + K_a(K_a + [A_0])\,[P_0]} \tag{50}$$

From Eq. (2) and noncompetitive product inhibition, a similar equation takes the form

$$(v_0)_{[P]=0} - (v_0)_{[P]>0} = \frac{K_a\{V_f(K_{iip}K_a + K_p[A_0])\,[A_0] + V_rK_{iip}K_a(K_a + [A_0])\}\,[P_0]}{K_pK_{iip}K_a(K_a + [A_0])^2 + K_a(K_a + [A]_0)(K_{iip}K_a + K_p[A_0])[P_0]} \tag{51}$$

Primary plots of $1/\{(v_0)_{[P]=0} - (v_0)_{[P]>0}\}$ versus $1/[P_0]$ at various concentrations of substrate result in lines intersecting the Y axis:

$$Y_{\text{competitive}} = -\frac{K_a}{K_p(K_a + [A_0])} \tag{52}$$

$$Y_{\text{noncompetitive}} = -\frac{K_a(K_{iip}K_a + K_p[A_0])}{K_pK_{iip}K_a(K_a + [A_0])} \tag{53}$$

Secondary replots of the reciprocal of the Y intercepts versus $[A_0]$ are linear if inhibition is competitive, and nonlinear if noncompetitive. To determine K_{iip}, Darvey[35] constructs a tertiary replot of $1/(-1/K_p$ intercept) versus $1/[A_0]$ to obtain

$$\text{Slope}_{\text{III}} = \frac{K_aK_pK_{iip}}{K_p - K_{iip}} \tag{54}$$

$$\text{Intercept}_{\text{III}} = \frac{K_pK_{iip}}{K_p - K_{iip}} \tag{55}$$

Finally, K_{iip} is calculated from the relationship

$$K_{iip} = \frac{\text{Slope}_{\text{III}}}{K_a + \text{Intercept}_{\text{III}}(K_a/K_p)} \tag{56}$$

A citation search suggests that Darvey plots have not been applied to experimental data. Nevertheless, this exercise in analytical geometry documents some interesting relationships between kinetic parameters and illustrates the difficulty of analyzing Iso mechanisms before the AGIRE system was developed.

Ray–Roscelli Plots

Plots of $[P_t]/[P_\infty]$ versus time at different concentrations of initial substrate produce curves which overlap in the presence of competitive product inhibition or, alternatively, curves which decrease as substrate concentration approaches and passes the value for K_{iip}.[45] In the latter case, K_{iip} may be calculated from the following equation:

$$K_{iip} = \frac{[P]_x[P]_y K_{eq}([P]_x - [P]_y)}{2V_f(t_x[P]_y - t_y[P]_x)(K_{eq}+1)} - \frac{[P]_x[P]_y K_{eq}([A_0]_x - [A_0]_y)\{\ln(1 - [P]/[P_\infty])/([P]/[P_\infty]) - 1\}}{V_f(t_x[P]_y - t_y[P]_x)(K_{eq}+1)^2} \quad (57)$$

where $[A_0]_x$ and $[A_0]_y$ are two initial concentrations of substrate. The reaction is allowed to proceed to a constant percentage of reaction, identified as times t_x and t_y, respectively. The concentrations of products formed at these times are $[P]_x$ and $[P]_y$, respectively, but the ratio $[P]/[P_\infty]$ is constant for both times. Ray and Roscelli[45] chose to calculate K_{iip} after 80% of the phosphoglucomutase reaction had been completed; however, other percentages of reaction may be more suitable for the study of other enzymes.

Cennamo Plots

If a substrate is highly soluble or if the K_m is extremely low, the presence of a normal intercept effect in classical inhibition patterns may be approximated directly, without extrapolation.[46,47] A plot of reciprocal initial velocities versus concentration of product at saturating substrate is a linear function with a slope of $1/K_{iip}$.

Product Inhibition Patterns for Multireactant Reactions

Traditional product inhibition studies can be used to detect isomerizations in higher order reactions. These reactions can be transformed into the general kinetic form of Uni Bi reactions by holding the second substrate at a high, fixed concentration. Furthermore, special variants such as Theorell–Chance kinetic mechanisms can be deduced from Uni Bi kinetic equations by raising values of individual rate constants or by dropping

[45] W. J. Ray, Jr., and G. A. Roscelli, *J. Biol. Chem.* **239,** 3935 (1964).
[46] C. Cennamo, *J. Theor. Biol.* **21,** 260 (1968).
[47] C. Cennamo, *J. Theor. Biol.* **23,** 53 (1969).

$$\mathrm{E} + \mathrm{A} \underset{k_2}{\overset{k_1}{\rightleftharpoons}} \mathrm{EA} \underset{k_4}{\overset{k_3}{\rightleftharpoons}} \mathrm{FPQ} \underset{k_6}{\overset{k_5}{\rightleftharpoons}} \mathrm{FQ} \underset{k_8}{\overset{k_7}{\rightleftharpoons}} \mathrm{F} \underset{k_{10}}{\overset{k_9}{\rightleftharpoons}} \mathrm{E}$$

SCHEME V

kinetic terms. Therefore, only Iso Uni Bi reactions, illustrated in Scheme V, will be examined in detail.

Reversible Uni Bi reactions are described by the rate equation

$$v = \frac{V_f V_r\left([\mathrm{A}] - \frac{[\mathrm{P}][\mathrm{Q}]}{K_{eq}}\right)}{V_r K_a + V_r[\mathrm{A}] + \frac{V_f K_q[\mathrm{P}]}{K_{eq}} + \frac{V_f K_p[\mathrm{Q}]}{K_{eq}} + \frac{V_r[\mathrm{A}][\mathrm{P}]}{K_{ip}} + \frac{V_f[\mathrm{P}][\mathrm{Q}]}{K_{eq}}} \tag{58}$$

They have two Haldane relationships:

$$K_{eq} = \frac{V_f K_{ip} K_q}{V_r K_{ia}} = \frac{V_f K_p K_{iq}}{V_r K_a} \tag{59}$$

The addition of an Iso segment as in Scheme V results in an equation with two additional terms in the denominator, and AQ term and an APQ term:

$$v = \frac{V_f V_r\left([\mathrm{A}] - \frac{[\mathrm{P}][\mathrm{Q}]}{K_{eq}}\right)}{\begin{array}{l} V_r K_a + V_r[\mathrm{A}] + \frac{V_f K_q[\mathrm{P}]}{K_{eq}} + \frac{V_f K_p[\mathrm{Q}]}{K_{eq}} + \frac{V_r[\mathrm{A}][\mathrm{P}]}{K_{ip}} \\ \quad + \frac{V_f[\mathrm{P}][\mathrm{Q}]}{K_{eq}} + \frac{V_r[\mathrm{A}][\mathrm{Q}]}{K_{iiq}} + \frac{V_r K_{iq}^2 K_p[\mathrm{A}][\mathrm{P}][\mathrm{Q}]}{K_{eq} K_a K_q K_{ip} K_{iiq}} \end{array}} \tag{60}$$

The kinetic constants are defined below, together with the King–Altman coefficients from which they were derived:

$$V_f = \frac{(\mathrm{num1})}{(\mathrm{coefA})} = \frac{k_3k_5k_7k_9}{k_3k_5k_7 + k_3k_5k_9 + k_3k_7k_9 + k_4k_7k_9 + k_5k_7k_9} \tag{61}$$

$$V_r = \frac{(\mathrm{num2})}{(\mathrm{coefPQ})} = \frac{k_2k_4k_{10}}{k_2k_4 + k_2k_{10} + k_4k_{10} + k_3k_{10}} \tag{62}$$

$$K_a = \frac{(\mathrm{constant})}{(\mathrm{coefA})} = \frac{k_7(k_9 + k_{10})(k_2k_4 + k_2k_5 + k_3k_5)}{k_1(k_3k_5k_7 + k_3k_5k_9 + k_3k_7k_9 + k_4k_7k_9 + k_5k_7k_9)} \tag{63}$$

$$K_p = \frac{(\mathrm{coefQ})}{(\mathrm{coefPQ})} = \frac{k_{10}(k_2k_4 + k_2k_5 + k_3k_5)}{k_6(k_2k_4 + k_2k_{10} + k_3k_{10} + k_4k_{10})} \tag{64}$$

$$K_q = \frac{(\text{coefP})}{(\text{coefPQ})} = \frac{k_2k_4(k_9 + k_{10})}{k_8(k_2k_4 + k_2k_{10} + k_3k_{10} + k_4k_{10})} \tag{65}$$

$$K_{iq} = \frac{(\text{constant})}{(\text{coefQ})} = \frac{k_7(k_9 + k_{10})}{k_8k_{10}} \tag{66}$$

$$K_{iiq} = \frac{(\text{coefA})}{(\text{coefAQ})} = \frac{k_3k_5k_7 + k_3k_5k_9 + k_3k_7k_9 + k_4k_7k_9 + k_5k_7k_9}{k_3k_5k_8} \tag{67}$$

Many Uni Bi reactions are irreversible and obey the equation

$$v = \frac{V_f[A]}{K_a + [A] + \frac{K_a[Q]}{K_{iq}} + \frac{[A][P]}{K_{ip}} + \frac{K_a[P][Q]}{K_pK_{iq}}} \tag{68}$$

Equation (68) describes competitive inhibition by Q and uncompetitive inhibition by P. In Iso Uni Bi reactions, the irreversibility can reside either in the catalytic segment or in the isomerization segment, and these give rise to different rate equations. Setting $k_4 = 0$ in Scheme V and Eqs. (60)–(66) reduces the rate equation to

$$v = \frac{V_f[A]}{K_a + [A] + \frac{K_a[Q]}{K_{iq}} + \frac{[A][P]}{K_{ip}} + \frac{K_a[P][Q]}{K_pK_{iq}} + \frac{[A][Q]}{K_{iiq}} + \frac{[A][P][Q]}{K_pK_{iiq}}} \tag{69}$$

Hence, inhibition by Q is changed to noncompetitive. In contrast, setting $k_{10} = 0$ gives

$$v = \frac{V_f[A]}{K_a + [A] + \frac{K_a[P]}{K_{iip}} + \frac{[A][P]}{K_{ip}} + \frac{K_a[P][Q]}{K_{iip}K_q} + \frac{[A][Q]}{K_{iiq}} + \frac{[A][P][Q]}{K_{ip}K_q}} \tag{70}$$

Now inhibition by P becomes noncompetitive. When both the catalytic segment and isomerization are irreversible, the rate equation becomes

$$v = \frac{V_f[A]}{K_a + [A] + \frac{[A][P]}{K_{ip}} + \frac{[A][Q]}{K_{iiq}} + \frac{[A][P][Q]}{K_pK_{iiq}}} \tag{71}$$

In this last case, inhibition by both P and Q becomes uncompetitive. The expected product inhibition patterns for Uni Bi reactions are summarized in Table I. The product inhibition patterns for more complex reactions, including those with more than one isomerization segment, as in Di-Iso Ping Pong, are summarized in Table II.[48]

[48] F. B. Rudolph, this series, Vol. 63, p. 411.

TABLE I
PRODUCT INHIBITION PATTERNS VERSUS 1/[A] FOR UNI BI AND ISO UNI BI REACTIONS

Reaction	Product P	Product Q
Reversible Uni Bi	NC[a]	C
Reversible Iso Uni Bi	NC	NC
Irreversible Uni Bi	UC	C
Catalytic Segment Irreversible Iso Uni Bi	UC	NC
Iso Segment Irreversible Iso Uni Bi	NC	UC
Iso and Catalytic Segments Irreversible Iso Uni Bi	UC	UC

[a] NC is noncompetitive, C is competitive, and UC is uncompetitive inhibition.

Britton's Induced Transport Experiment

Detection of noncompetitive inhibition does not in itself warrant the conclusion that free enzyme exists in two or more forms. Noncompetitive terms of a rate equation can result from the formation of a nonproductive ternary complex of enzyme, substrate, and inhibitor[45] or from the presence of two types of enzymatic sites.[49] Therefore, confirmatory data from alternative experimental designs are necessary. Fortunately, Britton and coworkers[21,50,51] invented a brilliant experiment to test for very fast isomerizations of free enzyme. His experimental design employs isotopic exchange during net catalytic turnover and is diagrammed below in terms of the fumarase reaction.

During a preincubation phase, radiolabeled substrate and product, [^{14}C]fumarate and [^{14}C]malate, are allowed to come to chemical and isotopic equilibrium in the presence of enzyme (Scheme VI). To this mixture is added a pulse of excess unlabeled substrate (Scheme VII, fumarate, large print) which combines with and depletes the E form (small print) of free enzyme. The newly formed EA unlabeled complex undergoes catalysis to yield FP, which generates nearly one enzyme equivalent of unlabeled product and more F form (Scheme VII, large print) of the free enzyme. Surplus F form of the enzyme combines with labeled product, [^{14}C]malate, unimpeded by competition with fumarate because of free enzyme isomerizations, and catalyzes the reverse synthesis of labeled substrate, [^{14}C]-fumarate. This newly synthesized [^{14}C]fumarate is prevented from recom-

[49] G. B. Kistiakowsky and A. J. Rosenberg, *J. Am. Chem. Soc.* **60**, 5020 (1952).
[50] H. J. Britton, *Nature (London)* **205**, 1323 (1965).
[51] H. G. Britton and J. B. Clarke, *Biochem. J.* **110**, 161 (1968).

TABLE II
PRODUCT INHIBITION PATTERNS FOR COMPLEX ISO MECHANISMS[a]

Kinetic mechanism	Product	Varied reactant A	B	C
Iso Ordered Bi Uni	P	N[b](C)[c]	N	
Mono-Iso Ordered Bi Bi	P	N	N	
	Q	N(C)	N	
Di-Iso Ordered Bi Bi	P	N	N	
	Q	C	N	
Mono-Iso Theorell–Chance	P	N	C	
	Q	N(C)	N	
Di-Iso Theorell–Chance	P	N	C	
	Q	C	N	
Mono-Iso Tetra Uni Ping Pong	P	N	C	
	Q	N(C)	N	
Di-Iso Tetra Uni Ping Pong	P	N	N(C)	
	Q	N(C)	N	
Mono-Iso Uni Uni Bi Bi Ping Pong	P	N	N(C)	N
	Q	U	N	N
	R	C	U	U
Di-Iso Bi Uni Uni Bi Ping Pong	P	N	N	N(C)
	Q	U	U	N
	R	N(C)	N	U
Tri-Iso Hexa Uni Ping Pong	P	N	N(C)	U
	Q	U	N	N(C)
	R	N(C)	U	N

[a] Adapted from Fromm[8] and Rudolph.[45]
[b] C is competitive, N is noncompetitive, and U is uncompetitive inhibition.
[c] In parentheses are the patterns for the analogous non-Iso mechanism.

bining with the small amount of free E because of competition with excess unlabeled fumarate (Scheme VIII). Consequently, only if an isomerization of free enzyme is present will the forward conversion of [^{12}C]fumarate to [^{12}C]malate be accompanied by a transient reverse flow of [^{14}C]malate to [^{14}C]fumarate, causing a demonstrable decrease in [^{14}C]malate concentra-

[^{14}C]fumarate   [^{14}C]malate
↓↑   ↓↑

E   EA ⇌ FP   F ↔ E

SCHEME VI

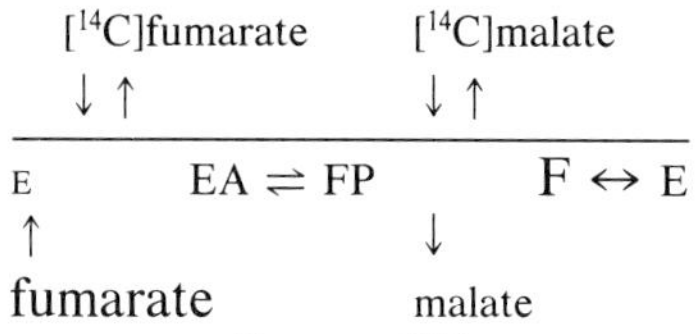

SCHEME VII

tion during steady-state catalysis. Data for such a transient caused by a pulse of fumarate are shown in Fig. 5, together with complementary data from a pulse of unlabeled malate. Note that the equilibrium favors malate more at high concentrations of reactants than at low, and that the perturbation is larger when pulsed by fumarate.

The general equation describing the perturbation of isotopic equilibrium as a function of unlabeled substrate was derived by Britton[21]:

$$\frac{[A^*]}{[A^*_\infty]} = \frac{[A]}{[A_\infty]} + \left(\frac{[A^*_0]}{[A^*_\infty]} - \frac{[A_0]}{[A_\infty]}\right)\left(\frac{[A] - [A_\infty]}{[A_0] - [A_\infty]}\right)^{(1+\alpha_{iso}[A_\infty])} \tag{72}$$

where $[A^*]$ represents the concentration of labeled substrate at any time; $[A^*_0]$ represents the concentration of labeled substrate at zero time; and $[A_0]$, $[A]$, and $[A_\infty]$ represent the total substrate concentration at zero time, at time t, and at the end of the reaction, respectively. Britton's α_{iso} may be extracted from the induced transport curve according to Eq. (72), or more simply from the maximum perturbation described by the relationship

$$\frac{[A^*]_{max} - [A^*_0]}{[A^*_\infty]} = \frac{\alpha_{iso}([A_0] - [A_\infty][A^*_0]/[A^*_\infty])}{(1 + \alpha_{iso}[A_\infty])^{(1+1/\alpha_{iso}[A_\infty])}} \tag{73}$$

If the concentration of labeled substrate is the same at the beginning and the end of the perturbation (i.e., if the equilibrium constant is not

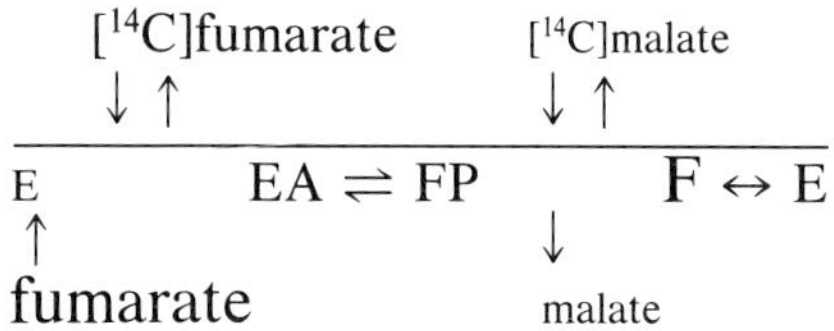

SCHEME VIII

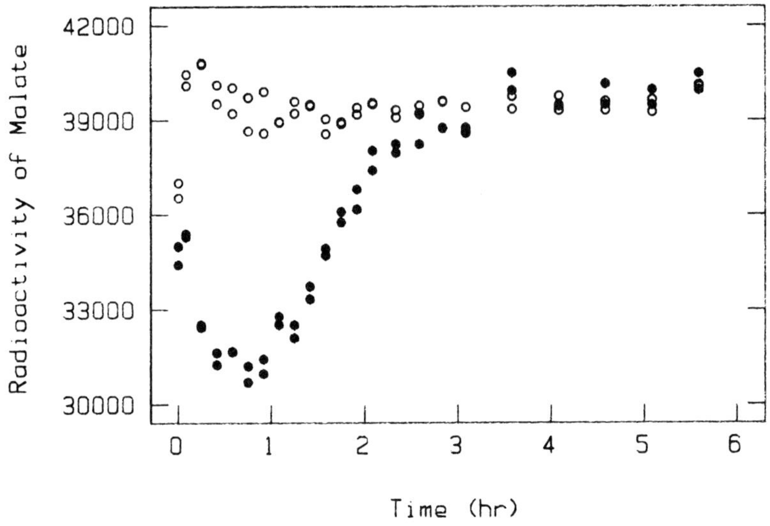

FIG. 5. Induced transport catalyzed by fumarase. Porcine heart fumarase (240 units/ml) was preincubated with [^{14}C] malate equilibrating with [^{14}C]fumarate (2.52×10^{-3} *M* total substrate and product, $K_{eq} = 4.4$). A pulse of cold fumarate ($[A_0] = 0.75$ *M*) was added, and radiolabeled malate was isolated and counted (●) as a function of time until the reaction came to equilibrium ($[A_0] = 0.1389$ *M*, $K_{eq} = 9.45$). At the point of maximum transport, [^{14}C]malate decreased and [^{14}C]fumarate increased ($[A^*] = 7.51 \times 10^{-5}$ *M*) from which a value of $\alpha_{iso} = 6.42$ was calculated using Eq. (73). Other data (○) show similar results with a pulse of cold malate, but with less induced transport because the equilibrium favors malate.

concentration independent as was the case for fumarase), the induced transport equations reduce to

$$\frac{[A^*]}{[A_0^*]} = 1 + \frac{[A] - [A_\infty]}{[A_\infty]}\left[1 - \left(\frac{[A] - [A_\infty]}{[A_0] - [A_\infty]}\right)^{\alpha_{iso}[A_\infty]}\right] \tag{74}$$

$$\frac{[A^*]_{max}}{[A_0^*]} = 1 + \frac{\alpha_{iso}([A_0] - [A_\infty])}{(1 + \alpha_{iso}[A_\infty])^{(1+1/\alpha_{iso}[A_\infty])}} \tag{75}$$

If, however, equilibrium is not achieved during the experiment, induced transport may be described by

$$\frac{[A^*]}{[A_\infty^*]} = \frac{[A]}{[A_\infty]} + \left(\frac{[A_0^*]}{[A_\infty^*]} - \frac{[A_0]}{[A_\infty]}\right)\left(\frac{[A] - [A_\infty]}{[A_0] - [A_\infty]}\right)^{(1+\alpha_{iso}[A_\infty])} \tag{76}$$

The value of α_{iso} is positively correlated with the equilibrium constant. As a result, the maximal perturbation is achieved by the reactant less favored at equilibrium. For example, for fumarase, $K_{eq} = 4.4 \pm 0.1 =$

[malate]/[fumarate][15]; hence, the displacement by unlabeled fumarate is about four times greater than that of unalabeled malate, as illustrated in Fig. 5. Note, however, that the fumarase equilibrium constant is concentration dependent, as is that of triose-phosphate isomerase.[11] Britton and others have calculated rate constants for isomerizations of a variety of mutases and isomerases based on α_{iso} obtained by induced transport, and their results are listed in Table III. The calculation is analogous to substituting α_{iso} and V/K into Eq. (37), but it requires the assumption that $k_{fiso} = k_{riso}$. Using data from dead-end or product inhibition and the relationships described in Eqs. (38)–(43), it is sometimes possible to calculate individual values for k_{fiso} and k_{riso}.

The induced transport experiment does not require knowledge of the concentration of enzyme, and partial inactivation of the enzyme during the experiment will not affect the results. The method is limited to detection of interconversion of enzyme forms with rate constants less than 10^8 sec^{-1} and assumes both that the maximal substrate association rate is 10^8 M^{-1} sec^{-1} and that the upper limit of substrate concentration used in the measurements is 1 M.[11]

TABLE III
RATE CONSTANTS OF ENZYME ISOMERIZATION DETERMINED BY INDUCED TRANSPORT

Enzyme	k_{cat} (sec^{-1})	k_{iso} (sec^{-1})	Ref.
Phosphoglucomutase (rabbit muscle)	750	$\sim 10^7$	*a*
Phosphoglycerate mutase (wheat germ)	270	$>10^6$	*b*
Phosphoglycerate mutase (yeast)	1600	$>10^6$	*c*
Phosphoglycerate mutase (rabbit muscle)	1900	$>4 \times 10^6$	*d*
Proline racemase (*Clostridium sticklandii*)	2000	$\sim 10^5$	*e*
Triose-phosphate isomerase (chicken muscle)	4300	$\sim 10^6$	*f*

[a] H. G. Britton and J. B. Clarke, *Biochem. J.* **110,** 161 (1968).
[b] H. G. Britton, J. Carreras, and S. Grisolia, *Biochemistry* **10,** 4522 (1971).
[c] H. G. Britton, J. Carreras, and S. Grisolia, *Biochemistry* **11,** 3008 (1972).
[d] H. G. Britton and J. B. Clarke, *Biochemistry* **130,** 397 (1972).
[e] L. M. Fisher, J. W. Albery, and J. R. Knowles, *Biochemistry* **25,** 2538 (1986).
[f] R. T. Raines and J. R. Knowles, *Biochemistry* **26,** 7014 (1987).

Substrate and Solvent Isotope Effects

A kinetic hydrogen isotope effect on an isomerization segment will not be expressed on V/K, but it may be expressed on either V or K_{iip}, or both, depending on how rate limiting proton transfers are in the isomerization and in a complete catalytic cycle. For the Iso mechanism in Scheme III, apparent isotope effects are described by the following equations:

$$^{D}V = \frac{k_7k_9[k_3k_5(k_7^* + k_8^* + k_9^*) + k_7^*k_9^*(k_3 + k_4 + k_5)]}{k_7^*k_9^*[k_3k_5(k_7 + k_8 + k_9) + k_7k_9(k_3 + k_4 + k_5)]} \quad (77)$$

$$^{D}K_{iip} = \frac{(k_8^* + k_9^*)[k_3k_5(k_7 + k_8 + k_9) + k_7k_9(k_3 + k_4 + k_5)]}{(k_8 + k_9)[k_3k_5(k_7^* + k_8^* + k_9^*) + k_7^*k_9^*(k_3 + k_4 + k_5)]} \quad (78)$$

where rate constants for deuterium are marked by an asterisk. The product of ^{D}V and $^{D}K_{iip}$ is

$$^{D}V^{D}K_{iip} = \frac{k_7k_9(k_8{}^* + k_9{}^*)}{k_7{}^*k_9{}^*(k_8 + k_9)} = \frac{k_7'}{k_7^{*\prime}} \quad (79)$$

where k_7' is the forward net rate constant for the multistep isomerization. If only one step in the Iso segment is sensitive to the isotope, then Eq. (79) can be rearranged to Eq. (80):

$$^{D}V^{D}K_{iip} = \frac{^{D}K + {}^{D}K_{eq}C_r}{1 + C_r} \quad (80)$$

where ^{D}k is the intrinsic isotope effect, C_r the reverse commitment to proton transfer in the isomerization segment, and $^{D}K_{eq}$ the equilibrium isotope effect.[31] For example, if $^{D}k = k_7/k_7^*$, then $C_r = k_8/k_9$ and $^{D}K_{eq} = k_7k_8^*/k_7^*k_8$. If C_r is small or there is but one step in the isomerization, then Eq. (80) reduces to

$$^{D}V^{D}K_{iip} = {}^{D}k \quad (81)$$

Equation (81) presents a curious relationship that is useful to think about. It states that if an isomerization is partially rate-limiting, D_2O will decrease V but increase K_{iip} by similar proportions: if very slow, the intrinsic isotope effect will be fully expressed on V but not at all on K_{iip}; and if relatively fast and therefore not significant to V, the isotope effect may be fully expressed only on K_{iip}. However, at both extremes K_{iip} itself becomes large (see Fig. 1) and may be difficult to detect. For example, pepsin hydrolysis of Leu-Ser-*p*-nitro-Phe-Nle-Ala-Leu-OMe gives $^{D}V = 1.5 \pm 0.02$ and $^{D}(V/K) = 0.84 \pm 0.21$, indicative of an isotopically sensitive Iso mechanism[29] but noncompetitive product inhibition by Leu-Ser-*p*-nitro-Phe is barely detectable with $K_{iip} = 5057 \pm 2700\ \mu M$ (versus a

$K_p = 36 \pm 10\ \mu M$ at pH 5 and 25°). Nevertheless, K_{iip} increases to $6350 \pm 4960\ \mu M$ in D_2O, consistent with the postulated Iso mechanism (K. L. Rebholz and D. B. Northrop (1993), unpublished results). When the numbers are standard errors are combined in Eq. (81) the limits of ${}^{D}k$ are calculated to be 1.5 to 2.3, which is surprisingly precise given the size of the errors. Similarly, Steiner *et al.*[52] observed solvent isotope effects on V but not on V/K of carbonate dehydratase, and these are consistent with an Iso mechanism. However, K_{iip} was ambiguous within initial velocity measurements[53] and undetectable in analysis of progress curves (F. B. Simpson and D.B. Northrop, unpublished results), indicative of a rate-limiting isomerization.

Conclusion

Methods for detection and analysis of enzymatic Iso mechanisms have been simplified and expanded. The simplification is not the result of simplifying assumptions, but rather of a conversion of kinetic terms into standard nomenclature together with the derivation of new equations. The expansion includes (a) new methods to fit progress curves to detect and quantify noncompetitive product inhibition, (b) a second method to detect α_{iso} based on noncompetitive dead-end inhibitors, and (c) an analysis of the expression of kinetic isotope effects from within an Iso segment, together with a means to calculate the intrinsic isotope effect. With less intimidating algebra than offered previously, enzymologists should find the consideration of an Iso mechanism more inviting, and, inevitably, more enzymes will be found to possess Iso mechanisms.

The results to date show that the addition and release of water and protons from an enzyme active site need not be in rapid equilibrium as generally has been assumed. This finding alone necessitates the reanalysis of published data for enzymes in which the movement of protons or water are involved, and this projected analysis most likely will produce many candidates for further study. Considerations of Iso mechanisms also will appeal to medicinal chemists seeking to avoid metabolic resistance, which occurs when concentrations of substrates become high and overcome inhibition caused by drugs which act as simple competitive inhibitors.[51] Metcalf *et al.*[52] have proposed that more effective drugs might be developed if forms of enzymes which accumulate along the reaction pathway

[52] H. Steiner, B.-H. Jonsson, and S. Lindskog, *Eur. J. Biochem.* **59,** 253 (1975).

[53] H. Steiner, B.-H. Jonsson, and S. Lindskog, *FEBS Lett.* **62,** 16 (1976).

[54] R. G. Duggleby and R. I. Christopherson, *Eur. J. Biochem.* **143,** 221 (1984).

[55] B. W. Metcalf, D. A. Holt, and M. A. Levy, *in* "Enzymes as Targets for Drug Design" (M. G. Palfreyman, P. P. McCann, W. Lovenberg, J. G. Temple, Jr., and A. Sjoerdsma, eds.), p. 85. Academic Press, San Diego, 1989.

were targeted for inhibitor binding, and the F form of an Iso mechanism is a most inviting candidate. Indeed, the archetype inhibitor of aspartic proteinases, pepstatin, which has served as a model for the design and synthesis of many drug candidates, binds faster to an F form of porcine pepsin than to free enzyme.[57] Whether it binds differently or more tightly are fascinating questions that remain to be addressed.

[57] Y.-K. Cho, K. L. Rebholz, and D. B. Northrop, *Biochemistry* **33,** 9637 (1994).

[10] Mechanism-Based Enzyme Inactivators

By RICHARD B. SILVERMAN

Terminology

An enzyme inactivator, in general, is a compound that produces irreversible inhibition of the enzyme; that is, it irreversibly prevents the enzyme from catalyzing its reaction.[1] Irreversible in this context, however, does not necessarily mean that the enzyme activity never returns, only that the enzyme becomes dysfunctional for an extended (but unspecified) period of time. A compound that leads to formation of a covalent bond to the enzyme often destroys enzyme activity indefinitely; however, there are many cases where the covalent bond that forms is reversible, and enzyme activity slowly returns. There are other cases of irreversible inhibition in which no covalent bond forms at all, but the compound binds so tightly to the active site of the enzyme that k_{off}, the rate constant for release of the compound from the enzyme, is exceedingly small. This, in effect, produces irreversible inhibition. Yet another type of irreversible inhibitor is one that is converted by the enzyme to a product that binds very tightly to the enzyme and, therefore, has a very slow off rate.

The name mechanism-based enzyme inactivator conjures up the idea of an enzyme inactivator that depends on the mechanism of the targeted enzyme. In the broadest sense of the term any of the inactivators described above that utilize the enzyme mechanism could be classified as mechanism-based enzyme inactivators. In fact, Pratt[2] and Krantz[3] have sug-

[1] R. B. Silverman, "The Organic Chemistry of Drug Design and Drug Action," Chap. 5. Academic Press, San Diego, 1992.
[2] R. F. Pratt, *BioMed. Chem. Lett.* **2,** 1323 (1992).
[3] A. Krantz, *BioMed. Chem. Lett.* **2,** 1327 (1992).

gested this general classification for mechanism-based enzyme inhibitors. However, there is a more narrow definition of mechanism-based enzyme inactivator that I have invoked previously[4] and which is used here. In this chapter a mechanism-based enzyme inactivator is defined as an unreactive compound whose structure resembles that of either the substrate or product of the target enzyme, and which undergoes a catalytic transformation by the enzyme to a species that, prior to release from the active site, inactivates the enzyme. By this definition the ultimate cause for inactivation can result from any of the above-cited covalent or noncovalent mechanisms, provided a catalytic step was required to convert the inactivator to the inactivating species and the species responsible for inactivation did not leave the active site prior to inactivation.

The inactivators as defined above can be differentiated from other types of enzyme inactivators, such as affinity labeling agents, transition state analogs, and slow, tight-binding inhibitors. Affinity labeling agents are generally reactive (electrophilic) compounds that alkylate or acylate enzyme nucleophiles. Often more than one enzyme nucleophile reacts with these compounds, and if a crude system containing more than one enzyme is being used, reactions with multiple enzymes can occur. The closer that the structure of the affinity labeling agent resembles that of the substrate for the target enzyme, the greater the specificity that will be attained. A less reactive variation of affinity labeling agents, termed quiescent affinity labeling agents, has been described.[3] These inactivators contain an electrophilic carbon that is stable in solution in the presence of external nucleophiles but which can react with an enzyme active site nucleophile involved in covalent catalysis.

Transition state analogs depend on the property of enzymes to bind substrates most tightly at the transition states of reactions. A transition state analog is a compound whose structure resembles that of the hypothetical transition state (or intermediate) structure. Because of this structural similarity to the transition state structure, the enzyme binds these compounds exceedingly tightly, leading to very small k_{off} values.

Slow, tight-binding inhibitors have relatively slow rates of binding (i.e., the k_{on} values are relatively small and, as a result, time-dependent inhibition is observed), and then they form complexes that are exceedingly stable (i.e., the k_{off} rate constants are even smaller). The causes for the slowness and tightness of binding of these inhibitors can be a conformational change in the enzyme during binding, a change in the protonation state of the enzyme, displacement of a water molecule at the active site, or

[4] R. B. Silverman, "Mechanism-Based Enzyme Inactivation: Chemistry and Enzymology," Vols. 1 and 2, CRC Press, Boca Raton, 1988.

$$\mathrm{E + ALA} \underset{k_{\mathrm{off}}}{\overset{k_{\mathrm{on}}}{\rightleftharpoons}} \mathrm{E \cdot ALA} \xrightarrow{k_2} \mathrm{E\text{-}ALA'}$$

SCHEME 1

reversible, covalent bond formation. Although these are often noncovalent inactivators, their affinities for the enzyme are so great that sometimes stoichiometric amounts of inactivator are sufficient to inactivate the enzyme.

A mechanism-based enzyme inactivator, by the definition used here, is a compound that is transformed by the catalytic machinery of the enzyme into a species that acts as an affinity labeling agent, a transition state analog, or a tight-binding inhibitor (either covalent or noncovalent) prior to release from the enzyme.

Basic Kinetics of Enzyme Inactivators

The different types of enzyme inactivators can be differentiated by their kinetic schemes. An affinity labeling agent (ALA) has a kinetic mechanism as shown in Scheme 1. Generally, the k_{on} and k_{off} rates are very fast, thereby establishing the equilibrium with E · ALA rapidly. A time-dependent reaction of an enzyme nucleophile with E · ALA leads to the covalent enzyme adduct depicted as E–ALA′. Because of the covalent nature of this type of inactivator, dialysis or gel filtration would not restore enzyme activity.

A transition state analog (TSA) generally is a tight-binding, noncovalent inactivator which displays the kinetic scheme shown in Scheme 2. In this case k_{on} is generally rapid, but k_{off} is slow. Because initial binding is rapid, inactivation occurs immediately, and there is no time dependence to the inactivation. Also, dialysis or gel filtration would restore the enzyme activity, because of the noncovalent nature of the interaction.

$$\mathrm{E + TSA} \underset{k_{\mathrm{off}}}{\overset{k_{\mathrm{on}}}{\rightleftharpoons}} \mathrm{E \cdot TSA}$$

SCHEME 2

$$E + STBI \underset{k_{off}}{\overset{k_{on}}{\rightleftharpoons}} E \cdot STBI$$

SCHEME 3

The slow, tight-binding inhibitor (STBI) has a relatively small k_{on} and an even smaller k_{off}, as shown in Scheme 3. These types of inactivators have been known to be both noncovalent and covalent in nature. Because of the slowness of the binding, a time-dependent loss of enzyme activity is observed. Dialysis or gel filtration can restore enzyme activity, but it may require longer dialysis times or a longer gel filtration column.

A mechanism-based enzyme inactivator (MBEI) requires a step to convert the compound to the inactivating species (k_2), as depicted in Scheme 4. This step, which is generally responsible for the observed time dependence of the enzyme inactivation, usually is irreversible and forms a new complex (E · MBEI′) which can have three fates: (1) if MBEI′ is not reactive, but forms a tight complex with the enzyme, then the inactivation may be the result of a noncovalent tight-binding complex (E · MBEI′); (2) if MBEI′ is a reactive species,[5] then a nucleophilic, electrophilic, or radical reaction with the enzyme may ensue (k_4) to give the covalent complex E–MBEI″; and (3) the species generated could be released from the enzyme as a product (k_3). Most often these inactivators result in covalent bond formation with the enzyme, and, therefore, dialysis or gel filtration does not restore enzyme activity.

The two principal kinetic constants that are useful in describing mechanism-based enzyme inactivators are k_{inact} and K_I. Based on Scheme 4, k_{inact} is a complex mixture of k_2, k_3, and k_4 [Eq. (1)], and K_I is a complex

$$k_{inact} = k_2k_4/(k_2 + k_3 + k_4) \quad (1)$$

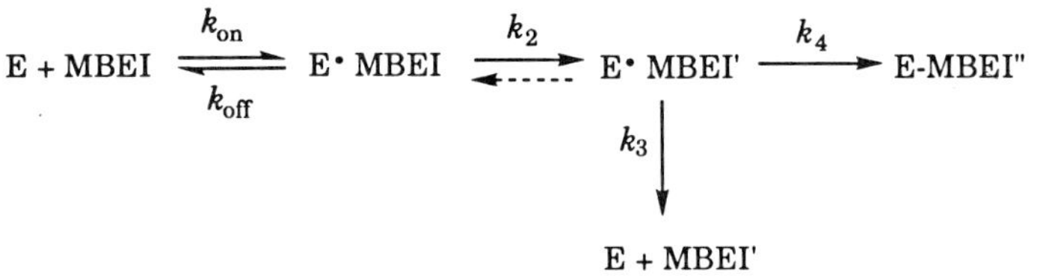

SCHEME 4

[5] The reactive species generated may be like an affinity labeling agent, in which case it could undergo a reaction with an active site nucleophile; or it could be radical in nature, in which case it could react with an enzyme radical; or it could be a nucleophile, in which case it could react with an electrophilic species, such as an oxidized cofactor on the enzyme.

mixture of k_{on}, k_{off}, k_2, k_3, and k_4 [Eq. (2)].[6] Only if k_2 is rate determining ($k_2 \ll k_4$) and k_3 is 0 (or nearly 0), will $k_{inact} = k_2$. Values given for k_{inact}

$$K_I = [(k_{off} + k_2)/k_{on}][(k_3 + k_4)/(k_2 + k_3 + k_4)] \tag{2}$$

are assumed to represent the inactivation rate constants at infinite concentrations of inactivator.

An expression can be derived[7] that relates the enzyme concentration (E), the inactivator concentration (I), k_{inact}, and K_I [Eq. (3)]. The half-life

$$\frac{\partial \ln E}{\partial t} = \frac{k_{inact} I}{K_I + I} \tag{3}$$

for inactivation ($t_{1/2}$), then, is described by Eq. (4). Extrapolation to infinite inactivator concentrations reduces Eq. (4) to $t_{1/2} = \ln 2/k_{inact}$. These are

$$t_{1/2} = \frac{\ln 2}{k_{inact}} + \frac{\ln 2\, K_I}{k_{inact}\, I} \tag{4}$$

the basic equations needed for typical mechanism-based enzyme inactivators; more complex discussions of kinetics related to mechanism-based enzyme inactivators can be found elsewhere.[8,8a,8b] The determination of these kinetic constants is described in the section entitled Experimental Protocols for Mechanism-Based Enzyme Inactivators.

The ratio of product release to inactivation is termed the partition ratio and represents the efficiency of the mechanism-based enzyme inactivator. When inactivation is the result of the formation of an E–MBEI″ adduct, the partition ratio is described by k_3/k_4. The partition ratio depends on the rate of diffusion of MBEI′ from the active site, its reactivity, and the proximity of an appropriate nucleophile, radical, or electrophile on the enzyme for covalent bond formation. It does not depend on the initial inactivator concentration. There are some cases where the partition ratio

[6] S. G. Waley, *Biochem. J.* **185,** 771 (1980); S. G. Waley, *Biochem. J.* **227,** 843 (1985).

[7] M. J. Jung and B. W. Metcalf, *Biochem. Biophys. Res. Commun.* **67,** 301 (1975).

[8] R. B. Silverman, "Mechanism-Based Enzyme Inactivation: Chemistry and Enzymology," Vol. 1, p. 12. CRC Press, Boca Raton, Florida, 1988; F. Garcia-Canovas, J. Tudela, R. Varon, and A. M. Vazquez, *J. Enzyme Inhibition* **3,** 81 (1989); J. Escribano, J. Tudela, F. Garcia-Carmona, and F. Garcia-Canovas, *Biochem. J.* **262,** 597 (1989); R. Varon, M. Garcia, F. Garcia-Canovas, and J. Tudela, *J. Mol. Catal.* **59,** 97 (1990).

[8a] T. Funaki, S. Ichihara, H. Fukazawa, and I. Kuruma, *Biochim. Biophys. Acta* **1118,** 21 (1991).

[8b] D. J. Kuo and F. Jordan, *Biochemistry* **22,** 3735 (1983).

has been shown to be 0, that is, every turnover of inactivator produces inactivated enzyme.[9]

For substrates and reversible competitive inhibitors, when k_{on} and k_{off} are very large, then the K_m and K_i represent the dissociation constants for the breakdown of the E · S and E · I complexes, respectively. The K_i value is obtained experimentally by determining the effect on the rate of conversion of subsaturating concentrations of substrate to product on addition of a constant amount of an inhibitor. The K_I value, a term used for mechanism-based enzyme inactivators, however, is obtained experimentally (see section entitled Experimental Protocols for Mechanism-Based Enzyme Inactivators) by determining the effect on the rate of inactivation by a change in the inactivator concentration. When k_{on} and k_{off} are very large and k_2 is rate determining, then K_i (for initial binding of the inactivator to the enzyme) and K_I should be the same for the same compound under identical conditions. When k_4 becomes partially rate determining, then $K_I \geq K_i$. Just as the K_m represents the concentration of substrate that gives one-half the maximal velocity, the K_I represents the concentration of the inactivator that produces one-half the maximal rate of inactivation.

Uses of Mechanism-Based Enzyme Inactivators

The two principal areas where mechanism-based enzyme inactivators have been most useful are in the study of enzyme mechanisms and in the design of new potential drugs.

Study of Enzyme Mechanisms

The reason mechanism-based enzyme inactivators are so important to the study of enzyme mechanisms is because they are really nothing more than modified substrates for the target enzymes. Once inside the active site of the enzymes, though, they are converted to products that inactivate the enzyme. However, the mechanisms by which these compounds are converted to the species that inactivate the enzymes proceed, at least initially, by the catalytic mechanisms for normal substrates of the enzymes. Therefore, whatever information can be obtained regarding the inactivation mechanism is directly related to the catalytic mechanism of the enzyme. One approach for the use of mechanism-based inactivators to elucidate enzyme mechanisms would involve formulating a hypothesis regarding the catalytic mechanism of the enzyme, then designing a com-

[9] R. B. Silverman and B. J. Invergo, *J. Am. Chem. Soc.* **25,** 6817 (1986).

pound that would require that mechanism to convert it to the activated form. If inactivation occurs, it supports the hypothetical mechanism. Of course, it does not prove the mechanism, because there may be other mechanisms that were not considered that also would lead to enzyme inactivation. An alternative approach arises when a time-dependent inactivator is serendipitously discovered, and then mechanisms are contemplated that might lead this compound to inactivate the enzyme.

Drug Design

Most enzyme inhibitor drugs (EID) are noncovalent, reversible inhibitors. The basis for their effectiveness is the tightness of their binding to the enzyme (i.e., the stability of E · EID and a small k_{off}), so that they compete with the binding of substrates (S) for the enzyme (Scheme 5). The efficacy of the drug continues as long as the enzyme is complexed with the drug (E · EID). Because the enzyme concentration is low and fixed, the equilibrium between E + EID and E · EID will depend on the concentrations of EID and S. When the concentration of EID diminishes because of metabolism, the concentration of E · EID diminishes and the concentration of E · S increases. To maintain the pharmacological effect of the drug, administration of the drug several times a day, then, becomes necessary. An effective mechanism-based enzyme inactivator, however, could form a covalent bond to the enzyme, thereby preventing the dissociation of the inactivator from the enzyme. This would mean that frequent administration of the drug would not be necessary. Inactivation of an enzyme, however, induces gene-encoded synthesis of that enzyme, but this could take hours to days before sufficient newly synthesized enzyme is present.

Although affinity labeling agents also could form a covalent bond to the enzyme, the reactivity of these compounds renders them generally unappealing for drug use because of the possibility that they could react with multiple enzymes or other biomolecules, thereby leading to toxicity and side effects. Mechanism-based enzyme inactivators, however, are

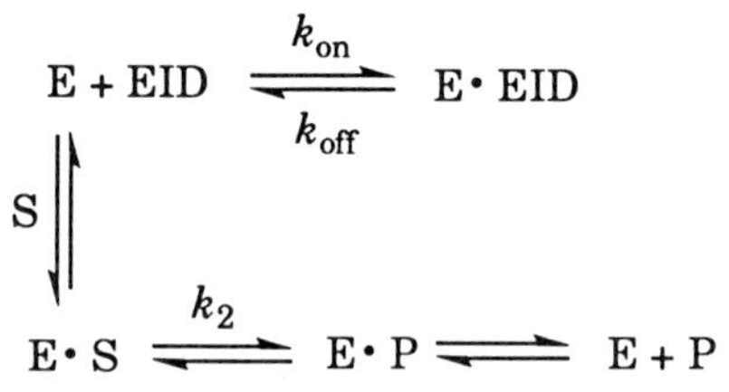

SCHEME 5

unreactive compounds, so nonspecific reactions with other biomolecules would not be a problem. Only enzymes that are capable of catalyzing the conversion of these compounds to the form that inactivates the enzyme, and also have an appropriately positioned active site group to form a covalent bond, would be susceptible to inactivation. With a sufficiently clever design it should be possible to minimize (to one?) the number of enzymes that would be affected. Provided the partition ratio is zero or a small number, in which case potentially toxic product release would not be important, then a mechanism-based enzyme inactivator could have the desirable drug properties of specificity and low toxicity.

Criteria for Mechanism-Based Enzyme Inactivators

The following experimental criteria have been established to characterize mechanism-based enzyme inactivators. The protocols to support these criteria are described in the next section.

Time Dependence of Inactivation

Following a rapid equilibrium between the enzyme and the mechanism-based enzyme inactivator to give the E · MBEI complex (see Scheme 4 above) there is a slower reaction that ensues to convert the inactivator to the form that actually inactivates the enzyme (k_2). This produces a time-dependent loss of enzyme activity in which the pH-inactivation rate profile should be consistent with the pH–rate profile for substrate conversion to product.

Saturation

Formation of the E · MBEI complex occurs rapidly, and the rate of inactivation is proportional to added inactivator until sufficient inactivator is added to saturate all of the enzyme molecules. Then there is no further increase in rate with additional inactivator, that is, saturation kinetics are observed.

Substrate Protection

Mechanism-based enzyme inactivators act as modified substrates for the target enzymes and bind to the active site. Therefore, addition of a substrate or competitive reversible inhibitor slows down the rate of enzyme inactivation. This is referred to as substrate protection of the enzyme.

Irreversibility

Most cases of mechanism-based inactivation result in the formation of covalent irreversible adducts. Consequently, dialysis or gel filtration does not restore enzyme activity. Some alternate substrates for enzymes give products that form tight, noncovalent complexes or weak covalent complexes with the enzyme. These enzyme · product complexes may dissociate during dialysis or gel filtration. It is not clear at what point in the continuum of enzyme–product complex stabilities that an alternate substrate should be reclassified as a mechanism-based enzyme inactivator.

Inactivator Stoichiometry

When a radioactively labeled inactivator is used, it is possible to measure the number of inactivator molecules attached per enzyme molecule. Because mechanism-based enzyme inactivators require the catalytic machinery at the active site of the enzyme to convert them to the form that inactivates the enzyme, at most one inactivator molecule should be attached per enzyme active site. In the case of multimeric enzymes not all of the subunits have functional active sites. Sometimes only one-half of the subunits are catalytically active, and, therefore, only one-half as much inactivator is incorporated per enzyme inactivated. This "half-sites" reactivity may occur because of negative cooperativity after one-half of the subunits are labeled.

Involvement of Catalytic Step

One of the most important aspects of mechanism-based enzyme inactivators is that they require the enzyme to convert them to the species that actually inactivates the enzyme. Therefore, it must be demonstrated that there is an enzyme-catalyzed reaction on the compound that is responsible for the activity of the inactivator.

Inactivation Prior to Release of Active Species

Another requirement for mechanism-based enzyme inactivators is that the species which is formed by enzyme catalysis inactivates the enzyme prior to release from the active site of the enzyme. It must be shown that a product which is released does not return to the enzyme to cause inactivation. Particularly if the activated product is highly electrophilic, release from the enzyme would be equivalent to the generation of an affinity labeling agent. A species that inactivates the enzyme after release may produce inactivation by attachment to a site other than the active site, or the compound may have undergone a rearrangement prior to its return to the enzyme, which may confuse the conclusion about the inactivation mechanism.

Experimental Protocols for Mechanism-Based Enzyme Inactivators

Each of the above-mentioned criteria for mechanism-based enzyme inactivators can be tested experimentally as described in this section.

Time Dependence of Inactivation

The experiment to determine time dependence of inactivation is as follows. All of the enzyme assay components except substrate are equilibrated at the desired temperature (generally 25° or 37°) in a preincubation tube. It is generally preferable to use a buffer at or near the pH optimum for the enzyme, because the maximum rate of inactivation should be observed at the pH optimum of the enzyme. Sufficient enzyme for 6–10 assays is added to give a solution of *at least* 50 times the concentration necessary for a standard assay of enzyme activity. When investigating a new inactivator, a series of arbitrary inactivator concentrations (e.g., 10 μM, 100 μM, 1 mM, and 10 mM final concentrations in the preincubation tube) are tried in separate preincubation tubes to determine the appropriate concentration range to monitor time-dependent inactivation. Sometimes it is more convenient or appropriate to add the enzyme last to initiate the reaction, particularly if the enzyme is somewhat unstable. These experiments are much quicker if a continuous enzyme assay (such as a spectrophotometric assay) is possible rather than a single time-point assay. With a continuous assay the relative rates can be observed immediately, and if complete inactivation has occurred at an early time point, the assay can be discontinued and a lower concentration of inactivator tried. Also, if no change in enzyme activity has occurred after an hour or so, the experiment can be aborted and a higher concentration of inactivator used. With a single-point assay, the rates generally are not known until the entire experiment has been completed. Of course, in a single-point assay, the time selected to make the assay measurement must be in the linear part of the assay rate.

As soon as all of the components of the preincubation tube are added, a stopwatch is begun to mark that time as time zero. Once the solution has been mixed, an aliquot is removed (preferably within 15 sec of inactivator addition) and is diluted *at least* 50-fold into the enzyme assay mixture containing the substrate at saturating concentration and preequilibrated at the desired temperature (it is most convenient to use 25° so that everything is at about room temperature); then the assay is carried out. This 50-fold dilution is why the enzyme in the preincubation tube must be at least 50 times more concentrated than is required for the assay. Even greater dilution would be desirable, because the purpose of the dilution is to quench the inactivation reaction by rapidly decreasing the inactivator concentration and rapidly increasing the substrate concentration so that the substrate is able to occupy the active sites of uninactivated enzyme molecules. If a 100-fold dilution is used, then 100 times more concentrated enzyme is used in the preincubation tube than is needed to monitor an assay. However, it should be noted that when the enzyme concentration is too concentrated in the preincubation tube, rates of inactivation decrease, possibly because of the increased viscosity or protein–protein interactions.

Periodically (seconds to hours, depending on the observed inactivation rate) identical aliquots are removed from the preincubation tube, diluted as above, and assayed. With the continuous assay it is easy to determine when to take aliquots, because the change in enzyme activity is observed immediately. Additional time points need to be taken with a single-point assay because the rates generally are not known until after the entire experiment is completed. Once a convenient concentration range is determined (one in which complete inactivation occurs within a couple of hours or less), the inactivation experiment is repeated with a series of concentrations (at least five different concentrations, but more is better) in that range spanning about an order of magnitude of concentrations. At least five time points, over at least two half-lives, should be taken; more time points will give more accurate data. Loss of enzyme activity should be monitored until no activity remains to determine if there is a change in kinetics with time. When a crude enzyme system is used or when a single-point assay is used, it is best to do duplicate or triplicate runs per concentration and average the results.

All inactivation experiments are monitored relative to a control sample which is done exactly as the experiment except without inactivator added (an equal volume of buffer is added instead). The enzyme activity in this sample is set to 100% at each time point. This control takes into account the normal loss of enzyme activity under the conditions of the experiment not as a result of the presence of the inactivator.

Once the data have been collected, the logarithm of the percentage of enzyme activity remaining (as measured by whatever means the enzyme activity is determined) relative to the uninactivated control is plotted against the time of preincubation (Fig. 1). Increasing inactivator concentrations should produce increased rates of inactivation. Often these plots exhibit pseudo-first-order kinetics because the inactivator is added in a large excess relative to the enzyme concentration. However, this is the ideal case. Equations have been derived[8a] showing that pseudo-first-order

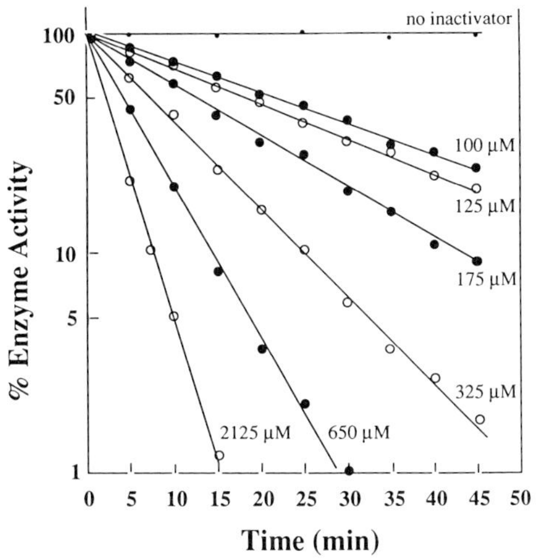

FIG. 1. Time- and concentration-dependent inactivation of an enzyme by a mechanism-based enzyme inactivator. The inactivator concentrations for each line are given.

kinetics are followed only when the ratio of the initial concentration of the mechanism-based inactivator to that of the enzyme is greater than the partition ratio. When the inactivator is very potent (i.e., has a low K_I and a large k_{inact}), it may not be possible to measure the inactivation rates fast enough unless the inactivator concentration is lowered to a point approaching the concentration of the enzyme; in that case, pseudo-first-order kinetics would not be observed. To slow down the inactivation rate, the use of nonideal conditions, such as lowering the temperature or changing the pH of the buffer from that needed to get optimum enzyme activity, can be used. However, not only will the rate change, but all kinetic constants (e.g., K_I and k_{inact}; see the next section on saturation) also will be altered. Because of the effects of temperature, pH, ionic strength, and buffer, kinetic constant comparisons cannot be made between inactivators measured under different conditions.

Another common problem in doing these experiments arises when the partition ratio for the inactivator is very high and the concentration of the inactivator diminishes during the time course of the inactivation experiment. This may lead to non-pseudo-first-order kinetics (Fig. 2). If this is suspected, then some type of analytical determination of the inactivator concentration during the time period of the experiment should be carried out, such as high-performance liquid chromatography (HPLC) analysis. If inactivator consumption is a problem, then the early time points would be more dependable than the later ones.

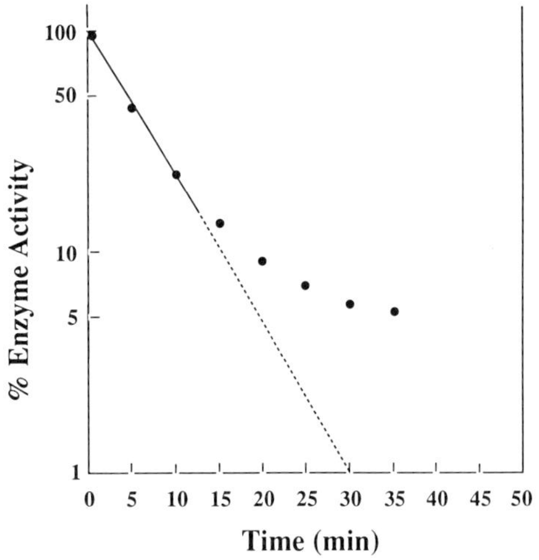

FIG. 2. Non-pseudo-first-order inactivation as a result of consumption of inactivator.

Non-pseudo-first-order kinetics also would be observed if a product generated is a good inhibitor of the enzyme. As the concentration of the metabolite increases, more inhibition occurs, and this prevents the inactivator from being effective. This problem is more prevalent with compounds having high partition ratios. In these cases the earlier time points and higher inactivator concentrations give more reliable data.

In the two cases above of non-pseudo-first-order kinetics, there is an upward deviation to the loss of enzyme activity (i.e., the rate decreases with time) relative to pseudo-first-order kinetics. In some cases the opposite may be observed, namely, there is an increase in the rate of inactivation with time (Fig. 3). The rate of increase of enzyme inactivation does not have to be as pronounced as is shown in Fig. 3. In general, this result suggests that the initial compound incubated with the enzyme is converted to another compound, which is the actual inactivator of the enzyme. As the concentration of the actual inactivator species increases, the rate of inactivation increases, until it reaches the maximum saturation rate. If a fresh aliquot of enzyme is then added to that same solution, which already is saturated with the actual inactivator species, then there will be no lag period to inactivation, and the maximum inactivation rate will be observed immediately.

There are several possible sources for the formation of the actual inactivator. It could be the result of a nonenzymatic conversion of the presumed inactivator into either a reactive species or a new mechanism-

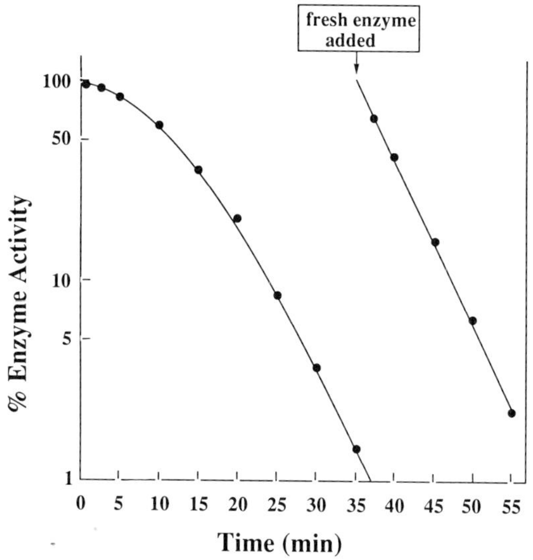

FIG. 3. Non-pseudo-first-order inactivation as a result of conversion of a presumed inactivator to the actual inactivating species.

based inactivator. Alternatively, the presumed inactivator may be a pro-mechanism-based inactivator; that is, the enzyme may convert the compound to a mechanism-based enzyme inactivator that is released into solution, and this new compound is the actual inactivator of the enzyme. Alternatively, the enzyme may convert the compound to a moderately reactive species, which escapes the active site, then, as it builds in concentration, returns to inactivate the enzyme. As described below (see the section *Inactivation Prior to Release of Activated Species*), a test for the generation of a reactive species is to add an electrophile trapping agent (such as a thiol) and see if this prevents inactivation. In any case, if the actual inactivator species is released from the enzyme, the compound is not classified as a mechanism-based enzyme inactivator.

Finally, there are times when biphasic kinetics are observed. This can arise when two or more inactivation mechanisms are occurring simultaneously, if the inactivated enzyme adduct is not stable and the breakdown of this adduct is rate determining, if there is negative cooperativity and attachment of the inactivator to one subunit causes an adjoining subunit to be less active, and if there is heterogeneity of subunit composition that results in nonequivalent binding to the subunits.

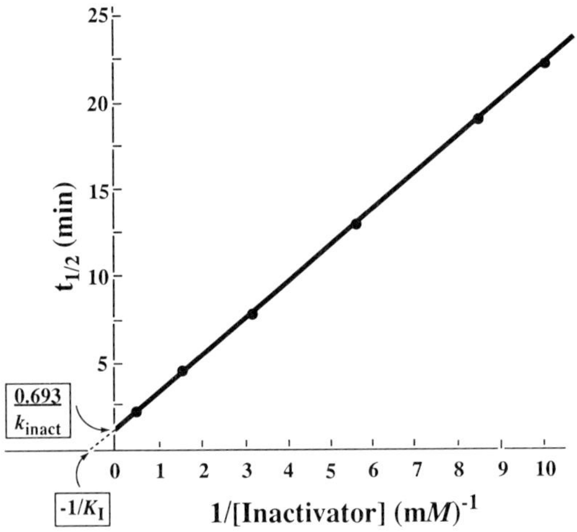

FIG. 4. Kitz and Wilson[10] replot of the half-life of enzyme inactivation as a function of the reciprocal of the mechanism-based inactivator concentration.

Saturation and Determination of K_I and k_{inact}

Saturation is demonstrated by carrying out the time-dependent experiment described above at several (at least five) different concentrations at a constant enzyme concentration. If little or no change in inactivation rate is observed with increasing concentration of inactivator, it suggests that the initial inactivator concentration selected is near saturation. In this case, lower concentrations should be used to produce sufficiently large differences in the rates of inactivation so that results are not ambiguous. The half-life for inactivation ($t_{1/2}$) at each inactivator concentration is plotted against 1/[inactivator concentration] (known as a Kitz and Wilson plot[10]). If inactivation proceeds with saturation, then a plot like that shown in Fig. 4 will be observed. Saturation is indicated by the intersection of the experimental line at the positive y axis. When that occurs, then there is a finite rate of inactivation at infinite inactivator concentration. The point of intersection at the y axis is ln $2/k_{inact}$, from which the value of k_{inact} can be determined. The extrapolated negative x axis intercept is $-1/K_I$. From Fig. 4 the k_{inact} is 0.46 min^{-1} and the K_I is 1.48 mM.

There are several kinetics and graphics programs that can be used to obtain these results instead of determining them manually. Inactivator concentrations should be selected so that the data points are not bunched at one end of the experimental line. This allows the points to be given

[10] R. Kitz and I. B. Wilson, *J. Biol. Chem.* **237,** 3245 (1962).

equal weighting in the analysis. Therefore, choose concentrations based on 1/[I], not [I]. If K_I is low and k_{inact} is large, all concentrations of the inactivator used may give the same rate of inactivation. This indicates that saturation has already been reached. If the inactivator concentration is lowered, nonlinear kinetics may result because the inactivator concentration may approach that of the enzyme. Carrying out the experiment at nonideal conditions, for example, at a lower temperature or at a nonideal pH, may permit subsaturating concentrations of inactivator to be used. Typically, the values for k_{inact} are quite small (on the order of 10^{-3}–1 min^{-1}) relative to values of k_{cat} for substrates (10–10^4 min^{-1}).

When the experimental line intersects at the origin, then saturation is not being observed (Fig. 5). This indicates that a bimolecular reaction is occurring and that k_{inact} is fast relative to formation of the E · MBEI complex. When this is observed, alteration of the experimental conditions, for example, lowering the temperature (to slow down k_{inact}) or changing the pH, may give saturation kinetics.

Intersection of the experimental line at the negative y axis has no physical meaning (a negative rate of inactivation). When this occurs, more concentrations should be used in different concentration ranges. If increasing the inactivator concentration leads to an upward deviation of the experimental line (larger half-lives, i.e., slower rates, at higher inactivation concentrations), then the rate of inactivation may be a function of $[I]^2$, that is, there may be two binding sites or two inactivator

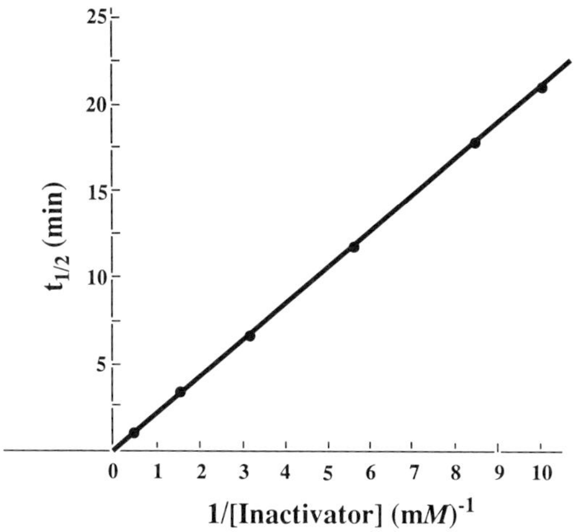

FIG. 5. Mechanism-based inactivation without saturation.

molecules involved in inactivation. Kinetics for two-site inactivation has been described.[8b]

Substrate (or Competitive Inhibitor) Protection

To show that inactivation is occurring at the active site of the enzyme, the inactivation experiment described by Fig. 1 is repeated under identical conditions in the presence and absence of a known substrate for the enzyme. Because both the mechanism-based enzyme inactivator and the substrate must bind at the active site, the presence of the substrate will prevent the binding of the inactivator, and, therefore, the rate of inactivation will decrease compared with the rate in the absence of substrate. It is best to do the experiment with several different substrate concentrations as is shown in Fig. 6. A competitive inhibitor of the enzyme also can be used to protect the enzyme from the inactivator.

Irreversibility

After inactivation is complete, the enzyme is dialyzed against several changes of buffer at 4° or subjected to gel filtration at 4° to remove any reversibly bound inactivator molecules. A noninactivated enzyme control should be run simultaneously and carried through the same operations as

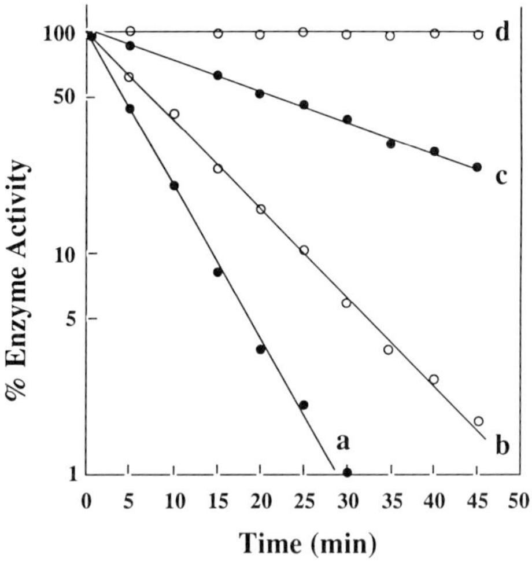

FIG. 6. Substrate protection from inactivation: (a) 650 μM inactivator with no substrate; (b) 650 μM inactivator with 100 μM substrate; (c) 650 μM inactivator with 500 μM substrate; (d) 650 μM inactivator with 2 mM substrate or no inactivator.

the inactivated enzyme. This control is required to show that under the conditions of the experiment the enzyme would still be active if it were not treated with the inactivator, and the control enzyme activity is set to 100% for comparison. If the inactivated enzyme complex is marginally stable, dialysis or gel filtration at room temperature instead of at 4° may result in partial or complete return of enzyme activity. If more than one adduct were formed, possibly only one of the adducts would decompose at room temperature with concomitant reactivation of that portion of the enzyme activity. Extended dialysis at room temperature may release more adduct. Arrhenius activation energies and free energies can be estimated if the rates of reactivation can be determined at various temperatures. Dialysis at higher or lower pH also may yield useful information about the adduct stability and structure.

Inactivator Stoichiometry

Often the most difficult or time-consuming part of the characterization of a mechanism-based inactivator is the chemical synthesis and purification of a high specific activity radioactively labeled analog. As a rule of thumb, a specific activity of 0.5 mCi/mmol [1.1×10^6 disintegrations/min (dpm)/μmol] is satisfactory for a ^{14}C-labeled compound, but higher specific activities (usually more easily done by tritium incorporation, but the tritium must be in a nonexchangeable position) will permit the use of less enzyme. Once the labeled inactivator has been obtained, it is incubated with the enzyme. If aliquots are removed during inactivation and each assayed for enzyme activity remaining and for radioactivity attached to the protein (either by acid precipitation or gel filtration followed by scintillation counting), there should be a correlation of loss of enzyme activity with gain of radioactivity; at 50% loss of enzyme activity there should be approximately 50% incorporation of total radioactivity (Fig. 7). When no activity remains, exhaustive dialysis or gel filtration is carried out to remove all of the unbound inactivator (a small volume dialysis can be carried out after several large volume dialyses in order to determine if additional radioactivity is being released). A protein assay and radioactivity determination are made, and from the specific activity of the inactivator the stoichiometry of inactivation (equivalents of compound bound per enzyme molecule) can be calculated.

This experiment is most effective if homogeneous enzyme is used, so that a direct protein and radioactivity determination can be made. However, gel electrophoresis of a labeled enzyme mixture can be done, the bands eluted, and protein and radioactivity determinations made on the eluted protein.

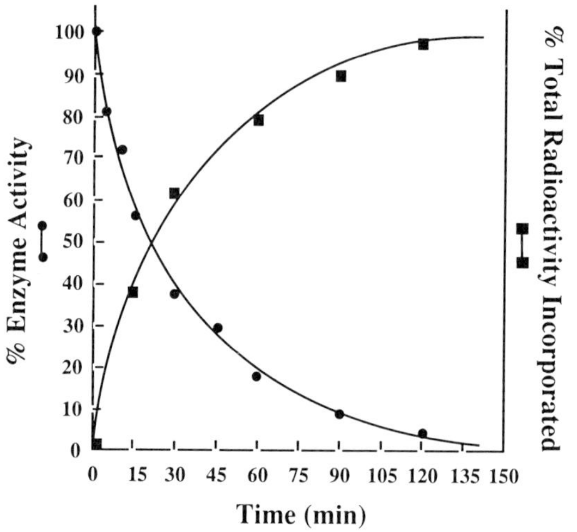

FIG. 7. Correspondence of the time-dependent inactivation of an enzyme with the incorporation of radioactivity from a radiolabeled mechanism-based inactivator. Periodically, aliquots are removed and assayed for remaining enzyme activity and for radioactivity covalently bound to the enzyme.

If some enzyme activity remains after dialysis, then only the radioactivity bound per inactivated enzyme should be calculated (multiply the total protein by the fraction inactivated). Generally, for a mechanism-based inactivator a 1:1 stoichiometry of binding is observed (1.0 ± 0.3 is usually considered to be one equivalent, given the errors in all of these determinations). If greater than a 1:1 stoichiometry is observed, it may indicate that incomplete dialysis or gel filtration was done or nonspecific covalent reactions took place, possibly as a result of an activated form of the inactivator being released during inactivation. An electrophile trapping agent, such as 2-mercaptoethanol or dithiothreitol, would be useful to have present during inactivation in order to prevent potent released electrophiles from labeling the periphery of the enzyme. Sometimes excess inactivator is tightly but reversibly bound to the protein. If the covalent adduct is stable, the enzyme can be denatured to release unstable adducts or excess reversibly bound radioactivity. Denaturants such as 8 *M* urea or 1% sodium dodecyl sulfate are very gentle means of denaturation, but trichloroacetic acid precipitation or heat denaturation also can be effective. When less than 1:1 stoichiometry is observed, it may indicate one or more of the following: (1) some of the enzyme used was inactive (possibly denatured) prior to treatment with the inactivator; (2) more than one adduct formed, but at least one of the adducts was unstable to dialysis or gel filtration; (3) in the case of multimeric enzymes, some kind of negative cooperativity was important. When a 0.5:1 stoichiometry is observed, often it indicates that half-site reactivity is occurring.

Involvement of Catalytic Step

Experiments to test the involvement of a catalytic step have the most variability of any of the experiments for mechanism-based enzyme inactivators because they depend on the specific reaction catalyzed by the particular enzyme. If an oxidation or reduction of a cofactor is involved, then this may be observed spectrophotometrically. If a C–H bond is cleaved, then either deuteration or tritiation of that hydrogen can be used; a deuterium isotope effect or release of tritium, respectively, can be monitored. If part of the inactivator is cleaved during inactivation, radioactive labeling of that part of the compound and demonstration that it is released during inactivation would be an appropriate experiment. The important aspect is to show that the enzyme is involved in the activation process.

Inactivation Prior to Release of Activated Species

There are several experiments that can be carried out to test whether inactivation occurs before or after an activated species is released from the enzyme. If inactivation occurs as a result of a species that is released, then the rate of inactivation may increase with time as the species increases in concentration outside of the enzyme. In this case pseudo-first-order kinetics would not be evident; rather, there would be an increase in the rate of inactivation over time (see Fig. 3 in the section on time dependence of inactivation). However, because the reactive species also may react with the buffer, buffer quenching may compete with the enzyme for the released species.

Another experiment to test this phenomenon is first to measure the rate of inactivation of the enzyme, then add a fresh aliquot of enzyme. If the inactivating species is building in concentration, then the rate of inactivation of the second aliquot of enzyme will be as fast as the maximum rate of inactivation with the first aliquot or faster than that for the first aliquot, if saturation had not yet been achieved with the first aliquot (see Fig. 3 and note that the lag period does not have to be as long as shown). With a mechanism-based enzyme inactivator there should be no difference in inactivation rates no matter how many successive aliquots are added (unless the concentration of the inactivator is depleted).

A third test is to have a trapping agent present in the preincubation solution to react with any electrophilic or radical species generated and released by the enzyme. If inactivation occurs subsequent to release of an activated species, then the trapping agent would prevent that species from inactivating the enzyme. In this case, there would be a considerable decrease in the inactivation rate in the presence of the trapping agent. In fact, it may totally prevent inactivation from occurring. The presence of the trapping agent should have no effect on the inactivation rate of a mechanism-based enzyme inactivator. Because thiols are both excellent nucleophiles and radical traps, 2-mercaptoethanol or dithiothreitol are commonly used in this experiment. It also is useful to use a bulky thiol, such as reduced glutathione, to minimize the possibility of the thiol entering the active site and trapping the reactive species prior to its release from the active site.

Determination of Partition Ratio

There are three common methods for the determination of the partition ratio. When a product is released, the rate constant that relates to the formation of a product is k_{cat}, the catalytic rate constant, determined from Michaelis–Menten kinetics. The rate constant that relates to inactivation is k_{inact}, determined as described above. The ratio k_{cat}/k_{inact}, then, defines the partition ratio. If k_{cat} can be measured easily, then this is the simplest procedure to follow. When a potent mechanism-based enzyme inactivator is used or if the partition ratio is small, inactivation may occur too rapidly for accurate k_{cat} measurements to be made.

Another method for partition ratio determination requires the synthesis of a radioactively labeled analog of the inactivator and is only effective for covalent inactivation. After complete inactivation occurs, the small molecules are separated from the enzyme by gel filtration, ultrafiltration, protein precipitation, or microdialysis. The latter technique involves suspension of the labeled enzyme solution contained in narrow dialysis tubing into a test tube containing a minimum volume of buffer (a dialysis clip at the top will hold it suspended; the bottom of the tubing is knotted). A flea stir bar is added to the test tube, and the buffer (3–5 ml) is stirred while dialysis takes place. The dialyzate is saved, a fresh aliquot of buffer is added, and dialysis is continued. After the second microdialysis is completed, the combined dialyzates are retained for further analyses of the small molecules. These two dialyses remove a large percentage of the radioactive small molecules. If, for example, 300 μl of enzyme solution is dialyzed twice in 3 ml each of buffer, then greater than 99% of the small molecules will be separated from the enzyme solution. The enzyme solution is then exhaustively dialyzed to remove the traces of remaining unbound radioactivity, and the stoichiometry of labeling is determined by protein and radioactivity assays. Because a large excess of inactivator is generally used for inactivation, the small molecules must be chromatographed (ion-exchange or reversed-phase silica gel chromatography) to separate the excess inactivator from the products. The amount of radiolabeled metabolites produced per radiolabeled enzyme can be determined to give the partition ratio directly.

Whereas standard gel filtration techniques dilute the small molecule fraction considerably and require more time to set up than microdialysis, a modified version of gel filtration, described by Penefsky,[11] is fast and results in much less dilution than the standard method. Dilution is not a problem with ultrafiltration, but protein has a tendency to adhere to the filtration membrane. A fast and convenient micro ultrafiltration tool, however, is the Centricon centifugal microconcentrator[12] which allows recovery of both filtrate and protein without dilution (generally, though, buffer is added to the concentrated protein to remove the last of the small molecules, so there may be some dilution).

A third approach for the determination of the partition ratio is to titrate the enzyme with the inactivator; this measures the number of inactivator molecules required to inactivate the enzyme completely. Increasing concentrations of inactivator are added to a fixed amount of enzyme, and each is incubated until no more enzyme activity is observed. After dialysis or gel filtration a plot is constructed of the enzyme activity remaining versus the ratio of amount of inactivator per enzyme active sites (Fig. 8). If the enzyme solutions of all of the samples are normalized to a constant protein concentration after dialysis or gel filtration, then direct comparisons can be made between samples.

In the ideal case, a straight line will be obtained from 100 to 0% enzyme activity remaining. The intercept with the x axis gives the number of inactivator molecules required to inactivate each enzyme molecule (the turnover number). This number includes the one molecule of inactivator required to inactivate the enzyme; consequently, the partition ratio is the turnover number minus one (assuming there is a 1:1 stoichiometry of inactivator and enzyme). Often the higher ratio data points deviate from linearity because of product inhibition or product protection of the enzyme from further inactivation (Fig. 9). In this case the linear portion (the lower inactivator concentrations) should be used for extrapolation to the turnover number. If the inactivated enzyme solution is not dialyzed or gel filtered prior to the determination of the enzyme activity remaining, and if a product formed is a potent reversible inhibitor, then each assay

[11] H. S. Penefsky, *J. Biol. Chem.* **252,** 2891 (1977); H. Penefsky, this series, Vol. 56, p. 527.

[12] Registered trademark of the Amicon Division of W. R. Grace & Co., Beverly, MA 01915-1065; telephone number 1-800-343-0696.

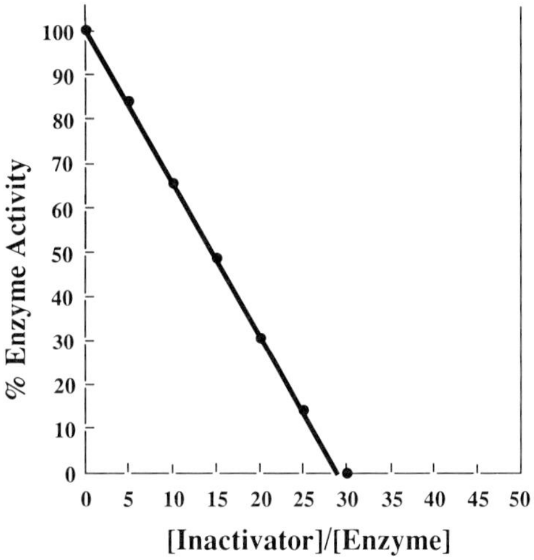

FIG. 8. Titration of an enzyme with a mechanism-based enzyme inactivator. The loss of enzyme activity is measured as a function of the ratio of the inactivator to enzyme concentration.

will show an artificially low enzyme activity, resulting in a falsely low partition ratio. This method is the least reliable because there are several ways in which falsely high and falsely low partition ratios can be obtained.

The point in the inactivation mechanism where product formation

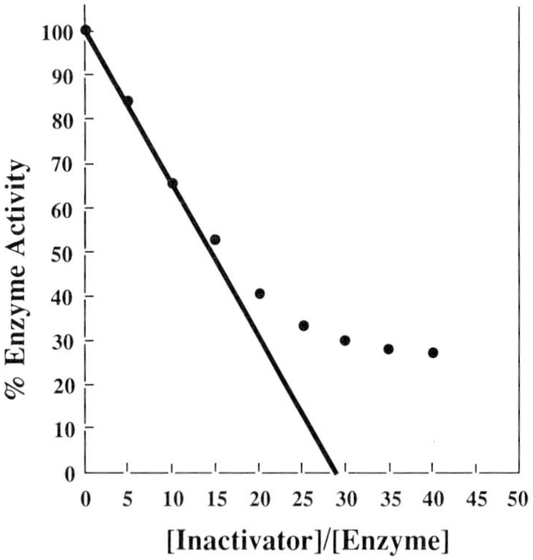

FIG. 9. Titration of an enzyme with a mechanism-based enzyme inactivator in which there is product inhibition.

branches from inactivation can be determined when C–H bond cleavage is involved.[13] In this case the deuterium isotope effect on the partition ratio, on the rate of product formation, and on the rate of inactivation should be measured. If there is no isotope effect on the partition ratio or on V_{max}, but the same isotope effect on k_{inact}/K_I and V_{max}/K_I, then inactivation is occurring from a species in which C–H bond cleavage has already occurred, and both product formation and inactivation occur from a common intermediate. If there is no isotope effect on the partition ratio, but a different isotope effect on V_{max} than on k_{inact}, the product formation and inactivation occur from different species, and both pathways involve C–H bond cleavage. If there is a normal deuterium isotope effect on the partition ratio, an inverse isotope effect on k_{inact}/K_I, and a normal isotope effect on V_{max}, then partitioning is occurring at the point of C–H bond cleavage, and this bond breakage only is involved in product formation, not in inactivation.

Applications

The remainder of this chapter deals with examples of the above-described methodologies for mechanism-based enzyme inactivators and approaches that can be taken to elucidate inactivator mechanisms. The experiments described for the given enzymes are general approaches that can be applied to other enzymes as well. The examples used come from the author's laboratory. It is not the intent of the author to lead the reader to believe that the only or best examples are from the author's laboratory; these are just the most convenient and well known to the author.

Mechanism-Based Inactivation of γ-Aminobutyric Acid Aminotransferase

γ-Aminobutyric acid (GABA) aminotransferase is a pyridoxal 5′-phosphate-dependent enzyme whose mechanism is shown in Scheme 6. 4-Amino-5-halopentanoic acids (**1**, X = F, Cl, Br; Scheme 7) are mechanism-based inactivators of GABA aminotransferase.[14] The standard experiments described above were carried out. Inactivation is time dependent, pH dependent, exhibits saturation, is blocked by the presence of substrate, and is not affected by the presence of the electrophile trapping agent 2-mercaptoethanol. Dialysis of the inactivated enzyme does not result in

[13] P. F. Fitzpatrick, D. R. Flory, Jr., and J. J. Villafranca, *Biochemistry* **24,** 2108 (1985); P. F. Fitzpatrick and J. J. Villafranca, *J. Biol. Chem.* **261,** 4510 (1986).

[14] R. B. Silverman and M. A. Levy, *Biochem. Biophys. Res. Commun.* **95,** 250 (1980).

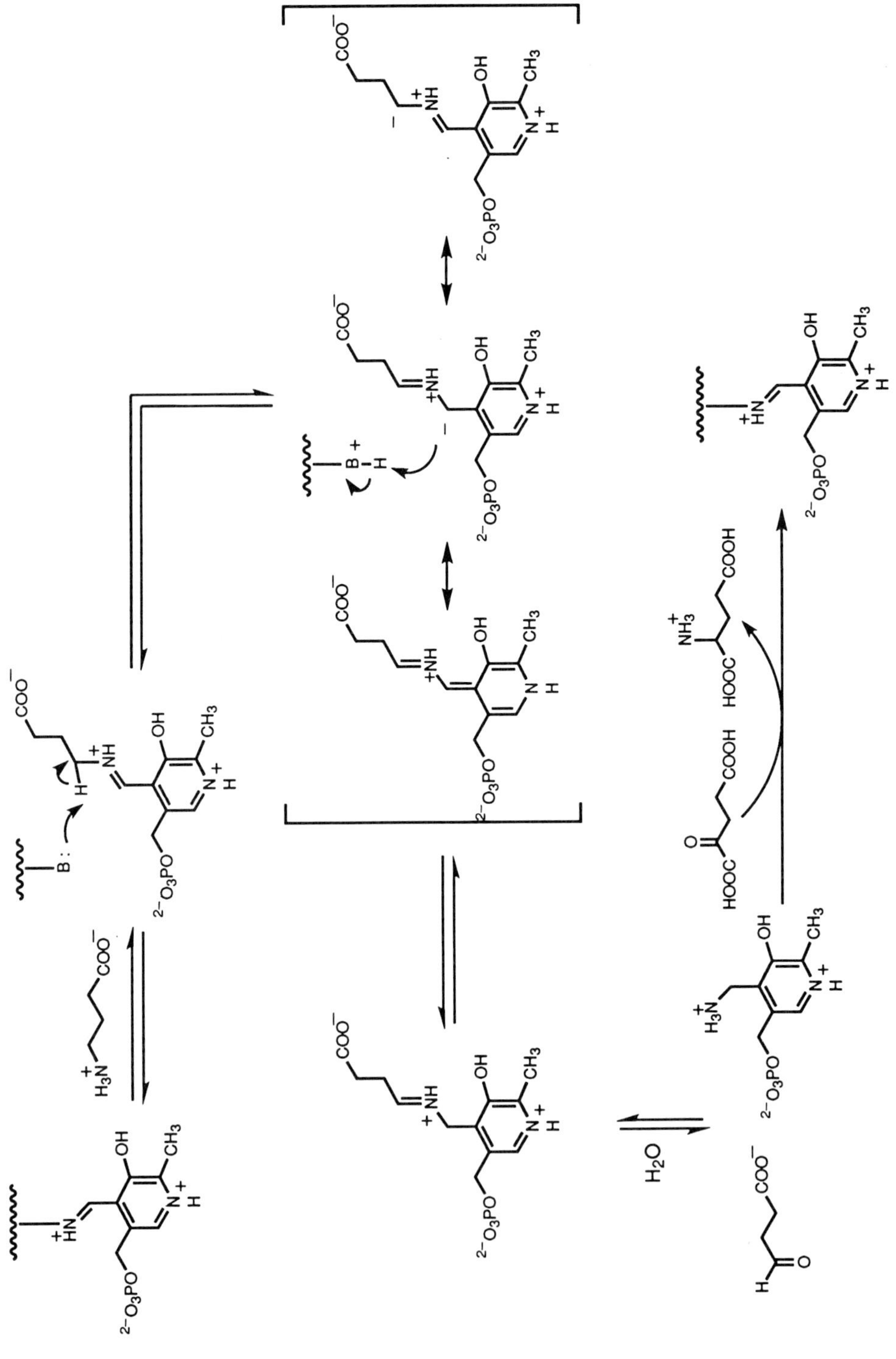

SCHEME 6

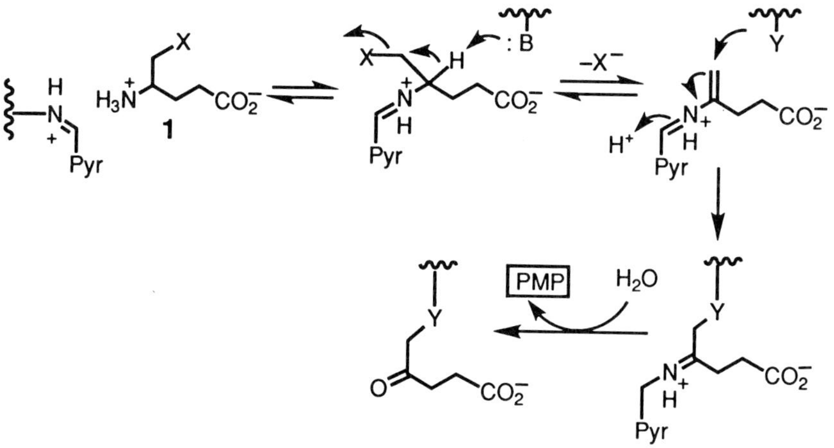

SCHEME 7

return of enzyme activity. Inactivation by (*S*)-[*U*-^{14}C]-4-amino-5-chloropentanoic acid leads to incorporation of 0.85–1.25 equivalents of radioactivity per active site.[14,15] No radioactivity was released on 8 *M* urea denaturation, indicating that a stable covalent bond was formed. Concomitant with enzyme inactivation there is a time-dependent change in the UV–visible spectrum of the pyridoxal 5′-phosphate (PLP) cofactor to a spectrum that resembles that of pyridoxamine 5′-phosphate (PMP).[15] The spectral change indicates that inactivation requires the catalytic action of the enzyme, an essential characteristic of a mechanism-based enzyme inactivator.

Another experiment to demonstrate that inactivation was enzyme catalyzed was to show that (*S*)-[4-^{2}H]-4-amino-5-chloropentanoic acid inactivated GABA aminotransferase with a deuterium isotope effect of 6.7. The fact that inactivation occurred with an isotope effect indicates that C–H bond cleavage is a rate-determining step in inactivation. When 4-amino-5-fluoropentanoic acid (**1**, X = F) was the inactivator, it was shown that one fluoride ion was released for every radioactive molecule incorporated into the enzyme. On the basis of these initial results, the inactivation mechanism shown in Scheme 7 was proposed.[14,15]

The partition ratio for this mechanism-based enzyme inactivator was determined by several methods to be zero. First, complete inactivation of the enzyme by these inactivators occurs in the absence of α-ketoglutarate, the second substrate required to convert any PMP back to PLP (see

[15] R. B. Silverman and M. A. Levy, *Biochemistry* **20,** 1197 (1981).

Scheme 6). When α-ketoglutarate is added to the inactivated enzyme, no activity is regenerated. If any of the inactivator molecules are turned over without inactivation, then some of the enzyme would have been converted to the PMP form, and active enzyme would be generated on incubation with α-ketoglutarate. Furthermore, when α-[*U*-^{14}C]ketoglutarate was present during inactivation with 4-amino-5-fluoropentanoic acid, no [*U*-^{14}C]glutamate, the product of transamination of PMP with α-ketoglutarate, was detected. Second, only one fluoride ion is released per inactivation event. If multiple turnovers were occurring, then more than one fluoride ion would be released. Third, when (*S*)-[*U*-^{14}C]-4-amino-5-chloropentanoic acid is used to inactivate the enzyme, essentially no radioactive nonamines are generated, indicating that no products are produced during inactivation.

All of the above results are consistent with the mechanism shown in Scheme 7, but additional labeling experiments proved otherwise. Apo-GABA aminotransferase was reconstituted with [4-^{3}H]PLP, inactivated with (*S*)-4-amino-5-fluoropentanoic acid, then denatured. According to the mechanism shown in Scheme 7, [^{3}H]PMP should be released; however, essentially no [^{3}H]PMP was detected. Instead, all of the radioactivity was released from the enzyme as **3** (Scheme 8), indicating that a different inactivation mechanism via an enamine intermediate occurs. This mecha-

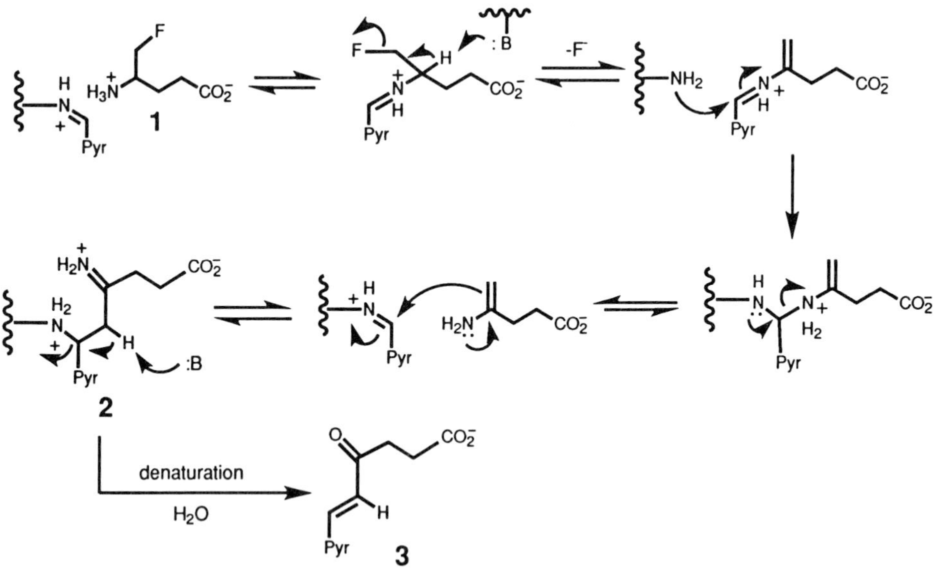

SCHEME 8

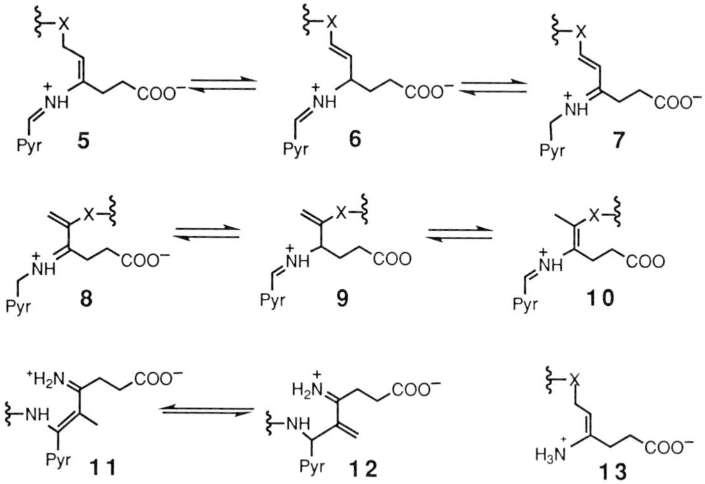

FIG. 10. Possible covalent adducts (**5–13**) produced during inactivation of GABA aminotransferase by γ-ethynyl-GABA.

nism is based on the earlier work of Metzler and co-workers.[16] Compound **3** is derived from the covalent adduct **2** by β-elimination. Part of the initial confusion about the inactivation mechanism was derived from the UV–visible spectrum of the inactivated enzyme, which appeared to that of PMP. Adduct **2** is similar in structure to PMP and would be expected to have a similar UV–visible spectrum.

Another example of how radioactive labeling can be used to differentiate inactivation mechanisms was described for the inactivation of GABA aminotransferase by γ-ethynyl-GABA (**4**).[17] At least nine different possible

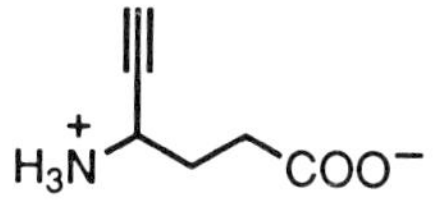

4

inactivation adducts, derived from three different inactivation mechanisms, can be imagined for this inactivation reaction (Fig. 10; see Burke and Silverman[17] for these mechanisms, if interested). Four are PLP adducts (**5**, **6**, **9**, and **10**), two are PMP adducts (**7** and **8**), two are adducts

[16] J. J. Likos, H. Ueno, R. W. Feldhaus, and D. E. Metzler, *Biochemistry* **21,** 4377 (1982); H. Ueno, J. H. Likos, and D. E. Metzler, *Biochemistry* **21,** 4387 (1982).

[17] J. R. Burke and R. B. Silverman, *J. Am. Chem. Soc.* **113,** 9329 (1991).

derived from an enamine mechanism as described above for 4-amino-5-fluoropentanoic acid (**11** and **12**), and the other is derived from an enamine, but by a mechanism different from that described above. This mechanism does not affect the PLP cofactor at all (**13**).

These possibilities were differentiated with the use of two radioactive compounds, [4-^{3}H]PLP and γ-ethynyl[2-^{3}H]-GABA. As described above, apo-GABA aminotransferase was reconstituted with [4-^{3}H]PLP, and the fate of the PLP was determined after inactivation. Greater than 95% of the tritium was released on denaturation as [4-^{3}H]PLP; no ^3H_2O was detected. These results exclude the two PMP adducts that could have been formed and the two enamine adducts. The four PLP adducts (**5**, **6**, **9**, and **10**) and the other enamine-derived product (**13**) were differentiated with γ-ethynyl[2-^{3}H]-GABA. Inactivation of GABA aminotransferase with γ-ethynyl[2-^{3}H]-GABA followed by exhaustive dialysis led to the incorporation of 1.0 equivalent of inactivator bound to the enzyme. Denaturation of the labeled enzyme with 6 *M* urea, however, resulted in the release of about one-half of the radioactivity, unless the denaturation was carried out at pH 7.4 or pH 9.5 and 4°, in which case no radioactivity was released. This suggests that either there are two different adducts formed, one stable to denaturation and the other unstable, or that one adduct can hydrolyze to give two different products, one stable to denaturation and one unstable. If the labeled enzyme is treated with sodium borohydride prior to denaturation, then, even at room temperature, little or no radioactivity is released, indicating that the unstable product is probably attached to the enzyme in the form of a Schiff base. To be consistent with that result, the linkage to the enzyme (X in **5–10** and **13**) must be an amino group, presumably from a lysine residue. The unstable product (that released on denaturation) was isolated and identified by HPLC comparison to a synthesized standard compound as **14**, which could be derived from either **9** or **10** as shown in Scheme 9.

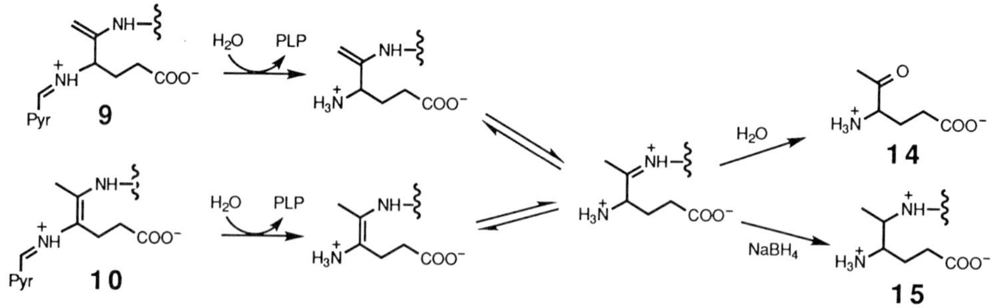

SCHEME 9

The stable adduct also must be derived from the PLP adducts (**5**, **6**, **9**, and **10**). Consider the products of hydrolysis of these adducts (structures **15–18**, Scheme 10). For **16** or **17** to be candidates for the stable adduct, X cannot be NH; otherwise, they would be unstable. If X is another heteroatom (O or S), then they could be stable. Treatment of the stable adduct under conditions known to hydrolyze vinyl ethers (X = O) and vinyl sulfides (X = S) did not release the radioactivity from the enzyme; therefore, **16** and **17** are not the stable adduct. Adducts **15** and **18** (X = NH) were differentiated by reduction with sodium borohydride followed by oxidation with sodium periodate (Scheme 11). If the adduct has structure **15** (X = lysine), then reduction with sodium borohydride would give adduct **19**, which would be stable to treatment with sodium periodate; consequently, all of the tritium would remain attached to the enzyme. If the

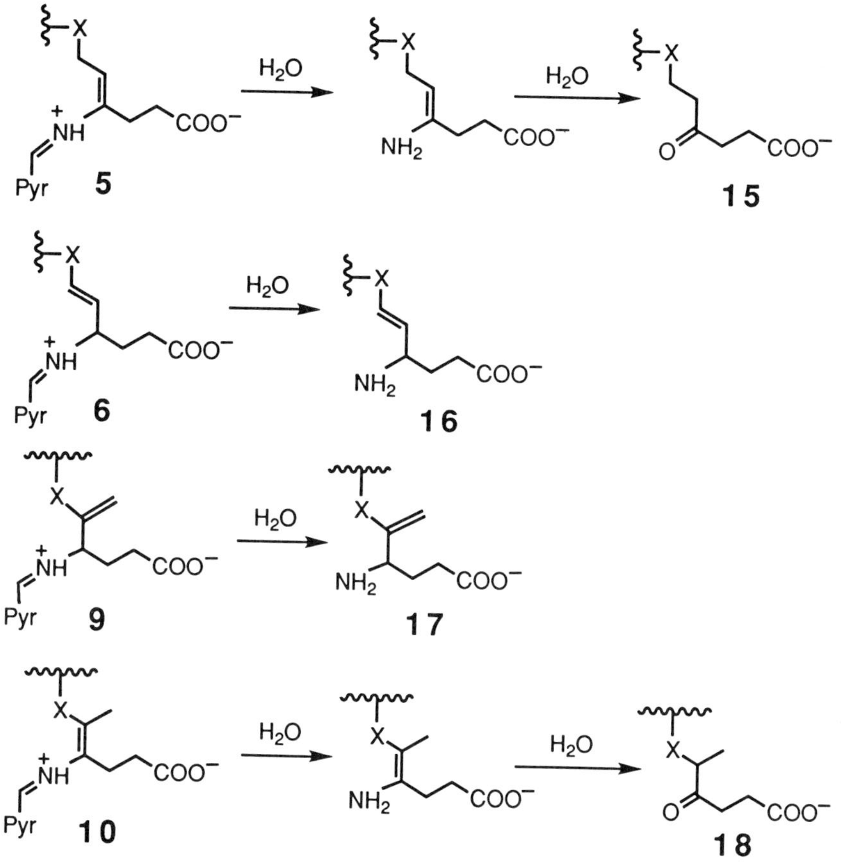

SCHEME 10

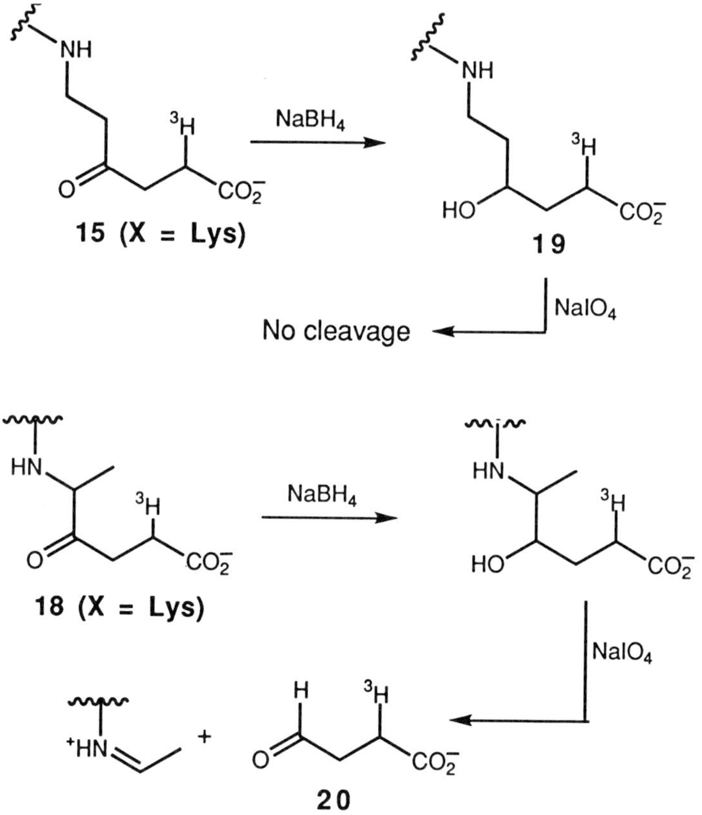

SCHEME 11

adduct has structure **18** (X = lysine), however, the sodium borohydride-reduced adduct would be oxidized by sodium periodate to give tritiated succinic semialdehyde (**20**). This experiment resulted in complete release of the tritium from the enzyme as succinic semialdehyde which was identified by HPLC. Therefore, the stable adduct has the structure **18** (X = lysine). Given that the unstable adduct could have come from **9** or **10** and the stable adduct is derived from **10**, it is most likely that both the unstable and the stable adducts are produced from **10** (Scheme 12). Hydrolysis of **10** gives **21** (Scheme 12), which can break down equally by pathways a and b to give the unstable adduct and release **14** or the stable adduct (**18**), respectively.

To determine the partition ratio and to complete the metabolic profile of inactivation of GABA aminotransferase by γ-ethynyl-GABA, turnover products formed during inactivation were isolated. Incubation of GABA

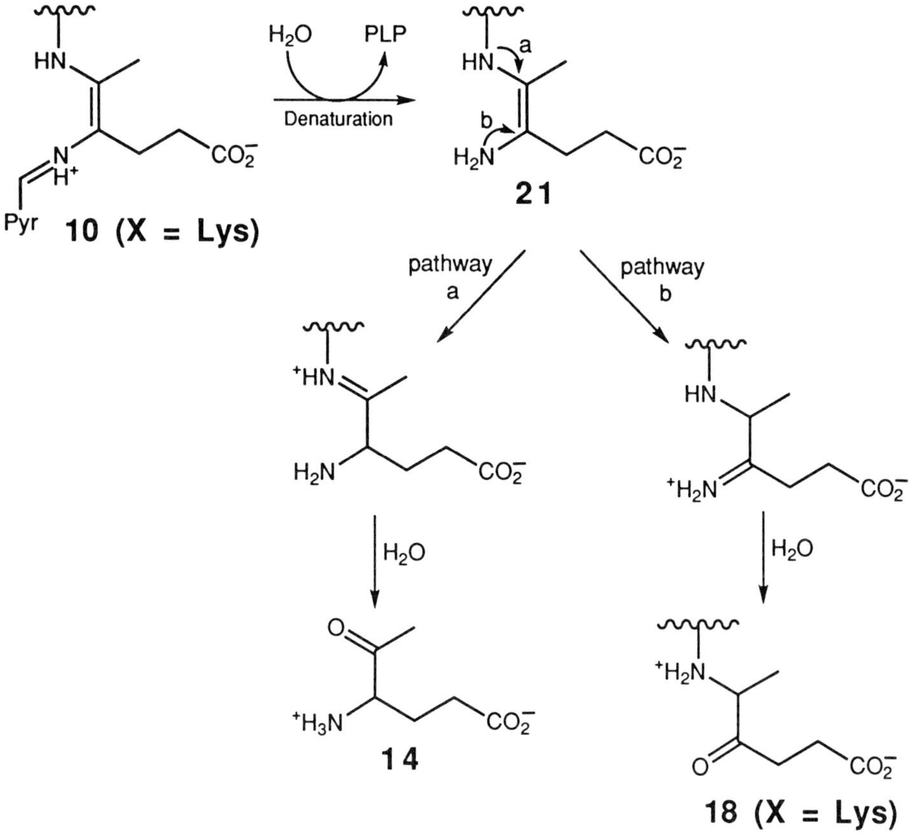

SCHEME 12

aminotransferase with γ-ethynyl-GABA in the presence of α-[5-^{14}C]ketoglutarate produced about 1 equivalent of [5-^{14}C]glutamate; consequently, in addition to inactivation there is one molecule of γ-ethynyl-GABA transaminated, presumably (but not identified) to 4-oxo-5-hexynoic acid (**22**, Scheme 13). Inactivation with γ-ethynyl[2-^{3}H]-GABA produced about 4 equivalents of radioactive nonamines, of which at least 2 equivalents were identified by HPLC to be 4-oxo-5-hexenoic acid (**23**, Scheme 13). Presumably 1 of these 4 equivalents of radioactive nonamines is derived from the transamination noted above to give **22**. About 8 equivalents of a radioactive amine, identified as **14**, the same product identified as coming from the unstable adduct, also were generated by incubation of GABA aminotransferase with γ-ethynyl[2-^{3}H] GABA. Therefore, the turnover number is 13, and the partition ratio is 12 (12 product molecules generated

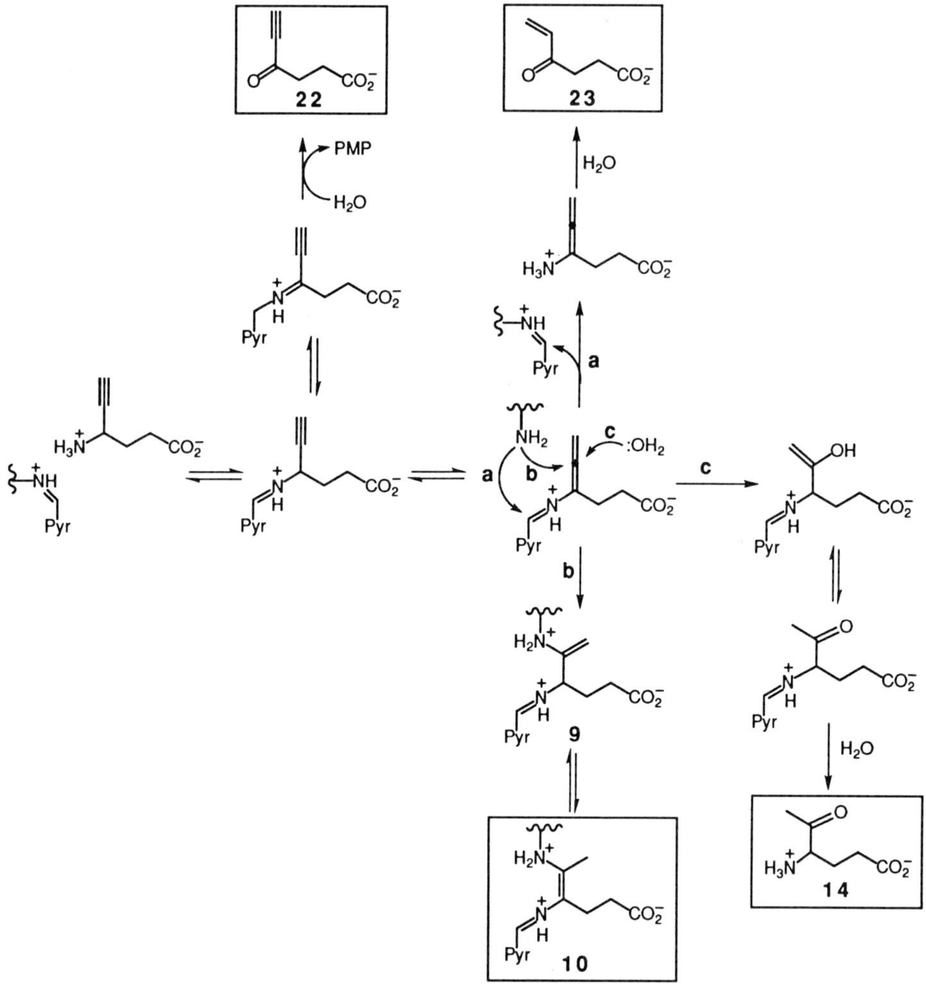

SCHEME 13

for each inactivation event). An overall set of pathways to accommodate these data is shown in Scheme 13.

Mechanism-Based Inactivation of Monoamine Oxidase

N-(1-Methylcyclopropyl)benzylamine (**24**)[18,19] and 1-phenylcyclopropylamine (**25**)[20] are mechanism-based inactivators of the flavoenzyme

[18] R. B. Silverman and S. J. Hoffman, *Biochem. Biophys. Res. Commun.* **101,** 1396 (1981).
[19] R. B. Silverman and R. B. Yamasaki, *Biochemistry* **23,** 1322 (1984).
[20] R. B. Silverman and P. A. Zieske, *Biochemistry* **24,** 2128 (1985).

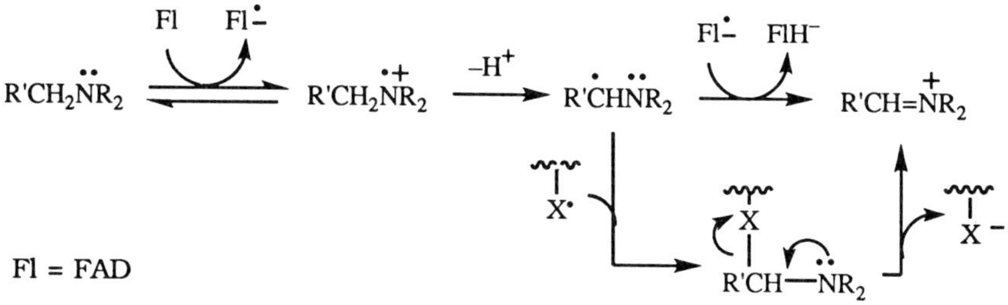

SCHEME 14

monoamine oxidase (MAO). The currently accepted mechanism for this enzyme, mostly derived from studies with mechanism-based inactivators,

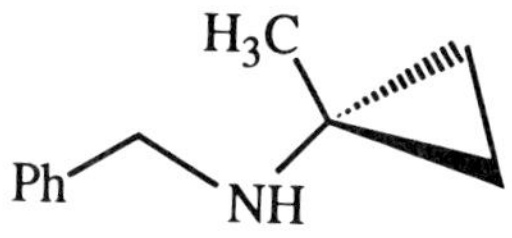

24

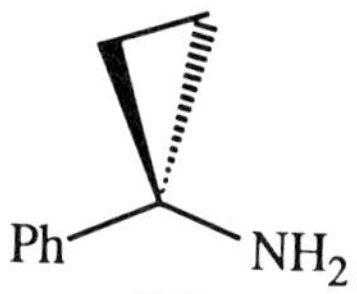

25

is shown in Scheme 14.[21] In addition to the criteria for mechanism-based inactivation described above for the inactivators of GABA aminotransferase, it was shown for **24** and **25** that there was no lag time for inactivation, and a second aliquot of MAO added to the inactivation solution with **24** was inactivated at the identical rate as was the first aliquot.[19] To demonstrate that enzyme catalysis accompanies inactivation, the flavin absorption spectrum was recorded during and after inactivation by **24** and **25**. In both cases[19,20] the flavin was shown to be reduced during inactivation and to remain reduced after denaturation of the inactivated enzyme. This indicates that two electrons are transferred during inactivation and that the inactivator becomes attached to the flavin.

1-Phenylcyclopropylamine (**25**, 1-PCPA) inactivates MAO via two different pathways; one pathway is irreversible and the other is reversible (Scheme 15). The observation that initially indicated that more than one pathway may be involved was that pseudo-first-order loss of enzyme activity occurred for the first two to three half-lives, then the rate of inactivation slowed. When a series of concentrations of 1-PCPA were used in ratios from 1 to 10 times the enzyme concentration, it was found that, over an extended period of time, the enzyme activity slowly returned but not completely (Fig. 11). This suggests that two adducts form, one

[21] R. B. Silverman, *in* Advances in Electron Transfer Chemistry'' (P. S. Mariano, ed.), Vol. 2, p. 177. JAI Press, Greenwich, Connecticut, 1992; R. B. Silverman, J. P. Zhou, and P. E. Eaton, *J. Am. Chem. Soc.* **115,** 8841 (1993).

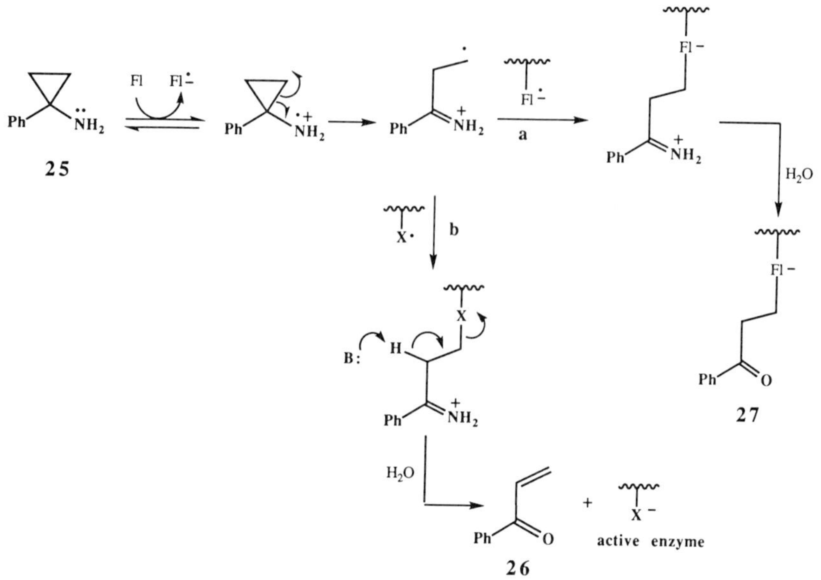

SCHEME 15

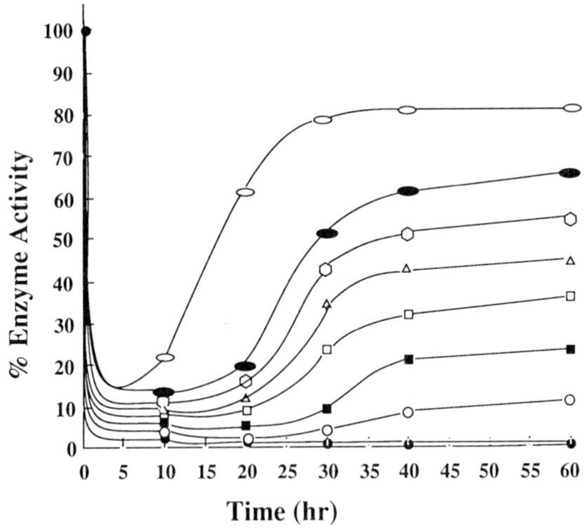

FIG. 11. Effect on enzyme activity by treatment of monoamine oxidase with varying concentrations of 1-phenylcyclopropylamine. The ratios of inactivator/enzyme used were as follows: open oval, 1 equivalent; closed oval, 2 equivalents; hexagon, 3 equivalents; triangle, 4 equivalents; open square, 5 equivalents; closed square, 6 equivalents; open circle, 7 equivalents; closed circle, 10 equivalents.

that is irreversible and one that is reversible. The higher the ratio of inactivator concentration to enzyme, the less enzyme activity returned (relative to a control that was not inactivated). When the final percentage of enzyme activity remaining was plotted against the inactivator/enzyme ratio (as described in Fig. 8), a straight line was obtained that intersected the x-axis at a ratio of eight inactivator molecules required to inactivate the enzyme completely, indicating a partition ratio of seven (seven product molecules plus one covalent adduct). Inactivation of MAO with 1-[*phenyl*-^{14}C]**25** resulted in the covalent attachment of 1.2 ± 0.2 equivalents of ^{14}C in addition to producing seven equivalents of radioactive acrylophenone (**26**). The mechanism shown in Scheme 15 accounts for both the attachment of one equivalent of radioactivity from 1-[*phenyl*-^{14}C]**25** and the formation of acrylophenone.

The structure of the irreversible adduct (**27**, after denaturation) was determined by carrying out three organic reactions on it (Scheme 16). Treatment of **27** with NaB^3H_4 resulted in the incorporation of 0.73 equivalent of tritium after subtracting out the noninactivated control, suggesting the presence of a ketone. Baeyer–Villiger oxidation of phenyl alkyl ketones with peroxytrifluoroacetic acid is known to give phenyl esters ex-

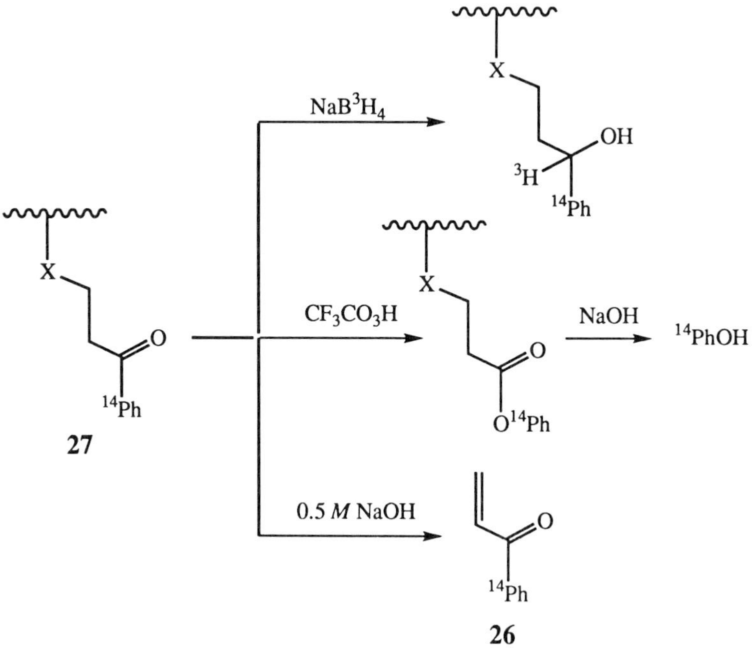

SCHEME 16

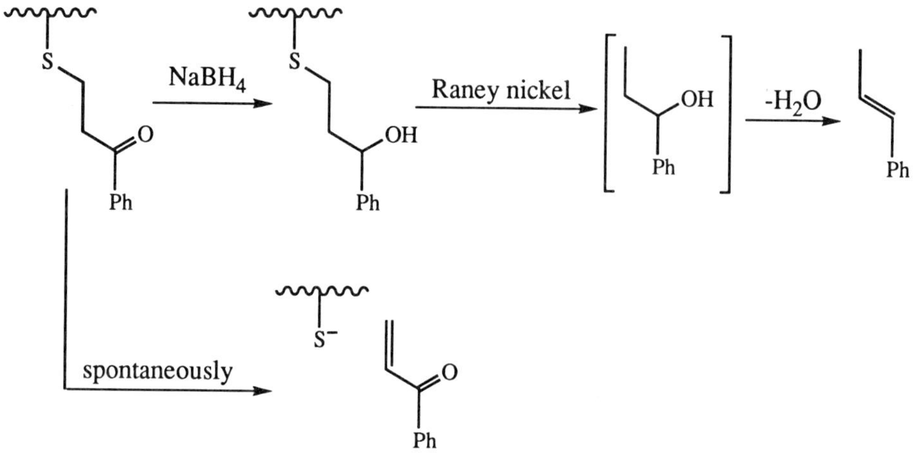

SCHEME 17

clusively; saponification of phenyl esters give phenol. When this reaction was carried out on the protease-digested **27**, 87% of the theoretical amount of [^{14}C]phenol was released from the enzyme, consistent with the presence of a [^{14}C]phenyl ketone in the adduct structure. Finally, hydroxide treatment resulted in the release of [^{14}C]acrylophenone, the expected product of a retro-Michael reaction on **27**. The identity of the X group in the irreversible adduct was determined in two ways. The absorption spectrum of the flavin is reduced on inactivation and remains reduced after denaturation, suggesting that the inactivator becomes attached to the reduced form of the flavin. Pronase digestion of the enzyme labeled with 1-[*phenyl*-^{14}C]**25** gives peptide fragments that contain the radioactivity from the inactivator and which have the absorption spectrum of the flavin.

The reversible adduct, which has a half-life of about 1 hr, is formed seven times more readily than is the irreversible adduct. On denaturation of this adduct the flavin spectrum becomes reoxidized, even though the radioactivity is still mostly bound to the enzyme. Therefore, this adduct is not attached to the flavin. The group on the enzyme to which the reversible adduct is attached was determined to be a cysteine residue.[22] Reduction of the adduct with sodium borohydride followed by treatment with Raney nickel gave *trans*-β-methylstyrene, the dehydration product of 1-phenyl-1-propanol, which is the expected reduction product of a (3-hydroxy-3-phenylpropyl)cysteine adduct (Scheme 17). This is consis-

[22] R. B. Silverman and P. A. Zieske, *Biochem. Biophys. Res. Commun.* **135,** 154 (1986).

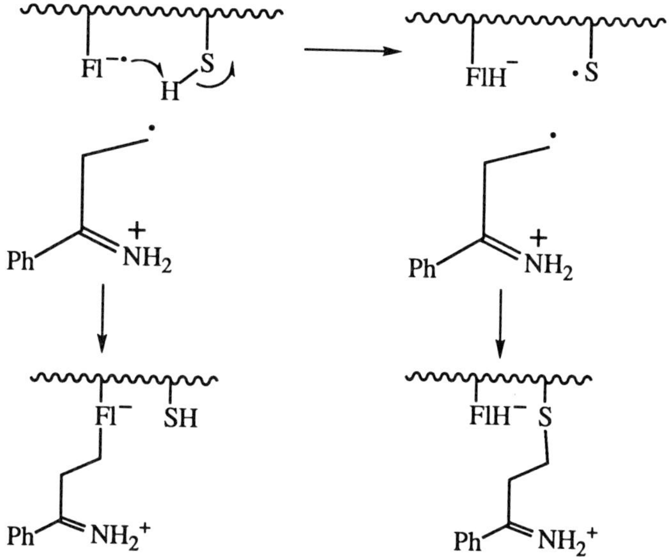

SCHEME 18

tent with the structure of the reversible adduct shown in Scheme 17; the instability of this reversible adduct could account for the fact that 7 equivalents of [^{14}C]acrylophenone are spontaneously produced from 1-[*phenyl*-^{14}C]**25** after complete irreversible inactivation.

Further support for a reversible cysteine adduct came from 5,5′-dithiobis(2-nitrobenzoic acid) titration of the native enzyme and of the reversibly inactivated and sodium borohydride-reduced enzyme. In this experiment it was shown that the native enzyme had one more cysteine than the inactivated enzyme, lending support to the hypothesis that the reversible adduct is a cysteine adduct. The difference in the stabilities of the cysteine and the flavin adducts could be in the leaving group ability of the group to which the inactivator is attached. Cysteine, the amino acid to which the reversible adduct is bound, would be a much better leaving group than the flavin cofactor, to which the irreversible adduct is attached. The rationale for the formation of a cysteine adduct is that a hydrogen atom transfer from an active-site cysteine residue to the flavin semiquinone would generate a thiyl radical, which could undergo radical combination with the inactivator radical in competition with flavin semiquinone combination to the inactivator radical (Scheme 18).

To obtain further support for a radical mechanism 1-phenylcyclobutylamine (**28**) was synthesized and its reaction with MAO studied (Scheme

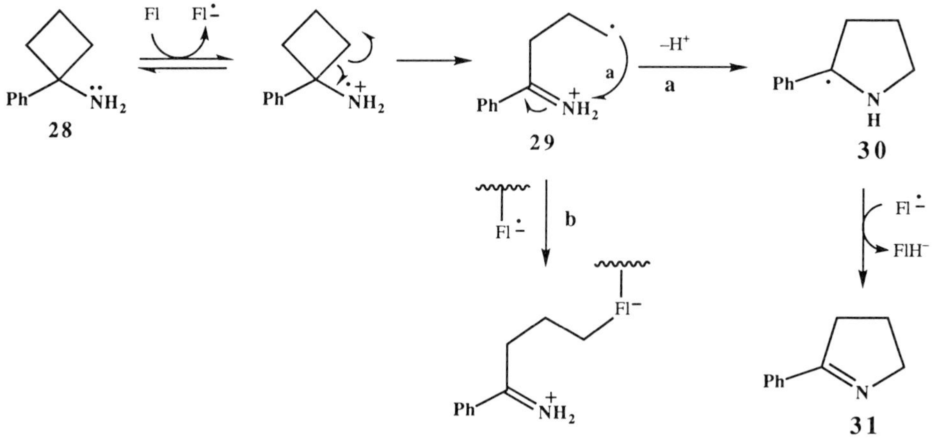

SCHEME 19

19).[23] If the mechanism proceeds by a one-electron oxidation mechanism, then, based on related nonenzymatic one-electron rearrangements,[24] **31** should be produced (pathway a in Scheme 19). Intermediate **29** is related to several intermediates known to undergo endocyclizations to intermediates related to **30**.[24] The alternative pathway from **29** is attachment to the flavin (pathway b, Scheme 19), leading to enzyme inactivation. In fact, both occur. Incubation of MAO with **28** leads to a slow loss of enzyme activity. Analysis by HPLC of aliquots removed periodically showed that concomitant with the consumption of **28** is the formation of **31**. Therefore, MAO catalyzes the oxidation and rearrangement of **28** to give the product expected from chemical studies known to proceed by one-electron pathways. Furthermore, if the MAO-catalyzed oxidation of **28** or of a related compound (**32**) is carried out in the presence of a nitrone spin trapping agent, such as α-phenyl-*N-tert*-butylnitrone (**33**) in an electron paramagnetic resonance (EPR) spectrometer, the expected triplet of doublets centered at a g value of 2.006 (an organic radical) is observed.[25]

Mass spectrometry of deuterated (or other isotopically labeled) metabolites can be a useful tool for the elucidation of inactivator mechanisms. (Aminomethyl)trimethylsilane (**34**) was shown to be a mechanism-based inactivator of MAO.[26] Two mechanisms were imagined (Scheme 20) based

[23] R. B. Silverman and P. A. Zieske, *Biochemistry* **25,** 341 (1986).

[24] D. D. Tanner and R. M. Rahimi, *J. Org. Chem.* **44,** 1674 (1979); L. W. Menapace and H. G. Kuivila, *J. Am. Chem. Soc.* **86,** 3047 (1964); J. W. Wilt, L. L. Maravetz, and J. F. Zawadzki, *J. Org. Chem.* **31,** 3018 (1966).

[25] K. Yelekci, X. Lu, and R. B. Silverman, *J. Am. Chem. Soc.* **111,** 1138 (1989).

[26] G. M. Banik and R. B. Silverman, *J. Am. Chem. Soc.* **112,** 4499 (1990).

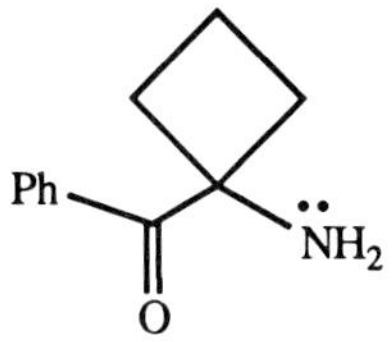

32

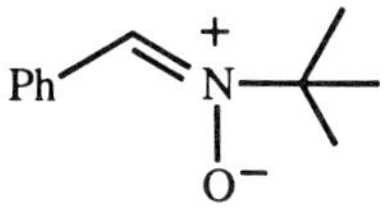

33

on the enzyme mechanism and the known reactivity of silicon that is β to an electron-deficient center.[27] Pathway a in Scheme 20 is the one based on the earlier work indicating that silicon β to an electron-deficient center is highly electrophilic.[27] In this case, an active site nucleophile reacts with the silicon atom, which is β to an electron-deficient nitrogen atom as a result of one-electron transfer of **34** to the flavin, cleaving the carbon–silicon bond and eventually producing formaldehyde. Pathway b (Scheme 20) also can produce formaldehyde by a reaction of water with released (formyl)trimethylsilane (via what is referred to as a Brook rearrangement[28]). If the inactivation is carried out with [1-$^2H_2$]**34** [$(CH_3)_3SiCD_2NH_2$], however, pathway a would produce dideuterioformaldehyde, but pathway b would produce monodeuterioformaldehyde because one deuterium would be removed from **35** to give **36**. Following inactivation of MAO with [1-$^2H_2$]**34**, treatment with 2,4-dinitrophenylhydrazine reagent to trap the formaldehyde(s) produced, extraction, and mass spectometry, it was found that both dideuterioformaldehyde and monodeuterioformaldehyde 2,4-dinitrophenylhydrazones were formed in the ratio of about 1:3.5. This indicates that both pathways are relevant. Furthermore, there is a deuterium isotope effect on the rate of inactivation ($k^H{}_{inact}/k^D{}_{inact}$) of 2.3 with no effect on the K_I. This kinetic isotope effect is almost identical with the deuterium isotope effects for MAO-catalyzed oxidation of tyramine[29] and dopamine.[30]

The observed deuterium isotope effect on inactivation indicates that C–H bond cleavage is a rate-determining step in inactivation, consistent with pathway b but not with pathway a (Scheme 20). Apparently, however, since dideuterioformaldehyde is generated, pathway a is a relevant pathway as well, but not relevant to enzyme inactivation. The converse experi-

[27] M. A. Brumfield, S. L. Quillen, U. C. Yoon, and P. S. Mariano, *J. Am. Chem. Soc.* **106,** 6855 (1984); A. J. Y. Lan, S. L. Quillen, R. O. Heuckeroth, and P. S. Mariano, *J. Am. Chem. Soc.* **106,** 6439 (1984); K. Ohga, Y. C. Yoon, and P. S. Mariano, *J. Org. Chem.* **49,** 213 (1984).

[28] A. G. Brook, *Acc. Chem. Res.* **7,** 77 (1974).

[29] B. Belleau and J. Moran, *Ann. N.Y. Acad. Sci.* **107,** 822 (1963).

[30] P. H. Yu, B. A. Bailey, D. A. Durden, A. A. Boulton, *Biochem. Pharmacol.* **35,** 1027 (1986).

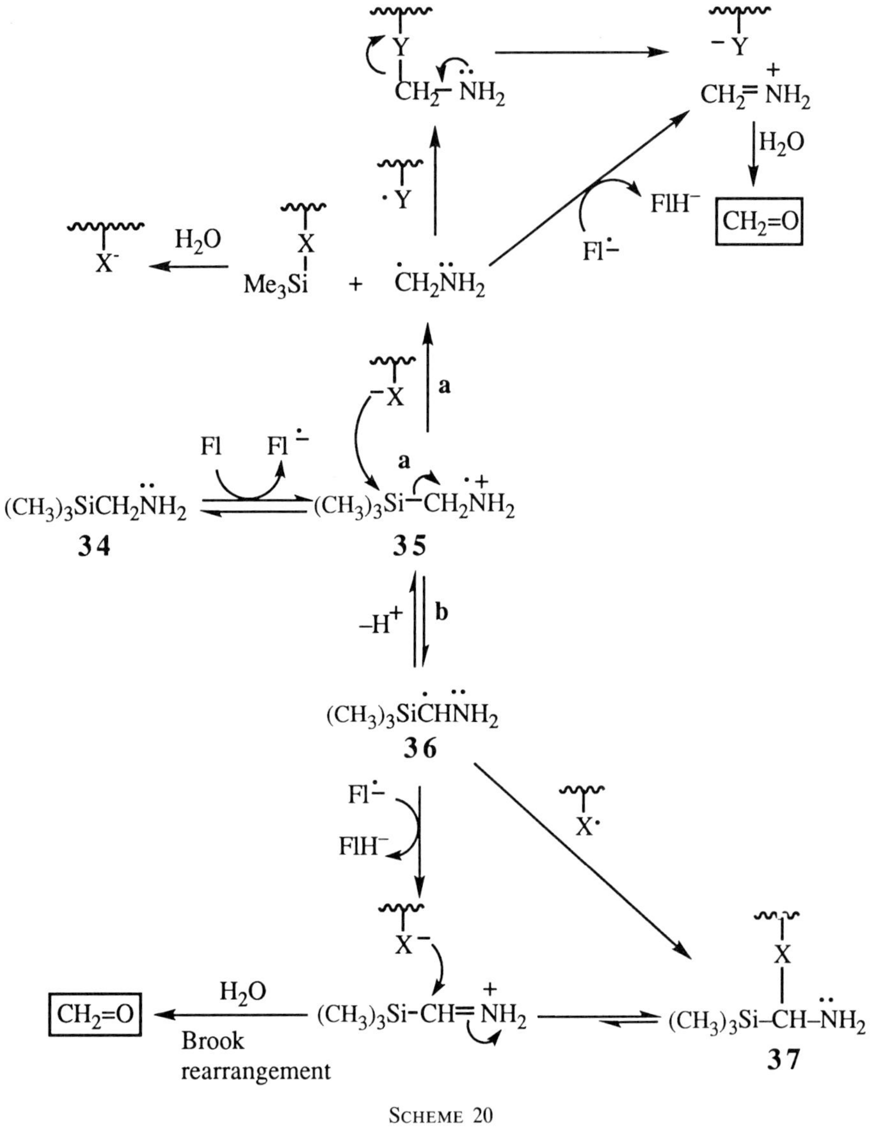

SCHEME 20

ment, namely, the inactivation of MAO with **34** in ^2H_2O, gave a mixture of undeuterated and monodeuterated formaldehyde, as expected for oxidation by pathways a and b, respectively. To confirm this conclusion, MAO was inactivated with [1-^{3}H]**34**. According to pathway a, no tritium would be incorporated into the enzyme; all would be released as [^{3}H]formaldehyde. Pathway b, however, would lead to incorporation of tritium into the enzyme (**37**) and, as a result of the release of (formyl)trimethylsilane and subsequent Brook rearrangement, also to formation of [^{3}H]formaldehyde. Incubation of MAO with [1-^{3}H]**34** resulted in the formation of [^{3}H] formaldehyde and in the incorporation of 1.2 equivalents of tritium into the enzyme. The stoichiometric incorporation of tritium into the enzyme is not consistent with pathway a but is consistent with pathway b. Given the 1:3.5 ratio of dideuterioformaldehyde to monodeuterioformaldehyde noted above and the 1 equivalent of tritium incorporated during inactivation, it suggests that for every 5.5 molecules of inactivator oxidized by MAO, one molecule goes via pathway a without inactivation and 4.5 go via pathway b, 3.5 of which are converted to formaldehyde and one becomes attached to the enzyme (**37**).

Isotopically labeled analogs have been used to clarify the pathway for inactivation of MAO by the anticonvulsant agent milacemide (**38a**).[31]

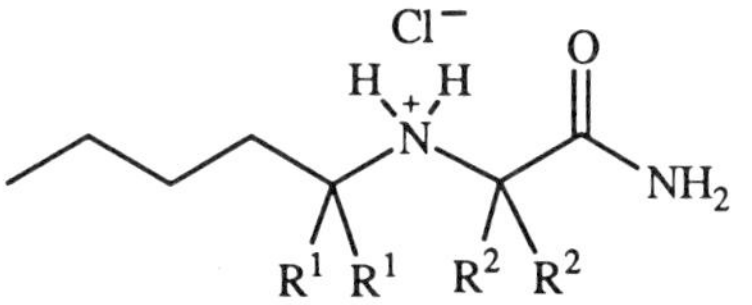

38

a; $R^1 = H, R^2 = H$
b; $R^1 = H, R^2 = D$
c; $R^1 = D, R^2 = H$

Milacemide is both a substrate and inactivator of MAO. Two pathways for oxidation and inactivation were envisioned subsequent to initial electron transfer to the flavin (Scheme 21). Pathway a involves the removal of the α-proton from the pentyl chain, leading to both inactivation (pathway c) and metabolite formation (pathway d). Pathway b proceeds via removal of the acetamido methylene proton, leading to inactivation (pathway e) and metabolite formation (pathway f). To test which pathway(s) was reasonable, the corresponding dideuterated analogs **38b** and **38c** were prepared. Analog **38b** inactivated MAO with a negligible isotope effect (1.25) on k_{inact}/K_I, but **38c** exhibited a large isotope effect (4.55) on k_{inact}/K_I. As a substrate **38b** showed no isotope effect (1.03) on k_{cat}/K_m, but **38c** had an isotope effect of 4.53. These results indicate that pathway a is the most reasonable one for both inactivation and oxidation of milacemide. The partition ratios for **38a–c** are similar (within a factor of 1.4), consistent

[31] R. B. Silverman, K. Nishimura, and X. Lu, *J. Am. Chem. Soc.* **115,** 4949 (1993).

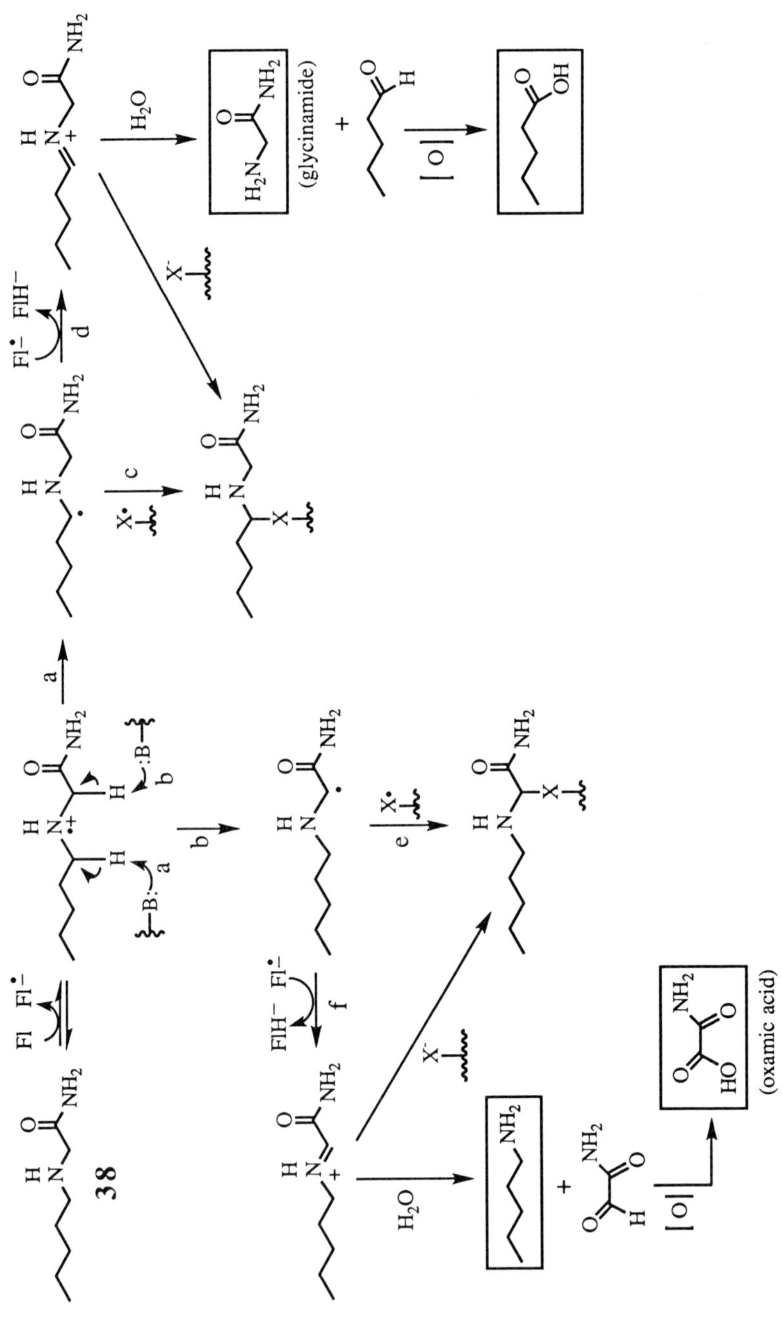

SCHEME 21

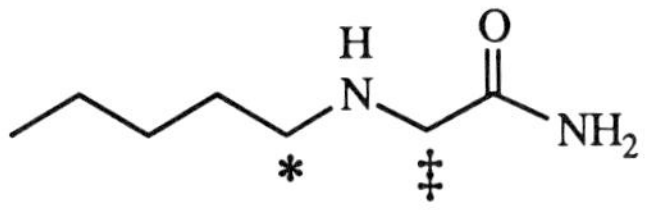

39

a, * = ^{12}C; ‡ = ^{14}C **b**, * = ^{14}C; ‡ = ^{12}C

with a mechanism in which partitioning between metabolite formation and inactivation occurs subsequent to the C–H bond cleavage step.

A slightly different picture emerged, however, when two ^{14}C-labeled analogs were used (**39**) and the radiolabeled metabolites were isolated. Incubation of MAO with **39a** gave both [^{14}C]glycinamide (from pathway a involving oxidation of the pentyl side chain) and [^{14}C]oxamic acid (from pathway b involving oxidation of the acetamide methylene to give 2-oxoacetamide, which was nonenzymatically oxidized). Inactivation of MAO with **39b** produced both [^{14}C]pentanoic acid (from pathway a involving oxidation of the pentyl side chain to give pentanal, which was nonenzymatically oxidized) and [^{14}C]pentylamine (from pathway b involving oxidation of the acetamide methylene). These results indicate that oxidation does occur at both methylenes adjacent to the amine nitrogen, but the deuterium isotope effect results support inactivation as resulting from pathway a.

Conclusion

Mechanism-based enzyme inactivation is a powerful tool for studies of enzyme mechanisms and mechanisms of enzyme inactivation by small molecules. Mechanistic hypotheses can be tested by appropriate molecular design, utilizing isotopically labeled analogs to permit the elucidation of structures of metabolites produced and to determine what portions of the mechanism-based inactivators become covalently attached to the target enzyme. This approach to studies of enzyme mechanisms is well suited for those who are geared more to the organic chemistry of enzyme-catalyzed reactions and who have insights into the chemical machinery of active sites of enzymes. The use of mechanism-based enzyme inactivators is yet another of the very important methods in enzymology.

[11] Transition State and Multisubstrate Analog Inhibitors

By ANNA RADZICKA and RICHARD WOLFENDEN

Transition State Stabilization by Enzymes

To lower the energy barrier that limits the rate of a reaction, a catalyst must bind the altered substrate in the transition state ($S^{\ddagger}$) more tightly than it binds the substrate in the ground state (S). In the moment, lasting perhaps 1 msec, during which the catalytic event occurs, binding is enhanced by a factor that equals or surpasses the factor by which the catalyst enhances the rate of the reaction.[1]

Scheme 1 illustrates this principle by comparing the rate of deamination of adenosine in the presence and absence of calf intestinal adenosine deaminase (Scheme 1a). The central postulate of transition state theory is that in the ground state the substrate S (in the nonenzymatic reaction), or the enzyme–substrate complex ES (in the enzyme-catalyzed reaction), exists in a state of equilibrium with a transition state ($S^{\ddagger}$ or $ES^{\ddagger}$, respectively) situated at the top of a potential energy barrier, from which its chances of going forward to products, or backward to substrates, are equal (Schemes 1b and 1c).

The rate of decomposition of any transition state to products is equal to kT/h, a universal rate constant composed of Planck's constant k, Boltzmann's constant h, and the absolute temperature T; and it has a value of approximately 0.62×10^{13} sec^{-1} at room temperature. At any given temperature, rates of reactions differ according to the difference between their equilibrium constants for reaching the transition state. That equilibrium constant is always unfavorable, even for the fastest enzyme reactions. Thus, for carbonic anhydrase, with a turnover number of roughly 10^6 sec^{-1},[2] the value of $K_{ES^{\ddagger}}$ is about 10^{-7}. For calf intestinal adenosine deaminase, a more conventional enzyme with a turnover number of 375 sec^{-1}, the value of $K_{ES^{\ddagger}}$ is approximately 6×10^{-11}. At neutral pH in water, adenosine is deaminated very slowly ($k_{non} = 1.8 \times 10^{-10}$ sec^{-1}; $t_{1/2} = 122$ years),[3] and the value of $K_{S^{\ddagger}}$ is approximately 3×10^{-23}. The ratio of $K_{ES^{\ddagger}}$ to $K_{S^{\ddagger}}$ matches the rate enhancement of 2×10^{12}-fold. Scheme 1b shows that K_S is related to K_{TX} by this same ratio, where K_S is the dissociation constant of the enzyme–substrate com-

[1] R. Wolfenden, *Nature* (*London*) **223,** 704 (1969).

[2] H. Steiner, B. H. Johnson, and S. Lindskog, *Eur. J. Biochem.* **59,** 253 (1975).

[3] L. Frick, J. P. Mac Neela, and R. Wolfenden, *Bioorg. Chem.* **15,** 100 (1987).

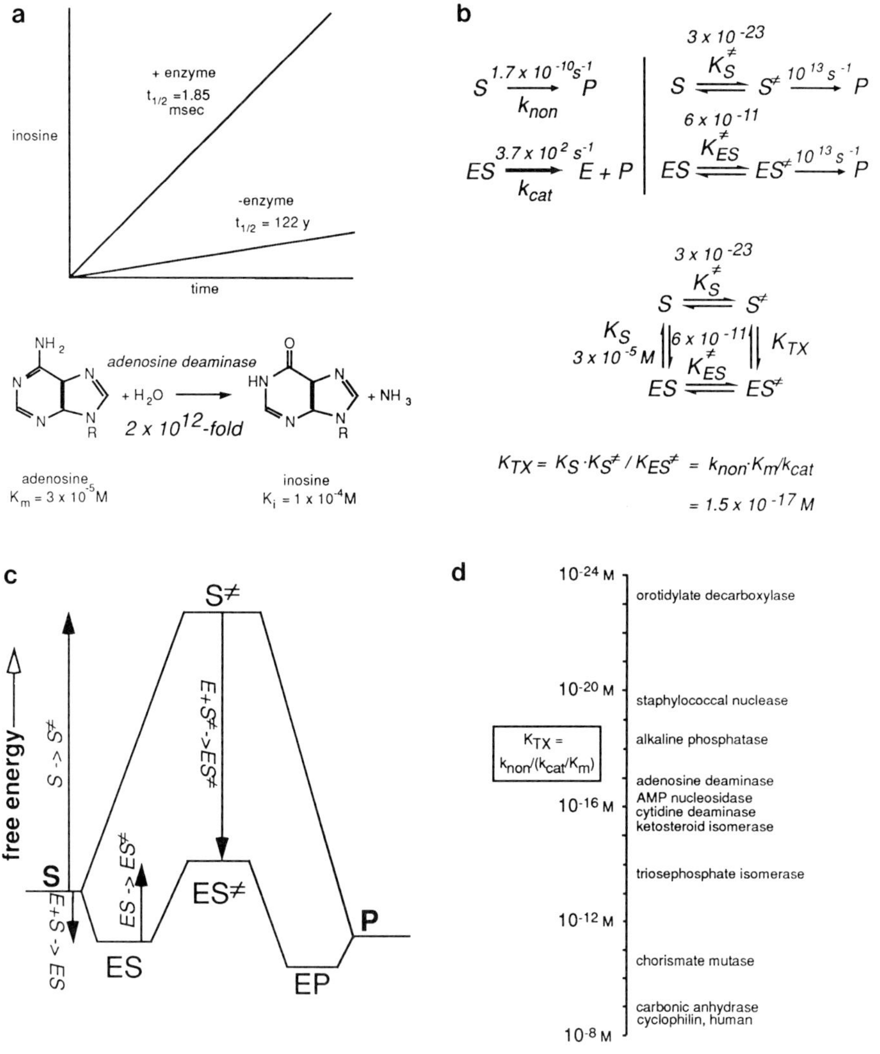

SCHEME 1

plex and K_{TX} is the dissociation constant of the enzyme–substrate complex in the transition state. Evidently the substrate, initially bound with a dissociation constant of roughly $3 \times 10^{-5}\ M$, is bound with a dissociation constant of approximately $1.5 \times 10^{-17}\ M$ in the transition state.

In the simple case described in Scheme 1, K_{TX} (expressed in moles/

liter) is equivalent to k_{non} (the first-order rate constant for the nonenzymatic reaction, usually expressed as sec^{-1}) divided by k_{cat}/K_m (the second-order rate constant for the enzyme reaction, usually expressed as $sec^{-1}\ M^{-1}$).

Second-order rate constants for enzyme reactions typically fall in the range between 10^5 and $10^7\ sec^{-1}\ M^{-1}$. Rate constants for the corresponding nonenzymatic reactions are distributed over a much wider range, with half-times that may be measured in tens of seconds or millions of years. Scheme 1d shows approximate values of K_{TX} for several enzymes, estimated from such measurements. The resulting affinities of $S^‡$ are seen to be high. In most cases, the observed rate enhancement probably provides a conservative estimate of the levels of binding affinity that are achieved in the transition state. As has been shown elsewhere,[4] transition state affinity may be underestimated, from simple comparison of rates, if (1) the enzymatic and nonenzymatic reactions differ mechanistically in some fundamental respect, so that their transition states are unrelated in structure; or (2) the chemical mechanisms are the same, but the rate-limiting transition state is reached at a different point on the reaction coordinate in the enzymatic and nonenzymatic reactions. In both cases, simple comparison of rates results in underestimation of the ability of the enzyme to stabilize the transition state for the step that limits the rate of the nonenzymatic reaction.

This view of catalysis, focusing attention on a concrete structure rather than a process, implies that the catalytic power of enzymes lies in their extremely high affinity for unstable intermediates in substrate transformation, as opposed to the substrate in the ground state. The rate of the nonenzymatic reaction can be enhanced only if this difference in binding affinities exists. This principle also furnishes a practical basis for designing powerful enzyme antagonists, in the form of stable analogs of $S^‡$. These inhibitors, usually termed "transition state analogs," can be designed to test alternative mechanisms by which the enzyme might act; that is, strong binding should be observed only for those inhibitors that resemble activated forms of the substrate that arise along the pathway of the enzyme reaction. Such inhibitors can be used to probe the source of the binding discrimination of the enzyme between the substrate in the ground state and that in the transition state, which lies at the heart of the catalytic process. Thus, exact structural observations on enzyme–inhibitor complexes should make it possible to identify the origins of the affinity of the enzyme for the inhibitor and, by inference, the interactions that stabilize

[4] For a review, see R. Wolfenden, *Annu. Rev. Biophys.* **5,** 271 (1976).

the actual transition state for the reaction, involving those amino acid residues of the enzyme that are directly involved in catalysis.

The transition state, by definition the highest point on the energy profile of a reaction, involves bond angles, bond distances, and electron distributions that can never be imitated precisely in any stable analog inhibitor. In addition, some compounds designed on this principle resemble reaction intermediates whose structures and energies may approach, but inevitably fall short of, that of the altered substrate in the transition state itself. For these reasons, the term "transition state analog" describes an ideal that will never be fully attained. Nevertheless, $S^{\ddagger}$ represents an ideal that is worth imitating, because a compound that shares even a few of the structural features that distinguish $S^{\ddagger}$ from S should be a very strong inhibitor, many orders of magnitude more strongly bound than the substrate or product. Potential transition state analog inhibitors have now been prepared against enzymes catalyzing reactions of every class (see Table I at the end of this chapter), and some show very high affinities. For example, 1,6-dihydroinosine (or nebularine 1,6-hydrate, described below) is bound by intestinal adenosine deaminase 3×10^8-fold more tightly than product inosine[5]; 3,4-dihydrouridine is bound by bacterial cytidine deaminase 2×10^9-fold more tightly than product uridine.[6]

A special kind of activation involves gathering of two or more substrates from dilute solution at the active site and their binding in an orientation appropriate for reaction. A multisubstrate analog inhibitor that incorporates binding determinants of two or more substrates within the same molecule may express a large entropic advantage in binding, as compared with the binding properties of analogs of the two substrates measured separately.[7] In principle[8] and in practice,[9] these effects can enhance reaction rates and inhibitor binding affinities by factors as large as 10^8 or more. Accordingly, given only that a compound is an exceptionally powerful inhibitor of a multisubstrate reaction, it may not be easy to decide whether its potency is due to some resemblance to the combined substrates, or to a chemically activated intermediate in which bonds are being made and broken. In some cases, such as that shown in Scheme 2, one inhibitor, tentatively identified as a multisubstrate analog inhibitor, is greatly sur-

[5] W. Jones, L. C. Kurz, and R. Wolfenden, *Biochemistry* **28,** 1242 (1989).
[6] L. Frick, C. Yang, V. E. Marquez, and R. Wolfenden, *Biochemistry* **28,** 9423 (1989).
[7] R. Wolfenden, *Acc. Chem. Res.* **5,** 10 (1972).
[8] M. I. Page and W. P. Jencks, *Proc. Natl. Acad. Sci. U.S.A.* **68,** 1678 (1971).
[9] W. M. Kati, S. A. Acheson, and R. Wolfenden, *Biochemistry* **31,** 7356 (1992).

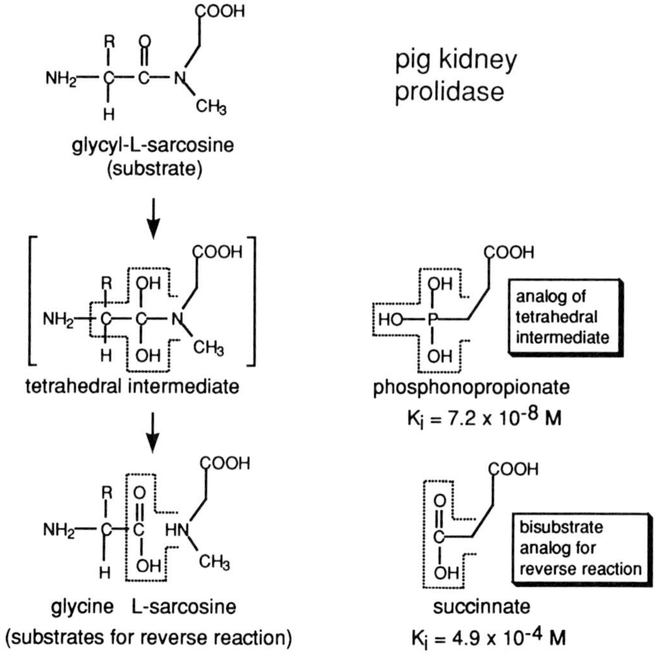

SCHEME 2

passed in potency by another inhibitor containing a bond arrangement that resembles a hydrated intermediate in peptide cleavage.[10] The first of these compounds might be considered a multisubstrate analog, whereas the second has some of the structural features expected of a transition state.

Designing Transition State Analogs

Transition state and multisubstrate inhibitors have been prepared against enzymes catalyzing reactions of every class (see Table I). Each of the various devices by which enzymes are able to enhance the rates of reactions that they catalyze, including catalysis by approximation, general acid–base catalysis, catalysis by desolvation, nucleophilic catalysis, and catalysis by distortion, now appears to have been exploited in several inhibitors (for a review, see Wolfenden and Frick[11]). In what follows, we

[10] A. Radzicka and R. Wolfenden, *Biochemistry* **30,** 4160 (1991).

[11] R. Wolfenden and L. Frick, *in* "Enzyme Mechanisms" (M. I. Page and A. Williams, eds.), p. 9. Royal Society of Chemistry, London, 1987.

consider a few such inhibitors that have been grouped into several classes that emphasize their charge, reactivity with nucleophiles, hydrophobicity, or multisubstrate character.

Analogs of Anionic Intermediates

Isotope exchange studies have shown that many enzymes abstract protons from substrates, generating carbanionic or oxyanionic intermediates that undergo subsequent addition or rearrangement reactions. Inhibitory compounds resembling these species can be prepared by incorporation of stable oxyanionic substituents such as carboxylate groups.

Triose-phosphate isomerase, for example, was found to be strongly inhibited by 2-phosphoglycolate, analogous in structure to a suspected ene-diolate intermediate in the enzyme reaction. Judging from the effect of pH on K_i, the inhibitor appeared to be bound as a dianion.[12] However, ^{13}C and ^{31}P nuclear magnetic resonance experiments showed that the inhibitor was bound as the trianion, requiring that the enzyme have taken up a proton as binding occurred.[13] It was inferred that the EI complex resembled a species on the reaction pathway in which a basic group on the enzyme, Glu-165, had abstracted a proton from the substrate to form an ene-diolate intermediate, and that interpretation has been confirmed by X-ray crystallography of the EI complex.[14] The true affinity of the enzyme for the inhibitor is therefore several orders of magnitude higher than the K_i value had suggested. A comparable approach has been used to examine the complex formed between carboxypeptidase A and the multisubstrate analog inhibitor L-benzylsuccinate. From the dependence of K_i on pH, the inhibitor appeared to be bound as one of two possible monoanionic species involving the two carboxylic acid groups of the inhibitor.[15] When these groups were substituted with ^{13}C, resonances of the bound inhibitor showed that it was bound instead as the dianionic species, with concomitant release of a hydroxide ion from the protein.[16]

Barbituric acid ribonucleoside 5′-phosphate is an excellent inhibitor of orotidylate decarboxylase from yeast, presumably because one of its canonical forms resembles a zwitterionic intermediate formed during decarboxylation.[17] The resonance of ^{13}C at C-5, as well as the UV difference

[12] R. Wolfenden, *Biochemistry* **9,** 3404 (1970).

[13] I. D. Campbell, R. B. Jones, P. A. Keiner, E. Richards, S. G. Waley, and R. Wolfenden, *Biochem. Biophys. Res. Commun.* **83,** 347 (1978).

[14] E. Lolis and G. A. Petsko, *Biochemistry* **29,** 6619 (1990).

[15] L. D. Byers and R. Wolfenden, *Biochemistry* **12,** 2070 (1973).

[16] A. R. Palmer, P. D. Ellis, and R. Wolfenden, *Biochemistry* **21,** 5056 (1982).

[17] H. L. Levine, R. S. Brody, and F. H. Westheimer, *Biochemistry* **19,** 4993 (1980).

spectrum of the bound inhibitor, shows that it is bound in zwitterionic form, not as a covalent 5,6-adduct that would have been expected if the enzyme had acted by an alternative mechanism involving addition and elimination of an enzyme nucleophile.[18]

Analogs of Cationic Intermediates

Carbonium ion intermediates are generated during acid-catalyzed hydrolysis of glycosides, and also during the action of most glycosidases. In the enzyme reactions, a covalent glycosyl–enzyme intermediate may also be formed, but carbonium ions intervene during its formation and breakdown. Presumably for this reason, several glycosidases are strongly inhibited by 1-amino sugars whose protonated forms may form ion pairs with carboxylate groups at the active site that normally serve as a source of protons to the leaving group.[17]

Sterol methyltransferases are believed to involve nucleophilic attack by the sterol on the methyl function of *S*-adenosylmethionine. The methylated intermediate contains a positive charge on an adjacent tertiary carbon atom, and when this atom is replaced by nitrogen potent inhibition results.[2] Squalene synthetase is believed to generate a carbonium ion adjacent to a cyclopropane ring, by elimination of a pyrophosphoryl group. Rearrangement of the intermediate, followed by hydride transfer, leads to squalene. An analogous azasterol serves as a strong inhibitor.[12]

Electrophilic Analogs

Many hydrolases and transferases act by a double displacement mechanism, in which an enzyme nucleophile displaces part of the substrate to form a covalently bound intermediate that undergoes hydrolysis or transfer in a second step. Enzymes involved in carboxyl (or phosphoryl) transfer reactions are often susceptible to inhibition by analogs that undergo the first stage of reaction, typically forming a tetrahedral (or trigonal bipyramidal) intermediate but stop at that stage because they lack an appropriate leaving group. Thus, aldehyde analogs of peptides and amides, in which a hydrogen atom replaces the normal leaving group, are powerful inhibitors of proteases with cysteine[19] or serine[20] nucleophiles at the active site, forming hemiacetals whose stability reflects (1) the ability of the enzyme to stabilize tetrahedral intermediates in substitution and (2) the unusually favorable equilibrium constant for addition of nucleophiles to aldehydes.

[18] S. A. Acheson, J. B. Bell, M. E. Jones, and R. Wolfenden, *Biochemistry* **29,** 3198 (1990).
[19] J. O. Westerik and R. Wolfenden, *J. Biol. Chem.* **247,** 8195 (1972).
[20] R. C. Thompson, *J. Biol. Chem.* **12,** 47 (1973).

When such an enzyme is also specific for the leaving group, an additional advantage can be gained by using a ketone in which one substituent represents the acyl group and the other substituent represents the leaving group. Equilibria of addition of nucleophiles to ketones are much less favorable than for addition to aldehydes, but this effect can be offset by incorporating fluoro groups to promote electrophilic character. In this way, inhibitors of remarkable potency were developed for acetylcholinesterase and several proteases.[21]

Instead of mediating a double displacement reaction, some hydrolases catalyze direct transfer to the acceptor water. Aldehydes and ketones can also inhibit reactions of these kinds, not by addition of an enzyme nucleophile, but by addition of water itself to form a *gem*-diol that resembles a tetrahedral intermediate in direct water attack on the peptide bond. Demonstrated first for pepsin inhibitors containing the unnatural amino acid statone,[22] this mode of binding has also been observed in complexes of fluoroketones with carboxypeptidase A.[23]

Enzymes of the latter type are extremely strongly inhibited by a special class of compounds in which a $-P(=O)O^{-}-NH_2-$ group replaces the peptide bond, allowing four substituents (including an oxygen anion) to interact with the active site in a manner resembling an oxyanionic intermediate in peptide hydrolysis,[24,25] and this mode of binding has been confirmed by X-ray crystallography.[26] This combination of features confers on these compounds very high binding affinities that may have been surpassed only by the complex formed between methionine sulfoximine phosphate and glutamine synthetase (glutamate–ammonia ligase).[27]

Analogs of Nonpolar Intermediates

Ethanol enhances the rate of decarboxylation of pyruvate by a factor of 10^4–10^5, approaching the value observed for pyruvate dehydrogenase catalyzing the same reaction.[28] This increase in the nonenzymatic rate was ascribed to the greater stability of the neutral resonance of the reactive ylide intermediate in the organic solvent, as compared with the charged

[21] M. H. Gelb, J. P. Svaren, and R. H. Abeles, *Biochemistry* **24,** 1813 (1985).
[22] D. H. Rich, A. S. Boparai, and M. S. Bernatowitz, *Biochem. Biophys. Res. Commun.* **104,** 3535 (1982).
[23] D. W. Christianson and W. N. Lipscomb, *J. Am. Chem. Soc.* **108,** 4998 (1986).
[24] P. A. Bartlett and C. K. Marlowe, *Biochemistry* **26,** 8553 (1987).
[25] A. P. Kaplan and P. A. Bartlett, *Biochemistry* **30,** 8165 (1991).
[26] B. W. Matthews, *Acc. Chem. Res.* **21,** 333 (1988).
[27] M. R. Maurizi and A. Ginsburg, *J. Biol. Chem.* **257,** 4271 (1982).
[28] J. Crosby and G. E. Lienhard, *J. Am. Chem. Soc.* **92,** 5707 (1970).

starting materials, and it was suggested that much of the transition state stabilization by pyruvate dehydrogenase was due to the hydrophobic nature of the active site. Keto or thioketo substitution at C-2 of the thiamine ring resulted in compounds with sp^3 hybridization at C-2, causing the ring nitrogen atom to lose its positive charge. These compounds proved to be extremely effective inhibitors or pyruvate dehydrogenase,[29] as did reduced TPP.[30]

Similarly, methyl transfer reactions involving positively charged *S*-adenosylmethionine probably involve charge dispersal in the transition state. Several nonpolar inhibitors of polyamine biosynthesis, synthesized by alkylation of thioamines with 5′-deoxy-5′-chloroadenosine, are considered to owe their effectiveness to hydrophobicity.[31]

Multisubstrate Analogs

Many enzymes play the role of a marriage broker, binding two or more substrates in a spatial relationship that is conducive to reaction. Multisubstrate analogs are single molecules that imitate the binding determinants in such a complex but save the enzyme the trouble of gathering the substrates from dilute solution. The first multisubstrate analog inhibitor ever prepared appears to have been pyridoxylalanine, a strong inhibitor of pyridoxamine–pyruvate transaminase.[32] Other early examples include L-benzylsuccinate, an inhibitor of carboxypeptidase A[33]; Ap_5A, a strong inhibitor of adenylate kinase[34]; and an inhibitor of ether lipid biosynthesis that incorporates the elements of both an attacking and a leaving group.[35] The number of such inhibitors is now very extensive, and, as discussed above, some may represent chemically activated species. One such analog, phosphorylated methionine sulfoximine, an inhibitor of glutamine synthetase,[36] appears to be bound with a K_d value in the range of 10^{-20} *M*.[27]

Practical Uses of Transition State Analogs

Transition state and multisubstrate analog inhibitors include a number of enzyme antagonists of practical importance. From a medicinal stand-

[29] J. A. Gutowski and G. E. Lienhard, *J. Biol. Chem.* **251,** 2863 (1976).

[30] P. N. Lowe, F. J. Leeper, and R. N. Perham, *Biochemistry* **22,** 150 (1983).

[31] K. C. Tang, A. E. Pegg, and J. K. Coward, *Biochem. Biophys. Res. Commun.* **96,** 1371 (1980).

[32] W. B. Dempsey and E. E. Snell, *Biochemistry* **2,** 1414 (1963).

[33] L. D. Byers and R. Wolfenden, *J. Biol. Chem.* **247,** 606 (1972).

[34] G. E. Lienhard and I. I. Secemski, *J. Biol. Chem.* **248,** 1121 (1973).

[35] S. Hixson and R. Wolfenden, *Biochem. Biophys. Res. Commun.* **101,** 1064 (1981).

[36] W. B. Rowe, R. A. Ronzio, and A. Meister, *Biochemistry* **8,** 2674 (1969).

point, these inhibitors represent attractive targets for drug design because the transition state tends to be uniquely characteristic of one kind of enzyme reaction, whereas substrates are usually shared by two or more enzymes. Most successful of these, from a clinical standpoint, have been antihypertensive inhibitors of the angiotensin-converting enzyme, namely, captopril[37] and enalapril,[38] whose design was based on benzylsuccinate, an early multisubstrate analog inhibitor of carboxypeptidase A.[33] A second major use of transition state analogs is as haptens in the production of catalytic antibodies, discussion of which is beyond the scope of this chapter (for reviews, see Lerner and Benkovic[39] and Shokat *et al.*[40]). Finally, transition state analogs constitute promising ligands for affinity chromatography and, more interestingly, as eluants from conventional substrate affinity columns. By using progressively increasing low concentrations of a transition state analog, enzymes can be eluted according to the molecular turnover numbers.[41]

Characterizing Transition State Analog Complexes

The affinities of transition state analogs for enzymes can be measured by methods conventionally used to characterize simple reversible inhibitors; however, when binding is extremely strong, the concentration of inhibitor present in kinetic experiments is so low that correction must be made for mutual depletion of the free enzyme and free inhibitor,[42] and for the time required for enzyme–inhibitor complexes to come to equilibrium. In cases of very tight binding, the best approach is to measure the "on" rate by determining the second-order rate constant for the onset of inhibition and the "off" rate by measuring the first-order rate constant for release of radiolabeled inhibitor in the presence of a large excess of unlabeled inhibitor. The quotient can be used to determine K_d values in the femtomolar range, which are inaccessible by other methods. Details of this method are given in papers describing applications to determining

[37] D. W. Cushman, H. S. Cheung, E. F. Sabo, and M. A. Ondetti, *Biochemistry* **16,** 5484 (1977).

[38] A. A. Patchett, E. Harris, E. W. Tristam, M. J. Wyvratt, M. T. Wu, D. Taub, E. R. Peterson, T. J. Ikeler, and J. ten Broeke, *Nature* (*London*) **288,** 280 (1980).

[39] R. A. Lerner and S. J. Benkovic, *Chemtracts—Org. Chem.* **3,** 1 (1990).

[40] K. M. Shokat, M. K. Ko, T. S. Scanlan, L. Kochersperger, S. Yonkovich, S. Thraisivongs, and P. G. Schultz, *Angew. Chem., Int. Ed. Engl.* **29,** 1296 (1990).

[41] L. Andersson and R. Wolfenden, *J. Biol. Chem.* **255,** 11106 (1980).

[42] J. L. Webb, *in* "Enzyme and Metabolic Inhibitors" Vol. 1, p. 184. Academic Press, New York, 1963.

the affinity of biotin for avidin[43] and to characterizing inhibitors of ribulose-bisphosphate carboxylase[44] and carboxypeptidase A.[25]

One implication of the theory described in Scheme 1 is that any alteration in structure of the substrate, or of the enzyme, which alters the value of k_{cat} or K_m should have a predictable effect on the binding affinity of an ideal transition state analog inhibitor that perfectly resembles $S^‡$. If a change in the structure of the substrate does not affect the rate of the nonenzymatic reaction, then any effect on k_{cat}/K_m should be matched by a change in the affinity of the enzyme for an ideal transition state analog inhibitor. This experimental test of analogy has been passed by inhibitors of papain,[19] elastase,[20] and thermolysin.[45] Alterations in the structure of the enzyme have been examined in the same way. After mutations at the active site, carboxypeptidase A[46] and cytidine deaminase[47] show changes in affinities for transition state analog inhibitors that are closely related to changes in k_{cat}/K_m.

The energetic consequences of stabilizing interactions can be analyzed individually, at least in principle, by deleting one of the interacting groups from either the inhibitor or the enzyme, then examining the thermodynamic consequences of alteration for binding affinity. Kinetic constants such as k_{cat}/K_m are sometimes open to ambiguities of interpretation, because the position of the transition state may vary along the reaction coordinate as alterations are made in the structure of an enzyme or substrate. In contrast, binding affinities of competitive inhibitors offer the advantage of being true dissociation constants that lend themselves to rigorous interpretation. This relationship has been tested for several enzymes by varying the inhibitor[9] and in two cases by varying the structure of the enzyme by site-directed mutagenesis.[46,47]

Examination of enzyme–inhibitor complexes by NMR, revealing states of ionization of both partners in enzyme complexes with transition state analog inhibitors, can provide important indications of the presence of acid–base catalysis as described in the above discussion of analogs of anionic reaction intermediates. States of covalent hydration of inhibitors can be equally revealing, as described in the discussion below of nucleoside deaminases.

[43] N. M. Green, *Biochem. J.* **89,** 585 (1963).
[44] J. V. Schloss, *J. Biol. Chem.* **263,** 4145 (1988).
[45] P. A. Bartlett and C. K. Marlowe, *Biochemistry* **22,** 4618 (1983).
[46] M. A. Phillips, A. P. Kaplan, W. J. Rutter, and P. A. Bartlett, *Biochemistry* **31,** 959 (1992).
[47] A. A. Smith, D. Carlow, R. Wolfenden, and S. A. Short, *Biochemistry* **33,** 6468 (1994).

Slow Binding

Strongly bound inhibitors are often bound slowly: lactate oxidase, for example, binds oxalate with a rate of onset of 80 M^{-1} sec^{-1}.[48] Several caveats should be borne in mind in considering whether these properties are likely to bear any functional relationship to transition state analogy. First, when an inhibitor is bound with high affinity, slow binding (when that is also present) tends to be obvious, simply because the behavior of the inhibitor must be examined at very low concentrations to determine the value of K_i. If, for example, an inhibitor is bound with a K_i value of 10^{-10} M, and the rate constant for formation of EI is 10^7 sec^{-1} M^{-1}, then a slow rate of onset of inhibition becomes obvious in any kinetic investigation intended to determine K_i, carried out over a period of a few minutes. At a concentration of 10^{-10} M, the pseudo-first-order rate constant for the onset of inhibition would be 10^{-3} sec^{-1}. However, if another inhibitor combined with the enzyme at the same rate, but was much less strongly bound ($K_i = 10^{-5}$ M, for example), then its rate of onset of inhibition would be very difficult to determine. For this reason, it remains to be demonstrated in most cases whether "slow" binding is any less common among conventional substrate analog inhibitors than among transition state and multisubstrate analog inhibitors that are much more strongly bound. Second, some transition state analog inhibitors combine rapidly with enzymes, whereas others combine very slowly.[11] If slow binding were an essential, rather than an accidental, feature of transition state analogy, then such variation might not have been expected. Thus, the relationship between transition state affinity and slow binding, if such a relationship exists, is inconsistent, and its origins seem likely to be complex.

Several possible reasons might be advanced to explain slow binding. First, strongly bound inhibitors tend to contain several binding determinants, each of which must be properly engaged for optimal binding. These various interactions might take effect in stages, with some adjustment of the configuration of the active site or the inhibitor that requires the elapse of time. It is also possible that some binding determinants may engage more rapidly than others, resulting in formation of a weak "abortive" complex; this complex must first dissociate before it becomes possible to form the final complex that engages all the appropriate binding determinants. Although rapid weak binding is often followed by slow tight binding, it has seldom if ever been determined whether the rapid weak complex lies on the pathway to the slow tight complex.

[48] S. Ghisla and V. Massey, *J. Biol. Chem.* **250,** 577 (1975).

Another explanation for slow binding of transition state analogs may be that the enzyme has not been prepared by natural selection to bind, rapidly, a molecule that resembles the altered substrate in the transition state, although its affinity for that species is extremely high. Ordinarily, the transition state develops from the bound substrate or product, which the enzyme has been prepared by natural selection to bind rapidly, permitting it to act at rates that often approach the limits imposed by encounter between the substrate and enzyme in solution. Subsequent reactions, within the ES complex, may involve topologically complex transformations, including the closing of a lid or flap, to maximize the attractive interactions that so greatly stabilize the altered substrate in the transition state.[49] New crystal structures show that bound transition state analog inhibitors are enveloped so completely that they become almost completely inaccessible to solvent water in the cases of triose-phosphate isomerase,[14] adenosine deaminase,[50] and cytidine deaminase.[51]

Mechanistic Uses of Transition State Analogs: Two Case Histories

Hydrolytic deamination of adenosine, catalyzed by fungal and mammalian enzymes, is strongly inhibited by analogs of an unstable hydrated intermediate formed by 1,6-addition of substrate water approaching from the front side of the adenosine ring as viewed in Scheme 3. Thus, 1,6-hydroxymethyl-1,6-dihydropurine ribonucleoside (HDHPR) and the antibiotics coformycin and 2′-deoxycoformycin are powerful competitive inhibitors. Modeling studies show that the critical hydroxyl group of the hydroxymethyl substituent of the active isomer of HDHPR can be superimposed on the ring hydroxyl group of the natural 8*R*-OH isomer of 2-deoxycoformycin, both compounds being similar in structure to the postulated intermediate in the catalytic process.

In a remarkable display of steric discrimination, adenosine deaminase binds the natural 8*R*-OH isomer of 2-deoxycoformycin more tightly than the synthetic 8*S* isomer by a factor of 10^7.[52] This difference in affinity might arise from strong attraction of the 8*R* isomer by the active site, from steric hindrance of binding of the 8*S* isomer, or from some combination of these effects. In the 8*S* isomer, the critical hydroxyl group projects from the back side of the ring, from which the leaving group is believed to depart during the catalytic process. The extreme lack of specificity of the

[49] R. Wolfenden, *Mol. Cell. Biochem.* **3,** 207 (1974).

[50] D. K. Wilson, F. B. Rudolph, and F. A. Quiocho, *Science* **252,** 1278 (1991).

[51] L. Betts, S. Xiang, S. A. Short, R. Wolfenden, and C. W. Carter, Jr., *J. Mol. Biol.* **235,** 635 (1994).

[52] V. L. Schramm and D. C. Baker, *Biochemistry* **24,** 641 (1985).

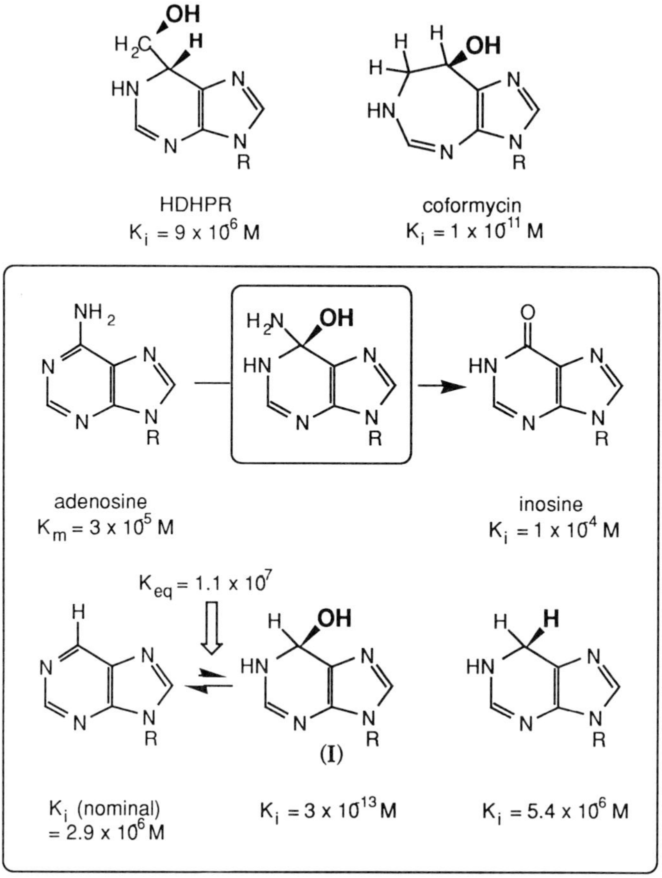

SCHEME 3

enzyme with respect to the leaving group (NH_2-, CH_3NH_2-, Cl-, and CH_3O- are similarly reactive) suggests that the active site of the enzyme appears to be "as big as a barn" on the leaving group side, so that steric hindrance is improbable, and the first of these explanations seems most likely to be correct.

Purine ribonucleoside resembles the substrate adenosine except for replacement of the leaving -NH_2 group by hydrogen, and was long considered to be bound by adenosine deaminase as a simple competitive inhibitor with an affinity similar to the apparent affinity of the substrate. That view became untenable when ^{13}C NMR studies revealed that purine ribonucleoside was bound by adenosine deaminase with a change of hybridization from sp^2 to sp^3 at C-6.[53] The NMR and UV spectra confirmed identification of enzyme-bound purine ribonucleoside as an oxygen adduct, presumably a 1,6-hydrate closely analogous in structure to the 1,6-hydrated intermediate in direct attack by water at the 6 position of adenosine.[5] In this structure, a hydrogen atom occupies the position presumed to be occupied by the leaving $-NH_2$ group in the normal reaction. Because the enzyme is nonspecific for leaving group (-Cl and $-NHCH_3$ are similarly reactive), it is also presumably indifferent to substitution by hydrogen at this position. If the apparent K_i value of purine ribonucleoside is combined with its extremely unfavorable equilibrium constant for hydration in free solution ($K_{eq} = 10^{-7}$), then the true K_i value of the more inhibitory of the two diastereomers of the 1,6-hydrate is found to be in the neighborhood of 3×10^{-13} M.[54]

From the rapid rate of onset of inhibition and the rarity of the hydrate in free solution, it was clear that inhibition normally occurs as a result of purine ribonucleoside binding, followed by hydration at the active site in a mockery of the normal catalytic process.[53] It could further be shown that the equilibrium of hydration is greatly enhanced at the active site of the enzyme, where the effective concentration of substrate water is approximately 10^{10} M.[9]

Cytidine deaminases from bacteria and mammals are strongly inhibited by 3,4,5,6-tetrahydrouridine, structurally analogous to a hypothetical intermediate formed by 3,4-addition of water to the alternate substrate 5,6-dihydrocytidine, shown in Scheme 4. The competitive inhibitors pyrimidin-2-one ribonucleoside [$K_{i(app)} = 3.6 \times 10^{-7}$ M] and 5-fluoropyrimidin-2-one ribonucleoside [$K_{i(app)} = 3.5 \times 10^{-8}$ M] exhibit UV absorption spectra, in their complexes with the enzyme, that are virtually identical with those of the products obtained when hydroxide ion combines with analogs quaternized at N-3.[6] These results indicate that the bound inhibitors are oxygen adducts and provide evidence in favor of binding as a covalent hydrate, not as an enzyme cysteine derivative that had been considered as an alternative possibility. The apparent K_i value of pyrimidin-2-one ribonucleoside as an inhibitor of bacterial cytidine deaminase, combined with its equilibrium constant for covalent hydration in free solution, indicates that $K_i = 1.2 \times 10^{-12}$ M for 3,4-dihydrouridine (the 3,4-hydrate of pyrimidin-2-one ribonucleoside).

Adenosine and cytidine deaminases are nonspecific for the leaving group in substrates, so that they are probably indifferent to replacement of the leaving group by hydrogen in analogs I and II and bind these

[53] L. Kurz and C. Frieden, *Biochemistry* **26,** 8450 (1987).

[54] W. Jones and R. Wolfenden, *J. Am. Chem. Soc.* **108,** 7444 (1986).

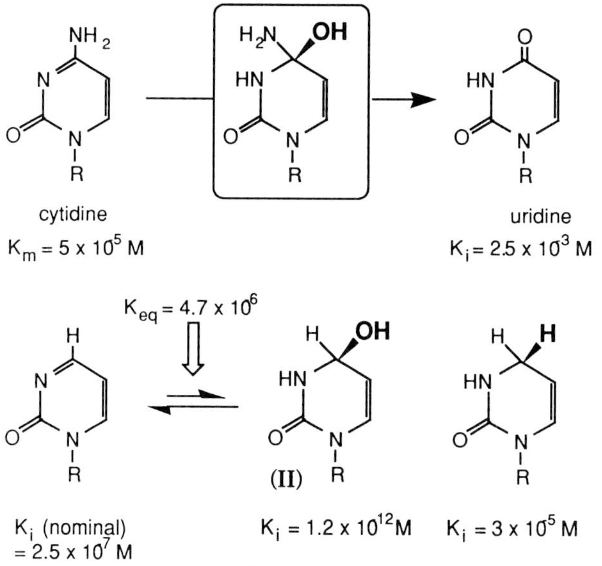

SCHEME 4

transition state analogs very tightly. Thus, the hydroxyl group at the sp^3-hybridized carbon atom probably offers one of the few structural features that could be used by either adenosine or cytidine deaminase to distinguish the altered substrate in the transition state for deamination from the substrate in the ground state (Schemes 3 and 4). To assess the contribution of this hydroxyl group to the binding of analogs I and II, we examined the results of replacement by hydrogen. 1,6-Dihydropurine ribonucleoside, prepared photochemically, was found to serve as a simple competitive inhibitor of adenosine deaminase, with a K_i value of 5.4×10^{-6} M. When this value was compared with the K_i value of the 1,6-hydrate of purine ribonucleoside (1.6×10^{-13} M), it became evident that the 6-hydroxyl group of the latter compound contributes −9.8 kcal/mol to the free energy of binding by calf intestinal adenosine deaminase (Scheme 3).[55,56]

Similar experiments on bacterial cytidine deaminase, performed with 3,4-dihydropyrimidin-2-one ribonucleoside ($K_i = 3.0 \times 10^{-5}$ M), showed that the 4-hydroxyl group of 3,4-dihydrouridine contributes −10.1 kcal/mol to the free energy of binding (Scheme 4).[6] Molecular orbital calculations suggest that the geometry and density of electrons are essentially identical at other positions in the hydrogen- and hydroxyl-substituted

[55] W. M. Kati and R. Wolfenden, *Science* **243,** 1591 (1989).

[56] W. M. Kati and R. Wolfenden, *Biochemistry* **28,** 7919 (1989).

ligands, so that these hydroxyl group contributions to binding affinity, approximately −10 kcal/mol, can be considered to result from simple replacement of -OH by -H.

Group Contributions and Role of Solvent Water

When bound by a protein, a ligand must normally be removed, at least in part, from solvent water. To compare the inherent affinities of the desolvated ligands for the active site, it would therefore be of interest to correct for the free energies of their prior removal from solvent water. (Binding also involves removal of the active site from previous contact with solvent water, but this is true in either case and does not contribute to the difference in affinities between the hydroxyl-containing and the hydrogen-containing ligands.) Free energies have now been determined for removal of many compounds of biological interest from solvent water, from their water-to-vapor or water-to-cyclohexane distribution coefficients. To a fair approximation, free energies of solvation of organic compounds are found to vary as an additive function of constituent groups, alcohols being solvated more strongly than the corresponding alkanes by a factor of roughly 10^5.[15] If a hydroxyl-containing ligand is more readily desolvated that the corresponding hydrogen-containing ligands by roughly 7 kcal/mol in free energy, then for both adenosine and cytidine deaminases the contribution of a desolvated hydroxyl group to the binding of a transition state analog inhibitor appears to be approximately −17 kcal/mol.

In arriving at this conclusion, we have assumed that solvent water has been stripped completely from ligands at critical points of contact with the enzyme. That assumption, although it seems plausible for the hydroxylated ligand whose high affinity implies a close fit to the active site, may not be appropriate in the case of the hydrogen-containing ligand. In the latter case a molecule of water may take the place of the missing hydroxyl group. This "trapping" of water would invalidate simple comparison of observed binding affinities as a measure of the contribution of the hydroxyl group to binding affinity. However, if water is trapped in this way, then the stability of the resulting "wet" complex of the hydrogen-containing ligand must presumably be greater than that of any hypothetical "dry" complex of the hydrogen-containing ligand, from which trapped water was absent. Otherwise, a "dry" complex, of the kind needed for direct comparison of binding affinities, would have been formed by the hydrogen-containing ligand. Under these circumstances, the observed difference in binding affinities would be less than the difference in "dry" binding affinities that is needed to determine the contribution of the hydroxyl group to ligand binding.

The meaning of our earlier estimate of the contribution of the critical hydroxyl group to binding, based on the difference in binding affinity between the two ligands, would also be clouded if the conformation of the enzyme were to change and, to a different extent, on binding of the different ligands. The strong affinity observed for the hydroxylated ligand suggests that the native conformation of the enzyme is already well adapted to tight binding of the hydroxyl-containing ligand. The hydrogen-containing ligand, being smaller, should be able to fit into any "native" structure that can accommodate the hydroxylated ligand. It would hardly be surprising, however, if the active site of the enzyme were to show some tendency to collapse around the hydrogen-containing ligand, forming a more compact structure than does the complex of the hydroxyl-containing ligand. Such a change in structure would invalidate simple comparison of binding affinities as a measure of hydroxyl group contribution to binding. If, however, the hydrogen-containing ligand were bound with such a change in conformation, then the stability of the resulting "collapsed" complex would necessarily be greater than that of any complex with the active site in the "native" configuration. Otherwise, the natively configured complex, being more stable, would have been the species actually observed at equilibrium. The contribution of the hydroxyl group to the stability of the complex of the hydroxyl-containing ligand in the native structure would again have been underestimated.

These considerations suggest that if "water trapping" or enzyme distortion accompany formation of the complex of the enzyme with the hydrogen-containing ligand, then either of these effects might be expected to exert a "leveling" influence on the relative affinities observed for the hydroxyl- and hydrogen-containing ligands, leading to underestimation of the contribution of the critical hydroxyl group to binding affinity. Evidently, the contributions of these hydroxyl groups to binding affinities of the desolvated ligands are probably at least as large, and could be larger, than values of approximately -17 kcal/mol suggested by the observed differences in K_i values.

Crystal structures have now been reported for the complexes formed between transition state analogs and adenosine[57] and cytidine[51] deaminases. The results confirm that these inhibitors are bound as the covalent hydrate almost completely removed from contact with solvent water, as in the complex that is formed between 2-phosphoglycolate and triosephosphate isomerase.[14] The implied conformation change, by maximizing the possibility of attractive interactions between the enzyme and substrate in the transition state, may help to answer the conflicting requirements

[57] D. K. Wilson, F. B. Rudolph, and F. A. Quiocho, *Science* **252,** 1278 (1991).

of transition state stabilization and rapid access of substrates and egress of products.[49] Several features of the new crystal structures are shown in Scheme 5. The critical hydroxyl group of the inhibitor, on which so much of the catalytic binding enhancement appears to depend, interacts with three groups, including a zinc atom and a carboxylate residue at the active site of the enzyme.

Conclusions

In summary, many of the structural features of a substrate remain unchanged as it passes from the ground state to the transition state. To enhance the rate of a reaction, an enzyme must therefore single out for chemical recognition those few features of a substrate that do change. We have considered the generation of hydrates I and II at the active sites of nucleoside deaminases as analogs of the process by which such enzymes generate intermediates in substrate transformation. In these compounds, a tetrahedrally oriented hydroxyl group is an obvious feature that distinguishes these compounds from the aromatic starting materials. Evidently a few polar interactions involving this group, arising fleetingly in the transition state, are capable of generating a large part of the added binding affinity that is needed to explain the rate enhancement (~10^{12}-fold) that an enzyme of this kind produces. Other interactions with the enzyme are obviously important in transition state stabilization and can be analyzed

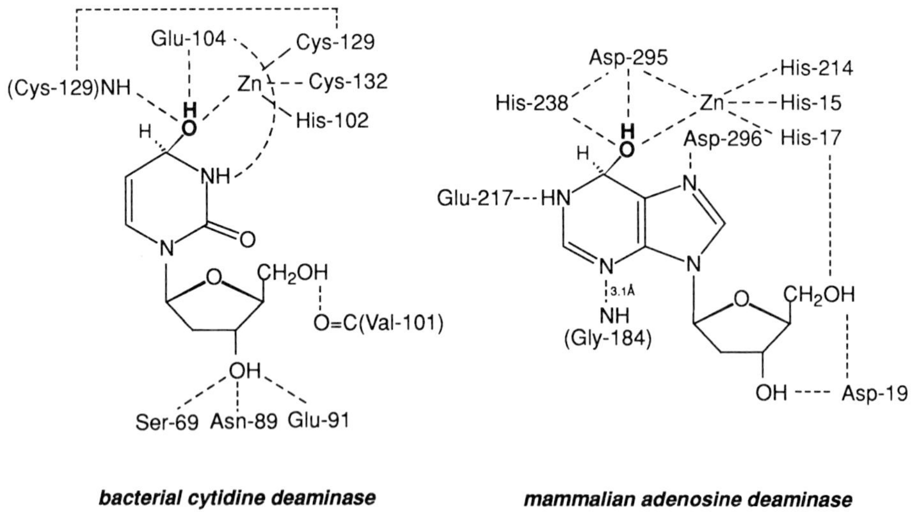

SCHEME 5

by similar methods, including active site directed and inhibitor directed mutagenesis. Particular mention should be made of the structural results, too extensive for present discussions, of crystallographic studies of other enzyme complexes with transition state and multisubstrate analog inhibitors, especially those of triose-phosphate isomerase,[14] thermolysin,[26] and carboxypeptidase A.[23]

Transition State and Multisubstrate Analogs (Table I)

The list of enzymes and inhibitors in Table I is organized according to EC classification, and an attempt has been made to cite the original reference in each case. A list of this kind is certain to contain errors of omission, for which the authors apologize.

TABLE I
TRANSITION STATE AND MULTISUBSTRATE ANALOGS

EC Number[a]	Enzyme[a]	Inhibitor[b]
1.1.1.1	Alcohol dehydrogenase	NAD^+ adduct[1]
1.1.1.27	L-Lactate dehydrogenase	NAD^+ adduct, oxalate,[2] oxalylethyl-NADH[3]
1.1.1.37	Malate dehydrogenase	NAD^+ adduct[1]
1.1.1.145	3β-Hydroxy-Δ^5-steroid dehydrogenase	4-Aza-4-methyl-5-pregnane-3,20-dione[4]
1.2.1.12	Glyceraldehyde-3-phosphate dehydrogenase	Threose-2,4-diphosphate[5]
1.2.7.1	Pyruvate synthase	Thiamine thiazolone and thiamine thiothiazolone pyrophosphates,[6] tetrahydrothiamine pyrophosphate,[7] acetylphosphonate[8]
1.4.1.1	Alanine dehydrogenase	Oxalylethyl-NADH[9]
1.4.1.2	Glutamate dehydrogenase	NAD^+ adduct[10]
1.5.1.3	Dihydrofolate reductase	Methotrexate[11]
1.5.1.X	Deoxyhypusine synthase	*N*-Guanyl-1,7-diaminoheptane[12]
1.11.1.9	Glutathione peroxidase	Mercaptosuccinate[13]
1.13.11.3	Protocatechuate 3,4-dioxygenase	2-Hydroxyisonicotinic acid *N*-oxide[14]
1.13.12.4	Lactate 2-monooxygenase	Oxalate,[15] malonate[16]
1.14.17.1	Dopamine β-monooxygenase	1-(4-Hydroxybenzyl)imidazole-2-thiol[17]
2.1.1.41	24-Sterol *C*-methyltransferase	25-Azacholesterol,[18] 24-(*R*,*S*)-epiminolanosterol,[19] 24-methyl-25-azacycloartanol[20]
2.1.1.45	Thymidylate synthase	1-(D-2′-Deoxyribofuranosyl)-8-azapurin-2-one 5′-monophosphate,[21] multisubstrate analog[22]
2.1.2.2	Phosphoribosylglycinamide formyltransferase	β-Thioglycinamide ribonucleotide dideazafolate[23]
2.1.3.1	Methylmalonyl-CoA carboxyltransferase	Oxalate[24]
2.1.3.2	Aspartate carbamoyltransferase	Phosphonoacetylaspartate[25]
2.1.3.3	Ornithine carbamoyltransferase	Phosphonoacetylornithine[26,27]
2.3.1.7	Carnitine *O*-acetyltransferase	*O*-[2-(*S*-Coenzyme A)acetyl]carnitine[28]
2.3.1.48	Histone acetyltransferase	*N*-[2-(*S*-Coenzyme A)acetyl]spermidine amide[29]
2.3.1.59	Gentamicin 2′-*N*-acetyltransferase	*N*-[2-(*S*-Coenzyme A)acetyl]gentamicin[30]
2.3.1.X	Succinyl-CoA: tetrahydrodipicolinate *N*-succinyltransferase	2-Hydroxytetrahydropyran-2,6-dicarboxylic acid[31]
2.3.2.2	γ-Glutamyltransferase	Serine–borate complex[32]
2.4.1.1	Glycogen phosphorylase	1,5-Gluconolactone,[33,34] 1-deoxy-α-D-*gluco*-heptulose 2-phosphate[35]
2.4.1.19	Cyclomaltodextrin glucanotransferase	Acarbose[36]

2.4.1.X	α(1 → 2)-Fucosyltransferase	2-*O*-(2-Guanosinophosphonoethyl)-1-*O*-(phenyl)-β-D-galactopyranoside[37]
2.4.2.1	Purine-nucleoside phosphorylase	8-Amino-2′-nordeoxyguanosine, 2′-nordeoxyguanosine diphosphate[38]
2.5.1.1	Geranylgeranyl diphosphate synthase	3-Azageranyl diphosphate[39]
2.5.1.6	Methionine adenosyltransferase	5′-(δ-Methylmethionine)-β,γ-imido-ATP[40]
2.5.1.9	Riboflavin synthase	6,7-Dioxolumazine[41]
2.5.1.16	Spermidine synthase	*S*-Adenosyl-3-thio-1,8-diaminooctane[42]
2.5.1.19	3-Phosphoshikimate 1-carboxyvinyltransferase	Carboxyallenyl phosphate, (*Z*)-3-fluorophosphoenolpyruvate[43]
2.5.1.21	Farnesyl-diphosphate farnesyltransferase	Ammonium analog of carbocation[44]
2.5.1.22	Spermine synthase	*S*-Adenosyl-1,12-diamino-3-thio-9-azadodecane[45]
2.5.1.26	Alkylglycerone-phosphate synthase	2(3)-Palmitoyl-1,2,3-trihydroxyeicosane 1-phosphate[46]
2.6.1.30	Pyridoxamine–pyruvate transaminase	Pyridoxylalanine[47]
2.7.1.1	Hexokinase	Chromium-ATP[48]
2.7.1.20	Adenosine kinase	Ap_4A[49]
2.7.1.21	Thymidine kinase	Ap_5T[48,50]
2.7.1.40	Pyruvate kinase	Oxalate[51]
2.7.3.2	Creatine kinase	Nitrate[52]
2.7.3.3	Arginine kinase	Nitrate[53,54]
2.7.4.3	Adenylate kinase	Ap_5A[55]
2.7.9.2	Pyruvate, water dikinase	Oxalate[56]
2.7.10.X	p70 S6 kinase	Rapamycin[57]
2.8.3.5	3-Oxoacid CoA-transferase	Succinic monohydroxamic acid[58]
3.1.1.1	Carboxylesterase	Benzil,[59] ethylphenyl glyoxalate[60]
3.1.1.4	Phospholipase A_2	1-Hexadecylthio-2-hexadecanoylamino-1,2-dideoxy-*sn*-glycero-3-phosphocholine, 1-hexadecylthio-1-deoxy-2-hexadecylphosphono-*sn*-glycero-3-phosphocholine,[61] 1-hexadecyl-3-trifluoroethylglycero-*sn*-2-phosphomethanol[62]
3.1.1.7	Acetylcholinesterase	Boronate,[63] fluoroketone[64,65]
3.1.1.8	Cholinesterase	Diphenylboric acid[66]
3.1.1.13	Sterol esterase	Phenyl-*n*-butylborinic acid,[67] chlorodifluoroacetophenone[68]
3.1.1.34	Lipoprotein lipase	Phenyl-*n*-butylborinic acid[69]
3.1.1.59	Juvenile-hormone esterase	Fluoroketones[70]
3.1.3.1	Alkaline phosphatase	Vanadate[71]
3.1.3.2	Acid phosphatase	Tungstate, molybdate[72]
3.1.6.1	Arylsulfatase	Sulfite[73]

(*continued*)

TABLE I (*continued*)

EC Number[a]	Enzyme[a]	Inhibitor[b]
3.1.27.5	Pancreatic ribonuclease	Uridine-vanadate[74]
3.2.1.1	α-Amylase	D-*malto*-Bionolactone[75]
3.2.1.10	*O*-Glycosyl glycosidase	Acarbose,[76] bis(hydroxymethyl)dihydroxypyrrolidine[77]
3.2.1.17	Lysozyme	Lactone, 2-acetamido-2-deoxyglucoseen[78]
3.2.1.18	Exo-α-sialidase	2-Deoxy-2,3-dehydro-*N*-acetylneuraminic acid[79]
3.2.1.21	β-Glucosidase	1-Aminoglucoside,[80] gluconolactone[81]
3.2.1.22	Galactosidase	1-Aminogalactoside,[82] galactal[83]
3.2.1.23	β-Galactosidase	1-Aminogalactoside,[82] galactal[83]
3.2.1.45	Glucosylceramidase	Glucosylamine[84]
3.2.1.48	Sucrose α-glucosidase	Castanospermine,[85] 2,6-Dideoxy-2,6-imino-7-*O*-(β-D-glucopyranosyl)-D-*glycero*-L-*gulo*-heptitol,[86]
3.2.1.52	β-*N*-Acetylhexosaminidase	NAG-lactone[87]
3.2.1.55	α-*N*-Arabinofuranosidase	L-Arabino-1,4-lactone[88]
3.2.2.4	AMP nucleosidase	Formycin 5′-phosphate[89]
3.2.2.X	Uracil-DNA-glycosylase	6-(*p*-*n*-Octylanilino)uracil[90]
3.4.11.1	Leucyl aminopeptidase	Aminoaldehydes,[91] bestatin,[92] amastatin,[93] aminohydroxamates[94]
3.4.15.1	Peptidyl-dipeptidase A	Captopril,[95] enalapril,[96] fluoroketone,[97] aminoalcohol,[98] 2-mercaptoacetyl dipeptides,[99] ketodimethyl peptide,[100] aminoketodimethyl peptides[101]
3.4.17.1	Carboxypeptidase A	Benzylsuccinate,[102] L-2-phosphoryloxy-3-phenylpropionic acid,[103] dipeptide phosphoramidates,[104] 2-benzyl-3-formylpropionate, 2-mercaptoacetyl dipeptides,[76] 3-phosphonopropionic acid,[105] (2-carboxy-3-phenylpropyl)methylsulfoximine and -sulfodiimine,[106] phosphonotetrapeptides[107]
3.4.17.2	Carboxypeptidase B	Benzylsuccinate[108]
3.4.17.3	Lysine carboxypeptidase	2-Mercaptomethyl-3-guanidinoethyl thiopropionate,[109] phosphono dipeptide[110]
3.4.19.3	Pyroglutamyl-peptidase I	Oxoprolinal[111]
3.4.21.1	Chymotrypsin	Aldehyde,[112] 1-acetamido-2-phenylethaneboronic acid,[113] peptidyl fluoromethyl ketones[114]
3.4.21.4	Trypsin	Peptide and nonpeptide borates[115]
3.4.21.36	Pancreatic elastase	Peptidyl fluoromethylketones,[116] acetyl-Pro-Ala-Pro-alaninal[117]
3.4.21.62	Subtilisin	Benzeneboronic acid[118]

3.4.22.2	Papain	Acetyl-Phe-glycinal,[119] *N*-acetyl-1-phenylalanylaminoacetonitrile[120]
3.4.22.3	Ficain	Acetyl-Phe-glycinal[121]
3.4.23.1	Pepsin A	Pepstatin,[122] methylpepstatin and statone,[123] fluoroketone,[124] phosphinic acid dipeptide[125]
3.4.23.15	Renin	Statine,[126] difluorostatine and difluorostatone,[127] reduced peptide,[128] dihydroxyethylene peptide analogs[129]
3.4.23.X	HIV protease	Hydroxyethylene peptide isostere[130]
3.4.24.3	Microbial collagenase	Isoamylphosphonyl peptide,[131] cinnamoyl-Phe-(CO)Gly-Pro-Pro ketone,[132] phosphonamide[133]
3.4.24.11	Neprilysin	*N*-Carboxymethyl peptides,[137] 2-mercaptoacetyl dipeptides[76]
3.4.24.27	Thermolysin	*N*-Carboxymethyl dipeptides,[134] phosphonamidate,[135] dipeptides,[76] hydroxamates[83,136]
3.5.1.1	Asparaginase	Aspartate semialdehyde[138]
3.5.1.4	Amidase	Acetaldehyde–ammonia[139]
3.5.2.3	Dihydroorotase	Borocarbamylethyl aspartate[140]
3.5.2.6	*β*-Lactamase	Arylmethyl phosphonate methyl ester[141]
3.5.4.3	Guanine deaminase	(1,2,6)-Thiadiazine-1,1-dioxides[142]
3.5.4.4	Adenosine deaminase	1,6-Dihydro-6-hydroxymethylpurine ribonucleoside,[143] coformycin,[144] 1,6-dihydroinosine[145]
3.5.4.5	Cytidine deaminase	Tetrahydrouridine,[146] phosphapyrimidine,[147] 1,3-diazepin-2-ol ribonucleoside,[148] 3,4-dihydrouridine[148a]
3.5.4.6	AMP deaminase	Coformycin 5′-phosphate[149]
3.5.4.12	dCMP deaminase	Tetrahydrouridine 5′-phosphate[150]
3.11.1.1	Phosphonoacetaldehyde hydrolase	Phosphite + acetaldehyde[151]
	Cholesterol 5,6-oxide hydrolase	5,6-Iminocholestanol[152]
4.1.1.3	Oxaloacetate decarboxylase	Oxalate[153]
4.1.1.4	Acetoacetate decarboxylase	Acetopyruvate,[154] acetoacetone[155]
4.1.1.23	Orotidine-5′-phosphate decarboxylase	1-(5′-Phospho-D-ribosyl)barbituric acid[156]
4.1.1.39	Ribulose-bisphosphate carboxylase	Carboxyribitol diphosphate,[157] carboxyarabinitol diphosphate[158]
4.1.2.13	Fructose-bisphosphate aldolase	Phosphoglycolohydroxamate[159]
4.1.3.1	Isocitrate lyase	3-nitropropionate[160]
4.1.3.7	Citrate (*si*)-synthase	Carboxymethyl-CoA and oxaloacetate,[161] *S*-acetonyl-CoA[162]
4.1.99.1	Tryptophanase	2,3-Dihydrotryptophan[159]
4.1.99.4	1-Aminocyclopropane-1-carboxylate deaminase	1-Aminocyclopropane phosphonate[163]

(continued)

TABLE I (*continued*)

EC Number[a]	Enzyme[a]	Inhibitor[b]
4.2.1.2	Fumarate hydratase	3-Nitro-2-hydroxypropionate,[164] *S*-2,3-dicarboxyaziridine[165]
4.2.1.3	Aconitate hydratase	Nitro analogs of citrate and isocitrate[166]
4.2.1.17	Enoyl-CoA hydratase	Acetoacetyl-CoA[167]
4.2.1.20	Tryptophan synthase	2,3-Dihydrotryptophan[168]
4.3.1.1	Aspartate ammonia-lyase	3-Nitro-2-aminopropionate[169]
4.3.1.5	Phenylalanine ammonia-lyase	L-2-Aminooxy-3-phenylpropionic acid[170]
4.3.2.2	Adenylosuccinate lyase	N^6-(L-2-Carboxyethyl-2-nitroethyl)-AMP[171]
4.4.1.5	Lactoylglutathione lyase	3-Hydroxy-2-methyl-4*H*-pyran-4-one, isoascorbate,[172] isopropyltropolon[173]
5.1.1.1	Alanine racemase	1-Aminocyclopropane phosphonate[164]
5.1.1.4	Proline racemase	Pyrrole,[174] pyrroline 2-carboxylates[175]
5.1.1.7	Diaminopimelate epimerase	3-Chlorodiaminopimelic acid[176]
5.3.1.1	Triose-phosphate isomerase	2-Phosphoglycolate,[177] 2-phosphoglycolohydroxamate[178]
5.3.1.6	Ribose-5-phosphate isomerase	4-Phosphoerythronate[179]
5.3.1.9	Glucose-6-phosphate isomerase	5-Phosphoarabinonate[180]
5.3.1.13	Arabinose-5-phosphate isomerase	4-Phosphoerythronate[181]
5.3.3.1	Steroid Δ-isomerase	17β-D-hydroequinalin[182]
5.3.3.2	Isopentenyl-diphosphate Δ-isomerase	2-(Dimethylamino)ethyl pyrophosphate[183]
5.4.2.2	Phosphoglucomutase	α-D-glucose 1-phosphate vanadate[184]
5.4.99.5	Chorismate mutase	Oxabicyclo[3.3.1]nonene,[185] 2-aza-2,3-dihydrosqualene[186]
6.3.1	Aminoacyl-tRNA ligases	Aminoalkyl adenylates,[187] aminophosphonyl adenylates[188]
6.3.1.2	Glutamate–ammonia ligase	Phosphinotricin,[189] methionine sulfoximine phosphate[190]
6.3.2.2	Glutamate–cysteine ligase	Buthionine-sulfoximine phosphate[191]
6.4.1.1	Pyruvate carboxylase	Oxalate[192]

[a] Classification and nomenclature of enzymes based on recommendations of the Nomenclature Committee of the International Union of Biochemistry and Molecular Biology, published in 1992 (E. C. Webb, ed., Academic Press).

[b] *Key to references:* (*1*) J. Everse, E. C. Zoll, L. Kahan, and N. O. Kaplan, *Biorg. Chem.* **1,** 207 (1971); (*2*) W. B. Novoa, A. D. Winer, A. J. Glaid, and G. W. Schwert, *J. Biol. Chem.* **234,** 1143 (1959); (*3*) H. Kapmeyer, G. Pfleiderer, and W. E. Trommer, *Biochemistry* **15,** 5024 (1976); (*4*) P. J. Bertics, C. F. Edman, and H. J. Karavolas, *J. Biol. Chem.* **259,** 107 (1984); (*5*) A. L. Fluharty and C. E. Ballou, *J. Biol. Chem.* **234,** 2517 (1958); (*6*) T. A. Gutowski and G. E. Lienhard, *J. Biol. Chem.* **251,** 2863 (1976); (*7*) P. N. Lowe, F. J. Leeper, and R. N. Perham, *Biochemistry* **22,** 150 (1983); (*8*) R. Kluger, *J. Am. Chem. Soc.* **99,** 4504 (1977); (*9*) H. Kapmeyer, G. Pfleiderer, and W. E. Trommer, *Biochemistry* **15,** 5024 (1976); (*10*) J. Everse, E. C. Zoll, L. Kahan, and N. O. Kaplan, *Bioorg. Chem.* **259,** 1043 (1984); (*11*) W. C. Werkheiser, *J. Biol. Chem.*

236, 888 (1961); (*12*) J. Jakus, E. C. Wolff, M. H. Park, and J. E. Folk, *J. Biol. Chem.* **268,** 13151 (1993); (*13*) J. Chaudiere, E. C. Wilhemsen, and A. L. Tappel, *J. Biol. Chem.* **259,** 1043 (1984); (*14*) S. W. May, C. D. Oldham, P. W. Mueller, S. R. Padgette, and A. L. Sowell, *J. Biol. Chem.* **257,** 12746 (1982); (*15*) S. Ghisla and V. Massey, *J. Biol. Chem.* **250,** 577 (1975); (*16*) S. Ghisla and V. Massey, *J. Biol. Chem.* **252,** 6729 (1977); (*17*) L. I. Kruse, W. E. DeWolf, Jr., P. A. Chambers, and P. J. Googhart, *Biochemistry* **25,** 7271 (1986); (*18*) A. C. Oehlschlager, R. H. Angus, A. M. Pierce, H. D. Pierce, Jr., and R. Srinivasan, *Biochemistry* **23,** 3582 (1984); (*19*) W. D. Nes, G. G. Jansen, R. A. Norton, M. Kalinowska, F. G. Crumley, B. Tal, A. Bergenstrahle, and L. Johsson, *Biochem. Biophys. Res. Commun.* **177,** 566 (1991); (*20*) A. Rahier, D. Benveniste, and F. Schubert, *J. Am. Chem. Soc.* **103,** 2408 (1981); (*21*) T. I. Kalman and D. Goldman, *Biochem. Biophys. Res. Commun.* **102,** 682 (1981); (*22*) A. Srinivasan, V. Armarnath, A. D. Broom, F. C. Zou, and Y.-C. Cheng, *J. Med. Chem.* **27,** 1710 (1984); (*23*) J. Inglese, R. A. Blatchy, and S. J. Benkovic, *J. Med. Chem.* **32,** 937 (1989); (*24*) D. Northrop and H. G. Wood, *J. Biol. Chem.* **244,** 5820 (1969); (*25*) K. D. Collins and G. R. Stark, *J. Biol. Chem.* **246,** 6599 (1971); (*26*) M. Mori, K. Aoyagi, M. Tatibana, T. Ishikawa, and H. Ishii, *Biochem. Biophys. Res. Commun.* **76,** 900 (1977); (*27*) N. J. Hoogenraad, *Arch. Biochem. Biophys.* **188,** 137 (1978); (*28*) J. F. A. Chase and P. K. Tubbs, *Biochem. J.* **111,** 225 (1969); (*29*) P. M. Cullis, R. Wolfenden, L. S. Cousens, and B. M. Alberts, *J. Biol. Chem.* **257,** 12165 (1982); (*30*) J. W. Williams and D. B. Northrop, *J. Antibiot.* **32,** 1147 (1979); (*31*) D. A. Berges, W. E. DeWolf, Jr., G. L. Dunn, D. J. Newman, S. J. Schmidt, J. J. Taggart, and C. Gilvarg, *J. Biol. Chem.* **261,** 6160 (1986); (*32*) S. S. Tate and A. Meister, *Proc. Natl. Acad. Sci. U.S.A.* **75,** 4806 (1978); (*33*) J. I. Tu, G. R. Jacobson, and D. J. Graves, *Biochemistry* **10,** 1229 (1971); (*34*) A. M. Gold, E. Legrand, and G. R. Sanchez, *J. Biol. Chem.* **246,** 5700 (1971); (*35*) H. W. Klein, M. J. Im, and D. Palm, *Eur. J. Biochem.* **157,** 107 (1986); (*36*) A. Nakamura, K. Haga, and K. Yamane, *Biochemistry* **32,** 6624 (1993); (*37*) M. M. Palcic, L. D. Heerze, O. P. Srivastava, and O. Hindsgaul, *J. Biol. Chem.* **264,** 17174 (1989); (*38*) J. M. Stein, J. D. Stoeckler, S.-Y. Li, R. L. Tolman, M. MacCoss, A. Chen, J. D. Karkas, W. T. Ashton, and R. E. Parks, Jr., *Biochem. Pharmacol.* **36,** 1237 (1987); (*39*) H. Sagami, T. Korenaga, K. Ogura, A. Steiger, H.-J. Pyun, and R. M. Coates, *Arch. Biochem. Biophys.* **297,** 314 (1992); (*40*) F. Kappler and A. Hampton, *J. Med. Chem.* **33,** 2545 (1990); (*41*) S. S. Al-Hassan, R. J. Kuilick, D. B. Livingstone, C. J. Suckling, and H. C. S. Wood, *J. Chem. Soc., Perkins Trans. 1* 2645 (1980); (*42*) K. C. Tang, A. E. Pegg, and J. K. Coward, *Biochem. Biophys. Res. Commun.* **96,** 1371 (1980); (*43*) M. C. Walker, J. E. Ream, R. D. Sammons, E. W. Logusch, M. H. O'Leary, R. L. Sommerville, and J. A. Sikorski, *BioMed. Chem. Lett.* **1,** 683 (1991); (*44*) R. M. Sandifer, M. D. Thompson, R. G. Gaughan, and C. D. Poulter, *J. Am. Chem. Soc.* **104,** 7376 (1982); (*45*) P. A. Woster, A. Y. Black, K. D. Duff, J. K. Coward, and A. E. Pegg, *J. Med. Chem.* **32,** 1300 (1989); (*46*) S. Hixson and R. Wolfenden, *Biochem. Biophys. Res. Commun.* **101,** 1064 (1981); (*47*) W. B. Dempsy and E. E. Snell, *Biochemistry* **2,** 1414 (1963); (*48*) K. D. Danenberg and W. W. Cleland, *Biochemistry* **14,** 28 (1975); (*49*) R. Bone, Y. C. Cheng, and R. Wolfenden, *J. Biol. Chem.* **261,** 5731 (1986); (*50*) S. Ikeda and D. H. Ives, *J. Biol. Chem.* **23,** 12659 (1985); (*51*) G. H. Reed and S. D. Morgan, *Biochemistry* **13,** 3537 (1974); (*52*) E. J. Milner-White and D. C. Watts, *Biochem. J.* **122,** 727 (1971); (*53*) D. H. Buttlaire and M. Cohn, *J. Biol. Chem.* **249,** 5733 (1974); (*54*) D. H. Buttlaire and M. Cohn, *J. Biol. Chem.* **249,** 5741 (1974); (*55*) G. E. Lienhard and I. I. Secemski, *J. Biol. Chem.* **248,** 1121 (1973); (*56*) S. Narindrasorasak and W. A. Bridger, *Can. J. Biochem.* **56,** 816 (1978); (*57*) C. J. Kuo, J. Chung, D. F. Fiorentino, W. M. Flanagan, J. Blenis, and G. R. Crabtree, *Nature* (*London*) **358,** 70 (1992); (*58*) C. M. Pickart and W. P. Jencks, *J. Biol. Chem.* **254,** 9120 (1979); (*59*) M. C. Berndt, J. de Jersey, and B. Zerner, *J. Am. Chem. Soc.* **99,** 8332 (1977); (*60*) M. C. Berndt, J. de Jersey, and B. Zerner, *J. Am. Chem. Soc.* **99,** 8334

(*continued*)

TABLE I (*continued*)

(1977); (*61*) L. Yu and E. A. Dennis, *Proc. Natl. Acad. Sci. U.S.A.* **88,** 9325 (1991); (*62*) M. K. Jain, W. Tao, J. Rogers, C. Arenson, H. Eibl, and B.-Z. Yu, *Biochemistry* **30,** 10256 (1991); (*63*) K. A. Koehler and G. P. Hess, *Biochemistry* **13,** 5345 (1974); (*64*) U. Brodbeck, K. Schweikert, R. Gentinetta, and M. Rottenberg, *Biochem. Biophys. Acta* **567,** 357 (1979); (*65*) M. H. Bleb, J. P. Svaren, and R. H. Abeles, *Biochemistry* **24,** 1813 (1985); (*66*) C. W. Garner, G. W. Little, and J. W. Pelley, *Biochim. Biophys. Acta* **790,** 91 (1984); (*67*) L. D. Sutton, J. S. Stout, L. Hosie, P. S. Spencer, and D. M. Quinn, *Biochem. Biophys. Res. Commun.* **134,** 386 (1986); (*68*) J. Sohl, L. D. Sutton, D. J. Burton, and D. M. Quinn, *Biochem. Biophys. Res. Commun.* **151,** 554 (1988); (*69*) L. D. Sutton, J. S. Stout, L. Hosie, P. S. Spencer, and D. M. Quinn, *Biochem. Biophys. Res. Commun.* **134,** 386 (1986); (*70*) B. D. Hammock, K. D. Wing, J. McLaughlin, V. M. Lovell, and T. C. Sparks, *Pestic. Biochem. Physiol.* **17,** 76 (1982); (*71*) V. Lopez, T. Stevens, and R. N. Lindquist, *Arch. Biochem. Biophys.* **175,** 31 (1976); (*72*) R. L. Van Etten, P. P. Waymack, and D. M. Rehkop, *J. Am. Chem. Soc.* **96,** 6782 (1974); (*73*) A. B. Roy, *Biochem. J.* **55,** 653 (1955); (*74*) R. N. Lindquist, J. L. Lynn, Jr., and G. E. Lienhard, *J. Am. Chem. Soc.* **95,** 8762 (1973); (*75*) E. Laszlo, J. Hollo, A. Hoschke, and G. Sarosi, *Carbohydr. Res.* **61,** 387 (1978); (*76*) D. Schmidt, W. Frommer, B. Junge, L. Muller, W. Wingender, E. Truscheit, and D. Schafer, *Naturwissenschaften* **64,** 535 (1977); (*77*) M. Schindler and N. Sharon, *J. Biol. Chem.* **251,** 4330 (1976); (*78*) P. J. Card and W. D. Hitz, *J. Org. Chem.* **50,** 891 (1985); (*79*) C. A. Miller, P. Wand, and M. Flashner, *Biochem. Biophys. Res. Commun.* **83,** 1479 (1978); (*80*) H. L. Lai and B. Axelrod, *Biochem. Biophys. Res. Commun.* **54,** 463 (1973); (*81*) C. D. Santos and W. R. Terra, *Biochim. Biophys. Acta* **831,** 179 (1985); (*82*) H. L. Lai and B. Axelrod, *Biochem. Biophys. Res. Commun.* **54,** 463 (1973); (*83*) D. F. Wentworth and R. Wolfenden, *Biochemistry* **13,** 4715 (1974); (*84*) K. M. Osiecki-Newman, D. Fabbro, G. Legler, R. J. Desnick, and G. A. Grabowski, *Biochim. Biophys. Acta* **915,** 87 (1987); (*85*) C. Danzin and A. Ehrhardt, *Arch. Biochem. Biophys.* **257,** 472 (1987); (*86*) B. L. Rhinehart, K. M. Robinson, P. S. Liu, A. J. Payne, M. E. Wheatley, and S. R. Wagner, *J. Pharmacol. Exp. Ther.* **241,** 915 (1987); (*87*) Y. C. Lee, *Biochem. Biophys. Res. Commun.* **35,** 161 (1969); (*88*) A. H. Fielding, M. L. Sinnott, M. A. Kelly, and D. Widows, *J. Chem. Soc., Perkin Trans. 1* 1013 (1981); (*89*) W. E. de Wolf, F. A. Fullin, and V. L. Schramm, *J. Biol. Chem.* **254,** 10868 (1979); (*90*) F. Focher, A. Verri, S. Spadari, R. Manservigi, J. Gambino, and G. E. Wright, *Biochem. J.* **292,** 883 (1993); (*91*) L. Andersson, T. C. Isley, and R. Wolfenden, *Biochemistry* **21,** 4177 (1982); (*92*) H. Umezawa, H. Aoyagi, H. Suda, M. Hamada, and T. Takeuchi, *J. Antibiot.* **29,** 97 (1976); (*93*) D. H. Rich, B. J. Moon, and S. Harbeson, *J. Med. Chem.* **27,** 417 (1984); (*94*) W. W. C. Chan, P. Dennis, W. Demmer, and K. Brand, *J. Biol. Chem.* **257,** 7955 (1982); (*95*) D. W. Cushman, H. S. Cheung, E. F. Sabo, and M. A. Ondetti, *Biochemistry* **16,** 5484 (1977); (*96*) A. A. Patchett *et al., Nature* (*London*) **288,** 280 (1980); A. A. Patchett and E. H. Cordes, *Adv. Enzymol. Relat. Areas Mol. Biol.* **57,** 1 (1985); (*97*) U. Brodbeck, K. Schweikert, R. Gentinetta, and M. Rottenberg, *Biochim. Biophys. Acta* **567,** 357 (1979); (*98*) E. M. Gordon, J. D. Godfrey, J. Pluscel, D. von Langen, and S. Natarajan, *Biochem. Biophys. Res. Commun.* **126,** 419 (1985); (*99*) B. Holmquist and B. L. Vallee, *Proc. Natl. Acad. Sci. U.S.A.* **76,** 6216 (1979); (*100*) E. M. Gordon, S. Natarajan, J. Pluscec, H. N. Weller, J. D. Godfrey, M. B. Rom, E. F. Sabo, J. Engebrecht, and D. W. Cushman, *Biochem. Biophys. Res. Commun.* **124,** 148 (1984); (*101*) E. M. Gordon, S. Natarajan, J. Pluscec, H. N. Weller, J. D. Godfrey, M. B. Rom, E. F. Sabo, J. Engebrecht, and D. W. Cushman, *Biochem. Biophys. Res. Commun.* **126,** 419 (1985); (*102*) L. D. Byers and R. Wolfenden, *J. Biol. Chem.* **247,** 606 (1972); (*103*) D. W. Cushman, H. S. Cheung, E. F. Sabo, and M. A. Ondetti, *Biochemistry* **16,** 5484 (1977); (*104*) N. E. Jacobsen and P. A. Bartlett, *J. Am. Chem. Soc.* **103,** 654 (1981); (*105*) D. Grobelny, U. B. Goli, and R. E. Galardy, *Biochem. J.* **232,** 15 (1985); (*106*) W. L. Mock and J.-T. Tsay, *J. Am. Chem. Soc.* **111,** 4467 (1989); (*107*) J. E. Hanson, A. P.

Kaplan, and P. A. Bartlett, *Biochemistry* **28,** 6294 (1989); (*108*) D. J. McKay and T. H. Plummer, *Biochemistry* **17,** 401 (1978); (*109*) T. H. Plummer and T. J. Ryan, *Biochem. Biophys. Res. Commun.* **98,** 4481 (1981); (*110*) K. Yamauchi, S. Ohtsuki, and M. Kinoshita, *Biochim. Biophys. Acta* **827,** 275 (1985); (*111*) T. C. Friedman, T. K. Kline, and S. Wilk, *Biochemistry* **24,** 3907 (1985); (*112*) E. J. Breaux and M. L. Bender, *FEBS Lett.* **56,** 81 (1975); (*113*) K. A. Koehler and G. E. Lienhard, *Biochemistry* **10,** 2477 (1971); (*114*) B. Imperiali and R. H. Abeles, *Biochemistry* **25,** 3760 (1986); (*115*) W. W. Bachovchin, W. Y. L. Wong, S. Farr-Jons, A. B. Shenvi, and C. A. Kettner, *Biochemistry* **27,** 12839 (1988); (*116*) B. Imperiali and R. H. Abeles, *Biochemistry* **25,** 3760 (1986); (*117*) R. C. Thompson, *Biochemistry* **12,** 47 (1973); (*118*) R. N. Lindquist and C. Terry, *Arch. Biochem. Biophys.* **160,** 135 (1974); (*119*) J. O. Westerik and R. Wolfenden, *J. Biol. Chem.* **247,** 8195 (1971); (*120*) T.-C. Liang and R. H. Abeles, *Arch. Biochem. Biophys.* **252,** 626 (1987); (*121*) C. A. Lewis, Jr., Ph.D. Thesis, University of North Carolina, Chapel Hill (1976); (*122*) H. Umezawa, T. Aoyagi, H. Morishima, M. Matzusaki, H. Hamada, and T. Takeuchi, *J. Antibot.* **23,** 259 (1970); (*123*) D. H. Rich, M. S. Bernatowicz, N. S. Agarwal, M. Kawai, and F. G. Salituro, *Biochemistry* **24,** 3165 (1985); (*124*) M. H. Gelb, J. P. Svaren, and R. H. Abeles, *Biochemistry* **24,** 1813 (1985); (*125*) P. H. Bartlett and W. B. Keyer, *J. Am. Chem. Soc.* **106,** 4282 (1984); (*126*) J. Boger, L. S. Payne, D. S. Perlow, N. S. Lohr, M. Poe, E. H. Blaine, E. H. Ulm, T. W. Schorn, B. I. LaMont, T.-Y. Lin, M. Kawai, D. H. Rich, and D. F. Veber, *J. Med. Chem.* **28,** 1779 (1985); (*127*) S. Thaisrivongs, D. T. Pals, W. Kati, S. R. Turner, L. M. Thomasco, and W. Watt, *J. Med. Chem.* **29,** 2080 (1986); (*128*) M. Szelke, B. Leckle, A. Hallett, D. M. Jones, J. Sueiras, B. Atrash, and A. F. Lever, *Nature* (*London*) **299,** 555 (1982); (*129*) J. R. Luly, N. BaMaung, J. Soderquist, A. K. L. Fung, H. Stein, H. D. Kleinert, P. A. Marcotte, D. A. Egan, B. Bopp, I. Merits, G. Bolis, J. Greer, T. J. Perun, and J. J. Plattner, *J. Med. Chem.* **31,** 2264 (1988); (*130*) G. B. Dreyer, B. W. Metcalf, T. A. Tomaszek, T. J. Carr, A. C. Chandler, III, L. Hyland, S. A. Fakhoury, V. W. Magaard, M. L. Moore, J. E. Strickler, C. Debouck, and T. D. Meek, *Proc. Natl. Acad. Sci. U.S.A.* **86,** 9752 (1989); (*131*) R. E. Galardy and D. Grobelny, *Biochemistry* **22,** 4556 (1983); (*132*) D. Grobelny, C. Teater, and R. E. Galardy, *Biochem. Biophys. Res. Commun.* **159,** 426 (1989); (*133*) V. Dive, A. Yiotakis, A. Nicolaou, and F. Toma, *Eur. J. Biochem.* **191,** 685 (1990); (*134*) A. L. Maycock, D. M. De Sousa, L. G. Payne, J. ten Broeke, M. T. Wu, and A. A. Patchett, *Biochem. Biophys. Res. Commun.* **102,** 963 (1981); (*135*) P. A. Bartlett and C. K. Marlowe, *Biochemistry* **22,** 4618 (1983); P. A. Bartlett and C. K. Marlowe, *Biochemistry* **26,** 8553 (1987); (*136*) N. Nishino and J. C. Powers, *Biochemistry* **13,** 2846 (1978); (*137*) S. Almenoff and M. Orlowski, *Biochemistry* **22,** 590 (1983); (*138*) J. O. Westerik and R. Wolfenden, *J. Biol. Chem.* **249,** 6351 (1974); (*139*) J. D. Findlater and B. A. Orsi, *FEBS Lett.* **35,** 109 (1973); (*140*) D. H. Kinder, S. K. Frank, and M. M. Ames, *J. Med. Chem.* **33,** 819 (1990); (*141*) J. Rahil and R. F. Pratt, *Biochemistry* **31,** 5869 (1992); (*142*) R. B. Meyer and E. B. Skibo, *J. Med. Chem.* **22,** 944 (1979); (*143*) B. Evans and R. Wolfenden, *J. Am. Chem. Soc.* **92,** 4751 (1970); (*144*) T. Sawa, Y. Fukagawa, I. Homma, T. Takeuchi, and H. Umezama, *J. Antibiot.* **204,** 227 (1967); (*145*) W. Jones, L. C. Kurz and R. Wolfenden, *Biochemistry* **28,** 1242 (1989); (*146*) R. M. Cohen and R. Wolfenden, *J. Biol. Chem.* **246,** 7561 (1971); (*147*) G. W. Ashley and P. A. Bartlett, *J. Biol. Chem.* **259,** 13621 (1984); (*148*) V. E. Marquez, P. S. Liu, J. A. Kelley, J. S. Driscoll, and J. J. McCormack, *J. Med. Chem.* **23,** 713 (1980); (*148a*) L. Frick, C. Yang, V. E. Marquez, and R. Wolfenden, *Biochemistry* **28,** 9423 (1989); (*149*) C. Frieden, H. R. Gilbert, W. H. Miller, and R. L. Miller, *Biochem. Biophys. Res. Commun.* **91,** 278 (1979); (*150*) F. Maley and G. F. Maley, *Arch. Biochem. Biophys.* **144,** 723 (1971); (*151*) J. M. Lanauze, H. Rosenberg, and D. C. Shaw, *Biochim. Biophys. Acta* **212,** 332 (1970); (*152*) N. T. Nashed, D. P. Michaud, W. Levin, and D. M. Jerina, *Arch. Biochem. Biophys.* **241,** 149 (1985); (*153*) A. Schmitt, I. Bottke, and G. Siebert, *Hoppe-Seyler's Z. Physiol. Chem.* **347,** 18 (1966); (*154*) W. Tagaki, J. P. Guthrie, and F. H. Westheimer, *Biochemistry* **7,** 905 (1968); (*155*) I. Fridovich, *J. Biol. Chem.* **243,** 1043 (1968); (*156*) B. W. Potvin, H. J. Stern, S. R. May, G. R. Lam, and R. S. Krooth, *Biochem. Pharmacol.* **27,** 655 (1978); (*157*) M.

(*continued*)

TABLE I (*continued*)

Wishnick, M. D. Lane, and M. C. Scrutton, *J. Biol. Chem.* **245,** 4939 (1970); (*158*) J. Pierce, N. E. Tolbert, and R. Barker, *Biochemistry* **19,** 934 (1980); J. V. Schloss, *J. Biol. Chem.* **263,** 4145 (1988); (*159*) K. D. Collins, *J. Biol. Chem.* **249,** 136 (1974); (*160*) J. V. Schloss and W. W. Cleland, *Biochemistry* **21,** 4420 (1982); (*161*) E. Bayer, B. Bauer, and H. Eggerer, *Eur. J. Biochem.* **120,** 155 (1981); (*162*) P. Rubenstein and R. Dryer, *J. Biol. Chem.* **255,** 7858 (1980); L. C. Kurz, S. Shah, B. C. Crane, L. J. Donald, H. W. Duckworth, and G. R. Drysdale, *Biochemistry* **31,** 7899 (1992); (*163*) M. D. Erion and C. T. Walsh, *Biochemistry* **26,** 3417 (1987); (*164*) D. J. T. Porter and H. J. Bright, *J. Biol. Chem.* **255,** 4772 (1980); (*165*) J. Greenhut, H. Umezawa, and F. B. Rudolph, *J. Biol. Chem.* **260,** 6684 (1985); (*166*) J. V. Schloss, D. J. T. Porter, H. J. Bright, and W. W. Cleland, *Biochemistry* **19,** 2358 (1980); (*167*) R. M. Waterson and R. L. Hill, *J. Biol. Chem.* **247,** 5258 (1972); (*168*) R. S. Phillips, E. W. Miles, and L. A. Cohen, *Biochemistry* **23,** 6228 (1984); (*169*) D. J. T. Porter and H. J. Bright, *J. Biol. Chem.* **255,** 4772 (1980); (*170*) K. R. Hanson, *Arch. Biochem. Biophys.* **211,** 575 (1981); (*171*) D. J. T. Porter, N. G. Rudie, and H. J. Bright, *Arch. Biochem. Biophys.* **225,** 157 (1983); (*172*) K. T. Douglas and I. N. Nadvi, *FEBS Lett.* **106,** 393 (1979); (*173*) J. F. Barnard and J. F. Honek, *Biochem. Biophys. Res. Commun.* **165,** 118 (1989); (*174*) G. J. Cardinale and R. H. Abeles, *Biochemistry* **7,** 3970 (1968); (*175*) M. V. Keenan and W. L. Alworth, *Biochem. Biophys. Res. Commun.* **57,** 500 (1974); (*176*) R. J. Baumann, E. H. Bohme, J. S. Wiseman, M. Vaal, and J. S. Nicols, *Antimicrob. Agents Chemother.* **32,** 1119 (1988); (*177*) R. Wolfenden, *Biochemistry* **9,** 3404 (1970); (*178*) K. D. Collins, *J. Biol. Chem.* **249,** 136 (1974); (*179*) W. W. Woodruff and R. Wolfenden, *J. Biol. Chem.* **254,** 5866 (1979); (*180*) J. M. Chirgwin and E. A. Noltmann, *Fed. Proc., Fed. Am. Soc. Exp. Biol.* **32,** 667 (1973); (*181*) E. C. Bigham, C. E. Gragg, W. R. Hall, J. E. Kelsey, W. R. Mallory, D. C. Richardson, C. Benedict, and P. H. Ray, *J. Med. Chem.* **27,** 717 (1984); (*182*) S. Wang, F. S. Kawahara, and D. Talalay, *J. Biol. Chem.* **238,** 576 (1963); (*183*) J. E. Reardon and R. H. Abeles, *Biochemistry* **25,** 5609 (1986); M. Muehlbacher and C. D. Poulter, *Biochemistry* **27,** 7315 (1988); (*184*) W. J. Ray and J. M. Puvathingal, *Biochemistry* **29,** 2790 (1990); (*185*) P. A. Bartlett and C. R. Johnson, *J. Am. Chem. Soc.* **107,** 7792 (1985); (*186*) A. Duriatti, P. Bouvier-Nave, P. Benveniste, F. Schuber, L. Delprino, G. Balliano, and L. Cattel, *Biochem. Pharmacol.* **34,** 2765 (1985); (*187*) D. Cassio, F. Le Moine, J. P. Waller, E. Sandrin, and R. A. Boissonas, *Biochemistry* **6,** 827 (1967); (*188*) A. I. Biryukov, B. Kh. Ishmuratov, and R. M. Khomutov, *FEBS Lett.* **91,** 249 (1978); (*189*) E. W. Logusch, D. M. Walker, J. F. McDonald, and J. E. Franz, *Biochemistry* **28,** 3043 (1989); (*190*) W. B. Rowe, R. A. Ronzio, and A. Meister, *Biochemistry* **8,** 2674 (1969); (*191*) O. W. Griffith, *J. Biol. Chem.* **257,** 13704 (1982); (*192*) A. S. Mildvan, M. C. Scrutton, and M. F. Utter, *J. Biol. Chem.* **241,** 3488 (1966).

Section III

Isotopic Probes of Enzyme Action

[12] Partition Analysis: Detecting Enzyme Reaction Cycle Intermediates

By IRWIN A. ROSE

Introduction

Many important questions regarding enzymology may be investigated by partition methods: Is a given complex competent for reaction and what is its rate of dissociation and formation? In complex formation, which is often much slower than diffusion, is the ligand interacting rapidly with a small subpopulation of enzyme (E) or with the major component of the population cooperatively? What is the order of addition of substrates and cofactors to form the central complex, and how can one distinguish between random addition to the same enzyme form and ordered additions to different enzyme forms that interconvert rapidly in the free enzyme state? What is the basis for noncompetitive inhibition by an inhibitor in a one substrate–one product reaction? What is the origin of positive and negative cooperativity? What is the location in the reaction sequence of a step causing an isotope effect? How do protons that are used in a reaction reach the active site from the solvent?

In partition analysis a reaction cycle intermediate is generated so that its distribution between alternative end products can be determined as a function of reaction conditions. Two procedures are in use to generate intermediate states for a partition study. The first is a pulse–chase method in which the enzyme under some set of conditions is chased into a diluting solution that contains a full complement of reactants. The products formed immediately after mixing reflect the character of the complex of the enzyme with the ligand(s) provided in the pulse. The ligand(s) may be a labeled substrate, a cofactor, or labeled hydrogens. This is usually a one turnover approach unless the measured aspect of the state of the enzyme in the pulse persists beyond one cycle of reaction. A second method for studying an intermediate of the reaction cycle is to generate the intermediate by action of the enzyme and to observe under steady-state conditions its distribution between group transfer versus dissociation or as a positional isotope exchange versus product formation. Nonliganded states, the free enzyme forms with which substrate, product, activator, or inhibitor interact, also undergo interconversions, and these are readily studied by new partition methods that are discussed in this chapter.

Partition analysis leads directly to a reaction rate constant that is specific to a particular intermediate. This is possible because the concentration of the partitioning intermediate cancels in the expression of the partition ratio, and one of the components of the ratio is usually a known steady-state parameter such as k_{cat} or k_{cat}/K_m (henceforth referred to as V/K_m). This is not to say that the rate constant determined may not be a composite of first-order rate constants, only that it is specific for one intermediate of the reaction cycle.

This chapter provides a comprehensive review of the subject of partition analysis as it has evolved since its first application 20 years ago and since a previous review in 1980.[1] The experiments used to describe the methods and concepts are taken from studies with yeast hexokinase and fumarase. With hexokinase the experiments are related mainly to the processing of the substrates. They are more or less typical of procedures used to study a wide variety of other enzymes by other workers. With fumarase we identify recycling, not product release, as the rate-limiting step and show how partition methods may be used to learn how protons travel between solvent and an active site.

Hexokinase Reaction Sequence

As studied by partition analysis[2,3] and the flux ratio method,[4] the sequence of substrate additions and product formation of yeast hexokinase and liver glucokinase[5] is fully ordered (Scheme 1).

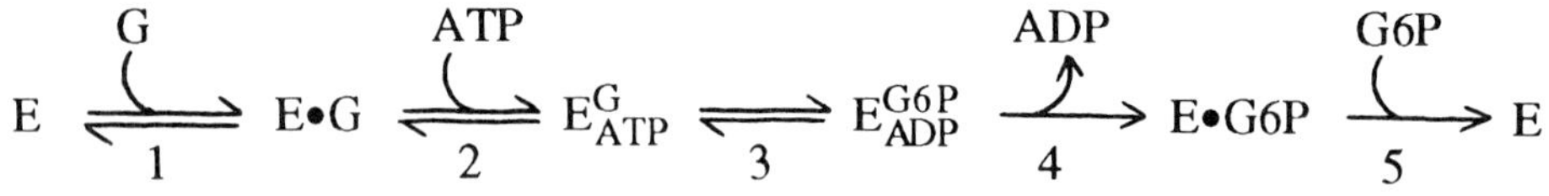

SCHEME 1

Enzyme–Glucose Complex

The pulse–chase method was designed as a way of demonstrating by isotope capture the catalytic competence of the noncovalent hexokinase · glucose (E · G) complex. Virtually all of the radioactive glucose

[1] I. A. Rose, this series, Vol. 64, p. 47.
[2] I. A. Rose, E. L. O'Connell, S. Litwin, and J. BarTana, *J. Biol. Chem.* **249,** 5163 (1974).
[3] R. E. Viola, F. M. Raushel, A. R. Rendina, and W. W. Cleland, *Biochemistry* **21,** 1295 (1982).
[4] H. G. Britton and J. B. Clarke, *Biochem. J.* **128,** 104P (1972).
[5] M. Gregoriore, I. P. Thayer, and A. Cornish-Bowden, *Biochemistry* **20,** 499 (1981).

known from rate of dialysis or dialysis equilibrium studies to be in complex with yeast hexokinase was converted to labeled glucose 6-phosphate (G*6P) when the pulse of E plus G* was mixed rapidly with unlabeled glucose and MgATP. The fraction of total bound glucose that was captured was a simple hyperbolic function of the MgATP concentration used in the chase. R_P, defined as the partition of EG* at any [ATP] between capture and dissociation, was a linear function of [ATP] from which the value of k_{-1} could be calculated according to

$$\begin{aligned} R_P &= (E \cdot G^* \rightarrow G^*6P)/(E \cdot G^* \rightarrow G^*) \\ &= (V/K_m)_{ATP}[ATP][E \cdot G^*]/k_{-1}[E \cdot G^*] \qquad (1) \\ k_{-1} &= (V/K_m)_{ATP}[ATP]/R_P. \end{aligned}$$

The concentration of ATP at which half of $E \cdot G^*$ is captured, $R_P = 1$, defines the capture constant $(K_c^{ATP})_{E \cdot G}$:

$$k_{-1} = (V/K_m)_{ATP} K_c^{ATP}. \qquad (2)$$

Using trypsin-treated hexokinase to increase its glucose affinity, the value calculated for k_{-1} was close to the k_{cat} of the reverse reaction and much slower than that for untreated enzyme.[3] Evidence that the pulse–chase result is relevant to the steady state is the agreement between k_1 calculated from k_1/K_1 and $(V/K_m)_{glucose}$[1] as expected in any ordered sequence.

The use of V/K_m of ATP in Eq. (2) assumes that $E \cdot G$ in the chase and the steady state are kinetically equivalent. If $E \cdot G$ complexes in the pulse represent an equilibrium of forms, some of which are not in the reaction cycle, these forms will also be present in the steady state in equilibrium with the form of $E \cdot G$ that is in the cycle. This equilibrium value comes into the definition of $(V/K_m)_{ATP}$ in the same way in the two experiments. For example, if in Scheme 1 the intermediate $E \cdot G$ were in equilibrium with a nonfunctional form, $E' \cdot G$, such that $K' = [E' \cdot G]/[E \cdot G]$, the value of $(V/K_m)_{ATP}$ would contain a factor $[E \cdot G]/([E \cdot G] + [E' \cdot G])$ or $(1 + K')^{-1}$. The rate of the $E' \cdot G \rightarrow E \cdot G$ transition influences K_c^{ATP} unless the $E' \cdot G^* \rightarrow G^*$ step is insignificant relative to k_{-1}. The fact that the kinetic component of V/K_m may be larger than the value determined by steady state becomes important if there are ways of generating $E \cdot G$ specifically in a pulse–chase experiment.

There are two cases in which the use of $(V/K_m)_{ATP}$ to calculate k_{-1} would be inappropriate, both of which result in the incomplete capture of $E \cdot G^*$. $E \cdot G^*$ of the pulse may be in a form that can dissociate G* but must undergo a hysteretic conversion to a form present in the steady state before capture can occur. This possibility can be evaluated by comparing the pulse–chase result with the partition of $E \cdot G$ generated in the reverse

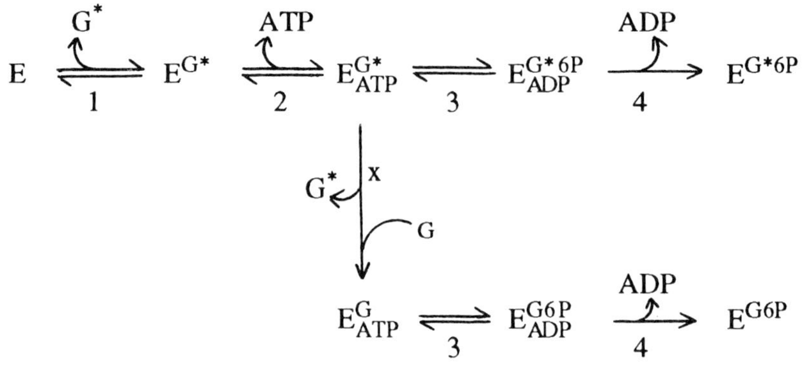

SCHEME 2

direction as an isotope exchange (*ATP → *ADP) versus G6P → G (which equals E · G → G).[3]

An unrelated cause of incomplete capture of E · G* depends on the dissociation of G* from the ternary complex E · G* · ATP (Scheme 2). This is not necessarily equivalent to a random order mechanism which requires formation of functional E · ATP from free E. The plot of $[P^*]^{-1}$ versus $[ATP]^{-1}$ would extrapolate to give $[P^*]_{max} < [E \cdot G^*]_{total}$ by the factor $\vec{k}_{3,4}/(k_x + k_{3,4})$ where $\vec{k}_{3,4} = k_3k_4/(k_{-3} + k_4)$. As shown by Cleland,[6] k_{-1} will lie within the range $(V/K_m)_{ATP}K'_c$ as the lower limit, where K'_c is the value of K_c when less than $[E \cdot G^*]_{total}$ is captured and the same value $x([E \cdot G^*]_{total}/[P^*]_{max})$ as the upper limit. This complexity derives from the fact that two intermediates of uncertain ratio are being partitioned.

A small difference between the G*6P that would be formed at $[ATP] \gg K_c$ and EG* present in the pulse, estimated from a binding constant, cannot be reliably used to distinguish between a small amount of dissociation of G* from E · G* · ATP and a complex in which G* is locked in. It would only be possible to rule out random order in this way if the fraction of E · G* that cannot be captured could be determined directly and could be distinguished from nonfunctional E · G*. A solution to this dilemma is to test the effect of a nonfunctional analog of ATP, in competition, on the value of K_c^{ATP}. When AMP ($K_{is} \approx 2.5$ mM) was present at 40 mM in the pulse and at 20 mM in the chase, there was no change in the capture of G* or in K_c^{ATP}, showing that G* was locked in by AMP and that the E · G* released from E · G* · AMP had the same partition ratio as E · G* itself (I. A. Rose, unpublished, 1979). Of course a longer chase time was used in the presence of AMP to ensure complete

[6] W. W. Cleland, *Biochemistry* **14,** 3220 (1975).

partitioning of E · G*. The fact that AMP and ATP lock in glucose makes unlikely an alternative mechanism for the reverse reaction in which MgADP replaces the first product, ATP, followed by glucose dissociation and glucose 6-phosphate binding for the next cycle. This leaves unresolved the interesting question of why MgADP fails to show dead-end inhibition by reaction with EG in the reverse direction.[3]

The two conclusions drawn for E · G, its competence as a substrate and its rate of dissociation, illustrate the way isotope capture may be used to interpret data obtained by other approaches. For example,[7] glucose is known to form a complex with brain hexokinase from the measured nuclear magnetic resonance line broadening of the glucose anomeric hydrogens by a single enzyme-bound Mn^{2+}. The minimum rate of glucose dissociation required to transfer the Mn^{2+} effect to free glucose was calculated to be around 1300 s^{-1}. An independent measure of the glucose dissociation rate by pulse–chase studies under the same conditions could be used to implicate the E · Mn^{2+} · G complex in the enzyme function.

Enzyme–ATP Complex

Although hexokinase has ATPase activity, no capture of [γ-^{32}P]ATP as G6P* could be obtained using molar levels of glucose in the chase. The implication of this is that ATP dissociation from the E^{G}_{ATP} state greatly exceeds the k_{cat}. However, because glucose does not dissociate from the ternary E^{G}_{ATP} complex (as was shown with AMP), it is not expected that E · ATP would ever be generated in the reverse direction and therefore (at equilibrium, by the principle of microscopic reversibility) could not be a first substrate to bind in the forward direction. If the ATPase activity is evidence for a dead-end complex, one should be aware that the concentrations of ATP used in the chase may prevent a normal reaction sequence with glucose.

When small, nonintegral amounts of isotope capture are found, it is important to be certain that the labeled product was derived from a binary complex present in the pulse solution and not the result of turnover prior to extensive dilution of isotope with unlabeled carrier of the chase solution. Inefficient mixing is usually not improved by using greater dilution by volume. The best magnetic mixing does not create enough turbulence for an injected pulse to diffuse adequately in competition with turnover rates greater than 100 s^{-1}. Since the distinction between capture and no capture is of primary importance, it will be desirable to use rapid flow mixing whenever possible.

[7] G. K. Jarori, S. R. Kasturi, and U. W. Kenkare, *Arch. Biochem. Biophys.* **211,** 258 (1981).

How Well Do Single Turnover Pulse–Chase Results Agree with Conclusions about the Steady State?

In most cases the discrepancy between the isotope capture and exchange results may be explained by assuming that one of the alternative binary complexes is too unstable to be captured. These enzymes will also be referred to as anomalously ordered to indicate that a difference exists between the first turnover result of the pulse–chase experiment and isotope exchange either at equilibrium or in the steady state.

Wiseman *et al.*[8] point out that "random order" is a qualitative characterization of a process that may simply be dependent on the concentrations of substrates used in the analysis. They suggest that in the case that alternative sequences are possible the ratio of the alternate substrates at which both pathways are equally likely be given. Having done so for the reaction of NAD^+ and cyclopropanone with aldehyde dehydrogenase, they find that NAD^+ is so much more reactive as to warrant the characterization of an ordered sequence for the enzyme.

Britton[9] has reported apparent contradictions between a strictly ordered sequence expected from pulse–chase studies with phosphofructokinase (PFK) and apparently random order when studied by the flux ratio method. For example, with PFK, capture of ATP by fructose 6-phosphate (F6P) was quantitative, whereas capture of F6P* by ATP was said to be less than 5% of that expected "if an E · F6P complex were formed that did not exchange with the solution." The authors take the position that their negative result means that E · F6P was not formed and that the conclusion from pulse–chase studies would be that the sequence of substrate addition is ordered. They then reviewed the flux ratio studies in which fructuse 1,6-bisphosphate (FDP) is converted to ATP plus F6P in the presence of [γ-^{32}P]ATP with F6P removed by coupling or in the presence of labeled F6P with ATP removed by coupling. Back-labeling of FDP was observed to increase at the expense of net reaction in both cases, suggesting that both E · ATP and E · F6P were functional complexes that could be captured when generated in the steady state. The concentration of F6P and ATP that gave half-maximum capture of E · ATP and E · F6P was 0.2 m*M*. These concentrations, being much less than the 10 m*M* used in each chase solution, mean that the possibility of incomplete capture of a functional E · F6P complex is ruled out. The only alternative seems to be that a functional complex was not formed to a significant extent at the 90 μM concentration of F6P* used in the pulse, although this is not far from the calculated binding constant at the high pH of the study. An explanation

[8] J. S. Wiseman, G. Tayrien, and R. H. Abeles, *Biochemistry* **19,** 4222 (1980).

[9] S. Merry and H. G. Britton, *Biochem. J.* **226,** 13 (1978).

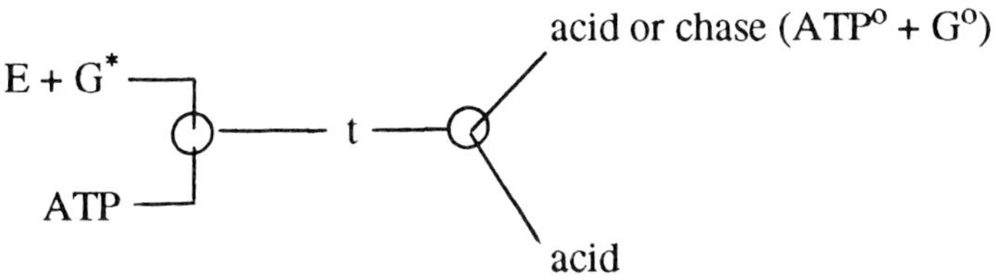

DIAGRAM 1

might be that in the absence of ATP, F6P binds the wrong way around at the FDP binding site. If this explanation is correct, PFK should be referred to as anomalously ordered. It is a realistic example of how an enzyme may be predominantly ordered in one direction and random in producing the same reactants in the reverse. (Microscopic reversibility applies to both directions only at equilibrium, of course.) Because of this dead-end form, F6P would not achieve its full rate potential when used with low (unphysiological) levels of ATP.

Enzyme · Glucose · ATP Complex: Partition on the Fly

A partition study of the E^{G}_{ATP} complex was originally undertaken[10] to determine whether the failure to capture ATP in pulse–chase experiments could be due to rapid exchange from the ternary complex. The steady state was generated in a continuous flow device as shown in Diagram 1 using G* and unlabeled ATP that are brought together in a first mixer and either quenched with acid or chased with a mixture of unlabeled substrates at high concentration after time t_1 in a second mixer, followed by acid after an interval sufficient to complete the partition of labeled intermediates.

Label in G6P in the acid-quenched sample indicates the following sum: free G6P formed + E^{G6P}_{ADP} + E · G6P. The [ATP] used to initiate the reaction was greater than its K_m in order to eliminate EG* from the steady state. The [ATP] used in the chase was sufficient to capture all G*-containing intermediates. The (chase minus acid quench) difference in *G6P captured therefore gives the level of the E^{G*}_{ATP} present in the steady state. This proved to be around 0.5 of $[E]_{total}$, with $[E]_{total}$ determined by pulse–chase experiments in which the pulse contained G* at much greater than K_1. The timed values for the acid-stopped samples extrapolated to about 0.5 $[E]_{total}$ representing E^{G6P}_{ADP} + E · G6P in the steady state. E · G6P must be low in amount since it follows the rate-limiting release of ADP. To determine how ATP partitions from E^{G}_{ATP} the same experiment was performed with [γ-^{32}P]ATP and tritiated glucose. Of the steady-state complexes, only

[10] K. D. Wilkinson and I. A. Rose, *J. Biol. Chem.* **254,** 12567 (1979).

E^{G*}_{ATP*} and E^{G*6P*}_{ADP} could contribute to the (chase minus acid) quench Δ depending on their partitions. The difference between the Δ value found for T and ^{32}P gives the fraction of the two complexes that dissociates to ATP* and fails to reach G6P* in the chase. From this result $k_{-2}/k_{cat} \approx 1/2$ and $k_{-2} \approx 200\ s^{-1}$, not fast enough to escape capture if the ternary substrates complex had formed from E_{ATP}. Partition "on the fly" experiments have been used by Benkovic and co-workers in studies of fructose-1,6-bisphosphate phosphatase[11] and DNA polymerase.[12]

Enzyme · Glucose 6-Phosphate · ADP Complex

The partition of the ternary products complex was determined by comparing the rate of G6P formation with positional exchange of the β–γ bridge oxygen of ATP.[13] The ratio 2 : 1 agrees with the partition found for E^{G}_{ATP}, which is consistent with rapid equilibration of the two species. Positional isotope exchange has been used to study the partition of ternary complex intermediates in UDPglucose pyrophosphorylase[14] and CTP synthetase.[15] In the latter case the absence of inhibition of positional isotope exchange (PIX) in ATP by high UTP was used to establish UTP as the first substrate to bind in the reaction sequence.

Enzyme · Glucose 6-Phosphate Complex

The complex was first suggested as an explanation for noncompetitive inhibition by ADP.[16] This was verified by showing back-labeling of ATP from labeled MgADP during the steady state using glucose 6-phosphate dehydrogenase to ensure irreversibility from free enzyme. As shown by Kim and Raushel[17] in a similar application with argininosuccinate lyase, Eq. (2) can be used to determine the dissociation rate constant of the last product complex.

Failure to capture MgADP with G6P by pulse–chase methods[3] cannot be attributed to rapid dissociation of ADP from the E · ADP · G6P complex since the partition ratio of the central complexes would have allowed capture of a significant fraction (~33%). Presumably a functional E · MgADP complex is not formed. On the other hand, failure to trap

[11] J. F. Rahil, M. M. de Maine, and S. J. Benkovic, *Biochemistry* **21,** 3358 (1982).
[12] M. E. Dahlberg and S. J. Benkovic, *Biochemistry* **30,** 4835 (1991).
[13] I. A. Rose, *Biochem. Biophys. Res. Commun.* **94,** 573 (1980).
[14] L. S. Hester and F. M. Raushel, *Biochemistry* **26,** 6465 (1987).
[15] W. von der Saal, P. M. Anderson, and J. J. Villafranca, *J. Biol. Chem.* **260,** 14993 (1985).
[16] D. P. Kosow and I. A. Rose, *J. Biol. Chem.* **245,** 198 (1970).
[17] S. C. Kim and F. M. Raushel, *J. Biol. Chem.* **261,** 8163 (1986).

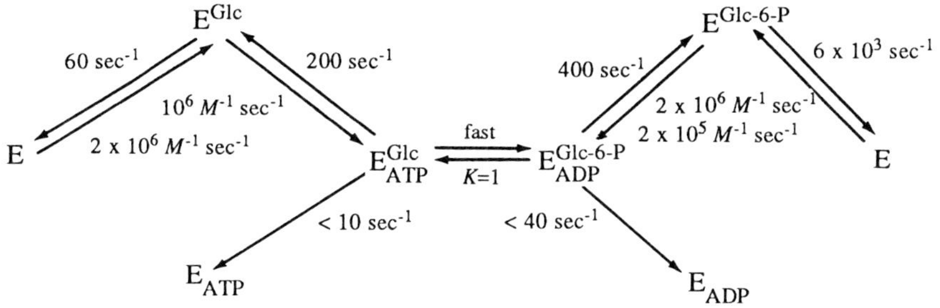

SCHEME 3

MgADP[18] under conditions that give capture of creatine phosphate (Cr-P)[19] on creatine kinase has been attributed to too rapid exchange of a functional binary complex.

Summary

A summary of the rate constants established by partition methods and the relevant dissociation constants is shown in Scheme 3 for yeast hexokinase form S1.[10] Using partition analysis as a major tool, similar kinetic road maps have been elaborated for many enzymes. Most of these were shown to be ordered: the hexokinases, aldehyde dehydrogenase,[8] glutamine synthetase (glutamate–ammonia ligase),[20] prenyl transferase (dimethylallyl transferase),[21] fructose-1,6-bisphosphate phosphatase,[11,22] UDPglucose pyrophosphorylase (UTP–glucose-1-phosphate uridylyltransferase),[14] hammerhead ribozyme-RNase,[23] CTP synthetase,[15,24] and dihydropteridine reductase.[25]

Some bireactant enzymes with which only one substrate could be trapped but which exchange studies identify as random are pyruvate kinase,[26] fructose diphosphatase,[9] fructokinase,[27] and creatine kinase.[18,19] Enzymes with which both of the substrates have been shown to give functional complexes by isotope capture (therefore random order) are the

[18] V. L. Pecoraro, J. Rawlings, and W. W. Cleland, *Biochemistry* **23,** 153 (1984).
[19] P. F. Cook, G. L. Kenyon, and W. W. Cleland, *Biochemistry* **20,** 1204 (1981).
[20] T. D. Meek, K. A. Johnson, and J. J. Villafranca, *Biochemistry* **21,** 2158 (1982).
[21] F. M. Laskovis and C. D. Poulter, *Biochemistry* **20,** 1893 (1981).
[22] C. A. Caperelli, W. A. Frey, and S. J. Benkovic, *Biochemistry* **17,** 1699 (1978).
[23] M. J. Fedor and O. C. Uhlenbeck, *Biochemistry* **31,** 12042 (1992).
[24] D. A. Lewis and J. J. Villafranca, *Biochemistry* **28,** 8454 (1989).
[25] S. Poddar and J. Henkin, *Biochemistry* **23,** 3143 (1984).
[26] E. G. Dann and H. G. Britton, *Biochem. J.* **169,** 39 (1978).
[27] F. M. Raushel and W. W. Cleland, *Biochemistry* **16,** 2176 (1979).

NAD^+-malic enzyme[28] of *Ascaris* worms and cAMP-dependent protein kinase.[29]

The pulse–chase method has been used to study the following: the rate of dissociation and functionality of tight E · S complexes in the ubiquitin–protein ligase system,[30,31] the binding of initiating ATP to the RNA polymerase · DNA–RNA complex,[32] the processing of an unstable intermediate between the enzyme responsible for its formation and the one using it in the 5-oxoprolinase system,[33] the binding of ATP to myosin subfragment as a two-step process,[34] and the order of addition in a ribozyme · RNA cleavage reaction.[35]

The Dark Side of the Reaction Cycle

The dark side of the reaction cycle refers to the changes that occur at the active site after products are formed so that substrate can bind functionally. It is well known, most especially from crystallographic studies, that enzymes are stabilized in different structural states due to the presence of specific ligands.[36,37] Very often the exchange rates of many backbone hydrogens of enzymes are found to change greatly in the presence of substrate or a substrate analog.[38] What does this say about the mobility of ligand binding sites in the absence of ligands? These are the regions of the protein in the crystal and in solution that by three-dimensional (3D) NMR often seem to be the least clearly defined. The common experience that specific ligand binding often stabilizes purified enzymes indicates that there may be cooperative interactions between the reversible fluctuations of the active site region and structural elements of the protein. Much has been written about the dynamic fluctuations of structural domains and loops of proteins,[39] but their importance in an enzyme reaction cycle at the level of the free enzyme has not received much attention.

[28] C.-Y. Chen, B. G. Harris, and P. F. Cook, *Biochemistry* **27,** 212 (1988).
[29] C. T. Kong and P. F. Cook, *Biochemistry* **27,** 4795 (1988).
[30] A. Hershko, H. Heller, E. Eytan, and Y. Reiss, *J. Biol. Chem.* **261,** 1992 (1986).
[31] S. Elias and A. Ciechanover, *J. Biol. Chem.* **265,** 15511 (1990).
[32] N. Shimamoto and C.-W. Wu, *Biochemistry* **19,** 842 (1980).
[33] A. P. Seddon and A. Meister, *J. Biol. Chem.* **261,** 11538 (1986).
[34] T. E. Barman, D. Hillaire, and F. Travers, *Biochemistry* **209,** 617 (1993).
[35] D. Herschlag and T. R. Cech, *Biochemistry* **29,** 10159 (1990).
[36] W. S. Bennett and R. Huber, *Crit. Rev. Biochem.* **15,** 291 (1983).
[37] R. Lumry, *in* "A Study of Enzymes" (E. S. A. Kuby, ed.), p. 3. CRC Press, Boca Raton, Florida, 1991.
[38] C. A. Browne and S. G. Waley, *Biochemistry* **14,** 753 (1974).
[39] C. L. Brooks III, M. Karplus, and B. M. Pettitt, *in* "Proteins: A Theoretical Perspective of Dynamics, Structure and Thermodynamics." Wiley, New York, 1988.

This is probably because they can be subsumed under a single rate constant in the usual steady-state representation, and all of the chemistry is performed after the substrate binds and before the product is released.

We now take the position that the conformational dynamics of the unliganded enzyme is not only to be rationalized in terms of the need to bring substrates, once they are bound, into a correct arrangement for the reaction or only to contribute to a stabilization of the transition state of that reaction. In addition, we propose that the rapid interconversion of detailed conformational states is necessary to provide structures with which the alternative substrates and products can react specifically. Steric limitations to the conformational states possible in the ensemble will determine whether a reaction sequence will be highly ordered as with the three substrate–two product reaction of glutamine synthetase[20] with only two reactant-specific states of the free enzyme, one for the first substrate in each direction, or random as for NAD^+-malic enzyme[28] with at least four states in "rapid" equilibrium.

There are three questions related to this proposition that are not receiving experimental attention because of difficulties in distinguishing actions that occur with different rapidly interconverted free enzyme species from those in which alternative ligands react with the same form. For example, an inhibitor that is shown to be competitive by steady-state rate measurements may overlap the substrate site, may compete sterically for the same enzyme species, or may bind exclusively to another conformation of the enzyme and appear to be competitive due to rapid interconversion of the two forms (kinetic competition). There is currently no simple method for distinguishing between these possibilities. In the same vein one cannot readily distinguish between two models for random order of substrate–enzyme association (Scheme 4). Both models are consistent with incomplete capture of the pulsed substrate due to its dissociation from the

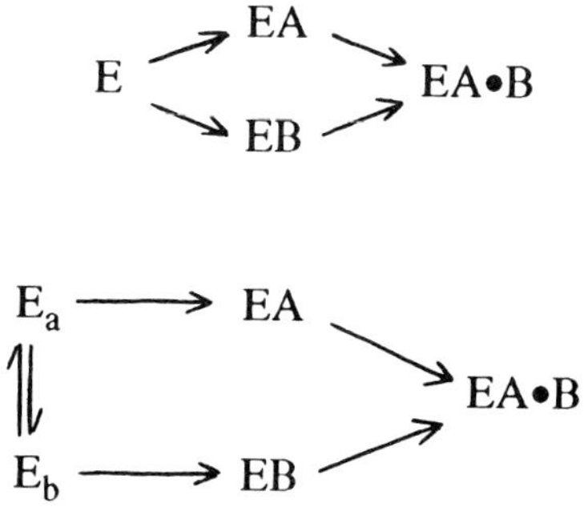

SCHEME 4

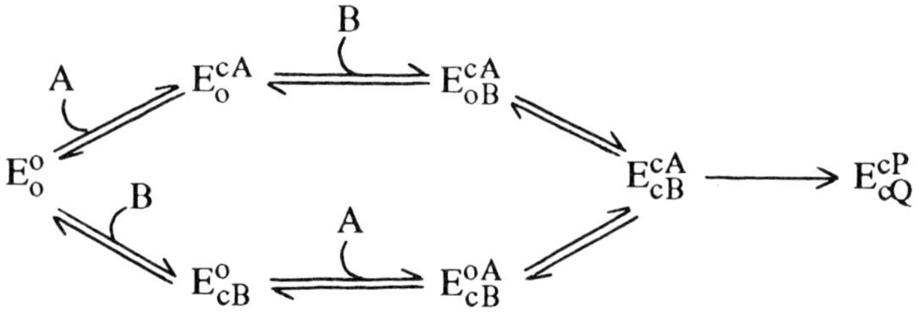

SCHEME 5

ternary complex. Experience has shown that a ternary complex is a relatively closed structure. It seems probable that reactions that show random dissociation of substrate from the ternary complex, such as the NAD^+-malic enzyme of *Ascaris,* do so from conformationally different, partially open isoforms and that reaction occurs from a truly committed third form (Scheme 5, where the symbols c and o refer to conformationally committed or "closed" and conformationally "open" conditions of a ligand binding domain.)[36]

A third problem not easily resolved is whether substrate and product react with the same free enzyme form. When different forms are required, E_s and E_p, their interconversion may occur through intermediates, E_{ps}, with dual specificity for substrate and product. Noncompetitive inhibition by product is then easily explained by interactions of P with E_p (K_{ii}) and E_{ps} (K_{is}). One does not require an allosteric site for the intercept effect and overlapping specificity for the slope effect.

Formation of Enzyme–Substrate Complexes

Enzyme–substrate complexes that by steady-state analysis are used for further processing are rarely formed at rates greater than a few percent of the diffusion limit. What are the mechanisms and limiting factors in forming binary and ternary complexes in the reaction cycle? The changes

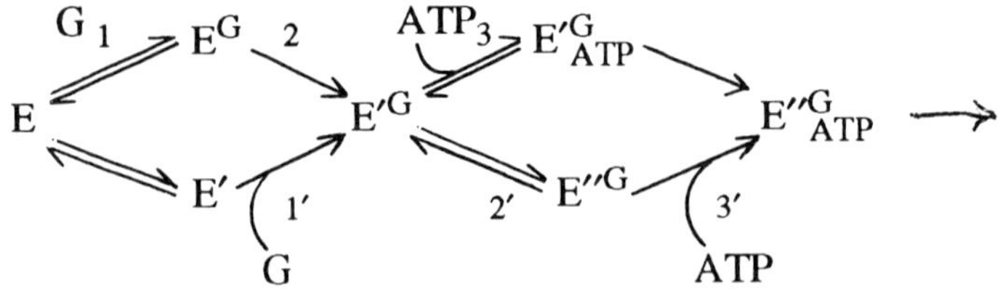

SCHEME 6

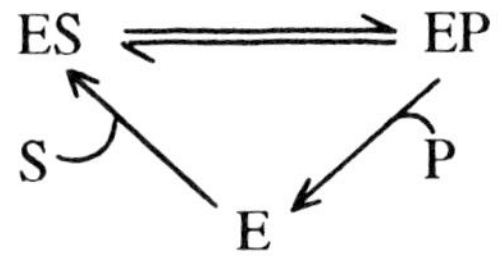

SCHEME 7

that occur in the process of complex formation must ensure that the complex has a sufficient thermal lability for dissociation to occur at a useful rate.

Two limiting mechanisms may be considered for forming the catalytically active ternary complex (Scheme 6). The upper sequence in Scheme 6 is an induced fit mechanism in which a weakly reversible interaction with glucose is followed by changes to form a structure with which ATP can react. In the lower sequence a minor component of the free enzyme and of enzyme–glucose complexes selectively interacts with the substrate in what is a lock and key mechanism in which the correct lock is a small fraction of the population of locks. The pulse–chase method may be useful for distinguishing between the two models if the interconversion of isoforms is not much more rapid than the mixing. For example, if only minor forms can react directly with ATP there should be conditions under which $E \cdot G^*$ will not be fully captured by ATP depending on k_{-1}/k_2 or $k_{-1'}/k_{2'}$. Failure to find such conditions would argue against both steps 2 and 2′ and in favor of the lower path for E'^{G} formation and upper path for ternary substrates complex formation.

On the other hand, an argument for multiple forms of E^G can be made from the fact that crystals grown in glucose are unstable in the presence of ATP.[36] Indeed, it is possible to prepare different crystalline forms of E^G depending on whether ADP is present during the crystallization.[40] Labeled glucose was incorporated into the two crystalline forms by exchange and excess G* and all nucleotide removed by washing the crystals with $(NH_4)_2SO_4$. Crystals grown in G plus ADP, but not in G alone, rapidly converted one-half of the bound G* to G*6P on addition of ATP in keeping with the solution equilibrium. Both types of crystals were dissolved with rapid mixing into a solution of unlabeled glucose and MgATP.[40] Crystals grown in G alone required 3 m*M* MgATP for 50% capture, the same K_c^{ATP} found with soluble enzyme. The value of K_c^{ATP} with crystals grown with glucose and ADP was significantly lower (~1 m*M*), suggesting an E''^{G} intermediate and a preequilibrium fluctuation model.

[40] K. D. Wilkinson and I. A. Rose, *J. Biol. Chem.* **255,** 7569 (1980).

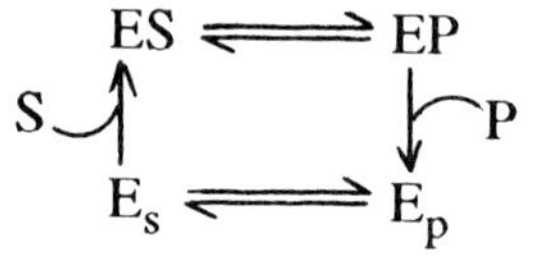

SCHEME 8

A more accessible method of generating binary complexes to test for alternative isoforms of E^G is to include different ATP analogs with G* in the pulse. As mentioned earlier, AMP locks in bound glucose but does not change K_c^{ATP} in the pulse–chase study compared with E · G. This suggests that the form of E · G* produced on dissociation of AMP in the chase is not further advanced toward the species that reacts with ATP than E · G itself, or else these isoforms of E^G are too rapidly interconverted to be observe in this way.

The question of how complexes are formed and the mechanism of recycling are immediately related. If an induced fit view of complex formation is used, the trinodal model of the enzyme cycle (Scheme 7) is noted. Any differences in the enzyme in the two forms ES and EP are absorbed in the differences in the binding constants of S and P to a common E and are assumed to occur during the binding process by virtue of the interactions of E with S and P. Such differences are usually such as to adjust the internal equilibrium to close to unity. This model ascribes all of the catalytic potential of the enzyme to the binding interactions. If we expand to a four-node model (Scheme 8), we relieve an important constraint on evolution of enzyme structure–function (i.e., we do not select better structures on the basis of the ratio of binding constants alone). Binding constants and rates may need to satisfy different selective pressures. Changes in the free enzymes themselves may contribute to adjusting the internal equilibrium to the enzyme-independent solution equilibrium in the four-node model.

It is important that the reality of the E_p/E_s model does not depend on its ability to be demonstrated kinetically. Fisher[41] has shown that the thermodynamic properties of the binding constants of ligands of glutamate dehydrogenase require a two-state model for the free enzyme. There are a few examples in which ligand-independent recycling steps are clearly kinetically important: carbonic anhydrase,[42] proline racemase,[43] and fumarase.[44] In the first two cases E_s and E_p are known to differ in their state of protonation, as shown by D_2O and buffer effects. A proton difference

[41] H. F. Fisher, *Adv. Enzymol.* **61,** 1 (1988).
[42] D. N. Silverman and S. H. Vincent, *Crit. Rev. Biochem.* **14,** 207 (1983).
[43] L. M. Fisher, W. J. Albery, and J. R. Knowles, *Biochemistry* **25,** 2538 (1986).
[44] I. A. Rose, J. V. B. Warms, and R. G. Yuan, *Biochemistry* **32,** 8504 (1993).

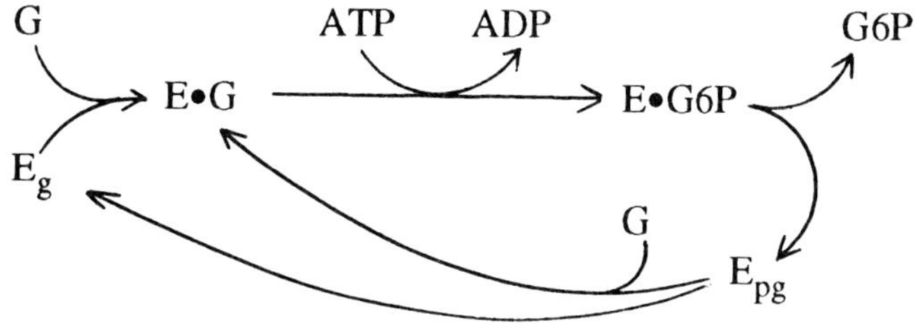

SCHEME 9

between E_s and E_p may not be sufficient to confer absolute substrate specificity as shown with aconitase and fumarase, neither of which forms dead-end complexes at high substrate. The species E^H–citrate[45,46] and E^H–malate[47] are functional, showing the ability to dissociate to E–citrate and E–malate. The mechanisms of these hydrogen dissociations are of interest in themselves, but the point to make here is that substrate-derived protonation does not confer specificity in these cases.

Perhaps the first kinetic observation to be explained by ligand-free isoforms with different specificities was made with the glucose rate dependence of a wheat germ hexokinase.[48] This enzyme showed activation by glucose to rates above those predicted from measurements made in the K_m concentration range, a phenomenon often given the operational name of negative cooperativity. In this case cooperativity was considered to be unlikely since the enzyme is a monomer at the concentration used. Nor was an allosteric site for glucose proposed. In the model suggested to explain activation by glucose (Scheme 9), a product form of the free enzyme, E_{pg}, was proposed that could interact functionally with glucose at high concentration as well as with G6P at low concentration. In the K_m range of glucose, the enzyme would be undersaturated in the nomenclature of Albery and Knowles,[49] and the step $E_{pg} \rightarrow E_g$ provides the necessary recycling that is rate determining for V_{max}. At high glucose this step and the following one are bypassed and are replaced by $E_{pg} + G \rightarrow E \cdot G$, which being faster explains the activation. Consistent with this, G6P was an uncompetitive inhibitor at saturating levels of glucose but competitive at high G where activation was seen. The possibility of an enzyme form that reacts only with G6P, E_p, was not considered. The fundamental

[45] I. A. Rose and E. L. O'Connell, *J. Biol. Chem.* **242,** 1870 (1967).
[46] D. J. Kuo and I. A. Rose, *Biochemistry* **26,** 7589 (1987).
[47] I. A. Rose, J. V. B. Warms, and D. J. Kuo, *Biochemistry* **31,** 9993 (1992).
[48] J.-C. Meunier, J. Buc, A. Navarro, and J. Ricard, *Eur. J. Biochem.* **49,** 209 (1974).
[49] W. J. Albery and J. R. Knowles, *J. Theor. Biol.* **124,** 137 (1987).

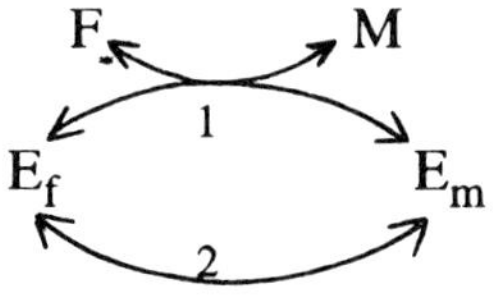

SCHEME 10

property of this explanation of substrate activation is slow recycling of ligand-free enzyme isoforms.

Perhaps the earliest report of substrate activation is the effect of high fumarate on the fumarase rate with enzyme from pig heart.[50] Activation greater than 5-fold has been reported with both fumarate and malate using yeast fumarate.[51] As is often found in cases of cooperative phenomena the homotropic effect can be modified by unrelated ligands. In the case of fumarase, inorganic phosphate (P_i) facilitates the activation by substrate.[50,51] Activation has usually been attributed in these cases not to heterogeneity of ligand-free enzymes as in the wheat germ model, but to effects at allosteric sites or to subunit interactions. The next section considers evidence for a recycling model for fumarase based on partition experiments.

Partition Studies of Nonliganded Enzyme Forms

Slow recycling in the fumarase reaction cycle was indicated using the substrate-induced countertransport model of Britton.[52] This is a critical test of whether net flux in one direction can significantly alter the ratio of E_p/E_s and depends on both isoforms being specific and not interconverted rapidly relative to k_{cat}. The condition of slow recycling is shown when a radioactive probe, initially a mixture of P* and S* at equilibrium, becomes enriched in S* when added to an S → P reaction. Following Scheme 10,[44] the equilibrium ratio $(E_m/E_f)_{eq} = k_{-2}/k_2$ at which the labeled maleate/fumarate (M*/F*) equilibrium was established may be exceeded in the F → M steady state if the rate constants of step 2 are not extremely large relative to k_1F and k_{-1}M of Eq. (3):

$$(E_m/E_f)_{ss} = (k_1F + k_{-2})/(k_{-1}M + k_2) > (E_m/E_f)_{eq} \tag{3}$$

[50] R. A. Alberty and R. M. Bock, *Proc. Natl. Acad. Sci. U.S.A.* **39,** 895 (1953).
[51] J. S. Keruchenko, I. D. Keruchenko, K. L. Gladilin, V. N. Zaitsev, and N. Y. Chirgadze, *Biochim. Biophys. Acta* **1122,** 85 (1992).
[52] H. G. Britton, *Biochem. J.* **133,** 255 (1973).

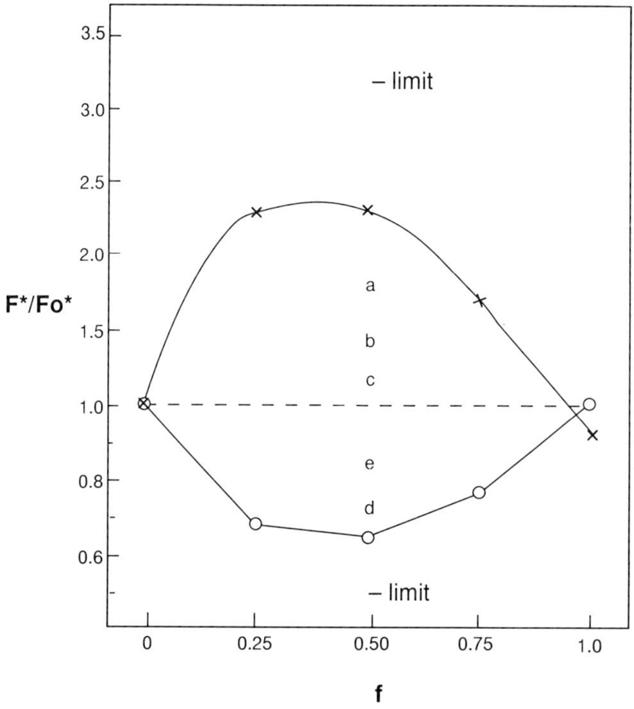

FIG. 1. Counterflow shown by fumarase. An equilibrium probe ($M^*/P^* = 4.4$) was added to 5 mM malate (○) or 3 mM fumarate (X). The reaction with yeast fumarase at pH 7.3 (Hepes, 25 mM) was quenched at f of equilibrium. Label in fumarate (F^*) was determined (isolated by repeated crystallization from acid). The decrease and increase in F^* reached ~70% of the theoretical limits (—) at $f = 0.50$. $K_\alpha^M/K_\alpha^F = 6.3$, close to the predicted value, 4.4. Sodium acetate at 50 (a, d), 100 (b), and 150 mM (c, e) are shown to decrease counterflow greatly.

As long as this inequality exists, M* → F* will exceed F* → M* resulting in $F^*/F_o^* > 1$. Equation (4) relates this ratio to the fraction $f = (P/S)/K_{eq}$ and a characteristic constant, α, after Britton,[52]

$$(S^*/S_o^*) - 1 = K_{eq}(1 - f)[1 - (1 - f)^{\alpha \cdot So/(K_{eq}+1)}], \qquad (4)$$

where $\alpha^{-1} = K_\alpha$ is a complex partition constant defined by

$$K_\alpha^S(V/K_m)_S = k_2k_{-2}/(k_2 + k_{-2}). \qquad (5)$$

The results of counterflow experiments with yeast fumarase[53] in both directions of mass flow are shown in Fig. 1, in which the data are plotted

[53] I. A. Rose, *Biochemistry*, submitted for publication.

as the ratio F^*/F_o^*. Note that the direction of displacement of this ratio depends on the direction of net flux and that the ratio returns to unity as equilibrium is approached, $f = 1$. Similar results showing that E_f and E_m can be displaced from the static equilibrium by the catalytic flux are obtained with enzyme from pig heart[44] and *Escherichia coli*.[53] Counterflow is measurable at initial concentrations of fumarate only a few fold greater than its K_m, which indicates that recycling partially determines the rate of the reaction cycle even in the so-called V/K_m range.

With a reliable method for measuring recycling, one may ask whether known effectors of k_{cat} act at step 1 or 2 of Scheme 10. Insensitivity to effectors of step 1 will be expected if step 2 is rate limiting for k_{cat}. Counterflow, however, should be sensitive to changes at either step since it depends on the ratio of their rates. The preferential activation of substrate interconversion will increase counterflow but may not change k_{cat}, whereas specific inhibitors of step 1 will decrease counterflow at lower concentrations than will be required to influence k_{cat}.

Monoanions such as acetate, chloride, and cacodylate are strong activators of yeast and pig heart fumarase, especially in the F → M direction. Concentrations that increase overall rate also suppress counterflow, shown for acetate in Fig. 1. It follows that anions owe their effect on k_{cat} to their effect on recycling. The influence of 100 m*M* acetate on K_α^F, a 4-fold increase, was much greater than its effect on F/M equilibrium exchange, a 1.4-fold increase.

Activation of initial rate in the 5–10 m*M* range of substrate, 100 × K_m, is probably not similar in origin to the effect of anions. At concentrations that give similar 3-fold activated rates, fumarate does not increase K_α significantly. A second indication is that the sensitivity of K_α to acetate is about the same at 0.5, 1, and 3 m*M* fumarate, showing that they act at kinetically different sites. An activator of recycling in the F → M direction will act on E_m, with which fumarate should not interact.

Although binding studies do not disclose a second site for substrate, such studies are not convincing in the m*M* concentration of substrate required for activation. An alternative possibility is suggested by the finding that despite the necessary exclusiveness of E_m and E_f, product inhibition studies clearly show a competitive element.[44] A nonspecific recycling intermediate, E_{mf}, with relatively low affinity for both substrates would explain both inhibition as a product and stimulation of recycling as a substrate. If $E_m \leftrightarrows E_{mf}$ were rate limiting for recycling, fumarate could activate by removing E_{mf} as soon as it forms, thus increasing the net flux through this step. Combination of either substrate with E_{mf} must function for product formation since neither forms inhibitory or dead-end complexes at any concentration. Therefore an alternative pathway to product

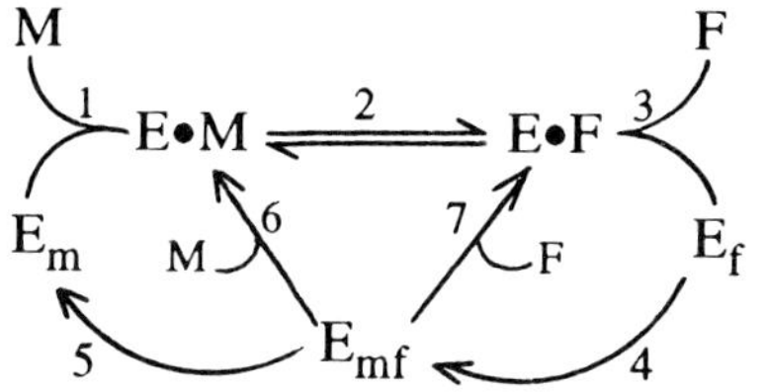

SCHEME 11

formation must be possible (Scheme 11). Activation by high substrate results whenever this alternative pathway is more rapid than the steps of recycling that it bypasses.

Fumarase is 2- to 3-fold less active in D_2O over most of the substrate concentration range studied in either direction. Counterflow was significantly increased in D_2O, suggestive of a primary isotope effect on the recycling process. A protonated residue of the active site would play the role of carrying and abstracting –OH that is performed by metal ions, Mg, Zn, or Fe, of other enzymes (Scheme 12).

Dissociation of the residue may be required to make way for an H_2O molecule to bind (Scheme 12A), or if H_2O were to bind directly, the excess proton would have to leave in a later step (Scheme 12B). The ability of E_f to bind –OH may be incompatible with the ability to bind malate. Also opposing Scheme 12B is a failure to detect any [^{18}O]hydroxyl group transfer between malate and fumarate, although based on the exchange studies of Hansen *et al.*[54] it is clear that –OH/water exchange occurs before free fumarate is formed from malate. [^{18}O]H_2O/malate exchanges before free fumarate is formed. Activating monoanions may stimulate recycling by facilitating the deprotonation step. This is supported by a significant decrease in the D_2O effect in the presence of activating concentrations of acetate.

Unfortunately, the valuable substrate-induced countertransport method of Britton cannot be used with irreversible reactions, and its use with reactions more complex than single substrate/single product interconversions may be complicated by isotope exchange pathways that occur prior to formation of free enzyme.

The Role of Proton Transfer in Recycling

A variety of physical and chemical changes may be required for recycling to occur. Hydrogen transfer(s) is likely to be important. Not only is

[54] J. N. Hansen, E. C. Dinovo, and P. D. Boyer, *J. Biol. Chem.* **244,** 6270 (1969).

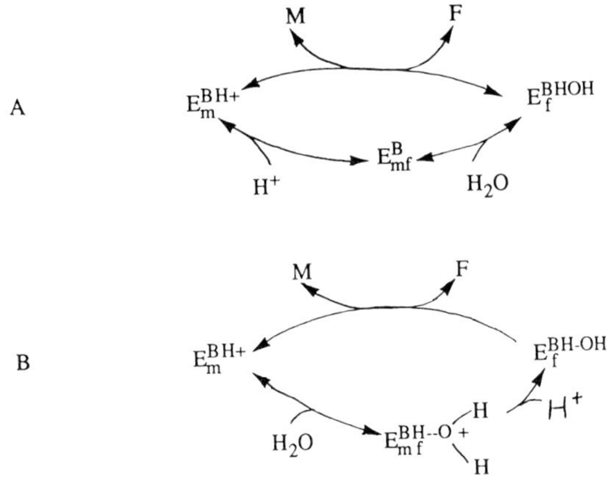

SCHEME 12

hydrogen transfer the most common of the transformations of proteins and enzymes, but hydrogen transfers occur in the rate range of enzyme catalysis and therefore may well be rate determining. It is possible that the most likely catalytic cycles to have slow recycling will be those in which the reaction that is catalyzed leaves the enzyme with a proton excess or deficit that must be corrected for recycling to be successful.

If a proton transfer to solvent is required in the recycling transition between E_p and E_s and has not yet occurred before $E_p \rightarrow E_{ps}$, it must occur after E_{ps}^{H} goes to $E^{H} \cdot SH$ or else SH will be inhibitory at high concentration. To avoid this requires a way to remove the proton of the previous cycle to allow the bound substrate to proceed.

A note of caution may be appropriate at this point: the habit of carrying out steady-state kinetic studies in the presence of 50–100 m*M* buffer may be useful for obtaining linear double reciprocal plots but may obscure a role of hydrogen transfer in recycling as well as in substrate activation. An activating effect of buffers on recycling in the carbonic anhydrase reaction[42] and the activating effect of acetate and P_i on fumarase are examples of each result.

Partition Studies of Free Enzyme as General Acid

The great mobility of exchangeable hydrogens of model compounds in aqueous solvent might well discourage the use of partition analysis to measure the lifetime of E–T where T (tritium) is at the donor site for a reaction that fixes T into an exchange stable position of the product. That this anticipation was exaggerated became evident from early studies with aconitase[45] showing the intermolecular transfer of T between 3T,2-methylisocitrate as donor and *cis*-aconitate as acceptor. With 2 m*M cis*-aconitate (16 times its K_m) about 14% of the T that was mobilized (i.e., 14% of E^T that was formed) was captured into isocitrate, and approximately half that much would be estimated to be in citrate. This partition would correspond to $k_{off} = 79/21 \cdot k_{cat}/K_m \cdot 2\ mM \cong 900\ sec^{-1}$ or $\sim 60 \times k_{cat}$. When given 2T-citrate, the enzyme forms the intermediate $E^T \cdot$ *cis*-aconitate that has three directions of partition: return to T-citrate, intramolecular transfer to 3T isocitrate, or dissociation to E^T + *cis*-aconitate. Exchange of T occurs only from E^T formed after dissociation of *cis*-aconitate and can be fully prevented in the limit of high *cis*-aconitate.[46] The second way to generate E^T was to pulse aconitase briefly in T-water and chase in normal water containing *cis*-aconitate.[46] Although we may judge, from the intra- and intermolecular transfer studies, the donor residue of aconitase to be monoprotonic and not to exchange with hydrogens of the enzyme at the $E^T \cdot$ *cis*-aconitate state, the pulse–chase experiment can show whether the hydrogen that regenerates the donor after the first cycle of hydration comes from the solvent or from the enzyme. The results of this experiment were startling, showing the capture of ~4 enzyme equivalents of T in the limit of high *cis*-aconitate, half-capture requiring only 131 μM *cis*-aconitate under the reaction conditions, at 0° where $k_{cat} \cong 6\ s^{-1}$. The donor residue of pig heart aconitase is a serine hydroxyl by crystallographic analysis of the enzyme–substrate complex.[55] Dissociation of hydrogen of this high pK_a residue would be very slow, much slower than 900 sec^{-1} found by intermolecular transfer without a hydrogen bonding partner. From the crystallographic analysis,[55] this role seems to be satisfied by the guanidinium group of an arginine that has been shown by site-directed mutagenesis to be essential for activity.[56] Such a group might have been anticipated as the source of the ~4T easily captured by *cis*-aconitate in the pulse–chase experiment. The donor hydrogen with its much greater exchange rate was clearly not captured at the concentration range of *cis*-aconitate used.

The low exchange rate observed by pulse–chase may correspond to the exchange between the hydrogens of the arginine and the serine hydroxyl group. Internal exchange, mixing of the two sets of protons, must be slow in the $E^D \cdot$ *cis*-aconitate complex as shown by the largely intramolecular nature of the conversion of 2D-citrate to 3D-isocitrate.[45] This interpretation points to the value of doing both intermolecular transfer and pulse/chase experiments. About 100% transfer between 3D-malate and

[55] H. Lauble, M. C. Kennedy, H. Beinert, and C. D. Stout, *Biochemistry* **31,** 2740 (1992).
[56] L. Zheng, M. C. Kennedy, H. Beinert, and H. Zalkin, *J. Biol. Chem.* **267,** 7895 (1992).

acetylene dicarboxylate by Fe–sulfur fumarase A was shown by Flint and McKay[57] in spite of a calculated exchange rate from E · D of ~10^5 sec^{-1}. This high exchange rate, which is in sharp contrast to the negligible rate from the $E^T \cdot F$ complex, is too rapid to verify by T^+-pulse–chase experiments. Due to the practical limitations of mixing two solutions, the initial E^T of the pulsed enzyme will be lost before turnover occurs. In addition, in case of rapid exchange it will not be possible to rule out capture of T that exchanged with solvent in the presence of substrate before dilution with H_2O of the chase is complete, a mixing artifact.[2]

The special value of the pulse–chase approach is to show the presence of additional hydrogens of the enzyme that have access to the donor heteroatom. The observation that unshared electron pairs are almost always satisfied by H-bonding in proteins[58] is supported by the frequency with which higher order values are found in so-called proton inventory investigations of solvent kinetic isotope effects.[59] To demonstrate the relevance of multiple hydrogen capture found in pulse–chase to the chemistry of proton transfer to C of the substrate, it is necessary to compare the time course of T-capture at $S > K_c$ with the rate $S \rightarrow P$. This comparison is unfortunately complicated by the likely occurrence of T discrimination which in the first turnover will favor H capture. This discrimination may effectively decrease in subsequent cycles of reaction as the donor site and the linked hydrogens become enriched in T in those H-bonding networks in which the succession of transfer of linked hydrogens is linear rather than random. Preliminary studies of multiple T capture with aconitase suggest a significant discrimination value since ~3 turnovers were required for capture of the first T (D. J. Kuo and I. A. Rose, unpublished observations 1989).

Hydrogen Transfer in the Fumarase Reaction

Intermolecular hydrogen transfer between the abstracted C_3-H of malate and fumarate was first proposed by Hansen *et al.*[54] to explain the inhibitory effect of high levels of reactants on the equilibrium exchange rate ratio: (T-malate → T-water)/([^{14}C]malate → [^{14}C]fumarate). This explanation was corroborated more recently showing ~50% T capture in the reaction (3*R*)-T-malate + fluorofumarate → fumarate + 3T,2-fluoromalate (isolated as a deriviative of 3T-oxolacetate) and ~90–95% T trans-

[57] D. H. Flint and R. G. McKay, *J. Am. Chem. Soc.* **116,** 5534 (1994).

[58] E. N. Baker and R. E. Hubbard, *Prog. Biophys. Mol. Biol.* **44,** 97 (1984).

[59] K. B. Schowen, *in* "Transition States of Biochemical Processes" (R. D. Gandour and R. L. Schowen, eds.), p. 225. Plenum Press, New York, 1978.

fer in the double-labeling identity reaction: [T,^{14}C]malate + fumarate → [^{14}C]fumarate + [T,^{12}C]malate.[47] From the concentration of fumarate required for 50% capture it was estimated, using Eq. (2), that T exchanged from the donor site at a rate not much greater than $k_{cat(M \rightarrow F)}$. On the basis of ease of capture by intermolecular transfer, the donor hydrogen of fumarase would be characterized as a sticky proton. In pulse–chase T-capture experiments with pig heart fumarase[60] K_c of fumarate was again not much greater than K_m of fumarate, showing again the similarity between T exchange and k_{cat}. Using enzyme rate as a measure of the amount of fumarase delivered into the 40-fold greater volume (H_2O) of the chase, it was estimated that ~2 enzyme equivalents of T were captured with this rate constant at 23°. When this experiment was repeated at ~4°, ~2 equivalents of T were again captured but the double reciprocal plot was biphasic with K_c values differing by ~10-fold: ~0.1 and ~1 m*M*. This could be explained by an approximately equal mixture of two kinetically different forms of the enzyme in the pulse or if the donor hydrogen and a linked hydrogen of a single enzyme had different exchange rates. The time course of T capture at 4° showed three phases (Fig. 2).[60] Very rapid capture of ~0.5 T was followed by ~1 T with a rate similar to the turnover rate of the enzyme after which there was a continuous capture of T, at the 41-fold slower rate due to dilution of T in the chase. The rapid capture of ~0.5 enzyme equivalent has been interpreted as an indication that approximately half of the equilibrium of enzyme forms in the pulse is in E_f^T + E_{mf}^T forms which could react with the high level of fumarate of the chase without passing through a rate-limiting recycling step required for capture of E_m^T.

Malate itself does not capture T from E^T. For this to occur in pulse–chase experiments the hydrogen derived from the malate would have to exchange with hydrogens of the enzyme. This does not occur as shown also by the absence of back exchange of T from tritiated water into malate under conditions that show exchange from $H_2{}^{18}O$.[59] Malate is far from passive with respect to E^T, as anticipated if some of the T is associated with E_m and E_{mf}. Malate, when included in the chase with fumarate, decreased the capture of T from ~2 equivalents to a limit ~0.4–0.5 equivalent, supposedly a measure of the fraction of enzyme in the E_f form in the pulse.[60]

The fumarase T-capture experiments have been interpreted in terms of a monoprotonic donor with hydrogen bonding to an additional T that has a greater exchange rate with solvent than does T in the donor position. The first turnover to occur at the steady-state rate after E_f^T is captured

[60] I. A. Rose and D. J. Kuo, *Biochemistry*, submitted for publication.

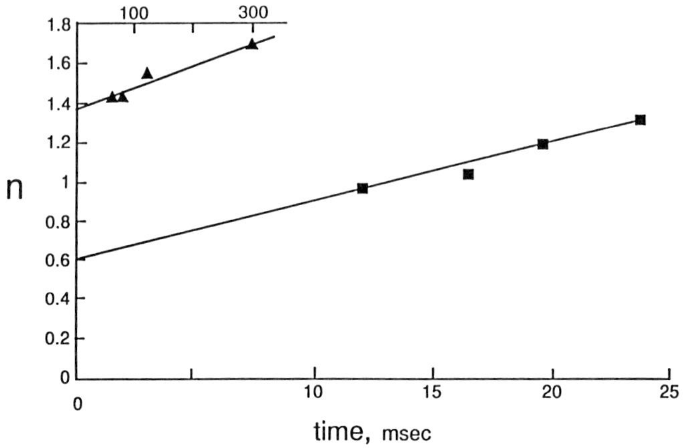

FIG. 2. Time course of T capture at 4°. A pulse solution in imidazole–acetate (10 m*M*, pH 7.0) and EDTA (0.5 m*M*) in tritiated water was machine-mixed with 40 vol of a chase containing 10 m*M* fumarate and the same buffer in normal water. The time (14–28 msec) before acid quenching in a second mixer was determined by the length of capillary between the two mixers. The data are plotted to compare the T equivalents in malate that were derived from E · T (■, within 25 msec) with the additional counts derived from the 41-fold diluted reaction solution (▲, from 50 to 300 msec) shown on a 41-fold compressed time axis. The extrapolated intercepts indicate 0.6 enzyme equivalents of T to be captured at a rate too rapid to resolve, with another 0.8 equivalents captured at the turnover rate of the enzyme as judged from the similar slopes of the two best lines.

utilizes the T originally hydrogen-bonded to the heteroatom of the donor group, signifying it to be on the pathway between solvent and the donor, the relay hydrogen. This model leads to the possibility that exchange of the donor hydrogen with solvent is slow in intermolecular transfer experiments (e.g., sticky) because the linked hydrogen dissociates more rapidly, thereby stabilizing the donor hydrogen electrostatically or as a consequence of its isolation from solvent.[61]

Newer Applications of the Pulse–Chase Method

Pulse–chase competition methods may be used to establish the presence of ligand-specific free enzyme forms in cases not suitable for steady-

[61] A particularly extreme example of a low exchange rate from a T-labeled free-enzyme intermediate is that of yeast flavocytochrome b_2–lactate dehydrogenase in which the C-2 hydrogen of lactate may be abstracted as a proton [P. Urban and F. Lederer, *J. Biol. Chem.* **260,** 11115 (1985)]. Hydrogen transfer (2T-lactate + 3F-pyruvate → pyruvate + 2T,3F-lactate) in competition with T exchange could be easily demonstrated with $K_c \sim 6K_m$ of 3F-pyruvate. However, k_{cat} for 5F-pyruvate is only 0.4 sec^{-1}. This leads to a rate constant for T exchange from E^T of only ~2 sec^{-1}.

state methods. For example, irreversible reactions cannot be studied by either the flux ratio or counterflow methods. The pulse–chase method can be used to study the effects of inhibitors and pH on the distribution of substrate-free isoforms and in favorable cases gives information as to their rate of interconversion. The approach is not one of determining the *partition* of the pulse enzyme since there is no way the nonliganded enzyme can be labeled to produce a labeled product (with the exception of a labeled hydrogen donor, or course). To determine the fraction of total enzyme in the E_s form, a *competition* is set up in the chase using a labeled substrate and a second component which may be either a product or an inhibitor believed to interact with E_p and E_{ps} but not E_s. The amount of P* that is produced rapidly after mixing should depend on whether the second component has been used in sufficient amount to trap E_p and E_{ps}. In the case of reversible reactions the use of unlabeled product(s) as the second component makes possible the undesirable regeneration of E_s by the reverse reaction, $(E_p + E_{ps}) + P \rightarrow E_s + S$ and therefore subsequent $S^* \rightarrow P^*$ conversion. The extent of this problem may be evaluated by a time-course study of P* production and extrapolation to zero time. This problem may be solved in uni–bi or bi–bi reactions by omitting one of the products or in general by electing to use as S* the substrate of the more rapid direction of reaction. If the second component is an inhibitor, the need for short interval quenching is lessened by the factor by which I/K_{ii} is greater than 1. The concentrations of S* and the second component, P or I_P (inhibitor-specific for E_p), must be sufficient to prevent the interconversion of their respective target isoforms. The plot of $S^* \rightarrow P^*$ against the concentration of I_p may be a step function that reflects the relative content of E_s, E_{ps}, and E_p. At low inhibitor more than one equivalent of P* will be found when enzyme turnover is in excess of 1 equivalent at the time of sampling. A plateau at $P^* = E_t$ may be found in the region of concentration at which these extra turnovers become minimal and assuming that $EP^* \rightarrow E + P^*$ is rate limiting. A second plateau would be reached at $E_s + E_{ps}$ if inhibitor binds tightly to E_P. A third would reflect the content of E_s as inhibitor binds only to $E_p + E_{ps}$. The concentration of S* required to capture $1/2E_s$, K_c, can be used to determine $k_{Es \rightarrow Eps}$ by $(V/K)_S \times (E_t/E_S) \times K_c$.

By using reaction rate instead of isotope capture the pulse–chase approach can be extended greatly to analyze both the functional state of an enzyme and the stability of that state through successive cycles of reaction. For example in a reaction that requires Mg^{2+} the possibility that $E \cdot Mg^{2+}$ forms a functional complex may be tested by including EDTA with the substrates in the chase and determining if an equivalent of product is formed at high concentration of substrates or more if the Mg^{2+} does not dissociate when the products are produced. By determining the effect of substrate concentration on product formation the rate of dissociation of $E \cdot Mg^{2+}$ can be determined. This same approach might be used to test the rate of response of free enzyme to a change in pH by using an unfavorable pH for the chase. A variant of this has been used with fumarase as follows: Excess enzyme and labeled malate are used as a pulse under conditions such that neither substrate is in the free form. The chase contains diluting amounts of malate and fumarate at the same pH as the pulse or at another pH, and the mixture is terminated with acid at an early time. With the chase and pulse at the same pH, the ratio of label in M and F indicates the partition of the central complexes to free products which relates the internal interconversion and off rates. The effect of changing the pH in the chase indicates how and whether the distribution of products is influenced by pH in the liganded forms of the enzyme.

With proton transfer reactions the transfer is frequently to an exchange-labile position, to an oxygen or nitrogen atom of the product, so that T capture cannot be used. However, pulse–chase methodology can be applied using rate of reaction to identify the donor hydrogen if the transfer step is responsible for a solvent isotope effect. For example, enzyme in a pulse of H_2O is chased with a solution of substrate(s) in H_2O or D_2O. The rate of product formation during the first turnover will be the same, the H_2O rate, if the enzyme uses the hydrogen that it acquired in the pulse. Otherwise, the rate in D_2O will be linear at the steady-state D_2O rate. Since we know that hydrogen exchange is often within the range that allows its capture at $K_c > K_m$, this "solvent shift" approach should be successful in some cases in demonstrating the site of origin of hydrogen responsible for the D_2O effect. By relating the result to rate, one may also be able to observe rate-determining hydrogen transfers that are regulatory. In this case internal transfer of a hydrogen derived from H_2O or D_2O may affect the reaction rate for several turnovers in the presence of high substrate to suppress exchange. One may use solvent inventory techniques to distinguish sustained effects of a single hydrogen from the participation of more than one hydrogen.

Acknowledgments

This work was supported by National Institutes of Health Grants GM20940 (IAR), CA06927, and RR05539 (ICR), and by an appropriation from the Commonwealth of Pennsylvania.

[13] Isotope Effects: Determination of Enzyme Transition State Structure

By W. W. CLELAND

Introduction

Enzymologists wish to determine transition state structures for enzymatic reactions for several reasons: (1) to compare the mechanisms of enzymatic and nonenzymatic reactions in order to understand better the factors involved in enzymatic catalysis and (2), with "transition state analogs" being used for drugs, to know what structure should be mimicked. One of the best tools for determining transition state structure is the use of isotope effects, since these are very sensitive to changes in transition state structure. Isotope effects have been used for many years to study nonenzymatic reactions.[1] For many enzymatic reactions the observed isotope effects can be reasonably interpreted, once the intrinsic values are known on the chemical steps. The problem is that the chemical step (or steps) of an enzymatic reaction is often not totally rate limiting, so that one has to either find a way to make this step rate limiting (by changing substrate, changing pH, or mutating the enzyme) or find a way to calculate the intrinsic isotope effect.

The basic equations that describe an isotope effect on an enzymatic reaction, as well as the method of Northrop[2] that compares deuterium and tritium isotope effects on V/K to estimate the intrinsic isotope effect, are given elsewhere in this series.[3] We do not repeat this material here, but rather review some basic definitions and then describe the more modern methods of measuring isotope effects and determining intrinsic isotope effects. After that we give examples of the determination of transition state structure and what these studies tell about enzymatic mechanisms. The reader may also find other references useful.[4]

[1] L. Melander and W. H. Saunders, Jr., "Reaction Rates of Isotopic Molecules." Wiley, New York, 1980; R. D. Gandour and R. L. Schowen (eds.), "Transition States for Biochemical Processes" Plenum, New York, 1978.

[2] D. B. Northrop, *Biochemistry* **14,** 2644 (1975).

[3] W. W. Cleland, this series, Vol. 87, p. 625.

[4] W. W. Cleland, *Crit. Rev. Biochem.* **13,** 385 (1982); W. W. Cleland, *Bioorg. Chem.* **15,** 283 (1987); W. W. Cleland, *in* "Isotopes in Organic Chemistry" (E. Buncel and C. C. Lee, eds.), Vol. 7, p. 61. Elsevier, Amsterdam, 1987; P. F. Cook (ed.), "Enzyme Mechanism from Isotope Effects" CRC Press, Boca Raton, Florida, 1991.

METHODS IN ENZYMOLOGY, VOL. 249

Definitions

Kinetic Isotope Effect. The kinetic isotope effect is defined as the ratio between the rate with an unlabeled molecule and the rate with a molecule containing a heavy isotope, and it is written ^{x}k, $^{x}(V/K)$, or ^{x}V for isotope effects on a rate constant, on V/K for a given substrate, or on the maximum velocity. The superscript x is D, T, 13, 14, 15, 17, or 18 for deuterium, tritium, ^{13}C, ^{14}C, ^{15}N, ^{17}O, or ^{18}O isotope effects. An isotope effect on a rate constant is an intrinsic isotope effect, whereas those on V/K or V are observed isotope effects and are often closer to unity than intrinsic isotope effects.

Equilibrium Isotope Effect. The ratio between the K_{eq} value with unlabeled reactant and K_{eq} with a reactant containing a heavy isotope is known as the equilibrium isotope effect. For an isotope-sensitive step, the ratio of intrinsic isotope effects in forward and reverse directions equals the equilibrium isotope effect on that step. Equilibrium isotope effects are either greater or less than unity, depending on whether the isotopic atom is more strongly bonded in the reactant or product, respectively. The equilibrium isotope effect in the reverse reaction is the reciprocal of that in the forward reaction. Equilibrium isotope effects are easily measured experimentally and are presented elsewhere in this series.[5]

Primary Isotope Effect. A primary isotope effect is an isotope effect where a bond is made or broken to the isotopic atom during the reaction. Primary kinetic isotope effects on enzymatic reactions will always be normal (i.e., greater than unity) because the molecular vibration which corresponds to motion in the reaction coordinate has a negative force constant at the transition state and an imaginary frequency. The isotopic atom is thus more weakly bonded in the transition state than in the reactant.

Secondary Isotope Effect. A secondary isotope effect is an isotope effect where no bonds are made or broken to the isotopic atom during the reaction, but where the strength of bonding changes between reactant and transition state (for kinetic isotope effects) or between reactant and product (for equilibrium isotope effects). Deuterium or tritium isotope effects are called α, β, or γ when the isotopic atom is attached to the carbon undergoing bond cleavage or formation or is attached to the first or second carbon removed from the reacting one. Secondary kinetic isotope effects can be either normal (greater than unity) or inverse (less than unity), depending on whether the bonding to the isotopic atom is weaker or stronger in the transition state than in the reactant.

[5] W. W. Cleland, this series, Vol. 64, p. 104.

Commitments. A commitment to catalysis reduces the size of the observed isotope effect on V/K below the intrinsic value according to Eq. (1):

$$^{x}(V/K) = \frac{^{x}k + c_f + c_r{}^{x}K_{eq}}{1 + c_f + c_r} \tag{1}$$

where ^{x}k is the intrinsic isotope effect on the isotope-sensitive step and $^{x}K_{eq}$ is the equilibrium isotope effect on the reaction. The forward commitment (c_f) is the ratio of the forward rate constant for the isotope-sensitive step to the net rate constant for release back off of the enzyme of (a) the variable substrate when the isotope effect is determined by direct comparision of V/K values for unlabeled and labeled substrates or (b) the labeled substrate when the isotope effect is determined by internal competition between labeled and unlabeled molecules. The reverse commitment (c_r) is the ratio of the reverse rate constant for the isotope-sensitive step to the net rate constant for release of the first product from the enzyme. To determine the intrinsic isotope effect one tries to minimize commitments by choosing substrates where the chemistry is slow, changing the pH to make the chemistry rate limiting, or mutating the enzyme to make the chemistry rate limiting.

Measurement of Isotope Effects

There are three methods for measuring isotope effects on enzymatic reactions. First, one compares the kinetics of labeled and unlabeled substrates. This direct comparison method is useful only for deuterium isotope effects that are at least 1.1, but this is the only way to determine isotope effects on V_{max}. The isotope effects on V/K and V are evaluated separately by comparing slopes or vertical intercepts of reciprocal plots with labeled and unlabeled substrates. It is not necessary to vary the concentration of the labeled substrate; one normally varies the concentration of the last substrate to add to the enzyme, or in a random mechanism the one that is released most rapidly from the enzyme, as this minimizes the commitments.

The second method of measuring isotope effects is equilibrium perturbation. This involves adding enzyme to reaction mixtures that are at chemical equilibrium, but with a labeled reactant and corresponding unlabeled product. The initially faster reaction in one direction leads to an apparent perturbation from the equilibrium position, but, as isotopic mixing proceeds, the reaction returns to chemical as well as isotopic equilibrium. The size of the perturbation is proportional to the isotope effect.

This method is described in detail elsewhere in this series.[5] It is capable of measuring isotope effects as low as 1.03.

The third method of measuring isotope effects is by internal competition. In this method the changes in isotopic ratio as the reaction proceeds are used to compute the isotope effect. The equations for doing this are presented elsewhere in this series.[6] This is the method one has to use with tritium or ^{14}C, since these are used as trace labels. By careful comparison of the specific activity of product or residual substrate with that of the initial substrate (or product after 100% reaction), one can calculate isotope effects to 1 or 2%.

The most precise method of measuring isotope effects, however, makes use of the natural abundance of ^{13}C (1.1%), ^{15}N (0.37%), or ^{18}O (0.2%) as trace labels and the isotope ratio mass spectrometer as the detection device. The machine reads mass ratios such as $^{13}C/^{12}C$ with errors of 0.0001 or less, but in practice works only with CO_2 or N_2, since the molecule being looked at must be a gas that does not fragment in the beam. Hydrogen can also be used; however, it is easier to determine deuterium isotope effects by direct comparison, and one usually does not need the precision of the isotope ratio mass spectrometer to determine a deuterium isotope effect. The low natural abundance of deuterium (0.015%) is also a drawback.

The isotope ratio mass spectrometer gives high precision only when reading natural abundance ratios. This is because the machine is designed that way, and also because this minimizes errors from contamination. A 1% contamination of CO_2 in a sample containing equal amounts of ^{13}C and ^{12}C would look like a 2% isotope effect, whereas in a sample with a near natural abundance ratio such contamination will have little effect. The use of the isotope ratio mass spectrometer for measuring CO_2 is described elsewhere in this series,[6] and in other places.[7]

If there is only one nitrogen atom in a molecule (as in *p*-nitrophenol), one uses combustion to convert the nitrogen to N_2. To a 25 cm quartz tube (7 mm ID, 9 mm OD) sealed at one end one adds material containing at least 30 μmol of nitrogen, either as solid or evaporated from ether in the tube. One then adds 2.5 cm of CuO wire (previously baked at 850° for 1 hr to remove contaminants) with a 1 mm layer of copper powder (previously heated at 550° for 15 min in an atmosphere of hydrogen and cooled to room temperature) on top plus a 4 mm^2 piece of silver foil to trap halogens. The tube is pumped to a high vacuum, sealed, and heated

[6] M. H. O'Leary, this series, Vol. 64, p. 83.

[7] P. M. Weiss, *in* "Enzyme Mechanism from Isotope Effects" (P. F. Cook, ed.), p. 291. CRC Press, Boca Raton, Florida, 1991.

at 850° for 2 hr, after which it is cooled over 1 hr to 550° and then held there for 8 hr. After the tube is cracked, the N_2 is separated on the vacuum line by using a 2-propanol–dry ice trap to remove water and a liquid N_2 trap to remove CO_2. The nitrogen is then adsorbed on 5 Å sieves in a tube sealed with a vacuum stopcock and cooled in liquid N_2 (the sieves must be previously heated for several hours under vacuum at 200° to free them from nitrogen or other adsorbed gases, and cooled and then stored under vacuum). The tube containing the sieves is then heated to 200° to release the nitrogen for analysis.

For amides and such molecules, the Kjeldahl procedure[8] can also be used to generate ammonia, which is then oxidized to N_2 by hypobromite prior to analysis. If ammonia is the product of the reaction, it can be isolated directly after pH adjustment by steam distillation.[9] In working with nitrogen, one must be careful to avoid having the final N_2 sample contaminated by any molecules fragmenting to give ethyl radicals (mass 29).

Remote Label Method

The limitation of using the internal competition method of determining isotope effects is that one may not have a practical way to isolate the isotopic atom as N_2 or CO_2 for measurement in the isotope ratio mass spectrometer. This is usually the case when there is more than one nitrogen in a molecule or when one is measuring ^{18}O or ^{13}C isotope effects on a reaction other than one producing CO_2.[10] In these cases the procedure used is the remote label method.[11]

In this procedure one uses a mixture of a double-labeled material and a depleted material. The double-labeled material has as high a percent label as possible both in the remote labeled position and in the discriminating position (the position where one wishes to know the isotope effect). The depleted material has a depleted content of the heavy atom in the remote labeled position but otherwise natural abundance content in other positions, including the discriminating one. The two species are mixed to restore the natural abundance level of the heavy atom in the remote label position. Discrimination between the two species is then caused by the product of labels in the remote label and discriminating positions. As a

[8] R. Ballentine, this series, Vol. 3, p. 984.

[9] J. D. Hermes, P. M. Weiss, and W. W. Cleland, *Biochemistry* **24,** 2959 (1985).

[10] For decarboxylations, however, one can determine ^{18}O isotope effects directly by running the reaction under high vacuum at pH values less than pH 6.5; CO_2 escapes from solution without exchanging its oxygens to an extent greater than 3%.

[11] M. H. O'Leary and J. F Marlier, *J. Am. Chem. Soc.* **101,** 3300 (1979).

control, one measures the isotope effect in the remote labeled position, using natural abundance material. The isotope effect in the discriminating position is then the ratio of the two values. The actual equation for this calculation, correcting for incomplete labeling and allowing for labels of more than one similar atom (as in -PO_3 units), is[12]

$$^{j}k \text{ per atom} - 1 = \frac{(\sqrt[i]{\ }) - 1}{1 - (\sqrt[i]{\ })\dfrac{(1 - y)}{i}}$$

where

$$(\sqrt[i]{\ }) = \sqrt[i]{\frac{P/R}{1 - Q(P/R - 1)}}$$

and $P = {}^{q,j}(V/K)$ is the observed isotope effect with remote label [i.e., ${}^{15,18,18}(V/K)$]; R is the isotope effect in the remote labeled position with natural abundance material; i is the number of discriminating atoms (1, 2, or 3), and j indicates the nature of the discriminating atoms (15, 18, etc.) and q the nature of the remote label (15, 18); $Q = (1 - b)z/bx \approx z/b$ is the degree to which light material in the remote labeled mixture is depleted below natural abundance, where b is the fraction double-labeled material in the remote labeled mixture, z the fraction heavy label present in the remote labeled position of light material used for the remote labeled mixture, x the fraction of heavy label in the remote labeled position of double-labeled material used for mixing, and y the fraction of heavy discriminating label in double-labeled material (or, if 2 or 3 discriminating atoms are present, fraction containing all atoms labeled; for example, 90% labeling with $i = 2$ gives $y = 0.81$).

Not only is this method useful for ^{13}C, ^{15}N, and ^{18}O isotope effects, but it also permits measurement of secondary deuterium isotope effects with a precision far greater than can be obtained by direct comparison of rates with deuterated and unlabeled substrates. Convenient remote labels that have been used are as follows: (1) the nitro group of *p*-nitrophenol, which is easily labeled with ^{14}N or ^{15}N, since these are available as nitrate salts, and one can readily nitrate triphenyl phosphate solely in the para position and hydrolyze it (*p*-nitrophenol is readily oxidized to give N_2; see above); (2) the exocyclic amino group of adenine, which is generated by reaction of chloropurine riboside with ammonia in ethanol solution in

[12] S. R. Caldwell, F. M. Raushel, P. M. Weiss, and W. W. Cleland, *Biochemistry* **30,** 7444 (1991).

a bomb (for analysis this group is removed by adenosine deaminase as ammonia, which can be oxidized to N_2); (3) amide groups, which can be readily synthesized from ammonia and later hydrolyzed to ammonia for analysis; (4) carbon-1 of glucose incorporated by cyanohydrin synthesis from arabinose and cyanide, which is available labeled with ^{12}C and ^{13}C (after phosphorylation to glucose 6-phosphate by MgATP and hexokinase, carbon-1 is removed as CO_2 by the action of $NADP^+$ and glucose-6-phosphate and 6-phosphogluconate dehydrogenases); and (5) formate, which can be obtained containing either ^{12}C or ^{13}C, and the carbon used as the remote label for measurement of ^{18}O or small deuterium isotope effects (when used to measure the ^{18}O isotope effect, the resulting CO_2 needs to be equilibrated with water to normalize the ^{17}O content prior to analysis).

The remote label method can be used with radioactive substrates to determine isotope effects that are hard to measure directly. For measurement of a ^{14}C isotope effect, one mixes a ^{14}C-labeled substrate and one labeled with tritium in a position where no isotope effect is expected. The change in the $^{14}C/^{3}H$ ratio then determines the isotope effect. Similarly, a tritium isotope effect can be measured the same way by having tritium in the discriminating position and ^{14}C in a position that does not give an isotope effect.

Determination of Intrinsic Isotope Effects

As noted earlier, one must determine the intrinsic isotope effects on a reaction in order to gain information on transition state structures. For enzymatic reactions, this requires some means of estimating commitments, if they cannot be eliminated, or of demonstrating that they are not present when the chemistry is thought to be rate limiting.

The first method for determining intrinsic deuterium and tritium isotope effects was that of Northrop,[2] which is based on the fact that tritium isotope effects are the 1.442 power of deuterium ones.[13] The deviation from this relationship can be used to calculate the forward commitment and the intrinsic isotope effects as long as (a) there is no reverse commitment or (b) the equilibrium isotope effect on the reaction is unity (in which case one gets the sum of the forward and reverse commitments). In cases where there are both forward and reverse commitments and $^{D}K_{eq}$ is not unity, however, one gets answers which are not exact. The true value, however, will lie between that calculated from $^{D}(V/K)$ and $^{T}(V/K)$ in the

[13] C. G. Swain, E. C. Stivers, D. F. Reuwer, Jr., and L. J. Schaad, *J. Am. Chem. Soc.* **80,** 5885 (1958).

forward reaction and that calculated from the comparable isotope effects in the reverse reaction (the latter do not need to be measured but can be computed by dividing the isotope effects in the forward reaction by the respective equilibrium isotope effects). This method is outlined elsewhere in this series,[14] and a computer program is listed to make the calculations.

When there are both deuterium and ^{13}C (or other heavy atom) isotope effects on the same step of a reaction (such as C–H cleavage), one can use the effect of deuteration on the ^{13}C isotope effect to calculate intrinsic deuterium and ^{13}C isotope effects, as well as commitments.[15] This gives very precise results when the isotope ratio mass spectrometer is used to measure the ^{13}C isotope effects. The deuterium isotope effect is given by Eq. (1) with the superscript being D, whereas the ^{13}C one with a nondeuterated substrate is given by Eq. (1) with the superscript being 13. The ^{13}C isotope effect with a deuterated substrate, however, is given by Eq. (2):

$$^{13}(V/K)_{\mathrm{D}} = \frac{^{13}k + c_{\mathrm{f}}/^{\mathrm{D}}k + c_{\mathrm{r}}\,^{13}K_{\mathrm{eq}}/(^{\mathrm{D}}k/^{\mathrm{D}}K_{\mathrm{eq}})}{1 + c_{\mathrm{f}}/^{\mathrm{D}}k + c_{\mathrm{r}}/(^{\mathrm{D}}k/^{\mathrm{D}}K_{\mathrm{eq}})} \tag{2}$$

Because deuteration slows the isotope-sensitive step by the size of the intrinsic deuterium isotope effect ($^{\mathrm{D}}k$ in the forward reaction and $^{\mathrm{D}}k/^{\mathrm{D}}K_{\mathrm{eq}}$ in the reverse reaction), while all of the other steps are unaltered, the commitments are divided in both numerator and denominator by the intrinsic deuterium isotope effects.

If either c_{f} or c_{r} is zero, or either $^{13}K_{\mathrm{eq}}$ or $^{\mathrm{D}}K_{\mathrm{eq}}$ is unity, the intrinsic isotope effects are given directly by

$$^{\mathrm{D}}k = \frac{[^{13}(V/K)_{\mathrm{D}} - d][^{\mathrm{D}}(V/K)]}{^{13}(V/K)_{\mathrm{H}} - d} \tag{3}$$

$$^{13}k = \frac{^{\mathrm{D}}(V/K)^{13}(V/K)_{\mathrm{D}} - {}^{13}(V/K)_{\mathrm{H}}c}{^{\mathrm{D}}(V/K) - c} \tag{4}$$

where

$$d = \frac{1 + (c_{\mathrm{r}}/c_{\mathrm{f}})^{13}K_{\mathrm{eq}}}{1 + (c_{\mathrm{r}}/c_{\mathrm{f}})} \tag{5}$$

$$c = \frac{1 + (c_{\mathrm{r}}/c_{\mathrm{f}})^{\mathrm{D}}K_{\mathrm{eq}}}{1 + (c_{\mathrm{r}}/c_{\mathrm{f}})} \tag{6}$$

[14] W. W. Cleland, this series, Vol. 87, p. 625.

[15] J. D. Hermes, C. A. Roeske, M. H. O'Leary, and W. W. Cleland, *Biochemistry* **21**, 5106 (1982).

In the general case where c_f and c_r are both finite and neither equilibrium isotope effect is unity, Eqs. (3)–(6) give approximate answers, and one can calculate only limits for the intrinsic isotope effects, although these are very narrow for ${}^{13}k$ and fairly narrow for ${}^{D}k$ in most cases. The procedure is to assume a ratio of c_r/c_f (values of 0, 0.3, 1.0, 3.0, and ∞ suffice) and then use the following equations to calculate ${}^{D}k$ and ${}^{13}k$:

$$ {}^{D}k = \frac{b + \sqrt{b^2 - 4\,{}^{D}(V/K)\,(yp)\,(zp)}}{2(yp)} \tag{7} $$

$$ {}^{13}k = {}^{13}(V/K)_H + [{}^{D}k - {}^{D}(V/K)](yp)/(xp) \tag{8} $$

where

$$ (xp) = {}^{D}(V/K)(1 + c_r/c_f) - [1 + (c_r/c_f){}^{D}K_{eq}] \tag{9} $$

$$ (yp) = {}^{13}(V/K)_H(1 + c_r/c_f) - [1 + (c_r/c_f){}^{13}K_{eq}] \tag{10} $$

$$ (zp) = {}^{13}(V/K)_D[1 + (c_r/c_f){}^{D}K_{eq}] - [1 + (c_r/c_f){}^{D}K_{eq}\,{}^{13}K_{eq}] \tag{11} $$

$$ b = {}^{D}(V/K)(yp) + [{}^{13}(V/K)_D - {}^{13}(V/K)_H](xp) + (zp) \tag{12} $$

The forward commitment is then

$$ c_f = [{}^{D}k - {}^{D}(V/K)]/(xp) \tag{13} $$

and the reverse commitment is calculated from the assumed ratio, except when c_f was assumed to be zero, in which case

$$ c_r = [{}^{D}k - {}^{D}(V/K)]/[{}^{D}(V/K) - {}^{D}K_{eq}] \tag{14} $$

A computer program which makes these calculations (ISOCALC4) appears at the end of this chapter in Appendix I.

Exact Solution for Intrinsic Isotope Effects

Measuring ${}^{13}C$ (or other heavy atom) isotope effects with a deuterated and a nondeuterated substrate will, as described above, give narrow ranges for the intrinsic isotope effects. The range is often narrowed further by the fact that some choices of the c_r/c_f ratio give negative values for one or more parameters, or totally ridiculous values for intrinsic isotope effects, and thus these solutions can be discarded. Two methods have been proposed for exact solution for the intrinsic isotope effects.

The first involves combining the tritium isotope effect with the three others [${}^{D}(V/K)$, ${}^{13}(V/K)_H$, and ${}^{13}(V/K)_D$] used above. Because one now has four equations in ${}^{D}k$, ${}^{13}k$, and the two commitments [${}^{T}k = ({}^{D}k)^{1.442}$, so ${}^{T}k$ is not a separate unknown], one can achieve a solution.[15] This method has yet to be applied, however, as measurement of ${}^{D}(V/K)$ and ${}^{T}(V/K)$ values is subject to larger errors than those from the heavy atom isotope effects, but it should work when there is no secondary deuterium isotope effect that can be measured. The equations for this method are given in Ref. 15, and the computer program ISOCALC4 (Appendix I) will make the calculations.

The method which has been used several times involves measurement of an α-secondary deuterium isotope effect on the reaction, as well as the ^{13}C isotope effect with an α-secondary deuterated substrate.[15–17] An example is determining the deuterium isotope effect at C-4 of the nicotinamide ring of NAD^+ or $NADP^+$ during a dehydrogenase reaction when C–H cleavage in the substrate contributes the primary deuterium and ^{13}C isotope effects.

The equation for the α-secondary deuterium isotope effect is given by Eq. (1) with α-D as the superscript. The equation for the ^{13}C isotope effect with α-secondary deuteration is given by Eq. (2) with D replaced by α-D. This situation results in five equations for the 5 isotope effects in terms of ${}^{D}k$, ${}^{13}k$, ${}^{\alpha\text{-D}}k$, c_f, and c_r. A computer program that has been used to solve these equations (ISOCALC5) is given at the end of this chapter in Appendix II. It assumes a range of values for c_r/c_f and uses the equations above to determine ${}^{D}k$, ${}^{13}k$, c_f, and c_r. These values are then used to calculate ${}^{\alpha\text{-D}}k$ from Eq. (1) with α-D superscripts, and ${}^{\alpha\text{-D}}k$ is then used to calculate the expected ${}^{13}(V/K)_{\alpha\text{-D}}$ from Eq. (2) with α-D superscripts. This value is compared with the experimental one and the partial derivative of ${}^{13}(V/K)_{\alpha\text{-D}}$ with respect to the c_r/c_f ratio used to calculate a new value for this ratio. Iterations are continued until calculated and experimental values of ${}^{13}(V/K)_{\alpha\text{-D}}$ converge.

Errors in the final intrinsic isotope effects and commitments are estimated by seeing the effects of 1% errors in each input parameter on the answers, and multiplying these effects by the actual percent errors in the input parameters. The percent error in each final answer is then the square root of the sum of the squares of the percent errors resulting from error in each input parameter. Because each input parameter comes from a separate experiment, the errors are independent, as this calculation assumes.

This method was used on glucose-6-phosphate dehydrogenase in H_2O and D_2O,[16] and with liver alcohol dehydrogenase with benzyl alcohol as substrate.[17] The method determines ${}^{D}k$ and ${}^{13}k$ with small errors, but c_f and c_r are less well known, although their sum is well determined. For example, with glucose-6-phosphate dehydrogenase in water, ${}^{13}k$ =

[16] J. D. Hermes and W. W. Cleland, *J. Am. Chem. Soc.* **106,** 7263 (1984).

[17] M. Scharschmidt, M. A. Fisher, and W. W. Cleland, *Biochemistry* **23,** 5471 (1984).

1.041 ± 0.002, ${}^{D}k = 5.3 \pm 0.3$, ${}^{\alpha\text{-}D}k = 1.054 \pm 0.035$, $c_f = 0.75 \pm 0.26$, and $c_r = 0.49 \pm 0.27$. The sum of c_f and c_r, however, was 1.24 ± 0.14.

Calculation of Intrinsic Isotope Effects in Stepwise Mechanisms

When a reaction is stepwise and deuterium-sensitive and ^{13}C-sensitive steps are not the same, the ^{13}C isotope effects with deuterated and unlabeled substrates are not independent, but are related by

$$\frac{{}^{13}(V/K)_H - 1}{{}^{13}(V/K)_D - 1} = \frac{{}^{D}(V/K)}{{}^{D}K_{eq}} \tag{15}$$

in the direction where the deuterium-sensitive step comes first, and by Eq. (16) in the reverse direction where the ^{13}C-sensitive step comes first:

$$\frac{{}^{13}(V/K)_H - {}^{13}K_{eq}}{{}^{13}(V/K)_D - {}^{13}K_{eq}} = {}^{D}(V/K) \tag{16}$$

Note that the parameters in Eqs. (15) and (16) relate to different directions of the reaction [i.e., ${}^{D}(V/K)$ in Eq. (16) equals ${}^{D}(V/K)/{}^{D}K_{eq}$ in Eq. (15), so that the equations are really the same]. When ${}^{D}K_{eq}$ is significantly different from unity, however, Eqs. (15) and (16) differ sufficiently that one can tell which one fits the experimental data in a given direction of the reaction. This defines which step comes first in the mechanism, and thus the nature of the intermediate.

Equations (15) and (16) do not permit determination of intrinsic isotope effects. One can of course look at multiple isotope effects that are on just one step of a stepwise mechanism, but when the intermediate is stable enough to add to the enzyme and partition in forward and reverse directions, the partition ratio can be used to help deconvolute the system. With malic enzyme, where dehydrogenation precedes decarboxylation and there is no reverse commitment for CO_2, this was done by adding oxaloacetate and NADPH and observing formation of both pyruvate and malate spectrophotometrically (disappearance of A_{340} for malate production and change at 281 nm, the isosbestic point of $NADP^+$ and NADPH, for disappearance of oxaloacetate).[18] If r_H is the ratio of pyruvate/malate formed initially (0.47 for malic enzyme), then

$${}^{13}k = {}^{13}(V/K)_H + r_H[{}^{13}(V/K)_H - 1] \tag{17}$$

where ${}^{13}k$ is for the decarboxylation step. An independent determination of ${}^{13}k$ was obtained by using A-side deuterated NADPH and combining

[18] C. B. Grissom and W. W. Cleland, *Biochemistry* **24,** 944 (1985).

the resulting r_D value (0.81) with $^{13}(V/K)_D$ by an equation analogous to Eq. (17). Values of 1.044 from r_H and 1.045 from r_D were obtained. The relationship between r_H and r_D is

$$^{D}(V/K)/^{D}K_{eq} = (r_D + 1)/(r_H + 1) \tag{18}$$

To get the intrinsic deuterium isotope effect on the dehydrogenase step, the deuterium and tritium isotope effects on V/K were used in Eqs. (19)–(21):

$$T_r = [^{T}K_{eq} - {}^{T}(V/K)]/r_H \tag{19}$$

$$D_r = [^{D}K_{eq} - {}^{D}(V/K)]/r_H \tag{20}$$

$$\frac{^{T}(V/K) - 1 - T_r}{^{D}(V/K) - 1 - D_r} = \frac{(^{D}k)^{1.442} - {}^{T}(V/K) + T_r}{^{D}k - {}^{D}(V/K) + D_r} \tag{21}$$

Equation (21) is solved for ^{D}k by Newton's method of successive approximations, and then

$$c_{fH} = \frac{^{D}k - {}^{D}(V/K) + D_r}{^{D}(V/K) - 1 - D_r} \tag{22}$$

where c_{fH} is the forward commitment for hydride transfer, and

$$P = r_H/(1 + c_{fH}) \tag{23}$$

where P is the ratio of the forward rate constant for decarboxylation to the reverse rate constant for hydride transfer (i.e., P is the immediate partition ratio of the intermediate). A computer program to make these calculations has been written by Dr. Charles Grissom (Department of Chemistry, University of Utah, Salt Lake City, Utah 84112).

Isotope Effects on More Than One Step

Until now we have assumed that each isotopic substitution affects only one step in an enzymatic mechanism. There are a number of mechanisms, however, where a given isotope will affect two or more steps. For example, when $^{15}NH_3$ is used with glutamate dehydrogenase, there will be a primary ^{15}N isotope effect on carbinolamine formation from α-ketoglutarate, followed by a primary ^{15}N isotope effect on proton removal as well as a secondary effect as water is eliminated to give iminoglutarate (the order of the steps is not clear). There will then be a secondary ^{15}N isotope effect on hydride transfer as $=NH_2^+$ becomes $—NH_2$. If the amino group is later protonated while still on the enzyme, there will be another primary ^{15}N isotope effect on that step.

In mechanisms like this the equations for $^{15}(V/K)$ with deuterated or unlabeled NADPH become complicated, and exact solutions for intrinsic isotope effects cannot be obtained. One can deduce quite a bit from such studies, however, and the reader is referred to experimental papers on alanine and glutamate dehydrogenases,[19] phenylalanine-ammonia lyase,[9] and adenosine deaminase.[20] Similar studies have also been reported on pyruvate carboxylase, where ^{13}C isotope effects were determined in D_2O and H_2O.[21]

Transition State Structures

An earlier chapter in this series on the use of isotope effects to deduce transition state structure[14] gives only a few examples, including catechol *O*-methyltransferase and several dehydrogenase-catalyzed reactions. The interesting effects of coupled hydrogen motions in the transition states for hydride transfer, which we now know involve tunneling, were mentioned, and the reader is referred to [14] of this volume for the latest developments in this area. We cannot discuss all of the work that has been done since the late 1980s on transition state structures of enzymatic reactions, but we give representative examples plus several examples of more recent work.

Hydride Transfer in Dehydrogenase-Catalyzed Reactions

The transition states for several reactions catalyzed by dehydrogenases have been studied in detail. The first enzyme studied was formate dehydrogenase, where changing the nucleotide from NAD^+ ($E^{\circ\prime}$, -0.32 V) to acetylpyridine-NAD^+ ($E^{\circ\prime}$, $-$ 0.258 V) caused the primary deuterium isotope effect to change from 2.2 to 3.1, the primary ^{13}C isotope effect to change from 1.042 to 1.036, the α-secondary deuterium isotope effect (in NAD^+) to drop from 1.32 to 1.06, and the secondary ^{18}O isotope effect to increase from 1.005 to 1.008 per oxygen.[22] Isotope effects with thio-NAD^+ ($E^{\circ\prime}$, $-$ 0.285 V) and pyridine aldehyde-NAD^+ ($E^{\circ\prime}$, $-$ 0.262 V) were intermediate, except for the ^{18}O isotope effect with the latter, which was slightly inverse.

The data were interpreted to mean that the transition state was rather late with NAD^+, with the C–H bond of formate nearly broken, and very

[19] P. M. Weiss, C.-Y. Chen, W. W. Cleland,and P. F. Cook, *Biochemistry* **27,** 4814 (1988).
[20] P. M. Weiss, P. F. Cook, J. D. Hermes, and W. W. Cleland, *Biochemistry* **26,** 7378 (1987).
[21] P. V. Attwood, P. A. Tipton, and W. W. Cleland, *Biochemistry* **25,** 8197 (1986).
[22] J. D. Hermes, S. W. Morrical, M. H. O'Leary, and W. W. Cleland, *Biochemistry* **23,** 5479 (1984).

tight coupling between the hydride approaching C-4 of NAD^+ and the secondary hydrogen at C-4 which was bending out of plane (this causes the normal secondary deuterium isotope effect for the reaction, where ${}^{\alpha\text{-D}}K_{eq}$ is 0.89). As the redox potential of the nucleotide becomes more positive, the transition state becomes earlier and there is less coupling of the motions of the hydride and the secondary hydrogen. The ^{18}O isotope effects are expected to be inverse, since the oxygens are more stiffly bonded in CO_2 than in formate (estimated ${}^{18}K_{eq}$ of 0.983), and the trend in values is in the expected direction, but all values except for that with pyridine aldehyde-NAD^+ are normal. Hermes *et al.*[22] proposed that dehydration of formate when it was bound by the enzyme produced a normal isotope effect that was larger than the inverse kinetic effect on dehydrogenation, and that the aberrant value for pyridine aldehyde-NAD^+ resulted from incomplete dehydration on binding (it is a very slow substrate, with a V_{max} 0.05% that of NAD^+).

The above data for the enzymatic reaction were compared with those for the oxidation of formate to CO_2 in dimethyl sulfoxide (DMSO), or DMSO–water mixtures by I_2.[22] In pure DMSO the transition state is early (and the rate is nearly 7 orders of magnitude faster than in water[23]), as evidenced by a ^{13}C isotope effect of only 1.0154, ^{18}O isotope effect of 0.994, and deuterium isotope effect of 2.2. As water was added, the transition state became later, with the ^{13}C and deuterium isotope effects going to 1.0362 and 3.8 in water. These values are very similar to those for the enzymatic reaction with acetylpyridine-NAD^+, and they probably represent a transition state that is nearly symmetrical.

Similar data were obtained for liver alcohol dehydrogenase with benzyl alcohol as substrate.[17] The primary ^{13}C isotope effect in benzyl alcohol varied from 1.0254 with NAD^+ to 1.0115 with acetylpyridine-NAD^+, whereas the deuterium isotope effects with dideuterobenzyl alcohol (this includes both primary and secondary isotope effects) increased from 4.0 to 6–7. Again this appears to represent a trend from late to earlier transition states.

It is interesting to note the differences in the size of primary ^{13}C isotope effects in dehydrogenase-catalyzed reactions. Formate dehydrogenase (and decarboxylations in general) shows a value of around 1.04,[22] and the same is true for glucose-6-phosphate dehydrogenase,[16] whereas the values for alcohol dehydrogenase with benzyl alcohol are about 1.02.[17] The difference presumably results from how much carbon motion is taking place in the transition state. In a decarboxylation (and in the formate dehydrogenase reaction), the carbon must move into a colinear arrangement with the two oxygens of the carboxyl group, and their mass is 32 while carbon

[23] F. W. Hiller and J. H. Krueger, *Inorg. Chem.* **6,** 528 (1967).

is only 12. Thus considerable carbon motion is taking place in the transition state. Carbon-1 of glucose 6-phosphate has only one hydrogen attached. Thus when this carbon changes from tetrahedral to trigonal during hydride transfer it must move back into the plane of the two oxygens and one carbon it is attached to, and carbon motion is occurring in the transition state. With a primary alcohol such as benzyl alcohol, however, there is a second hydrogen attached, and the change from tetrahedral to trigonal geometry can be accomplished by having this second hydrogen move. Thus carbon-1 does not have to move appreciably, and the primary ^{13}C isotope effect is only one-half that in the other two cases.

Three predictions come from this analysis: (1) there should be a sizable normal secondary deuterium isotope effect on benzyl alcohol oxidation as the result of the required motion of this hydrogen, (2) the primary ^{13}C isotope effect for oxidation of secondary alcohols by alcohol dehydrogenase should approach or exceed 4%, and (3) the primary ^{13}C isotope effects at carbon-4 of NAD^+ should not greatly exceed 2%, since motion of the hydrogen at carbon-4 will permit geometry changes from trigonal to tetrahedral. These predictions have so far not been tested, except for a preliminary value of 1.007 for the ^{13}C isotope effect at C-4 of NAD^+ in the formate dehydrogenase reaction obtained by D. Kiick in our laboratory (unpublished results, 1988).

In this connection it is useful to look at the isotope effects on prephenate dehydrogenase.[24] The ^{13}C isotope effect for the decarboxylation was 1.0103 with deuterated substrate but only 1.0033 with unlabeled substrate, whereas the deuterium isotope effect on V/K was 2.34. The increase in the ^{13}C value on deuteration shows that the mechanism is concerted, not stepwise, and application of Eqs. (3)–(6) above allowed calculation of the intrinsic deuterium isotope effect as 7.3, the intrinsic ^{13}C one as 1.0155, and the forward commitment as 3.7 (the reverse commitment was assumed to be zero for CO_2 release). These values suggest a transition state where the hydride that is being transferred to NAD^+ is midway along the reaction coordinate, but where C–C cleavage is only just beginning. Thus the transition state is concerted but not synchronous.

Transition States for Phosphoryl Transfer

Heavy atom isotope effects can yield considerable information on the transition states for phosphoryl transfer.[25] A 1–2% primary ^{18}O isotope effect shows that P–O cleavage is occurring in the rate-limiting transition

[24] J. D. Hermes, P. A. Tipton, M. A. Fisher, M. H. O'Leary, J. F. Morrison, and W. W. Cleland, *Biochemistry* **23,** 6263 (1984).

[25] W. W. Cleland, *FASEB J.* **4,** 2899 (1990).

state, whereas the secondary ^{18}O isotope effect in the nonbridge oxygen or oxygens indicates changes in bond order, and thus actual transition state structure. When *p*-nitrophenol is the leaving group, the size of the secondary ^{15}N isotope effect tells the degree of electron delocalization into the nitro group and is thus a measure of P–O cleavage. The effect is small ($^{15}K_{eq}$ is 1.0023 for ionization of *p*-nitrophenol[26]) but readily measured with the isotope ratio mass spectrometer after combustion of *p*-nitrophenol to N_2. The effect comes from an increased contribution of the quinonoid resonance form:

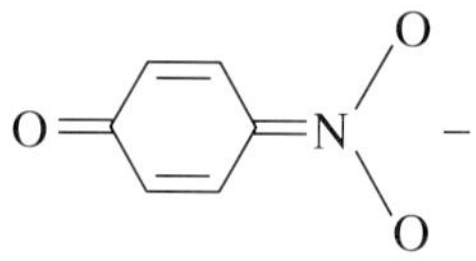

where the negative charge is in the nitro group and stems from the fact that in the quinonoid form the nitrogen is double-bonded to carbon and single-bonded to each oxygen, rather than single-bonded to carbon and 1.5-bonded to each oxygen. Because bonds to oxygen are more stiffening than bonds to carbon, ^{15}N enriches in the protonated or phosphorylated species where the quinonoid resonance form (which gives the color to the ion) contributes less than in the ion.

The ^{15}N isotope effect when *p*-nitrophenoxide is the leaving group was discovered when using this nitrogen as the remote label for measuring primary and secondary ^{18}O isotope effects on phosphoryl transfer, since the control for such experiments is to measure the ^{15}N isotope effect with natural abundance substrates.

Triesters

Triester hydrolysis has been studied with diethyl-*p*-nitrophenyl phosphate or diethyl-*p*-carbamoylphenyl phosphate as substrates and using either hydroxide or a phosphotriesterase as catalyst.[12] The primary ^{18}O isotope effects with hydroxide were 1.006 and 1.027 for *p*-nitrophenol and *p*-carbamoylphenol, respectively, showing that these reactions are concerted, and do not involve five-coordinate phosphorane intermediates. Such an intermediate would partition forward to release *p*-nitrophenol (p*K* 7) or *p*-carbamoylphenol (p*K* 8.6) rather than decompose to release hydroxide (p*K* 15.7), and thus any primary ^{18}O isotope effect would be suppressed. The enzymatic reaction shows primary ^{18}O isotope effects of 1.002 and 1.036, respectively, for the two substrates, showing that these

[26] A. C. Hengge and W. W. Cleland, *J. Am. Chem. Soc.* **112,** 7421 (1990).

reactions also are concerted. The value with diethyl-*p*-nitrophenyl phosphate is not the intrinsic value, since the chemistry is not totally rate limiting with this substrate. The lower primary isotope effects with *p*-nitrophenol as the leaving group presumably result from the delocalization of electrons into the nitro group, with the resulting increase in double bond character for the bond between the phenolic oxygen and the ring. Such delocalization cannot occur with *p*-carbamoylphenol, so a larger ^{18}O isotope effect is observed.

The secondary ^{18}O isotope effects for the nonbridge oxygen were 1.0063 and 1.025 with hydroxide as the attacking group, and 1.002 and 1.018 for the enzymatic reactions for the substrates with *p*-nitrophenol and *p*-carbamoylphenol as leaving groups. These values reflect the decreased bond order in the transition state relative to the triester substrate, and show that the reactions, although concerted, have associative transition states in which the sum of the bond orders along the axis of the trigonal bipyramidal transition state is greater than unity. The transition state with *p*-carbamoylphenol as leaving group is more associative than that with *p*-nitrophenol, as one might expect from the higher pK of *p*-carbamoylphenol. What is not yet clear is whether reactions of phosphate triesters will remain concerted when the pK of the leaving group reaches 16 (as with ethanol), or whether phosphorane intermediates will occur in such cases. Further isotope effect studies should help answer this question, although the fact that ^{18}O exchange with the nonbridge oxygen during hydrolysis is not observed with such triesters argues that the reaction will remain concerted. There is much better evidence for phosphorane formation and ^{18}O exchange under acid conditions.

For diethyl-*p*-nitrophenyl phosphate, the size of the secondary ^{15}N isotope effect (1.0007) can be used to estimate the bond order to the leaving group in the transition state, and the secondary ^{18}O isotope effect gives evidence of the bond order to the nonbridge oxygen. Assuming that the bond order to phosphorus is 5 in the reactant and will remain 5 in the transition state, the bond order to the nonbridge oxygen is 1.85, and those to the entering OH and leaving *p*-nitrophenol are 0.40 and 0.75, respectively.[12] The sum of axial bond orders of 1.15 is greater than unity, but only slightly, showing that this reaction, although associative, is not extremely so.

Phosphodiesters

Secondary ^{18}O and ^{15}N isotope effects have been measured for phosphoryl transfer reactions of 3,3-dimethylbutyl-*p*-nitrophenyl phosphate catalyzed by acid, hydroxide, β-cyclodextrin, and snake venom phosphodiesterase.[27] All reactions were concerted, as evidenced by ^{15}N isotope effects ranging from 1.0009 in acid to 1.0013 to 1.0017 with the other catalysts. The smaller value in acid shows that protonation of the nonbridge oxygen converts the reaction to a more associative one, with less cleavage to the leaving group in the transition state. The other values suggest 50–65% cleavage of the P–O bond in the transition state.

The secondary ^{18}O isotope effect from the nonbridge oxygens was slightly inverse in alkaline hydrolysis (0.995) and slightly normal (1.001) for the reaction with β-cyclodextrin, in which an ionized secondary hydroxyl of a sugar unit attacks the phosphorus of the adsorbed substrate (the dimethylbutyl group forms an inclusion complex), with release of *p*-nitrophenol. Both reactions are basically S_N2 ones, but the alkaline hydrolysis is slightly dissociative because of the higher pK of the attacking hydroxide (15.7) compared with the estimated first pK of 12 (including the statistical effect of there being 7 of them) for the sugar hydroxyl.

The surprising value was the secondary ^{18}O isotope effect of 0.984 for phosphodiesterase hydrolysis. This value suggests that the enzyme protonates a nonbridge oxygen during the reaction, since this inverse value of 1.6% equals the equilibrium isotope effect for protonation of a phosphate oxygen. Interestingly, the transition state does not appear to be as associative as when the hydrolysis is carried out in acid, since the ^{15}N isotope effect is 1.0017 versus 1.0009 in acid. Presumably, efficient general base catalysis by the enzyme is responsible for the difference (the enzyme catalyzes a Ping Pong mechanism, with a threonine on the enzyme acting as the initial attacking group).

The secondary ^{18}O isotope effect for acid hydrolysis of dimethylbutyl-*p*-nitrophenyl phosphate, after correction to full protonation of one nonbridge oxygen, was 1.0139.[27] This normal value shows that the transition state is quite associative, with the reaction resembling those of triesters.

p-(*tert*-Butyl)phenyl-*p*-nitrophenyl phosphate has been used as a substrate for mono- and bisimidazoyl-β-cyclodextrin, and also hydrolyzed by acid or base.[28] The 1,2- and 1,3-bisimidazoylcyclodextrins were made by Breslow as mimics of ribonuclease.[29] These reactions are all concerted, since they show secondary ^{15}N isotope effects of 1.0007 to 1.0016 and primary ^{18}O isotope effects of 0.5 to 3.5%. The exact transition state structures await determination of secondary ^{18}O isotope effects.

Uridylyl-3′-*p*-nitrophenyl phosphate is a substrate for ribonuclease. So far only secondary ^{15}N isotope effects have been measured, but these

[27] A. C. Hengge and W. W. Cleland, *J. Am. Chem. Soc.* **113,** 5835 (1991).

[28] A. E. Tobin and A. C. Hengge, unpublished experiments (1994).

[29] E. Anslyn and R. Breslow, *J. Am. Chem. Soc.* **111,** 5972 (1989).

all suggest that the reactions of this molecule, which produce *p*-nitrophenol and uridylyl-3′,5′-cyclic phosphate, are concerted, regardless of whether the reaction is catalyzed by ribonuclease, acid, imidazole buffer, or hydroxide.[28] Catalysis by metal ions at pH 6.5 is also concerted, with transition states showing a high degree of electron delocalization into the nitro group.[30] Primary and secondary ^{18}O isotope effects should be available shortly to help determine the exact nature of the transition states.

Phosphate Monoesters

A number of reactions of *p*-nitrophenyl phosphate have been studied. Hydrolysis of the monoanion has a primary ^{18}O isotope effect of 1.0087, an ^{18}O isotope effect in the nonbridge oxygens of 1.019, and a secondary ^{15}N isotope effect of 1.0004.[26,31] The isotope effect for the nonbridge oxygens reflects the fact that O–H cleavage is part of the reaction, and thus this is mainly a primary isotope effect. Hydrogen transfer to the leaving group appears to be concerted with P–O cleavage, but lags slightly behind it, since there is a small secondary ^{15}N isotope effect; if the proton shift occurred in a preequilibrium step one would not expect any ^{15}N isotope effect.

As the dianion, *p*-nitrophenyl phosphate hydrolyzes in water with a primary ^{18}O isotope effect of 1.019, a secondary ^{18}O isotope effect for the nonbridge oxygens of 0.999, and a secondary ^{15}N isotope effect of 1.0028.[26,31] The large value of the latter indicates a considerable degree of electron delocalization into the nitro group and is consistent with the expected highly dissociative character of the reaction. When the dianion is placed in *tert*-butanol, it reacts under mild conditions (30°) to give *tert*-butyl phosphate with racemization of configuration, and thus presumably via a metaphosphate intermediate (reaction in water involves inversion).[32] The primary ^{18}O isotope effect of 1.021 and the secondary ^{15}N isotope effect of 1.0039 are consistent with this picture, with the very large secondary ^{15}N isotope effect reflecting a high degree of electron delocalization into the nitro group as a result of the negatively charged PO_3 group in intimate contact with the phenolic oxygen, and the low dielectric constant of *tert*-butanol.[31] This is, in fact, the largest secondary ^{15}N isotope effect reported for *p*-nitrophenol.

The preliminary value of the secondary ^{18}O isotope effect of 1.000 is a puzzle. One would predict that this isotope effect would be inverse because the P–O bond order would increase from $\frac{4}{3}$ to $\frac{5}{3}$ as one goes from

[30] M. A. Rishavy and A. C. Hengge, unpublished experiments (1994).
[31] A. C. Hengge, W. A. Edens, and H. Elsing, *J. Am. Chem. Soc.* **116,** 5045 (1994).
[32] J. M. Friedman, S. Freeman, and J. R. Knowles, *J. Am. Chem. Soc.* **110,** 1268 (1988).

a phosphate monoester to metaphosphate. At the same time, however, one is losing both bending vibrations with the oxygens of the leaving group and torsional vibrations of the O–P–O–C type. Possibly, in a fully dissociative transition state the loss of these vibrations overrides the stiffening of the stretching and in-plane bending modes and gives the effect observed. Further experimental and theoretical work is needed on this problem.

Primary and secondary ^{18}O and secondary ^{15}N isotope effects have been measured for the hydrolysis of *p*-nitrophenyl phosphate by alkaline phosphatase, but since the primary ^{18}O isotope effect was only 1.0003 at pH 8 and 1.0005 at pH 6 (where catalysis should be more rate limiting), it is clear that there are large forward commitments with this substrate.[31] With glucose 6-phosphate, the secondary ^{18}O isotope effect was 0.9994 at pH 8 and 0.9982 at pH 6, and the slightly more inverse value at pH 6 was taken as evidence that the transition state was dissociative, as expected from model chemical studies on phosphate monoesters.[34] Presumably, the forward commitment is less with this substrate than with *p*-nitrophenyl phosphate because the free energy of hydrolysis is less, and thus the phosphoryl transfer step less favorable. The primary ^{18}O isotope effect needs to be determined with glucose 6-phosphate, however, before this conclusion is established.

An interesting series of isotope effects have been measured with *cis*-hydroxo-*p*-nitrophenylphosphatobisethylenediaminecobalt(III).[35] The complex undergoes competing reactions in which (1) attack by the cis-coordinated hydroxide leads to formation of a bidentate phosphato complex with liberation of *p*-nitrophenol (84%) and (2) dissociation of *p*-nitrophenyl phosphate occurs (16%).[36] The ^{18}O isotope effect in the bridge oxygen (primary isotope effect) was 1.022 on reaction (1), showing that the reaction is a concerted one. The secondary ^{18}O isotope effect in the nonbridge oxygens was 1.0007, which suggests a nearly S_N2 type of reaction, and the secondary ^{15}N isotope effect in the nitro group of *p*-nitrophenyl phosphate of 1.0021 also argues for considerable P–O cleavage in the transition state.

What was surprising were the isotope effects on reaction (2). The isotope effect in the nonbridge oxygens (which involves Co–O cleavage) was 1.0107, suggesting that the inert Co–O bond has sufficient covalent character to lead to a primary isotope effect on its cleavage. Even more

[33] Deleted in proof.
[34] P. M. Weiss and W. W. Cleland, *J. Am. Chem. Soc.* **111,** 1928 (1989).
[35] J. Rawlings and A. C. Hengge, unpublished experiments (1994).
[36] D. R. Jones, L. F. Lindoy, and A. M. Sargeson, *J. Am. Chem. Soc.* **105,** 7327 (1983).

surprising was the bridge ^{18}O isotope effect of 1.006, which is a secondary isotope effect in this case. There are two possible causes for this effect. First, the P–O bond in the coordinated *p*-nitrophenyl phosphate is probably not a pure single bond, but may have a bond order of approximately 1.15.[37] When the coordinated oxygen is released from cobalt, it has a chance to increase its bond order by entering into resonance with the other two nonbridge oxygens, and this will permit the phosphorus to bridge oxygen bond to lower its bond order to about 1.09.[37] This would cause an isotope effect of approximately 1.002, since calculations suggest that a change in bond order of 1.0 for P–O bonds corresponds to an equilibrium ^{18}O isotope effect of 3–4%.

The second possible cause for the observed isotope effect is the loss of torsional vibrations as the coordination bond to cobalt is broken. The bridge oxygen is involved in a torsional vibration of the Co–O–P–O type, and because oxygen is lighter than cobalt the frequency of the vibration will be highly sensitive to ^{18}O substitution. Further work is needed on the cause of this interesting isotope effect.

Secondary ^{18}O isotope effects in the nonbridge oxygens of the γ-phosphate have been determined for reaction of MgATP with glucose or 1,5-anhydro-D-glucitol (a slow substrate) catalyzed by yeast hexokinase.[38] The value was 0.996 at pH 8.2 and 0.990 at pH 5.3 with glucose, whereas with anhydroglucitol it was 0.993 at pH 8.2. The more inverse values with a slow substrate, or at a pH where catalysis should be more rate limiting, were interpreted to mean that the transition state was dissociative, as expected for reaction of phosphate monoesters. Primary ^{18}O isotope effects have not been determined in this system, however, so the size of any commitments is not known.

Transition States for Acyl Transfer

Isotope effects have been determined for a number of reactions of *p*-nitrophenyl acetate.[39,40] By using the nitrogen of *p*-nitrophenol as the remote label, accurate primary and secondary ^{18}O isotope effects in the carboxyl group and β-deuterium isotope effects in the methyl group of acetate have been determined. Attacking nucleophiles have been hydroxide (p*K* 15.7), phenolate (p*K* 9.9), hexafluoro-2-propanoxide (p*K* 9.3), mercaptoethanol (p*K* 9.5), Ethyl-2-mercaptopropionate (p*K* 9.3), and methoxyethylamine (p*K* 9.7), and the hydrolysis catalyzed by chymotrypsin

[37] W. J. Ray, Jr., J. W. Burgner II, H. Deng, and R. Callender, *Biochemistry* **32,** 12977 (1993).

[38] J. P. Jones, P. M. Weiss, and W. W. Cleland, *Biochemistry* **30,** 3634 (1991).

[39] A. C. Hengge, *J. Am. Chem. Soc.* **114,** 6575 (1992).

[40] A. C. Hengge and R. A. Hess, *J. Am. Chem. Soc.* **116,** in press (1994).

at pH 7 has also been studied. In all cases the primary ^{18}O isotope effects were between 1.3% (with hydroxide) and 3.3% (with the amine), with the other reactants showing approximately 2% isotope effects. These values show that the reactions are concerted, and do not involve a tetrahedral intermediate, since any such intermediate would partition forward to release *p*-nitrophenol (pK 7) much faster than it would release the attacking group with higher pK, and thus show no primary ^{18}O isotope effect. The secondary ^{18}O isotope effects in the carbonyl oxygen were about 0.5% for all oxygen nucleophiles, including chymotrypsin, but were 1.2% with the thiolates. Along with this the secondary ^{15}N isotope effects, which were approximately 1.001 with oxygen nucleophiles (again including chymotrypsin) other than hydroxide (where the low value of 1.0002 as well as the small primary ^{18}O isotope effect presumably reflect an early transition state), were less than 1.0003 with the thiolates. The β-secondary deuterium isotope effects were 5% inverse with oxygen nucleophiles but only 2.5% inverse with thiolates.

These data point to a fundamental difference in transition state structures between oxygen and sulfur as attacking nucleophiles. With sulfur the C–O bond order to the carbonyl oxygen decreases twice as much, and the smaller β-secondary deuterium isotope effect indicates that there is less loss of hyperconjugation in going to the transition state. This in turn argues for some residual positive charge on the carbonyl carbon in the transition state, which is consistent with the small secondary ^{15}N isotope effects, since such a positive charge would prevent electron delocalization into the nitro group.

The amine used for these experiments gave the largest primary ^{18}O isotope effect, as well as a 3.3% inverse β-secondary deuterium isotope effect, a secondary ^{18}O isotope effect of 1.0066, and a secondary ^{15}N isotope effect of 1.0011. It is not clear at this point whether the data represent a concerted reaction in which a proton from the attacking amine is lost during the reaction, or whether a zwitterionic tetrahedral intermediate is formed in a preequilibrium step, which breaks down with rate-limiting C–O cleavage. These isotope effects need to be measured as a function of pH before definitive conclusions are reached concerning what happens with amines, where, in contrast with the situation with oxygen or sulfur as the attacking nucleophile, proton loss must accompany the reaction.

Mechanism of Aspartate Transcarbamylase

The enzyme aspartate transcarbamylase (aspartate carbamoyltransferase) catalyzes transfer of a carbamoyl group from carbamoyl phosphate to aspartate to give carbamoylaspartate. It has long been assumed that the reaction involved attack of aspartate on carbamoyl phosphate to form a tetrahedral intermediate, which decomposed after proton transfer to phosphate and carbamoylaspartate. The nonenzymatic breakdown of carbamoyl phosphate does not involve such chemistry, however, with the dianion decomposing by internal general acid catalysis with C–O cleavage to phosphate and cyanic acid (H—N=C=O), and the monoanion losing a PO_3 group to water by P–O cleavage to leave carbamic acid as the other product.[41] Thus there have been suggestions that the enzymatic reaction might involve breakdown of carbamoyl phosphate to cyanic acid, followed by attack of aspartate on the electrophilic intermediate. To answer this question, ^{15}N isotope effects were measured in carbamoyl phosphate for the nonenzymatic breakdown as well as the enzymatic reaction.[42] The monoanion gave 1.0028, which is consistent with this being a small secondary isotope effect associated with changes in the amide resonance, or possibly loss of N–C–O–P torsional vibrations. The dianion gave 1.0105, which corresponds to a primary ^{15}N isotope effect for N–H cleavage during formation of cyanic acid. This value is not larger because most of the motion in the transition state is that of the hydrogen. By contrast, the ^{13}C isotope effect on this cleavage is 4–5%, since the carbon must move to become colinear with the nitrogen and oxygen in the product.

The enzymatic reaction was studied with three combinations of enzyme and substrate.[42] A His134Ala mutant with 5% of wild-type activity with aspartate as substrate gave a ^{15}N isotope effect of 1.0027, whereas the wild-type enzyme with cysteine sulfinate as substrate (V_{max} 2% of that with aspartate) gave 1.0024. Both combinations give the full intrinsic ^{13}C isotope effect on the reaction of 4%, and these ^{15}N isotope effects are presumably also intrinsic. The wild-type enzyme with aspartate gives a ^{13}C isotope effect of 2.2%, so there is some commitment present. It gave a ^{15}N isotope effect of 1.0014, which appears to be suppressed to the same extent as the ^{13}C isotope effect. These intrinsic ^{15}N isotope effects of about 0.25% were interpreted to indicate that the mechanism did not involve a cyanic acid intermediate, but rather involved a tetrahedral intermediate, with the small ^{15}N isotope effect reflecting loss of amide resonance in forming the tetrahedral intermediate. This is an excellent example of how isotope effects can determine transition state structures, and thus mechanism.

Glycosyltransferases

α-Secondary deuterium (or tritium) isotope effects have been determined for a number of glycosyltransferases, since the size of this isotope

[41] C. M. Allen and M. E. Jones, *Biochemistry* **3,** 1238 (1964).
[42] G. L. Waldrop, J. L. Urbauer, and W. W. Cleland, *J. Am. Chem. Soc.* **114,** 5941 (1992).

effect gives information on the degree of oxycarbonium ion character at C-1 in the transition state.[43] The method was first applied to lysozyme, where values of 1.11 to 1.14 for the deuterium isotope effect were taken as indication of an oxycarbonium ion intermediate.[44] Such an intermediate is too unstable to exist, however, and these data really show that the transition state for the transfer of the glycosyl unit to the aspartate on the enzyme is highly dissociative, with little axial bond order to the entering and leaving groups. All work since then suggests that glycosyltransferase reactions involve such dissociative transition states.

Other isotope effects have also been used to help determine transition state structures for these reactions. Primary ^{18}O isotope effects in the leaving group have ranged from 1.025 for acid catalysis reactions where the leaving group is protonated to 1.047 where it is not.[45,46] The primary ^{13}C isotope effects at the glycosyl carbon are only about 1% in acid hydrolysis of methyl glucosides[45] but are about 2% in the hydrolyses of AMP studied by Schramm and co-workers (these were determined as ^{14}C isotope effects, which are 1.9 times as large as ^{13}C ones).[43] For this C–N cleavage, the ^{15}N isotope effect in the base was approximately 3%.

In these reactions the β-secondary deuterium isotope effect at C-2 is normal as the result of hyperconjugation with the oxycarbonium ion-like transition state structure, although the exact value depends on the ring pucker and thus the angle between the C–H bond at C-2 and the C–O bond in the ring. The ring ^{18}O isotope effects were inverse (0.991–0.996) in the one case where they were measured (acid hydrolysis of α- and β-methylglucoside), as expected from the dissociative transition state and partial double bond character between C-1 and the ring oxygen.[45]

By a comparison of measured isotope effects and calculated ones, Schramm has determined transition state structures for several hydrolytic reactions of AMP. The transition states generally have 10–20% bond order to the leaving group (adenine) and 3–4% bond order to the attacking nucleophile (water). The C-1 to ring oxygen bond has a bond order of approximately 1.8.[43]

Conclusion

Methods are now in hand for measuring both deuterium (or tritium) and heavy atom isotope effects on enzymatic reactions and then determining

[43] V. L. Schramm, *in* "Enzyme Mechanism from Isotope Effects" (P. F. Cook, ed.), p. 367. CRC Press, Boca Raton, Florida, 1991.

[44] F. W. Dahlquist, T. Rand-Meir, and M. A. Raftery, *Biochemistry* **8,** 4214 (1969): L. E. H. Smith, L. H. Mohr, and M. A. Raftery, *J. Am. Chem. Soc.* **95,** 7497 (1973).

[45] A. J. Bennet and M. L. Sinnott, *J. Am. Chem. Soc.* **108,** 7287 (1986).

[46] S. Rosenberg and J. F. Kirsch, *Biochemistry* **20,** 3196 (1981).

intrinsic isotope effects. With intrinsic isotope effects in hand for all atoms that change their bonding during the reaction, it is then possible to deduce reasonable transition states for the reactions. It should be possible, in fact, to determine transition state structures more accurately for enzymatic reactions than for nonenzymatic ones, since the geometry is under precise control, the reactions are fast and normally without side reactions, and one does not have to worry about the structure of or interactions with the solvent. With X-ray determination of enzyme structures becoming more efficient all the time, the geometry of the active site is readily determined, and thus the transition state structure can be modeled with some confidence. The next decade should see considerable expansion in our knowledge of transition state structures for enzymatic reactions.

Computer Programs in FORTRAN

The appendix at the end of this chapter presents the computer programs for ISOCALC4 and ISOCALC5.

ISOCALC4 Program

The ISOCALC4 program calculates a range of intrinsic deuterium and ^{13}C isotope effects from experimental values of deuterium (on V/K) and ^{13}C isotope effects (the latter both with deuterated and unlabeled substrates) on the same step of a reaction. If a tritium isotope effect is available, it is used to attempt an exact solution for the intrinsic isotope effects.

Data input is as follows. The first line, columns 1–5, gives the number of r values (we have used 10), I format. The rest of the line has the chosen r values in I5 fields. The first must be 0, and we have used 0.1, 0.3, 0.5, 0.75, 1.0, 1.5, 2.0, 3.0, and 10.0 for the rest. Here, r is the ratio of c_r/c_f. The second line contains, in F10.5 fields, $^{13}K_{eq}$, $^{13}(V/K)_H$, $^{13}(V/K)_D$, $^{D}K_{eq}$, $^{D}(V/K)$, and $^{T}(V/K)$. If $^{T}(V/K)$ is not known, leave the field blank. As many sets of data as desired may be placed on subsequent lines; a line with a blank second field stops the program.

Data output consists of the following: (1) The input data are listed. (2) A table of c_r/c_f, c_f, c_r, ^{D}k, ^{13}k, and calculated $^{T}(V/K)$ values is printed for each r value chosen. (3) When an experimental value of $^{T}(V/K)$ was supplied, it is printed out for comparison, and an exact solution is attempted. (4) Six lines of iteration should show convergence. The columns tabulate r, calculated $^{T}(V/K)$, ^{D}k, c_f, c_r, and the derivatives with respect to r of calculated $^{T}(V/K)$, c_r, c_f, ^{D}k, and ^{13}k. If the experimental value of $^{T}(V/K)$ lies outside the range of calculated $^{T}(V/K)$ values, "SOLUTION IMPOSSIBLE" is printed instead of the iterations. (5) Printout of final values of ${}^{D}k$, ${}^{13}k$, c_f, c_r, their derivatives with respect to calculated ${}^{T}(V/K)$, and the coefficients of variation of $({}^{D}k - 1)$, $({}^{13}k - 1)$, c_f and c_r divided by the coefficient of variation of $[{}^{T}(V/K) - 1]$. The latter values indicate the degree of confidence one can have in the fitted answers, with values above 5 indicating great uncertainty. Thus if ${}^{T}(V/K) - 1$ can only be determined to 5% accuracy, a parameter with a ratio of 5 is known only to 25% accuracy.

ISOCALC5 Program

The ISOCALC5 program calculates intrinsic isotope effects and standard errors from experimental values of primary deuterium, secondary deuterium, and ^{13}C isotope effects (the latter determined with primary or secondary deuterated or unlabeled substrates) on the same step of a reaction. All isotope effects must be ones on V/K. One also needs values for the equilibrium primary and secondary deuterium and ^{13}C isotope effects. Provision is made for lack of full deuteration of the substrate in the secondary position. If the fractional labeling is less than 1.00, change the statement on line 2 that says FI = 1. to give the actual degree of labeling.

Data input is as follows. The first line is the same as that for ISOCALC4. The second line, in F10.5 fields, contains ${}^{13}K_{eq}$, ${}^{13}(V/K)_{H}$, ${}^{13}(V/K)_{\text{pri-D}}$, ${}^{\text{pri-D}}K_{eq}$, ${}^{\text{pri-D}}(V/K)$, ${}^{\text{sec-D}}K_{eq}$, ${}^{\text{sec-D}}(V/K)$, and ${}^{13}(V/K)_{\text{sec-D}}$. The third line gives, in F10.5 fields, standard errors of the corresponding parameters on the second line. As many sets of second and third lines as desired may be included; each data set is processed separately. A blank second field on a line that should have second line data stops the program.

Data output consists of one page of iteration similar to that given by ISOCALC4, followed by a second page with two tables. The first gives fractional change in output values for a 1% change in each input parameter (parameter − 1 for ^{13}C isotope effects, and likewise ${}^{13}k - 1$ for the output parameter). The standard errors of the input parameters are then used to compute the second table, which shows the actual fractional changes in answers caused by the actual errors in input values. Net coefficient of variation (assuming all input errors are independent), standard error, and the values are then printed. The standard error and value are for ${}^{13}k$, and not ${}^{13}k - 1$. Note that the sum of c_f and c_r is generally well determined, but that the ratio is not.

Appendix I: Computer Code for ISOCALC4

```
c      Program ISOCALC4
       DIMENSION R(15),CW(16)
     1 FORMAT(I5,15F5.2)
     2 FORMAT(F10.2,3F10.3,F10.4,F10.3)
     3 FORMAT(10H  INFINITE3F10.3,F10.4,F10.3 /)
     4 FORMAT(34X,16H EXPTL T(V/K) =  F10.3 /  )
     5 FORMAT(9H 13KEQ = F7.4,14H   13(V/K)H = F7.4,14H   13(V/K)D = F7.4
      1,10H   DKEQ = F6.3,12H   D(V/K) = F7.3/)
     6 FORMAT(62H     CR/CF        CF         CR         DK       13K    CALC
      1T(V/K)  /)
     7 FORMAT(1H1)
       write(6,7)
       READ(5,1) NR, R
    14 READ(5,2) C13KEQ, Y, Z, DKEQ, X, W
       M = 1
       IF(Y) 20,20,21
    21 IF(W) 22,22,23
    22 M = 2
    23 write(6,5) C13KEQ, Y, Z, DKEQ, X
       write(6,6)
       DO 10 I = 1,NR
       XP = X*(1.+R(I))-1.-DKEQ*R(I)
       YP = Y*(1.+R(I))-1.-C13KEQ*R(I)
       ZP = Z*(1.+DKEQ*R(I))-1.-DKEQ*C13KEQ*R(I)
       B = X*YP+(Z-Y)*XP + ZP
       T = B**2 - 4.*YP*X*ZP
       IF(T) 40,41,41
    40 write(6,43)
    43 FORMAT(18H B**2-4AC NEGATIVE  )
       C = (1.+R(I)*DKEQ)/(1.+R(I))
       D = (1.+R(I)*C13KEQ)/(1.+R(I))
       DK = (Z-D)*X/(Y-D)
       C13K = (X*Z - Y*C)/(X-C)
       CF = (DK-X)/((1.+R(I))*X - 1. - R(I)*DKEQ)
       GO TO 42
    41 DK = (B+SQRT(T))/2./YP
       C13K = Y+(DK-X)*YP/XP
       CF = (DK-X)/XP
    42 CR = R(I)*CF
       IF(DK)24,25,25
    24 CW(I) = 0
       GO TO 26
    25 CW(I)=(DK**1.442+ CF + CR*DKEQ**1.442)/(1.+CF+CR)
    26 write(6,2) R(I), CF, CR, DK, C13K, CW(I)
    10 CONTINUE
       C = DKEQ
       D = C13KEQ
       DK = (Z-D)*X/(Y-D)
       C13K = (X*Z - Y*C)/(X-C)
       CF = 0
       CR = (DK-X)/(X-DKEQ)
       IF(DK)27,28,28
    27 CW(NR+1) = 0
       GO TO 29
    28 CW(NR+1) = (DK**1.442 + CR*DKEQ**1.442)/(1.+CR)
    29 write(6,3) CF, CR, DK, C13K, CW(NR+1)
       GO TO (12,14),M
    12 write(6,4) W
       NR1 = NR + 1
       DO 62 I = 1, NR1
       IF(CW(I))62,62,67
    67 IF(W-CW(I))62,60,60
    62 CONTINUE
       write(6,69)
    69 FORMAT(20H SOLUTION IMPOSSIBLE  ///)
       GO TO 14
```

```
60 IF(I-1)64,64,65
64 DO 63 I = 1,NR1
68 IF(W-CW(I))65,65,63
63 CONTINUE
   write(6,69)
   GO TO 14
65 write(6,80)
80 FORMAT(100H        R        CW         DK         CF         CR       D
  1W/DR    DCR/DR    DCF/DR  D(DK)/DR  D(13K)/DR       )
   YC = Y - C13KEQ
   XD = X - DKEQ
   ZC = Z - C13KEQ
   PBR = X*YC + (Z-Y)*XD + DKEQ*ZC
   RFIN = R(I)
   DO 70 J = 1,6
   XP = X*(1.+RFIN)-1.-DKEQ*RFIN
   YP = Y*(1.+RFIN)-1.-C13KEQ*RFIN
   ZP = Z*(1.+DKEQ*RFIN)-1.-DKEQ*C13KEQ*RFIN
   B = X*YP+(Z-Y)*XP + ZP
   T = B**2 - 4.*YP*X*ZP
   DK = (B+SQRT(T))/2./YP
   CF = (DK-X)/XP
   CR = RFIN*CF
   CW(I)=(DK**1.442+ CF + CR*DKEQ**1.442)/(1.+CF+CR)
   PDKR=(PBR-DK*2.*YC)/2./YP+ (B*PBR-2.*X*(YP*DKEQ*ZC+ZP*YC))/2./YP
  1 /SQRT(T)
   P13K = ((DK-X)*(XP*YC-YP*XD)+PDKR*XP*YP)/XP**2
   PCF = (PDKR- CF*XD    )/XP
   PCR = RFIN*PCF + CF
   PNUM = 1.442*DK**.442*PDKR + PCF + DKEQ**1.442*PCR
   D = 1. + CF + CR
   PW = PNUM/D-(DK**1.442+CF+DKEQ**1.442*CR)*(PCF+PCR)/D**2
71 FORMAT(12F10.5)
   write(6,71)RFIN, CW(I), DK, CF, CR ,PW,PCR,PCF,PDKR,P13K
70 RFIN = RFIN + (W-CW(I))/PW
   C13K = Y+(DK-X)*YP/XP
   PDKW = PDKR/PW
   P13KW = P13K/PW
   PCFW = PCF/PW
   PCRW = PCR/PW
   CVDKW = PDKW*(W-1.)/(DK-1.)
   CV13KW = P13KW*(W-1.)/(C13K-1.)
   CVCFW = PCFW*(W-1.)/(CF)
   CVCRW = PCRW*(W-1.)/(CR)
   write(6,72) DK ,PDKW,CVDKW
   write(6,73) C13K   ,P13KW,CV13KW
   write(6,74) CF   ,PCFW,CVCFW
   write(6,75) CR,PCRW,CVCRW
   GO TO 14
72 FORMAT(/8H   DK = F7.3,16H     D(DK)/DW = F10.5,23H    CV(DK-1)/CV
  1(W-1) = F10.5)
73 FORMAT(8H C13K = F8.4,15H   D(13K)/DW = F10.5,23H   CV(13K-1)/CV(W
  1-1) = F10.5)
74 FORMAT(8H   CF = F7.3,16H     D(CF)/DW = F10.5,23H      CV(CF)/CV(
  1W-1) = F10.5)
75 FORMAT(8H   CR = F7.3,16H     D(CR)/DW = F10.5,23H      CV(CR)/CV(
  1W-1) = F10.5///)
20 STOP
   END
```

Appendix II: Computer Code for ISOCALC5

```
c     Program ISOCALC5
      DIMENSION R(15),CW(16),CU(16),Q(8,7),QM(8,7),CV(7)
      FI = 1.
    1 FORMAT(I5,15F5.2)
    2 FORMAT(F10.2,3F10.3,F10.4,F10.3,F10.4,F10.4)
    3 FORMAT(10H  INFINITE3F10.3,F10.4,F10.3,F10.4,F10.4/)
    4 FORMAT(39X,'  EXPTL 13(V/K)A-D = ',F10.4/)
    5 FORMAT(9H 13KEQ = F7.4,14H    13(V/K)H = F7.4,14H     13(V/K)D = F7.4
     1,10H    DKEQ = F6.3,12H    D(V/K) = F7.3 )
    6 FORMAT(  '       CR/CF        CF          CR          DK        13K        A
     1-DK    CALC 13(V/K)A-D   CALC T(V/K)'/)
    7 FORMAT(1H1)
   21 FORMAT(' ALPHA-D KEQ = ',F6.3,'    A-D(V/K) = ',F6.3,'     13(V/K)A-D
     1 = ',F7.4/)
   43 FORMAT(18H B**2-4AC NEGATIVE   )
   69 FORMAT(20H SOLUTION IMPOSSIBLE   ///)
   80 FORMAT(120H          R         CU          DK          CF          CR         D
     1U/DR      DCR/DR      DCF/DR     D(DK)/DR   D(13K)/DR   D(AK)/DR CF+CR/DR)
   71 FORMAT(12F10.5)
   72 FORMAT(/8H    DK = F7.3,16H        D(DK)/DU = F10.5,23H       CV(DK-1)/CV
     1(U-1) = F10.5)
   76 FORMAT( 8H    AK = F7.3,16H        D(AK)/DU = F10.5,23H       CV(AK-1)/CV
     1(U-1) = F10.5)
   73 FORMAT(8H C13K = F8.4,15H     D(13K)/DU = F10.5,23H    CV(13K-1)/CV(U
     1-1) = F10.5)
   74 FORMAT(8H    CF = F7.3,16H       D(CF)/DU = F10.5,23H          CV(CF)/CV(
     1U-1) = F10.5)
   75 FORMAT(8H    CR = F7.3,16H       D(CR)/DU = F10.5,23H          CV(CR)/CV(
     1U-1) = F10.5///)
   46 FORMAT('1FRACTIONAL CHANGE IN ANSWERS FOR A 1 % CHANGE IN EACH INP
     1UT PARAMETER'/)
   48 FORMAT('                          DK          CF          CR          AK
     1       13K-1      CF+CR      CF/CR'/)
   49 FORMAT(' 13KEQ - 1      ',7F10.5)
   50 FORMAT(' 13(V/K)H-1     ',7F10.5)
   51 FORMAT(' 13(V/K)D-1     ',7F10.5)
   52 FORMAT(' DKEQ           ',7F10.5)
   53 FORMAT(' D(V/K)         ',7F10.5)
   54 FORMAT(' AKEQ           ',7F10.5)
   55 FORMAT(' A(V/K)         ',7F10.5)
   56 FORMAT(' 13(V/K)A-1     ',7F10.5)
   57 FORMAT(//' FRACTIONAL CHANGE IN ANSWERS RESULTING FROM ACTUAL STAN
     1DARD ERRORS OF INPUT PARAMETERS'/)
   66 FORMAT(//' NET CV          ',7F10.5/)
   77 FORMAT(' S. E.          '7F10.5/)
   78 FORMAT(' VALUE          '7F10.5)
   81 FORMAT(' STANDARD ERRORS OF  ABOVE DATA')
   90 FORMAT('                          DK          CF          CR          AK
     1       13K       CF+CR      CF/CR'/)
      READ(5,1) NR, R
   14 write(6,7)
      READ(5,2) C13KEQ, Y, Z, DKEQ, X, AKEQ, V, U
      IF(Y) 20,20,23
   23 write(6,5) C13KEQ, Y, Z, DKEQ, X
      write(6,21) AKEQ, V, U
      READ(5,2) SE13KQ,SEY,SEZ,SEDKQ,SEX,SEAKQ,SEV,SEU
      write(6,81)
      write(6,5) SE13KQ, SEY, SEZ, SEDKQ, SEX
      write(6,21) SEAKQ, SEV, SEU
      CV13KQ  = SE13KQ/(C13KEQ-1.)
      CVY0 = SEY/(Y-1.)
      CVZ0 = SEZ/(Z-1.)
      CVDKQ = SEDKQ/DKEQ
      CVX = SEX/X
      CVAKQ = SEAKQ/AKEQ
      CVV = SEV/V
```

```
      CVU = SEU/(U-1.)
      M = 1
      write(6,6)
      DO 10 I = 1,NR
      XP = X*(1.+R(I))-1.-DKEQ*R(I)
      YP = Y*(1.+R(I))-1.-C13KEQ*R(I)
      ZP = Z*(1.+DKEQ*R(I))-1.-DKEQ*C13KEQ*R(I)
      B = X*YP+(Z-Y)*XP + ZP
      T = B**2 - 4.*YP*X*ZP
      IF(T) 40,41,41
   40 write(6,43)
      C = (1.+R(I)*DKEQ)/(1.+R(I))
      D = (1.+R(I)*C13KEQ)/(1.+R(I))
      DK = (Z-D)*X/(Y-D)
      C13K = (X*Z - Y*C)/(X-C)
      CF = (DK-X)/((1.+R(I))*X - 1. - R(I)*DKEQ)
      GO TO 42
   41 DK = (B+SQRT(T))/2./YP
      C13K = Y+(DK-X)*YP/XP
      CF = (DK-X)/XP
   42 CR = R(I)*CF
      VP = V*(1.+R(I))-1.-R(I)*AKEQ
      AK = V + CF*VP
      PAKF = (AK-1.)*FI + 1.
      PAKR = (AK/AKEQ-1.)*FI + 1.
      IF(DK)24,25,25
   24 CW(I) = 0
      GO TO 22
   25 CW(I)=(DK**1.442+ CF + CR*DKEQ**1.442)/(1.+CF+CR)
   22 IF(C13K)19,19,18
   19 CU(I) = 0
      GO TO 26
   18 CU(I)=(C13K+CF/PAKF+CR*C13KEQ/PAKR)/(1.+CF/PAKF+CR/PAKR)
   26 write(6,2) R(I), CF, CR, DK, C13K, AK, CU(I), CW(I)
   10 CONTINUE
      C = DKEQ
      D = C13KEQ
      DK = (Z-D)*X/(Y-D)
      C13K = (X*Z - Y*C)/(X-C)
      CF = 0
      CR = (DK-X)/(X-DKEQ)
      AK = V*(1.+CR)-CR*AKEQ
      PAKF = (AK-1.)*FI + 1.
      PAKR = (AK/AKEQ-1.)*FI + 1.
      IF(DK)27,28,28
   27 CW(NR+1) = 0
      GO TO 30
   28 CW(NR+1) = (DK**1.442 + CR*DKEQ**1.442)/(1.+CR)
   30 IF(C13K)31,31,32
   31 CU(NR+1) = 0
      GO TO 29
   32 CU(NR+1)=(C13K+CR*C13KEQ/PAKR)/(1.+CR/PAKR)
   29 write(6,3) CF, CR, DK, C13K, AK, CU(NR+1), CW(NR+1)
      write(6,4) U
      NR1 = NR + 1
      DO 62 I = 1, NR1
      IF(CU(I))62,62,67
   67 IF(U-CU(I))62,60,60
   62 CONTINUE
      write(6,69)
      GO TO 14
   60 IF(I-1)64,64,65
   64 DO 63 I = 1,NR1
   68 IF(U-CU(I))65,65,63
   63 CONTINUE
      write(6,69)
```

```
      GO TO 14
   65 write(6,80)
      YC = Y - C13KEQ
      XD = X - DKEQ
      ZC = Z - C13KEQ
      PBR = X*YC + (Z-Y)*XD + DKEQ*ZC
      RFIN = R(I)
   47 DO 70 J = 1,6
      XP = X*(1.+RFIN)-1.-DKEQ*RFIN
      YP = Y*(1.+RFIN)-1.-C13KEQ*RFIN
      ZP = Z*(1.+DKEQ*RFIN)-1.-DKEQ*C13KEQ*RFIN
      B = X*YP+(Z-Y)*XP + ZP
      T = B**2 - 4.*YP*X*ZP
      DK = (B+SQRT(T))/2./YP
      CF = (DK-X)/XP
      CR = RFIN*CF
      CFPCR = CF+CR
      CFCR = CF/CR
      VP = V*(1.+RFIN)-1.-RFIN*AKEQ
      AK = V + CF*VP
      C13K = Y+(DK-X)*YP/XP
      PAKF = (AK-1.)*FI + 1.
      PAKR = (AK/AKEQ-1.)*FI + 1.
      CU(I)=(C13K+CF/PAKF+CR*C13KEQ/PAKR)/(1.+CF/PAKF+CR/PAKR)
      PDKR=(PBR-DK*2.*YC)/2./YP+ (B*PBR-2.*X*(YP*DKEQ*ZC+ZP*YC))/2./YP
     1 /SQRT(T)
      P13K = ((DK-X)*(XP*YC-YP*XD)+PDKR*XP*YP)/XP**2
      PCF = (PDKR- CF*XD    )/XP
      PCR = RFIN*PCF + CF
      PCFCR = PCF + PCR
      PAK = (V-1.)*PCF + (V-AKEQ)*PCR
      PP1 = (AK*PCF-CF*PAK)/AK**2
      PP2 = AKEQ*(AK*PCR-CR*PAK)/AK**2
      PNUM = P13K + PP1 + C13KEQ*PP2
      PD = PP1 + PP2
      D = 1. + CF/AK + CR*AKEQ/AK
      PU = PNUM/D-(C13K+CF/AK+C13KEQ*AKEQ*CR/AK)*PD/D**2
      IF(M-1)44,44,70
   44 write(6,71)RFIN, CU(I), DK, CF, CR ,PU,PCR,PCF,PDKR,P13K,PAK,PCFCR
   70 RFIN = RFIN + (U-CU(I))/PU
      IF(M-1)45,45,79
   45 PDKU = PDKR/PU
      P13KU = P13K/PU
      PCFU = PCF/PU
      PCRU = PCR/PU
      PAKU = PAK/PU
      CVDKU = PDKU*(U-1.)/(DK-1.)
      CV13KU = P13KU*(U-1.)/(C13K-1.)
      CVCFU = PCFU*(U-1.)/(CF)
      CVCRU = PCRU*(U-1.)/(CR)
      CVAKU = PAKU*(U-1.)/(AK-1.)
      write(6,72) DK ,PDKU,CVDKU
      write(6,73) C13K   ,P13KU,CV13KU
      write(6,76) AK, PAKU, CVAKU
      write(6,74) CF   ,PCFU,CVCFU
      write(6,75) CR,PCRU,CVCRU
      M = 2
      RDK = DK
      RCF = CF
      RCR = CR
      RAK = AK
      R13K = C13K
      RCFPCR = CF+CR
      RCFCR = CF/CR
      R13KQ = C13KEQ
      C13KEQ = C13KEQ*1.05-.05
```

```
      GO TO 47
   79 Q(M-1,1)=(DK-RDK)/RDK
      Q(M-1,2)=(CF-RCF)/RCF
      Q(M-1,3)=(CR-RCR)/RCR
      Q(M-1,4)=(AK-RAK)/RAK
      Q(M-1,5)=(C13K-R13K)/(R13K-1.)
      Q(M-1,6)=(CFPCR-RCFPCR)/RCFPCR
      Q(M-1,7)=(CFCR-RCFCR)/RCFCR
      GO TO(45,8,9,11,12,13,15,16,17),M
    8 DO 33 I = 1,7
   33 QM(1,I) = Q(1,I)*CV13KQ /.05
      C13KEQ = R13KQ
      RY = Y
      Y = Y*1.05 - .05
      M = 3
      GO TO 47
    9 DO 34 I = 1,7
   34 QM(2,I) = Q(2,I)*CVY0/.05
      Y = RY
      RZ = Z
      Z = Z*1.05 -.05
      M = 4
      GO TO 47
   11 DO 35 I = 1,7
   35 QM(3,I) = Q(3,I)*CVZ0/.05
      Z = RZ
      RDKQ = DKEQ
      DKEQ = DKEQ*1.01
      M = 5
      GO TO 47
   12 DO 36 I = 1,7
   36 QM(4,I) = Q(4,I) *CVDKQ/.01
      DKEQ = RDKQ
      RX = X
      X = X*1.01
      M = 6
      GO TO 47
   13 DO 37 I = 1,7
   37 QM(5,I) = Q(5,I)*CVX/.01
      X = RX
      RAKQ = AKEQ
      AKEQ = AKEQ*1.01
      M = 7
      GO TO 47
   15 DO 38 I = 1,7
   38 QM(6,I) = Q(6,I)*CVAKQ/.01
      AKEQ = RAKQ
      RV = V
      V = V*1.01
      M = 8
      GO TO 47
   16 DO 39 I = 1,7
   39 QM(7,I) = Q(7,I)*CVV/.01
      V = RV
      RU = U
      U = U*1.05-.05
      M = 9
      GO TO 47
   17 DO 61 I = 1,7
   61 QM(8,I) = Q(8,I)*CVU/.05
      U = RU
      write(6,46)
      write(6,48)
      write(6,49)(Q(1,I),I=1,7)
      write(6,50)(Q(2,I),I=1,7)
      write(6,51)(Q(3,I),I=1,7)
```

```
      write(6,56)(Q(8,I),I=1,7)
      write(6,52)(Q(4,I),I=1,7)
      write(6,53)(Q(5,I),I=1,7)
      write(6,54)(Q(6,I),I=1,7)
      write(6,55)(Q(7,I),I=1,7)
      write(6,57)
      write(6,48)
      write(6,49)(QM(1,I),I=1,7)
      write(6,50)(QM(2,I),I=1,7)
      write(6,51)(QM(3,I),I=1,7)
      write(6,56)(QM(8,I),I=1,7)
      write(6,52)(QM(4,I),I=1,7)
      write(6,53)(QM(5,I),I=1,7)
      write(6,54)(QM(6,I),I=1,7)
      write(6,55)(QM(7,I),I=1,7)
      DO 87 I = 1,7
   87 CV(I) = 0
      DO 58 I = 1,7
      DO 58 J = 1,8
   58 CV(I) = CV(I) + QM(J,I)**2
      DO 59 I = 1,7
   59 CV(I) = SQRT(CV(I))
      write(6,66) CV
      CV(1) = CV(1)*RDK
      CV(2) = CV(2)*RCF
      CV(3) = CV(3)*RCR
      CV(4) = CV(4)*RAK
      CV(5) = CV(5)*(R13K-1.)
      CV(6) = CV(6)*RCFPCR
      CV(7) = CV(7)*RCFCR
      write(6,90)
      write(6,78)RDK,RCF,RCR,RAK,R13K,RCFPCR,RCFCR
      write(6,77) CV
      GO TO 14
   20 STOP
      END
```

[14] Hydrogen Tunneling in Enzyme Catalysis

By BRIAN J. BAHNSON and JUDITH P. KLINMAN

Introduction

An essential part of our understanding of enzyme catalysis is an experimental description of the mechanism, transition state structure, and potential energy surface of the reaction. A semiclassical picture of the reaction can be described from primary and secondary deuterium kinetic isotope effects. However, the isotopes of hydrogen [protium (H), deuterium (D), and tritium (T)] have de Broglie wavelengths that are similar to the distances they must typically travel during a hydrogen transfer reaction (Table I). This property has led to the recognition that the behavior of hydrogen is poised between classical and quantum mechanical realms. To describe enzyme-catalyzed hydrogen transfers rigorously, these reactions need to be examined for quantum effects. Hydrogen tunneling has been shown

TABLE I
PARTICLE UNCERTAINTY SHOWN BY DE BROGLIE WAVELENGTHS[a]

	Particle[b]				
Parameter	e^-	H	D	T	C
Mass (amu)	0.0006	1	2	3	12
λ (Å)	27	0.63	0.45	0.36	0.18

[a] Calculated using the de Broglie equation, $\lambda = h/(2mE)^{1/2}$, where h is Planck's constant, m is the mass of the particle, and E = 20 kJ/mol.
[b] The isotopes of hydrogen are protium, deuterium, and tritium and are abbreviated H, D, and T.

to be a general feature of a variety of enzyme systems that have been examined.

Some features of a chemical reaction that affect the tunneling probability,[1] and which have been observed to be altered in enzyme-catalyzed reactions, are the degree of participation of solvent reorganization,[2] the thermodynamic relationship between substrates and products,[3,4] and the height and width of the reaction coordinate energy barrier.[5] The optimization of enzyme catalysis may entail the evolutionary implementation of these chemical strategies that increase the probability of tunneling, and thereby accelerate the reaction rate. Additionally, the exploration of variations in the extent of hydrogen tunneling offers a powerful new approach in determining how protein structure is linked to the optimization of enzyme catalysis.

Methods to Demonstrate Hydrogen Tunneling

Historically, kinetic isotope effects have been measured for a variety of organic and enzyme-catalyzed reactions, and many books and reviews

[1] R. P. Bell, "The Tunnel Effect in Chemistry." Chapman & Hall, New York, 1980.
[2] P. A. Bartlett and C. K. Marlowe, *Biochemistry* **26,** 8553 (1987).
[3] W. J. Albery and J. R. Knowles, *Biochemistry* **15,** 5631 (1976).
[4] K. P. Nambiar, D. M. Stauffer, P. A. Kolodziej, and S. A. Benner, *J. Am. Chem. Soc.* **105,** 5886 (1983).
[5] J. Rodgers, D. A. Femec, and R. L. Schowen, *J. Am. Chem. Soc.* **104,** 3263 (1982).

have been written.[6–9] In some instances, the magnitude of these isotope effects is larger than expected based on a semiclassical reaction barrier,[1] and is suggestive of hydrogen tunneling. However, a more detailed comparison of the reaction rates for H-, D-, and T-labeled substrates at a single temperature and also across the range of experimentally accessible temperatures can more rigorously probe for quantum effects. A concern that is common to each approach is whether the enzyme-catalyzed chemical step is rate limiting. If this is not the case, the evidence of tunneling may be masked.

Rule of the Geometric Mean Breakdown

Large secondary kinetic isotope effects, which exceed the corresponding equilibrium isotope effect value, are suggestive of tunneling and coupled motion.[10] For systems that have coupled motion between the transferred hydrogen and the secondary hydrogen, it may be possible to probe for tunneling by demonstrating a breakdown of the rule of the geometric mean. The latter states that isotope effects are multiplicative or, in other words, that there are no isotope effects on isotope effects.[11]

In a hypothetical reaction that has tunneling and coupled motion between the transferred hydrogen (primary position) and the secondary hydrogen, a comparison of the secondary isotope effect with either protium or deuterium in the primary position will probe for a breakdown of the rule of the geometric mean. Compared to deuterium, the protium transfer is characterized by a greater extent of tunneling. The magnitude of the secondary isotope effect is greatly influenced by the primary hydrogen tunneling, and this influence is realized through the coupled motion. A secondary isotope effect that is significantly reduced on replacement of the primary protium by deuterium is evidence for hydrogen tunneling in the reaction.

Care must be taken when exploring for a breakdown in the rule of the geometric mean in reactions where it is not clear whether the primary and secondary isotope effects arise in a concerted fashion from the same

[6] W. W. Cleland, M. H. O'Leary, and D. B. Northrop (eds.), "Isotope Effects on Enzyme-Catalyzed Reactions." University Park Press, Baltimore, Maryland, 1977.

[7] R. D. Gandour and R. L. Schowen (eds.), "Transition States of Biochemical Processes." Plenum, New York, 1978.

[8] P. F. Cook (ed.), "Enzyme Mechanism from Isotope Effects." CRC Press, Boca Raton, Florida, 1991.

[9] W. W. Cleland, this volume [13].

[10] J. P. Klinman, *in* "Enzyme Mechanism from Isotope Effects" (P. F. Cook, ed.), p. 127. CRC Press, Boca Raton, Florida, 1991.

[11] J. Bigeleisen, *J. Chem. Phys.* **23,** 2264 (1955).

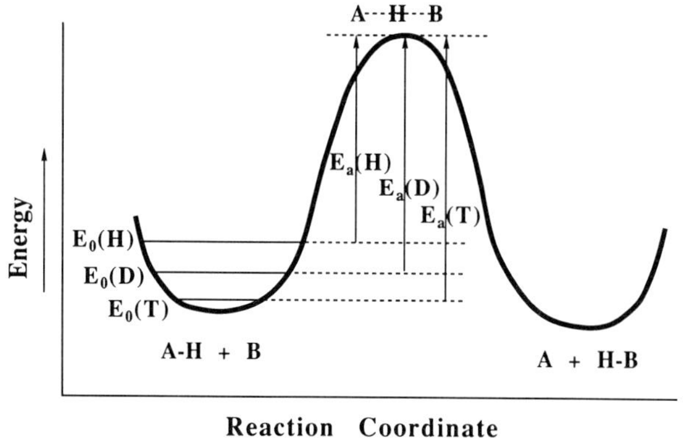

FIG. 1. Hydrogen transfer reaction coordinate diagram without any tunneling. Kinetic isotope effects arise from differences of the activation energies E_a(H), E_a(D), and E_a(T), which arise from differences of the zero point energies E_0(H), E_0(D), and E_0(T) for H-, D-, and T-labeled substrates.

isotopically sensitive step. A misleading result could occur in a two-step mechanism where the primary and secondary isotope effects result from different steps. In this case, primary deuteration could reduce the rate limitation of the secondary isotope sensitive step, thereby decreasing the value of the secondary isotope effect.

Exponential Breakdown

Semiclassically, a primary kinetic isotope effect will have its maximum value for a reaction coordinate that is linear and symmetrical, and has no restoring force in the transition state (identical transition state energies for H-, D-, and T-labeled substrates).[12] In this case the primary isotope effect arises from differences in the zero point energies of the H-, D-, or T-labeled substrates, causing the activation energies to increase in the order $E_a(H) < E_a(D) < E_a(T)$ and the rates to decrease in the order $k_H > k_D > k_T$ as shown in Fig. 1. The Swain-Schaad relationship [Eq. (1)] was derived to predict the semiclassical relationship between k_H/k_D and k_H/k_T isotope effects.[13] Another form of the Swain–Schaad relationship interrelates the k_D/k_T and k_H/k_T isotope effects [Eq. (2)]. The exponent value

$$k_H/k_T = (k_H/k_D)^{1.44} \tag{1}$$

[12] J. P. Klinman, *in* "Transition States of Biochemical Processes" (R. D. Gandour and R. L. Schowen, eds.), p. 165. Plenum, New York, 1978.

[13] C. G. Swain, E. C. Stivers, J. F. Reuwer, Jr., and L. J. Schaad, *J. Am. Chem. Soc.* **80,** 5885 (1958).

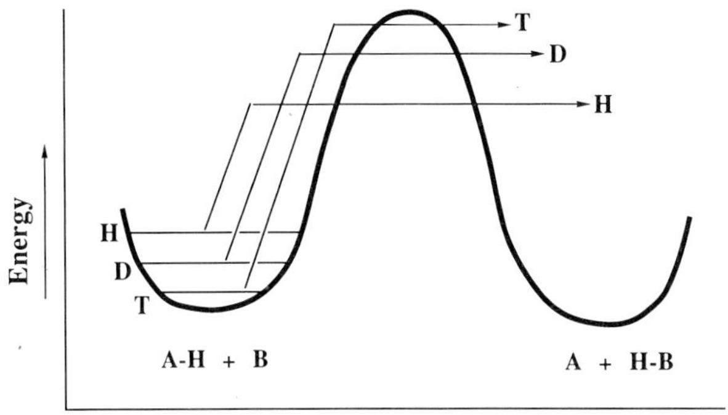

FIG. 2. Hydrogen transfer reaction coordinate diagram with a moderate degree of tunneling. Owing to the greater uncertainty of protium relative to deuterium and tritium, the protium-labeled substrate tunnels at a point lower in energy relative to the semiclassical transition state.

$$k_H/k_T = (k_D/k_T)^{3.26-3.34} \quad (2)$$

of 3.26 in Eq. (2) is derived assuming that the hydrogen atom is transferred from a donor of infinite mass. The exponent value of 3.34 is an upper limit of the semiclassical relationship when reduced mass considerations are included, and when a model is used which assumes that the hydrogen is transferred from a lone carbon atom that is not bonded to or interacting with other atoms.[14]

Based on model calculations for a reaction with a moderate amount of hydrogen tunneling, Saunders[14] proposed that a comparative study of k_D/k_T and k_H/k_T isotope effects would show a breakdown of the semiclassical relationship [Eq. (2)]. This scenario is described intuitively in Fig. 2, which is a diagram of a hypothetical potential energy barrier for a reaction with hydrogen tunneling. When the hydrogen transfer has a quantum mechanical component to it, and the hydrogen tunnels through the semiclassical potential energy barrier, it is expected that the protium-labeled substrate will tunnel below the range of energies where the deuterium- and tritium-labeled species tunnel. It is this type of differential barrier penetration that a comparative study of k_D/k_T and k_H/k_T isotope effects is able to detect, by showing an exponential breakdown signature of hydrogen tunneling according to Eqs. (3) and (4):

$$k_H/k_T > (k_D/k_T)^{3.26-3.34} \quad (3)$$

$$k_H/k_T = (k_D/k_T)^{\text{exponent}}; \qquad \text{exponent} > 3.34 \quad (4)$$

[14] W. H. Saunders, Jr., *J. Am. Chem. Soc.* **107,** 164 (1985).

As illustrated in Eq. (4), hydrogen tunneling is demonstrated when the exponent which relates the experimentally measured k_H/k_T and k_D/k_T isotope effects is larger than 3.34.

The experimental demonstration of hydrogen tunneling is complicated in many enzyme-catalyzed reactions because the chemical step is not completely rate limiting. When steps other than the chemical step contribute to the overall rate of the reaction, it is said to be kinetically complex. Kinetic complexity causes the masking of the intrinsic isotope effect, which is the isotope effect on the isotopically sensitive step. For a simple reaction, $E + S \rightleftharpoons ES \rightarrow E + P$, the influence of kinetic complexity on the magnitude of the observed primary tritium isotope effects is described by Eqs. (5) and (6):

$$k_H/k_T = [(k_H/k_T)_{int} + C_H]/(1 + C_H) \qquad (5)$$
$$k_D/k_T = [(k_D/k_T)_{int} + C_D]/(1 + C_D) \qquad (6)$$

where $(k_H/k_T)_{int}$ and $(k_D/k_T)_{int}$ are the intrinsic tritium kinetic isotope effects and C_H and C_D, the commitments to catalysis, are the ratio of rate constants for the chemical step versus dissociation of substrate from the enzyme.[15] For primary kinetic isotope effects, kinetic complexity in a reaction leads to a greater reduction in the observed k_H/k_T isotope effect relative to its intrinsic value than in the observed k_D/k_T isotope effect relative to its intrinsic value. Kinetic complexity leads to a breakdown in the Swain–Schaad relationship that is opposite to that predicted for tunneling, as shown in Eqs. (7) and (8):

$$k_H/k_T < (k_D/k_T)^{3.26-3.34} \qquad (7)$$
$$k_H/k_T = (k_D/k_T)^{\text{exponent}}; \qquad \text{exponent} < 3.26 \qquad (8)$$

For the case in which both kinetic complexity and hydrogen tunneling exist in a reaction, it is possible that a canceling of effects on the exponential breakdown takes place, such that one fortuitously observes the semiclassical relationship [Eq. (2)].

The effect of kinetic complexity on the exponential relationship between secondary k_H/k_T and k_D/k_T isotope effects can give a false indication of hydrogen tunneling for reactions with a secondary kinetic isotope effect significantly less than the corresponding equilibrium isotope effect. This necessitates the need to examine primary and secondary isotope effects simultaneously, or to demonstrate that the secondary kinetic isotope effect is greater than or equal to the equilibrium isotope effect.[16]

[15] D. B. Northrop, *in* "Isotope Effects on Enzyme-Catalyzed Reactions" (W. W. Cleland, M. H. O'Leary, and D. B. Northrop, eds.), p. 122. University Park Press, Baltimore, Maryland, 1977.

[16] K. L. Grant and J. P. Klinman, *Bioorg. Chem.* **20,** 1 (1992).

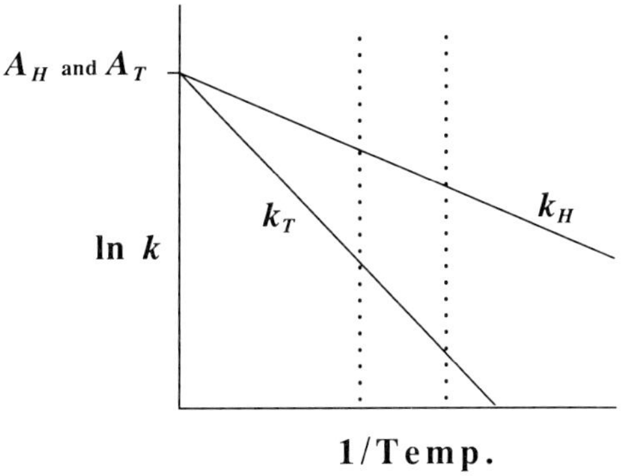

FIG. 3. Arrhenius plot of ln(k) versus $1/T$ for a hydrogen transfer reaction with no tunneling. Over the experimental temperature region (between the dotted lines) the rates for H- and T-labeled substrates extrapolate to the same Arrhenius prefactor ($A_H/A_T = 1$).

Temperature Dependence of Kinetic Isotope Effects

The measurement of k_H/k_D isotope effects across an experimentally achievable temperature range has been widely used as a method to demonstrate hydrogen tunneling in organic model reactions.[1,17,18] This traditional method has had success in demonstrating hydrogen tunneling in enzyme-catalyzed reactions by measuring the temperature dependence of k_H/k_T and k_D/k_T isotope effects.[19,20]

The Arrhenius equation is shown for an H-labeled substrate in Eq. (9):

$$\ln k_H = \ln A_H + E_A(H)/RT \qquad (9)$$

where A_H is the Arrhenius prefactor and E_a(H) is the activation energy for the protium transfer. Figure 3 shows an Arrhenius plot for a hypothetical reaction that has no hydrogen tunneling. In this case, the Arrhenius equation predicts that an Arrhenius plot of H- and T-labeled substrates will be linear across the measurable temperature range. Extrapolation to infinite temperature is expected to yield values of A_H and A_T that are approximately equal.

For a reaction in which hydrogen tunneling occurs, the Arrhenius plots of H-, D-, and T-labeled substrates are expected to exhibit a degree of

[17] D. Devault, "Quantum Mechanical Tunneling in Biological Systems," 2nd Ed. Cambridge Univ. Press, Cambridge, 1984.

[18] H. Kwart, *Acc. Chem. Res.* **15,** 401 (1982).

[19] K. L. Grant and J. P. Klinman, *Biochemistry* **28,** 6597 (1989).

[20] B. J. Bahnson, D.-H. Park, K. Kim, B. V. Plapp, and J. P. Klinman, *Biochemistry* **32,** 5503 (1993).

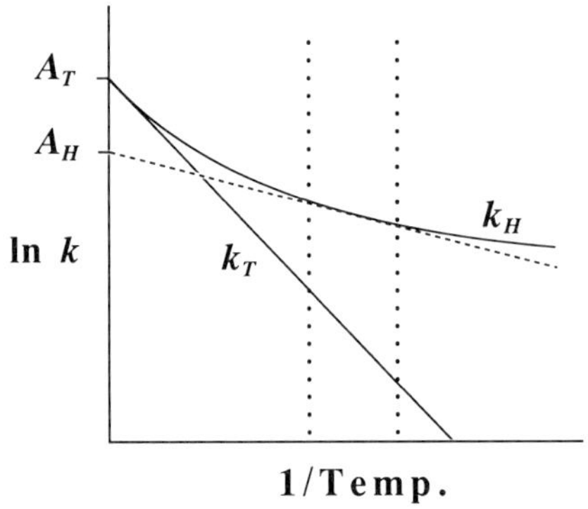

FIG. 4. Arrhenius plot of ln(k) versus $1/T$ for a hydrogen transfer reaction with significant protium tunneling. Over the experimental temperature region (between the dotted lines) the rates for H- and T-labeled substrates extrapolate to very different Arrhenius prefactors ($A_H/A_t < 1$). In this hypothetical reaction an extrapolation of the k_H and k_T rates from a lower temperature would yield $A_H/A_T \ll 1$ and from a temperature region approaching infinity, $A_H/A_T = 1$.

curvature that is related to the extent of hydrogen tunneling in the order H > D > T. Figure 4 shows an Arrhenius plot for a hypothetical reaction in which protium tunnels while tritium does not. The curvature of the Arrhenius plot for hydrogen transfer is related to its tunneling contribution; for a simple model in which the tunneling rate is temperature independent, the extent of reaction which proceeds by tunneling will rise with decreasing temperature. In this situation, a linear fit of the reaction rate for H- and T-labeled substrates over the experimental temperature range produces crossing lines, which in the limit of infinite temperature extrapolate to $A_H < A_T$.

Experimentally, Arrhenius plots of isotope effects are fit to a ratio form of the Arrhenius equation which is shown in Eq. (10) for a k_H/k_T isotope effect:

$$\ln(k_H/k_T) = \ln(A_H/A_T) + [E_a(T) - E_a(H)]/RT \qquad (10)$$

Evidence of hydrogen tunneling is obtained when isotope effects on experimental Arrhenius prefactors are below a conservatively low semiclassical lower limit of $A_H/A_T < 0.6$, $A_H/A_D < 0.7$, and $A_D/A_T < 0.9$[21] and the differences in activation energies are above a conservatively high semiclassical limit of $E_a(T) - E_a(H) > 8.3$ kJ/mol, $E_a(D) - E_a(H) > 5.8$ kJ/mol, and $E_a(T) - E_a(D) > 2.5$ kJ/mol.[1] Kim and Kreevoy[22] have argued that

[21] M. E. Schneider and M. J. Stern, *J. Am. Chem. Soc.* **94,** 1517 (1972).
[22] Y. Kim and M. M. Kreevoy, *J. Am. Chem. Soc.* **114,** 7116 (1992).

the case for tunneling is quite strong when there is significant deviation of A_H/A_D below a value of unity, combined with $E_a(D) - E_a(H) > 5.0$ kJ/mol.

The k_H/k_T and k_D/k_T isotope effects are much more informative than the k_H/k_D isotope effect when measuring isotope effects across a temperature range. The value of A_H/A_T will show the greatest deviation below the semiclassical lower limit, and it can also be measured with the greatest precision. Although k_D/k_T is the smallest among the three ratios, an Arrhenius plot of k_D/k_T isotope effects becomes an especially powerful probe in the event of extensive quantum behavior, where deuterium as well as protium is tunneling. From a mechanistic perspective, the measured value of k_D/k_T is closer to the intrinsic k_D/k_T than either k_H/k_D or k_H/k_T are to their intrinsic isotope effects. This makes the k_D/k_T isotope effect the most dependable tool to describe the transition state structure of a reaction that may have some kinetic complexity.

As discussed by Koch and Dahlberg, it is important to restrict this approach to reactions that are nearly kinetically limited in rate by the isotopically sensitive chemical step, since changes in the rate-limiting step can give artifactual curvature in Arrhenius plots.[23] This is especially true in the case of A_H/A_T measurements. As summarized in Table II, for a reaction in which kinetic complexity increases with a decrease in temperature, the evidence of H tunneling can be masked because of an inflated value of A_H/A_T. In contrast, if the kinetic complexity of the reaction decreases significantly with decreasing temperature, an artifactually low value of A_H/A_T could give a false indication of H tunneling. Finally, an Arrhenius plot of a reaction with a constant degree of kinetic complexity across the experimental temperature range will accurately represent the extent of hydrogen tunneling. The lack of a temperature-dependent commitment can, in favorable circumstances, be shown from a constant exponential relationship between k_H/k_T and k_D/k_T isotope effects across the experimental temperature range. It should be noted that A_D/A_T is expected to be consistently below a value of 0.9 when tunneling is occurring, even in the event of a modest amount of kinetic complexity in the k_H/k_T measurements. Thus, the simultaneous measurement of Arrhenius plots of k_H/k_T and k_D/k_T isotope effects provides an important control in establishing tunneling in a hydrogen transfer reaction.

As has been discussed for each experimental approach, kinetic complexity in an enzyme mechanism can mask the experimental demonstration of hydrogen tunneling. It is then necessary either to chose a system for study that already has a rate-limiting chemical step or to find conditions

[23] H. F. Koch and D. B. Dahlberg, *J. Am. Chem. Soc.* **102,** 6102 (1980).

TABLE II
EFFECT OF TUNNELING AND KINETIC COMPLEXITY ON INTERCEPTS OF ARRHENIUS PLOTS OF ISOTOPE EFFECTS

Case and effect	Diagnostic
I. No tunneling[a]	$A_H/A_T > 0.6$ and $A_D/A_T > 0.9$
II. H and D tunneling[a]	$A_H/A_T < 0.6$ and $A_D/A_T < 0.9$
III. H and D tunneling with kinetic complexity[b]	
(a) Increasing kinetic complexity with decreasing temperature	$A_H/A_T \approx A_D/A_T$ and $A_D/A_T < 0.9$
(b) Decreasing kinetic complexity with decreasing temperature	$A_H/A_T \ll A_D/A_T$ and $A_D/A_T < 0.9$
(c) Kinetic complexity constant with temperature	$A_H/A_T < 0.6$ and $A_D/A_T < 0.9$
IV. Kinetic complexity[b] and no tunneling	
(a) Increasing kinetic complexity with decreasing temperature	$A_H/A_T \approx A_D/A_T$ and $A_D/A_T > 0.9$
(b) Decreasing kinetic complexity with decreasing temperature	$A_H/A_T \ll A_D/A_T$ and $A_D/A_T > 0.9$
(c) Kinetic complexity constant with temperature	$A_H/A_T > 0.6$ and $A_D/A_T > 0.9$

[a] The semiclassical lower limit for the Arrhenius prefactor ratios in a reaction without tunneling was conservatively estimated to be $A_H/A_T = 0.6$ and $A_D/A_T = 0.9$ [M. E. Schneider and M. J. Stern, *J. Am. Chem. Soc.* **94,** 1517 (1972)]. It has been argued that any deviation below unity of A_H/A_T and A_D/A_T is indicative of tunneling [Y. Kim and M. M. Kreevoy, *J. Am. Chem. Soc.* **114,** 7116 (1992)].

[b] Assuming that kinetic complexity influences the observed k_H/k_T but not the k_D/k_T isotope effect.

that reduce the kinetic complexity. Some approaches that have been used successfully to reduce kinetic complexity include varying the reaction conditions such as pH, using alternate slow substrates, or using site-directed mutagenesis of the enzyme to increase the substrate and product dissociation rates.

Competitive Comparison of k_H, k_D, and k_T

The best approach to obtain convincing evidence of hydrogen tunneling is to measure the relative rates of H-, D-, and T-labeled substrates in a competitive fashion. Ideally an experiment could be designed in which all three labeled substrate reaction rates would be compared directly. In many systems, however, it is far more practical to compare the rates of two isotopes at a time. A commonly used technique,[24] which provides

[24] D. W. Parkin, *in* "Enzyme Mechanism from Isotope Effects" (P. F. Cook, ed.), p. 269. CRC Press, Boca Raton, Florida, 1991.

very accurate results, is to compare the rate of tritium transfer from a T-labeled substrate to the rate of a H- or D-labeled substrate, which is remotely labeled with ^{14}C. To compare the rates of H-, D-, and T-labeled species, the following combination of isotope effect experiments could be measured using the remove label approach: k_H/k_D versus k_H/k_T, k_H/k_D versus k_D/k_T, or k_D/k_T versus k_H/k_T. There has been some disagreement over which pair of isotope effects will give the most convincing evidence of tunneling.[16,25] Experimentally the k_D/k_T versus k_H/k_T pair has the following advantages. (i) For a reaction with a moderate degree of tunneling, the exponential breakdown has been predicted to be largest for the k_D/k_T versus k_H/k_T pair.[14,16] (ii) It is the only pair in which independent primary and secondary isotope effects can be measured simultaneously. The other two pairs each require the measurement of k_H/k_D isotope effect, which would involve remote labeling both the H- and D-substrate for the measurement. Separate substrates would need to be synthesized, as well as separate experiments run, to obtain primary and secondary isotope effects. (iii) As described above, the k_H/k_T and k_D/k_T isotope effects are much more informative than the k_H/k_D isotope effect when measuring isotope effects across a temperature range.

The accurate measurement of k_D/k_T isotope effects requires the synthesis of deuterated substrate which has a ^{14}C remote label. It is essential to minimize the protium contamination in the deuterated substrate, because it will artifactually inflate the value of the k_D/k_T isotope effect. The k_D/k_T isotope effect can be algebraically corrected[19,26] following the determination of H contamination in the deuterated substrate by mass spectrometry or some other means. Alternatively, for systems that have rather large isotope effects, the H-contaminated substrate will be depleted more rapidly than the deuterium-labeled substrate, and the isotope effect versus fractional conversion can be fit with an H-contamination value that yields no trend in the isotope effect.[27]

Measuring Isotope Effects: Alcohol Dehydrogenase Reaction

The following k_D/k_T versus k_H/k_T isotope effect strategy was used to study the reactions of yeast and horse liver alcohol dehydrogenase (YADH and LADH) with benzyl alcohol as a substrate.[20,26] The alcohol dehydrogenase system will be used to describe how these experiments are routinely

[25] D. B. Northrop and R. G. Duggleby, *Bioorg. Chem.* **18,** 435 (1990).

[26] Y. Cha, C. J. Murray, and J. P. Klinman, *Science* **243,** 1325 (1989).

[27] J. Rucker, Y. Cha, T. Jonsson, K. L. Grant, and J. P. Klinman, *Biochemistry* **31,** 11489 (1992).

run to achieve the accuracy needed to characterize the extent of hydrogen tunneling in an enzyme-catalyzed hydrogen transfer reaction.

The LADH-catalyzed oxidation of benzyl alcohol by NAD^+ produces benzaldehyde and NADH as shown in Fig. 5. The enzyme-catalyzed production of benzaldehyde is reversible so it was coupled with semicarbazide which forms benzaldehyde semicarbazone as a final product. The coupled formation of the semicarbazone makes the enzyme reaction irreversible at the benzaldehyde product release step. Under the conditions of the experiments, irreversibility was tested by measuring isotope effects at varying enzyme activity, while keeping the concentration of semicarbazide hydrochloride constant.

The synthesis of four labeled forms of benzyl alcohol was required in order to measure k_H/k_T and k_D/k_T isotope effects. These substrates, shown in Fig. 6, enabled the simultaneous determination of the primary and secondary kinetic isotope effects. Because the k_D/k_T isotope effects were measured using the dideuterated radiolabeled substrates, the primary and secondary k_D/k_T isotope effects were measured with deuterium in the secondary and primary positions, respectively. By contrast, the primary and secondary k_H/k_T isotope effects were measured with protium in the opposing position. This pattern of labeling is sensitive to a breakdown of the rule of the geometric mean as will be described below.

The deuterium content of the [1,1-$^2H_2$-*ring*-^{14}C(U)]benzyl alcohol was determined by mass spectrometry of the urethane derivative to be

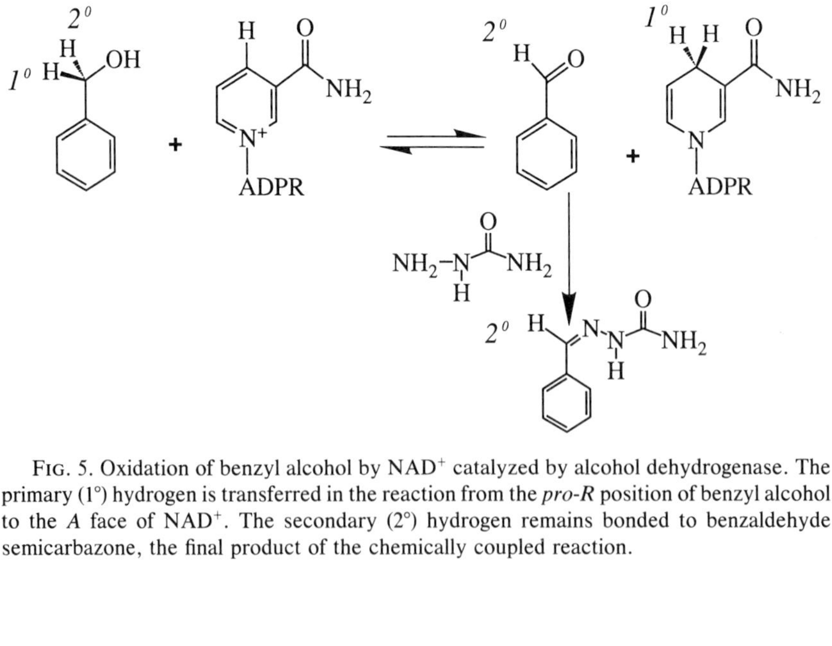

FIG. 5. Oxidation of benzyl alcohol by NAD^+ catalyzed by alcohol dehydrogenase. The primary (1°) hydrogen is transferred in the reaction from the *pro-R* position of benzyl alcohol to the *A* face of NAD^+. The secondary (2°) hydrogen remains bonded to benzaldehyde semicarbazone, the final product of the chemically coupled reaction.

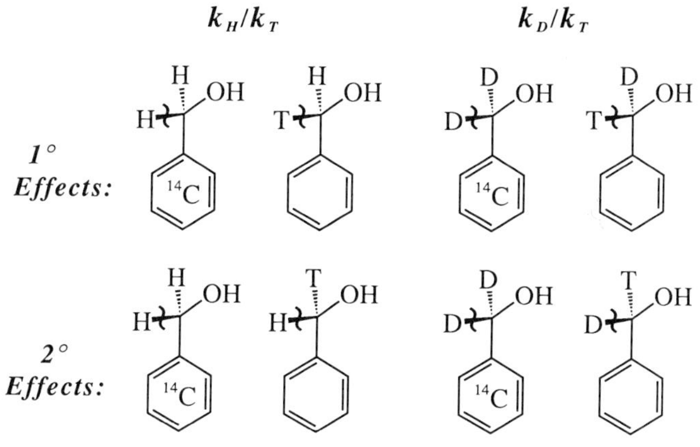

1 2 3 4

FIG. 6. The k_H/k_T isotope effects were measured for the alcohol dehydrogenase reaction with [*ring*-^{14}C(U)]benzyl alcohol (**1**) and [1-^{3}H]benzyl alcohol (**2**), which is randomly tritiated. The k_D/k_T isotope effects were measured with [1,1-$^2H_2$-*ring*-^{14}C(U)]benzyl alcohol (**3**) and [1-^{2}H,1-^{3}H]benzyl alcohol (**4**), which is randomly tritiated.

98.7 ± 0.3% at C-1. An H-contamination level of 1.3% would influence the value of the k_D/k_T isotope effect and required correction. The [1-^{2}H,1-^{3}H]benzyl alcohol was synthesized from [1-^{2}H]benzaldehyde which was deuterated to a level exceeding 99.8%, as determined by mass spectral analysis. Within experimental error, the k_D/k_T isotope effect would not be influenced by this low level of H-contamination, and no correction was necessary.

Isotope effects were measured at pH 7.0 in 300 m*M* potassium phosphate, 300 m*M* semicarbazide hydrochloride, 10 m*M* NAD$^+$, and the temperature was thermostatted to ±0.1° in a water bath. Typically, a reaction mixture of 6 ml was made, from which three aliquots (0.6 ml) were removed prior to the addition of LADH for the measurement of the ^{3}H to ^{14}C ratio in the starting benzyl alcohol. Radiolabeled benzyl alcohol was added so that a 0.5 ml high-performance liquid chromatography (HPLC) injection had approximately 5 × 10^5 dpm (disintegrations per minute) ^{3}H and 5 × 10^4 dpm ^{14}C. The amount of enzyme added was adjusted to ensure proper semicarbazide coupling. Reaction aliquots of 0.6 ml were removed at fractional conversions of 5–30%, then quenched by the addition of $HgCl_2$, and the samples were frozen overnight (−20°). The aliquots of unreacted and reacted substrates were separated by HPLC with a C_{18} reversed-phase column (20.0 × 0.46 cm) eluted isocratically at 1 ml/min with methanol/acetonitrile/water (12:12:76, v/v/v). The product

NADH, substrate benzyl alcohol, and product benzaldehyde semicarbazone eluted at 2–6, 12–15, and 24–28 mm, respectively. Fractions (1.5 ml) were collected and mixed with 12 ml of scintillation cocktail, and the $^3H/^{14}C$ ratio was estimated by counting for 2 min with a dual $^3H/^{14}C$ program by a liquid scintillation counter (LKB Model 1209 Rackbeta).

Primary isotope effects were determined from the comparison of the ratio of ^3H in product NADH to ^{14}C in product benzaldehyde semicarbazone (primary R_f) at a fractional conversion versus the ^3H to ^{14}C ratio of the initial benzyl alcohol substrate (R_0). Secondary isotope effects were determined by comparison of the ^3H to ^{14}C ratio of the product benzaldehyde semicarbazone (secondary R_f) versus the R_0. The k_H/k_T isotope effects were calculated from Eq. (11):

$$k_H/k_T = \ln(1 - f)/\ln[1 - f(R_f/0.5\,R_0)] \tag{11}$$

where f is the fractional conversion as determined by the ^{14}C in product versus substrate.[28] The ratios in unreacted substrate of ^3H in the primary or secondary position at C-1 to ^{14}C (ring label) are each equal to one-half of the R_0 of the initial benzyl alcohol substrate, since the benzyl alcohol is randomly tritiated at C-1. Controls were routinely run that compared the R_0 of benzyl alcohol to the ^3H to ^{14}C of the products, following the complete oxidation to [*ring*-^{14}C(U)]benzoic acid and NAD^{3}H with LADH and aldehyde dehydrogenase. The k_D/k_T isotope effects were corrected for 1.3% H contamination in [1,1-$^2H_2$-*ring*-^{14}C(U)]benzyl alcohol by using an integrated expression of products arising from H-contaminated substrate.[19,26]

Verification of Hydrogen Tunneling

Coupled Motion and Hydrogen Tunneling

Hydrogen tunneling was first suggested in enzyme-catalyzed reactions following the measurement of anomalously large secondary kinetic isotope effects that exceeded the equilibrium isotope effect, as shown in Table III. In the absence of tunneling and coupled motion a secondary kinetic isotope effect is expected to be intermediate between unity and the value of the equilibrium isotope effect, and its value thereby gives information

[28] R. G. Duggleby and D. B. Northrop, *Bioorg. Chem.* **17,** 177 (1989).

TABLE III
ANOMALOUSLY LARGE SECONDARY KINETIC ISOTOPE EFFECTS

Reaction	Isotope Effect		Ref.
	2° Kinetic	Equilibrium	
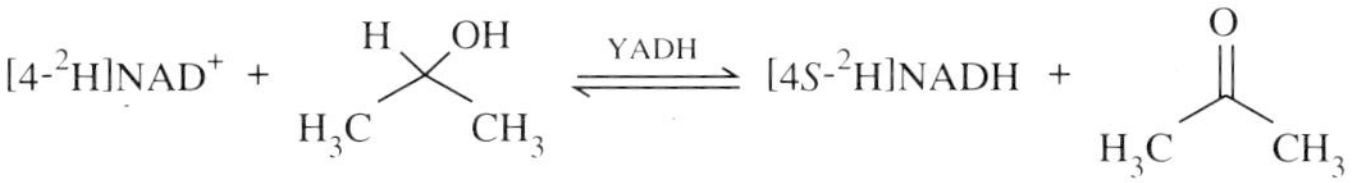	1.22	0.89	*a*
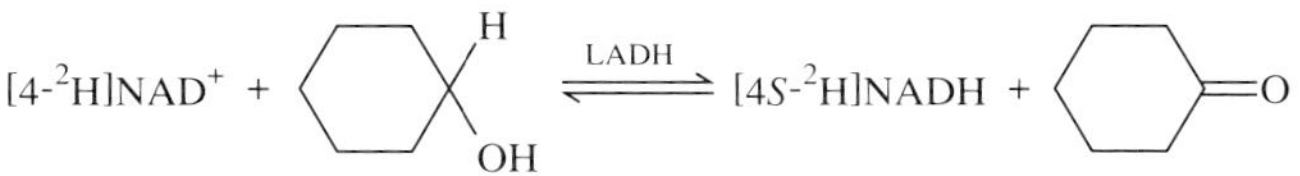	1.34	0.89	*a*
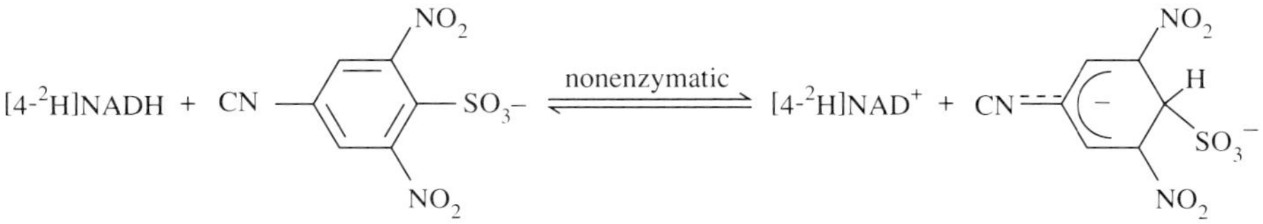	1.15	1.03	*b*

[a] P. F. Cook, N. J. Oppenheimer, and W. W. Cleland, *Biochemistry* **20,** 1817 (1981).
[b] L. C. Kurz and C. Frieden, *J. Am. Chem. Soc.* **102,** 4198 (1980).

regarding the nature of the transition state.[29–31] It was postulated by Kurz and Frieden[32] and Cook *et al.*[33] that the anomalously large secondary kinetic isotope effects were the result of a transition state coupling of the motion of the primary hydrogen, which is transferred in the reaction, to the secondary hydrogen. This is shown below for the reduction of NAD^+ by a secondary alcohol (as alkoxide):

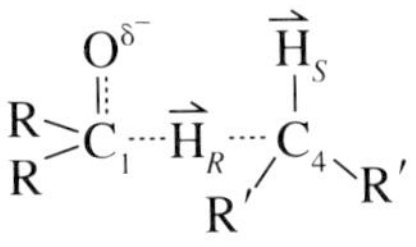

Huskey and Schowen[34] performed model vibrational-analysis calculations to explain the observed anomalies in the secondary kinetic isotope effects. It was found that merely invoking coupled motion between the primary and secondary hydrogens for the hydrogen transfers shown in Table III was not sufficient to model experimentally observed values. The combination of coupled motion and hydrogen tunneling was required in order to reproduce the experimental values.

Breakdown of Rule of Geometric Mean

The calculations of Huskey and Schowen[34] support the prediction that secondary isotope effects, which are coupled to the reaction coordinate, are dependent on the mass of the transferred particle for a reaction with hydrogen tunneling. Increasing the mass of a transferred hydrogen (H $\rightarrow$ D) is expected to reduce the tunnel correction and thereby reduce the magnitude of the coupled secondary isotope effect.

Breakdown in the rule of the geometric mean has been experimentally measured for the NAD^+-dependent formate dehydrogenase reaction,[35] in which the α-deuterium secondary kinetic isotope effect value of 1.22 for protium transfer was reduced to a value of 1.07 when deuterium was transferred in the reaction coordinate. Interestingly, by using alternate

[29] J. F. Kirsch, *in* "Sixth Steenbock Symposium on Isotope Effects on Enzyme Catalyzed Reactions" (W. W. Cleland, M. H. O'Leary, and P. B. Northrop, eds.), p. 100. University Park Press, Baltimore, Maryland, 1977.

[30] J. L. Hogg, *in* "Transition States of Biochemical Processes" (R. D. Gandour and R. L. Schowen, eds.), p. 201. Plenum, New York, 1978.

[31] J. P. Klinman, *Adv. Enzymol. Relat. Areas Mol. Biol.* **46,** 415 (1978).

[32] L. C. Kurz and C. Frieden, *J. Am. Chem. Soc.* **102,** 4198 (1980).

[33] P. F. Cook, N. J. Oppenheimer, and W. W. Cleland, *Biochemistry* **20,** 1817 (1981).

[34] W. P. Huskey and R. L. Schowen, *J. Am. Chem. Soc.* **105,** 5704 (1983).

[35] J. D. Hermes, S. W. Morrical, M. H. O'Leary, and W. W. Cleland, *Biochemistry* **23,** 5479 (1984).

NAD^+ analogs, it was found that the effect of primary deuteration on the magnitude of the secondary isotope effects decreased from a maximum with the natural NAD^+ cofactor to a smaller effect as the unnatural NAD^+ analog gave rise to an earlier transition state. These results raise an interesting speculation that one of the catalytic strategies of formate dehydrogenase is to modulate the internal thermodynamics of the enzyme-bound substrates and products in order to utilize an increased tunneling probability that will accelerate the rate.

Using a variation of the approach described above, Hermes and Cleland[36] demonstrated a breakdown in the rule of the geometric mean for the reaction of glucose-6-phosphate dehydrogenase. They reported a decrease of the hydride transfer primary isotope effect from a value of 5.3 to 3.7 on replacement of H_2O by D_2O. The replacement of deuterium on the alcoholic position of glucose 6-phosphate is believed to cause a reduction in tunneling of the O—H and C—H bond cleavages owing to an increased mass of the reaction coordinate. To come to this conclusion it was necessary to demonstrate the concerted nature of the C—H and O—H bond cleavages; this was achieved by the measurement of identical heavy atom ^{13}C isotope effects for the reaction in H_2O and D_2O.[36]

Alcohol Dehydrogenase

Early studies of the YADH-catalyzed oxidation of benzyl alcohol indicated an α-secondary kinetic isotope effect as large as the equilibrium isotope effect, suggesting that the transition state was late and resembled product.[37] This result contrasted with previous structure–reactivity correlations, which had indicated an early transition state resembling alcohol. These anomalous results, together with those of Cook *et al.*,[33] strongly suggested that the alcohol dehydrogenase hydride transfer occurs with a significant amount of hydrogen tunneling and coupled motion between secondary and primary positions. The YADH system was ideal for exploring for additional proof of hydrogen tunneling because of the lack of any kinetic complexity, as it has a rate-limiting H-transfer step with benzyl alcohol as a substrate.[38] As seen from Table IV, the measured primary and secondary exponents of 3.58 and 10.2, respectively, indicated a significant breakdown from the semiclassical upper limit, thus providing the first rigorous demonstration of room temperature hydrogen tunneling in an enzyme-catalyzed reaction.[26] The Arrhenius plots of isotope effects did

[36] J. D. Hermes and W. W. Cleland, *J. Am. Chem. Soc.* **106,** 7263 (1984).
[37] K. M. Welsh, D. J. Creighton, and J. P. Klinman, *Biochemistry* **19,** 2005 (1980).
[38] J. P. Klinman, *Biochemistry* **15,** 2018 (1976).

TABLE IV

KINETIC CONSTANTS, EXPONENTIAL RELATIONSHIP BETWEEN k_D/k_T AND k_H/k_T ISOTOPE EFFECTS, AND PRIMARY ARRHENIUS PARAMETERS FOR ENZYME-CATALYZED OXIDATION OF BENZYL ALCOHOL

Parameter	Wild-type YADH[a]	Wild-type LADH[b]	Larger alcohol pocket[b] ESE	Larger alcohol pocket[b] $Leu^{57} \rightarrow Val$	Smaller alcohol pocket[b] $Leu^{57} \rightarrow Phe$	Smaller alcohol pocket[b] $Phe^{93} \rightarrow Trp$
V_{max} (sec^{-1})	0.83[d]	0.32	1.7	0.095	0.24	0.13
V_{max}/K_m (mM^{-1} sec^{-1})[c]	0.17[d]	8.8	3.3	3.5	8.6	4.7
Primary exponent[e]	3.58	3.08	3.23	3.14	3.30	3.31
Secondary exponent[e]	10.2	4.10	3.96	4.55	8.50	6.13
Primary A_H/A_T[f]	0.91	g	0.58	g	g	0.49
Primary A_D/A_T[f]	0.84	g	0.61	g	g	0.86

[a] Y. Cha, C. J. Murray, and J. P. Klinman, *Science* **243,** 1325 (1989). Measurements at pH 8.5.

[b] B. J. Bahnson, D.-H. Park, K. Kim, B. V. Plapp, and J. P. Klinman, *Biochemistry* **32,** 5503 (1993). Measurements were performed at pH 7.0, kinetic parameters were measured at 25°.

[c] The Michaelis constant K_m is for the substrate benzyl alcohol.

[d] J. P. Klinman, *Biochemistry* **15,** 2018 (1976). Kinetic parameters were measured at 25°, pH 8.5.

[e] The exponent relating $(k_D/k_T)^{exponent} = k_H/k_T$ for primary and secondary isotope effects at 25°.

[f] The semiclassical lower limit for a reaction without tunneling is $A_H/A_T = 0.6$ and $A_D/A_T = 0.9$. M. E. Schneider and M. J. Sterne, *J. Am. Chem. Soc.* **94,** 1517 (1972).

[g] Not determined.

not provide evidence for tunneling in YADH. This is most likely due to some kinetic complexity at low temperatures that is elevating the Arrhenius prefactors and, as a result, masking tunneling.

Unlike the oxidation of benzyl alcohol catalyzed by YADH, the LADH reaction is partially limited in rate by product benzaldehyde dissociation.[39] It was then not surprising that the primary and secondary exponential relationships shown in Table IV for the wild-type form of LADH did not deviate greatly from the semiclassical range of 3.26–3.34. This result most likely reflects opposing influences of tunneling and kinetic complexity. With the goal of reducing this kinetic complexity, site-directed mutagenesis of LADH was undertaken in order to increase the dissociation rate of bound alcohol and aldehyde, which would thereby unmask tunneling. The two approaches explored were to increase or decrease the size of the alcohol binding pocket by substitutions of hydrophobic amino acids that contact the bound alcohol.

[39] V. C. Sekhar and B. V. Plapp, *Biochemistry* **29,** 4289 (1990).

Using the crystallographic structure of the productive ternary complex of *p*-bromobenzyl alcohol · NAD^+ · LADH,[40] the residues that were selected for mutagenesis are shown in Fig. 7. Substitutions that increased the size of the alcohol binding pocket had only a slight decrease of kinetic complexity and no significant change in isotope effects (data not shown)[20] or exponential relationships (Table IV). By contrast, mutations that have a smaller alcohol binding pocket provided a clear demonstration of protium tunneling from the breakdown of the exponential relationship of secondary isotope effects, which are significantly elevated relative to the semiclassical upper limit of 3.34. The value of the $A_H/A_T = 0.49$ from the temperature dependence of the primary isotope effects of the Phe[93] → Trp mutant strongly suggests protium tunneling. The unmasking of tunneling was achieved without a significant alteration in either V_{max} or V_{max}/K_m, suggesting subtle changes in rate constants for the hydride transfer step relative to the product release step.

In this case, site-directed mutagenesis was able to unmask tunneling by decreasing the kinetic complexity of the reaction. However, the Leu[57] → Phe and Phe[93] → Trp mutants each have smaller k_D/k_T isotope effects compared to the wild-type enzyme (data not shown),[20] which suggests an altered hydride transfer step. Additionally, the A_D/A_T value of the Phe[93] → Trp enzyme, which is at the semiclassical lower limit, has changed considerably from the A_D/A_T value of 0.61 for the ESE enzyme, which suggests that deuterium as well as protium is tunneling in the hydrogen transfer catalzyed by the ESE enzyme. The ESE enzyme has residue 115 deleted, which is believed to shift Leu[116] away from the bound alcohol. This enzyme form was chosen as a model for studying the temperature dependence of isotope effects for the wild-type enzyme because it has less kinetic complexity and identical intrinsic isotope effects (data not shown)[20] to the wild-type enzyme. These results suggest that in addition to unmasking H-tunneling, mutagenesis to a smaller alcohol binding pocket has altered the enzyme in such a way as to decrease the extent of H tunneling.

The pattern of labeling of substrates used to measure k_D/k_T and k_H/k_T isotope effects for the ADH reaction (Fig. 6) is expected to influence a breakdown of the rule of the geometric mean.[41] If the k_D/k_T isotope effects had been measured with protium instead of deuterium in the opposing positions, our expectation is that the exponents relating k_D/k_T and k_H/k_T isotope effects would change in the following manner. (i) The secondary

[40] H. Eklund, B. V. Plapp, J.-P. Samama, and C.-I. Branden, *J. Biol. Chem.* **257,** 14349 (1982).

[41] W. P. Huskey, *Phys. Org. Chem.* **4,** 361 (1991).

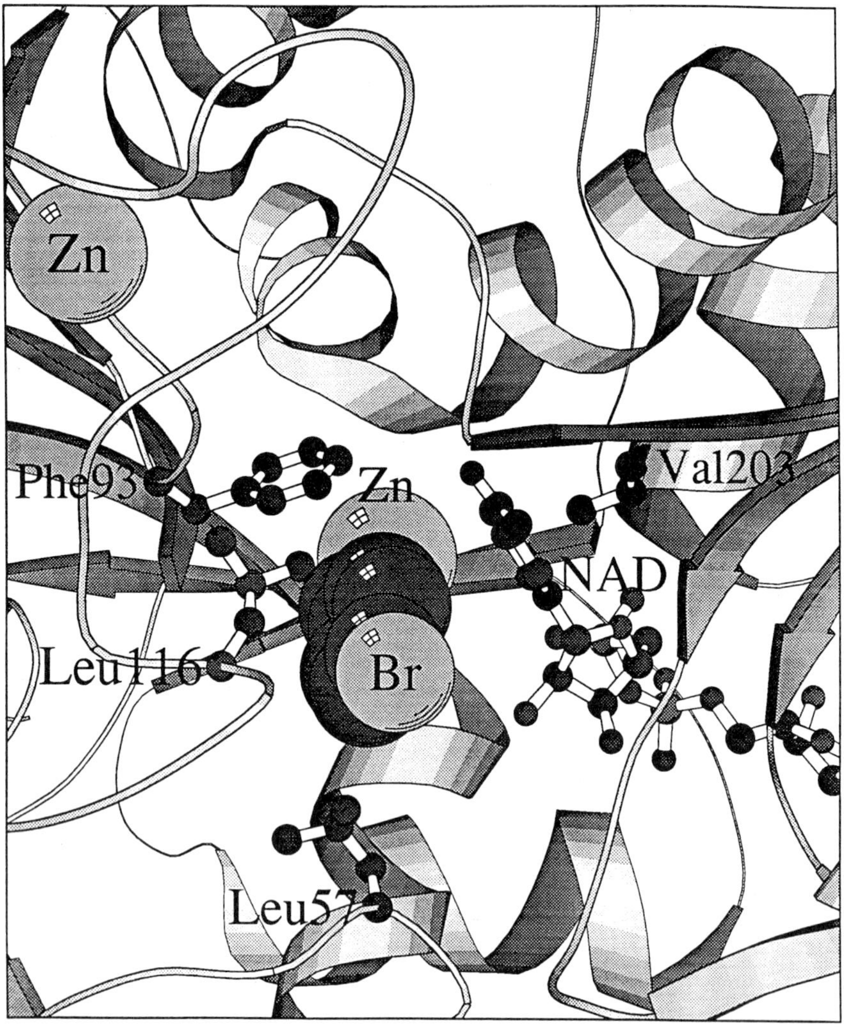

FIG. 7. Ribbon structure of LADH displaying the positions of Leu^{57}, Phe^{93}, Leu^{116}, and Val^{203} relative to the bound *p*-bromobenzyl alcohol, the cofactor NAD^+, and the structural and catalytic Zn atoms of the productive ternary complex crystallographic structure. Residue 115 is deleted in the ESE enzyme, and this change is believed to shift Leu^{116} away from the bound alcohol.

k_D/k_T isotope effect measured with protium instead of deuterium in the primary position would be larger owing to more hydrogen tunneling and mass-dependent coupled motion as has been seen with formate dehydrogenase.[35] This would give rise to a smaller secondary exponent, which is closer to the semiclassical limit. According to this prediction, the substrate labeling scheme used previously,[20,26] which is sensitive to a breakdown of the rule of the geometric mean, is optimal to detect tunneling from secondary isotope effects. (ii) The primary k_D/k_T isotope effect measured with protium instead of deuterium in the secondary position may be smaller owing to an increase in coupled motion, which decreases the magnitude of primary kinetic isotope effects.[10,34] However, if there were a significant degree of deuterium tunneling, the k_D/k_T isotope effect would be increased by substitution of protium for deuterium in the secondary position owing to a lower reduced mass for the H transfer and therefore more tunneling.[1] These compensating effects make it difficult to predict how the exponential relationship between primary k_D/k_T and k_H/k_T isotope effects will change for the alternate labeling scheme.

Huskey has predicted that the exponential breakdown indicative of tunneling [Eq. (4)] will never be experimentally observed in a reaction with hydrogen tunneling, unless there is a coupling of secondary and primary hydrogens and the k_H/k_T and k_D/k_T isotope effects are measured in such a way as to be sensitive to a breakdown in the rule of the geometric mean.[41] Several approaches are currently being explored to test the Huskey prediction. In one approach, primary and secondary isotope effects will be measured separately for ADH using stereospecifically deuterated benzyl alcohol substrates. In this way, k_D/k_T isotope effects can be measured with protium in the opposing position, as discussed above. In an alternate approach, the general use of an exponential breakdown of k_H/k_T and k_D/k_T isotope effects as a probe of tunneling is being explored in enzyme systems with substrates that lack a secondary hydrogen.

Bovine Serum Amine Oxidase

In addition to dehydrogenases, the existence of hydrogen tunneling is suggested in many enzyme systems that have large primary kinetic isotope effects. One such system is the topa-quinone and copper-containing enzyme bovine serum amine oxidase (BSAO).[19] The rate-limiting step of the BSAO-catalyzed reaction is a proton abstraction from an amine substrate to give an imine intermediate as shown in Fig. 8 for benzylamine. This isotopically sensitive chemical step was shown to be free of kinetic complexity from a comparison of isotope effects measured under steady state

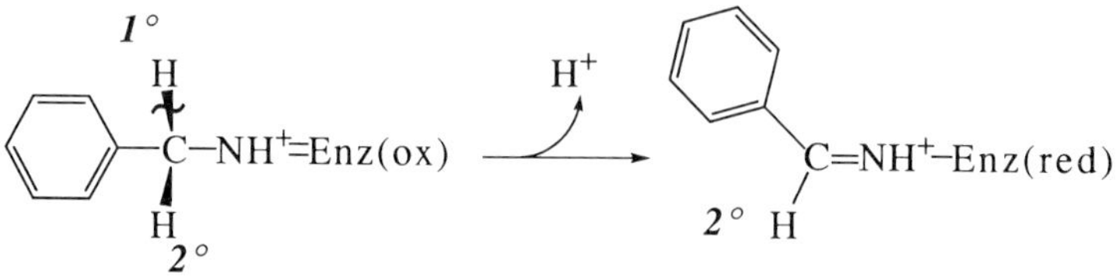

FIG. 8. Oxidation of benzylamine to an imine intermediate catalyzed by BSAO.

and stopped flow conditions,[19] thus making BSAO an ideal system to explore for hydrogen tunneling.

The k_D/k_T and k_H/k_T isotope effects and their temperature dependence for the BSAO-catalyzed oxidation of benzyl amine were measured using substrates labeled in a manner analogous to Fig. 6. The Arrhenius plot of the primary isotope effects is shown in Fig. 9. The value of A_H/A_T = 0.12 is well below the semiclassical limit, thus providing very convincing evidence that protium is tunneling. Surprisingly, the A_D/A_T value of 0.51 is also significantly below the semiclassical limit of 0.9, showing that deuterium as well as protium is tunneling in this enzyme-catalyzed hydrogen transfer.[19]

Unlike the ADH system, the exponential relationships between both primary and secondary k_D/k_T and k_H/k_T isotope effects for the BSAO reaction do not break down from the semiclassical range. In light of the evidence supporting the lack of kinetic complexity, this result appears to contrast the very convincing evidence of hydrogen tunneling from the temperature dependence of isotope effects. One possible explanation involves the observation that coupling between the primary and secondary hydrogens in the BSAO reaction appears to be minimal.[19] As discussed above, Huskey[41] has argued that the breakdown of the exponential relationship between k_D/k_T and k_H/k_T isotope effects will only occur when a reaction is characterized by a significant degree of coupled motion and breakdown in the rule of the geometric mean. An alternative explanation is that for systems with extensive tunneling where both protium and deuterium tunnel, the predictions made by Saunders[14] do not hold. This idea is supported in a theoretical paper by Grant and Klinman, in which they used the expanded Bell correction for tunneling to show that the exponent relating k_D/k_T and k_H/k_T isotope effects is a function of the reaction coordinate frequency, reaching a maximum value at 1000 i cm^{-1} and then declining to a value approximating the semiclassical exponent value of 3.26.[16]

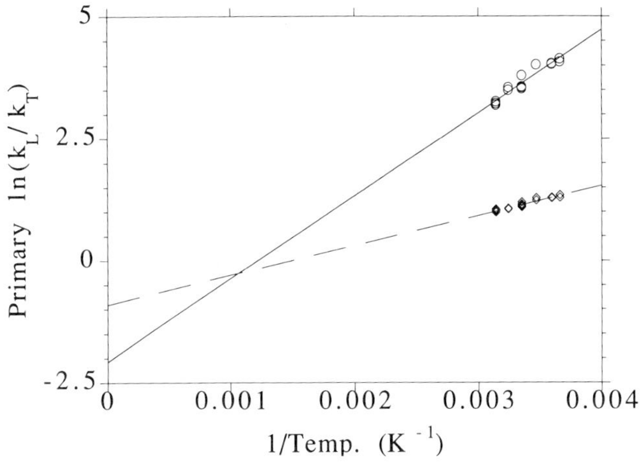

FIG. 9. Arrhenius plot for the temperature dependence of primary k_H/k_T (○—○) and k_D/k_T (◇--◇) isotope effects for the reactions catalyzed by BSAO over the temperature range of 0°–45°. The points are fit to the Arrhenius equation [Eq. (10)] to obtain the Arrhenius parameters A_H/A_T and A_D/A_T from the intercept at infinite temperature and the values of $E_a(T) - E_a(H)$ and $E_a(T) - E_a(D)$ from the slope. [Data from K. L. Grant and J. P. Klinman, *Biochemistry* **28,** 6597 (1989).]

Monoamine Oxidase B

Strong evidence of hydrogen tunneling has been determined with the flavoprotein monoamine oxidase B from the temperature dependence of k_H/k_T and k_D/k_T isotope effects.[42] This extends the observation of hydrogen tunneling to an enzyme-catalyzed reaction that utilizes the redox cofactor flavin. Unlike the ADH hydride transfer or the BSAO proton abstraction, experiments in the monoamine oxidase-catalyzed reaction support the existence of radical intermediates,[43] thus extending the demonstration of tunneling into a system that may involve hydrogen atom abstraction.[44]

Factors Influencing Hydrogen Tunneling in Enzyme Catalysis

Despite the difficulties associated with kinetic complexity, hydrogen tunneling has been shown to be a general feature of enzyme catalysis. Along with the demonstration of hydrogen tunneling, some results ob-

[42] T. Jonsson, D. E. Edmondson, and J. P. Klinman, *in* "Proceedings of the 11th International Symposium on Flavins and Flavoproteins," (K. Yagi, ed.) in press. de Gruyter, New York, 1994.

[43] M. C. Walker, Ph.D. Thesis, Emory University, Atlanta, Georgia (1987).

[44] M. C. Walker and D. E. Edmondson, *in* "Proceedings of the 11th International Symposium on Flavins and Flavoproteins," (K. Yagi, ed.) in press. de Gruyter, New York, 1994.

tained suggest that the extent of tunneling is greater in the enzyme-catalyzed reaction compared to the corresponding nonenzymatic reaction. All of the anomalously large secondary kinetic isotope effects shown in Table III are indicative of coupled motion and tunneling. However, the magnitudes of the secondary kinetic isotope effects for the LADH- and YADH-catalyzed reactions[33] exceed the equilibrium limit to a greater extent than the nonenzymatic model reaction,[32] thus suggesting a greater degree of tunneling in the enzyme-catalyzed reaction. In a different enzyme system, the unnatural NAD^+ analog dependence on the breakdown of the rule of the geometric mean for formate dehydrogenase[35] is suggestive that the enzyme has the greatest extent of hydrogen tunneling with the natural NAD^+ cofactor. As the NAD^+ analogs are varied to an earlier transition state and lower catalytic efficiency, the extent of hydrogen tunneling is reduced. A final example that suggests that the enzymatic reaction is characterized by a greater degree of tunneling is from the mutagenesis studies of LADH.[20] Here mutagenesis to a smaller alcohol binding pocket has apparently altered the enzyme in such a way as to decrease the extent of H tunneling that appears to exist in the wild-type enzyme. These examples suggest that along with the classical rate acceleration[45] of some enzyme-catalyzed hydrogen transfers, the tunneling probability increases as well. This stresses the importance of describing quantum effects in order to understand more fully the strategies of enzyme catalysis. In particular, it becomes important to explore what features of enzyme catalysis increase the tunneling probability of a hydrogen transfer reaction.

Internal Thermodynamics

The importance of the equalization of the energies of enzyme-bound substrates and products on the achievement of evolutionary perfection has received considerable attention.[3,4] It has also been predicted that the probability of hydrogen tunneling increases as the energy of reactant and products approach one another.[46] In the interest of examining the correlation of internal thermodynamics and the extent of hydrogen tunneling, Rucker *et al.*[27] extended the initial work on YADH with benzyl alcohol[26] to include para-substituted benzyl alcohols and aldehydes that show a range of internal equilibrium constants in a ternary complex with YADH. The maximal rate, V_{max}, was measured in the forward and reverse direction across a temperature range in order to obtain values of ΔG°, ΔH°, and $T\Delta S^\circ$ for a series of para-substituted substrates. These values

[45] J. Kraut, *Science* **242,** 533 (1988).
[46] J. R. De La Vega, *Acc. Chem. Res.* **15,** 185 (1982).

were compared to the intrinsic k_D/k_T isotope effects, and a conclusion was reached that the degree of tunneling appeared to increase as the value of ΔH° decreased toward zero. Similar studies on additional enzyme systems should indicate whether this is a general trend.

Static and Dynamic Protein Structure

Hydrogen tunneling is particularly dependent on the ratio of the height and width of the reaction barrier.[1] The compression of a reaction coordinate through the binding of substrates in close proximity at the active site of an enzyme may explain the apparent increase in tunneling seen in some enzyme systems. If compression of the reaction coordinate is important for an increase of hydrogen tunneling, does this arise as a static feature of the active site geometry of the enzyme or a dynamic feature utilizing protein thermal motions to compress the donor and acceptor atoms of the hydrogen transfer? Bruno and Bialek have successfully modeled the temperature dependence of isotope effects for the BSAO reaction[19] using a model of a fluctuating potential of protein motion combined with the premise that H motion in the reaction coordinate is purely quantum mechanical.[47] In their view of catalysis, protein motion is required to bring the substrates within a tunneling distance. As discussed by Cha *et al.*[26] the crystallographically determined distances of about 3.5 Å between the donor and acceptor atoms of the hydride transfer catalyzed by LADH and dihydrofolate reductase are unreasonably long for either a semiclassical hydrogen transfer or a reaction in which hydrogen tunnels. They suggest that protein breathing motions may provide the narrowing of the reaction coordinate required for hydrogen to tunnel through the barrier. In a recent finding with LADH,[48] modifications of the size of a hydrophobic residue (Val^{203} shown in Fig. 7), which resides on the opposite face of NAD^+ from the bound alcohol, is found to cause significant changes in the degree of tunneling. Although these experiments cannot distinguish between a static or a dynamic model for reaction coordinate optimization, they indicate the importance of geometric constraints in bringing about efficient H tunneling.

[47] W. J. Bruno and W. Bialek, *Biophys. J.* **63**, 689 (1992).

[48] J. P. Klinman, 4th European Symposium on Organic Reactivity, Royal Soc. of Chem., Perkin Div., in press.

[15] Positional Isotope Exchange as Probe of Enzyme Action

By LEISHA S. MULLINS and FRANK M. RAUSHEL

Introduction

The application and development of new mechanistic probes for enzyme-catalyzed reactions have significantly expanded our knowledge of the molecular details occurring at the active sites of many enzymes. Enzyme kinetic techniques have progressed from a simple determination of K_m and V_{max} values through a discrimination between sequential and ping-pong kinetic mechanisms to detailed evaluations of transition state structures by measurement of the very small differences in rate on isotopic substitution at reaction centers. The positional isotope exchange (PIX) technique, originally described by Midelfort and Rose,[1] is one of the newer techniques that has made a significant contribution to the elucidation of enzyme reaction mechanisms over the last decade.

The PIX technique can be used as a mechanistic probe in any enzymatic reaction where the individual atoms of a functional group within a substrate, intermediate, or product become torsionally equivalent during the course of the reaction. This criterion is best illustrated with the example first presented by Midelfort and Rose.[1] Glutamine synthetase (glutamate–ammonia ligase) catalyzes the formation of glutamine via the overall reaction presented in Eq. (1).

$$\text{MgATP} + \text{glutamate} + \text{NH}_3 \rightleftharpoons \text{MgADP} + \text{glutamine} + \text{P}_i \quad (1)$$

The questions addressed by Midelfort and Rose concerned whether γ-glutamyl phosphate was an obligatory intermediate in the reaction mechanism and if this intermediate was synthesized at a kinetically significant rate on mixing of ATP, enzyme, and glutamate in the absence of ammonia. The putative reaction mechanism is illustrated in Scheme I.

In the first step (Scheme I) glutamate is phosphorylated by MgATP at the terminal carboxylate group to form enzyme-bound MgADP and γ-glutamyl phosphate. In the second step ammonia displaces the activating phosphate group to form the amide functional group of the product glutamine. Previous experiments with this enzyme had failed to detect an equilibrium isotope exchange reaction between ATP and ADP in the presence of enzyme and glutamate.[2] Thus, if the γ-glutamyl phosphate interme-

[1] C. G. Midelfort and I. A. Rose, *J. Biol. Chem.* **251,** 5881 (1976).
[2] F. C. Wedler and P. D. Boyer, *J. Biol. Chem.* **247,** 984 (1972).

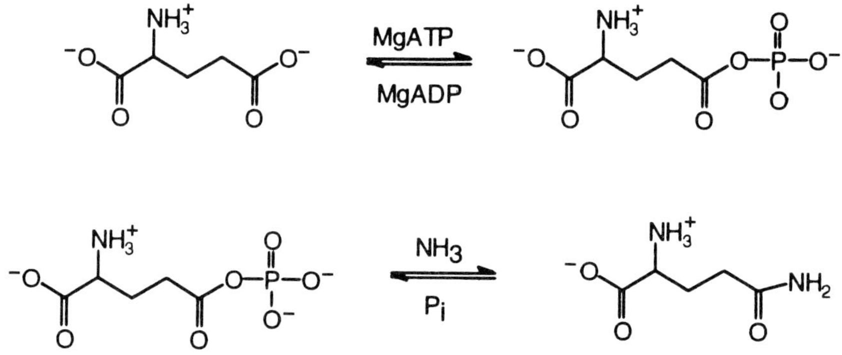

SCHEME I

diate was formed at the active site then the release of MgADP from the E · MgADP · intermediate complex must be very slow. The formation of the intermediate can be confirmed, however, when the oxygen atom between the β- and γ-phosphoryl groups in the substrate ATP is labeled with oxygen-18.

If the γ-glutamyl phosphate intermediate is formed and the β-phosphoryl group of the bound ADP is free to rotate then an isotopic label that was originally in the β, γ-bridge position will eventually be found 67% of the time in one of the two equivalent β-nonbridge positions of ATP as illustrated in Scheme II. This migration of the isotopic label can only occur if the γ-phosphoryl group of ATP is transferred to some acceptor and if the β-phosphoryl group of the enzyme-bound ADP is free to rotate torsionally. Midelfort and Rose found a PIX reaction with glutamine synthetase, and thus γ-glutamyl phosphate is an obligatory intermediate in the enzymatic synthesis of glutamine.

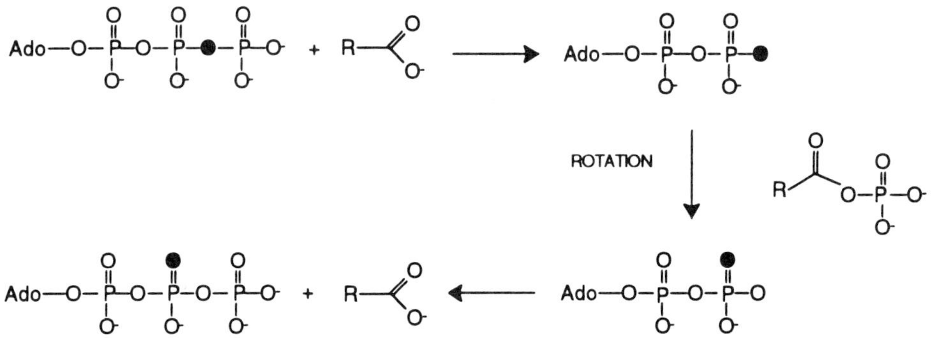

SCHEME II

Functional Groups for Positional Isotope Exchange

A variety of functional groups common to many substrates and products found in enzyme-catalyzed reactions are suitable for positional isotope exchange analysis. Some of the generic examples are illustrated in Diagram 1. The two most common functional groups that have been uti-

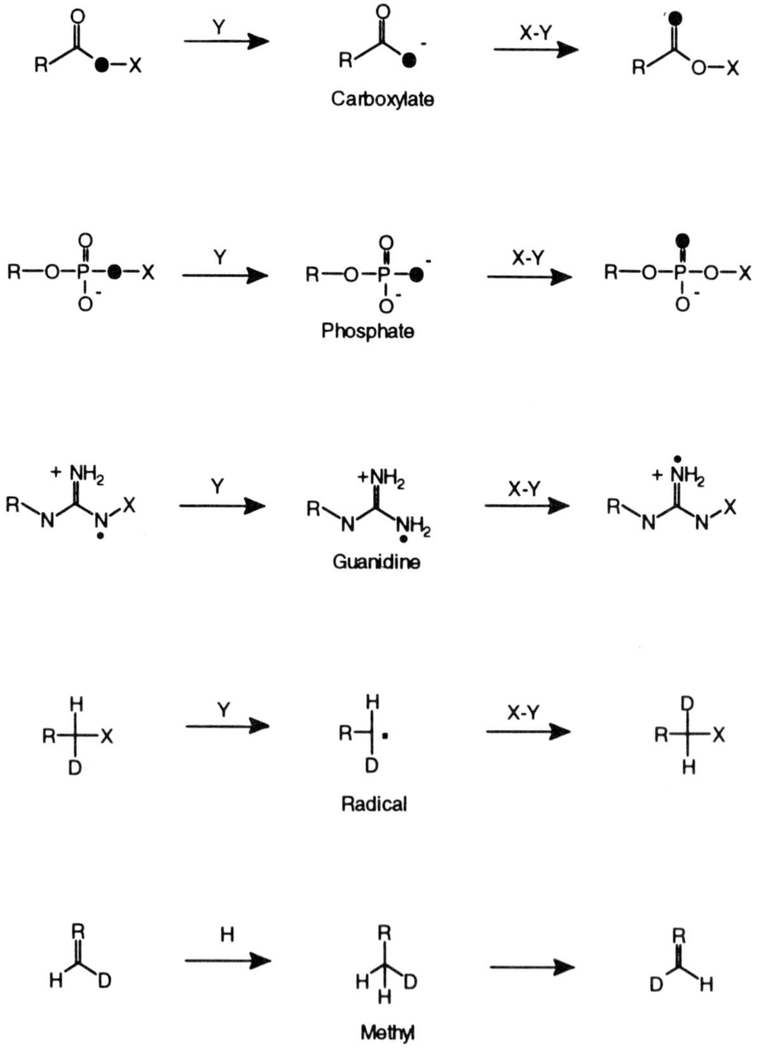

DIAGRAM 1. Functional groups for positional isotope exchange analysis.

lized for PIX studies are esters of substituted carboxylic and phosphoric acids. Thus, all reactions involving nucleophilic attack at either the β- or γ-phosphoryl groups of nucleotide triphosphates are amenable to PIX analysis. Moreover, all reactions utilizing UDP-sugars for complex sugar biosynthesis can use these PIX techniques for mechanistic evaluation. Less common are examples that involve reactions at a substituted guanidino functional group. Specific examples include creatine kinase[3] and argininosuccinate lyase.[4] With these enzymes positional isotope exchange of the two amino groups within arginine or creatine can be monitored by labeling with ^{14}N and ^{15}N and following the reaction with ^{15}N nuclear magnetic resonance (NMR) spectroscopy. Other cases include formation of radical or cationic centers at methylene carbons. Production of achiral acetaldehyde from chiral ethanolamine has been explained by intermediate formation of a C-2 radical during the course of the reaction.[5] Loss of stereochemistry in sp^2 centers can be exploited as a mechanistic probe when a methyl group is formed as an intermediate or product as for the case in the reaction catalyzed by pyruvate kinase.

Qualitative and Quantitative Approaches

The technique of positional isotope exchange can be used for two interrelated probes of enzyme-catalyzed reactions. In the first, as exemplified by the example with glutamine synthetase, the experimenter is primarily interested in determining whether a particular intermediate is formed during the course of the reaction. In the second approach the primary interest is in a quantitative determination of the partition ratio of an enzyme–ligand complex. The partition ratio is defined here as the fraction of the enzyme–ligand complex that proceeds forward to form unbound products versus the fraction of the enzyme complex in question that returns to unbound substrate and free enzyme.

When one is interested in whether a specific intermediate or complex is formed the experiment is generally conducted with one of the substrates absent from the reaction mixture. Take, for example, the simplified case of an enzyme reaction where two substrates are converted to two products via the covalent transfer of a portion of one substrate to the second substrate. This type of reaction is commonly found in many kinase reactions where ATP is used to phosphorylate an acceptor substrate. The generalized scheme is presented in Eq. (2).

[3] R. E. Reddick and G. L. Kenyon, *J. Am. Chem. Soc.* **109,** 4380 (1987).

[4] F. M. Raushel and L. J. Garrard, *Biochemistry* **23,** 1791 (1984).

[5] J. Retey, C. J. Suckling, D. Arigoni, and B. Babior, *J. Biol. Chem.* **249,** 6359 (1974).

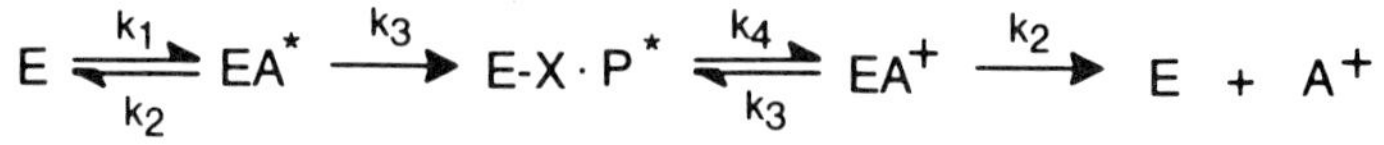

SCHEME III

$$A + B \rightleftharpoons P + Q \tag{2}$$

If the chemical mechanism involves the phosphorylation of the enzyme by ATP (substrate A) then a positional isotope exchange reaction may proceed in the absence of the acceptor substrate (B) if the acceptor B is not required to be bound to the enzyme in order for the phosphorylation to occur. Because substrate B is not required to be bound to the protein in order for the torsional equilibration of the isotopic label to occur, then all that is required for the PIX reaction to be observed is the incubation of enzyme and the isotopically labeled substrate (A^*). In such cases the velocity of the PIX reaction can be derived from the model presented in Scheme III where A^* represents the substrate as originally synthesized with the isotopic label and A^+ represents that fraction of A where the isotopic label is positionally equilibrated. It can be demonstrated that in such mechanisms the maximum value for the PIX reaction (V_{ex}) will be given by Eq. (3).

$$V_{ex} = (k_2k_3k_4)/[(k_2 + k_3)(k_3 + k_4)] \tag{3}$$

Rose has also demonstrated that in such mechanisms the minimum value for the PIX reaction is given by Eq. (4)[6]:

$$V_{ex} \geq (V_1V_2)/(V_1 + V_2) \tag{4}$$

where V_1 and V_2 are the maximal velocities of the steady-state reaction in the forward and reverse directions, respectively. This approach can be utilized with great utility to identify activated intermediates in many synthetase-type reactions where ATP is used to activate a second substrate molecule prior to condensation with a third substrate. This can be done either by phosphorylation or adenylylation of the second substrate. Equation (4) can be utilized to demonstrate that the presumed intermediate is formed rapidly enough to be kinetically competent.

When one is more interested in a quantitative analysis of the partition ratio of an enzyme–product complex the PIX methodology can easily provide that information. Take, for example, a case where a substrate, A, is converted to a single product, P:

$$A \rightleftharpoons P \tag{5}$$

[6] I. A. Rose, *Adv. Enzymol. Relat. Areas Mol. Biol.* **50,** 361 (1979).

In this example we will assume that the isotopic label originally in A becomes torsionally equivalent in the product, P.

The simplified kinetic scheme can be diagrammed as shown in Scheme IV. In this case the bonds are broken and the isotopes positionally equilibrated in the EP complex. This complex can partition in one of two ways. Either the product P can dissociate irreversibly from the EP complex with a rate constant k_5, or it can reverse the reaction and form free enzyme and substrate A with a net rate constant of k_4'. The fraction of EP that partitions forward can be quantitated by the amount of P that is formed per unit time, whereas the fraction of EP that partitions backward toward free enzyme and unbound A can be quantitated by the rate of positional equilibration of the isotope labels within the pool of substrate A.

In the steady state the partition forward is given by

$$[\mathrm{EP}]k_5 = V_{\mathrm{chem}} \tag{6}$$

whereas the partition backward is given by

$$[\mathrm{EP}]k_4' = [\mathrm{EP}](k_2k_4)/(k_2 + k_3) = V_{\mathrm{ex}} \tag{7}$$

and thus the ratio of $V_{\mathrm{ex}}/V_{\mathrm{chem}}$ is

$$V_{\mathrm{ex}}/V_{\mathrm{chem}} = (k_2k_4)/[k_5(k_2 + k_3)] \tag{8}$$

Because the maximal velocity in the reverse direction, V_2/E_t, is given by

$$V_2 = (k_2k_4)/(k_2 + k_3 + k_4) \tag{9}$$

then

$$k_2/k_5 \geq V_{\mathrm{ex}}/V_{\mathrm{chem}} \geq (V_2/E_t)k_5 \tag{10}$$

Therefore, the lower limit for the experimentally determined partition ratio is given by the maximal velocity in the reverse direction divided by k_5 (the dissociation rate constant for the product P). The upper limit for the partition ratio is given by the relative magnitude for the off-rate constants for A and P from the enzyme. These values would be difficult to obtain directly by any other method.

Experimentally, the positional isotope exchange rate, V_{ex}, is determined by measuring the fraction of positional isotope scrambling equilibrium (F) as a function of time (t) where $V_{\mathrm{ex}} = ([\mathrm{A}]/t)\ln(1 - F)^{-1}$. If

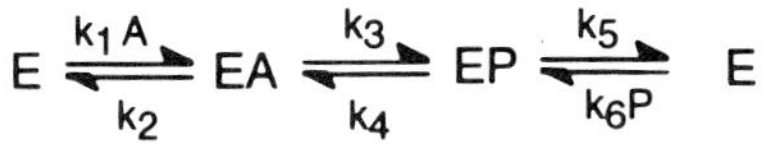

SCHEME IV

the isotopically labeled substrate is being chemically depleted during the course of the PIX analysis then the corrected PIX rate is calculated from $V_{ex} = [(X)/\ln(1 - X)]\,(A_0/t)\ln(1 - F)^{-1}$, where X is the fraction of substrate lost at time t and A_0 is the initial concentration of the labeled substrate. In the next section we examine the quantitative effects induced by variation of the concentration of the unlabeled substrates and products on the positional isotope exchange rates.

Variation of Nonlabeled Substrates and Products

The PIX technique can be used to obtain information about the partitioning of enzyme complexes and the order of substrate addition and product release. This information is obtained by measuring the PIX rate in the enzyme-catalyzed reaction relative to the overall chemical transformation rate in the presence of variable amounts of added substrates or products. In many cases, it is possible with these methods to determine the microscopic rate constants for the release of substrates and products from the enzyme–ligand complexes. The following provides a presentation for how the PIX technique can be applied to the analysis of sequential and ping-pong kinetic reaction mechanisms.

Sequential Mechanisms

In sequential mechanisms, substrate addition and/or product release can be either ordered or random or a combination of various pathways. The PIX reaction can be utilized to distinguish between the possible kinetic mechanisms and to determine the net reaction flux through the various kinetic pathways. The effect of product addition can lead to an *enhancement* of the PIX reaction relative to overall rate (PIXE), whereas the variation of unlabeled substrates may reduce or *inhibit* the PIX reaction relative to overall chemical rate (PIXI).

Positional Isotope Exchange Enhancement

The simplest kinetic mechanism that can be written for an enzyme reaction with multiple products is a Uni Bi mechanism. In this mechanism the release of the products can be either random (Uni Bi Random) or ordered (Uni Bi Ordered). A general model of a Uni Bi Random kinetic mechanism is shown in Scheme V, where A is designated as the substrate with the positionally labeled isotopic atoms and Q is the product in which the positional exchange occurs.

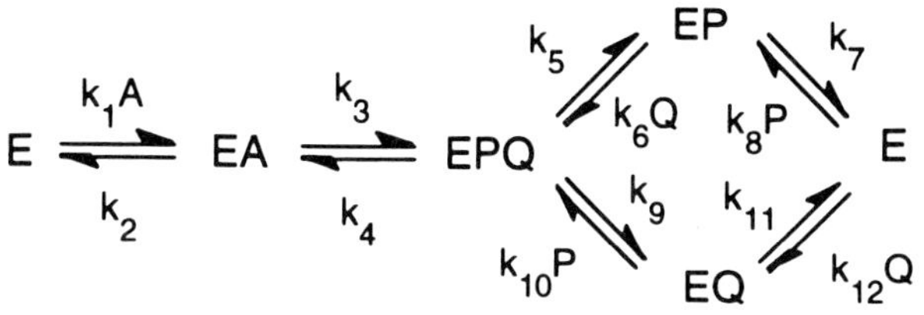

SCHEME V

In Scheme V the ratio of the PIX rate relative to the net chemical rate (V_{ex}/V_{chem}) is determined by the partitioning of the EPQ complex. Using the method of net rate constants, the partitioning of EPQ can be written as[7]

$$\frac{V_{ex}}{V_{chem}} = \frac{(k_2k_4)/(k_2 + k_3)}{k_5 + (k_9k_{11})/(k_{11} + k_{10}[P])} \tag{11}$$

As can be seen in Eq. (11), the ratio V_{ex}/V_{chem} is dependent on the amount of P added to the reaction mixture. In a random mechanism, the ratio V_{ex}/V_{chem} increases as a function of [P] because the addition of P inhibits the flux through the lower pathway and then the ratio plateaus at a level that is determined by the net flux through the upper pathway. The net flux through each pathway can be determined by the ratio V_{ex}/V_{chem} as a function of [P]. At [P] = 0, Eq. (11) becomes

$$V_{ex}/V_{chem} = [k_2k_4/(k_2 + k_3)]/(k_5 + k_9) \tag{12}$$

and at [P] = ∞

$$V_{ex}/V_{chem} = [k_2k_4/(k_2 + k_3)]/(k_5) \tag{13}$$

Thus the ratio of k_5 and k_9 can be determined by measurement of the PIX ratio at zero and saturating P.

Figure 1 also illustrates the change in the ratio V_{ex}/V_{chem} as a function of [P] as a sequential mechanism changes from random to ordered release of products. The ordered release of the products simplifies Eq. (11). If P is released first ($k_5 = 0$), then the equation becomes

$$\frac{V_{ex}}{V_{chem}} = \frac{k_2k_4/(k_2 + k_3)}{k_9k_{11}/(k_{11} + k_{10}[P])} \tag{14}$$

The ratio V_{ex}/V_{chem} is now linearly dependent on the concentration of added P. The intercept at [P] = 0 becomes

$$V_{ex}/V_{chem} = [k_2k_4/(k_2 + k_3)]/k_9 \tag{15}$$

[7] W. W. Cleland, *Biochemistry* **14,** 3220 (1975).

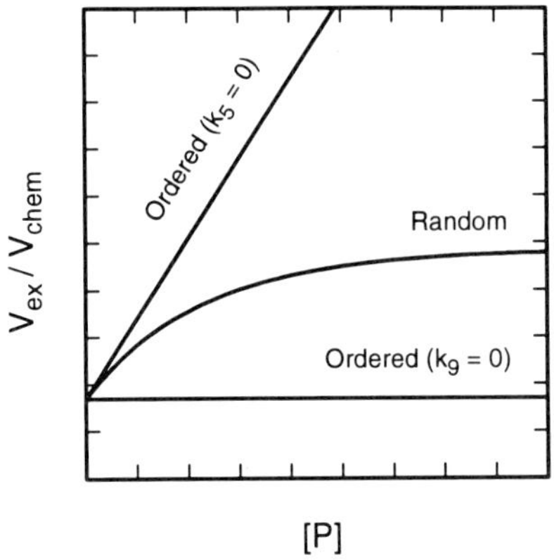

FIG. 1. Enhancement of the ratio of the positional isotope exchange rate and the net rate of chemical turnover as a function of added product inhibitor.

It is possible to determine the lower limits for the off-rate constant from the ternary complex EPQ (k_9) and the binary complex EQ (k_{11}) relative to the turnover number in the reverse direction (V_2/E_t) as shown in Eqs. (16) and (17) since $V_2/E_t = (k_2k_4)/(k_2 + k_3 + k_4)$:

$$k_9/(V_2/E_t) = (V_{chem}/V_{ex}) + (k_9/k_2) \tag{16}$$

$$k_{11}/(V_2/E_t) = (1/K_p)[\text{intercept} + 1)/\text{slope}] \tag{17}$$

where K_p is the Michaelis constant for the product P.

If Q is the first product to be released ($k_9 = 0$) in an ordered mechanism, then there would be no dependence on V_{ex}/V_{chem} as [P] is varied, as apparent in Eq. (18):

$$V_{ex}/V_{chem} = k_2k_4/(k_2 + k_4) \tag{18}$$

Again, the lower limit for the release of the product from the ternary complex can be determined relative to the turnover number in the reverse reaction:

$$k_5/(V_2/E_t) = (V_{chem}/V_{ex}) + (k_5/k_2) \tag{19}$$

Positional isotope exchange enhancement experiments can provide valuable information about enzyme kinetic mechanisms. Utilizing the PIXE technique, a simple inspection of a plot of V_{ex}/V_{chem} as a function of [P] readily identifies the kinetic mechanism (see Fig. 1). The PIXE experiments also allow the determination of the microscopic rate constants for product release from both the ternary and binary complexes. An often overlooked advantage of the PIXE experiment is that it can identify a

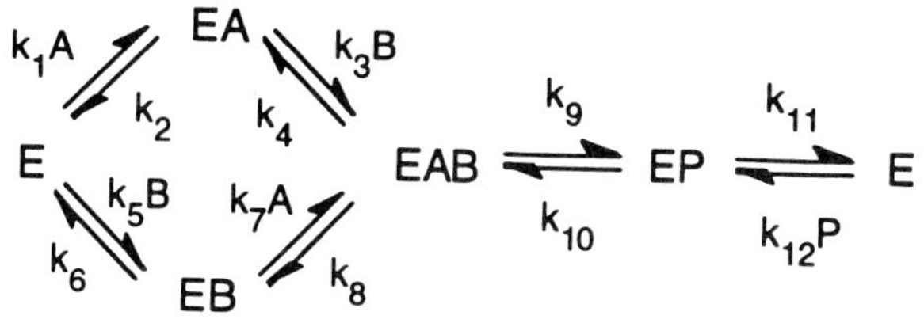

SCHEME VI

"leaky" product in what otherwise would appear as an ordered product release.

Positional Isotope Exchange Inhibition

The concentration of unlabeled substrate also affects the ratio of the PIX rate relative to the net chemical rate.[8] An enzyme kinetic mechanism with random addition of substrates is shown in Scheme VI, where A is the substrate with the positionally labeled atoms and P is the product in which the positional exchange occurs.

If A is the first substrate to bind in an ordered kinetic mechanism ($k_8 = 0$), then increasing concentrations of B will inhibit the PIX rate relative to the net chemical rate (V_{ex}/V_{chem}) because saturation with B prevents A from dissociating from the enzyme. In contrast, if B must bind first ($k_4 = 0$), then its concentration will have no effect on V_{ex}/V_{chem}. This is because saturation with B would not affect the rate of dissociation of A from the EAB complex. If addition of the substrates to the enzyme is random, then saturation with B will reduce the value of V_{ex}/V_{chem} but not eliminate it entirely. A plot of V_{ex}/V_{chem} as a function of [B] will plateau at a value of V_{ex}/V_{chem} that equals the flux through the lower pathway and thus enables the determination of the ratio of k_4 and k_8. Figure 2 illustrates the plot of V_{ex}/V_{chem} as a function of added substrate. Substrate inhibition by the nonlabeled substrate can be quite diagnostic of the particular reaction mechanism.

In special cases, the investigation of substrate inhibition of PIX reactions can be used to determine the microscopic rate constants for the release of substrates and products from the enzyme complexes. A Bi Bi Ordered mechanism can be used to illustrate this application. The simplest mechanism that can be written for a Bi Bi Ordered reaction is shown in Scheme VII, where A is the substrate with the positional label and P is the product undergoing rotational exchange. For the forward reaction,

[8] F. M. Raushel and J. J. Villafranca, *Crit. Rev. Biochem.* **23,** 1 (1988).

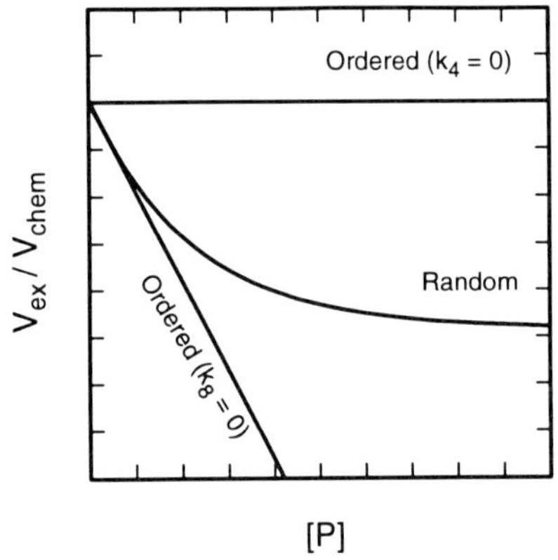

FIG. 2. Inhibition of the ratio of the positional isotope exchange rate and the net rate of chemical turnover as a function of added substrate.

the partitioning of the EPQ complex determines the PIX rate relative to the rate of net product formation. The partitioning of the EPQ complex can be written as

$$\frac{V_{\text{chem}}}{V_{\text{ex}}} = \frac{k_7(k_2k_4 + k_2k_5 + k_3k_5[\text{B}])}{k_2k_4k_6} \tag{20}$$

The plot of $V_{\text{chem}}/V_{\text{ex}}$ as a function of the concentration of B yields a straight line. An analysis of the plot gives the following information:

$$\text{Intercept}_{[\text{EPQ}]} = k_7(k_4 + k_5)/(k_4k_6) \tag{21}$$
$$\text{Slope}_{[\text{EPQ}]} = (k_3k_5k_7)/(k_2k_4k_6) \tag{22}$$

If a PIX reaction can be followed for the reverse reaction, then analogous equations can be derived for the partitioning of the EAB complex:

$$\frac{V_{\text{chem}}}{V_{\text{ex}}} = \frac{k_4(k_6k_9 + k_7k_9 + k_6k_8[\text{P}])}{k_5k_7k_9} \tag{23}$$

where the slope$_{[\text{EAB}]}$ is equal to $(k_4k_6k_8)/(k_5k_7k_9)$ and the intercept$_{[\text{EAB}]}$ is $k_4(k_6 + k_7)/(k_5k_7)$.

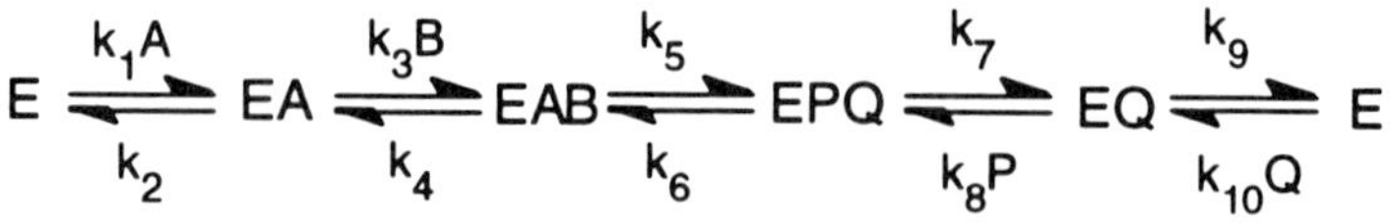

SCHEME VII

It can be shown that the microscopic rate constants for the release of the A and Q from the binary enzyme complexes can be determined from a combination of the respective slopes, intercepts, and the Michaelis constants as shown in Eqs. (24) and (25)[8]:

$$k_2/(V_1/E_t) = [(1 + \text{intercept}_{[EPQ]})/(\text{slope}_{[EPQ]})](1/K_b) \quad (24)$$
$$k_9/(V_2/E_t) = [(1 + \text{intercept}_{[EAB]})/(\text{slope}_{[EAB]})](1/K_q) \quad (25)$$

The lower limits for the release of B and P from the two ternary complexes can also be obtained:

$$k_7/(V_2/E_t) = (k_7/k_2) + (k_7/k_4) + \text{intercept}_{[EPQ]} \quad (26)$$
$$k_4/(V_1/E_t) = (k_4/k_9) + (k_4/k_7) + \text{intercept}_{[EAB]} \quad (27)$$

Therefore, the rate constants k_2 and k_9 relative to the net substrate turnover in the forward and reverse reactions are known from Eqs. (24) and (25), leaving two unknown rate constants, k_4 and k_7. Because there are two independent equations [Eqs. (26) and (27)] that contain both k_4 and k_7, both of the rate constants can be determined. Using this analysis of the PIX reaction, the rate constants for the release of all substrates and products from the enzyme complexes can be determined. If these results are combined with the steady-state kinetics parameters, V_1/E_t, V_2/E_t, K_a, K_b, K_p, and K_q, and the thermodynamic parameter, K_{eq}, then it is possible to estimate all 10 microscopic rate constants for the minimal Bi Bi Ordered kinetic mechanism shown in Scheme VII.

The analysis of PIX reactions as a function of added substrates and products in ordered kinetic mechanisms provides information that would otherwise be difficult to determine. It can also be used to corroborate classic steady-state experiments as well as to give information on the relative "stickiness" of a substrate or product.

Ping-Pong Mechanisms

In ping-pong reaction mechanisms, the first substrate binds and reacts covalently with the enzyme in the total absence of the second substrate to form a stable but modified enzyme–product complex. The second substrate binds and reacts with the modified enzyme to form the second product and regenerate the starting enzyme form. A simple Bi Bi Ping Pong reaction is shown in Scheme VIII.

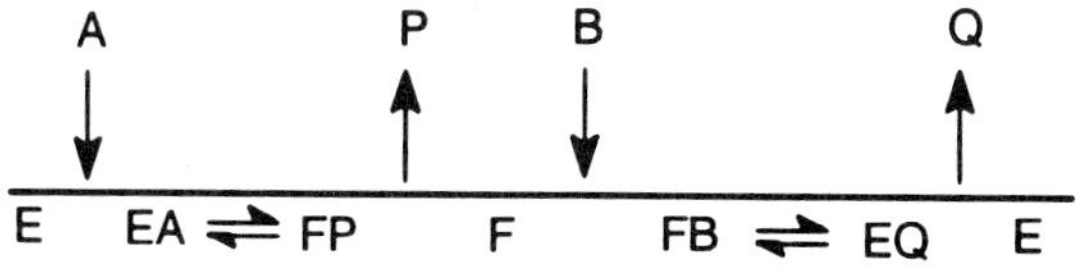

SCHEME VIII

$$\mathrm{E} \underset{k_2}{\overset{k_1\mathrm{A}}{\rightleftharpoons}} \mathrm{EA} \underset{k_4}{\overset{k_3}{\rightleftharpoons}} \mathrm{FP} \underset{k_6\mathrm{P}}{\overset{k_5}{\rightleftharpoons}} \mathrm{F} \underset{k_8}{\overset{k_7}{\rightleftharpoons}} \mathrm{EQ} \underset{k_{10}\mathrm{Q}}{\overset{k_9}{\rightleftharpoons}} \mathrm{E}$$

SCHEME IX

Because the first substrate can bind and react with the free enzyme in the absence of the second substrate, it is often assumed that the ping-pong reaction mechanism would be ideal for analysis by the PIX kinetic technique. The product bound to the modified enzyme could undergo torsional scrambling, and then reformation of the first substrate would yield a positional isotope exchange reaction. However, this is a misconception because in reality no significant PIX reaction is expected to be observed because of the rapid dissociation of the first product from the modified enzyme.[9] Essentially 1 enzyme equivalent of product is produced, leaving the enzyme in the modified form unable to process more substrate.

The minimal kinetic mechanism for a ping-pong reaction is shown in Scheme IX, where E is enzyme; A, substrate with the positional isotope label; P, product undergoing torsional scrambling; and F, stable modified enzyme. An analysis of the PIX reaction would give the partitioning of the FP complex shown in Eq. (28):

$$V_{\text{chem}}/V_{\text{ex}} = [k_5(k_2 + k_3)]/k_2k_4 \tag{28}$$

However, for a PIX reaction to be observed, a method for reconverting the modified enzyme form F back to free enzyme E is required. This can be achieved with the addition of a large excess of unlabeled product P to the reaction mixture. The large excess of P returns the modified enzyme form F back to E, producing unlabeled A. This process provides a measure of the partitioning of the FP and F complexes. The excess level of P also dilutes the concentration of labeled P. This is important since the purpose of the PIX reaction is to determine the partitioning of the FP complex between F and E. To obtain an accurate determination for the partitioning of the FP complex, the formation of positionally exchanged substrate from the true PIX reaction must be distinguished from a pseudo-PIX reaction (caused by product dissociation and reassociation with the modified enzyme).

The ping-pong PIX experiment is initiated by addition of enzyme to labeled substrate A and unlabeled product P. The first-order equilibration of the positional label between substrate and product gives the partitioning

[9] L. S. Hester and F. M. Raushel, *J. Biol. Chem.* **262,** 12092 (1987).

between FP and F. Statistical considerations permit the determination of how much of the labeled product that is formed and released into solution will partition back to either the original labeled substrate or the positionally exchanged substrate. The rate of the pseudo-PIX mechanism can be calculated from the exchange rate of the label (positionally exchanged and nonexchanged) between product P and total substrate. The rate of the pseudo-PIX reaction can be used to correct the overall rate of formation of the positionally exchanged substrate to yield the true PIX rate for the interconversion of positionally labeled substrate. The correction factor will decrease as the ratio of unlabeled product relative to labeled substrate increases.

An analysis of the ping-pong PIX experiment is shown in Scheme X, where M, N, and O represent the positionally labeled, unlabeled, and positionally exchanged substrates, respectively. The equilibration of labeled and unlabeled substrates is represented by the interconversion of M → N → O. The formation of the positionally exchanged substrate O by this pathway would be a pseudo-PIX reaction because of the disassociation and reassociation of the product. The interconversion of M → O without the formation of unbound labeled product provides the true PIX rate.

The rate constants in Scheme X can be defined as follows:

$$k_a = xy(1 + w) \quad (29)$$

$$k_b = x \quad (30)$$

$$k_c = xw \quad (31)$$

$$k_d = xy(1 + w) \quad (32)$$

$$k_e = xz \quad (33)$$

$$k_f = xzw \quad (34)$$

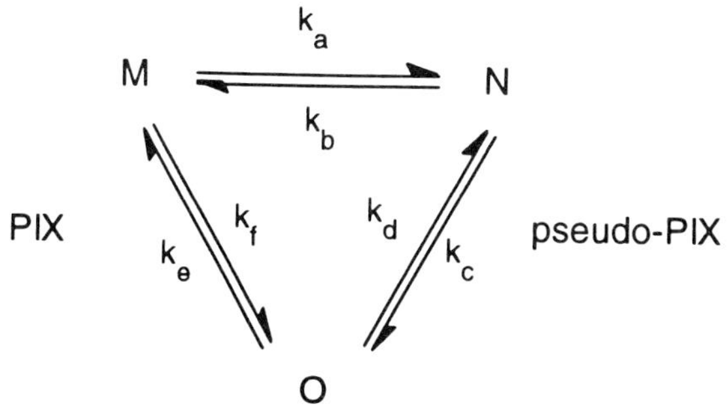

SCHEME X

where x is proportional to the amount of enzyme used and therefore affects each step equally, y is the ratio of the initial concentration of unlabeled product P and labeled substrate A, w is the equilibrium ratio of positionally exchanged substrate and labeled substrate, and z is proportional to the true PIX reaction.

The factor x can be determined from an analysis of the plot of ([M] + [O])/([M] + [N] + [O]) versus time since the values of y and w are known. Once x is determined then z can be obtained by an analysis of the plot of ([O])/([M] + [O]) versus time. A numerical solution for the determination of the rate constants k_a through k_f is thus possible.

The time course for the equilibration of the labeled substrate and the positionally exchanged substrate, represented as a plot of [O]/([M] + [O]) versus time, is dependent on the ratio of the PIX and the pseudo-PIX pathways. The flux through the pseudo-PIX pathway can be diminished by increasing the ratio of unlabeled product to labeled substrate. This is represented in Fig. 3, where the curves are simulated with increasing values of z from 0 to 100. In the absence of a pathway for the direct interconversion of M $\rightarrow$ O, there is a noticeable lag in the appearance of the positionally exchanged substrate.

To determine the partitioning of FP, the ratio V_{chem}/V_{ex} [Eq. (28)] must be expressed in terms of the rate constants k_a through k_f. The chemical rate for the conversion of substrate to product, V_{chem}, is $[X_0]k_a$, and the rate of positional isotope exchange, V_{ex}, is $[X_0](k_e + k_f)$. The ratio V_{chem}/V_{ex} can now be expressed as

$$V_{chem}/V_{ex} = y/z \tag{35}$$

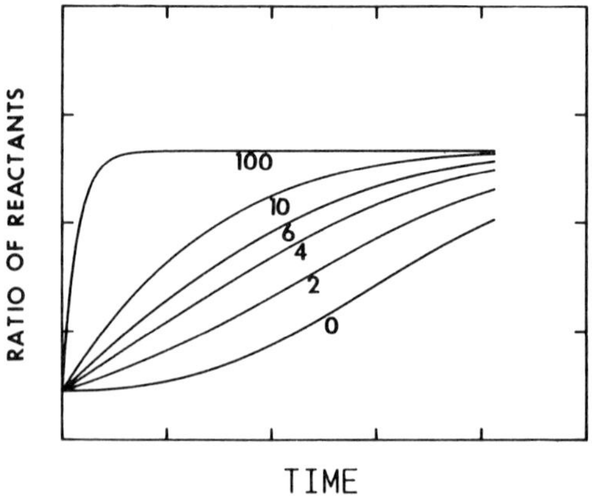

FIG. 3. Simulations for the interconversion of the labeled substrate (M) and the positionally exchanged substrate (O) as a function of time.

The lower limit for the conversion of FP to F, k_5, relative to the maximal velocity in the reverse direction, V_2/E_t, can be determined since V_2/E_t is less than or equal to $(k_2k_4)/(k_2 + k_5)$:

$$k_5/(V_2/E_t) = V_{chem}/V_{ex} \tag{36}$$

Specific Examples

Presented below are examples of five enzyme-catalyzed reactions that have been successfully studied by the positional isotope exchange technique. Three of these enzymes, argininosuccinate lyase, UDPG pyrophosphorylase (UTP–glucose-1-phosphate uridylyltransferase), and galactose-1-phosphate uridylyltransferase (UTP–hexose-1-phosphate uridylyltransferase) give representative examples of the type of information available from a quantitative analysis of enzyme–ligand dissociation rates using positional isotope exchange. The last two examples, D-alanine–D-alanine ligase and carbamoyl-phosphate synthase, are presented to illustrate how the PIX technique can be best utilized to identify reaction intermediates.

Argininosuccinate Lyase

The effect of added products on the observed PIX rate was first investigated by Raushel and Garrard on the reaction catalyzed by argininosuccinate lyase.[10] Argininosuccinate lyase catalyzes the cleavage of argininosuccinate to arginine and fumarate. The effect of fumarate addition on the $^{15}N/^{14}N$ positional exchange reaction within argininosuccinate was measured by ^{15}N NMR spectroscopy. Scheme XI shows the positional isotope exchange reaction that was followed with the enzyme. The two external nitrogens in the guanidino moiety of the product arginine are torsionally equivalent. This structural feature enables the scrambling of the ^{15}N and ^{14}N labels within argininosuccinate when the guanidino group is free to rotate about the C–N bond in the enzyme–arginine–fumarate complex.

Raushel and Garrard demonstrated that at zero fumarate there was no observable PIX reaction relative to the net formation of product ($V_{ex}/V_{chem} < 0.15$).[10] However, at higher levels of added fumarate the ratio V_{ex}/V_{chem} increased until a plateau of 1.8 was reached. These results indicate quite clearly that the release of products from the enzyme–arginine–fumarate complex in the argininosuccinate lyase-catalyzed reaction is kinetically random. Because no exchange was observed at low fumarate,

[10] F. M. Raushel and L. J. Garrard, *Biochemistry* **23,** 1791 (1984).

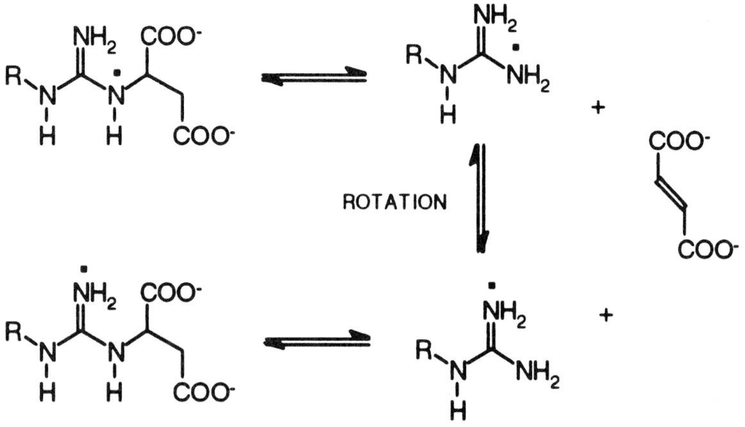

SCHEME XI

it can also be concluded that the release of at least one of the two products must be very fast. Moreover, the release of arginine from the ternary complex, relative to V_2/E_t, must be greater than or equal to 0.5 (based on the limiting value of 1.8 at saturating fumarate). Therefore, the release of fumarate from the ternary complex (see Scheme V) is very fast relative to the maximal velocity in the reverse direction $[k_9/(V_2/E_t) > 6]$, and thus the release of fumarate from the enzyme–arginine–fumarate complex is at least 10 times faster than arginine release.

UDPglucose Pyrophosphorylase

Hester and Raushel investigated the effects of substrate addition on the PIX reaction catalyzed by UDPG pyrophosphorylase.[11] UDPG pyrophosphorylase catalyzes the conversion of UTP and glucose 1-phosphate to UDP-glucose and pyrophosphate. The reaction proceeds by the nucleophilic attack of glucose 1-phosphate on the α-phosphate of UTP. The kinetic mechanism has been previously shown to be ordered with UTP as the first substrate to bind and pyrophosphate the first product to be released (see Scheme VII).[12]

Hester and Raushel showed that it was possible to follow a PIX reaction in both the forward and reverse reactions of the enzyme by utilizing the

[11] L. S. Hester and F. M. Raushel, *Biochemistry* **26,** 6465 (1987).
[12] K. K. Tsuboi, K. Fukunaga, and J. C. Petricciani, *J. Biol. Chem.* **244,** 1008 (1969).

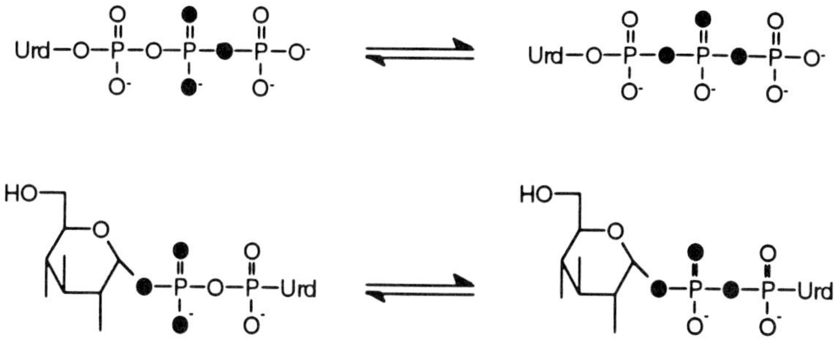

SCHEME XII

positionally labeled substrates shown in Scheme XII.[11] The PIX reactions in the forward and reverse directions were able to be suppressed by increasing the concentration of the second substrate, thus confirming the strictly ordered nature of the reaction mechanism (see Fig. 4).

By combining the quantitative information from the PIX reactions with the steady-state kinetic and thermodynamic parameters, Hester and Raushel were able to obtain estimates for all of the microscopic rate constants in the UDPG pyrophosphorylase reaction. The values calculated for the rate constants (shown in Scheme XIII) reveal that the release of UDPglucose is 3 times slower than the release of pyrophosphate and that the release of glucose 1-phosphate is 5 times slower than the release to UTP. The back-calculated kinetic constants are in excellent agreement

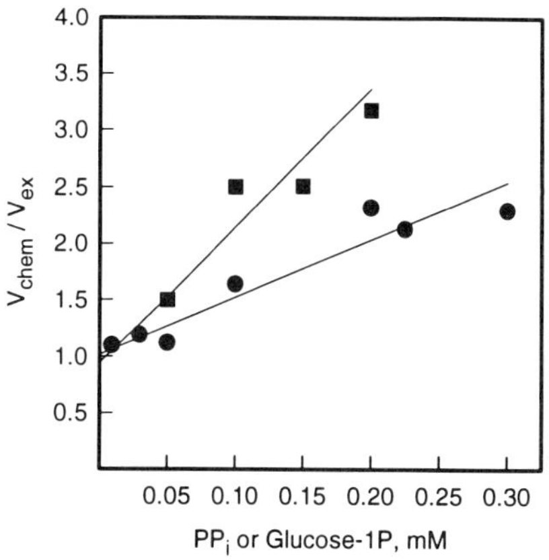

FIG. 4. Plot of the ratio of the net chemical turnover rate and the positional isotope exchange rate as a function of the concentration of added glucose 1-phosphate (●) or pyrophosphate (PP_i) (■).

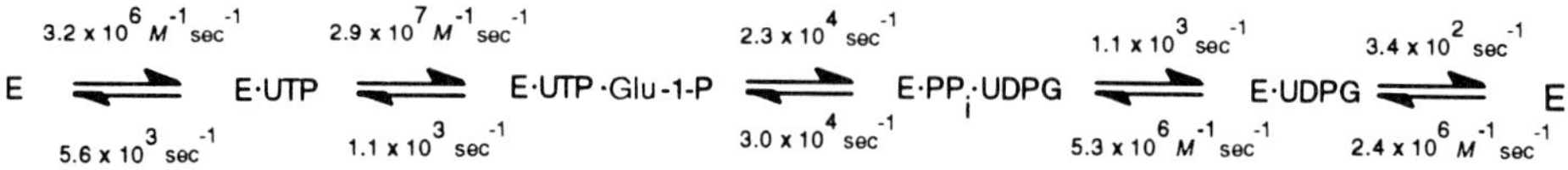

SCHEME XIII

with the experimental values.[11] This example demonstrates quite clearly the potential for obtaining quantitative information about the binding and release of substrates and products from a thorough analysis of PIX reactions.

Galactose-1-Phosphate Uridylyltransferase

The PIX technique has rarely been applied to a ping-pong reaction mechanism.[13] However, Hester and Raushel have investigated the reaction catalyzed by galactose-1-phosphate uridylyltransferase. Galactose-1-phosphate uridylyltransferase catalyzes the transfer of the uridylyl group from UDPglucose to galactose-1-phosphate. A covalent enzyme–uridylyl adduct is an intermediate in the reaction as determined by stereochemical and steady-state kinetic analysis.[14] The PIX experiment monitored only the first half-reaction of galactose-1-phosphate uridylyltransferase as shown in Scheme XIV.

Unlabeled glucose-1-phosphate was included in the reaction mixture to recycle the enzyme so that the partitioning of the uridylyl enzyme–

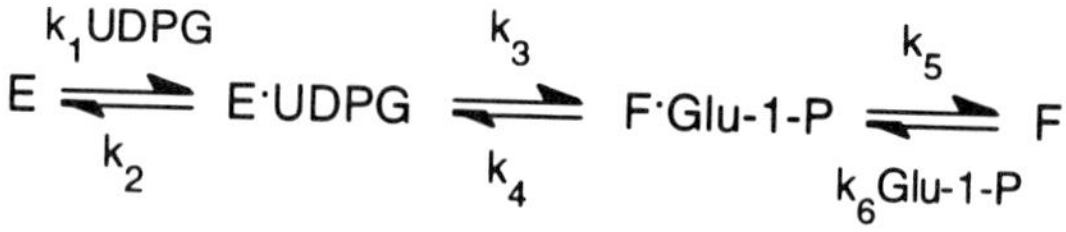

SCHEME XIV

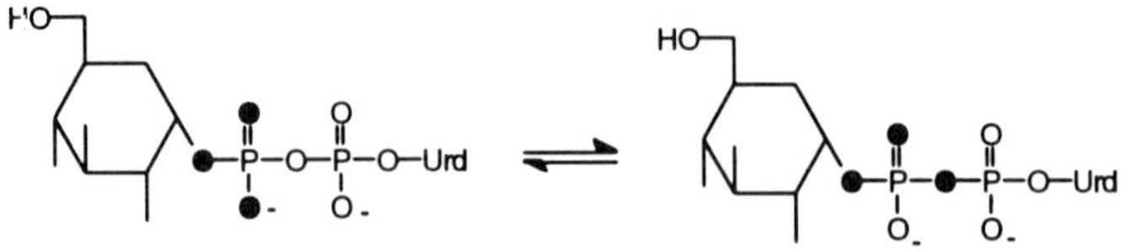

SCHEME XV

[13] L. S. Hester and F. M. Raushel, *J. Biol. Chem.* **262,** 12092 (1987).
[14] K. R. Sheu and P. A. Frey, *J. Biol. Chem.* **253,** 3378 (1978).

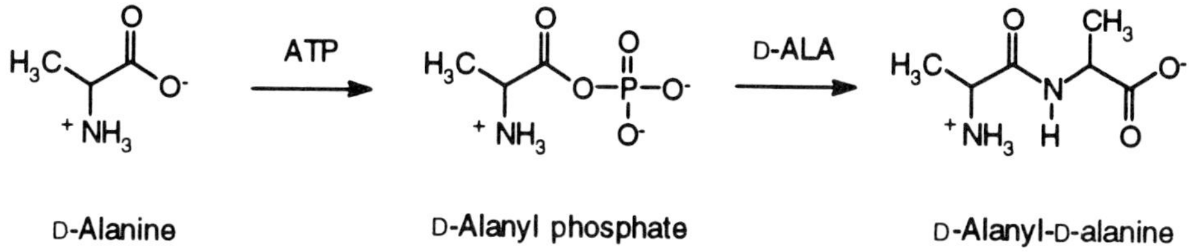

SCHEME XVI

glucose-1-phosphate complex could be determined. The PIX reaction followed the torsional scrambling of $^{18}O_3$ within [β-$^{18}O_3$]UDPglucose (Scheme XV) in the presence of variable amounts of unlabeled glucose-1-phosphate as a function of time. No PIX reaction was observed in the absence of glucose 1-phosphate as a function of time. No PIX reaction was observed in the absence of glucose-1-phosphate as expected for a ping-pong reaction. The partitioning of the uridylylenzyme–glucose-1-phosphate (V_{chem}/V_{ex}) was determined to be 3.4. Therefore, the release of glucose-1-phosphate from the uridylyl enzyme complex is 3.4 times faster than the maximal velocity in the reverse direction.

D-Alanine–D-Alanine Ligase

Mullins *et al.* utilized the PIX technique in order to obtain kinetic evidence for the intermediate formation of D-alanyl phosphate in the reaction catalyzed by D-alanine–D-alanine ligase.[15] The D-alanine–D-alanine ligase reaction is a difficult reaction to study mechanistically because it utilizes the same substrate twice. D-Alanine–D-alanine ligase catalyzes the formation of D-alanyl-D-alanine from ATP and two molecules of D-alanine. The reaction has been proposed to proceed through an acylphos-

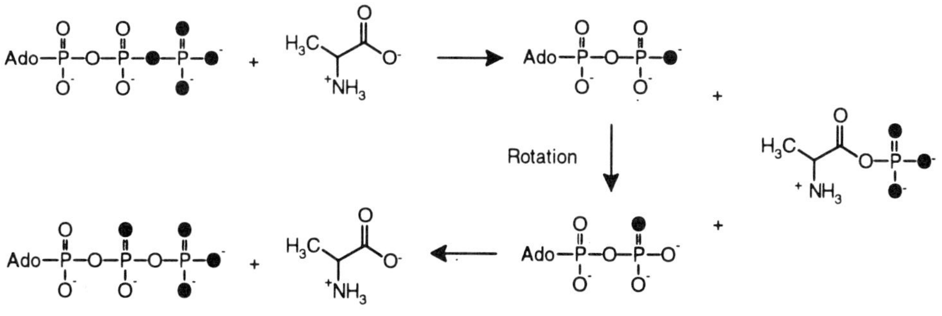

SCHEME XVII

[15] L. S. Mullins, L. E. Zawaszke, C. T. Walsh, and F. M. Raushel, *J. Biol. Chem.* **265,** 8993 (1990).

phate intermediate formed via the phosphorylation of the carboxyl group of the first D-alanine by the γ-phosphate of ATP (Scheme XVI).[16]

Mullins *et al.* followed the PIX reaction shown in Scheme XVII. In D-alanine–D-alanine ligase, the difficulty of investigating the kinetic mechanism is increased because the reaction can not be artificially stopped after the addition of the first two substrates to probe for the existence of the intermediate. A PIX reaction is observed only in the presence of D-alanine, and this is consistent with the direct attack of the carboxyl group of one D-alanine on the γ-phosphate of ATP to give an acyl phosphate intermediate. Cleavage of the γ-phosphate of ATP could occur either after the first D-alanine binds or, alternatively, only after the binding of all three substrates. However, the observed positional isotope exchange rate is diminished relative to the net substrate turnover as the concentration of D-alanine is increased. This is consistent with an ordered kinetic mechanism. In addition, the ratio of the PIX rate relative to the net chemical turnover of substrate (V_{ex}/V_{chem}) approaches a value of 1.4 as the concentration of D-alanine becomes very small. This ratio is 100 times larger than the ratio of the maximal reverse and forward chemical reaction velocities (V_2/V_1). This situation is only possible when the reaction mechanism in the forward direction proceeds in two distinct steps and the first step is much faster than the second step. Thus it appears that formation of the acyl-phosphate intermediate is faster than amide bond formation.

Carbamoyl-Phosphate Synthetase

The PIX technique has been used to identify two reactive intermediates in the reaction catalyzed by carbamoyl-phosphate synthetase (CPS). CPS catalyzes the following reaction:

$$2\,\text{MgATP} + \text{HCO}_3^- + \text{glutamine} + \text{H}_2\text{O} \rightarrow \text{carbamoyl phosphate} + 2\,\text{MgADP} + \text{P}_i + \text{glutamine} \quad (37)$$

The protein is heterodimeric consisting of a large (120 kDa) and a small (42 kDa) subunit.[17] The proposed mechanism for the synthesis of carbamoyl phosphate (shown in Scheme XVIII) is composed of at least four individual steps.[18] Carboxyphosphate and carbamate have been postulated as the key intermediates in the overall reaction mechanism. The hydrolysis of glutamine to glutamate ocurs on the small subunit while the remaining

[16] K. Duncan and C. T. Walsh, *Biochemistry* **27,** 3709 (1988).

[17] P. P. Trotta, M. E. Burt, R. H. Haschemeyer, and A. Meister, *Proc. Natl. Acad. Sci. U.S.A.* **68,** 2599 (1971).

[18] P. M. Anderson and A. Meister, *Biochemistry* **4,** 2803 (1965).

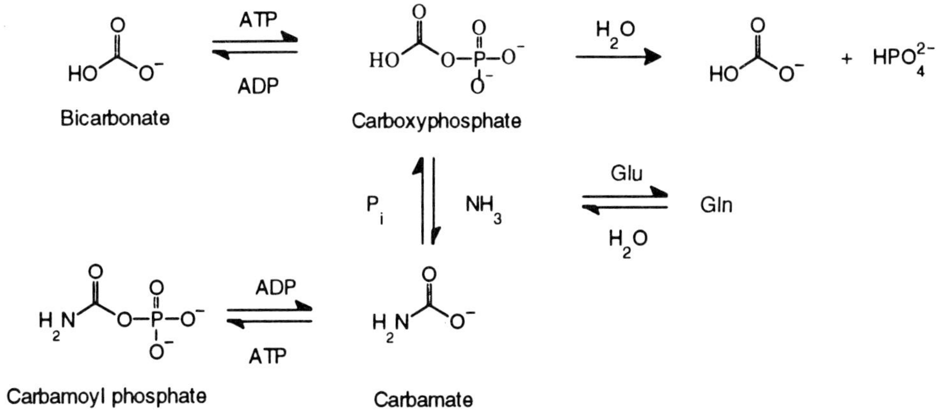

SCHEME XVIII

reactions occur on the large subunit. Therefore, the ammonia nitrogen must be transferred from the small subunit to the active site of the large subunit. Free ammonia can be utilized as the nitrogen source by the large subunit in the absence of an active glutaminase reaction from the small subunit. A partial reaction catalyzed by the large subunit is the bicarbonate-dependent hydrolysis of ATP. This reaction is thought to result from the slow hydrolysis of the carboxyphosphate intermediate in the absence of a nitrogen source. In the reverse direction the enzyme can utilize carbamoyl phosphate to phosphorylate MgADP. The other two products of the partial back reaction are CO_2 and NH_3, which probably arise from the decomposition of carbamate.

Wimmer *et al.* utilized ^{18}O-labeled ATP to probe for a positional isotope exchange reaction that would support the formation of carboxyphosphate as a reactive intermediate (see Scheme XIX).[19] These workers found that in the presence of enzyme and bicarbonate CPS catalyzed a positional isotope exchange within ATP which was 1.7 times as fast as the net hydrolysis of ATP. Therefore the E · ADP · carboxyphosphate complex releases a product into solution about as fast as ATP is resynthesized and released into the bulk solution. These results have been confirmed by three other research groups.[20–22]

The second intermediate, carbamate, was probed by following the

[19] M. J. Wimmer, I. A. Rose, S. G. Powers, and A. Meister, *J. Biol. Chem.* **254,** 1854 (1979).
[20] F. M. Raushel and J. J. Villafranca, *Biochemistry* **19,** 3174 (1980).
[21] T. D. Meek, W. E. Karsten, and C. W. DeBrosse, *Biochemistry* **26,** 2584 (1987).
[22] M. A. Reynolds, N. J. Oppenheimer, and G. L. Kenyon, *J. Am. Chem. Soc.* **105,** 6663 (1983).

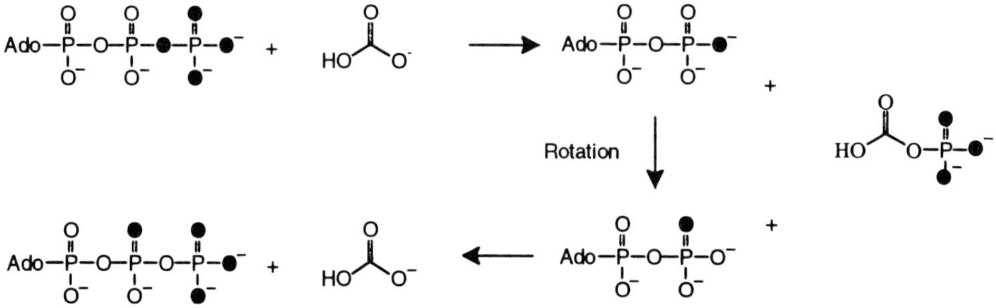

SCHEME XIX

positional exchange of an oxygen-18 label within carbamoyl phosphate.[20] Scheme XX shows that if carbamate is formed and stabilized at the active site on mixing of ADP and carbamoyl phosphate, then a positional isotope exchange reaction is possible via the torsional rotation of the carboxyl group of carbamate. The positional isotope exchange rate was found to be 4 times faster than the net synthesis of ATP, and thus carbamate is a kinetically significant reactive intermediate in the carbamoyl-phosphate synthase reaction.

Work from the Meister laboratory has provided evidence that the glutaminase reaction catalyzed by the small subunit requires an essential cysteine residue.[23] It was noted that when the glutaminase reaction was destroyed by incubation with a chloroketone analog of glutamine, the bicarbonate-dependent ATPase reaction was stimulated. Rubino *et al.* later identified the critical cysteine residue by site-directed mutagenesis to be Cys-269.[24] The two mutants (C269G and C269S) made by Rubino *et al.* not only lost all glutaminase activity, but more interestingly showed a significant enhancement of the bicarbonate-dependent ATPase reaction relative to wild-type enzyme. The bicarbonate-dependent hydrolysis of ATP normally occurs at approximately 5–10% of the rate of carbamoyl phosphate synthesis. The slow hydrolysis rate represents the protection of the unstable carboxyphosphate intermediate from water. Therefore, the stabilization of carboxyphosphate by the enzyme results from either exclusion of water from the active site or the very slow release of the reactive intermediate from the active site. Apparently, the two mutant enzymes (C269G and C269S) cannot adequately protect carboxyphosphate from water, thus permitting the partial hydrolysis reaction to compete with the overall synthesis of carbamoyl phosphate.

[23] L. M. Pinkus and A. Meister, *J. Biol. Chem.* **247,** 6119 (1972).

[24] S. D. Rubino, H. Nyunoya, and C. J. Lusty, *J. Biol. Chem.* **262,** 4382 (1987).

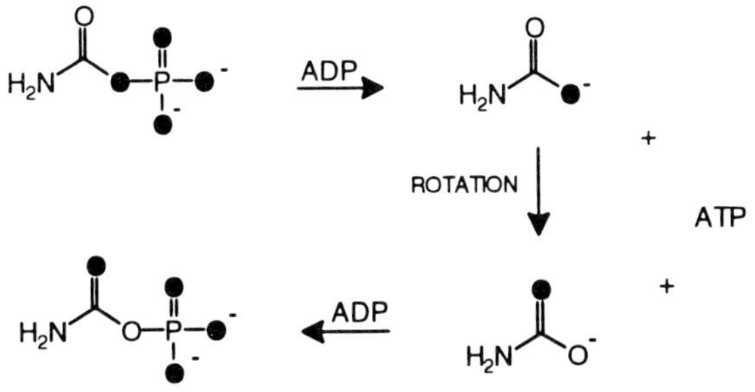

SCHEME XX

The energetics of the bicarbonate-dependent ATPase reaction catalyzed by the wild-type enzyme have been examined by Raushel and Villafranca using rapid quench and PIX experiments.[25] Scheme XXI shows the kinetic mechanism for the bicarbonate-dependent ATPase reaction along with the kinetic barrier diagram for that mechanism, where ES represents the enzyme–HCO_3^-–ATP complex and EI represents the enzyme–carboxyphosphate–ADP complex. At saturating levels of substrates the rate of the reaction shown in Scheme XXI is governed by Eq. (38), and the ratio of the PIX rate relative to the rate of net turnover of ATP (V_{ex}/V_{chem}) is given by the ratio of rate constants presented in Eq. (39):

$$V_{max} = (k_3k_5)/(k_3 + k_4 + k_5) \tag{38}$$

$$V_{ex}/V_{chem} = (k_2k_4)/[k_5(k_2 + k_3)] \tag{39}$$

The pre-steady-state time course exhibited "burst" kinetics with a rapid formation of acid-labile phosphate followed by a slower steady-state rate. Raushel and Villafranca measured the values of k_3, k_4, and k_5 as 4.2, 0.10, and 0.21 sec^{-1}, respectively, from the rapid quench and previous PIX data. They also showed that k_2 is much greater than 3.1 sec^{-1}. Therefore, the formation of carboxyphosphate is very fast, with the rate-limiting steps for net ATP hydrolysis involving either product release or hydrolysis of the intermediate in the active site.

Mullins *et al.* used the PIX technique to examine the changes in the reaction energetics of the ATPase reaction in the C269G and C269S mutants and the isolated large subunit.[26] The PIX reaction they followed has been previously shown in Scheme XIX. They found the bicarbonate-

[25] F. M. Raushel and J. J. Villafranca, *Biochemistry* **18,** 3424 (1979).

[26] L. S. Mullins, C. J. Lusty, and F. M. Raushel, *J. Biol. Chem.* **266,** 8236 (1991).

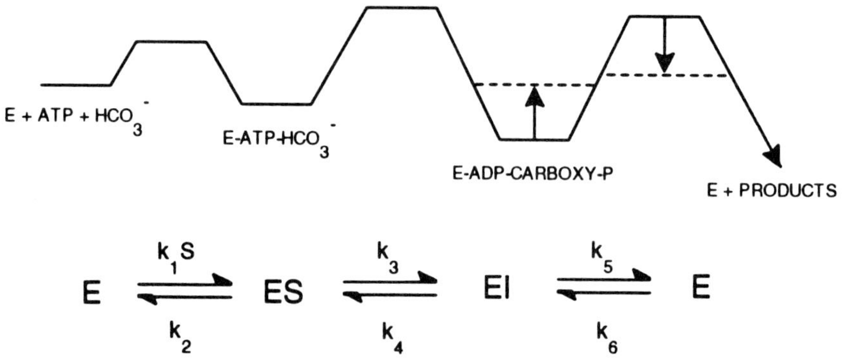

SCHEME XXI

dependent ATPase reaction for all three enzymes to be 2 to 3-fold faster than the wild-type enzyme, but 4- to 10-fold faster if glutamine is present. As no increase is observed in the NH_3-dependent carbamoyl phosphate synthesis rate, k_3 is not changed in the mutants. The significant increase in the bicarbonate-dependent ATPase reaction of the mutants can be the consequence of two possible alterations in the kinetic barrier diagram. It could reflect the stabilization of the transition state for the reaction of carboxyphosphate with water. However, this stabilization would only affect k_5 and k_6 while leaving k_4 unchanged. Alternatively, the ground state for the enzyme-bound carboxyphosphate may be destabilized, resulting in an increase in k_4 and k_5 by the same factor. Mullins *et al.* showed that the two cases could be distinguished by measuring the partitioning of the E–ADP–carboxyphosphate complex. The ratio V_{ex}/V_{chem} simplifies to k_4/k_5 since $k_2 \gg k_3$ [Eq. (39)]. Therefore, if the ground state of the enzyme–carboxyphosphate complex is destabilized, then the ratio V_{ex}/V_{chem} will be identical for the mutants and the wild-type enzyme. If the transition state for the hydrolysis of carboxyphosphate is stabilized, however, the ratio V_{ex}/V_{chem} will be reduced in the mutants relative to the wild-type enzyme.

Mullins *et al.* observed that the ratio of the PIX rate relative to ATP turnover was identical for the wild type, C269G, C269S, and the isolated large subunit.[26] Therefore, the alteration in the energetics is due to the destabilization of the ground state for the enzyme-bound carboxyphosphate–ADP complex. In the presence of NH_3, however, the value of V_{ex}/V_{chem} was reduced over 20-fold. Thus the increased turnover in the presence of NH_3 results from the substantial stabilization of the transition state for the reaction of ammonia with carboxyphosphate relative to the reaction with water.

Other Examples

A number of other enzymes have been successfully analyzed using positional isotope exchange methodology. Bass *et al.* have probed the reaction catalyzed by adenylosuccinate synthetase,[27] which involves the synthesis of adenylosuccinate from GTP, IMP, and aspartate. Incubation of [β,γ-^{18}O]GTP, IMP, and enzyme resulted in scrambling of the label from the $\beta\gamma$-bridge position to the β-nonbridge positions. No scrambling was observed in the absence of IMP, and the addition of aspartate was not required for positional isotope exchange to occur. This result has been interpreted to support a two-step mechanism for the synthesis of adenylosuccinate where the GTP phosphorylates the carbonyl oxygen of IMP to form a phosphorylated intermediate. The phosphorylated intermediate subsequently reacts with the α-amino group of aspartate to generate the final product, adenylosuccinate.

The Lowe laboratory has examined the reactions catalyzed by the aminoacyl-tRNA synthetases using the PIX technique.[28–30] The enzymes are responsible for the condensation of amino acids to the cognate tRNA. The reaction mechanism has been proposed to involve the formation of an aminoacyl adenylate intermediate from an amino acid and ATP. In support of this mechanism Lowe *et al.* have shown that the isoleucyl-, tryrosyl-, and methionyl-tRNA synthetases all catalyze a positional isotope exchange reaction from the β-nonbridge position of ATP to the $\alpha\beta$-bridge position in the presence of the required amino acid. No exchange was observed in the absence of the amino acid, nor was any PIX reaction observed in the presence of the dead-end alcohol analogs of the amino acids.

The enzyme CTP synthase catalyzes the formation of CTP from glutamine, ATP, and UTP. The two most reasonable reaction mechanisms involve either the attack of ammonia at C-4 of UTP to form the carbinolamine or, alternatively, the phosphorylation of the carbonyl oxygen of UTP by the ATP. von der Saal *et al.* have shown that the enzyme will catalyze the exchange of label from the $\beta\gamma$-bridge position to the β-nonbridge position in the presence of UTP.[31] No ammonia or glutamine is

[27] M. B. Bass, H. J. Fromm, and F. B. Rudolf, *J. Biol. Chem.* **259,** 12330 (1984).
[28] G. Lowe, B. S. Sproat, G. Tansley, and P. M. Cullis, *Biochemistry* **22,** 1229 (1983).
[29] G. Lowe, B. S. Sproat, and G. Tansley, *Eur. J. Biochem.* **130,** 341 (1983).
[30] G. Lowe and G. Tansley, *Tetrahedron* **40,** 113 (1984).
[31] W. von der Saal, P. M. Anderson, and J. J. Villafranca, *J. Biol. Chem.* **260,** 14993 (1985).

required for the reaction to be observed. This result is consistent only with the phosphorylated UTP intermediate, and thus the carbinolamine intermediate can be discarded.

The enzymatic synthesis of GMP follows a reaction scheme that is analogous to the synthesis of CTP. The enzyme GMP synthetase utilizes ATP, XMP, and glutamine to construct the final bond in the formation of GMP. The proposed reaction mechanism involves the adenylation the carbonyl oxygen of XMP by ATP. von der Saal *et al.* were able to demonstrate that on incubation of ^{18}O-labeled ATP and XMP a PIX reaction occurred which did not require the presence of glutamine or ammonia.[32]

The enzyme pyruvate-phosphate dikinase catalyzes the formation of phosphoenolpyruvate (PEP) from ATP, phosphate, and pyruvate. The other two products of the reaction are AMP and pyrophosphate. The proposed reaction mechanism is thought to involve at least three separate reactions. ATP pyrophosphorylates the enzyme in the first reaction, and this intermediate subsequently phosphorylates phosphate to produce pyrophosphate and a phosphorylated enzyme intermediate. In the last step the phosphorylated enzyme transfers the phosphoryl group to pyruvate to form the ultimate product, PEP. Wang *et al.* used a variety of ^{18}O-labeled ATP molecules to demonstrate that the observed PIX reactions were consistent with the proposed reaction mechanism.[33]

The enzyme PEP carboxykinase catalyzes the formation of PEP and CO_2 from oxaloacetate and GTP. Chen *et al.* have utilized PIX methodology to examine the partitioning of enzyme–product complexes.[34] No positional isotope exchange was observed within the labeled GTP under initial velocity conditions when enzyme was mixed with oxaloacetate and [β, γ-^{18}O]GTP. These results have been interpreted to indicate that at least one of the products dissociates rapidly from the enzyme–GDP–PEP–CO_2 complex relative to the net rate of GTP formation from the complex.

Summary

The positional isotope exchange technique has been found to be quite useful for the identification of reaction intermediates in enzyme-catalyzed reactions. For reactions where intermediates are not expected the method can be used with great utility for the quantitative determination of the partitioning of enzyme–product complexes. However, it must be remem-

[32] W. von der Saal, C.S. Crysler, and J. J. Villafranca, *Biochemistry* **24,** 5343 (1985).

[33] H. Ch. Wang, L. Ciskanik, D. Dunaway-Mariano, W. von der Saal, and J. J. Villafranca, *Biochemistry* **27,** 625 (1988).

[34] C. Y. Chen, Y. Sato, and V. L. Schramm, *Biochemistry* **30,** 4143 (1991).

bered that it has been explicitly assumed that the functional group undergoing positional exchange is free to rotate. This assumption is not always valid since examples have been discovered where the functional group rotation is indeed hindered. For instance, in the reaction catalyzed by argininosuccinate synthetase a PIX reaction was not observed on incubation of ATP and citrulline even though a citrulline–adenylate complex has been identified from rapid quench experiments.[35]

Acknowledgments

The authors are grateful for financial support from the National Institutes of Health (DK30343, GM33894, and GM49706).

[35] L. W. Hilscher, C. D. Hanson, D. H. Russel, and F. M. Rauschel, *Biochemistry* **24,** 5888 (1985).

[16] Manipulating Phosphorus Stereospecificity of Adenylate Kinase by Site-Directed Mutagenesis

By MING-DAW TSAI, RU-TAI JIANG, TERRI DAHNKE, and ZHENGTAO SHI

Introduction

Nucleoside phosphorothioates[1,2] occupy an important niche in the fields of molecular biology, biochemistry, and mechanistic enzymology. The applications in molecular biology, which include DNA sequencing and oligonucleotide-directed mutagenesis, have been reviewed.[3] The applications of nucleoside phosphorothioates to various biochemical problems have also been reviewed by Eckstein.[4] Many applications take advantage of the stereospecificity of enzymes toward specific isomer(s) of nucleoside phosphorothioates, a property uncovered by mechanistic enzy-

[1] ADP, Adenosine 5′-diphosphate; ADPαS, adenosine 5′-(1-thiodiphosphate); AK, adenylate kinase; AMP, adenosine 5′-monophosphate; AMPS, adenosine 5′-monothiophosphate; AP_5A, P^1,P^5-bis(5′-adenosyl)pentaphosphate; ATP, adenosine 5′-triphosphate; ATPαS, adenosine 5-(1-thiotriphosphate); ATPβS, adenosine 5′-(2-thiotriphosphate); EDTA, ethylenediaminetetraacetic acid; Tris, 2-amino-2-(hydroxymethyl)-1,3-propanediol.

[2] D. A. Usher, E. S. Erenrich, and F. Eckstein, *Proc. Natl. Acad. Sci. U.S.A.* **69,** 115 (1972).

[3] F. Eckstein and G. Gish, *Trends Biochem. Sci.* **14,** 97 (1989).

[4] F. Eckstein, *Annu. Rev. Biochem.* **54,** 367 (1985).

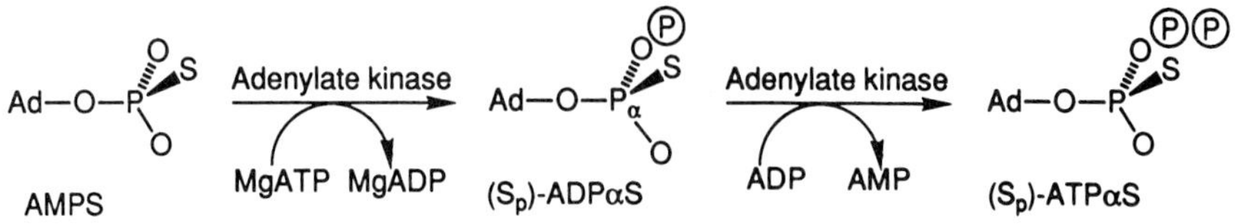

FIG. 1. Scheme showing the stereospecificity of wild-type adenylate kinase at the AMP site (AMPS to ADPαS) and ATP site (ADPαS to ATPαS).

mologists and subsequently exploited to study mechanisms of various enzymatic phosphoryl and nucleotidyl transfer reactions.[5-9]

Adenylate kinase (AK) was one of the first enzymes studied by the use of nucleoside phosphorothioates. Using the methods of coupled enzymatic reactions and ^{31}P nuclear magnetic resonance (NMR) spectroscopy, Sheu and Frey[10] demonstrated that the phosphorylation of AMPS catalyzed by rabbit muscle AK is highly stereospecific and yields predominantly "isomer A" of ADPαS. Later work established the phosphorus configuration of this isomer to be S_p.[6] The specificity toward MgADPαS or MgATPαS at the MgATP site has also been examined in several types of adenylate kinase. Without exception, the S_p isomer is preferred over R_p in AK from rabbit muscle,[11] porcine muscle,[12] and yeast.[13] Figure 1 shows the stereospecificity of both reactions (AMPS to ADPαS and ADPαS to ATPαS).

Although a vast amount of work has been reported on the stereochemical properties of the enzyme at the substrate level, little is known about the structural basis of the observed stereospecificity at the enzyme level. In general terms, the specific isomer preferred by the enzyme is the isomer that can fit into the active site and cause the least "pain" (owing to replacing an O with S) to the enzyme during the catalytic process. The stereospecificity, therefore, is a consequence of the balance of the enzyme–substrate interactions at the active site. A perturbation of such interactions can then be expected to lead to a perturbation in the stereo-

[5] P. A. Frey, J. P. Richard, H.-T. Ho, R. S. Brody, R. D. Sammons, and K.-F. Sheu, this series, Vol. 87, p. 213.
[6] F. Eckstein, P. J. Romaniuk, and B. A. Connolly, this series, Vol. 87, p. 197.
[7] M. Cohn, *Acc. Chem. Res.* **15,** 326 (1982).
[8] M.-D. Tsai, this series, Vol. 87, p. 235.
[9] P. A. Frey, *Adv. Enzymol. Relat. Areas Mol. Biol.* **49,** 119 (1989).
[10] K.-F. R. Sheu and P. A. Frey, *J. Biol. Chem.* **252,** 4445 (1977).
[11] F. Eckstein and R. S. Goody, *Biochemistry* **15,** 1685 (1976).
[12] A. G. Tomasselli and L. H. Noda, *Fed. Proc.* **40,** 1864 (1981).
[13] H. R. Kalbitzer, R. Marquetant, B. A. Connolly, and R. S. Goody, *Eur. J. Biochem.* **133,** 221 (1983).

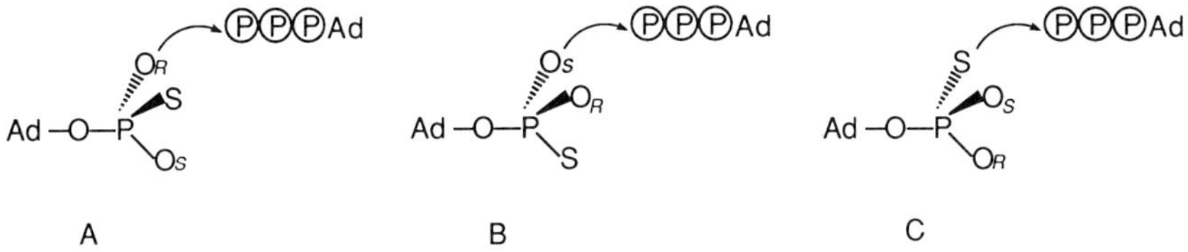

FIG. 2. Possible conformations of the P_α of AMPS with (A) *pro-R* oxygen, (B) *pro-S* oxygen, and (C) sulfur as the attacking atom for the γ-phosphate of ATP.

specificity. As shown by the elegant experiments of Jaffe and Cohn,[14] a reversal in stereospecificity can be achieved for E · metal · ATP complexes by substitution of cadmium, which preferentially coordinates sulfur over oxygen, for magnesium, which favors oxygen over sulfur. A reversal of stereospecificity from such experiments provides strong evidence that the metal ion coordinates directly to the particular phosphorothioate group during catalysis. By a complementary approach, one would predict that site-specific mutants of key residues will perturb the balance of interactions and thus the stereospecificity if these residues interact directly with the phosphoryl moiety. There is always a possibility that a perturbation can also be caused by indirect interactions or by a change in the conformation of the enzyme. However, the indirect interactions are likely to have smaller and nonspecific effects.

The stereospecificity pertaining to AMPS requires additional explanation. Here the choice is between two diastereotopic oxygens, not between two diastereomers. The preference of wild-type AK in phosphorylating the *pro-R* oxygen can be explained (from the substrate point of view) by a preferred conformation (conformer A in Fig. 2) in which the *pro-R* oxygen is positioned to attack the γ-phosphate of ATP. Such a preferred conformation is again a consequence of the balance of all interactions between the thiophosphoryl group and the active site residues. Perturbations of such interactions could lead to rotation of the O—PSO_2 bond (to find a new conformation least painful to the enzyme) and thus a perturbation in the observed stereospecificity.

The purpose of this chapter is to explain how to use site-directed mutagenesis to manipulate the stereospecificity of adenylate kinase in the conversion of AMPS to ADPαS (at the AMP site) and the conversion of ADPαS to ATPαS (at the MgATP site), by showing examples obtained from our laboratory. The stereospecificity toward R_p and S_p isomers of ATPβS was not examined owing to the very low activity of this substrate.

[14] E. K. Jaffe and M. Cohn, *J. Biol. Chem.* **253,** 4823 (1978).

Materials and Methods

Synthesis of Phosphorothioate Analogs

The single isomers (R_p)- and (S_p)-ATPαS, as well as the mixture (R_p + S_p)-ADPαS, are synthesized by the methods of Eckstein and Goody.[11] The single isomers (R_p)- and (S_p)-ADPαS are synthesized as described by Sheu and Frey.[10] The compounds are characterized by the ^{31}P chemical shifts[10,15] and high-performance liquid chromatography (HPLC) retention factors.[16]

Construction and Purification of Enzymes

The *Escherichia coli* expression system for chicken muscle AK was kindly provided by Nakazawa and co-workers.[17] The AK mutants are constructed according to the method of Eckstein and co-workers[18,19] by using the mutagenesis kit from Amersham (Arlington Heights, IL). The full-length AK gene is sequenced by the dideoxy nucleotide method of Sanger to ensure that no undesirable mutations have occurred. The method used to purify wild-type AK as described by Tian *et al.*[20] is modified as required for purifying the mutants. The purity of the enzyme is examined by sodium dodecyl sulfate–polyacrylamide gel electrophoresis (SDS–PAGE) with silver staining on a PhastSystem.

Steady-State Kinetics

The kinetic experiments are carried out at 30° and pH 8.0 by monitoring ADP formation with pyruvate kinase/lactate dehydrogenase as the coupling system.[21] The values of k_{cat}, K (Michaelis constant), and K_i (dissociation constant) are obtained by varying both MgATP and AMP concentrations followed by analyses according to the equation of Cleland[22] for a random Bi Bi system. The details have been described previously.[20]

[15] E. K. Jaffe and M. Cohn, *Biochemistry* **17,** 652 (1978).

[16] R. D. Sammons, Ph.D. Dissertation, The Ohio State University, Columbus (1982).

[17] Y. Tanizawa, F. Kishi, T. Kaneko, and A. Nakazawa, *J. Biochem. (Tokyo)* **101,** 1289 (1987).

[18] J. W. Taylor, W. Schmidt, R. Cosstick, A. Okruszek, and F. Eckstein, *Nucleic Acids Res.* **13,** 8749 (1985).

[19] J. W. Taylor, J. Ott, and F. Eckstein, *Nucleic Acids Res.* **13,** 8765 (1985).

[20] G. Tian, H. Yan, R.-T. Jiang, F. Kishi, A. Nakazawa, and M.-D. Tsai, *Biochemistry* **29,** 4296 (1990).

[21] D. G. Rhoads and J. M. Lowenstein, *J. Biol. Chem.* **243,** 3963 (1968).

[22] W. W. Cleland, *in* "Investigation of Rates and Mechanisms of Reactions, Part 1" (C. F. Bernasconi, ed.), p. 791. Wiley, New York, 1986.

Phosphorus-31 Nuclear Magnetic Resonance Methods

All ^{31}P NMR experiments are performed on Bruker AM-250 and AM-300 NMR spectrometers at 30°. All chemical shifts are referenced to external 85% H_3PO_4. All spectra are broadband decoupled with the WALTZ sequence. The spectral width is 75 ppm, and 16–32 K data points are recorded for each spectrum in the quadrature detection mode. A 45° pulse and a 0–1.0 sec relaxation delay are used. Acquisition times range from 1.5 to 2.0 sec, repetition times are 2.0 to 2.5 sec, and 1000–23,000 transients are obtained. A 0.5–2.0 Hz line broadening is applied to the time domain data prior to Fourier transformation. Unless otherwise noted, only the regions of the P_α resonances are shown in the spectra.

Assignment and Quantification of Isomers

The assignment of diastereomers within the reaction mixture is accomplished by addition of known isomers to the sample, whose relative chemical shifts are in agreement with those reported previously.[10,15,23] It should be noted that the absolute values of thiophosphate resonances are extremely sensitive to pH and magnesium ion concentration.[15] As a result, differences in chemical shifts for the same species may occur as the sample conditions differ from reaction to reaction. This fact should be kept in mind in comparing the chemical shifts of the same species in different figures in this chapter. The intensities of various components have been measured by cutting and weighing from greatly expanded spectra.

Manipulation of Stereospecificity toward Adenosine 5′-Monothiophosphate

Confirmation of Wild-Type Adenylate Kinase Stereospecificity

As the first step, the specificity at both binding sites of chicken muscle AK used in our work is established and compared to that of other variants of AK reported previously. The reaction of wild-type AK with MgATP and AMPS is followed by ^{31}P NMR spectroscopy, and a representative spectrum is shown in Fig. 3A. This reaction clearly affords one isomer of ADPαS, which is subsequently assigned as S_p by the addition of known isomers. The result in Fig. 3A also indicates that the newly formed (S_p)-ADPαS is readily converted to (S_p)-ATPαS at the MgATP site, suggesting that the preferred configuration at the P_α position of MgATP is also S_p.

[23] M.-D. Tsai, *in* "Phosphorus-31 NMR: Principles and Applications" (D. G. Gorenstein, ed.), p. 175. Academic Press, New York, 1984.

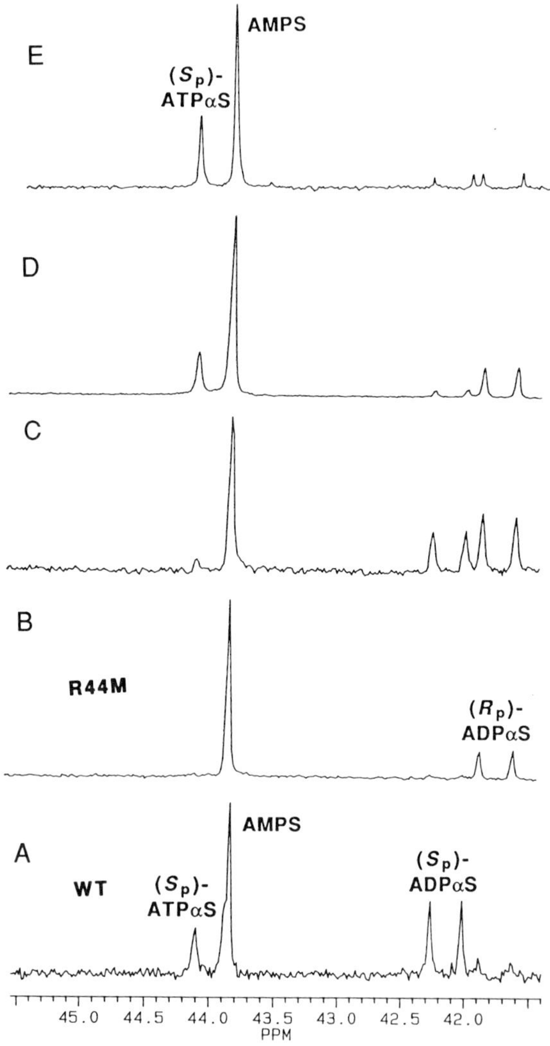

FIG. 3. Comparison by ^{31}P NMR of wild-type and R44M stereospecificity. (A) Wild-type AK, at 4 hr (midpoint of acquisition) after addition of the enzyme; (B) R44M, 2 hr; (C) addition of ADPαS ($R_p/S_p = \frac{1}{2}$) to B, 4 hr; (D) continuation of C, 9 hr; (E) continuation of D, 70 hr. The right half of the doublet of (S_p)-ATPαS overlaps with the singlet of AMPS. The position of (R_p)-ATPαS should be upfield from AMPS. The starting reaction mixture (2 ml) consisted of 22 m*M* AMPS, 75 m*M* ATP, 45 m*M* $Mg(NO_3)_2$, and about 40 μg of wild-type enzyme or 400 μg of R44M, in a 50 m*M* Tris buffer containing 50 m*M* KCl and 2.5 m*M* EDTA, pH 7.8. [Reprinted with permission from R.-T. Jiang, T. Dahnke, and M.-D. Tsai, *J. Am. Chem. Soc.* **113,** 5485 (1991). Copyright 1991 American Chemical Society.]

In conclusion, the stereospecificity of chicken muscle AK at both sites is consistent with previously established stereospecificities of AK from other sources.[10-13]

It is important to note that the degree of stereospecificity revealed by ^{31}P NMR is only qualitative and depends on the extent of reaction. Although the observed stereospecificity arises from differences in the thermodynamics of the interactions between active site residues and the two isomers (or conformers), there is little difference in the free energies of the isomers outside of the active site. Thus the observed stereospecificity is a kinetic event, and the ratio of S_p/R_p isomers should eventually reach the equilibrium value of about 1 on prolonged reaction.

Reversal of AMPS Stereospecificity with R44M Mutant

Our site-specific mutagenesis studies have determined that Arg-44 interacts with AMP starting with the binary complex.[24] This conclusion is based on selective 20- to 30-fold increases in the dissociation and Michaelis constants of AMP for the R44M mutant AK (Table I) and on the observation that the conformation of the mutant is unperturbed. Although the results of kinetic analysis do not specify how Arg-44 interacts AMP, the positively charged arginine side chain most likely interacts with the negatively charged phosphate group of AMP. In the crystal structure of the yeast AK complex with $MgAP_5A$,[25] Arg-53 (family numbering, corresponding to Arg-44 in muscle AK) is in contact with a phosphate group; in the crystal structure of the AMP complex of type 3 AK (from mammalian mitochondrial matrix),[26,27] Arg-41 (also corresponding to Arg-44 in muscle AK) has also been shown to interact with the phosphate group of AMP. This evidence led us to predict that the stereospecificity in the conversion of AMPS to ADPαS could be perturbed in R44M.

The reaction of R44M with AMPS and MgATP is shown in Fig. 3B. The spectrum clearly shows that R44M reverses the stereospecificity of the reaction, since the R_p isomer of ADPαS, instead of S_p as in the case of wild-type AK shown in Fig. 3A, is the predominant product. Furthermore, the fact that no ATPαS is found in Fig. 3B suggests that the stereospecificity at the P_α of the MgATP site has not been altered; that is, the change in stereospecificity appears to be a localized effect as predicted from the kinetic data. Figure 3C confirms the assignment by addition of $(R_p + S_p)$-ADPαS to the reaction mixture of B, and Fig. 3D confirms that

[24] H. Yan, T. Dahnke, B. Zhou, A. Nakazawa, and M.-D. Tsai, *Biochemistry* **29,** 10956 (1990).
[25] U. Egner, A. G. Tomasselli, and G. E. Schulz, *J. Mol. Biol.* **195,** 649 (1987).
[26] K. Diederichs and G. E. Schulz, *Biochemistry* **29,** 8138 (1990).
[27] K. Diederichs and G. E. Schulz, *J. Mol. Biol.* **217,** 541 (1991).

TABLE I
STEADY-STATE KINETIC DATA OF WILD-TYPE ADENYLATE KINASE AND VARIOUS MUTANT FORMS[a]

Parameter	Unit	Wild type[b]	R44M[d]	R97M[c]	R128A[d]	T23A[e]
k_{cat}	sec^{-1}	650	210	22	36	755
$K_{(MgATP)}$	mM	0.042	0.048	0.083	0.65	0.31
$K_{(AMP)}$	mM	0.098	3.53	2.78	1.38	0.60
$k_{cat}/K_{(MgATP)}$	sec^{-1} M^{-1}	1.55×10^7	0.44×10^6	0.26×10^6	5.5×10^4	0.24×10^7
$k_{cat}/K_{(AMP)}$	sec^{-1} M^{-1}	0.66×10^7	5.9×10^4	7.9×10^3	2.6×10^4	0.13×10^7
$K_{i(MgATP)}$	mM	0.16	0.11	0.066	1.63	0.76
$k_{i(AMP)}$	mM	0.37	8.25	2.22	3.48	1.47

[a] The kinetic data were obtained by varying concentrations of both substrates.
[b] Data from Tian *et al.*[20]
[c] Data from Dahnke *et al.*[28]
[d] Data from Yan *et al.*[24]
[e] Data from Shi *et al.*[28a]

the S_p isomer of ADPαS is indeed preferentially converted to ATPαS at the MgATP site. Figure 3E demonstrates that on prolonged reaction both isomers can be converted to ATPαS, since the stereospecificity is a kinetic effect. We suggest that this occurs via the back reaction of (R_p)-ADPαS to AMPS, with ephemeral formation of (S_p)-ADPαS, since direct formation of (S_p)-ATPαS by the phosphorylation of (R_p)-ADPαS cannot occur.[28]

Enhancement of AMPS Stereospecificity with R97M Mutant

Like the case with Arg-44, the corresponding mutation to methionine at position 97 yields selective perturbation of AMP binding, beginning at the stage of the binary complex (Table I), with no significant structural aberrations introduced by the site-specific change.[28] The crystal structure of the type 3 AK · AMP complex[26,27] also indicates that the guanidinium group of Arg-92 (Arg-97 in muscle AK) points toward the phosphoryl group. We therefore predicted that the stereospecificity of R97M should also be perturbed; because the effect cannot be identical to that of R44M, it is likely to enhance the stereospecificity of the wild-type reaction.

Because the ratio of R_p/S_p in the product ADPαS depends on the extent of reaction of AMPS as noted earlier, verification of an enhanced stereospecificity requires the demonstration that the ratio R_p/S_p or the percentage of R_p in the total phosphorothioate species in the reaction

[28] T. Dahnke, Z. Shi, H. Yan, and M.-D. Tsai, *Biochemistry* **31,** 6318 (1992).
[28a] Z. Shi, I.-J. L. Byeon, R.-T. Jiang, and M.-D. Tsai, *Biochemistry* **32,** 6450 (1993).

mixture, is smaller at a later stage of the reaction catalyzed by R97M than that at an earlier stage of reaction catalyzed by the wild type. The reaction mixtures of the wild type after 9, 17, and 29%, respectively, of AMPS have been converted to products are shown in Fig. 4A–C. The spectrum of R97M at 27% conversion is seen in Fig. 4D. The percent R_p in the total phosphorothioate species in the reaction mixture present for the wild-type reactions is approximately 0.27, 0.54, and 1.7% for points A–C (Fig. 4), respectively, whereas in spectrum D (R_p)-ADPαS is undetectable. It should be noted that the value of 1.7% for spectrum C could be misleadingly high, since equilibrium has already been established. In the absence of a change in stereospecificity, the theoretical percentage of R_p for R97M at 27% conversion will be considerably larger than 0.54% but less than 1.7%. If we select 1% as a conservative estimate, the amount of (R_p)-ADPαS in spectrum D should be 3–4 times the amount present in spectrum A (0.27%) were there no change in stereospecificity. Because the overall signal-to-noise ratios (S/N) of all spectra are similar and the S/N for (R_p)-ADPαS in spectrum A is about 3, the actual amount of (R_p)-ADPαS in spectrum D has decreased by at least 10-fold (3×3.5). Therefore, the stereospecificity of R97M has been enhanced by at least 10-fold.[28]

Stereospecificity at P_α of MgATP

Unperturbed ATPαS Stereospecificity with R44M and R97M Mutants

The results described above already suggest that the stereospecificity exhibited by the R44M and R97M mutants is not perturbed at the MgATP site. Another experiment using (R_p + S_p)-ADPαS provides additional support for the observed P_α stereospecificity of the MgATP site of wild type, R44M, and R97M.[28] A time course of the reaction of R97M and the ADPαS diastereomers in Fig. 5 illustrates the steady removal of (S_p)-ADPαS for the selective coupling of the S_p component to form (S_p)-ATPαS and AMPS (Fig. 5B–D). By comparison, the R_p isomer remains largely unreacted. Identical results are achieved in the wild-type reaction (Fig. 6A).

An analysis of the reaction of R44M with (R_p + S_p)-ADPαS as the sole substrates in Fig. 6B results in the favorable coupling of (R_p)-ADPαS (at the AMP site) with (S_p)-ADPαS (at the ATP site) to form AMPS and (S_p)-ATPαS, respectively, since both isomers have been consumed. This result not only reaffirms the S_p stereospecificity of the MgATP binding site for R44M, but strengthens the conclusion for a reversal of stereospecificity of the AMP site.

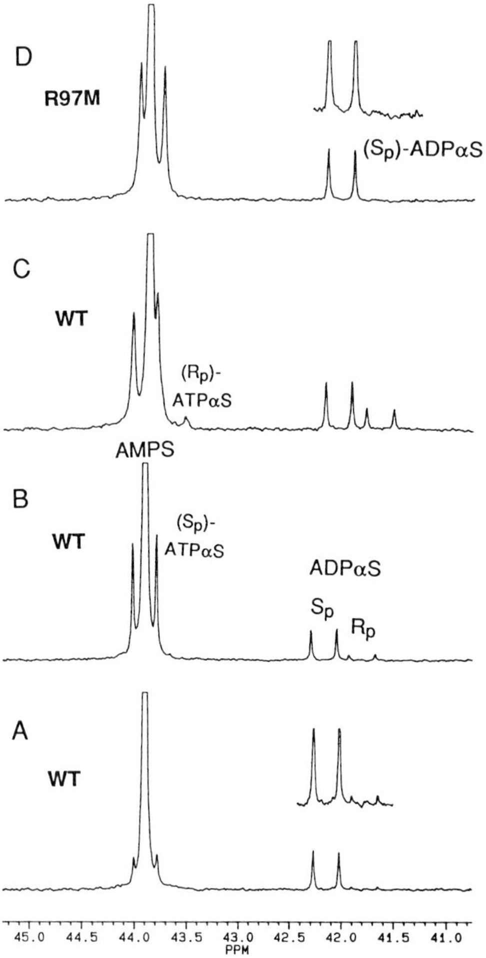

FIG. 4. The ^{31}P NMR spectra showing the conversion of AMPS to ADPαS (and subsequent conversion to ATPαS). (A–C) Wild-type AK, after 9, 17, and 29%, respectively, of AMPS has reacted; (D) R97M, at 27% conversion. Except for sample C, no (R_p)-ATPαS was detectable even under another set of conditions optimized for the detection of ATPαS (spectra not shown). The starting reaction mixture (600 μl) was similar to that of Fig. 3, except for the addition of D_2O (final concentration 15%) to the buffer. Spectra A–C were obtained after the addition of 2, 25, and 10 μg of wild-type AK, whereas 20 μg of R97M was added in D. EDTA and triethylamine were added to optimize conditions for the detection of ADPαS. [Reprinted with permission from T. Dahnke, R.-T. Jiang, and M.-D. Tsai, *J. Am. Chem. Soc.* **113,** 9388 (1991). Copyright 1991 American Chemical Society.]

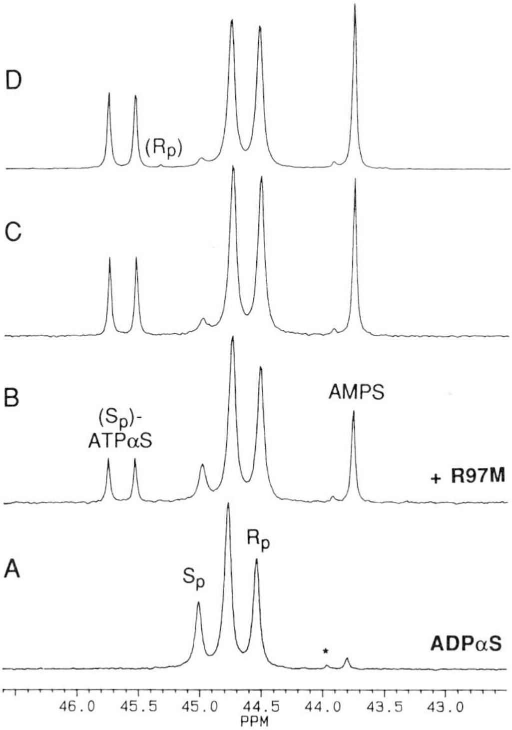

FIG. 5. Time course of the reaction of R97M and (R_p + S_p)-ADPαS. (A) (R_p + S_p)-ADPαS ($R_p/S_p \approx \frac{4}{3}$); (B) 30 min after the addition of R97M; (C) continuation of B, 1.5 hr; (D) continuation of C, 72 hr. A small amount of (R_p)-ATPαS can be detected after prolonged reaction in D. Sample conditions for the reverse reaction were similar to those of Fig. 4, except for the replacement of ATP and AMPS with 22 m*M* ADPαS. Spectra were obtained after the addition of 25 μg of R97M. The asterisk (*) marks a small impurity introduced during the synthesis of the phosphorothioate analogs. (Reprinted with permission from Dahnke *et al.*[28] Copyright 1992 American Chemical Society.)

Relaxation of ATPαS Stereospecificity with R128A Mutant

Arginine-128 does not play a clear structural or functional role since almost every kinetic (Table I) and structural property of the R128A mutant has been perturbed to a small extent.[24] Examination of the stereospecificity of R128A may shed some light on the contribution of this residue to substrate binding.

The two standard experiments were first performed with the R128A mutant. Reaction of AMPS with MgATP gave results (spectrum not shown) essentially the same as that of the wild type shown in Fig. 3A. In

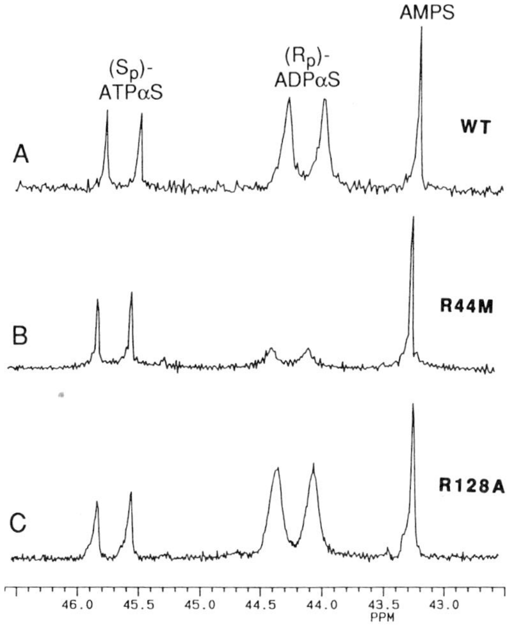

FIG. 6. Reaction of (R_p + S_p)-ADPαS with (A) wild-type AK, (B) R44M, and (C) R128A. Sample conditions for the reactions were similar to those of Fig. 5. The spectra were obtained after the addition of around 20–50 μg of enzyme. (A and B partially reprinted with permission from Dahnke *et al.*[28] Copyright 1992 American Chemical Society.)

the presence of (R_p + S_p)-ADPαS as sole substrates, the products shown in Fig. 6C are also essentially the same as those for the wild type (Fig. 6A). Based on the results of the two experiments, one might prematurely conclude that the stereospecificity of wild-type AK and R128A is identical. However, this was disproved on the basis of a third experiment. Figure 7 shows the products of the reactions of ADP and the single diastereomer (R_p)-ADPαS catalyzed by the wild type and R128A. The reaction mixture of wild-type AK consists of (R_p)-ADPαS, AMPS, and (S_p)-ATPαS, whereas that of R128A consists of (R_p)-ADPαS, AMPS, and (R_p)-ATPαS. The formation of (S_p)-ATPαS with the wild type can be explained by the (R_p)-ADPαS being dephosphorylated to form AMPS at the AMP site (allowed since the AMP site stereospecificity is not absolute), and further combined with ATP to form (S_p)-ADPαS (not detectable in Fig. 7) and subsequently (S_p)-ATPαS. The formation of (R_p)-ATPαS with R128A, however, can only come from direct phosphorylation of (R_p)-ADPαS at the MgATP site. This behavior indicates a relaxation of stereospecificity at the ATP site. The degree of this relaxation may not be large judging

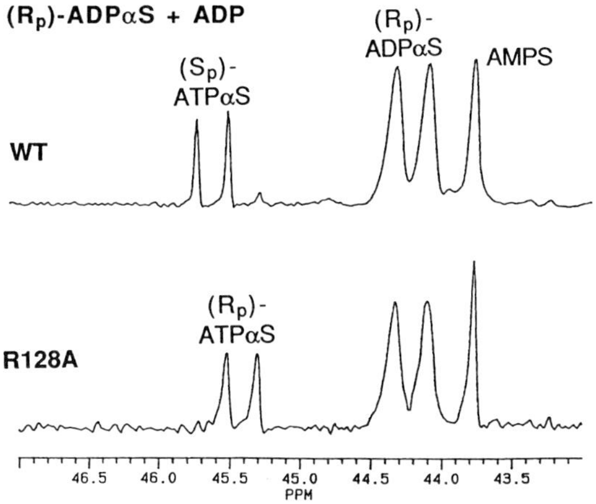

FIG. 7. Reaction of (R_p)-ADPαS and ADP with wild-type AK and R128A. Sample conditions for the reactions were similar to those of Fig. 5, except for the replacement of (R_p + S_p)-ADPαS with 22 m*M* (R_p)-ADPαS and 22 m*M* ADP. The spectra were obtained after the addition of around 20–50 μg of enzyme.

from the lack of perturbations (relative to wild type) in the two reactions starting with AMPS plus ATP and with (R_p + S_p)-ADPαS.

Because there is a relaxation of stereospecificity at the P_α site of ATP, a tempting interpretation is that the guanidinium group of Arg-128 interacts with the α-phosphate of ATP. However, because the degree of change in stereospecificity with the mutant is quite small, we can at best say that the results are consistent with such an interpretation. Further evidence needs to be sought to confirm the functional roles of this residue.

ATPαS Stereospecificity of T23A Mutant is Significantly Perturbed

The T23A mutant presents a case where kinetic data of the mutants are only minorly perturbed, but structural results suggest that the residue is likely to interact with the substrate.[29] The ^{31}P NMR analysis of the reaction between AMPS and MgATP catalyzed by T23A is illustrated in Fig. 8, which shows a noticeable difference from the result of wild-type AK in that there is a considerable amount of the R_p isomer of ATPαS (20% relative to S_p) in the product of T23A. This indicates a perturbation in the stereospecificity at the α-phosphate site of MgATP. The result, however, gives no indication as to the extent of perturbation in stereospecificity since the formation of (R_p)-ATPαS is limited by the availability of

[29] Z. Shi, I.-J. L. Byeon, R.-T. Jiang, and M.-D. Tsai, *Biochemistry* **32,** 6450 (1993).

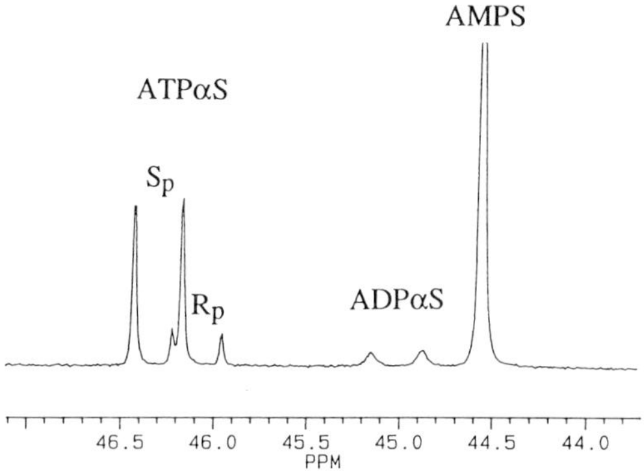

FIG. 8. Analysis by ^{31}P NMR of the reaction of AMPS and MgATP catalyzed by T23A showing the presence of (R_p)-ATPαS. The starting reaction mixture (0.6 ml) consisted of 22 m*M* AMPS, 75 m*M* ATP, 45 m*M* $Mg(NO_3)_2$, and around 100 μg of T23A in 50 m*M* Tris buffer containing 50 m*M* KCl and 2.5 m*M* EDTA, pH 7.8. The spectrum was obtained in the absence of the deuterium lock. Data acquisition started right after the addition of T23A. For better resolution, samples were adjusted to pH 14 before the spectra were taken. Notice that the peak of AMPS is shifted upfield under this condition. (Reprinted with permission from Shi *et al.*[29] Copyright 1993 American Chemical Society.)

(R_p)-ADPαS, which is controlled by the stereospecificity of the AMP site. The result also gives no indication of the stereospecificity of the AMP site since (R_p)-ADPαS does not accumulate.

Quantitative results were obtained by using (R_p)-ADPαS and (S_p)-ADPαS as substrates. In the case of the wild type, this experiment produces exclusively (S_p)-ATPαS (and AMPS) at the expense of exclusively (S_p)-ADPαS (at both sites), leaving (R_p)-ADPαS unreacted (Fig. 6A). The results with T23A, as shown in Fig. 9, indicate that a mixture of ATPαS with $R_p/S_p = 0.37:1$ is formed (spectrum B). This ratio should represent the stereochemical preference at the α-phosphorus of the MgATP site since the enzyme has ample supply of either isomer of ADPαS (spectrum A; an excess of S_p isomer was used). Because the ratio (R_p)-ATPαS/(S_p)-ATPαS is less than 0.02 (to the limit of detection) for the wild type,[28] the ratio R_p/S_p has been enhanced by more than 20-fold at the P_α of the MgATP site for T23A.

Before the above results can be used to suggest that the side chain of Thr-23 interacts directly with the α-phosphate of ATP during the catalysis by wild-type AK, it is necessary to address two issues. The first issue is whether the perturbation in stereospecificity is specific to the MgATP site. If the stereospecificity at the AMP site is also relaxed, it may be argued that the stereospecificity is relaxed nonspecifically owing to a

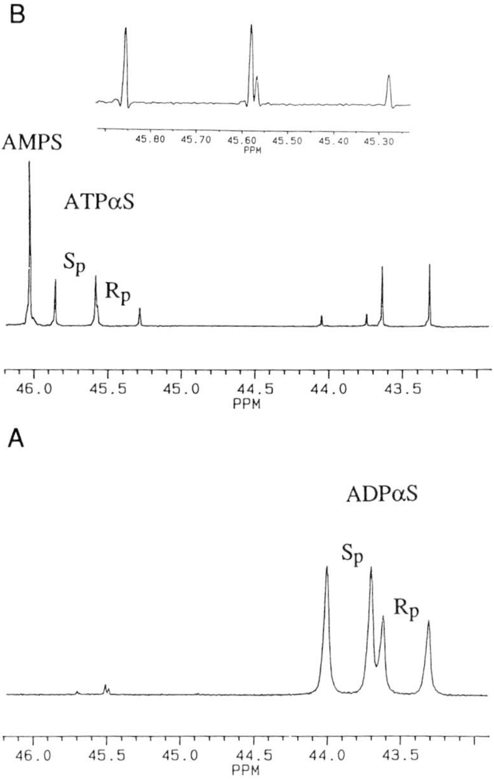

FIG. 9. Analysis by ^{31}P NMR of the reaction of (R_p)-ADPαS and (S_p)-ADPαS catalyzed by T23A. (A) Starting substrates: (R_p + S_p)-ADPαS (S_p/R_p = 1.6). (B) Reaction mixture, obtained at 30 min after the addition of T23A; the inset is a resolution-enhanced analysis of the product (R_p + S_p)-ATPαS. Sample conditions were similar to those of Fig. 6, except for the replacement of ATP and AMPS with 22 m*M* ADPαS. For better resolution, samples were adjusted to pH 5.6 before the spectra were taken. Notice that the peak of AMPS is shifted downfield under this condition.

perturbation in conformation or mechanism. This issue should be reflected in the R_p/S_p ratio of ADPαS consumed, and thus can be addressed by quantitative analyses of the results of Fig. 9. If the stereospecificity at the AMP site is strictly S_p, then the amount of R_p isomer of ADPαS consumed should be equal to the amount of R_p isomer of ATPαS formed, and AMPS should come exclusively from (S_p)-ADPαS. In other words, the ratio of [AMPS] + [(S_p)-ATPαS] + [(S_p)-ADPαS remaining] to [(R_p)-ATPαS] + [(R_p)-ADPαS remaining], defined as *R*, should be equal to the S_p/R_p ratio of the starting ADPαS (1.6). On the other hand, a relaxation in the stereospecificity of the AMP site should result in a larger R value since some of the (R_p)-ADPαS will be converted to AMPS at the AMP site, in addition to being converted to ATPαS at the MgATP site. The observed R value, on the basis of both peak height as well as integration of Fig. 9B, is 1.8; after correcting for the fact that the stereospecificity at the AMP site is around 95% instead of 100%, the observed R value becomes 1.65, which agrees with the predicted value of 1.6 within experimental errors. Thus, the stereospecificity at the AMP site of T23A is the same as that of the wild type.

The second issue is to ensure that the stereospecificity (at the MgATP site) is a "preequilibrium effect," that is, a kinetic effect. Because equilibration between R_p and S_p isomers must occur via AMPS as an intermediate, it should occur for ADPαS before ATPαS. This is not the case as shown in Fig. 9B. In addition, such equilibration should also make the observed R value deviate from the predicted value, which is not the case. Most importantly, the result is similar at an earlier stage of reaction (not shown).

The results of analysis of the T23A mutant allow us to interpret the stereochemical results: the side chain of Thr-23 interacts directly with the α-phosphate of ATP during the catalysis by wild-type AK, most likely via hydrogen bonding.[29]

Interpretation of Stereochemical Results

In Terms of Microscopic Rates

A scheme summarizing the differences in stereospecificities for wild-type AK and mutants is shown in Fig. 10. For the wild-type enzyme, the small arrows between AMPS and (R_p)-ADPαS indicate the unfavorable formation of the R_p isomer relative to the S_p isomer at the AMP site. The conversion of (R_p)-ADPαS to (R_p)-ATPαS is indicated as "unallowed" by a cross; however, it only means the rate is very slow. For R44M, the stereospecificity at the AMP site has been reversed, as indicated by the relative lengths of the arrows. For R97M, the stereospecificity at the AMP site has been enhanced; the conversion of AMPS to (R_p)-ADPαS is crossed out to indicate the low rate of this path. For R128A, the cross between (R_p)-ADPαS and (R_p)-ATPαS is replaced by small arrows to indicate relaxation of this path. In a scenario where the enzyme is given a "choice" of diastereomeric substrates, R128A still favors the S_p isomer at the ATP site (and the AMP site), as evidenced by the result of Fig. 6C. However, in the absence of (S_p)-ADPαS, the acceptance of the R_p isomer at the ATP site can be manifested as shown in Fig. 7B. For T23A, the R_p isomer

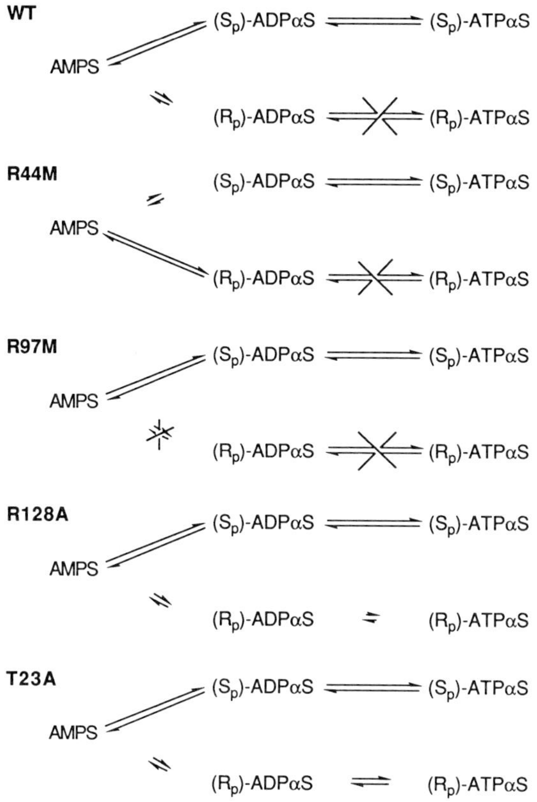

FIG. 10. Schemes summarizing the differences in stereospecificities for wild-type AK and R44M, R97M, R128A, and T23A mutants.

at the ATP site is acceptable to an extent nearly comparable to that for the S_p isomer, as indicated by the relative lengths of the arrows.

In Terms of Active Site Conformations

As shown schematically in Fig. 11, the stereospecificity of wild-type AK at the AMP site can be explained by an equilibrium between a major conformer A and a minor conformer B. The equilibrium is shifted toward conformer B in R44M, and toward conformer A in R97M, as indicated by the arrows. We speculate that Arg-44 is more important than Arg-97 in positioning the phosphoryl group during catalysis, since the major conformer is perturbed in the R44M mutant.

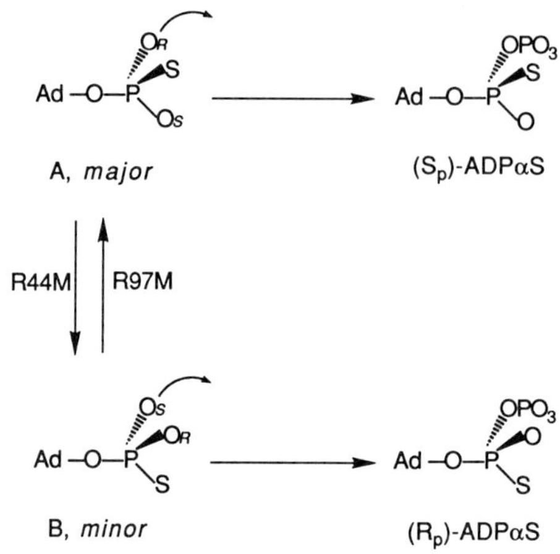

FIG. 11. Schemes showing the major and minor conformers of AMPS at the active site of wild-type AK and the conversion of these conformers to ADPαS. The equilibrium is shifted to conformer B on R44M mutation and to conformer A on R97M mutation. [Reprinted with permission from T. Dahnke, R.-T. Jiang, and M.-D. Tsai, *J. Am. Chem. Soc.* **113,** 9388 (1991). Copyright 1991 American Chemical Society.]

The role of conformer C in Fig. 2 is unclear and thus is not represented in Fig. 11. This conformer could lead to the formation of ADP with a bridging sulfur between P_β and P_α. Because such a product has never been detected, it may be tempting to consider this as a minor or nonproductive conformer. However, the finding of the high chemical reactivity of the bridging-monothio analog of pyrophosphate[30] suggests that even if a product is formed from conformer C it could be hydrolyzed back to AMPS (chemically) immediately and thus escape detection.

Potential Applications to Other Systems

The implications and possible applications of our work are many. First of all, a significant perturbation in stereospecificity is strong evidence that the mutated residue interacts with the phosphorothioate group. Our results unequivocally establish that both Arg-44 and Arg-97 interact with the phosphoryl group of AMP, and Thr-23 interacts with the P_α group of ATP, during the catalytic reaction of adenylate kinase. This method of assessing the functional role of residues is a step forward from direct interpretation of the static crystal structure or the kinetic data of mutant enzymes and

[30] C. J. Halkides and P. A. Frey, *J. Am. Chem. Soc.* **113,** 9843 (1991).

can enhance our understanding of the chemical basis of enzymatic catalysis and the biological effects of phosphorothioates. Furthermore, a combination of site-directed mutagenesis and changes in metal ions could be used to manipulate phosphorus stereospecificity so that enzymes may either accept or produce alternative diastereomers of phosphorothioate analogs. Potential candidates for manipulation of substrate stereospecificity include DNA and RNA polymerases which, without exception, all prefer the S_p isomer of dNTPαS or NTPαS.[3] With respect to immediate application, R44M provides a more direct route to the synthesis of pure (R_p)-ADPαS from AMPS, which has been previously achieved by the chemical synthesis of (R_p + S_p)-ADPαS followed by enzymatic removal of the S_p isomer.[11]

Acknowledgments

This work was supported by a grant from National Science Foundation (DMB8904727) and a grant from National Institutes of Health (GM43268).

[17] Equilibrium Isotope Exchange in Enzyme Catalysis

By FREDERICK C. WEDLER

Background and General Properties

Theoretical Basis

Isotope exchange kinetics at chemical equilibrium has been recognized as an important, if underdeveloped, subcategory of steady-state kinetics for over three decades. The fundamental concepts were pioneered and delineated by Boyer,[1] Boyer and Silverstein,[2] and Alberty *et al.*[3] Later contributions to theoretical concepts were made by Yagil and Hoberman,[4] Yagil,[5] Britton,[6,7] Britton and Dann,[8] Morales *et al.*,[9] Flossdorf and Kula,[10]

[1] P. D. Boyer, *Arch. Biochem. Biophys.* **82,** 387 (1959).
[2] P. D. Boyer and E. Silverstein, *Acta Chim. Scand.* **17,** S195 (1963).
[3] R. A. Alberty, V. Bloomfield, L. Peller, and E. L. King, *J. Am. Chem. Soc.* **84,** 4381 (1962).
[4] G. Yagil and H. D. Hoberman, *Biochemistry* **8,** 352 (1969).
[5] G. Yagil, *J. Theor. Biol.* **61,** 73 (1976).
[6] H. G. Britton, *Arch. Biochem. Biophys.* **177,** 167 (1966).
[7] H. G. Britton, *Tech. Life Sci.: Biochem.* **B115,** 1 (1978).
[8] H. G. Britton and L. G. Dann, *Biochem. J.* **169,** 29 (1978).
[9] M. F. Morales, M. Horovitz, and J. Botts, *Arch. Biochem. Biophys.* **99,** 258 (1962).
[10] J. Flossdorf and M. R. Kula, *Eur. J. Biochem.* **30,** 325 (1972).

Schultz and Fisher,[11] and Wong and Hanes.[12] Ainsworth[13] and Darvey[14] have derived equations for isotope exchanges with simple systems. Despite using different initial assumptions and a variety of theoretical and algebraic approaches, the derived equations are functionally consistent and compatible with one another. Reviews of this area include those by Boyer[15] and Purich and Allison.[16] The reader is also directed to summaries in monographs, including those by Segal,[17] Dixon and Webb,[18] Piszkiewicz,[19] and Wong.[20]

Isotope exchange kinetics fall into two subclasses: those carried out at true chemical equilibrium and those occurring as back-exchange from labeled product during net flux from reactant to product. The latter approach is especially useful for probing for central complexes and rates of product dissociation with enzymes catalyzing reactions with K_{eq} values very far from unity. A review of steady-state kinetics by Cleland[21] provides an overview of isotope exchange, including back-exchange during net catalysis. This chapter is restricted to exchanges at chemical equilibrium. As a background, we first consider some basic properties, strengths, and limitations of equilibrium isotope exchange kinetics (designated with the acronym EIEK).

Methodology

To obtain saturation curves for an isotopic exchange reaction while maintaining chemical equilibrium, one must (as a minimum) vary one reactant and one product in constant ratio. Variation of structurally related reactant–product pairs yields information about substrate binding order and the relative rates of substrate association (at subsaturating conditions) and dissociation (at saturation). Variation of structurally dissimilar pairs can lead to formation of abortive (dead-end) complexes, for example,

[11] A. R. Schulz and D. D. Fisher, *Can. J. Biochem.* **48,** 922 (1970).
[12] J. T.-F. Wong and C. S. Hanes, *Nature (London)* **203,** 492 (1964).
[13] S. Ainsworth, *J. Theor. Biol.* **64,** 381 (1977).
[14] I. G. Darvey, *J. Theor. Biol.* **42,** 55 (1973).
[15] P. D. Boyer, *Acc. Chem. Res.* **11,** 218 (1978).
[16] D. L. Purich and D. R. Allison, this series, Vol. 64, p. 3.
[17] I. H. Segel, "Enzyme Kinetics," p. 864. Wiley (Interscience), New York, 1975.
[18] M. Dixon and E. C. Webb, "Enzymes" 3rd Ed., p. 111. Academic Press, New York, 1979.
[19] D. Piszkiewicz, "Kinetics of Chemical and Enzyme-Catalyzed Reactions," p. 140. Oxford Univ. Press, New York, 1977.
[20] J. T.-F. Wong, "Kinetics of Enzyme Mechanisms," p. 171. Academic Press, New York, 1975.
[21] W. W. Cleland, *in* "The Enzymes" (D. S. Sigman and P. D. Boyer, eds.), 3rd Ed., Vol. 19, Mechanisms of Catalysis, p. 122. Academic Press, New York, 1990.

with lactate dehydrogenase varying the NAD^+/pyruvate or NADH/lactate pairs. For this reason, the first experiment now typically performed in a systematic EIEK study of a Bi Bi enzyme is variation of all four reactants in constant ratio at equilibrium, which tends to minimize dead-end complex formation.[22] Varying a reactant–product pair results in changing more than one variable at once, so that clear-cut interpretation of any observed effect is not possible from a single experiment. In a complete study all possible combinations of reactant–product pairs are varied, which allows one to deduce logically which component causes a specific kinetic effect.

Cleland[21] has suggested an alternative procedure, namely, increasing the concentration of one reactant while simultaneously lowering the concentration of the coreactant, without changing [P] or [Q]. Although not yet tested in practice, this approach should probably only be used along with the standard method of varying reactant–product pairs in constant ratio. *A priori,* increasing the degree of saturation with one reactant while decreasing saturation with the coreactant could overlook substrate-induced enzyme isomerization or synergistic effects, or it could produce inhibition effects arising from ordered substrate binding or dead-end complex formation involving reactant A that do not occur with high or increasing concentrations of B.

A typical experimental protocol for EIEK experiments is outlined as follows. One must first determine accurately the value of K_{eq} for the reaction catalyzed, usually by an approach-to-equilibrium method, using isotopes or spectrophotometry. In addition, one must have good values for all substrate K_m, at the pH and temperature of interest. A series of reactions are then set up in which the reactant–product pairs of interest are varied from a minimum (zero) to a maximum (saturating) value, with nonvaried cosubstrates held constant, usually near or slightly above half-saturation. To minimize pipetting errors, two stock solutions A and B are made up, A containing the full complement of buffer, salt, and reaction components and B containing identical concentrations of all components except those being varied. Then A and B are simply mixed together in different ratios to the same final volume to create the reaction series. Solutions A and B can be prepared with all components at twice the final concentrations of each, to allow for addition of enzyme, labeled material, and other components such as modifiers. Enzyme is then added, and the reactions are incubated for sufficient time to allow exact adjustment to chemical equilibrium. The chemical reaction is now poised at chemical equilibrium so that no net flow occurs into or out from any component pool, but a dynamic exchange does occur between all reactant–product

[22] F. C. Wedler and P. D. Boyer, *J. Biol. Chem.* **247,** 984 (1972).

pools, typically at a rate 10- to 100-fold slower than initial velocity. The isotopic exchange reaction of interest is then initiated by addition of a labeled component, negligible (in micromoles) relative to the size of the pool to which it is added, to avoid perturbing the equilibrium condition. The number of different labeled components that can be used is limited only by the structures of the exchanging species, their commercial availability, and the ingenuity and energy of the investigator. With glutamine synthetase (glutamate–ammonia ligase),[22,23] a total of ten different exchange reactions have been observed:

L-[^{14}C]glutamate $\rightleftharpoons$ L-glutamine, [^{14}C]ADP $\rightleftharpoons$ ATP,
$^{15}NH_3 \rightleftharpoons$ L-glutamine, [^{32}P]$P_i \rightleftharpoons$ ATP

plus various ^{18}O exchanges, including

L-Glu $\rightleftharpoons P_i$, L-Gln $\rightleftharpoons P_i$, L-Glu $\rightleftharpoons$ L-Gln,
ATP $\rightleftharpoons$ L-Glu, ATP $\rightleftharpoons$ L-Gln, ATP $\rightleftharpoons P_i$

Exchange is usually allowed to occur for a time sufficient to attain 10–50% of isotopic equilibrium, although percentages up to 90% are feasible. The reactions then are quenched by a method that does not cause degradation or fragmentation of labeled components, most frequently by addition of mild acid or base to shift the pH to a range in which enzyme is inactive. This may not be appropriate with some systems; with dehydrogenases, for example, owing to instability of NAD(P)$^+$ cofactors outside the range of pH 6–8 or on slow freezing at $-20°$, flash-freezing on dry ice/2-propanol or liquid nitrogen is the preferred quench-and-hold method. Another method is to remove a component essential for enzyme activity, such as divalent cation with EDTA, as long as this does not interfere with subsequent separation of labeled components. Once the labeled components have been separated, the isotopic content of each is determined by liquid scintillation counting or mass spectroscopy, and these values are used to calculate the rates of exchange (micromoles/minute/milligram enzyme) for each reaction, as described below.

Criteria and Limitations

A primary requirement for being able to study a given enzyme by EIEK methods is that the enzyme-catalyzed reaction have a K_{eq} that is near enough to unity that the concentrations of all reactants and products can be varied from well below to well above the respective K_m values.

[23] B. O. Stokes and P. D. Boyer, *J. Biol. Chem.* **251,** 5558 (1976).

This eliminates a number of systems with potentially interesting regulatory properties, for example, kinases such as phosphofructokinase and pyruvate kinase, as well as decarboxylation-driven dehydrogenases such as pyruvate, isocitrate, and α-ketoglutarate dehydrogenases.

A second requirement is chemical stability of reaction components, labeled and unlabeled, throughout all steps of the experimental protocol: preincubation, exchange, quench, and separation. Breakdown of a nonlabeled component during preincubation or exchange can cause a continuous nonequilibrium flow of material (and label) from one side of the reaction to the other. This can be especially serious if it involves the pool that is limiting in size. Fragmentation of a labeled component at any stage can lead to serious errors in the calculation and interpretation of exchange data, depending on how the subfragments migrate in the selected separation procedure. The experimentalist must prove unambiguously either that such degradation does not occur or that it is small relative to the overall micromoles exchanged and can be corrected for systematically. Typically, "zero-time" control reactions, in which enzyme or labeled compound is added after quenching, are carried out in parallel with the main reaction series. Other basic requirements are that the enzyme concentration be much less than the concentration of any substrate or modifier present, and that kinetic isotope effects are negligible, which could be significant for dehydrogenases with ^2H or ^3H label in the atom being transferred.

Advantages

Equilibrium isotope exchange kinetics methods offer the experimentalist a number of features that are useful for gaining unique insights to the dynamic properties of enzymes. First, in terms of the inherent kinetic properties of EIEK, the advantages include the following: (a) Variety of information: the dynamic flux between isotopically labeled reactant–product pools ($^*X \rightleftharpoons Y$) involves both the association–dissociation steps for each plus catalytic interconversion of the central complexes. Thus, by observing multiple exchange reactions one can obtain information about both the rate-limiting and non-rate-limiting steps simultaneously. Furthermore, EIEK data allow one to distinguish between certain kinetic mechanisms that initial velocity methods cannot, for example, Random Bi Bi versus rapid equilibrium random with dead-end complexes.[24] (b) Multiple observables: using multiple exchange rates serves to refine the

[24] C. Frieden, *Biochem. Biophys. Res. Commun.* **68,** 914 (1976).

mechanistic interpretation of the data, for instance, exactly which rate constants are altered by modifiers or site-specific mutations. (c) Mechanistic relevance: the observed isotopic exchanges relate directly to the chemical intermediates and enzyme–substrate complexes formed along the main reaction pathway. In addition, with simple systems, it has been argued that half-saturation values obtained from EIEK experiments very closely correspond to enzyme–substrate dissociation constants.[2,14] The "stickiness" and "commitment factors" for the first substrate to bind can produce a lack of agreement between these constants,[21] as observed with as $NAD(P)^+$-dependent dehydrogenases. (d) Sensitivity: using radioisotopes, the EIEK method can easily reach into the nanomolar range for any component of the system. (e) Probing "modified" enzymes: these inherent properties make EIEK ideal for defining in detail the kinetic consequences of "modifying" an enzyme, either with a bound effector ligand or by a site-specific mutation, as described below.

Second, in terms of practical methodology, EIEK studies provide the following advantages: (a) Simple equipment: the experimentalist needs only inexpensive and basic chromatography devices and a scintillation counter. If stable isotopes are used, access to a mass or nuclear magnetic resonance (NMR) spectrometer will be required. (b) Experimental ease: given improved methods for component separations and data analysis (described below), EIEK is now no more labor-intensive than other currently used kinetic methods (e.g., isotope effects).[25,26] (c) Multilevel data analysis: the EIEK approach offers both qualitative and quantitative data analysis. (i) Phase I, qualitative: In virtually all EIEK studies prior to the mid-1980s, the interpretation of data was first carried out by the simple, direct, nonesoteric logic set forth in the early 1960s[1,2]: if multiple exchanges occur at different rates, then catalysis cannot be rate limiting; the most rapid exchange must occur between the first components released from the central complexes; and inhibition effects are due either to ordered sequential substrate binding or dead-end complex formation, or both. As with most kinetics, the most powerful logic is negative, eliminating all possibilities but one or a limited few. The differences in kinetic behavior that allow one to distinguish between random versus ordered (compulsory) sequential kinetic mechanisms [Eqs. (1A) and (1B)] have been discussed at length.[15,16] As a brief summary, inhibition effects occur for ordered binding systems that are not observed with random ones. This occurs because of conversion of complexes EA and EP (the only ones from which

[25] W. W. Cleland, *Crit. Rev. Biochem.* **13,** 385 (1983).
[26] M. H. O'Leary, *Annu. Rev. Biochem.* **58,** 377 (1989).

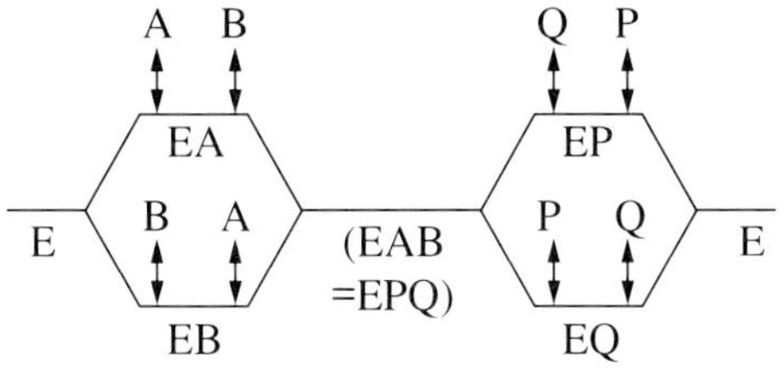

Random (1A)

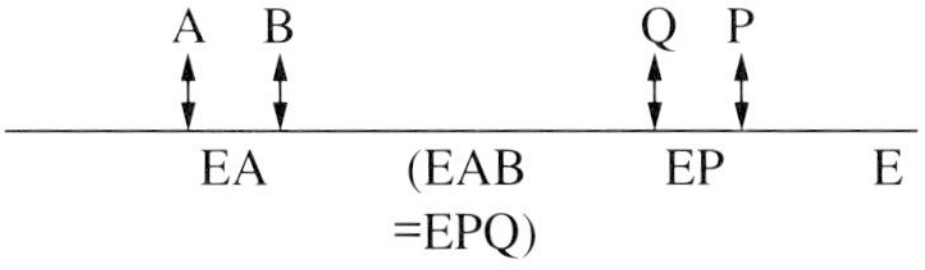

Compulsory (1B)

A and P can escape in the compulsory scheme) into the central complexes EAB and EPQ, which causes inhibition of any exchanges involving A or P. (ii) Phase II, quantitative: Simulation of EIEK data using computer programs (ISOBI, ISOTER), described below, avoids altogether the need to derive complex algebraic equations. This approach allows one to derive explicit rate constants not only for the "outer" (binary) complexes but also the "inner" (ternary, central) ones, which are often much more difficult to obtain by other kinetic approaches. These programs include such realistic features as dead-end complexes and cooperative substrate binding, and they allow one to define "confidence limits" (allowed limits of variation) for each constant.

Equations

Bi Bi Sequential Systems. For a Random Bi Bi enzyme system catalyzing the reaction

$$A + B \rightleftharpoons P + Q \qquad (2)$$

the expressions for isotope exchange at equilibrium have been derived.[2] For the exchange between components A and P ($A \rightleftharpoons P$) these can be written as

$$R = E_t \Big/ \left[1 + \frac{k'(k_{-6} + k_5 Q)}{k_{-5}k_{-6} + k_{-7}(k_{-6} + k_5 Q)} + \frac{k(k_{-1} + k_2 B)}{k_{-1}k_{-2} + k_{-4}(k_{-1} + k_2 B)} \right] Z \qquad (3)$$

and for the exchange between components B and Q (B $\rightleftharpoons$ Q) as

$$R' = E_t \Big/ \left[1 + \frac{k'(k_{-8} + k_7P)}{k_{-7}k_{-8} + k_{-5}(k_{-8} + k_7P)} + \frac{k(k_{-3} + k_4A)}{k_{-3}k_{-4} + k_{-2}(k_3 + k_4A)}\right] Z \quad (4)$$

where the term Z is defined as

$$Z = \left[\frac{1}{k'}\left(1 + \frac{K_5}{Q} + \frac{K_7}{P}\right) + \frac{1}{k}\left(1 + \frac{K_2}{B} + \frac{K_4}{A} + \frac{K_1K_2}{AB}\right)\right] \quad (5)$$

These equations differ from those presented in the original Boyer and Silverstein paper[2] in that components P and Q are substituted for components C and D. For a compulsory (ordered) sequential pathway, these equations simplify to the following for (A $\rightleftharpoons$ P)

$$R = E_t \Big/ \left[1 + \frac{k'(k_{-6} + k_5Q)}{k_{-5}k_{-6}} + \frac{k(k_{-1} + k_2B)}{k_{-1}k_{-2}}\right] Z \quad (6)$$

and for (B $\rightleftharpoons$ Q)

$$R' = E_t \Big/ \left[1 + \frac{k'}{k_{-5}} + \frac{k}{k_{-2}}\right] Z \quad (7)$$

where

$$Z = \left[\frac{1}{k'}\left(1 + \frac{K_5}{Q}\right) + \frac{1}{k}\left(1 + \frac{K_2}{B} + \frac{K_1K_2}{AB}\right)\right] \quad (8)$$

The rate steps for the fully random sequential mechanism can be represented as in Scheme 1. In the compulsory order system, the rate constants for one branch of the random pathway become vanishingly small. Assuming simple values for rate constants, for example, $k_{\pm A,P} = 5$, $k_{\pm B,Q} = 2$,

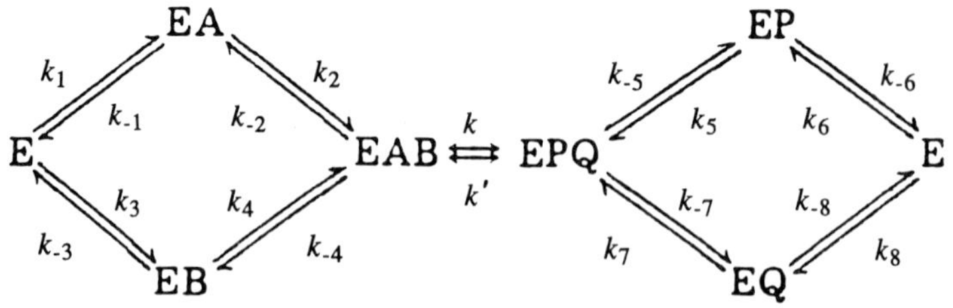

SCHEME 1. Fully random Bi Bi sequential mechanism with individual rate steps (see text). (Reprinted from Boyer[15] with permission.)

and $k = k' = 100$, on variation of the concentration of all substrates with [A] = [B] = [P] = [Q] to a maximum of $[S]/K_d = 20$, one obtains the saturation curves for the A $\rightleftharpoons$ P and B $\rightleftharpoons$ Q exchanges shown in Fig. 1.

For the random system, the more rapid on–off rates for substrates A and P compared to B and Q make the A $\rightleftharpoons$ P exchange more rapid than that for B $\rightleftharpoons$ Q under all conditions (Fig. 1, top). In addition, both exchanges rise smoothly to a maximum without inhibition effects. If, however, the rate constants for the pathways E–EB–EAB and E–EQ–EPQ decrease to values that are zero or insignificantly small, the random binding scheme converts to compulsory order. Using identical values for substrate on–off rates, the kinetic consequences of these changes are shown in Fig. 1 (bottom). Obligatory release of B or Q prior to A or P, respectively, dictates that the B $\rightleftharpoons$ Q exchange be equal to or more rapid than A $\rightleftharpoons$ P under all conditions. This is true, despite the fact that the on–off rates for A and P are more than twice as rapid as those for B and Q, since the most rapid exchange occurs between the first components released from the central complexes. In addition, exchange between A and P (now released only from EA and EP) exhibits a strong inhibition effect above $[S]/K_d$(app) ratios of around 2, as the concentrations of EA and EP decrease owing to conversion to EAB and EPQ.

In practice, a major shortcoming of these explicit algebraic equations, based on Scheme 1, is the failure to incorporate steps involving formation of dead-end complexes, cooperative substrate binding, enzyme isomerization, or other features commonly encountered with real enzymes. This problem has been addressed, in part, by development of new computer programs that can take these features into account (described in later sections).

Bi Bi Ping Pong Systems. Enzyme-catalyzed reactions that include discrete covalent intermediates frequently exhibit isotopic exchange reactions that occur even with incomplete (partial) sets of substrates. In other words, A $\rightleftharpoons$ P exchange may occur in the absence of B or Q, and B $\rightleftharpoons$ Q exchange occurs without A or P. Release of one product occurs prior to binding of the second reactant, and covalent chemistry occurs in the absence of ternary central complexes, as shown in Eq. (9). Partial ex-

$$\begin{array}{ccccccc} & \text{A} & & \text{P} & & \text{B} & & \text{Q} \\ & \updownarrow & & \updownarrow & & \updownarrow & & \updownarrow \\ \hline \text{E} & & (\text{EA} = \text{EP}) & & (\text{E–X}) & & (\text{EB} = \text{EQ}) & & \text{E} \end{array} \qquad (9)$$

change reactions have been used for over 40 years to detect covalent intermediates. In the E–X intermediate, X may be covalently attached or very tightly bound to E. Various synthetases, mutases, and transferases

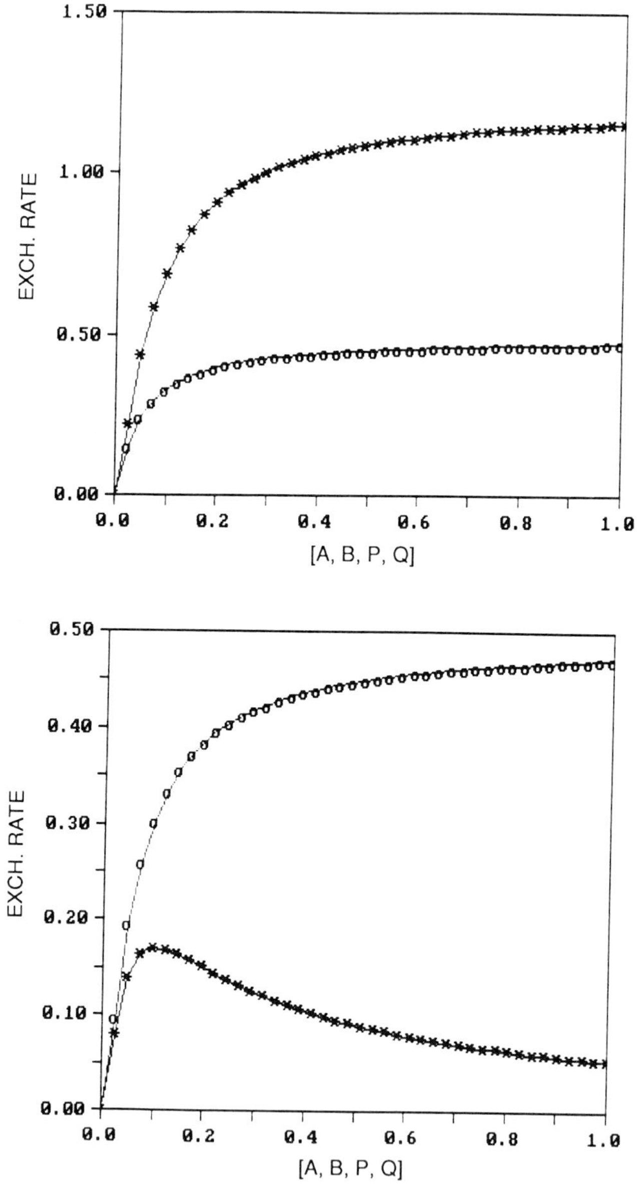

FIG. 1. Saturation curves generated on varying the concentration of all substrates in constant ratio at equilibrium for (top) random order Bi Bi kinetic mechanism (see Scheme 1) and (bottom) compulsory order kinetic mechanism. Curves were calculated and plotted with the ISOBI-HS program. See text for rate constants and substrate concentrations. (*) A–P; (○) B–Q.

have been identified that catalyze such partial reactions, as cited in previous reviews.[15,16]

Failure to observe partial reactions does not eliminate the possibility of formation of a covalent intermediate. A classic example is glutamine synthetase (glutamate–ammonia ligase), for which no isotopic exchange reactions were observed in the absence of a complete reaction system,[22] but for which evidence suggesting a γ-glutamyl phosphate intermediate was provided by borohydride trapping,[27] isolation of a labeled complex,[28] and reaction with a rigid analog of L-glutamate with ATP.[29] Ultimately, one must prove the kinetic competence of the postulated intermediate. For glutamine synthetase, this was finally accomplished by the elegant positional isotope exchange (PIX) experiments of Midelfort and Rose using [β,γ-^{18}O]ATP.[30]

Conversely, even though an intermediate may form on the main reaction pathway, the enzyme may not operate via a Ping Pong reaction. This occurs if all substrates must be bound before covalent chemistry occurs at an appreciable rate, or if once all substrates are bound the covalent conversion steps are much more rapid than release of the first product. An example of the first possibility is *Escherichia coli* glutamine synthetase. For the latter, a classic example is *E. coli* succinyl-CoA synthetase [succinate–CoA ligase (ADP-forming)], which uses ATP and succinate to form two intermediates, E-His-P and an E(succinyl-P),[31] neither of which is kinetically competent in the absence of cosubstrates. The enzyme exhibits strong synergistic effects of catalysis by bound substrates on catalytic steps.[15,31] That is, intermediate covalent species that form do not release ADP at a rate that compares to catalysis to the final succinyl-CoA product.[32] Consequently, the complex E-P(succinate)(ADP)(CoA) is kinetically competent, but free E(His)-P and E(succ-P) are not.

Bi Bi Rapid Equilibrium Random Systems. If catalysis is rate limiting, this creates a kinetic bottleneck at the one step common to all species in the kinetic scheme and forces all exchanges to occur at equal rates. A few well-known examples include phosphorylase and creatine kinase.[15,16] In this case, EIEK methods provide no insight to relative rates of substrate dissociation but under subsaturating conditions may provide clues to relative association rates. Other interesting effects are suppressed or masked

[27] J. A. Todhunter and D. L. Purich, *J. Biol. Chem.* **250,** 3505 (1975).

[28] P. R. Krishnaswamy, V. Pamiljans, and A. Meister, *J. Biol. Chem.* **237,** 2932 (1962).

[29] J. D. Gass and A. Meister, *Biochemistry* **9,** 842 (1970).

[30] C. A. Midelfort and I. A. Rose, *J. Biol. Chem.* **251,** 5881 (1976).

[31] W. A. Bridger, *in* "Enzymes" (P. D. Boyer, ed.), 3rd Ed., Vol. 10, p. 581. Academic Press, New York, 1974.

[32] F. J. Moffet and W. A. Bridger, *J. Biol. Chem.* **245,** 2758 (1970).

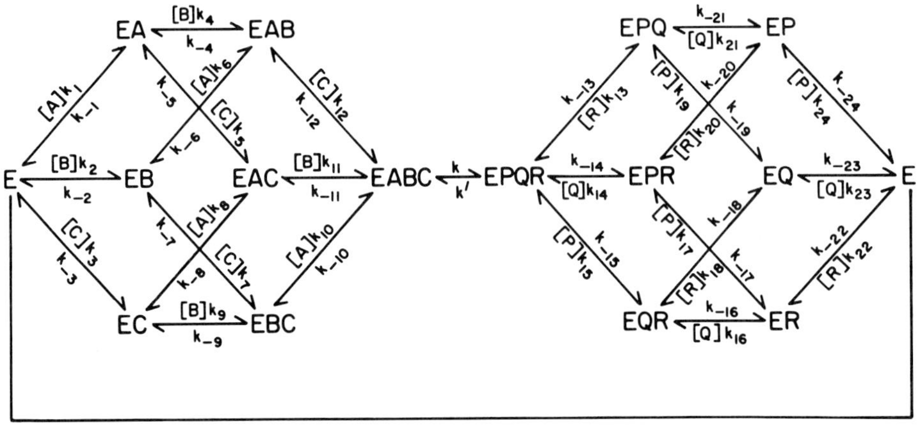

SCHEME 2. Kinetic steps for a Random Ter Ter enzyme system modeled by the ISOTER program (see text). [Reprinted from F. C. Wedler and R. W. Barkley, *Anal. Biochem.* **177**, 268 (1989).]

by the k and k' steps (Scheme 1) being rate determining, including modifier action. These shortcomings may be overcome if one can find an experimental condition (pH, temperature, activating cation, etc.) at which catalysis is not definitively slow.

Ter Ter Systems. A limited number of equations for initial velocity with enzymes involving three reactants/products have been derived.[33-36] The number of closed cycles for a fully Random Ter Ter system makes derivation of algebraic expression even for initial velocity prohibitive by classic King–Altman methods. When one considers exchange of isotope at equilibrium, including rate steps on both sides of the reaction, this problem is magnified by at least an order of magnitude, as illustrated in Scheme 2. This fully Random Ter Ter scheme includes a minimum of 15 unique enzyme–substrate complexes, interconnected by 50 rate constants. Algebraic expressions for equilibrium isotope exchange, if derived, would contain thousands of terms in the denominator, making such equations impractical to use. Nonetheless, a number of Ter Ter enzymes are mechanistically important, including various ATP-dependent systems such as aminoacyl-tRNA synthetases, glutamine and asparagine synthetases, and succinyl-CoA synthetase. Other intriguing enzymes involve Ter Bi or Bi Ter mechanisms. To approach these complex enzymes, computer pro-

[33] R. E. Viola and W. W. Cleland, this series, Vol. 87, p. 353.
[34] K. Dalziel, *Biochem. J.* **114,** 514 (1969).
[35] F. B. Rudolph and H. J. Fromm, *J. Theor. Biol.* **39,** 363 (1973).
[36] K. R. F. Elliott and K. F. Tipton, *Biochem. J.* **141,** 789 (1974).

grams for modeling EIEK data have been developed that circumvent the problem of deriving explicit (but uselessly complex) algebraic expressions for exchange rates.

Advances in Methods

Lack of access to a solid theoretical basis, compared to that available for other kinetic methods, has hindered application of the powerful and versatile EIEK technique to full potential. Laborious multiple separations of labeled components in sets of EIEK reactions has also been an ongoing challenge.[21] Several breakthroughs in methodology serve to overcome many of these problems. The key advances on the theoretical problem have come from the revolution in microcomputer technology and software, along with identifying the best mathematical methods for solving arrays of differential equations for complex kinetic schemes. Specifically, simulation programs for the fully Random Bi Bi and Ter Ter kinetic schemes have been developed that allow anyone to fit systematically EIEK experimental data and obtain an explicit set of constants. Spreadsheet programs also greatly facilitate EIEK rate calculations. Regarding separation methods, some advances and how these may be optimized are discussed below.

Simulation Programs

ISOBI, ISOMOD, ISO-COOP. The first attempts to design a computer program to simulate experimental saturation curves from EIEK experiments with Bi Bi systems involved writing a FORTRAN program that used algebraic expressions [Eqs. (3)–(8)] to calculate R (A $\rightleftharpoons$ P) and R' (B $\rightleftharpoons$ Q), based on the kinetic model in Scheme 1. Once this was applied to a complex enzyme, it became immediately apparent that this model was too simple. The goal then became to model the fully Random Bi Bi scheme, including 12 dead-end complexes and cooperative substrate binding, as shown in Scheme 3.

After writing differential equations for all enzyme–substrate complexes in Scheme 3, various methods were considered for solving for the unknowns in the equations. The first approach used difference calculus, but the resulting equations (in the early 1980s) required a mini- or mainframe computer.[37] With the advent of the IBM PC, our goal became to develop equations that could run on microcomputers to make these methods readily available to individual laboratories. Clearly, a better mathematical approach was needed, which was eventually provided by matrix algebra

[37] W. H. Shalongo, M.S. Thesis, The Pennsylvania State University, University Park (1982).

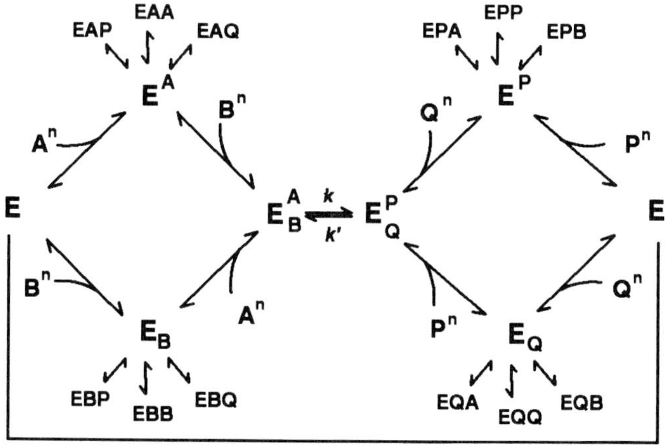

SCHEME 3. Kinetic steps for a Random Bi Bi enzyme system modeled by the ISOBI-HS program, including all productive pathways, 12 dead-end complexes, and cooperative substrate binding (see text).

used with numerical methods.[38] The general expression for the matrices was $\mathbf{A} \cdot \mathbf{X} - \mathbf{V} = \mathbf{0}$, where the matrix $\mathbf{A}$ is defined as

$$\mathbf{A} = \begin{vmatrix} A_{11} & A_{12} & A_{13} & \cdots & A_{1n} \\ A_{21} & A_{22} & A_{23} & \cdots & A_{2n} \\ \vdots & \vdots & \vdots & \vdots & \vdots \\ A_{n1} & A_{n2} & A_{n3} & \cdots & A_{nn} \end{vmatrix}$$

consisting of the coefficients A_{11} to A_{nn} of the system and the vectors

$$\mathbf{X} = \begin{vmatrix} x_1 \\ x_2 \\ x_3 \\ \vdots \\ x_n \end{vmatrix}, \qquad \mathbf{V} = \begin{vmatrix} B_1 \\ B_2 \\ B_3 \\ \vdots \\ B_n \end{vmatrix}, \qquad \mathbf{0} = \begin{vmatrix} 0 \\ 0 \\ 0 \\ \vdots \\ 0 \end{vmatrix}$$

define the unknowns, constants, and zero vector, respectively.

The best numerical method for simplifying the matrices was Gaussian elimination, in which all nonzero terms are placed above or on a diagonal of the matrix. Based on these operations, a FORTRAN program was written to solve the appropriate differential equations. This first version,

[38] P. Henrici, "The Essentials of Numerical Analysis." Wiley, New York, 1982.

called ISOBI,[39,40] was designed to run on the original IBM PC with an 8088 central processing unit (CPU), two floppy disk drives, an 8087 math coprocessor chip, and color graphics adapter (CGA) graphics. The major problems were slow calculation and plotting speeds and tedious procedures for making reiterative changes to rate constants. To calculate and plot the saturation curves for two exchanges at 50 different substrate concentrations required 45 sec, and changing the value of a rate constant required multiple strokes of both function and cursor keys. In addition, the split-screen CGA graphics used made comparisons to experimental data somewhat difficult. Despite these problems, with patience, ISOBI provided a way to fit the experimental data and obtain a set of explicit values for each rate constant in the kinetic model of Scheme 3. This was carried out with both wild-type and several mutant forms of aspartate transcarbamylase (ATCase; aspartate carbamoyl transferase), as described in sections that follow.

An interest in using EIEK to identify modes of action of modifiers with key regulatory enzymes such as ATCase led to development of programs to model changes in exchange rates as a function of the concentration of modifier.[15,16] This produced a subroutine, ISOMOD, run in conjunction with ISOBI.[40] To model the concerted M–W–C model for cooperative substrate binding a second subroutine, ISOCOOP, was also developed.[40] This allows one to manipulate directly the true allosteric parameters, L and substrate K_d values, rather than simply mathematically simulating cooperativity by using Hill numbers as exponents.

ISOBI-HS. A much more user-friendly version of ISOBI has been developed to overcome the problems cited above with the original ISOBI, namely, speed and ease of use. The calculational program has been reprogrammed with Microsoft FORTRAN 5.1 for use with 80386 or faster CPUs. Using an 80386 math chip, this reduces calculational and plotting times for 100 data points and two exchanges to under 3 sec. Second, a "front-end" package (HiScreen ProII, Softway, San Francisco, CA) has been interfaced with the calculation program, which allows custom-design of mouse-interactive screens. Thus, the user can now click on mouse "buttons" to facilitate making rapid, reiterative changes to rate constants or other parameters. The format for entering a rate constant is very simple, for example 1.85×10^{-4} is typed as 1.85e − 4. The main screens for ISOBI-HS, are shown in Fig. 2.

To initiate a simulation session, two types of parameters must first be entered: (1) the experimental conditions (EXPT) to be simulated, usually those used in practice with a real enzyme in the laboratory, namely, the

[39] Y. Hsuanyu and F. C. Wedler, *Arch. Biochem. Biophys.* **259,** 316 (1987).
[40] Y. Hsuanyu and F. C. Wedler, *J. Biol. Chem.* **263,** 4172 (1988).

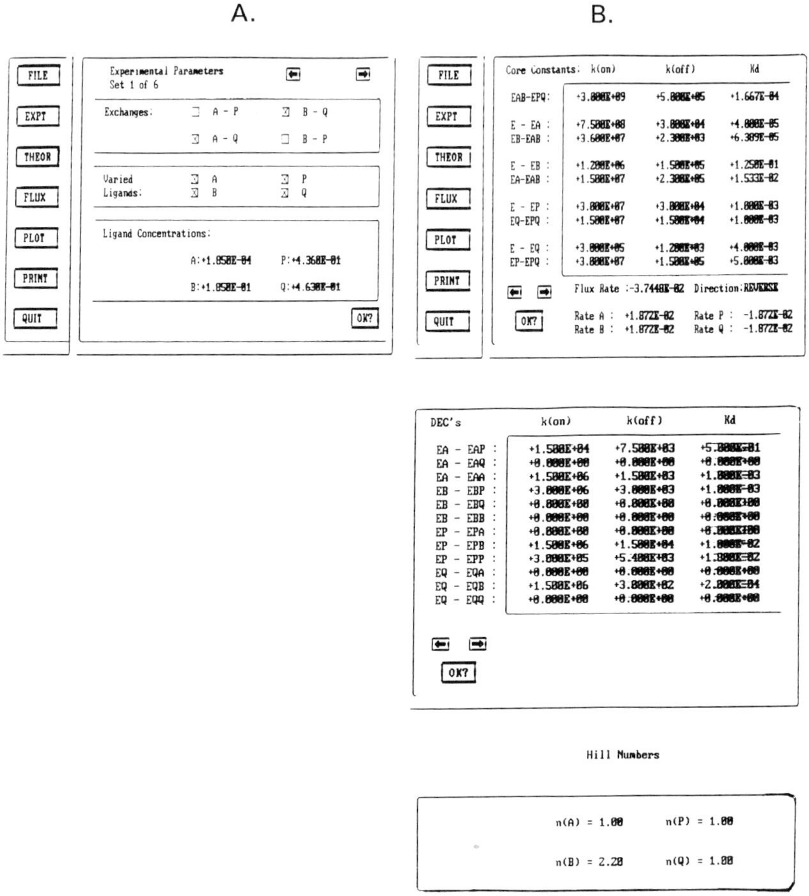

FIG. 2. Screens from the ISOBI-HS program (Scheme 3) for simulation and fitting of EIEK data. Values listed are those used to simulate curves for aspartate transcarbamylase.[39] (A) EXPT screen containing experimental parameters: exchanges observed, ligands (substrates) varied or held constant, with maximum (or constant) substrate concentrations. (B) THEOR screen containing theoretical (rate) constants: (top) covalent interconversion rates (k, k') plus association–dissociation of substrates for the productive pathways of Scheme 3; (middle) association–dissociation rates for formation of nonproductive (dead-end) complexes; (bottom) Hill numbers used for mathematical simulation of cooperative substrate binding.

exchanges to be observed and the ligands (substrates) to be varied, along with the maximal concentrations (see Fig. 2A); and (b) rate constants (THEOR) for the productive-pathway core of Scheme 3 (see Fig. 2B). In addition, although not required, one may use separate screens to specify binding of substrates in nonproductive (dead-end) complexes or in a cooperative manner, using Hill numbers. The program will automatically calculate a "FLUX rate" (net flow of material from one side of the reaction to the other), which is used as an index of nearness to equilibrium. To judge whether equilibrium has been significantly altered, FLUX can be recalculated after each change in rate constants by selecting (clicking on) this button in the menu bar to the left.

Although HiScreen Pro II (Softway, Inc.; San Francisco, CA) incorporates a video graphics array (VGA) plotting package, obtaining a hard copy of calculated plots from the video screen requires a screen-capture utility (e.g., PizazzPlus, Application Techniques, Pepperell, MA). Sample plots from ISOBI-HS are shown in Fig. 1. A useful feature of this program is the ability to overlay successive saturation curves, allowing one to compare the effects of changing particular rate constants. The program will also print out a list of rate constants currently being used, along with calculated steady-state concentrations of each complex on the productive pathways at any specified percent saturation. Entire sets of rate constants and experimental conditions can be stored and recalled as separate files.

ISOTER. The EIEK data for *E. coli* glutamine synthetase[22,23] were difficult to interpret by extrapolation of qualitative principles from Bi Bi systems,[1,2,15] which led us to develop a program to simulate kinetic behavior for the fully Random Ter Ter system of Scheme 2. The kinetic behavior of this complex scheme was envisioned essentially as flow through a network of pipes or wires, governed by specified rates between junctions (complexes) within the network, a classic problem in chemical engineering for which mathematical methods are well established. Solving the differential equations (involving a 15×15 matrix) by difference calculus methods, although successful,[37] were soon abandoned as too cumbersome. As with ISOBI matrix algebra and numerical methods proved to be the best approach.[38] Each component (A, B, C, P, Q, or R) that could be involved in an exchange reaction was first considered individually in terms of the specific enzyme complexes in which it could be involved. Dissected out, these species were rewritten as subnetworks specific to each component. The expanded version of Scheme 2 that resulted from this process is shown as Scheme 4. After writing differential equations for each complex in each subnetwork, the appropriate set of equations for each reactant was paired with those for each product to produce nine sets of equations, corresponding to each of the nine possible exchange reactions:

$$A \rightleftharpoons P, \quad A \rightleftharpoons Q, \quad A \rightleftharpoons R, \quad B \rightleftharpoons P, \quad B \rightleftharpoons Q, \quad B \rightleftharpoons R,$$
$$C \rightleftharpoons P, \quad C \rightleftharpoons Q, \quad C \rightleftharpoons R$$

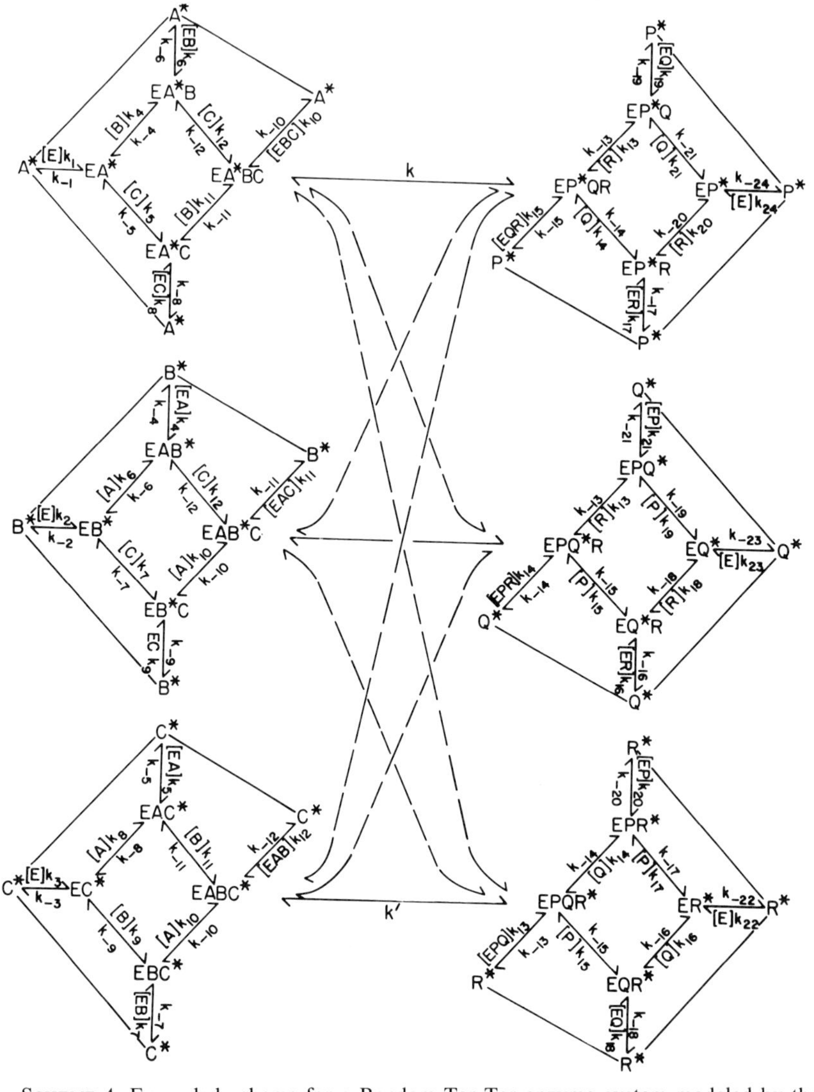

SCHEME 4. Expanded scheme for a Random Ter Ter enzyme system modeled by the ISOTER program, showing complexes specific to each labeled component (see text). [Reprinted from F. C. Wedler and R. W. Barkley, *Anal. Biochem.* **177,** 268 (1989).]

The ISOTER program solves each set of equations (by matrix algebra/Gaussian elimination) to calculate an exchange rate at a specified set of substrate concentrations.[41]

[41] F. C. Wedler and R. W. Barkley, *Anal. Biochem.* **177,** 268 (1989).

The ability to model equilibrium exchange kinetics for the fully Random Ter Ter system constitutes a major breakthrough. No other method offers such a simple and versatile approach to this complex problem.[33–36] In terms of number of rate constants that can be considered, ISOTER exceeds even the KINSIM program developed by Barshop *et al.*[42] for modeling pre-steady-state, initial velocity, and full time course experiments. Although KINSIM is more versatile, allowing *in situ* custom design of kinetic schemes, it requires use of a minicomputer. Modeling a Random Ter Ter system has the major drawback of requiring one to specify many rate constants, many of which cannot even be estimated initially with accuracy. Obviously, prior measurement of a number of k_{on} or k_{off} values by rapid kinetic methods, along with having measured substrate K_d values by fluorescence or UV difference titrations, is invaluable in overcoming this problem.

Knowledge of several accurately measured values of k_{on}, k_{off}, or K_d has a "ripple" effect on the data simulation, fitting, and validation process. Once one has fixed the values of several rate constants to agree with measured values, the relative magnitudes of the exchange rates at saturation dictate the range of values that are possible for substrate k_{off} values that have not been determined. Except for rapid equilibrium random (RER) mechanisms, catalytic rates are simply set to be faster than all substrate k_{off} values.

Validation Procedures. Ultimately, one must be convinced that the set of rate constants that provide optimal fit to the EIEK data are unique and valid. In studies to date, the following criteria have been applied: (a) agreement between all measured substrate association or dissociation rates, as well as $K_d = (k_{off}/k_{on})$ values; (b) determining "confidence limits" (allowed range of variation) for each rate constant; (c) proving that the constants predict the correct initial velocities in both directions (using KINSIM[42]); and (d) proving that the constants provide thermodynamic balance in all closed cycles of the scheme being modeled.

Data Handling

The equation[2,15] used to calculate equilibrium exchange rates for the reaction $^*X \rightleftharpoons Y$ is

$$R = -\{[X][Y]/([X] + [Y])\} \ln(1 - F) \quad (10)$$

[42] B. A. Barshop, R. F. Wrenn, and C. Frieden, *Anal. Biochem.* **130,** 134 (1983).

where F is the fraction of isotopic equilibrium, defined as $F = ([X] + [Y]/[Y])\{y/(x + y)\}$, in which x and y are disintegrations per minute (dpm) of radiolabel in the X and Y pools. Based on these expressions, hand calculation of EIEK rates might typically involve the following 14 items, listed as columns on a data sheet:

> Reaction no., f (the fraction of maximal saturation)—if appropriate, [X] (μmol), [Y] (μmol), ([X] + [Y])/[Y], [X][Y]/([X] + [Y]), x, y, $y/(x + y)$, F, $(1 - F)$, $\ln(1 - F)$, R (μmol), and R' (μmol/min/mg)

The parameter R can also be expressed as a pseudo-first-order process with units of reciprocal seconds or reciprocal minutes, either of which allows comparisons to turnover rates from initial velocity. Where $F <$ 10%, Eq. (10) simplifies to the expression $R = -\{[X][Y]/([X] + [Y])\}F$. Data error is reduced, however, by running exchanges so that F lies in the range of 20–70%.

Such repetitive, sequential calculations are laborious to perform by hand, even for today's programmable calculators. Currently available spreadsheet programs allow one to simplify this chore. As an example, using Lotus 1-2-3, one can quickly set up a calculation routine as follows:

⟨Row 1⟩	Set up columns A–J and label them as x, y, C, X, Y, A, B, F, R, and R′	
⟨Row 2⟩	(Leave blank)	
⟨Row 3⟩	Column A: x = dpm of X (final)	F: (D + E)/E {=(X + Y)/Y}
	B: y = dpm of Y (final)	G: (D * E)/(D + E) {=XY/(X + Y)}
	C: B/(A + B) {= y/(x + y)}	H: "F" = C * F
	D: X = μmol X in rxn	I: R = −@log(1 − H) * (2.3) * G
	E: Y = μmol Y in rxn	J: R′ = I * (1000/[μgE * t])
	[where the Lotus-specific symbols are * (times) and @ ("do" command)]	
⟨Row 4⟩	Insert values for [X], [Y], x, and y into columns A, B, D, and E	

The program then calculates values for "cells" in each row, using formulas in row 3. Obviously, other spreadsheet programs can be used for facilitating such calculations.

Separation Methods

As with any kinetic method, to obtain reliable data one must prove reproducibility by means of multiple determinations of rate at each substrate concentration, typically in triplicate. In EIEK studies a minimum of 8–10 rates are determined over the range of substrate–product concentrations. When observing two different exchange rates, this means that a single experiment requires that one perform $8 \times 2 \times 3 = 48$ separations. Twice this number is not unusual, including control reactions. Obviously, the carrying out so many separations can be laborious and time-consuming unless rapid, inexpensive, facile methods are devised. The experimentalist is advised that a serious initial investment of time and effort on this problem is essential.

To avoid isotopic spillover, one must prove the method selected provides quantitative, baseline separation of labeled components, especially if the pool sizes are appreciably different. Furthermore, it is crucial to know whether the labeled (and other) components are stable under the conditions of preincubation, reaction, quenching, and separation. To correct for chemical degradation or fragmentation of the exchanging species, "zero time" control reactions (either enzyme or the labeled component added after the quench step) should be carried out routinely as part of each reaction series.

Although modern high-performance liquid chromatography (HPLC) methods offer rapid, automated separation of a variety of reaction components on both micro- and preparative scales, this approach has the disadvantage that samples are processed "linearly," one at a time. If breakdown of any component after quench occurs, this can be a serious problem. Methods that process an entire set of samples in parallel are far preferable. More "primitive" techniques such as those using multiple ion-exchange or adsorption chromatography columns or thin-layer chromatography (TLC) plates satisfy this requirement. Inexpensive minicolumns and column racks made to deposit samples directly into scintillation vials are now commercially available at reasonable cost (e.g., from Bio-Rad, Richmond, CA). In addition, a number of fairly sophisticated chromatographic media, previously available only in the form of resin beads (e.g., DEAE- or CM-cellulose), are now offered in the form of TLC plates or sheets. Media for performing separations based on chiral differences are also available. The TLC sheet media are also offered as thick-film formulations that permit processing of larger samples. Moreover, EIEK reactions can be easily run with total volumes of 0.1 ml to minimize sample size and total micromoles that must be separated, which reduces column size and time required for separation. One may use only 25–50% of a reaction sample, reserving the remainder for later use.

Several general caveats can be given. (a) On variation of the concentration of all substrates, fairly large changes in ionic strength may occur, so one must be sure that the separation method operates identically throughout the reaction series. A background level of KCl or other salt (0.1–0.2 *M*) may be added to dampen such effects. (b) It is highly recommended to visualize directly each labeled component in each reaction sample separation, rather than just removing the same portion of a set of chromatograms. Otherwise, serious errors in observed dpm can occur, especially if some fragmentation of components occurs.

Applications

This section summarizes EIEK investigations that have been performed since the reviews by Boyer[15] and Purich and Allison[16] in the early 1980s. These were selected not to constitute an exhaustive list, but rather to illustrate two points: first, the major strides made in terms of methodology, especially computer programs for EIEK data analysis; and, second, the depth of understanding of complex enzyme dynamics that can now be gained by these approaches.

Kinetic Mechanism

In the first layer of defining enzyme mechanism, a distinction needs to be made between substrate binding orders that are kinetically determined versus those that result from structural or steric factors or protein conformational changes, simply defined as follows: (a) kinetic factors: in an inherently random system (Scheme 1), substrate A binds to free E much faster than does B, and/or B is released from the central complex EAB much faster than A; (b) structural, steric, and protein conformational factors: substrate A binds to free E, which induces a conformational change of the active site that, without which, enzyme is incapable of recognizing and binding B. Alternatively, substrate A binds into a cleft which is overlaid by substrate B, so owing to steric factors A cannot escape from the central complex EAB, only from EA. Because structural factors give rise to kinetic effects, the distinction between these is not always clear-cut. To distinguish between these purely from kinetics may be impossible. Because preferred rather than compulsory order kinetic mechanisms are frequently observed,[15,16] along with variations in the degree of preference for one branch over the other with pH or temperature, alternative branches of the random scheme apparently can often operate at appreciable rates. This points to kinetic rather than structural effects being predominant with a considerable number of enzymes. Furthermore, the relative paucity of identified rapid equilibrium random kinetic mechanisms argues that catalysis is not often the rate-limiting step in net turnover.

Glutamine Synthetases. As part of developing computer programs for quantitative analysis of EIEK data, ISOTER[41] was used to simulate the saturation curves observed for glutamine synthetases from mammalian, plant, and bacterial sources.[21,43] The best fit was obtained with the *E. coli* enzyme, for which extensive initial velocity, equilibrium binding, and rapid kinetic data were also available.

[43] F. C. Wedler, *J. Biol. Chem.* **249**, 5080 (1974).

An additional concern was differences between observed kinetics with the mammalian enzyme. Wedler[43] had reported inhibition effects in ADP $\rightleftharpoons$ ATP and $P_i \rightleftharpoons$ ATP exchanges on variation of substrate pairs involving L-Glu, L-Gln, or NH_3 at pH 6.5, 37°, interpreted as preferred order binding of NH_3 and L-Glu after ATP, plus release of Gln and P_i prior to ADP. In contrast, Allison *et al.*[44] failed to observe such inhibition effects at the same pH and temperature, which was interpreted in favor of a random order mechanism. To explain these differences, use of non-equilibrium conditions was suggested.[45] This possibility was tested with ISOTER, using rate constants derived for the brain enzyme to simulate exchange kinetics at both true and nonequilibrium conditions.[41] This revealed that no significant change in the shapes of the saturation curves occurred with the system shifted 20% away from equilibrium. Thus, the brain enzyme is relatively insensitive to nonequilibrium conditions. Some other explanation must be found for differences in the EIEK data.

γ-Glutamylcysteine Synthetase. The enzyme γ-glutamylcysteine synthetase (GCS; glutamate–cysteine ligase) is coupled to a second synthetase that forms glutathione. Various kinetic mechanisms have been suggested for GCS from pig liver or toad bladder (Ping Pong) and rat kidney and liver (sequential).[16] Partial reactions reported by Meister and co-workers[46,47] with rat kidney enzyme included the following: ATP → ADP + P_i and ATP + P_i → ADP + PP_i [inhibited by L-Glu, activated by L-α-aminobutyrate (αAB)], and L-Glu + ADP → 5-oxoproline + ADP + P_i (inhibited by L-α-aminobutyrate). The reaction αAB → γGlu–αAB is faster than Glu → γGlu–αAB in the presence of ADP and P_i. No ATP $\rightleftharpoons$ ADP or ATP $\rightleftharpoons$ P_i exchange occurs in the absence of other substrates. The data are consistent with the Random BC, Random RQ mechanism derived by Schandle and Rudolph from EIEK studies using multiple exchanges as a function of the concentration of all substrates: Glu $\rightleftharpoons$ γGlu–αAB, αAB $\rightleftharpoons$ γGlu–αAB, ATP $\rightleftharpoons$ ADP, and $P_i \rightleftharpoons$ ATP.[48] This kinetic mechanism resembles that deduced for *E. coli* glutamine synthetase,[21] in which nucleotide (ATP, ADP) binds first to free enzyme, followed by random binding of Glu and the third cosubstrate (Cys or NH_3). Earlier kinetic results of Yip and Rudolph[49] involved use of GCS that may have contained adenylate kinase activity as a contaminant.

[44] R. D. Allison, J. A. Todhunter, and D. L. Purich, *J. Biol. Chem.* **252,** 6046 (1977).
[45] R. D. Allison and D. L. Purich, *Biochem. Biophys. Res. Commun.* **115,** 220 (1983).
[46] M. Orlowski and A. Meister, *J. Biol. Chem.* **246,** 7095 (1971).
[47] R. Sekura and A. Meister, *J. Biol. Chem.* **252,** 2599 (1977).
[48] V. B. Schandle and F. B. Rudolph, *J. Biol. Chem.* **256,** 7590 (1981).
[49] B. Yip and F. B. Rudolph, *J. Biol. Chem.* **251,** 3563 (1976).

Adenylosuccinate Synthase. Cooper *et al.*[50] investigated the kinetic mechanism of the following reaction catalyzed by adenylosuccinate synthase:

$$\text{GTP} + \text{L-aspartate} + \text{IMP} \rightleftharpoons \text{adenylosuccinate (Ad-succ)} + \text{GDP} + \text{P}_i$$

by observing the IMP $\rightleftharpoons$ Ad-succ, Asp $\rightleftharpoons$ Ad-succ, GTP $\rightleftharpoons$ GDP, and GTP $\rightleftharpoons$ P_i exchanges, varying the concentration of all substrates in constant ratio at equilibrium ($K_{eq} = 10$). The kinetic patterns were consistent with a random sequential mechanism, with a preference in the forward direction for L-Asp binding to the E–IMP–GTP complex. Binding of L-Asp, Ad-succ, GDP, and P_i is rapid equilibrium (faster than catalysis), whereas IMP and GTP bind in a "steady state" manner. No partial isotope exchange reaction, Ad-succ $\rightleftharpoons$ Asp, occurred (with or without added P_i) that was indicative of a phosphoryl-IMP covalent intermediate. Simultaneous commitment of GTP and IMP to product formation was consistent with an interaction prior to commitment of L-Asp. This was proposed to be transfer of the C-6 oxygen of IMP to the γ-phosphoryl group of GTP prior to attack by L-Asp, in accord with the previously proposed chemical mechanism.[51]

Carbon Monoxide Dehydrogenase. The enzyme carbon monoxide dehydrogenase (CODH) catalyzes the key reaction in the autotrophic acetyl-CoA pathway for CO or CO_2 fixation, synthesis of acetyl-CoA from the methylated corrinoid/iron–sulfur protein and CoA (called the Wood pathway): CH_3—[C/Fe—SP] + CO + CoA—SH—→ CH_3—CO—S—CoA + [C/Fe—SP]. Three isotopic exchange reactions are observable[52–55]:

(a) CH_3—^{14}CO—SCoA + CO $\rightleftharpoons$ CH_3—CO—SCoA + ^{14}CO
(b) $^{14}CH_3$—E + CH_3—CO—SCoA $\rightleftharpoons$ $^{14}CH_3$—CO—SCoA + CH_3—E
(c) *CoA—SH + Ac—S—CoA $\rightleftharpoons$ Ac—S—CoA* + CoA—SH

Lu and Ragsdale probed the final steps in the overall reaction by observing CoA $\rightleftharpoons$ acetyl-CoA exchange as a function of redox potential.[56] The CoA $\rightleftharpoons$ acetyl-CoA rate was activated by a one-electron reduction process with a half-maximum at −486 mV, with overall stimulation by 2000-fold

[50] B. F. Cooper, H. J. Fromm, and F. B. Rudolph, *Biochemistry* **25,** 7323 (1985).
[51] M. R. Webb, B. F. Cooper, F. B. Rudolph, and G. H. Reed, *J. Biol. Chem.* **259,** 3044 (1984).
[52] E. Pezacka and H. G. Wood, *J. Biol. Chem.* **261,** 1609 (1986).
[53] W. E. Ramer, S. A. Raybuck, W. H. Orme-Johnson, and C. T. Walsh, *Biochemistry* **28,** 4675 (1989).
[54] S.-I. Hu, H. L. Drake, and H. G. Wood, *J. Bacteriol.* **149,** 440 (1982).
[55] S. W. Ragsdale and H. G. Wood, *J. Biol. Chem.* **260,** 3970 (1985).
[56] W.-P. Lu and S. W. Ragsdale, *J. Biol. Chem.* **266,** 3554 (1991).

as the redox potential was varied from −80 to −575 mV. This was not caused by perturbations in CoA binding, but rather was proposed to result from redox changes at the novel Ni/Fe_{3-4} site of the enzyme, at which the CH_3—corrinoid/Fe-S protein and CO bind and where methyl migration occurs. Carbon monoxide performs a dual role: it acts as a cosubstrate to activate the CoA ⇌ acetyl-CoA exchange and as a noncompetitive inhibitor, both of which roles could also be played by nitrous oxide (NO).

Creatine Kinase. Using the NMR saturation transfer method as a means for providing information on the position of rate determining step(s) in enzyme-catalyzed reactions, Kupriyanov *et al.*[57] studied creatine kinase and compared radioisotope exchange rates, [γ-^{32}P]ATP ⇌ P—Cr (where P—Cr represents phosphocreatine) and [^{3}H]MgADP ⇌ MgATP, to those measured by ^{31}P NMR magnetization transfer, P—Cr ⇌ [γ-^{32}P]ATP and [β-^{32}P]MgADP ⇌ [β-^{32}P]MgATP. Because all exchange rates occurred with near-equality, it was concluded that phosphoryl group transfer (catalysis) is rate determining at 30°–37°. At temperatures below 25°, at high [P-Cr]/[creatine] ratios, or with [ADP] < 20 μM, the rate-limiting steps became dissociation of ADP and ATP.

Hexokinases. Some of the earliest isotope exchange studies were carried out on yeast hexokinase.[58] The reaction ATP ⇌ ADP was observed to be faster than Glc ⇌ Glc6P (where Glc and Glc6P stand for glucose and glucose 6-phosphate, respectively) by a factor of two, and neither exchange showed inhibition effects on variation of [ADP, ATP] or [Glc, Glc6P]. This was interpreted as indicative of a random mechanism, with covalent interconversion being faster than substrate dissociation, in agreement with the majority of currently available kinetic evidence.[59,60] The basis for a claim by Noat and Richard[61] that glucose must bind prior to ATP appears to have been experimentally flawed because of changes in [ATP]/[ADP] ratios (nonconstant pool sizes) as saturation curves were observed, which may have caused net (nonequilibrium) flux for some data points.[16]

Because glucose binding to the active site induces large conformational changes in hexokinase,[62,63] preferred order binding of glucose prior to ATP is possible but not proved to be obligatory. With rat liver hexokinase,

[57] V. V. Kupriyanov, R. S. Balaban, N. V. Lyulina, A. Ya. Steinscheider, and V. A. Saks, *Biochim. Biophys. Acta* **1020,** 290 (1990).

[58] H. J. Fromm, E. Silverstein, and P. D. Boyer, *J. Biol. Chem.* **239,** 3645 (1964).

[59] D. L. Purich, H. J. Fromm, and F. B. Rudolph, *Adv. Enzymol.* **39,** 249 (1974).

[60] W. W. Cleland, *Adv. Enzymol.* **45,** 273 (1977).

[61] G. Noat and J. Richard, *Eur. J. Biochem.* **5,** 71 (1968).

[62] W. S. Bennett, Jr., and T. A. Steitz, *Proc. Natl. Acad. Sci. U.S.A.* **75,** 4848 (1978).

[63] C. M. Anderson, F. H. Zucker, and T. A. Steitz, *Science* **204,** 375 (1979).

isotope exchange data obtained by Gregoriou *et al.*[64] was interpreted in favor of an ordered mechanism (Glc binding before ATP, ADP being released before G6P). Plots of flux rate ratios versus substrate concentrations were used as a diagnostic method: $F(\text{G6P} \rightleftharpoons \text{ATP})/F(\text{G6P} \rightleftharpoons \text{Glc})$ and $F(\text{ATP} \rightleftharpoons \text{ADP})/F(\text{ATP} \rightleftharpoons \text{G6P})$ increased as a function of [ATP] or [ADP], respectively, whereas $F(\text{G6P} \rightleftharpoons \text{Glc})/F(\text{G6P} \rightleftharpoons \text{ATP})$ and $F(\text{ATP} \rightleftharpoons \text{G6P})/F(\text{ATP} \rightleftharpoons \text{ADP})$ were constant versus [Glc] and [G6P], respectively. Steady-state trapping occurred for E(Glc) and E(G6P) complexes, but not E(ATP) or E(ADP). A "mnemonical" (memory) effect on enzyme conformation, involving a slow (sugar-induced) conformational change, was postulated.

Kinetic data for rat muscle hexokinase have been interpreted in differing ways. Gregoriou *et al.*[65] used the observation that the $F(\text{G6P} \rightleftharpoons \text{ATP})/F(\text{G6P} \rightleftharpoons \text{Glc})$ ratio increased with [ATP], but was independent of [Glc], to argue that Glc binds before ATP. In addition, allosteric sites for both G6P and ADP were proposed, based on G6P induction of nonlinear responses to ATP and other effects.[65,66] A different mechanism was derived by Ganson and Fromm[67] from initial rate studies in the presence of competitive inhibitors of ADP and G6P, interpreted in favor of a kinetic mechanism with rapid equilibrium characteristics. The EIEK data[67] in fact suggest a "partially ordered" or preferred order random pathway, with Glc and G6P binding to free enzyme prior to nucleotides favored by 2 : 1. A critical analysis by Bass and Fromm[67,68] of all available kinetic data on muscle hexokinase was used to argue against ATP acting as an allosteric modifier.

Aspartate Kinase–Homoserine Dehydrogenase I. In *E. coli* several isoforms of aspartate kinase–homoserine dehydrogenase I (AK/HD-I) catalyze the first and third reactions of the aspartate kinase (AK) pathway, which ultimately leads to L-homoserine, L-threonine, L-lysine, L-isoleucine, and L-methionine.[69] The kinetic mechanism for AK is difficult to study at equilibrium because of a very low K_{eq} (3.5×10^{-4}), substrate K_m values relative to K_{eq}, and chemical instability of β-phospho-L-aspartate (L-Asp-β-P). Attempts to vary substrates in constant ratio for EIEK experiments with AK-III were only moderately successful; kinetic data were supplemented by observation of substrate stabilization against ther-

[64] M. Gregoriou, I. P. Trayer, and A. Cornish-Bowden, *Biochemistry* **20,** 499 (1981).
[65] M. Gregoriou, I. P. Trayer, and A. Cornish-Bowden, *Eur. J. Biochem.* **134,** 238 (1983).
[66] M. Gregoriou, I. P. Trayer, and A. Cornish-Bowden, *Eur. J. Biochem.* **161,** 171 (1986).
[67] N. J. Ganson and H. J. Fromm, *J. Biol. Chem.* **260,** 12099 (1985).
[68] M. B. Bass and H. J. Fromm, *Arch. Biochem. Biophys.* **256,** 708 (1987).
[69] P. Truffa-Bachi, M. Veron, and G. N. Cohen, *Crit. Rev. Biochem.* **2,** 379 (1974).

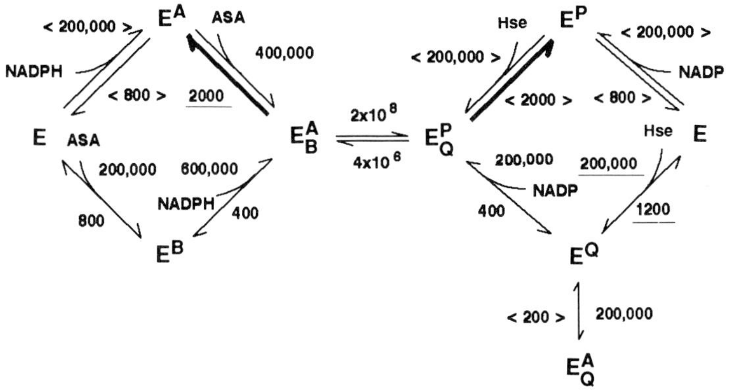

FIG. 3. Kinetic mechanism of *E. coli* homoserine dehydrogenase I (from Wedler and Ley,[73] with permission).

mal denaturation.[70] Initial velocity studies of AK-III (Lys-sensitive) and AK-I (Thr-sensitive) indicated random Asp and ATP binding but ordered release of ADP prior to β-Asp-P, determined by very slow dissociation of β-Asp-P.[70,71]

The kinetic mechanism of homoserine dehydrogenase-I (HD-I; part of AK/HD-I from *E. coli*) has been investigated by initial velocity[71] and EIEK methods.[72,73] The former approach suggested a "fully ordered" (compulsory) mechanism, with NADP(H) cofactors binding prior to L-aspartate β-semialdehyde (L-Asp-β-CHO; ASA) or L-Hse at pH 7.0.[71] The EIEK data at pH 9.0 indicated a similar but preferred order scheme.[72,73]The Hse ⇌ ASA rate was 2- to 3-fold faster than $NADP^+$ ⇌ NADPH, and variation of substrate pairs involving Hse or ASA caused inhibition effects in saturation curves for $NADP^+$ ⇌ NADPH. Analysis of the EIEK data with ISOBI-HS produced an optimal set of rate constants that explained the preferred order kinetic mechanism in terms of faster dissociation of ASA and Hse than NADP(H) cofactors from the central complexes, not faster association of cofactor than amino acids to free enzyme (see Fig. 3). Rate limitation in both directions is by cofactor dissociation rates rather than catalytic interconversion. A strong E(NADPH)(Hse) dead-end complex also forms.

Simulation and fitting of EIEK data for HD-I were carried out with

[70] J. Shaw and W. G. Smith, *J. Biol. Chem.* **252,** 5304 (1977).

[71] T. S. Angeles and R. E. Viola, *Arch. Biochem. Biophys.* **283,** 96 (1990).

[72] F. C. Wedler, B. W. Ley, S. L. Shames, S. J. Rembish, and D. L. Kushmaul, *Biochim. Biophys. Acta* **1119,** 247 (1992).

[73] F. C. Wedler and B. W. Ley, *J. Biol. Chem.* **268,** 4880 (1993).

ISOBI-HS without measured k_{on} or K_d values for substrates. Although the data fitting was adequate to eliminate alternative explanations, the confidence limits for cofactor on–off rates were less well-defined than is desirable. This illustrates the advantage of having measured rate and binding constants available prior to or in conjunction with computer simulations.

Aspartate Transcarbamylase. Previous studies of *E. coli* ATCase by initial velocity and EIEK methods at pH 7.8–8.3, 28°–37°, indicated compulsory order binding of carbamoyl phosphate (CP) prior to L-Asp and carbamoyl-Asp (CAsp) release prior to P_i, plus several dead-end complexes.[74,75] This early work also elucidated a temperature-dependent change in kinetic mechanism to rapid equilibrium random below 15°.

One problem with EIEK studies on ATCase near pH 8 is the high K_{eq} value, which necessitates using different reactant–product ratios in different saturation experiments.[75] For this reason, the ATCase kinetic mechanism was redetermined at pH 7.0, 30°, conditions at which K_{eq} is 10-fold lower. The ISOBI program was used for EIEK data simulation to derive a set of explicit rate constants.[39] These constants definitively proved that a preferred order kinetic scheme operated under these conditions, altered from the compulsory scheme derived near pH 8. These findings further argue against the rapid equilibrium random mechanism proposed from earlier initial velocity studies by Heyde *et al.*[76]

Modifier Action

A unique advantage of EIEK is the ability to observe both the fast and slow steps in the forward and reverse directions simultaneously. Thus, EIEK has the potential to define exactly which rate constants are altered by bound feedback modifiers, demonstrated in previous preliminary studies in theory and practice.[77–79] It is important to recognize that changes in initial velocity macroscopic constants (V_{max}, K_m) often may have a nonobvious basis in terms of the microscopic rate constants that are actually altered. Without knowledge of the kinetic mechanism this is difficult to predict. A prime example is the feedback inhibition of *E. coli* glutamine synthetase by AMP and GDP, shown to be V_{max} (noncompetitive) inhibitors, but found by EIEK studies to alter release of product

[74] R. W. Porter, M. O. Modebe, and G. R. Stark, *J. Biol. Chem.* **244,** 1846 (1969).
[75] F. C. Wedler and F. J. Gasser, *Arch. Biochem. Biophys.* **163,** 57 (1974).
[76] E. Heyde, A. Nagabhushanam, and J. F. Morrison, *Biochemistry* **12,** 4718 (1973).
[77] F. C. Wedler and P. D. Boyer, *J. Biol. Chem.* **247,** 993 (1972).
[78] F. C. Wedler and P. D. Boyer, *J. Theor. Biol.* **38,** 539 (1973).
[79] F. C. Wedler and W. H. Shalongo, this series, Vol. 87, p. 647.

ADP, which was the slow step in net turnover (not catalysis). Bound modifiers produced strong inhibition of the ATP $\rightleftharpoons$ P_i and ATP $\rightleftharpoons$ ADP exchanges, whereas a "cryptic" Glu $\rightleftharpoons$ Gln exchange occurred unimpaired.[77] Similar, nonobvious modes of action have been defined for other systems.

Aspartate Transcarbamylase. The primary allosteric modifiers of *E. coli* ATCase are CTP and ATP, causing the $S_{0.5}$ for L-aspartate to shift to higher or lower values, respectively. Previous EIEK data, interpreted without benefit of analytical computer programs, suggested that ATP and CTP may operate by separate and distinctly different mechanisms.[80] In opposition to this hypothesis is a macroscopic, monolithic model in which modifiers simply perturb the allosteric T–R transition in opposite directions and are inextricably linked to it.[81,82] The majority of current evidence suggests that in wild-type holoenzyme the action of ATP or CTP is linked to the T–R transition but can be independently altered or deleted by chemical modification or site-specific mutagenesis.[40,80]

An in-depth systematic EIEK study of the action of CTP and ATP on ATCase has been carried out, using computer-based simulation and fitting to define specific rate steps altered.[40] Modifiers were varied with substrates at fixed levels, and substrates were varied with fixed concentrations of each modifier. ATP, CTP, and the bisubstrate analog *N*-phosphonacetyl-L-aspartate (PALA) were observed to alter Asp $\rightleftharpoons$ CAsp more strongly than CP $\rightleftharpoons$ P_i, perturbing the observed R_{max} and $S_{0.5}$ values for each exchange in very different fashions. Computer simulation studies indicated that optimal modeling of these effects occurred when modifiers perturbed the T–R transition only indirectly, differentially altering the rates for L-Asp association greater than those for dissociation. The process of L-Asp binding immediately follows and is coupled in an equilibrium manner to the T–R transition, as shown in Fig. 4.

Modeling with ISOCOOP and ISOMOD was used to demonstrate that this action is an energetically more efficient way to alter enzyme properties than direclty altering the T–R transition step. Modifier binding to *r*-chains involves 7–8 kcal of energy, and only a small fraction can be transmitted over 60 Å to active sites on *c*-chains. Thus, the mode of highest efficiency is considered the most likely.

Homoserine Dehydrogenase I. (*E. coli*) The *E. coli* HD-I isoform is feedback inhibited by L-Thr and differentially activated by monovalent cations ($K^+ \gg Na^+ >$ others).[71] A systematic EIEK study, including

[80] F. C. Wedler and F. J. Gasser, *Arch. Biochem. Biophys.* **163,** 69 (1974).
[81] G. G. Hammes and C.-W. Wu, *Biochemistry* **10,** 1051 (1971).
[82] H. K. Schachman, *J. Biol. Chem.* **263,** 18583 (1988).

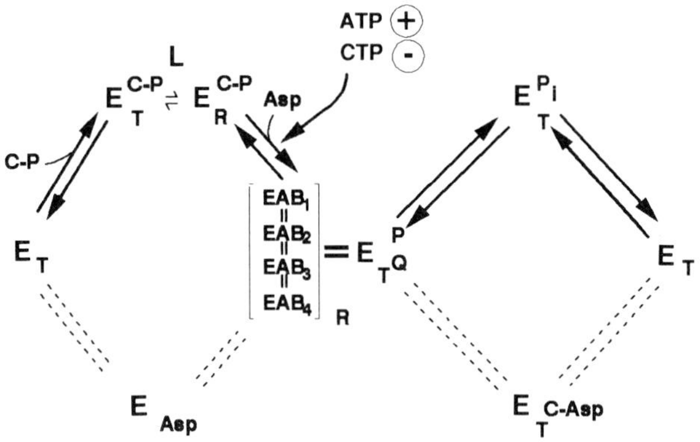

FIG. 4. Mode of action of allosteric feedback modifiers on *E. coli* aspartate transcarbamylase. (From Hsuanyu and Wedler,[40] with permission.)

extensive simulation studies with ISOBI, revealed that L-Thr inhibits catalytic interconversion of the central complexes without altering substrate association–dissociation rates.[73] Bound L-Thr produced an apparent change in kinetic mechanism toward a rapid equilibrium random scheme, making the ASA ⇌ Hse and $NADP^+$ ⇌ NADPH rates nearly equal, and suppressed the inhibition effects observed with $NADP^+$ ⇌ NADPH. These changes were best modeled with ISOBI-HS by making catalysis (k and k') definitively rate limiting. Decreasing the values for k and k', it should be noted, is not equivalent to removal of active enzyme from solution. This occurs without altering the substrate on–off rates that inherently define a preferred order kinetic mechanism, as shown in Fig. 5.

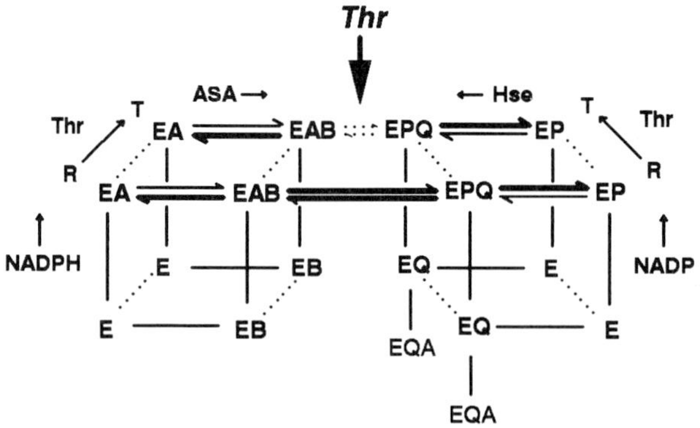

FIG. 5. Mode of action of L-Thr on *E. coli* homoserine dehydrogenase I. (From Wedler and Ley,[73] with permission.)

Comparison of kinetic properties by computer modeling of EIEK data was also performed to define which steps of the HD-I kinetic mechanism are altered on substitution of Na^+ for K^+ ion into the structure of AK/HD-I.[83] Whether these ions play a structural or a catalytic role with HD and AK is not yet well defined. Interestingly, although HD-I has a 10-fold higher affinity for Na^+ than for K^+, the Na^+-activated enzyme exhibits a V_{max} that is an order of magnitude lower and substrate K_m values that are 2- to 3-fold higher than with K^+. Saturation curves for the ASA $\rightleftharpoons$ Hse and $NADP^+ \rightleftharpoons$ NADPH exchanges were compared in the presence of K^+ versus Na^+, varying all possible combinations of substrate pairs in constant ratio. In contrast to the preferred order random scheme for K^+-activated HD-I, the Na^+-activated enzyme exhibited a fully random scheme, with significant rate limitation at catalysis and the dissociation of amino acid substrates, relative to the K^+-activated enzyme (see Fig. 6).

Site-Specific Mutations

Substituting one amino acid residue for another is an ideal method for dissecting factors related to the kinetics and regulation of complex enzyme systems.[84] One problem in such work is the fact that a point mutation can have far-reaching kinetic and structural consequences, that is, can alter an ensemble of properties.[85] This demands that changes in kinetic behavior be carried to the level of microscopic rate constants, not macroscopic K_m and V_{max} values. Such detailed analysis has been provided for only a few enzyme systems, primarily by EIEK, pre-steady-state, and isotope effect kinetics, plus unfolding–refolding studies. Examples of these include dihydrofolate reductase, *Staphylococcus* nuclease, thymidylate synthase, subtilisin, and ribonuclease. Mutant enzymes investigated and characterized by EIEK methods are discussed below.

Aminoacyl-tRNA Synthetases. With Tyr-tRNA synthetase, Wells *et al.*[86] used the ATP $\rightleftharpoons$ PP_i exchange to measure true rate constants and K_d values with wild-type and engineered mutants. Knowledge of such values has allowed a detailed accounting of energy changes for intermediates and transition states in the overall reaction pathway for these enzymes caused by specific mutations.[87]

[83] F. C. Wedler and B. W. Ley, *Arch. Biochem. Biophys.* **301,** 416 (1993).

[84] K. A. Johnson and S. J. Benkovic, *in* "The Enzymes" (D. S. Sigman and P. D. Boyer, eds.), 3rd Ed., Vol. 19, Mechanisms of Catalysis, p. 159. Academic Press, New York, 1990.

[85] D. W. Hibler, N. J. Stolowich, M. A. Reynolds, J. A. Gerlt, J. A. Wilde, and P. H. Bolton, *Biochemistry* **26,** 6278 (1987).

[86] T. N. C. Wells, J. W. Knill-Jones, T. E. Gray, and A. R. Fersht, *Biochemistry* **30,** 5151 (1991).

[87] A. R. Fersht, R. J. Leatherbarrow, and T. N. C. Wells, *Biochemistry* **26,** 6030 (1987).

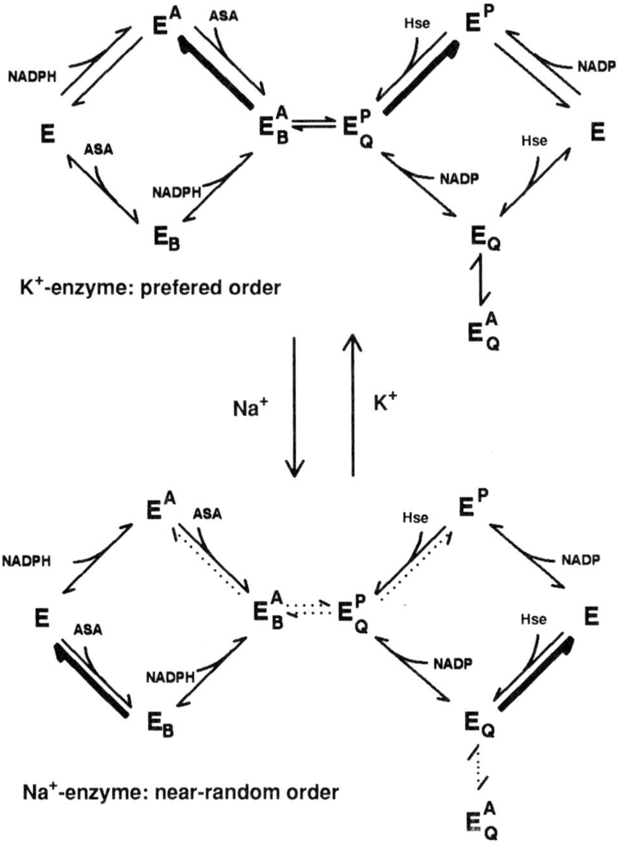

FIG. 6. Differences in kinetic mechanism between K^+- and Na^+-activated *E. coli* homoserine dehydrogenase I. (From Wedler and Ley.[83])

Aspartate Transcarbamylase. Stevens *et al.*[88] summarized the properties of over 50 mutant forms of ATCase characterized as of 1991, examples of which are illustrated in Fig. 7.[89] Some of the first mutants involved residues located in the so-called 240's loop of the *c*-chains, a structure that undergoes a major conformational change as part of the T–R transition. Residue Tyr-240 forms an intrachain hydrogen bond to Asp-271 in the T-state which is disrupted on conversion to the R-state. Loss of this bond on mutating Tyr to Phe is expected to destabilize the T-state. In the T-state

[88] R. C. Stevens, Y. M. Chook, C. Y. Cho, W. N. Lipscomb, and E. R. Kantrowitz, *Protein Eng.* **4,** 391 (1991).

[89] N. J. Dembowski, C. J. Newton, and E. R. Kantrowitz, *Biochemistry* **29,** 3716 (1990).

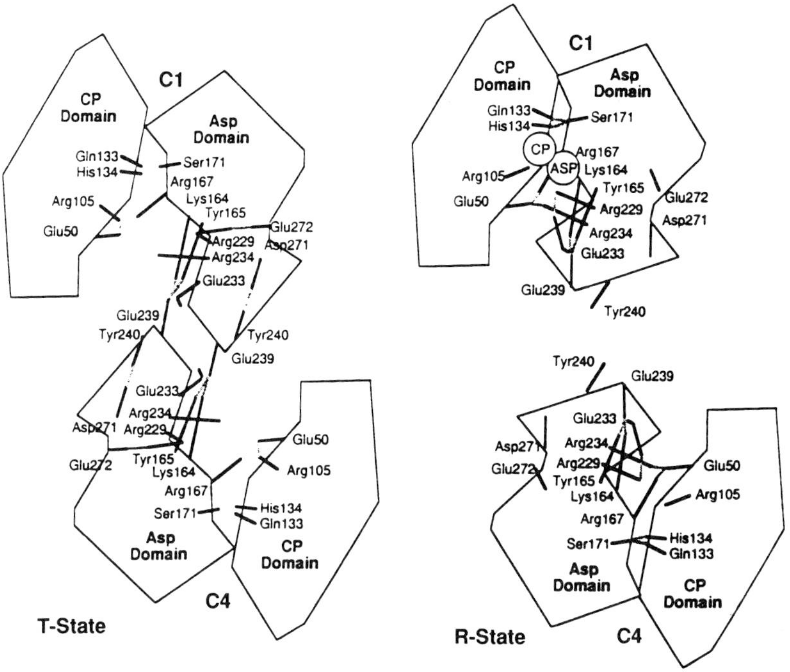

FIG. 7. Residues in *E. coli* aspartate transcarbamylase identified by site-specific mutagenesis as critical for substrate binding, catalysis, and the allosteric T–R transition. (Reprinted with permission from Dembowski *et al.*[89] Copyright 1990 American Chemical Society.)

enzyme Glu-239 forms multiple interchain bonds with Lys-164 and Tyr-165, and in the R-state Glu-239 reforms multiple bonds with these same groups within the same chains, making the kinetic and thermodynamic consequences of the Glu-239 to Gln mutation less predictable.

Wedler, Kantrowitz, and co-workers[90–92] carried out in-depth analysis of rate constants altered by these mutations in ATCase, using EIEK methods plus simulations with the ISOBI program. The studies revealed that the Tyr-240 to Phe mutation alters the kinetic mechanism[88] in opposite ways on each side of the reaction: the binding of CP prior to Asp shifted from preferred order random to fully compulsory order, whereas the order for CAsp and P_i shifted from preferred order to fully random. These

[90] Y. Hsuanyu, F. C. Wedler, E. R. Kantrowitz, and S. A. Middleton, *J. Biol. Chem.* **264,** 17259 (1989).

[91] Y. Hsuanyu and F. C. Wedler, *Biochim. Biophys. Acta* **957,** 455 (1988).

[92] Y. Hsuanyu, F. C. Wedler, S. A. Middleton, and E. R. Kantrowitz, *Biochim. Biophys. Acta* **995,** 54 (1989).

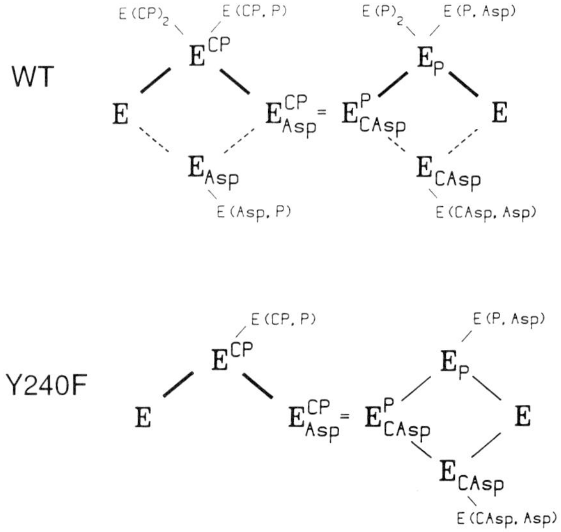

FIG. 8. Consequences of the Tyr-240 to Phe mutation on the kinetic mechanism of *E. coli* aspartate transcarbamylase. (From Hsuanyu *et al.*,[90] with permission.)

changes are illustrated in Fig. 8. It seems likely that these differences may relate to the fact that the T–R transition occurs at different positions in the kinetic scheme relative to substrate binding steps. Note that it is after CP but before Asp binding on the reactant side, but occurs just after catalysis but before release of either product. Although the unliganded Y240F enzyme retains a T-state conformation, the Tyr-240 to Phe mutation makes the R-state more kinetically and thermodynamically accessible, decreasing the energy barrier between the T- and R-states by destabilizing the T-state.

The EIEK studies of kinetic effects caused by the Glu-239 to Gln mutation showed saturation patterns that resemble catalytic subunits (c_3), which behaves as a high affinity R-state enzyme.[91] Cooperativity is lost, but the kinetic mechanism remains strongly preferred order: CP binding before Asp, CAsp released before P_i.[92] In contrast to effects seen with Y240F in which only the T-state is destabilized, data for the E239Q enzyme indicated destabilization for both the T- and R-state enzymes (see Fig. 9).

Changes in the regulatory properties of the Y240F mutant were also investigated.[93] Whereas with wild-type enzyme CTP strongly inhibits both exchange reactions, with Y240F CTP inhibits Asp $\rightleftharpoons$ CAsp but activates

[93] F. C. Wedler, Y. Hsuanyu, E. R. Kantrowitz, and S. A. Middleton, *J. Biol. Chem.* **264,** 17266 (1989).

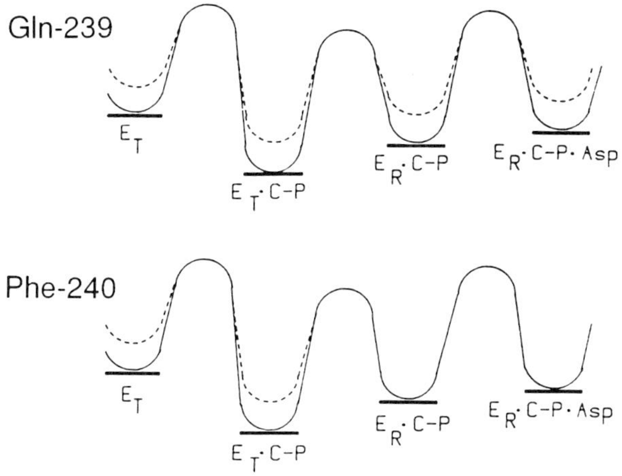

FIG. 9. Comparison of the kinetic and energetic consequences of the Glu-239 to Gln and Tyr-240 to Phe mutations on *E. coli* aspartate transcarbamylase. (From Hsuanyu and Wedler,[91] with permission.)

$CP \rightleftharpoons P_i$. Similarly, PALA activation of $CP \rightleftharpoons P_i$ is selectively lost, but activation of Asp $\rightleftharpoons$ CAsp is retained. In contrast, ATP activates both exchanges with both the wild type and Y240F. These data clearly indicate a change in the mode of action for CTP that does not occur for ATP, offering further evidence of separate (and separable) modes of action for the two modifiers. Modifier-induced kinetic effects with Y240F were analyzed by modeling with the ISOBI and ISOMOD programs, leading to the following conclusions: ATP increases the association and dissociation rates for both CP and L-Asp in a proportional manner, whereas CTP differentially increases the rate for CP association over dissociation, while decreasing the rates for L-Asp in equal proportion. These different modes of action are illustrated in Fig. 10.

Energy-Transducing Systems

Over the past several decades, an array of diverse, elegant studies on the enzyme systems involved with the ATP utilization and production processes of photosynthesis, muscle proteins, mitochondria, and other systems have been carried out. Isotope exchange methods have provided detailed molecular level, chemical insights to events in the active sites of these complex systems. In particular, exchange methods utilizing ^{18}O-labeled components have proved especially useful for defining intermediates, characterizing conformational states, and probing the rates of forma-

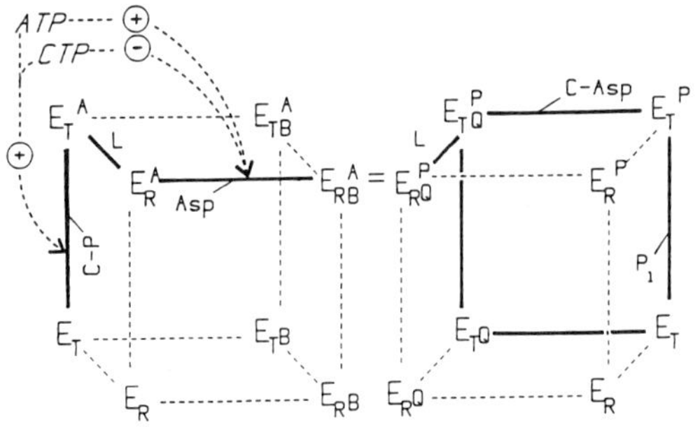

FIG. 10. Consequences of the Tyr-240 to Phe mutation on the sites of action of allosteric feedback modifiers of *E. coli* aspartate transcarbamylase. (From Hsuanyu *et al.*,[92] with permission.)

tion and breakdown of transient species in these investigations. These and other relevant topics have been discussed critically in a review on ATP synthase by Boyer,[15,94] which includes a detailed discussion of the so-called binding change mechanism.

Concluding Remarks

In 1980 Purich and Allison[16] observed that the EIEK method was "undergoing rapid expansion in theory and experiment." At that time, in the absence of convenient methods for modeling enzyme kinetic behavior at equilibrium, the theoretical base for EIEK was badly underdeveloped compared to what existed for initial velocity and other methods. With the development of the ISOBI and ISOMOD programs, described above, many of these problems have been overcome. A primary objective of this chapter has been to document the breakthroughs in methodology resulting from the development of computer programs for quantitative simulation and analysis of EIEK data. These new approaches offer the nonexpert instant access to a versatile theoretical basis for almost any complex Bi Bi or Ter Ter kinetic scheme, unfettered by the requirement to derive complex algebraic equations, in many cases an impossible task. Without these new methods, EIEK data analysis would still be done by the qualitative logic set forth by Boyer and Silverstein in the early 1960s.[2] Understandably, some may hestitate to use simulation methods, owing to a preconception that with enough (or too many) constants one can fit anything without ever being sure that a unique solution has been achieved. This is only true if each rate constant in a complex network is assumed

[94] P. D. Boyer, *Biochim. Biophys. Acta* **1140,** 215 (1993).

to be infinitely variable. In practice, this is not done. Rather, it is clear that a few substrate on or off rates or K_d values must be measured to "pin down" the system (i.e., reduce the allowed degrees of freedom). Once this is done, the limits of variation allowed for other constants are restricted.

Overall, these advances should convince a greater number of researchers concerned with enzyme mechanism of the power and insightfulness of this kinetic approach. Equilibrium isotope exchange kinetics has matured to the point of application to its full potential and scope in the field. For example, it is anticipated that EIEK will be used more extensively on mutant enzymes that have been engineered by the latest methods in molecular biology for specific or interesting properties.

[18] Proton Transfer in Carbonic Anhydrase Measured by Equilibrium Isotope Exchange

By DAVID N. SILVERMAN

Introduction

The isozymes of carbonic anhydrase (EC 4.2.1.1, carbonate dehydratase) have revealed many properties of proton transfer accompanying the catalyzed hydration of CO_2 to form bicarbonate and a proton [Eq. (1)].

$$CO_2 + H_2O \rightleftharpoons HCO_3^- + H^+ \quad (1)$$

The catalytic pathway contains inter- and intramolecular proton transfer steps that, under various conditions, can contribute to the overall maximal velocity. From the initial concerns about the role of proton transfer in the catalytic pathway[1] to the present-day studies of site-specific mutants, the carbonic anhydrases have been used to understand the properties of proton transfers and their role in enzymatic catalysis. In this chapter, the many studies of this catalysis are integrated into an overall view of the role of the various proton transfers and the role of active-site water including zinc-bound water in the carbonic anhydrases. Previous reviews have emphasized proton transfer in the catalysis.[2–5]

[1] H. DeVoe and G. B. Kistiakowski, *J. Am. Chem. Soc.* **83,** 274 (1961).

[2] S. Lindskog, *in* "Zinc Enzymes" (T. G. Spiro, ed.), p. 78. Wiley, New York, 1983.

[3] D. N. Silverman and S. H. Vincent, *Crit. Rev. Biochem.* **14,** 207 (1983).

[4] D. N. Silverman and S. Lindskog, *Acc. Chem. Res.* **21,** 30 (1988).

[5] R. G. Khalifah and D. N. Silverman, *in* "The Carbonic Anhydrases" (S. J. Dodgson, R. E. Tashian, G. Gros, and N. D. Carter, eds.), p. 49. Plenum, New York, 1991.

All carbonic anhydrases sequenced to date can be classified in two broad groups according to similarities in the primary structures: those similar in amino acid sequence with the animal and human carbonic anhydrases and those related to the plant carbonic anhydrases. These two groups have no apparent homologies in amino acid sequence although both contain zinc, and the crystal structure of a plant carbonic anhydrase has not yet been determined. The first class includes the seven known isozymes of human carbonic anhydrase which have amino acid sequence identities from 28 to 59%.[6] Nearly all mechanistic and structural studies on carbonic anhydrases have been done on three isozymes, I, II, and III, some steady-state properties of which appear in Table I. Among these are the most and least efficient of the carbonic anhydrases, isozymes II and III. Carbonic anhydrases I and II appear in red cells, with isozyme II appearing also in secretory tissues.[7] Carbonic anhydrase III (CA III) is an abundant cytosolic protein in skeletal muscle [8] and adipose tissues,[9] and its function is under investigation.[8] A membrane-bound form of mammalian carbonic anhydrase, carbonic anhydrase IV, is found in many tissues including lung (capillary endothelium) and kidney (brush border), where it is anchored to membranes at least in part by phosphatidylinositol–glycan linkages.[10] Carbonic anhydrase V has been identified in mitochondria, and isozymes VI and VII appear in the salivary gland and are secreted in saliva.[6]

The only reaction of physiological significance known to be catalyzed by the carbonic anhydrases is hydration of CO_2 to give HCO_3^- and a proton and the reverse dehydration. Carbonic anhydrases carry out many physiological functions based on two fundamental processes: (1) rapid liberation of CO_2 stored in the form of HCO_3^-, as in red cells and in photosynthesis, and in the facilitation of diffusion of CO_2 across membranes; and (2) production of HCO_3^- in formation of secretory fluids such as ocular and cerebrospinal fluids, and production of H^+ in renal acidification of urine and gastric acid secretion. Many references on function can be found in two volumes devoted to the carbonic anhydrases.[11,12]

[6] R. E. Tashian, *BioEssays* **10,** 186 (1989).

[7] T. H. Maren, *Physiol. Rev.* **47,** 595 (1987).

[8] C. Geers and G. Gros, *in* "The Carbonic Anhydrases" (S. J. Dodgson, R. E. Tashian, G. Gros, and N. D. Carter, eds.), p. 227. Plenum, New York, 1991.

[9] C. J. Lynch, W. A. Brennan, Jr., T. G. Vary, N. Carter, and S. J. Dodgson, *Am. J. Physiol.* **264,** E621 (1993).

[10] X. L. Zhu and W. S. Sly, *J. Biol. Chem.* **265,** 8795 (1990).

[11] S. J. Dodgson, R. E. Tashian, G. Gros, and N. D. Carter (eds.), "The Carbonic Anhydrases." Plenum, New York, 1991.

[12] F. Botre, G. Botre, and B. T. Storey (eds.), "Carbonic Anhydrase." VCH, Weinheim, 1991.

Structures of Active Sites of Carbonic Anhydrases

Carbonic anhydrases II and III are monomeric, have a molecular mass of 30 kDa, and contain one zinc per molecule. Refined crystal structures to 2.0 Å resolution are reported for human CA II[13] and for bovine CA III.[14] With 56% amino acid identity, the isozymes have backbone structures that are quite similar; human CA II (HCA II) and bovine CA III have a root mean squared deviation in backbone atoms of 0.92 Å.[14] These structures have a foundation of twisted β-pleated sheet with very little helix content. The single zinc per molecule is coordinated to three histidine residues at the bottom of a conical cavity (Fig. 1). The fourth ligand of the zinc is a water molecule which ionizes to the zinc-bound hydroxide.

The active-site cavity of HCA II is conical with a hydrophobic side (Val-121, Val-143, Leu-198, Val-207, Trp-209) and a hydrophilic side (Tyr-7, Asn-62, Asn-67, Gln-92, Glu-106). Threonine-199 is situated adjacent to the metal and acts as a hydrogen-bond acceptor for the zinc-bound water or hydroxide; in turn, the carboxylate of glutamate-106 is a hydrogen-bond acceptor for the hydroxyl side chain of threonine-199[13] as shown in Fig. 1. This interaction is suggested to provide an orientation of the lone pair electrons of the oxygen of the zinc-bound hydroxide in the direction appropriate for attack on CO_2 and to decrease the entropic penalty entering the transition state.[15] The active site of HCA III also contains the hydrogen-bonded network of glutamate-106, threonine-199, and the zinc-bound water. The major differences compared with isozyme II are lysine-64, arginine-67, and phenylalanine-198; thus, the active site of HCA II is more positively charged and more sterically constrained (Fig. 1).[14]

Although there is no firm picture of the binding site for CO_2 in the catalysis, current studies are giving a clearer picture of this site. The hydrophobic side of the cavity appears to have only an indirect effect in any specific orienting of CO_2 at its binding site. Site-directed mutagenesis of isozyme II in this region, with hydrophobic residues Val-121 and Val-143 replaced by residues of equivalent or smaller size, resulted in very small effects on both k_{cat} and k_{cat}/K_m for hydration of CO_2.[16,17] However, there was great decrease in activity with substitution of residues of greater size (Tyr, Phe). Krebs *et al.*[18] have identified an asymmetric stretching

[13] A. E. Eriksson, T. A. Jones, and A. Liljas, *Proteins: Struct. Funct. Genet.* **4,** 274 (1988).

[14] A. E. Eriksson and A. Liljas, *Proteins: Struct. Funct. Genet.* **16,** 29 (1993).

[15] K. M. Merz, Jr., *J. Mol. Biol.* **214,** 799 (1990).

[16] S. K. Nair, T. L. Calderone, D. W. Christianson, and C. A. Fierke, *J. Biol. Chem.* **266,** 17320 (1991).

[17] C. A. Fierke, T. L. Calderone, and J. F. Krebs, *Biochemistry* **30,** 11054 (1991).

[18] J. F. Krebs, F. Rana, R. A. Dluhy, and C. A. Fierke, *Biochemistry* **32,** 4496 (1993).

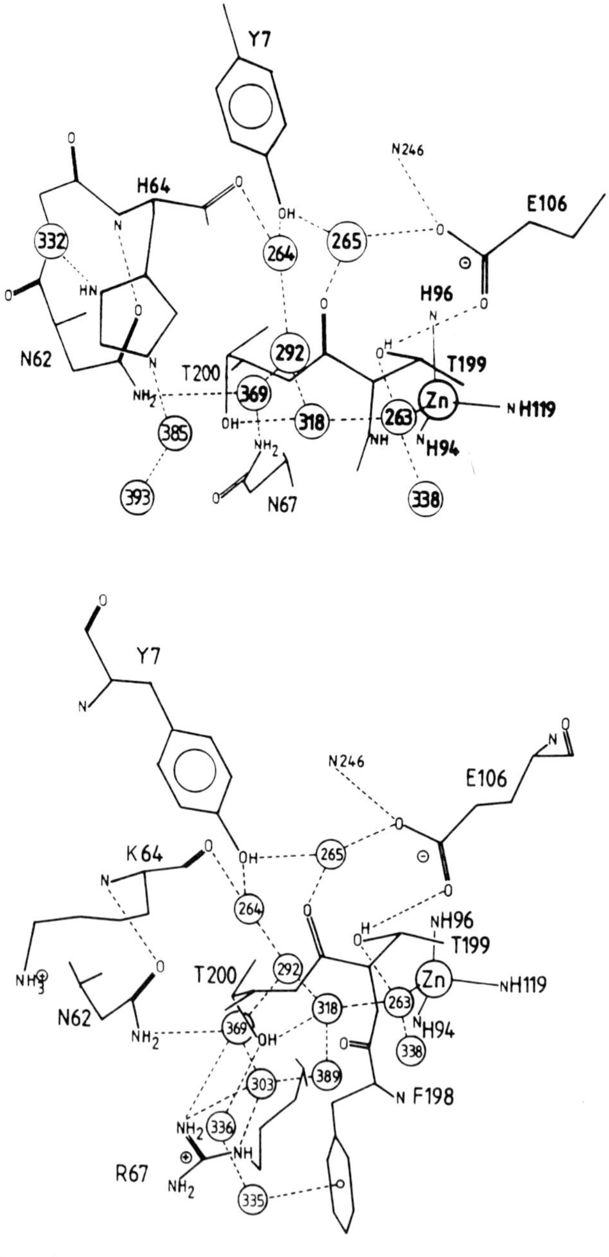

FIG. 1. (Top) Selected residues in the active-site cavity of human carbonic anhydrase II from the published crystal structure [A. E. Eriksson, T. A. Jones, and A. Liljas, *Proteins: Struct. Funct. Genet.* **4,** 274 (1988)]. The circles represent water molecules identified in the cavity. (Bottom) Residues near the active-site cavity of human carbonic anhydrase III from the published crystal structure [A. E. Eriksson and A. Liljas, *Proteins: Struct. Funct. Genet.* **16,** 29 (1993).

vibration of CO_2 in the presence of HCA II that is shifted in the infrared spectrum compared with free CO_2 in solution. This vibrational frequency has been identified with CO_2 bound in the active site of HCA II by use of specific inhibitors of carbonic anhydrase, reviving a method first applied to carbonic anhydrase by Reipe and Wang.[19] The results of this approach estimate an equilibrium dissociation constant near 100 m*M* for the binding of CO_2 to its catalytic site in a nonpolar region of isozyme II. Such a value is consistent with measurements of the catalyzed interconversion of CO_2 and HCO_3^- by equilibrium methods, ^{13}C nuclear magnetic resonance (NMR),[20] and ^{18}O exchange between CO_2 and water,[21] which indicate very weak substrate binding.

Another indication of the binding site for CO_2 has been obtained by Lindahl *et al.*,[22] who have located the binding site in HCA II of cyanate and cyanide. These inhibitors do not bind directly to the zinc and do not displace the water ligand of the zinc, according to the X-ray diffraction patterns, but they appear to form a hydrogen bond with the backbone NH of residue Thr-199. This hydrogen bond may also be formed in an analogous manner in the of binding of CO_2. These data are complemented by the structure, determined crystallographically, of HCO_3^- bound to the zinc in a mutant of HCA II containing the replacement of Thr-200 by histidine.[23]

Overview of Catalytic Mechanism

There is a wide body of evidence supporting a two-stage catalysis (Ping Pong) for both isozymes II and III.[4] The first stage, shown in Eq. (2), is the interconversion of CO_2 and HCO_3^- by the enzyme (E). One role of the metal in carbonic anhydrase is to act as a Lewis acid to activate the zinc-bound water molecule by reducing its pK_a to near 7 in isozymes I and II, and to near 5 for isozyme III. This provides the zinc-bound hydroxide as a prominent species at physiological pH; it is the zinc-bound hydroxide that is the catalytic entity in the hydration of CO_2, as shown in Eq. (2). Another role for the zinc is to lower the energy of the transition state by delocalizing developing negative charge.

$$CO_2 + EZnOH^- \rightleftharpoons EZnHCO_3^- \xrightleftharpoons{H_2O} EZnH_2O + HCO_3^- \quad (2)$$

[19] M. E. Riepe and J. H. Wang, *J. Biol. Chem.* **243,** 2779 (1968).
[20] I. Simonsson, B. H. Jonsson, and S. Lindskog, *Eur. J. Biochem.* **93,** 409 (1979).
[21] D. N. Silverman, C. K. Tu, S. Lindskog, and G. C. Wynns, *J. Am. Chem. Soc.* **101,** 6734 (1979).
[22] M. Lindahl, L. A. Svensson, and A. Liljas, *Proteins: Struct. Funct. Genet.* **15,** 177 (1993).
[23] Y. Xue, J. Vidgren, L. A. Svensson, A. Liljas, B. H. Jonsson, and S. Lindskog, *Proteins: Struct. Funct. Genet.* **15,** 80 (1993).

The proposed direct nucleophilic attack of the zinc-bound hydroxide on CO_2 relies on structural evidence and the solvent hydrogen isotope effect (SHIE) on catalysis. The structural evidence is the position of the metal-bound bicarbonate in the enzyme–product complex[23,24] as well as ^{13}C NMR measurements of distances from the cobalt in Co(II)-substituted CA II.[25] The SHIE determined for the steps of Eq. (2) is unity, measured by ^{13}C NMR[20] and by ^{18}O exchange,[21] suggesting that the pathway for hydration proceeds by direct nucleophilic attack with no proton transfer rather than in a general base mechanism involving proton transfer. The dissociation of bicarbonate from the metal leaves the active site with a zinc-bound water.

The second stage of catalysis in hydration is the proton transfer sequence that converts the enzyme back to the active zinc hydroxide form, as shown in Eq. (3). In Eq. (3) B is buffer in solution or possibly water,

$$EZnH_2O + B \rightleftharpoons H^+EZnOH^- + B \rightleftharpoons EZnOH^- + BH^+ \qquad (3)$$

and H^+ written before E indicates a protonated shuttle residue. Intermolecular proton transfer to buffers in solution is essential for maximal activity of the carbonic anhydrases in the hydration and dehydration directions.[26,27]

There are many experiments giving indirect evidence that His-64 shuttles protons between buffer in solution and the zinc-bound water in CA II,[4] a hypothesis based initially on observation of solvent hydrogen isotope effects.[28] This hypothesis is further supported by site-directed mutagenesis as described below. The imidazole of His-64 is too far from the zinc for direct proton transfer (Fig. 1), and solvent hydrogen isotope effects are consistent with proton transfer through intervening hydrogen-bonded water bridges.[29] HCA III has lysine at position 64, and no such intramolecular shuttle has been detected in this isozyme at physiological pH.

There are many unanswered questions about the catalytic pathway of the carbonic anhydrases. Among these are why the intramolecular proton transfer, which could possibly be as fast as a molecular vibration at 10^{12} sec^{-1}, is no faster than 10^6 sec^{-1}. Is this proton transfer rate, which is the limiting step for maximal velocity, sensitive to the properties of nearby residues? What is the property of the active site that gives a zinc-bound

[24] K. Hakansson and A. Wehnert, *J. Mol. Biol.* **228,** 1212 (1992).

[25] T. J. Williams and R. W. Henkens, *Biochemistry* **24,** 2459 (1985).

[26] R. S. Rowlett and D. N. Silverman, *J. Am. Chem. Soc.* **104,** 6737 (1982).

[27] Y. Pocker, N. Janjic, and C. H. Miao, *in* "Zinc Enzymes" (I. Bertini, C. Luchinat, W. Maret, and M. Zeppezauer, eds.), p. 341. Birkhauser, Boston, 1986.

[28] H. Steiner, B. H. Jonsson, and S. Lindskog, *Eur. J. Biochem.* **59,** 253 (1975).

[29] K. S. Venkatasubban and D. N. Silverman, *Biochemistry* **19,** 4984 (1980).

water with a pK below 6 a significant nucleophilicity? What are the dynamic properties and role of active-site water in the proton transfer pathways? What are the functions in catalysis of the hydrophilic and hydrophobic sides of the active site cavity?

Experimental Approaches

Cloning, Expression, and Site-Specific Mutagenesis of Human Carbonic Anhydrases II and III

The studies described below examining the catalytic mechanism of carbonic anhydrase required large amounts of enzyme with specific amino acid modifications. Successful expression (up to 60 mg of carbonic anhydrase per liter) has been achieved using the T7 RNA polymerase–T7 promoter PET vector system.[30] Because efficient production of multiple mutants is often necessary, the coding segments of both HCA II and HCA III have been inserted into constructs which contain the T7 promoter along with a phage f1 origin (for single-strand DNA synthesis) in a PUC12 backbone.[31] This allows mutation, sequencing, and expression of both CA isozymes without subcloning steps; in addition, the vector is suitable for efficient cassette mutagenesis. Tanhauser *et al.*[31] constructed two HCA III vectors with unique *Afl*II and *Cla*I sites bounding the 60–70 region and *Nru*I and *Stu*I sites bounding the 189–200 region. These cassettes allow rapid mutation (and multiple site mutation) by ligation of a synthetic double-strand oligonucleotide into the unique sites.

Once a mutant CA vector has been isolated (based on specific marker restriction sites introduced during the mutagenesis) and confirmed by DNA sequencing, it is transformed into BL21(DE3)pLysS[30] and protein expressed. Expression volumes range from 1 liter (flasks) to 15 liters (fermenter), depending on the activity of the mutant. The process is efficient; new cassette mutations are often expressed within 1 week of receipt of the double-strand oligonucleotides, with single point mutants elsewhere in the molecule taking little longer. Mutations at many positions in HCA II and III have been constructed using either mutating single-strand oligonucleotides or double-strand oligonucleotides inserted into the cassette sites.

[30] F. W. Studier, A. H. Rosenberg, J. J. Dunn, and J. W. Dubendorff, this series, Vol. 185, p. 80.

[31] S. M. Tanhauser, D. A. Jewell, C. K. Tu, D. N. Silverman, and P. J. Laipis, *Gene* **117,** 113 (1992).

Measurement of Catalysis at Steady State

Initial velocities of CO_2 hydration and HCO_3^- dehydration are measured by stopped-flow spectrophotometry following the change in absorbance of a pH indicator.[26,32] Solutions of CO_2 are prepared by bubbling CO_2 into water under controlled conditions to obtain solutions saturated in CO_2; dilutions are made using syringes with gas-tight connections.

Oxygen-18 Exchange Kinetics

The ^{18}O exchange method is carried out at chemical equilibrium and is based on initial studies of the uncatalyzed rate of hydration of CO_2 by Mills and Urey in 1940.[33] It has been extended to measure the catalysis by carbonic anhydrase.[34] This method relies on two measurements: (1) the rate of depletion of ^{18}O from CO_2 caused by the hydration–dehydration cycle and the appearance of the label in $H_2{}^{18}O$ where it is greatly diluted by $H_2{}^{16}O$ [Eqs. (4) and (5)], and (2) the rate of exchange of ^{18}O from $^{12}CO_2$

$$HCOO^{18}O^- + EZnH_2O \rightleftharpoons EZn^{18}OH^- + CO_2 + H_2O \quad (4)$$

to $^{13}CO_2$ caused by the transitory labeling of the active site of carbonic anhydrase with ^{18}O. To make these measurements, the reaction solution is in contact with a membrane permeable to gases; CO_2 passing across the membrane enters a mass spectrometer providing a continuous measure of isotopic content of CO_2. The ^{18}O exchange method is carried out at chemical equilibrium and can therefore be performed without buffers, an advantage in studying intramolecular proton transfer. In Eq. (5), BH^+ can be buffer in solution, water, or a side chain of the enzyme such as His-64.

$$EZn^{18}OH^- + BH^+ \rightleftharpoons EZn^{18}OH_2 + B \xrightleftharpoons{H_2O} EZnH_2O + H_2{}^{18}O + B \quad (5)$$

Two independent rates at chemical equilibrium can be determined: R_1, the rate of interconversion of CO_2 and HCO_3^- as in Eq. (4); and R_{H_2O}, the rate of release from the enzyme of $H_2{}^{18}O$ as in Eq. (5). The rate-limiting step for R_{H_2O} involves a proton transfer to the zinc-bound hydroxide, forming a zinc-bound water which is readily exchangeable with solvent water. The rate-limiting nature of this proton transfer has

[32] R. G. Khalifah, *J. Biol. Chem.* **246,** 2561 (1971).

[33] G. A. Mills and H. C. Urey, *J. Am. Chem. Soc.* **62,** 1019 (1940).

[34] D. N. Silverman, this series, Vol. 87, p. 732.

been confirmed by solvent hydrogen isotope effects and pH profiles,[35] buffer enhancement of R_{H_2O},[34,36] and simulations of catalysis.[37,38]

Effect of Site-Directed Mutagenesis on Proton Transfer Rates

Histidine-64 as Proton Shuttle in Catalytic Pathway

Catalysis by the site-specific mutant of HCA II with the putative shuttle residue His-64 replaced with alanine (H64A HCA II) is consistent with the proton shuttle hypothesis.[39] The rate of the proton-transfer dependent release of $H_2{}^{18}O$ from the enzyme [Eq. (5)] was decreased as much as 20-fold for the mutant H64A HCA II, as shown in Fig. 2, whereas the rate of interconversion of CO_2 and HCO_3^- [Eq. (4)] was the same for the mutant and wild-type enzymes. This conclusion was further supported in the same report by the steady-state results; the mutant had k_{cat} decreased as much as 25-fold, but k_{cat}/K_m was unchanged. The turnover number for hydration of CO_2, k_{cat}, contains rate constants for all forward steps in the catalysis from the enzyme–substrate complex and is dominated by proton transfer for both isozymes II and III. On the other hand, k_{cat}/K_m contains rate constants only for the steps of Eq. (4).[4]

Catalysis by the mutant H64A HCA II can be enhanced by certain buffers in solution; the buffer imidazole activated the mutant H64A HCA II in a saturable manner to a maximal level of activity about equal to that of the wild type.[39] The buffer MOPS [3-(*N*-morpholino)propanesulfonic acid] did not activate the mutant presumably because the buffer molecule is too large or has inappropriate charge to reach into the active-site cavity. The still sizable activity of the H64A mutant in the absence of buffer indicates a proton transfer rate as large as 10^4 sec^{-1} that is rather independent of pH in the range of these measurements (Fig. 2). This supports the role of proton transfer in the catalysis and is one of the first examples among site-specific mutants of replacing the catalytic function of an amino acid residue with a molecule from solution; another is the work of Toney and Kirsch with aspartate aminotransferase.[40]

[35] C. K. Tu and D. N. Silverman, *Biochemistry* **24,** 6353 (1982).

[36] C. K. Tu, S. R. Paranawithana, D. A. Jewell, S. M. Tanhauser, P. V. LoGrasso, G. C. Wynns, P. J. Laipis, and D. N. Silverman, *Biochemistry* **29,** 6400 (1990).

[37] R. S. Rowlett, *J. Protein Chem.* **3,** 369 (1984).

[38] S. Linkskog, *J. Mol. Catal.* **23,** 357 (1984).

[39] C. K. Tu, D. N. Silverman, C. Forsman, B.-H. Jonsson, and S. Lindskog, *Biochemistry* **28,** 7913 (1989).

[40] M. D. Toney and J. F. Kirsch, *Science* **243,** 1485 (1989).

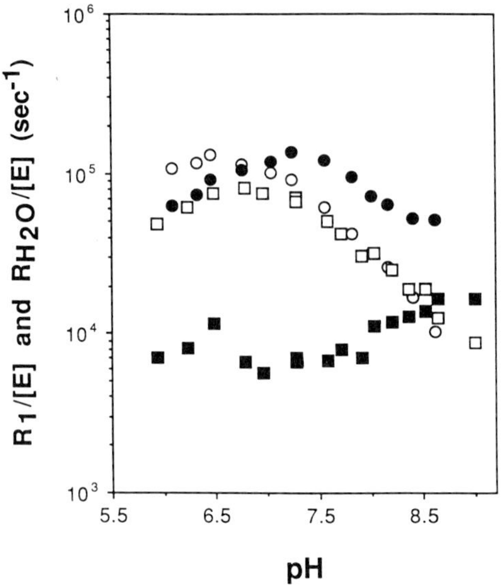

FIG. 2. Dependence of R_1/[E] and R_{H_2O}/[E] on pH for wild-type HCA II and H64A HCA II, the mutant of human carbonic anhydrase II with histidine-64 replaced with alanine. Solutions contained 25 m*M* total concentration of CO_2 and HCO_3^-, and no buffers were used. Experiments were carried out at 10° with the total ionic strength maintained at 0.2 *M* by adding the appropriate amounts of Na_2SO_4. The rates for the wild-type human carbonic anhydrase II are (○) R_1/[E] and (●) R_{H_2O}/[E]. The rates for the H64A mutant are (□) R_1/[E] and (■) R_{H_2O}/[E]. The concentration [E] of wild-type enzyme and H64A mutant was 6 n*M*. (Reprinted with permission from Tu *et al.*[39] Copyright 1989 American Chemical Society.)

Chemical modification of His-64 in isozyme II is also consistent with the proton shuttle hypothesis. HCA II was reacted with acrolein under conditions reported by Pocker and Janjic[41] to result in maximal formylethylation of exposed histidine and lysine residues. Oxygen-18 exchange established that the alkylation had decreased the proton transfer component of the catalysis.[42] Two experiments strongly suggested that the alkylation of His-64 is the main cause: (1) The ^{18}O exchange rates and pH profiles catalyzed by the formylethylated HCA II closely resembled the rates catalyzed by the site-specific mutant H64A HCA II; (2) Cu^{2+} and Hg^{2+}, known to inhibit proton transfer by binding to His-64,[43] had no effect on catalysis by formylethylated HCA II.

[41] Y. Pocker and N. Janjic, *J. Biol. Chem.* **263,** 6169 (1988).

[42] C. K. Tu, G. C. Wynns, and D. N. Silverman, *J. Biol. Chem.* **264,** 12389 (1989).

[43] A. E. Eriksson, T. A. Jones, and A. Liljas, *in* "Zinc Enzymes" (I. Bertini, C. Luchinat, W. Maret, and M. Zeppezauer, eds.), p. 317. Birkhauser, Boston, 1986.

TABLE I
STEADY-STATE CONSTANTS FOR HYDRATION OF CO_2 CATALYZED BY ISOZYMES OF HUMAN CARBONIC ANHYDRASE[a]

Isozyme	k_{cat} (sec^{-1})	k_{cat}/K_m (M^{-1} sec^{-1})	Ref.
I	2×10^5	5×10^7	*b*
II	1.4×10^6	1.5×10^8	*b*
III	1×10^4	3×10^5	*c*

[a] Maximal values at 25°.
[b] R. G. Khalifah, *J. Biol. Chem.* **246,** 2561 (1971).
[c] D. A. Jewell, C. K. Tu, S. R. Paranawithana, S. M. Tanhauser, P. V. LoGrasso, P. J. Laipis, and D. N. Silverman, *Biochemistry* **30,** 1481 (1991).

Further support for the proton shuttle hypothesis was obtained by showing that His-64 is a proton shuttle when inserted into the least efficient carbonic anhydrase, isozyme III.[44] HCA III occurs naturally with a lysine at position 64 and has k_{cat} and k_{cat}/K_m values 100- to 500-fold smaller than those for HCA II (Table I). The mutant with histidine at position 64, K64H HCA III, had the proton-transfer dependent rates of $H_2{}^{18}O$ release from the enzyme enhanced about 7-fold compared with wild-type HCA III (Fig. 3). This activation describes a titration curve of pK_a 7.5 with a maximum at low pH, consistent with proton transfer from His-64 to the zinc-bound hydroxide. Further support for this interpretation was obtained from the activation of HCA III by a large concentration (150 m*M*) of imidazole, which resulted in kinetics nearly identical to that of the K64H mutant. Again, neither the mutation of Lys-64 to His-64 nor the addition of imidazole had any significant effects on the steps of the interconversion of CO_2 and HCO_3^- [Eq. (4)].

Effect of Other Active-Site Residues on Proton Transfer and Catalysis

The most straightforward way to determine whether other residues of the active-site cavity participate as proton transfer groups is to replace His-64 in isozyme II with a residue that is not capable of proton transfer and measure the effect on the kinetic parameters. In approaching this question it is best to use the ^{18}O-exchange method which at chemical equilibrium can be carried out without external buffers. The mutant of

[44] D. A. Jewell, C. K. Tu, S. R. Paranawithana, S. M. Tanhauser, P. V. LoGrasso, P. J. Laipis, and D. N. Silverman, *Biochemistry* **30,** 1484 (1991).

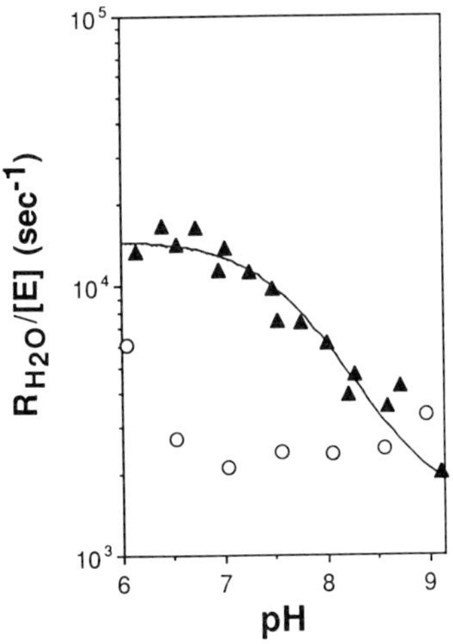

FIG. 3. Variation with pH of R_{H_2O}/[E], the rate constant for release from enzyme of water bearing substrate oxygen, catalyzed by (○) wild-type human carbonic anhydrase III and (▲) K64H HCA III. Data were obtained at 25° with solutions containing 100 m*M* of all species of CO_2 and with the total ionic strength of solution maintained at 0.2 *M* with Na_2SO_4. No buffers were added. The solid line is a least-squares fit of a titration curve for activation of the enzyme by the protonated form of a group with a pK_a of 7.5 ± 0.2. (Reprinted with permission from Jewell *et al.*[44] Copyright 1991 American Chemical Society.)

isozyme II with histidine 64 replaced with alanine, H64A HCA II, has the proton transfer-dependent release of $H_2{}^{18}O$ [Eq. (5)] near 1×10^4 sec^{-1} (10°) and does not vary much in the range of pH 6 to 9 (Fig. 2). This is about 5% of the $H_2{}^{18}O$ release rate of the wild-type enzyme at pH 7.5. There are no other significant proton donors in the active-site cavity with pK_a near or less than 7; these would show a pH dependence of R_{H_2O} such as that observed for the wild type. Other poor proton donor groups in the active-site cavity of high pK_a such as Tyr-7 could be involved. This is consistent with the observation that the mutant of HCA II with Tyr-7 replaced by Phe has maximal velocity 60 to 80% that of the wild-type HCA II in the hydration of CO_2.[45] There is the consideration of substrate bicarbonate itself as a proton donor in the dehydration direction; buffers, especially of small size such as imidazole, are able to enhance the rate of catalysis of H64A HCA II by acting as proton transfer agents.[39] The rate of release of $H_2{}^{18}O$ from H64A HCA II does not depend on the concentration of bicarbonate in the range from 5 to 100 m*M*; either this

[45] Z. Liang, Y. Xue, G. Behravan, B.-H. Jonsson, and S. Lindskog, *Eur. J. Biochem.* **211,** 821 (1993).

is not a prominent pathway for proton transfer in this enzyme, or it reaches its saturation level before 5 m*M* bicarbonate.

There is no evidence from the ^{18}O-exchange experiment that proton transfer from H_3O^+ to the zinc-bound hydroxide is a prominent pathway for H64A HCA II (or HCA III). Proton transfer from H_3O^+ to the zinc-bound hydroxide at the active site is not consistent with the observed pH independence of the water off-rate in the region above pH 6. In addition, proton transfer from water to the active site is probably not rapid enough to account for observations, assuming a maximal value of 10^{10} M^{-1} sec^{-1} for diffusion-controlled transfer to OH^- in the opposite direction.

It is useful to ask at this point whether a histidine residue placed at other locations in mutants of HCA II is effective in intramolecular proton transfer. Liang *et al.*[46] have prepared four double mutants of HCA II each with His-64 replaced by Ala; in addition, a histidine residue replaced Asn-62, Ala-65, Asn-67, and Thr-200 in the four mutants, each of these being a site in the active-site cavity. The maximal velocities of the mutants in CO_2 hydration varied slightly depending on the buffers used; however, on average the proton transfer efficiency at each of the alternate sites was about 5% that of His-64. Thus, His-64 seems well positioned in HCA II for its proton transfer role.

The effect on proton transfer of mutations at other sites near the zinc is under current study, and a consistent picture has not yet emerged. Amino acid substitutions in the hydrophobic site 121 have little or no effect on k_{cat} for hydration catalyzed by mutants of HCA II.[17] Substitutions at position 198 also on the hydrophobic side of the cavity can both increase (Leu-198 → Ala) and decrease (Leu-198 → Arg) proton transfer.[18] In mutants of HCA III, k_{cat} for hydration had a low-pH plateau showing a dependence on hydrophobicity, with the more hydrophobic residues at position 198 showing larger values of k_{cat} (correlation 0.79 with free energy of transfer from nonaqueous phase to water).[47] As hydrophobicity is believed to be related to water orientation around nonpolar groups, this correlation suggests that solvent structure near the active site enhances proton transfer pathways.

Rate–Equilibria Relationships

Brønsted Plots for Proton Transfer Rates

Correlations between the rate of proton transfers and the acidity of the reactants is an important endeavor of physical organic chemistry as

[46] Z. Liang, B.-H. Jonsson, and S. Lindskog, *Biochim. Biophys. Acta* **1203,** 142 (1993).
[47] P. V. LoGrasso, Doctoral Dissertation, University of Florida, Gainesville (1992).

a means of categorizing data, making predictions, and determining reaction mechanisms. Such a free energy relationship is the basis of the Brønsted plot from which information is obtained about transition state structure. The Brønsted relation correlates rate constants for proton transfer k_B with the difference in acid or base strength of the proton donor and acceptor[48,49]:

$$\log(k_B) = \beta[pK_a(\text{acceptor}) - pK_a(\text{donor})] + \text{constant}$$

The applications of the Brønsted plot to the catalytic reactions of enzymes have been few because of the difficulty of varying the properties of the donor or acceptor group within the structural and chemical constraints of the active site. Removing some of the constraints of the active site by using chemical rescue, Toney and Kirsch[40] replaced a lysine at the active site of aspartate aminotransferase with alanine and showed that enhancement of catalysis by external amines follows a Brønsted relation. Rowlett and Silverman[26] and Pocker *et al.*[27] were able to obtain a Brønsted plot for intermolecular proton transfer between different buffers in solution as proton acceptors and HCA II as proton donor in the catalyzed hydration of CO_2. In this mechanism, it is the proton shuttle residue His-64 near the surface of the enzyme that acts as the major proton donor.

Mutants of HCA III can be used to determine a Brønsted plot for the intramolecular proton transfer between zinc-bound hydroxide and the histidine residue at position 64.[50] The mutants of HCA III allow the pK_a of the zinc-bound water to be varied from a value near pK_a 5 for the wild type to pK_a 9.2 for the mutant with an aspartate residue at 198. The use of the ^{18}O-exchange method allows the measurement of catalysis without external buffers since pH control is not needed in the techniques carried out at chemical equilibrium. In this approach, each mutant contains His-64 as intramolecular proton donor and R_{H_2O}, the proton transfer-dependent rate of release of $H_2{}^{18}O$ from the enzyme, is measured.

The pH profile for R_{H_2O} is adequately fit by Eq. (6), which describes the release of ^{18}O-labeled water from the active site as limited in rate by transfer of a proton from a donor group, the imidazolium group of His-64, to the (labeled) zinc hydroxide.[34,51] In Eq. (6) K_E is the ionization

$$R_{H_2O}/[E] = k_B/\{(1 + K_B/[H^+])(1 + [H^+]/K_E)\} \quad (6)$$

[48] J. N. Brønsted and K. J. Pederson, *Z. Phys. Chem. Stoechiom. Verwandtschaftsl.* **108,** 185 (1924).

[49] A. J. Kresge, *Acc. Chem. Res.* **8,** 354 (1975).

[50] D. N. Silverman, C. K. Tu, X. Chen, S. M. Tanhauser, A. J. Kresge, and P. J. Laipis, *Biochemistry* **32,** 10757 (1993).

[51] C. K. Tu and D. N. Silverman, *Biochemistry* **24,** 5881 (1985).

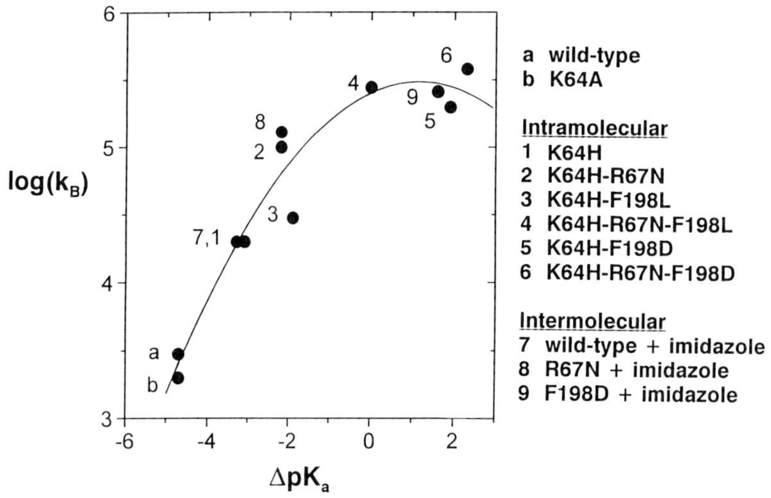

FIG. 4. Brønsted plot of the logarithm of the rate constant for proton transfer k_B (sec^{-1}) versus ΔpK_a [pK_a (zinc-bound water) − pK_a (donor group)]. The data points represent wild type and mutants of carbonic anhydrase III as indicated. Values of k_B and pK_a were obtained by a nonlinear least-squares fit of Eq. (6) to data for R_{H_2O}/[E] taken at 25°. The solid line is a least-squares fit of the Marcus equation [Eq. (7)] to all of the data, resulting in $\Delta G_0^{\ddagger}$ = 1.4 ± 0.3 kcal/mol, w^r = 10.0 ± 0.2 kcal/mol, and w^p = 5.9 ± 1.1 kcal/mol. (Reprinted with permission from Silverman *et al.*[50] Copyright 1993 American Chemical Society.)

constant of the zinc-bound water and K_B is that of the proton shuttle group, [E] is total enzyme concentration, and k_B is the rate constant for the proton transfer. Previous work shows the application of Eq. (6) to HCA II[51] and to mutants of HCA III containing His-64 (K64H HCA III and K64H–R67N HCA III[44]; K64H–R67N–F198L HCA III[50]). The constants obtained by a least-squares fit of Eq. (6) for these and additional variants of HCA III containing His-64 are shown in Fig. 4. In the fitting of Eq. (6), the value of K_E was fixed to that obtained from observation of the pH dependence of k_{cat}/K_m for hydration of CO_2. Individual values of pK_a are given by Silverman *et al.*[50] who also explain the procedure for rough estimates of values of pK_a for zinc-bound water less than 6. These could not be determined directly because of denaturation of the enzyme below pH 5.5.

Also included in the Brønsted plot of Fig. 4 are data obtained by chemical rescue of mutants of HCA III without a histidine residue as proton donor. At imidazole concentrations saturating proton transfer, the pH profile of R_{H_2O} was also adequately described by Eq. (6), with data given for HCA III and two variants in Fig. 4.

Figure 4 has the expected form for a Brønsted plot between oxygen and nitrogen acids and bases in which there is a transition from unit to zero slope in a rather narrow interval of ΔpK_a.[49] Interpretation of the Brønsted plot is obscured somewhat because of an uncertainty in the pK_a of the zinc-bound water for the forms of HCA III with slower proton transfer, as described above. Also, statistical corrections have not been made to account for multiple proton donation sites on the imidazolium ring; these may be nonequivalent sites because of preferred orientations of the histidine side chain. Nevertheless, the values of pK_a are not grossly in error, and the overall curvature of the Brønsted plot is confirmed by the maximum in solvent hydrogen isotope effect for k_B occurring at a ΔpK_a of -0.5.[50] This maximum is expected for proton transfer when the values of pK_a for the donor and acceptor group are nearly equal.[52–54] In this case the maximum occurs for the triple mutant K64H–R67N–F198L HCA III, the mutant in which the pK_a of zinc water is near 7, matching approximately the pK_a of the imidazolium ring of His-64.

Interaction between Mutation Sites

It is necessary to consider whether there are structural, noncovalent, or solvent-mediated interactions that affect the donor and acceptor sites and thereby the Brønsted plot; that is, to what extent the Brønsted plot is affected by intramolecular interactions in the enzyme. The consideration of double mutant cycles has been used to reflect on this issue. This approach, developed by Carter *et al.*,[55] analyzes quantitatively the effect of a second mutation on a first mutation. The largest rate constant for proton transfer observed by ^{18}O exchange catalyzed by HCA III is in the mutants with the replacement of Phe-198 with Asp. The double mutant cycle of Fig. 5 presents the rate constants for proton transfer k_B from His-64 to the zinc-bound hydroxide determined by ^{18}O exchange and the corresponding changes in the free energies of activation. Figure 5 shows an additive interaction, meaning that the effect of the replacements Ala-64 → His and Phe-198 → Asp are noninteracting; the introduction of His-64 causes the same decrease in the activation energy, about 1.4 kcal/mol, independent of whether Phe or Asp is at position 198.[56] This is useful in interpretation

[52] N. A. Bergman, Y. Chiang, and A. J. Kresge, *J. Am. Chem. Soc.* **100,** 5954 (1978).
[53] M. M. Cox and W. P. Jencks, *J. Am. Chem. Soc.* **100,** 5956 (1978).
[54] F. H. Westheimer, *Chem. Res.* **61,** 265 (1961).
[55] P. J. Carter, G. Winter, A. J. Wilkinson, and A. R. Fersht, *Cell (Cambridge, Mass.)* **38,** 835 (1984).
[56] X. Chen, C. K. Tu, X. Ren, P. J. Laipis, and D. N. Silverman, (1993). Unpublished.

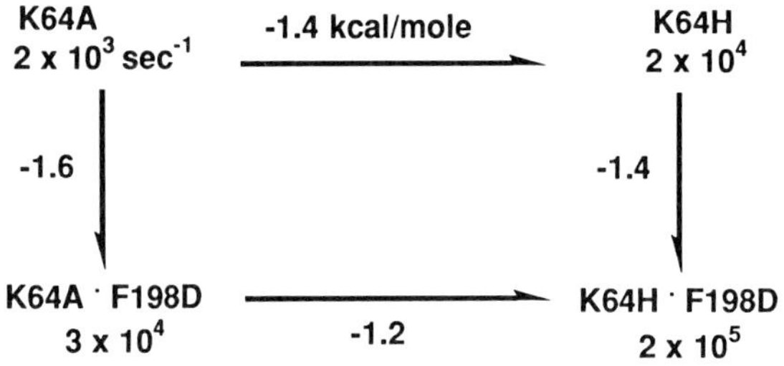

FIG. 5. Comparison of the values of k_B (sec^{-1}), the rate constant for proton transfer to the zinc-bound hydroxide, for variants of HCA III obtained by site-directed mutagenesis. Values of k_B appear beneath each designated mutant. The values adjacent to the arrows are the changes in free energy barriers (kcal/mol) for the catalysis corresponding to the designated mutations. Free energy changes were determined using $\Delta\Delta G = -RT \ln[(k_B)_{mut2}/(k_B)_{mut1}]$.

of the subsequent Brønsted plot; we do not have to account for other interactions between these sites and can emphasize the proton transfer itself.

The introduction of Leu at residue 198 is not so straightforward. The interaction between this residue and His-64 is best described as partially additive (using the nomenclature of Mildvan *et al.*[57]). Here the replacement Ala-64 → His lowers the activation energy by 1.4 kcal/mol with Phe at site 198, and lowers it by 0.5 kcal/mol with Leu at 198. Among other possibilities, there may be a structural effect of Leu-198 that reduces the ability of His-64 to transfer a proton to the zinc-bound hydroxide. Although not always clearly noninteracting, the mutations made by introduction of histidine at residue 64 in HCA III *enhance* catalysis by introduction of a proton transfer group. This is easier to interpret than mutations that cause a partial decrease but do not abolish catalytic activity. In interpretation of the Brønsted plot, it is assumed that the enhancements observed by introduction of histidine-64 are due to proton transfer and not steric effects or conformational changes.[50]

Application of Marcus Rate Theory

The purpose of introducing the Marcus rate theory[58] is to provide a quantitative analysis of the Brønsted plot in terms of the proton transfer illustrated in Scheme I. In Scheme I, w^r is a work term giving the energy required for solvent reorganization and side-chain conformational change required prior to the proton transfer. Structural data[13,14] and isotope ef-

[57] A. J. Mildvan, D. J. Weber, and A. Kuliopulos, *Arch. Biochem. Biophys.* **294,** 327 (1992).
[58] R. A. Marcus, *J. Phys. Chem.* **72,** 891 (1968).

fects[29] for carbonic anhydrase suggest that proton transfer between His-64 and the zinc-bound hydroxide proceeds through hydrogen-bonded water molecules in a proton relay mechanism. The side chain of His-64 in HCA II cannot extend close enough to the zinc-bound hydroxide for direct proton transfer.[13] These features are a likely possibility for CA III as well, especially since the backbone conformations of the two isozymes are very similar.[14] The work term w^r is then the energy required to align acceptor, donor, and intervening hydrogen-bonded water for facile proton transfer. The energy w^p is this work term for the corresponding reorganization for the reverse process (Scheme I). The standard free energy of reaction with the required active-site conformation is ΔG°_R, with the measured overall free energy for the reaction given by $\Delta G^\circ = w^r + \Delta G^\circ_R - w^p$.

In Marcus theory, the observed overall activation barrier for the proton transfer $\Delta G^\ddagger$ is given by Eq. (7), which relates this to an intrinsic energy barrier $\Delta G^\ddagger_0$, which is the value of $\Delta G^\ddagger$ when $\Delta G^\circ_R = 0$:[58]

$$\Delta G^\ddagger = w^r + \{1 + \Delta G^\circ_R/4\Delta G^\ddagger_0\}^2 \Delta G^\ddagger_0 \qquad (7)$$

The slope β of a Brønsted plot is given by Marcus theory as $d\Delta G^\ddagger/d\Delta G^\circ_R$. Brønsted plots for intermolecular proton transfer between nitrogen and oxygen acids and bases show sharp curvature, changing slope from zero to one over the span of ΔpK_a −2 to 2. Analysis by Marcus theory shows that such proton transfers have a low intrinsic energy barrier $\Delta G^\ddagger_0$, near 2 kcal/mol, with a work term w^r also near 2 kcal/mol representing energy required to form the reaction complex.[49]

Figure 4 represents the Brønsted plot for proton transfer between a nitrogen acid and oxygen base in carbonic anhydrase, and it displays the sharp curvature characteristic of such processes between small molecules. However, it is unusual because the overall rate of proton transfer is relatively slow, less than 10^6 sec^{-1}, and the magnitude of the overall energy barrier for the proton transfer is large, near 10 kcal/mol. A least-squares

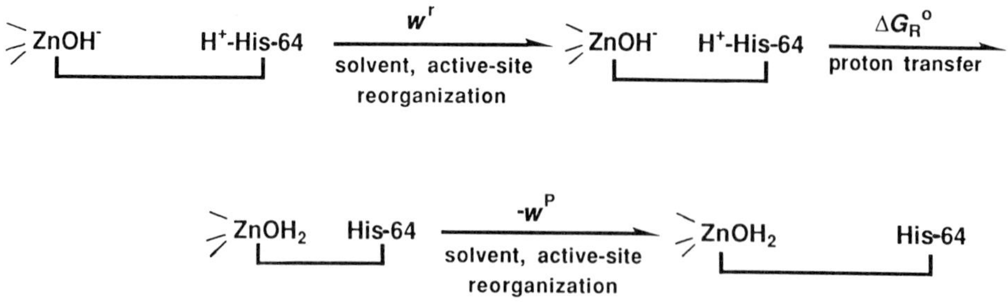

SCHEME I

fit of the Marcus equation [Eq. (7)] to all of the data of Fig. 4 gives an intrinsic energy barrier for the proton transfer of $\Delta G_0^{\ddagger} = 1.4 \pm 0.3$ kcal/mol with work terms $w^r = 10.0 \pm 0.2$ kcal/mol and $w^p = 5.9 \pm 1.1$ kcal/mol.[50] {For this calculation, $\Delta G^{\ddagger} = -RT \ln(hk_B/kT)$ and $\Delta G^\circ = RT \ln[(K_a)_{ZnH_2O}/(K_a)_{donor}]$.} In other words, the data are consistent with a rather low intrinsic energy barrier for the transfer accompanied by substantial work functions.

Points a and b of Fig. 4 (wild-type and K64A HCA III) were included in the fit although it is clear that an imidazolium group is not the proton donor in these cases. For these enzymes the donor group (or groups) is uncertain, but the pH dependence of R_{H_2O} in both cases shows an apparent pK_a near 9 and could possibly involve HCO_3^- which under our solution conditions has a pK_a of 9.8 for the equilibrium between HCO_3^- and CO_3^{2-}. Comparison of wild type with K64A HCA III and C66A HCA III shows that Lys-64 and Cys-66 are not predominant proton donor groups. The values of $\Delta G_0^{\ddagger}$ and the work terms are not significantly altered by omitting the intermolecular processes (points 7–9 of Fig. 4) ($\Delta G_0^{\ddagger} = 1.6 \pm 0.5$, $w^r = 10.0 \pm 0.2$, and $w^p = 5.6 \pm 1.6$ kcal/mol), nor are they significantly altered (although uncertainties are greater) by omitting the points 7–9 for examples of intermolecular proton transfer and points a and b which do not involve imidazole as proton donors ($\Delta G_0^{\ddagger} = 1.9 \pm 1.8$, $w^r = 10.0 \pm 0.4$, and $w^p = 5.0 \pm 5.2$ kcal/mol). These values are further supported by the analysis of the solvent hydrogen isotope effects on k_B. Kresge *et al.*[59] used Marcus theory to obtain an expression for the dependence of the solvent hydrogen isotope effect on ΔG_R°. The fit of this expression to the data (see Silverman *et al.*[50]) gave an intrinsic energy barrier for the proton transfer of $\Delta G_0^{\ddagger} = 1.3 \pm 0.3$ kcal/mol with $[(k_B)_{H_2O}/(k_B)_{D_2O}]_{max} = 4.7 \pm 0.7$ and work terms $w^r - w^p = 0.6 \pm 0.5$ kcal/mol. This latter value is different from the value of $w^r - w^p$ near 4 kcal/mol determined from the data of Fig. 4. The reasons for this difference are uncertain but may reflect experimental errors in the solvent hydrogen isotope effect.

The observation that the work terms for proton transfer in the hydration (w^p) and dehydration (w^r) directions may be unequal can be rationalized in a qualitative manner by noting that in the direction of HCO_3^- dehydration (Scheme I) the proton transfer occurs from a charged imidazolium group to a zinc-bound hydroxide, whereas in the reverse direction the proton transfer is from zinc-bound water to an uncharged imidazole group. Electrostatic effects in the active-site cavity could make significant contributions to these different work terms. In principle a Brønsted plot should also be observed for intramolecular proton transfer from His-64 to the

[59] A. J. Kresge, D. S. Sagatys, and H. L. Chen, *J. Am. Chem. Soc.* **99,** 7228 (1977).

zinc-bound hydroxide of HCA II; isozyme III was attempted first because a wider range of pK_a values for the zinc-bound water was attainable through mutagenesis.

Why Is Proton Transfer in Carbonic Anhydrase so Slow?

The hydration of CO_2 catalyzed by carbonic anhydrase with a maximal turnover of 10^6 sec^{-1} is one of the most rapid enzymatic reactions known. It is then ironic to ask why the intramolecular proton transfer between His-64 and the zinc-bound water, which is the rate-limiting step in the turnover, is so slow. In principle and in experiment, an intramolecular proton transfer can occur as fast as a molecular vibration; values near 10^{12} sec^{-1} are observed in electronically excited states.[60] It is a clue that in many examples of intramolecular proton transfer at 10^{12} sec^{-1} there is little solvent involvement. The work terms in the interpretation of the Brønsted plot (Fig. 4) represent the energy required for conformational changes and solvent reorganization to align the active site for subsequent facile proton transfer. It is these work terms that make intramolecular proton transfer in carbonic anhydrase III much slower than the maximal values of a molecular vibration. The source of these work terms is not clear, but presumably they contain the energy required to align the donor and acceptor residues and the intervening water molecules through which the proton transfer occurs. Such reorganization is not expected to be accompanied by a sizable solvent hydrogen isotope effect, although if many hydrogens change bonding even slightly the accumulated effect could be significant.[61]

Proton transfer is a specific process involving defined pathways in which facile transfer occurs when donor and acceptor are aligned within severely restricted distances and angles.[62] Moreover, we can envisage an approximate pathway for the proton transfer in HCA II from the crystal structure, which demonstrates that two water molecules form a bridge between His-64 and the zinc-bound water.[13] The nitrogen assignments of His-64 depend on the positions of the water molecules, and it appears that Nε2 of His-64 is the ring nitrogen closest to the zinc with a distance between the Nε2 atom and the zinc-bound hydroxide of 7.5 Å. However, this water bridge apparently does not form a hydrogen bond to the imidazole of His-64 in the crystal structure. The water molecules between this residue and the zinc come no closer than 3.4 Å to the imidazole[13]; thus, a completed, hydrogen-bonded water bridge for proton transfer is not

[60] P. F. Barbara, P. K. Walsh, and L. E. Brus, *J. Phys. Chem.* **93,** 29 (1989).

[61] A. J. Kresge, *J. Am. Chem. Soc.* **95,** 3065 (1973).

[62] S. Scheiner, this series, Vol. 127, p. 86.

found in this structure. Moreover, the distance between His-64 and the zinc means that desolvation energy is not involved in the intramolecular proton transfer. It is interesting that this water structure is disrupted in the mutant Thr-200 → His HCA II but the maximal value of k_{cat} is still very large, near 2×10^5 sec^{-1}.[23,63] This brings into question the significance of the ordered water structure observed in the crystal as it pertains to the kinetics of proton transfer.

The proton transfer accompanying catalysis in carbonic anhydrase requires that the imidazole side chain of His-64 rotate as it alternatively accepts a proton from the zinc-bound water and delivers it to buffer in solution. It appears unlikely that the rotation of this side chain requires much energy. In NMR studies, the proton line widths have been interpreted as a very mobile imidazole ring in isozyme II.[64] Moreover, there is abundant evidence from the crystal structures of CA II and mutants that there are two or more orientations of the side chain of His-64 that depend on the pH of crystallization,[65] the presence of inhibitors,[66] and the identity of the residue at position 200.[67]

It was also observed that in catalysis by H64A HCA II in the presence of large concentrations of imidazole, the magnitude of k_{cat}, its solvent hydrogen isotope effect, and the dependence of this isotope effect on the deuterium content of water are nearly identical with the corresponding experiments using wild-type CA II.[68] This is accompanied by the knowledge that a significant fraction of the proton transfer in HCA II proceeds through His-64, but that in H64A HCA II nearly all of the proton transfer proceeds through imidazole.[39] Also the steady-state data of Tu *et al.*[39] comparing k_{cat} with large buffers (MOPS, TAPS) and small buffers (imidazole) lead to the same conclusion.

These facts are consistent with the suggestion that the proton transfer is not significantly dependent on the orientation of the proton acceptor imidazole in the active site or on the difference between the mobility of the free imidazole buffer in contrast with the bound imidazole side chain of histidine. In fact, as demonstrated for intermolecular proton transfer in HCA II[26] and intramolecular proton transfer in mutants of HCA III,[50] a major determinant for the magnitude of these rate constants is the

[63] G. Behravan, B.-H. Jonsson, and S. Lindskog, *Eur. J. Biochem.* **190,** 351 (1990).

[64] I. D. Campbell, S. Lindskog, and A. I. White, *J. Mol. Biol.* **98,** 597 (1975).

[65] S. K. Nair and D. W. Christianson, *J. Am. Chem. Soc.* **113,** 9455 (1991).

[66] G. M. Smith, R. S. Alexander, D. W. Christianson, B. M. McKeever, G. S. Ponticello, J. P. Springer, W. C. Randall, J. J. Baldwin, and C. N. Habecker, *Protein Sci.* **3,** 118 (1994).

[67] J. F. Krebs, C. A. Fierke, R. S. Alexander, and D. W. Christianson, *Biochemistry* **30,** 9153 (1991).

[68] S. Taoka, C. K. Tu, and D. N. Silverman, (1993). Unpublished results.

difference in pK_a between donor and acceptor groups since these rates can be described by linear free energy relationships such as the Brønsted plot. Moreover, the nearly equivalent dependence of the solvent hydrogen isotope effect on the atom fraction of deuterium in solvent water for HCA II and H64A HCA II with imidazole suggests that in both enzymes a mechanism utilizing intervening water molecules is involved in the proton transfer; this means that imidazole buffer does not participate appreciably in direct transfer with the zinc-bound water but also uses the intervening water structures.

These features of the proton transfer are consistent with the calculations of Liang and Lipscomb[69] for HCA II and of Åqvist and Warshel[70] for HCA I, demonstrating that in proton transfer from the zinc-bound water to solution the first proton transfer from the zinc-bound water to the next water in the chain has the most significant energy barrier and is rate limiting. Subsequent proton transfers between H_3O^+ and H_2O have much smaller energy barriers provided the orientation of the donor and acceptor are fully optimized. In these calculations, the number of water molecules in the proton relay does not affect the overall energy barrier for the proton transfer. This same approach explains why intramolecular proton transfer in a series of mutants of HCA III containing His-64 follows a Brønsted plot[50]; the rate of intramolecular proton transfer is relatively independent of the predominant conformation of the histidine side chain.

The intervening water molecules through which the proton transfer proceeds must also rotate and form hydrogen bonds prior to each cycle of the proton transfer. This process as it occurs in carbonic anhydrase must be very similar to the processes of proton conduction in inorganic models and biological systems in a description by Williams[71] which requires continuous reorientations of water molecules and formation of hydrogen bonds. In the protonation of zinc-bound hydroxide in the direction of HCO_3^- dehydration, three hydrogen bonds are formed for an estimated total free energy near 6 kcal/mol. According to the temperature dependence of catalysis by isozyme II,[72] the entropic contribution to the proton-transfer dependent turnover in hydration is quite small at $-T\Delta S = 1.2$ kcal/mol; however, this may represent a balance between larger electrostatic and nonelectrostatic contributions to the turnover in hydration. These estimates near 7 kcal/mol are entirely consistent with the proton transfer rate of isozyme II. These numbers are in the range of

[69] J.-Y. Liang and W. N. Lipscomb, *Biochemistry* **27,** 8676 (1988).

[70] J. Åqvist and A. Warshel, *J. Mol. Biol.* **224,** 7 (1992).

[71] R. J. P. Williams, *Annu. Rev. Biophys. Chem.* **17,** 71 (1988).

[72] A. F. Ghannam, W. Tsen, and R. S. Rowlett, *J. Biol. Chem.* **261,** 1164 (1986).

values listed by Williams[71] for migration of protons in inorganic systems and are consistent with the transfer of protons through structural chains of water and not migration of H_3O^+.

Relation to Proton Transfer in Other Enzymatic Reactions

It is to be expected that intermolecular proton transfer between small molecules follows a Brønsted relationship; however, it is less predictable because of electrostatic and steric features that this relationship will hold when one of the donor–acceptor pair is a residue at the active site of an enzyme. The first studies that hold promise for the application of these methods applied to intermolecular proton transfer to enzymes are the work with aspartate aminotransferase[40] and carbonic anhydrase[26,50] mentioned above. In each case a Brønsted plot was possible with a variety of small molecule proton acceptors that, by chemical rescue, enhance the catalysis by mimicking a residue replaced at the active site through mutagenesis. The significance of such observations is the characterization of the catalytic mechanism that is then possible.

Extension of this approach to other enzyme systems will be useful. Likely candidates are enzymes, like carbonic anhydrase, for which the rate-limiting event is not the conversion of substrate to product but the subsequent rate-contributing isomerization by which the active form of the enzyme is regenerated. One enzyme in this category is proline racemase from a strain of *Clostridium* in which the transfer of protons between two residues of the enzyme, both most likely cysteines, occurs at a rate near 10^5 sec^{-1} and limits the maximal rate of catalysis.[73] Moreover, like carbonic anhydrase, the catalysis by proline racemase is activated by large concentrations (>100 m*M*) of buffers of small size, which has allowed points on a Brønsted plot to be observed,[73] another feature of the rate-contributing proton transfer.

Considerations of the Brønsted relationship are inherent in many of the interpretations of kinetic, spectroscopic, and structural studies from which we surmise a mechanism. One example from many is the mechanism of Δ^5-3-ketosteroid isomerase from *Pseudomonas* species which catalyzes the allylic isomerization of $\Delta^{5,6}$- and $\Delta^{5,10}$-3-oxo-steroids to the conjugated $\Delta^{4,5}$ isomers through a dienol or dienolate intermediate. Site-directed mutagenesis has indicated that Asp-38 in the active site has a significant catalytic role,[74] and work with affinity reagents has suggested that this residue is

[73] J. G. Belasco, T. W. Bruice, L. M. Fisher, W. J. Albery, and J. R. Knowles, *Biochemistry* **25,** 2564 (1986).

[74] L. A. Xue, A. Kuliopulo, A. S. Mildvan, and P. Talalay, *Biochemistry* **30,** 4991 (1991).

the base with a pK_a of 4.9 which abstracts a proton in the isomerization.[75] As the pK_a of the C-4 hydrogen of the (unbound) substrate to be abstracted is 12.6, the aspartate at position 38 would be inefficient in proton abstraction unless its basicity were greatly increased or the acidity of the C-4 hydrogen were greatly increased. Such simple implications of the Brønsted relationship suggest that another residue of the enzyme (Tyr-14) is an electrophilic group that lowers the pK_a of the C-4 hydrogen either by protonating or hydrogen bonding the adjacent C-3 carbonyl.[76] A similar consideration is reported for triose-phosphate isomerase for which an unprotonated imidazole side chain of histidine acts as an electrophilic catalyst through hydrogen bonding to a carbonyl oxygen of substrate.[77]

Others[78,79] have extended these concepts to explain stabilization of transition states in abstraction of α-protons from carbon acids, acyl-transfer reactions, and displacement reactions of phosphodiesters. They suggest that one (or more) general acid catalyst acts as proton donor to stabilize intermediates by forming short and strong hydrogen bonds and by delocalizing negative charge without requiring significant structural reorganization. This effect is maximized when the values of pK_a of the protonated intermediate and the general acid catalyst are equal.

Conclusions

Kinetic and structural studies demonstrate the pathways of inter- and intramolecular proton transfer in the hydration of CO_2 catalyzed by carbonic anhydrase. A histidine residue at position 64 in the most efficient of the carbonic anhydrases, carbonic anhydrase II, acts as a transfer group shuttling protons between buffer in solution and the zinc-bound water at the active site. This conclusion was confirmed by mutations to replace His-64, which resulted in inactivation of catalysis. Mutations to introduce His-64 in the less efficient carbonic anhydrase III resulted in activation, with the mutant showing proton transfer properties similar to the efficient form of the enzyme. In a series of site-specific mutants of isozyme III, the rate constants for proton transfer obeyed a Brønsted correlation and showed sharp curvature characteristic of facile proton transfers. Application of Marcus rate theory showed that this proton transfer has the small intrinsic energy barrier (near 1.5 kcal/mol) characteristic of rapid bimolec-

[75] P. L. Bounds and R. M. Pollack, *Biochemistry* **26,** 2263 (1987).
[76] L. A. Xue, P. Talalay, and A. S. Mildvan, *Biochemistry* **30,** 10858 (1991).
[77] P. J. Lodi and J. R. Knowles, *Biochemistry* **30,** 6948 (1991).
[78] J. A. Gerlt and P. G. Gassman, *Biochemistry* **32,** 11943 (1993).
[79] W. W. Cleland and M. M. Kreevoy, *Science* **264,** 1887 (1994).

ular proton transfer between nitrogen and oxygen acids and bases in solution. The observed overall energy barrier (near 10 kcal/mol) for proton transfer in the catalysis indicates the involvement of accompanying, energy-requiring processes such as solvent reorganization or conformational change. Some of this barrier is likely to involve the formation of hydrogen-bonded water in the active-site cavity through which intramolecular proton transfer occurs.

Acknowledgments

The unpublished work from the author's laboratory given in this report was obtained with support from the National Institutes of Health (GM25154). The author thanks Dr. Thomas H. Maren and Dr. A. Jerry Kresge for helpful discussions and comments.

Section IV

Kinetics of Specialized Systems

[19] Expression of Properly Folded Catalytic Antibodies in *Escherichia coli*

By JON D. STEWART, IRENE LEE, BRUCE A. POSNER, and STEPHEN J. BENKOVIC

Introduction

The field of catalytic antibodies has advanced along with the technology to produce and engineer the proteins. The notion of catalytic antibodies rests on the recognition, first articulated clearly by Linus Pauling, that enzymes owe a large portion of their catalytic power to their ability to bind selectively and thereby stabilize the transition state of a reaction to a greater extent than either the reactants or the products.[1] This theory also explained why antibodies should lack such catalytic power, despite the fact that they can strongly and selectively bind virtually any molecule, since they selectively bind to the ground states of molecules. However, if an antibody could be found that selectively bound the transition state for a given reaction, that antibody would possess catalytic power.

Although the basic idea for using antibodies as catalytic entities was articulated by Jencks over two decades ago,[2] two major technical advances were required before the idea was reduced to practice in 1986. Since the transition state of a reaction has a negligible lifetime, it was clear that an antibody would have to be raised against a stable molecule that approximated its geometric and electronic characteristics (i.e., a transition state analog). Thus, advances in transition state analog design and synthesis were required, and these were motivated mainly by the need to produce potent inhibitors for medically important enzymes, particularly proteases.[3] The second advance was the ability to create monoclonal antibodies that allowed one to study a homogeneous catalyst with clearly defined properties, rather than earlier work that relied on an undefined, polyclonal mixture of antibodies.[4–6]

[1] L. Pauling, *Am. Sci.* **36,** 51 (1948).
[2] W. P. Jencks, "Catalysis in Chemistry and Enzymology." McGraw-Hill, New York, 1969.
[3] R. Wolfenden, *Acc. Chem. Res.* **5,** 10 (1972).
[4] L. I. Slobin, *Biochemistry* **5,** 2836 (1966).
[5] F. Kohen, J.-B. Kim, G. Barnard, and H. R. Lindner, *Biochim. Biophys. Acta* **629,** 328 (1980).
[6] V. Raso and B. D. Stollar, *Biochemistry* **14,** 591 (1975).

METHODS IN ENZYMOLOGY, VOL. 249

Since the initial, simultaneous reports of catalytic antibodies by two groups working independently,[7,8] the repertoire of catalytic antibodies has grown to some fifty reactions, encompassing a diverse range of chemistry including hydrolysis, pericyclic reactions, and a peptide rearrangement. Previous reviews have documented many aspects of this field including surveys of antibody-catalyzed reactions[9,10] and mechanistic studies.[11–13] Thus, to avoid repetition, this review focuses on the technology to produce and modify these catalysts in bacterial hosts with an emphasis on producing antibodies in functional form. Antibody expression in bacteria has been discussed elsewhere in this series by Better and Horwitz[14] and Plückthun and Skerra,[15] and the reader is directed to these excellent descriptions of other expression methods.

With a single exception,[16] all catalytic antibodies produced to date have been isolated as hybridoma-derived monoclonal antibodies. This method of producing antibodies has a number of advantages, especially the ability to produce large amounts of pure antibody from ascites fluid. Large quantities are especially useful for mechanistic studies of catalytic antibodies, since high protein concentrations are often required for the experiments. The hybridoma method also has advantages for applications of antibodies that require effector functions. However, for the study of catalysis, the hybridoma method has the distinct disadvantage that the protein sequence cannot be easily altered. This precludes the use of site-directed mutagenesis experiments in order to improve the catalytic activity of an existing antibody or to probe its mechanism. The desire to imbue antibodies with desirable characteristics has led investigators to explore the production of antibodies in a number of heterologous hosts, particularly bacteria. Antibodies (or antibody fragments) have been successfully produced in bacteria (*Escherichia coli*[17] and *Bacillus subtilis*[18]), yeast,[19]

[7] S. J. Pollack, J. W. Jacobs, and P. G. Schultz, *Science* **234,** 1570 (1986).

[8] A. Tramontano, K. D. Janda, and R. A. Lerner, *Science* **234,** 1566 (1986).

[9] S. J. Benkovic, *Annu. Rev. Biochem.* **61,** 29 (1992).

[10] R. A. Lerner, S. J. Benkovic, and P. G. Schultz, *Science* **252,** 659 (1991).

[11] J. D. Stewart, L. J. Liotta, and S. J. Benkovic, *Acc. Chem. Res.* **26,** 396 (1993).

[12] J. D. Stewart and S. J. Benkovic, *Chem. Soc. Rev.* **22,** 213 (1993).

[13] D. Hilvert, *Pure Appl. Chem.* **64,** 1103 (1992).

[14] M. Better and A. H. Horwitz, this series, Vol. 178, p. 476.

[15] A. Plückthun and A. Skerra, this series, Vol. 178, p. 497.

[16] Y.-C. J. Chen, T. Danon, L. Sastry, M. Mubaraki, K. D. Janda, and R. A. Lerner, *J. Am. Chem. Soc.* **115,** 357 (1993).

[17] A. Plückthun, *Bio/Technology* **9,** 545 (1991).

[18] X.-C. Wu, S.-C. Ng, R. I. Near, and S.-L. Wong, *Bio/Technology* **11,** 71 (1993).

[19] K. Bowdish, Y. Tang, J. B. Hicks, and D. Hilvert, *J. Biol. Chem.* **266,** 11901 (1991).

insect,[20] and mammalian cells.[21,22] This review focuses on the production of antibodies in *E. coli* since expression in this host is intimately connected with our efforts to construct and use combinatorial libraries of antibody genes in bacteriophage λ for isolating catalytic antibodies.[23]

To construct such libraries, mRNA from the spleen cells of mice immunized with a transition state analog is converted to cDNA by reverse transcriptase, and the expressed antibody genes are amplified by the polymerase chain reaction using antibody-specific primers.[23] The amplified genes coding for heavy and light chains are cloned into suitable λ bacteriophage expression vectors, and these are randomly combined to produce a combinatorial library that aims to reconstruct the expressed antibody repertoire of the immunized mouse. The first task is to screen the library for antibodies that bind to the transition state analog. This group of binders represents the pool of potential catalytic antibodies. Each of these candidates must then be individually screened for catalytic activity. Finally, once a catalyst has been identified, relatively large quantities (on the order of several milligrams) must be prepared for mechanistic studies. The production of antibodies in *E. coli* is important in each of these three steps: screening for binding, screening for catalysis, and large-scale production.

Choice of Antibody Fragment

The modular nature of the antibody structure allows them to be produced in several functional forms (Fig. 1). As noted above, the full monoclonal antibody is best suited for applications requiring the effector functions to allow interactions with other components of the immune system. Despite intensive efforts, it has not been possible to produce complete monoclonal antibodies in bacteria in significant yield. However, several fragments of the antibody molecule have been successfully expressed in bacteria, and in general these retain essentially all of the antigen binding affinity of the parent monoclonal antibody.

The Fab fragment, which consists of the entire light chain and the *N*-terminal one-third of the heavy chain, has several advantages, especially for use in combinatorial libraries of antibodies. First, several groups have reported the expression of functional Fab fragments in *E. coli*, a

[20] C. A. Hasemann and J. D. Capra, *Methods (San Diego)* **2,** 146 (1991).
[21] C. R. Bebbington, *Methods (San Diego)* **2,** 136 (1991).
[22] A. Wright and S.-U. Shin, *Methods (San Diego)* **2,** 125 (1991).
[23] W. D. Huse, L. Sastry, S. A. Iverson, A. S. Kang, M. Alting-Mees, D. R. Burton, S. J. Benkovic, and R. A. Lerner, *Science* **246,** 1275 (1989).

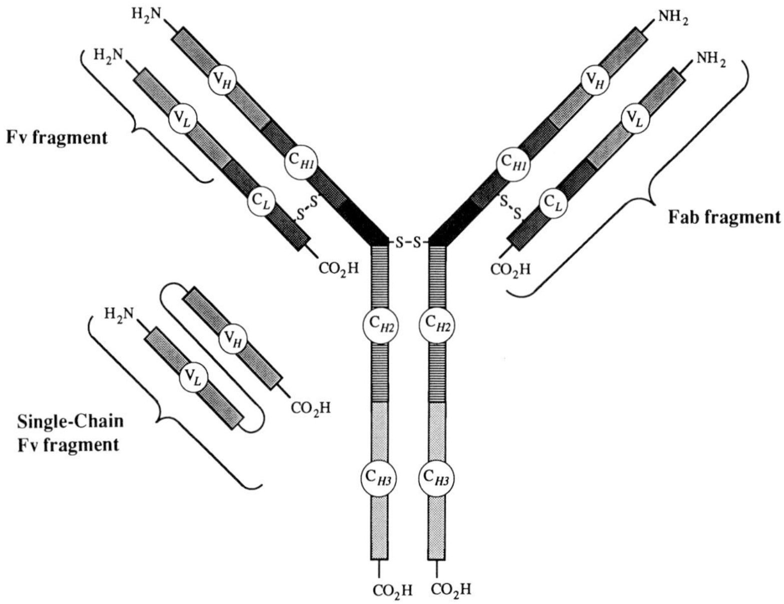

FIG. 1. Schematic diagram of an immunoglobulin G (IgG) antibody. The antibody consists of two identical heavy chains and two identical light chains. These chains are composed a variable region (V_H or V_L) that contains the complementarity-determining regions as well as constant regions that provide a scaffolding for the antibody structure. There are a number of intra- and interchain disulfide bonds, only some of which are shown. The regions corresponding to the Fab, Fv, and scFv fragments are also indicated.

method pioneered by Better and co-workers.[24] Clearly, the expression of a properly folded antibody is essential for screening antibody libraries for binding affinity. Second, the Fab fragment is monovalent, which simplifies binding screens by removing avidity effects. Finally, the presence of the heavy and light chains can be verified by commercially available antisera directed against the constant regions of the two chains. The same antisera can also be used to quantitate the level of Fab by enzyme-linked immunosorbent assay (ELISA) methods. Unfortunately, there are two main drawbacks to producing antibodies in the form of Fab fragments. First, the size of Fab fragments (~50 kDa) makes them too large to study by certain physical techniques, particularly nuclear magnetic resonance (NMR) spectroscopy. Also, the requirement to express simultaneously comparable levels of the two chains, as well as the requirement for their assembly

[24] M. Better, C. P. Chang, R. R. Robinson, and A. H. Horwitz, *Science* **240,** 1041 (1988).

under oxidizing conditions, often results in relatively low yields of the functional protein.

The Fv fragment, consisting of only the variable regions of the heavy and light chains, solves some of these problems.[25] Like the Fab fragment, the Fv fragment is monovalent, obviating avidity effects. In addition, its small size (~25 kDa) facilitates physical studies as well as cloning and site-directed mutagenesis. The major disadvantage is that the Fv fragment lacks a disulfide link between the two polypeptides. This often leads to dissociation of the chains at low protein concentrations and aggregation at high protein concentrations. These problems have been solved by joining the two polypeptide chains of the Fv fragment into a single polypeptide (scFv) by interposing a 14–20 amino acid linker peptide between the two chains. Constructions in which the chains are arranged as V_L–linker–V_H as well as V_H–linker–V_L have been reported, and the order of the two chains does not appear to be critical. The design of such single-chain antibodies has been summarized by Whitlow and Filpula.[26] The main disadvantage of the Fv and scFv fragments is the difficulty in quantitating expression levels since they lack the constant regions recognized by the commercially available antisera used to quantitate Fab fragments. In fact, it is this inability to detect Fv and scFv fragments at the plaque level that led to the choice of Fab fragments for combinatorial antibody libraries, despite the generally lower expression levels.

Catalytic Antibody 43C9 as Model System

We have chosen antibody 43C9 as a model system for antibody expression in *E. coli*. The antibody, which cleaves an activated amide as well as a series of related esters, was raised against a phosphonamidate transition state analog designed to mimic the geometric and electronic characteristics of the transition state or high-energy tetrahedral intermediate formed during amide hydrolysis (Scheme 1).[27] Antibody 43C9 was chosen as our test system both because it is a model for antibodies that can cleave peptide bonds and because it is one of the most efficient catalytic antibodies ever isolated, accelerating amide hydrolysis by a factor of 10^6 over the background rate.

Kinetic investigations of the hydrolysis catalyzed by the monoclonal antibody form of 43C9 revealed that it follows a multistep pathway involving an antibody-bound intermediate, even though the antibody was raised

[25] A. Skerra and A. Pluckthun, *Science* **240,** 1038 (1988).
[26] M. Whitlow and D. Filpula, *Methods (San Diego)* **2,** 97 (1991).
[27] K. D. Janda, D. Schloeder, S. J. Benkovic, and R. A. Lerner, *Science* **241,** 1188 (1988).

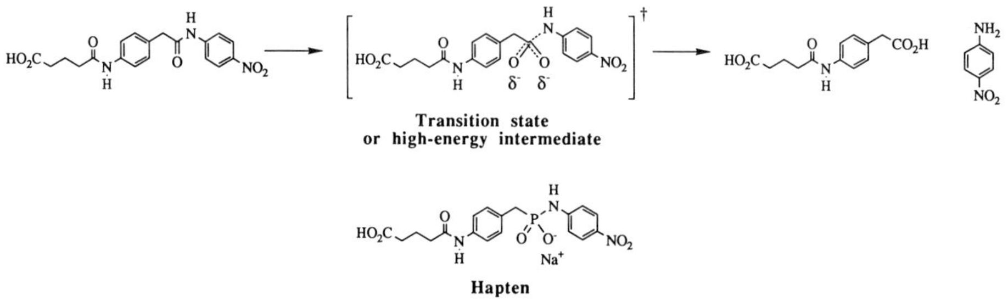

SCHEME 1

against a single transition state analog (Scheme 2).[28] The genes encoding the 43C9 Fab fragment have been cloned from the corresponding hybridoma,[29] and these have been used to construct the scFv form.[30] Antibody 43C9 thus represents one of the few cases for which a single antibody is available in multiple forms [monoclonal antibody (MAb), Fab, and scFv]. As expected, the catalytic parameters are the same for each of these forms.[30,31] A computer model of the 43C9 Fv fragment was constructed based on homologies with antibodies of known three-dimensional structure, and the phosphonamidate hapten was docked into a likely binding site.[32] This model has been used to guide site-directed mutagenesis studies that have tested some of the predictions arising from the model structure as well as the nature of the antibody-bound intermediate.[33]

Our current efforts focus on improving the catalytic activity of 43C9, and we are proceeding along two pathways. The first takes advantage of our ability to clone and screen antibody combinatorial libraries. Hybridoma technology allows access to only a small fraction ($\leq$100) of the estimated 10^8 binding specificities encoded by the murine immune system. This means that efficient catalysts may be missed if the hybridoma method is used to isolate potential catalysts. On the other hand, the combinatorial library approach yields library sizes exceeding 10^6 members, providing a

[28] S. J. Benkovic, J. A. Adams, C. L. Borders, Jr., K. D. Janda, and R. A. Lerner, *Science* **250,** 1135 (1990).

[29] L. Sastry, M. Mubaraki, K. D. Janda, S. J. Benkovic, and R. A. Lerner, *Catalytic Antibodies* (*Ciba Foundation Symp.* **159**), 145 (1991).

[30] R. A. Gibbs, B. A. Posner, D. R. Filpula, S. W. Dodd, M. A. J. Finkelman, T. K. Lee, M. Wroble, M. Whitlow, and S. J. Benkovic, *Proc. Natl. Acad. Sci. U.S.A.* **88,** 4001 (1991).

[31] B. Posner, I. Lee, T. Itoh, J. Pyati, R. Graff, G. B. Thornton, R. La Polla, and S. J. Benkovic, *Gene* **128,** 111 (1993).

[32] V. A. Roberts, J. D. Stewart, S. J. Benkovic, and E. D. Getzoff, *J. Mol. Biol.* **235,** 1098 (1993).

[33] J. D. Stewart, V. A. Roberts, N. Thomas, E. D. Getzoff, and S. J. Benkovic, *Biochemistry* **33,** (1994).

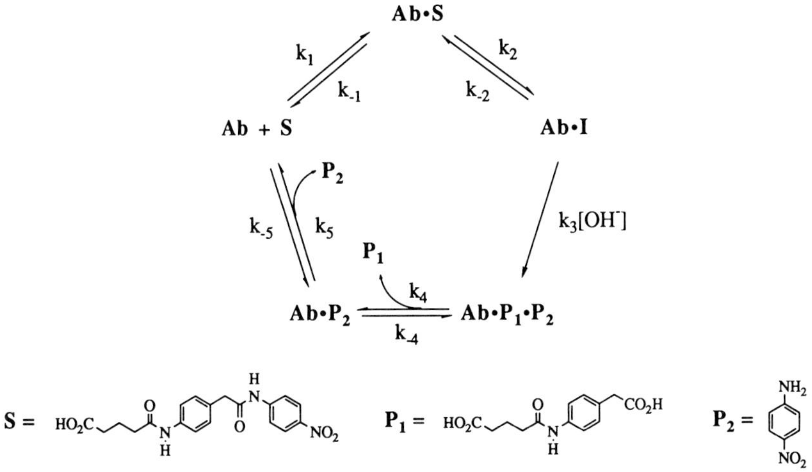

$$k_{cat}\,(\mathrm{pH} < 9.0) = \frac{k_1 \cdot k_3 \cdot [\mathrm{OH}^-]}{k_2 + k_{-2}} \qquad k_{cat}\,(\mathrm{pH} > 9.0) = \frac{k_2 \cdot k_4 \cdot k_5}{k_4 \cdot k_5 + k_2(k_4 + k_5)}$$

SCHEME 2

much larger pool of potential catalysts. To test this approach, a combinatorial library has been constructed from a mouse immunized with the 43C9 hapten, and this should allow us to ask whether a better catalyst than 43C9 exists in the immune response to the hapten. The second method has employed rational site-directed mutagenesis using mechanistic information in an effort to accelerate the rate-limiting steps of catalysis.[33] The first approach relies on the ability to produce functional Fab fragments of 43C9 in *E. coli,* whereas the second study has utilized the scFv form of the antibody. Antibodies vary widely in expression level in *E. coli,* and 43C9 represents a rather pooly expressed clone. We therefore summarize these results as a case study in antibody expression in *E. coli* with the expectation that the results will have general applicability for other antibodies.

Expression of 43C9 Fab Fragment

As noted above, the ability to screen λ phage plaques for antibodies that bind a given antigen depends on the ability to express functional antibodies within the plaques. The original expression vectors described

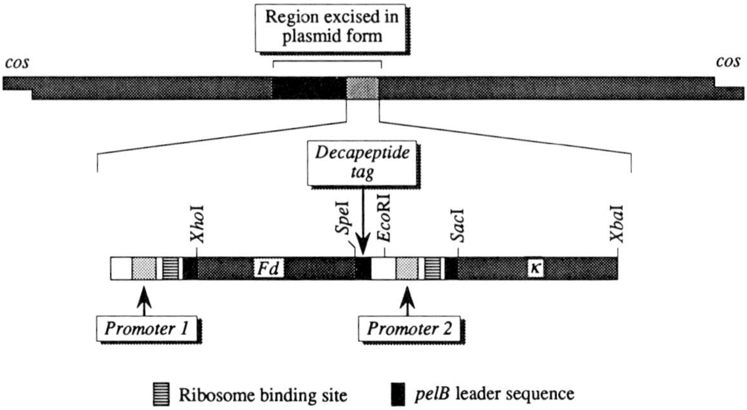

FIG. 2. Schematic diagram of the vectors used to express 43C9 Fab fragments. The bacteriophage λ vector region is identical in each of the vectors, and they differ principally in the region shown at the bottom. The positions of the promoters, the ribosome binding site, *pelB* leader sequence, and antibody genes (*Fd* or κ) are shown. In the single-*lac* vector, *Promoter 1* is a *lac* promoter and *Promoter 2* is absent. In the double-*lac* vector, both *Promoter 1* and *2* are *lac* promoters. Finally, in the T3 vector, *Promoter 1* is a *lac* promoter and *Promoter 2* is absent. The excised plasmid forms have the general structure of pBP105 (Fig. 3).

by Huse *et al.* incorporated a number of features designed to aid in antibody screening and overproduction.[23] For transcriptional control, a single *lac* promoter was used in front of the heavy (*Fd*) and light chain (κ) genes, creating a dicistronic operon (Fig. 2). Both the heavy and light chain genes were fused to the leader sequence from the *Erwinia carotovora* pectate lyase gene (*pelB*) to direct transport of the nascent polypeptides into the periplasmic space for folding and assembly under oxidizing conditions. In addition, a 12-amino acid peptide was fused to the C terminus of the heavy chain gene. This "decapeptide tag" was designed to serve two purposes: to aid in screening plaques for heavy chain expression since a monoclonal antibody against this sequence was available and to allow for a one-step immunopurification of the Fab fragments using the same monoclonal antibody immobilized onto a column.[34] Finally, the combinatorial λ vector was designed to yield a plasmid form of the cloned antibody genes for protein overproduction after superinfection with M13 helper phage. One of the major problems of the original λ library system was the difficulty in screening the plaques for binders, which was due to poor

[34] J. Field, J.-I. Nikawa, D. Broek, B. MacDonald, L. Rodgers, I. A. Wilson, R. A. Lerner, and M. Wigler, *Mol. Cell. Biol.* **8,** 2159 (1988).

expression of the Fab fragments at the plaque level and to technical difficulties with the radioactive detection method.

Several elements of this expression system have been reconsidered with the aim of improving antibody expression at the plaque level as well as at the level of soluble Fab fragments.[31] As it would be extremely difficult to assess directly the expression level of Fab fragments in plaques, a series of plasmids was first constructed and the expression levels of functional 43C9 Fab measured in liquid culture by an ELISA method using immobilized hapten. The plasmids corresponded to the excised portion of the λ vector. Because both chains of the Fab fragment are fused to a *pelB* leader sequence, the functional Fab was found in the periplasmic space of the *E. coli* host. Using the original single-*lac* expression system, the functional 43C9 Fab fragment represented approximately 0.10% of the total periplasmic protein.

Because comparable levels of both the heavy and light chain must be synthesized and exported to the periplasmic space, the dicistronic nature of the original expression system was a concern since the light chain gene was distal to the *lac* promoter. A derivative was therefore constructed in which a second *lac* promoter was introduced immediately upstream from the light chain gene (double-*lac*). Unfortunately, in liquid culture, the addition of the second *lac* promoter had no effect on the expression level of the Fab fragment. To test the effects of a stronger promoter, the *lac* promoter in the original construct was replaced with that from bacteriophage T3. In liquid culture, this promoter produced approximately 10-fold more functional Fab fragment (1.0% of the total periplasmic protein) than the original, single-*lac* vector. The T3 RNA polymerase was supplied from a second plasmid (pTG119) in which gene 1.0 was under the control of the *lacUV5* promoter (Fig. 3).

Finally, the effect of the decapeptide tag was probed by deleting it from each of the above constructs (single-*lac*, double-*lac*, and T3). In each case, the constructs lacking the decapeptide tag expressed approximately 2-fold more functional Fab than those retaining the tag. The conclusion from these studies was that the optimal expression system for Fab fragments in liquid culture used the bacteriophage T3 promoter without the decapeptide tag fused to the heavy chain. This expression plasmid was designated pBP108 (Fig. 3). Using this system, the 43C9 Fab fragment represented approximately 2.1% of the periplasmic protein.

Although the expression level of the Fab fragment is an important parameter in screening λ plaques for binders, other factors also play a role. To determine which vector system would be most useful at the plaque level, three λ expression vectors were constructed containing either a single *lac* promoter, two *lac* promoters (upstream from both the heavy-

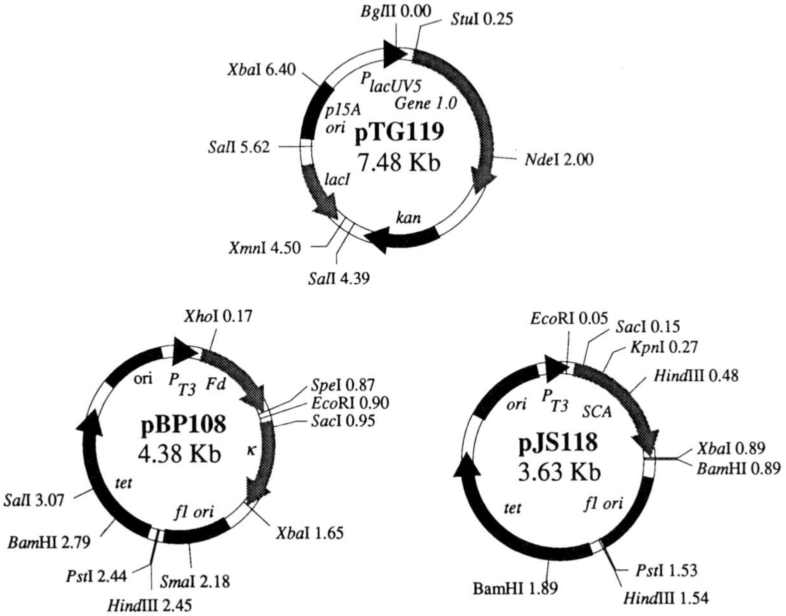

FIG. 3. Structures of plasmids used to express the 43C9 Fab and scFv fragments. Plasmid pTG119, a derivative of pACYC177, contains the gene for the T3 RNA polymerase (*gene 1.0*) under control of the *lacUV5* promoter. Plasmid pBP105 contains the 43C9 *Fd* and κ genes under control of the T3 promoter without the decapeptide tag. Plasmid pJS118 contains the scFv gene under control of the T3 promoter, also without the decapeptide tag.

and light-chain genes) or a single bacteriophage T3 promoter. These λ vectors are analogous to the plasmid vectors described above. All lacked a decapeptide tag fused to the heavy chain. The 43C9 Fab was then expressed from each of the vectors, and plaque lifts were screened for binding to the phosphonamidate hapten by a nonradioactive method employing an insoluble chromogenic alkaline phosphatase substrate. Unfortunately, the plaques produced by the λ vector containing the T3 promoter were extremely small and were not visible in the binding screen, probably owing to interference in DNA replication caused by transcription by the T3 RNA polymerase. However, the two *lac*-based vectors did give normal-sized plaques. Furthermore, the signal intensity of the positives was greater for the λ vector that contained *lac* promoters in front of both the heavy and light chains.

These studies indicate that a single promoter system is not adequate for antibody production in both plaques and liquid cultures. We therefore recommend the double-*lac* system for expression from λ phage and the

T3 system for Fab overexpression. Because the experimental protocols for the purification of the 43C9 Fab and scFv fragments are the same, this procedure is presented in the next section.

Expression of 43C9 scFv

For our mutagenesis study of 43C9, we had several reasons for avoiding the formation of inclusion bodies. First, the requirement for refolding introduces the possibility that the final product might be contaminated with improperly folded material that could be difficult to remove chromatographically. In addition, selection of catalysts by complementation of an *E. coli* mutation or other *in vivo* selections require the production of active catalysts. Finally, we wished to extend the λ library method to include scFv fragments since these are generally expressed in higher yields than Fab fragments. For these reasons, our goal was the production of the 43C9 scFv protein in functional form, rather than in the form of inclusion bodies, as had been reported previously.[30]

After exploring a number of alternative promoters and leader sequences for expression of the 43C9 scFv, we selected the bacteriophage T3 system described above for the 43C9 Fab fragment. Restriction sites flanking the scFv gene were altered to match those of the plasmid designed to express the Fab light chain, and the final expression construct was designated pJS118 (Fig. 3).[33] For protein overexpression, the *E. coli* host (BL21, F^-, *ompT, hsdS*$_B$) was cotransformed with pJS118 and pTG119, which supplies the T3 RNA polymerase under control of the *lacUV5* promoter.

The scFv gene was fused to the *pelB* leader sequence to direct secretion of the protein into the periplasmic space. However, we found that following induction, most of the protein was found in the growth media. Because other proteins normally found in the periplasm were also detected in the growth medium by electrophoresis, it seems likely that the proteins are released by a leaky outer membrane. The same phenomenon was observed for the 43C9 Fab fragment. This situation considerably simplifies antibody purification by removing the need for cell lysis, allowing the protein to be recovered by ultrafiltration after removing the cells by low-speed centrifugation. Our current protocol for antibody purification is given below. We have found that buffering the medium with MOPS at pH 7.5 significantly increases the yield of antibody over that obtained in the absence of added buffer, probably by providing an environment more conducive to protein folding and assembly.

For expression of the Fab or scFv fragment, BL21 cotransformed with pTG119 and the appropriate antibody expression plasmid (pBP108

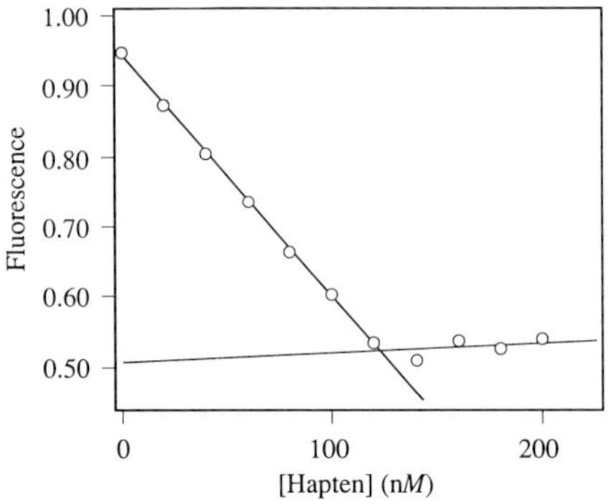

FIG. 4. Typical active site titration for purified 43C9 scFv. The intrinsic antibody fluorescence is plotted as a function of added hapten. In this case, the active antibody concentration is 122 n*M*.

or pJS118 for the Fab or scFv fragment, respectively) is precultured overnight in 25 ml of Luria–Bertani (LB) medium containing 10 μg/ml tetracycline and 30 μg/ml kanamycin at 37°.[35] Ten-milliliter portions of the preculture are diluted into two 1-liter portions of LB medium containing 10 μg/ml tetracycline, 30 μg/ml kanamycin, and 10 g/liter of MOPS, buffered to pH 7.5 before autoclaving. The cultures are grown in nonbaffled flasks at 37° to an OD_{600} of 1.0, then cooled to room temperature for 15 min. Isopropyl-β-D-thiogalactoside (IPTG) is then added to a final concentration of 2 m*M*, and the cultures are shaken at room temperature for 18 hr. Cells are removed by centrifuging at 7000 *g* for 30 min (4°), and the supernatant is concentrated by ultrafiltration to approximately 10 ml (4°). The concentration is usually performed by a tangential flow apparatus (Minitan, Millipore Corp., Bedford, MA) equipped with 10,000 NMWL membranes.

The concentrated material is dialyzed against three 1-liter changes of 20 m*M* sodium MOPS, 0.4 m*M* calcium acetate, pH 6.3 (4°). The antibody is then purified by chromatography on a 21 × 250 mm polyaspartic acid cation-exchange column (PolyCAT A, Nest Group) using a Pharmacia (Piscataway, NJ) FPLC (fast protein liquid chromatography) system. The column is initially equilibrated with 40 m*M* sodium MOPS, 1 m*M* calcium acetate, pH 6.3, and, after loading, the antibody is eluted with a linear gradient (total of 750 ml) of the starting buffer and 40 m*M* sodium MOPS, 25 m*M* calcium acetate, pH 7.5. The flow rate is 4 ml/min throughout the gradient. Fractions containing the antibody are identified by gel electro-

[35] J. Sambrook, E. F. Fritsch, and T. Maniatis, "Molecular Cloning: A Laboratory Manual," 2nd Ed. Cold Spring Harbor Laboratory, Cold Spring Harbor, New York, 1989.

phoresis, and these are combined and concentrated by ultrafiltration (Amicon, Danvers, MA, YM10 membrane), then dialyzed into 100 m*M* sodium HEPES, 50 m*M* NaCl (4°). Purified antibody fragments should be stored at 4° since they are unstable when subjected to multiple freeze–thaw cycles. The expected yield for the 43C9 Fab fragment from this procedure is 500 μg per liter of culture, and that for the scFv is 700 μg per liter of culture.

The antibody concentration is assayed by fluorescence active site titration with the hapten. Antibodies have large intrinsic fluorescence owing to the relatively high proportion of Trp residues, and ligand binding often leads to a quenching of the fluorescence.[36] A solution of antibody fragment in buffer (typically 100 m*M* sodium HEPES, 50 m*M* NaCl, pH 7.5) is placed in a fluorescence cuvette with the excitation wavelength set at 280 nm and the emission wavelength at 340 nm. Known quantities of hapten (in a volume $\leq$5 μl) are added to the antibody solution, and the fluorescence is recorded as a function of added ligand. If the antibody concentration is more than approximately 10-fold greater than the K_D value for hapten binding, a monotonic decrease in fluorescence will be observed that levels off at the active antibody concentration. The fluorescence values obtained for the initial and final regions of the titration are separately fit to lines using least-squares methods, and these equations are solved simultaneously to obtain the antibody concentration (Fig. 4).

Acknowledgments

Work in this laboratory was supported by a grant from the Office of Naval Research (N00014-91-J-1593) and by a postdoctoral fellowship from the Helen Hay Whitney Foundation to J.D.S.

[36] S. F. Velick, C. W. Parker, and H. N. Eisen, *Proc. Natl. Acad. Sci. U.S.A.* **46,** 1470 (1960).

[20] Cooperativity in Enzyme Function: Equilibrium and Kinetic Aspects

By KENNETH E. NEET

Introduction

Cooperativity in protein or enzyme systems commonly refers to ligand binding with the type of positive cooperativity that gives rise to the sigmoid curves of oxygen binding to hemoglobin. However, the concept and the practical applications are much more encompassing than that single interpretation. Cooperativity in its broadest sense includes protein folding–unfolding reactions (e.g., denaturation, helix–coil transitions), macromolecular assembly (e.g., tubulin, sickling of hemoglobin S), binding of proteins as ligands to DNA (e.g., transcriptional factors, single-stranded DNA binding proteins), and binding of proteins as ligands to membrane receptors (e.g., hormones, growth factors). Cooperativity, in general, is any process in which the initial event (hydrogen bonding, protein–protein interaction, ligand binding) affects subsequent similar events, in the cases cited via communication through intra- or intermolecular protein interactions. Protein structures are uniquely suited for the ability to communicate the information needed for cooperativity through conformational changes and intersubunit binding affinities. This chapter mainly deals with substrate and ligand binding to enzymes but also briefly discusses the extension to proteins as ligands binding to other macromolecular systems. The current discussion of cooperativity, which is an update of an earlier chapter in this series,[1] describes the basics of enzyme cooperativity, emphasizes advances since the early 1980s, and utilizes examples of several systems to describe how cooperativity is currently studied and analyzed. The reader is referred to previous reviews[1,2] for historical perspectives, more details of kinetic analysis of some systems, and a description of mechanisms not described here.

Cooperativity in enzyme systems is initially phenomenological, that is, it usually begins with the observation of non-Michaelis–Menten (nonhyperbolic) kinetic assays and later develops into a mechanistic explanation in the particular enzyme. Thus, a variety of both equilibrium (e.g., site–site models, oligomer association) and kinetic (e.g., slow transition, random order) mechanisms can give essentially identical velocity (v) versus substrate concentration ([S]) plots. The form of the cooperative curves and their fundamental analyses are basic to understanding the physiological relevance and the underlying mechanism of the cooperativity. The shape of positive, negative, and noncooperative curves is shown in Fig. 1 in six standard plotting formats (discussed in the next section). The investigator must eliminate potential sources of artifacts leading to the appearance of cooperativity such as nonrelevant adsorption of substrate at low concentration, impure enzymes, complex formation (e.g., MgATP), and substrate inhibition (discussed by Cornish-Bowden and Cardenas[3]).

Homotropic cooperativity is the effect of the interaction of a ligand

[1] K. E. Neet, this series, Vol. 64, p. 139.

[2] K. E. Neet and G. R. Ainslie, Jr., this series, Vol. 64, p. 192.

[3] A. Cornish-Bowden and M. L. Cardenas, *J. Theor. Biol.* **124,** 1 (1987).

on subsequent binding or reaction of the same ligand (e.g., the first molecule of O_2 binding to hemoglobin affects the binding of the next three O_2 molecules, giving rise to the now classic sigmoidal binding curve). Homotropic cooperativity can be either positive or negative. In positive cooperativity the initial binding makes it easier for subsequent molecules to interact (e.g., O_2 enhances the binding of the next O_2 to hemoglobin). With positive cooperativity in its most general sense, the logical intermediate does not accumulate, that is, the initial and the final states are the most predominant at equilibrium. In negative cooperativity, the initial interaction makes it more difficult for subsequent molecules to bind or interact (e.g., NAD^+ binding to glyceraldehyde phosphate dehydrogenase[4]). In the extreme, negative cooperativity can give rise to half-the-sites reactivity [e.g., three molecules of carbamoyl phosphate[5] binding to the aspartate transcarbamylase (ATCase; aspartate carbamoyltransferase) hexamer].

Heterotropic cooperativity is the effect of a different ligand (or the same ligand acting at a quite different site on the protein) on the binding or reaction of the first molecule. Negative heterotropic effector, inhibitor, and antagonist are synonyms for the same type of ligand, producing a decrease in binding or reaction at the site (catalytic) of interest. Similarly, positive heterotropic effector is synonymous with activator, producing an increase in binding or reaction at the initial site. Heterotropic responses are the classic allosteric effectors, at least when the action is at a site different from the active site. Several examples of isosteric regulation have also been observed which produce similar kinetic effects (see below).

Kinetic effects in enzyme catalysis can alter existing cooperativity. The most comprehensive extension of cooperative binding to kinetic analysis has been developed by Ricard and co-workers[6] as "structural kinetics." Alternatively, purely kinetic mechanisms can give rise to phenomenological "cooperativity" that appears only in kinetic assays and, thus, can be assessed by comparison of equilibrium binding and kinetic measurements with the same ligand. Hysteresis or ligand-induced slow transitions produced by enzyme isomerizations that are on the same order of magnitude as the catalytic cycle can produce nonhyperbolic kinetics.[2,7,8] A steady-state random addition of substrates, with the appropriate combination of rate constants, may produce the appearance of cooperativity in enzyme

[4] A. Conway and D. E. Koshland, Jr., *Biochemistry* **7,** 4011 (1968).

[5] P. Suter and J. P. Rosenbusch, *J. Biol. Chem.* **251,** 5986 (1976).

[6] J. Ricard, *in* "Organized Multienzyme Systems" (G. R. Welch, ed.), p. 177. Academic Press, New York, 1985.

[7] G. R. Ainslie, Jr., J. P. Shill, and K. E. Neet, *J. Biol. Chem.* **247,** 7088 (1972).

[8] J. Ricard, J. C. Meunier, and J. Buc, *Eur. J. Biochem.* **49,** 195 (1974).

catalysis.[9] These kinetic consequences are also discussed in this chapter. Finally, certain structural features and thermodynamic considerations of cooperative enzymes are emerging and are presented.

Graphical Representation and Evaluation of Extent of Cooperativity

Six standard ways of graphing kinetic or binding data have been developed (Fig. 1) based mainly on rearrangements of the Michaelis–Menten, multisite binding isotherm, or Hill equations. Examples of different degrees of both positive and negative cooperativity are shown in Fig. 1 to contrast to the noncooperative (Michaelis–Menten) case.

The straight-up plot of response versus ligand concentration (Fig. 1A) gives rise to the 'sigmoidal' curve of positive cooperativity but does not readily distinguish the hyperbola from the near-hyperbola of negative cooperativity. The log–response curve (Fig. 1B), commonly used in pharmacological studies, makes all common binding patterns sigmoidal. This plot provides an easy way to assess cooperativity visually from the slope of the graph or to calculate quickly the R_s, that is, the ratio of ligand concentration at 90% response to ligand concentration at 10% response ($R_s = 81$ for no cooperativity; $R_s < 81$ for positive cooperativity; $R_s >$ 81 for negative cooperativity). The value of R_s is easily related to the Hill coefficient ($R_s = 81^{1/n_H}$).

The double-reciprocal plot (Lineweaver–Burke) (Fig. 1C) is standardly used for enzyme kinetics, and its merits and faults for obtaining good estimates of V_{max} and K_m have been extensively discussed.[10] The various forms of cooperativity are clearly distinct in the double-reciprocal plot by their deviations from linearity. Note that the asymptote to the positive cooperativity curve at low concentrations would intersect the Y axis at a negative value, distinguishing it from substrate inhibition, which would have a positive Y intercept.

The Hill plot (Fig. 1D) linearizes all common binding patterns (over the range of 10 to 90% saturation) and provides the value of the Hill coefficient, n_H, from the slope of the graph. The n_H is the standard and most common means of evaluating or comparing cooperativity ($n_H = 1$ for no cooperativity; $n_H < 1$ for negative cooperativity; $n_H > 1$ for positive cooperativity). Deviations from linearity in the central region of the Hill plot are indicative of mixed cooperativity or more complex binding patterns.[11] The value for V_{max} must be estimated to calculate the values for

[9] W. Ferdinand, *Biochem. J.* **98,** 278 (1966).

[10] F. B. Rudolph and H. J. Fromm, this series, Vol. 63, p. 138.

[11] A. Cornish-Bowden and D. E. Koshland, Jr., *J. Mol. Biol.* **95,** 201 (1975).

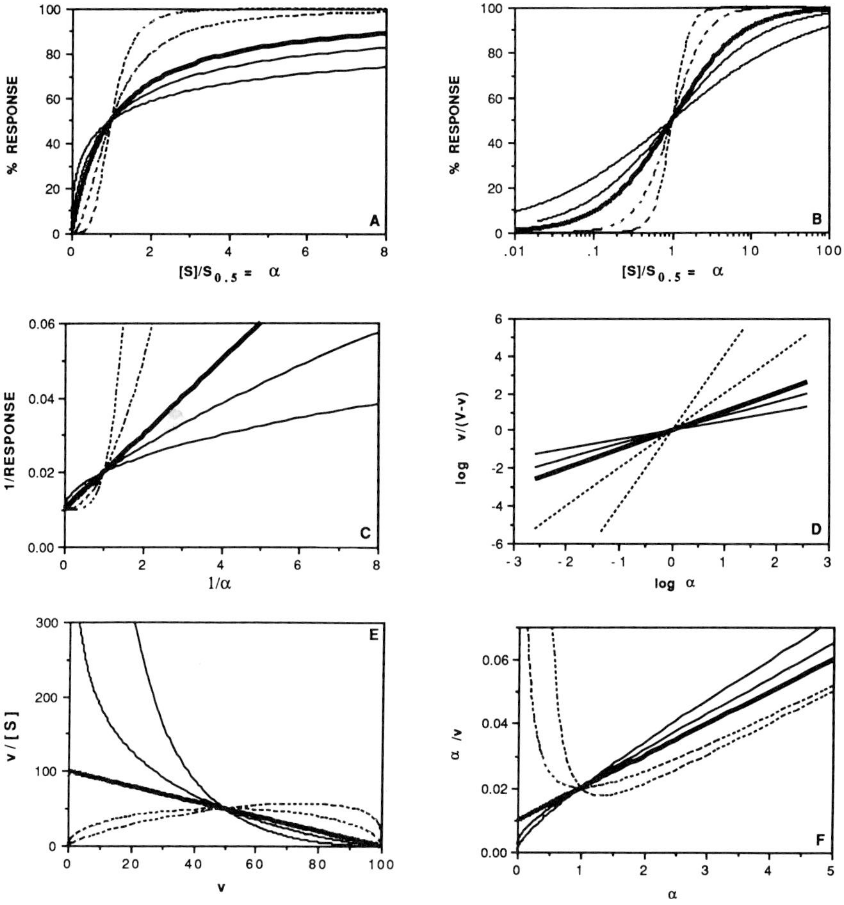

FIG. 1. Comparison of six ways of plotting kinetic or binding data to analyze the cooperativity. In each graph the noncooperative (Michaelis–Menten) case is the heavy line (n_H = 1), the two positively cooperative cases are dashed lines (n_H = 4, 2), and the two negatively cooperative cases are thin solid lines (n_H = 0.75, 0.5). Note that heterogeneous binding sites would appear the same as "negative cooperativity" in each of the plots. The % RESPONSE in (A) and (B) could represent enzymatic velocity or moles of bound ligand. The response is given as v in (C)–(F). The values of [S] are normalized to the $S_{0.5}$ value (α) for comparison purposes and would represent free ligand concentration in binding experiments.

the Y axis in the Hill plot. For positive cooperativity this estimate is easily obtained from the log–response or the double-reciprocal plot. For negative cooperativity the estimate of V_{max} is more difficult because of the slow asymptotic approach (in the v versus [S], the log–response, or the Scatchard plot) or the steep curvature at high concentrations (in the double-reciprocal plot). Fortunately, the minimum slope of the Hill plot in this case is not very sensitive to the value of the V_{max}. However, the shape of the Hill plot does change with wrong choices of V_{max}. Incorrect choice of a V_{max} that is 75 or 85% of the true value produces a curved Hill plot with a linear segment below the $S_{0.5}$, whose slope is about 10% higher than the true value, then breaks sharply upward with a slope approaching infinity.

The Scatchard–Eadie plot (or Hofstee–Augustinsson when rotated 90°) of v/[S] versus v (Fig. 1E) is commonly used in ligand binding analyses and has proponents for enzyme kinetics. The advantage of using the v/[S] versus v plot is that the data (or the curvature) are spread more evenly with retention of the distinction between the different forms of cooperativity. The position of the maximum in the positive cooperativity curves is easily related to the Hill coefficient; thus, $v/V_{max} = (1 - 1/n_H)$ at the optimum in the v/[S] curve.

Finally, the Hanes plot of [S]/v versus [S] (Fig. 1F) has similar characteristics to the Scatchard plot but is infrequently utilized. Note that, again, the [S] position of the minimum in the [S]/v curve is related to the degree of positive cooperativity.

Unfortunately, true negative cooperativity gives rise to the same shape of velocity or binding curve as the curve from a mixture of enzymes or binding sites within an enzyme.[3] Thus, the curves termed "negative cooperativity" in Fig. 1 are indistinguishable from a preexisting heterogeneity of binding sites. The shape of the curve is a result of nonidentical binding after ligand binding and does not depend on whether the protein initially had identical (negative cooperativity) or nonidentical (heterogeneous) sites. Therefore, demonstration of the purity of a protein preparation, including the possibility of partially denatured components, is essential for the study of negative cooperativity. Because ligand binding or kinetics cannot discriminate between these two cases, attempts to resolve such issues normally depend on physical methods to demonstrate or eliminate preexisting asymmetry or heterogeneity of ligand binding sites.

Although several alternate means of assessing the degree of cooperativity have been proposed (see Neet[1] for a comparison), the Hill coefficient remains the most generally used and conveys the most immediate, conceptual meaning to practicing biochemists. The n_H for positive cooperativity is equal to or less than the number of ligand binding sites, n; how different n_H is from n is a qualitative estimate of the strength of the site–site interactions. The n_H for negative cooperativity is not indicative of, nor related to, the number of sites. The important consideration in evaluating cooperativity is to obtain good data over a wide range of ligand concentration that is appropriately chosen for the graphical means of assessing the data. The use of several graphical methods is recommended in order to extract the most information, but such plots should only be used for visualization and/or presentation. An unbiased computer fitting algorithm of the raw data (v and [S]) to the equation for the appropriate model is the final and best means to judge the type and extent of cooperativity and obtain relevant parameters.

Estimation of Parameters by Curve Fitting

The estimation of cooperativity parameters and the evaluation of models by data fitting have been greatly facilitated with the general availability of computers, digital data acquisition, and appropriate software for analysis. The graph has become a means of presentation and "eyeballing" results, rather than a means of obtaining parameter values. An extensive discussion of statistical analysis of enzyme kinetic data and nonlinear least squares fitting procedures has been previously presented in this series[1,12,13] and is not elaborated on here. Those discussions tend to deal with Michaelian enzymes, but the approaches are still generally valid.

The appropriate cooperativity equation for the model to be tested should be chosen from those discussed below and the data fitted with a nonlinear regression procedure. The Marquardt algorithm is most commonly used. Even simple estimations of the Hill coefficient are more easily done with the Hill equation in a nonlinear fitting program that estimates V_{max} as well as K_m and n_H. Numerous programs for both MacIntosh and IBM-compatible computers are now commercially available that can handle higher order rate (or binding) equations for cooperativity and exponential functions for hysteretic transient analyses either intrinsic to the application or by writing the desired equation into the program. Typical commercial programs include LIGAND[14] (Lunden Software, Biosoft), ENZFITTER (Sigma, St. Louis, MO), GRAFIT (Erithacus Software Ltd.), ULTRAFIT (Biosoft), EZFIT[15] (Prentice Hall), JMP (SAS Institute, Inc.), KALEIDOGRAPH (Synergy Software), and MATHEMATICA (Wolfram Research Inc.). Simulations of data with chosen parameters can also be done with many of the programs and can be helpful in the design of experiments, for example, to determine the best concentration range for a ligand in the presence of several other ligands. No software, however,

[12] B. Mannervik, this series, Vol. 87, p. 370.

[13] W. W. Cleland, this series, Vol. 63, p. 103.

[14] P. J. Munson, this series, Vol. 92, p. 543.

[15] J. H. Noggle, "Practical Curve Fitting and Data Analysis." Prentice-Hall, Englewood Cliffs, New Jersey, 1993.

can substitute for precise data spread over an adequate concentration range and application of good common sense.

Mechanisms for Enzyme Cooperativity

Both the concerted[16] model and the sequential[17] model are well described in biochemistry textbooks such as Stryer,[18] Lehninger *et al.*[19] or Mathews and van Holde.[20] This section emphasizes the differences in the models, the basis for the resultant cooperativity, and the means for distinguishing the models. A later section describes studies that have chosen one model as the most appropriate for a particular system.

General Binding Model of Adair

$$n\overline{Y} = \frac{\Psi_1[S] + 2\Psi_2[S]^2 + 3\Psi_3[S]^3 + 4\Psi_4[S]^4}{1 + \Psi_1[S] + \Psi_2[S]^2 + \Psi_3[S]^3 + \Psi_4[S]^4} \quad (1)$$

The general Adair[21] binding equation [Eq. (1)] defines the ligand saturation function, $\overline{Y}$, in terms of the number of binding sites n (usually subunits) and n coefficients (Ψ_i), where $\Psi_i = \Pi_{j=1}^{i} K_j$. The Adair coefficients are readily related to the simple stoichiometric, K_i, or intrinsic, K_i', ligand association constants (see Table I for the tetramer case).

Stoichiometric constants are obtained by counting all liganded species with a certain stoichiometry, namely, $K_i = [PS_i]/[S][PS_{i-1}]$, whereas intrinsic constants are treated as if each liganded subunit were measured in isolation and therefore take into account the probability of binding to individual sites. An alternative definition of binding focuses on individual, discrete sites of binding on the oligomer and results in different, but related, equations utilizing these site binding constants.[22] The statistical relationship[1] between stoichiometric, K_i, and intrinsic K_i', constants for the ith site, is given by Eq. (2). In the case of a noninteracting oligomer, successive intrinsic constants will be identical.

$$K_i = \frac{(n + 1 - i)}{i} K_i' = \frac{\text{\# free sites on oligomer before binding}}{\text{\# occupied sites after binding}} K_i' \quad (2)$$

[16] J. Monod, J. Wyman, and J.-P. Changeux, *J. Mol. Biol.* **12,** 88 (1965).

[17] D. E. Koshland, Jr., G. Nemethy, and P. Filmer, *Biochemistry* **5,** 365 (1966).

[18] L. Stryer, "Biochemistry." Freeman, New York, 1988.

[19] A. L. Lehninger, D. L. Nelson, and M. M. Cox, "Principles of Biochemistry." Worth, New York, 1993.

[20] C. K. Mathews and K. E. van Holde, "Biochemistry" Benjamin/Cummings, Redwood City, California, 1990.

[21] G. S. Adair, *J. Biol. Chem.* **63,** 529 (1925).

[22] I. M. Klotz, *Acc. Chem. Res.* **7,** 162 (1974).

TABLE I
COEFFICIENTS OF ADAIR EQUATION FOR COOPERATIVE TETRAMER MODELS

	Adair	Intrinsic	KNF Tetrahedral	KNF Square	MWC Concerted
Ψ_1	K_1	$4K'_1$	$4K^3_{AB}K_SK_T$	$4K^2_{AB}K_SK_T$	$\frac{4(1+Lc)}{(1+L)K_R}$
Ψ_2	K_1K_2	$6K'_1K'_2$	$6K^4_{AB}K_{BB}K^2_SK^2_T$	$(2K^4_{AB}+4K_{BB}K^2_{AB})K^2_SK^2_T$	$\frac{6(1+Lc^2)}{(1+L)K^2_R}$
Ψ_3	$K_1K_2K_3$	$4K'_1K'_2K'_3$	$4K^3_{AB}K^3_{BB}K^3_SK^3_T$	$4K^2_{AB}K^2_{BB}K^3_SK^3_T$	$\frac{4(1+Lc^3)}{(1+L)K^3_R}$
Ψ_4	$K_1K_2K_3K_4$	$K'_1K'_2K'_3K'_4$	$K^6_{BB}K^4_SK^4_T$	$K^4_{BB}K^4_SK^4_T$	$\frac{(1+Lc^4)}{(1+L)K^4_R}$
$\frac{\Psi_1}{4}$		K'_1	$K^3_{AB}K_SK_T$	$K^2_{AB}K_SK_T$	$\frac{(1+Lc)}{(1+L)K_R}$
$\frac{3\Psi_2}{2\Psi_1}$		K'_2	$K_{AB}K_{BB}K_SK_T$	$(K^2_{AB}+2K_{BB})K_SK_T$	$\frac{(1+Lc^2)}{(1+Lc)K_R}$
$\frac{2\Psi_3}{3\Psi_2}$		K'_3	$\frac{K^2_{BB}K_SK_T}{K_{AB}}$	$\frac{K^2_{BB}K_SK_T}{(K^2_{AB}+2K_{BB})}$	$\frac{(1+Lc^3)}{(1+Lc^2)K_R}$
$\frac{4\Psi_4}{\Psi_3}$		K'_4	$\frac{K^3_{BB}K_SK_T}{K^3_{AB}}$	$\frac{K^2_{BB}K_SK_T}{K^2_{AB}}$	$\frac{(1+Lc^4)}{(1+Lc^3)K_R}$
		K'_2/K'_1	$\frac{K_{BB}}{K^2_{AB}}$	$\frac{(K^2_{AB}+2K_{BB})}{K^2_{AB}}$	$\frac{(1+Lc^2)(1+L)}{(1+Lc)^2}$
		K'_3/K'_2	$\frac{K_{BB}}{K^2_{AB}}$	$\frac{K^2_{BB}}{(K^2_{AB}+2K_{BB})^2}$	$\frac{(1+Lc^3)(1+Lc)}{(1+Lc^2)^2}$
		K'_4/K'_3	$\frac{K_{BB}}{K^2_{AB}}$	$\frac{(K^2_{AB}+2K_{BB})}{K^2_{AB}}$	$\frac{(1+Lc^4)(1+Lc^2)}{(1+Lc^3)^2}$

Intrinsic binding constants are frequently utilized instead of stoichiometric constants when the derivation of a binding (or rate) equation is made by considering the distribution of forms at equilibrium.[23,24] In this case, the saturation function may be presented in a slightly different form, where the numerical coefficients are the binomial expansion of $(n - 1)$ in the numerator and n in the denominator [Eq. (3)] and $\Psi'_i = \Pi^i_{j=1} K'_j$.

$$\bar{Y} = \frac{\Psi'_1[S] + 3\Psi'_2[S]^2 + 3\Psi'_3[S]^3 + \Psi'_4[S]^4}{1 + 4\Psi'_1[S] + 6\Psi'_2[S]^2 + 4\Psi'_3[S]^3 + \Psi'_4[S]^4} \quad (3)$$

These constants, K_i or K'_i, are phenomenological (or observed) constants and do not directly relate to molecular properties of the protein such as conformational changes or subunit interactions. However, they serve a useful purpose for initially fitting primary binding data and for comparison of molecular models.

Concerted, Symmetric Model of Monod–Wyman–Changeux

The concerted or Monod–Wyman–Changeux (MWC) model[16] provides the simplest explanation for cooperativity in ligand binding to enzymes and has been widely applied to many enzymes. This model occupies a historical place in enzymology, since it was the first general explanation for cooperativity and served as the basis for defining K systems (those effectors that primarily alter K_m or $S_{0.5}$), V systems (those effectors that primarily alter V_{max}), homotropic and heterotropic interactions, and the linkage between these interactions. The main advantages of the MWC model are that it is conceptually straightforward and requires few parameters to fit the experimental data. The main disadvantage is that the MWC model cannot account for negative cooperativity in equilibrium binding (or rapid equilibrium kinetic) systems.

The MWC model (Fig. 2) is based on a concerted change of an oligomeric structure (shown here as a tetramer) from the T-state (taut) to the thermodynamically more favored R-state (relaxed). The MWC model is "concerted" because symmetry is maintained as all subunits (or protomers) undergo the same conformational change concomitantly. The oligomeric conformational equilibrium is shifted by binding of substrate or activator with higher affinity to the R-state. The cooperativity arises because the shift in the number of high-affinity R-subunit conformations (squares in Fig. 2) is greater than the amount of ligand binding to trigger the shift, providing nonoccupied high-affinity sites to which subsequent

[23] J. Ricard and A. Cornish-Bowden, *Eur. J. Biochem.* **166,** 255 (1987).
[24] I. H. Segal, "Enzyme Kinetics." Wiley (Interscience), New York, 1975.

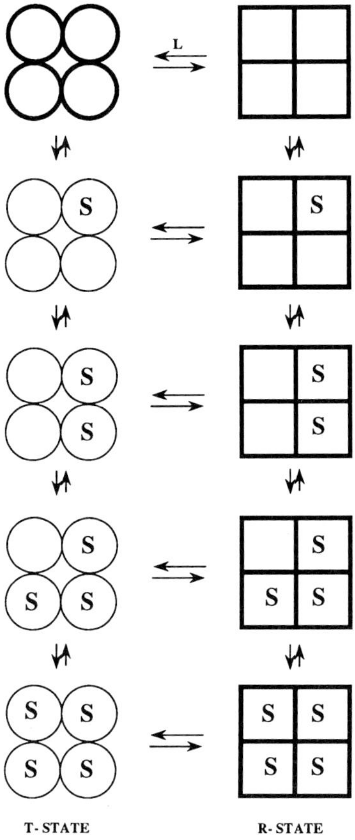

FIG. 2. Concerted Monod–Wyman–Changeux (MWC) model for cooperativity of a tetramer. The circles (T) and squares (R) represent the two symmetrical conformations of the tetramer. Ligand (S) binding shifts the equilibrium to the energetically favored R-state. The forms in heavy lines represent the original exclusive binding MWC model, and the full figure represents the complete nonexclusive binding model.

ligand can bind. The original proposal for the MWC model[16] presented exclusive binding of substrate (shown in bold, Fig. 2) but was subsequently extended to nonexclusive binding[25] (Fig. 2, all forms). Because of the linkage between equilibria, a negative heterotropic effector is predicted to increase the homotropic cooperativity in addition to raising the apparent $S_{0.5}$ value; a positive heterotropic effector is predicted to lower or eliminate the homotropic cooperativity with a lower apparent $S_{0.5}$. These relationships are frequently found with enzymes, for example ATCase.

[25] M. M. Rubin and J.-P. Changeux, *J. Mol. Biol.* **21,** 265 (1966).

The cooperativity derived from the nonexclusive binding MWC model can be described by the number of binding sites, n, and three parameters: the equilibrium constant between the two unliganded oligomer conformations, L; and the intrinsic dissociation constants for ligand dissociation from each conformer, K_R and K_T [Eq. (4)]. In Eq. (4), $L = [T_0]/[R_0]$; $K_R = [S][R]/[SR]$; $K_T = [S][T]/[ST]$; and $c = K_R/K_T$. The parameter α is simply a ligand normalization term equal to $[S]/K_R$. The concentrations of ligand, R-subunit, T-subunit, ligand bound to R-subunit, and ligand bound to T-subunit are given by [S], [R], [T], [SR], and [ST], respectively. Also shown in the right-hand side of Eq. (4) is the equivalent relationship using terms similar to those for the KNF model (see below) with $c = K_A/K_B$, $\alpha = [S]K_B$, $K_T = 1/K_A$, and $K_R = 1/K_B$.

$$\bar{Y} = \frac{Lc\alpha(1 + c\alpha)^{n-1} + \alpha(1 + \alpha)^{n-1}}{L(1 + c\alpha)^n + (1 + \alpha)^n}$$

$$= \frac{LK_A[S](1 + K_A[S])^{n-1} + K_B[S](1 + K_B[S])^{n-1}}{L(1 + K_A[S])^n + (1 + K_B[S])^n} \quad (4)$$

The degree of cooperativity in the MWC model is a function of the equilibrium between the R- and T-states, L, and the differential binding of ligand to the two states, c. The nature of the cooperativity and the lack of intermediates in a positively cooperative situation are clearly seen when typical concentrations of intermediate forms are examined. The conformational (or state) saturation, $\bar{R}$, occurs at lower ligand concentration than the ligand saturation, $\bar{Y}$, and is a characteristic structural feature of the MWC model. The concentration of enzyme forms with intermediate degrees of bound ligand, $\bar{Y}_1$, $\bar{Y}_2$, and $\bar{Y}_3$, never become large.

Sequential Model of Koshland–Nemethy–Filmer

The sequential or Koshland–Nemethy–Filmer (KNF) model[17] provides a more general explanation for cooperativity but, as a consequence, has more molecular parameters. The model is termed a sequential mechanism because of the sequential changes in subunit conformation within an oligomer. The advantages of the KNF model are that it can adequately describe negative (as well as positive) cooperativity of binding, which is frequently seen with enzymes, and can account for activators or inhibitors that do not produce the predicted results of the MWC model. The main disadvantage is that since more parameters are initially available for fitting (see below), more extensive data may be required to attain the same certainty in the final values. No independent means of assessing some of the molecular constants have been developed.

The KNF model[17] is based on ligand-induced conformational changes that consequently change subunit–subunit interactions in an oligomeric structure without (necessarily) affecting the conformation of the neighboring subunit (Fig. 3). The interaction constant (or implicitly the affinity) between the A-conformation (circles) and the B-conformation (squares), K_{AB}, would be different than the A–A interaction, K_{AA}, or the B–B interaction, K_{BB}. The cooperativity arises because of the alteration of the energetics of the subsequent ligand-induced conformational change of the subunit from A to B. A different number of K_{AA}, K_{BB}, and K_{AB} interactions modify the isomerization of the neighboring subunit in the complex. Be-

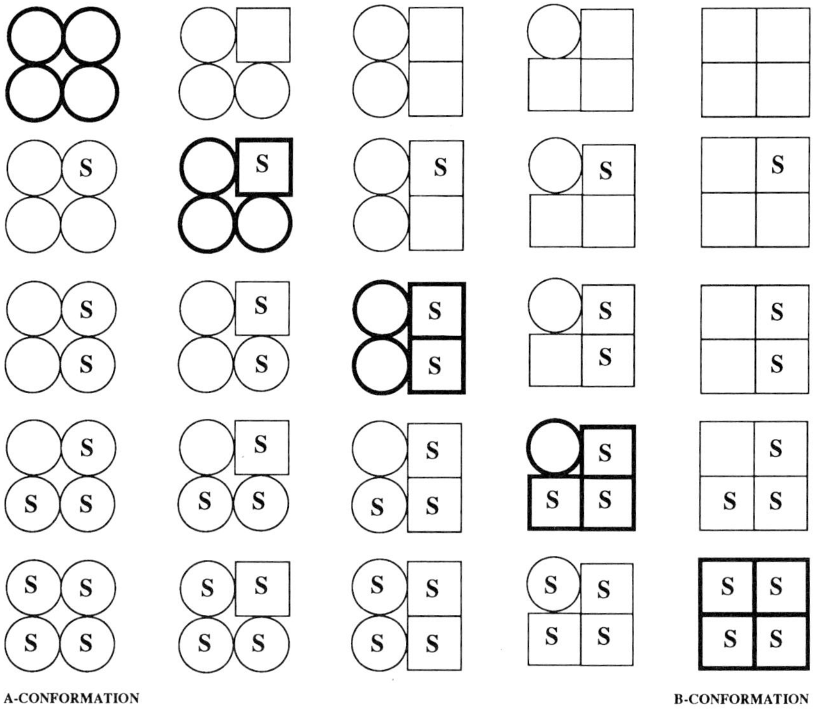

FIG. 3. Sequential Koshland–Nemethy–Filmer (KNF) model for cooperativity of a tetramer. The circles (A) and squares (B) represent two conformations of the subunit. Ligand (S) binds preferentially to the B-conformation and shifts the equilibria to the right. The diagonal forms in heavy lines represent the strict, ligand-induced KNF model with only unligated circles and ligated squares. The "square" geometry is implied in the drawing but actually depends on how the interactions between subunits are mathematically defined. The full figure is the most general model with nonexclusive binding to both A and B conformers and unligated hybrids containing A and B conformers. The MWC model is contained within the first and fifth columns that have no hybrids (cf. Fig. 2).

cause K_{AB} can be either greater or less than K_{BB}, the next ligand molecule can either bind more easily ($K_{BB} > K_{AB}$, positive cooperativity) or more difficultly ($K_{BB} < K_{AB}$, negative cooperativity), and either form of cooperativity can be generated. The original proposal of the sequential model was the most restrictive with an exact correspondence (Fig. 3, bold forms in diagonal sequence) between the bound ligand and each new subunit conformation, B. Generalization to include unliganded hybrid species and nonexclusive binding to A and B forms[4,26] and to allosteric modifiers[27] was subsequently made (Fig. 3, lighter forms).

The cooperativity derived from the restricted KNF model can be quantitatively described by n and three molecular parameters: the relative interaction constant between the oligomer conformations in their quaternary context, K_{AB} and K_{BB}, and the constant for binding of ligand to the A-conformation with the concomitant isomerization, K_tK_S. Here, K_{AB} = [AB][A]/[B][AA], K_{BB} = $[BB][A]^2/[AA][B]^2$, K_t = [B]/[A], and K_S = [SB]/[S][B]. Because the interaction constants are defined relative to the A–A interaction ($K_{AA} = [AA]/[A]^2$) and K_{AA} is taken to be equal to unity, the actual parameters simply represent subunit affinities or association constants for formation of the AB or BB dimeric components: K_{AB} = [AB]/[B][A] and $K_{BB} = [BB]/[B]^2$. The concentrations of ligand, A-subunit, B-subunit, ligand bound to A-subunit, and ligand bound to B-subunit are given by [S], [A], [B], [SA], [SB], respectively. The product of the equilibrium constant for the isolated protomer conformational transition and the intrinsic association constant for ligand with the B-conformer are never separated in the binding expression of the KNF model and thus represent a single fittable term, K_tK_S.

The KNF molecular constants (K_{AB}, K_{BB}, and K_tK_S) are determined from the Adair Ψ_i values fit to the data, based on n and the molecular geometry of the oligomer.[17] The "geometry" enters ($n > 2$) because interactions may exist or not exist with each other subunit. The Ψ parameters for the square and tetrahedral tetramer cases are given in Table I. In the original paper,[17] the "concerted" case was also considered in which the interactions of the oligomer were so tightly coupled that all subunits underwent the conformational transition simultaneously, conceptually the same as that of the MWC model.

Cooperativity is dependent on the relative subunit interactions, as can be seen by the ratio of successive intrinsic constants for the tetrahedral model which is simply equal to K_{BB}/K_{AB}^2. Either positive or negative cooperativity can be generated with the restricted KNF model with the

[26] J. E. Haber and D. E. Koshland, Jr., *Proc. Natl. Acad. Sci. U.S.A.* **58,** 2087 (1967).
[27] M. E. Kirtley and D. E. Koshland, Jr., *J. Biol. Chem.* **242,** 4192 (1967).

appropriate relationship of K_{AB} to K_{BB} (Fig. 3). Again, only low concentrations of intermediate liganded forms, $\overline{Y}_1$, $\overline{Y}_2$, and $\overline{Y}_3$ occur in the positively cooperative case when concentrations of intermediate forms are examined. Conversely, in a negatively cooperative case the intermediates accumulate, since full binding of ligand becomes more difficult. For either cooperativity, the conformational saturation, $\overline{R}$, of the restricted KNF model occurs concomitantly with ligand saturation, $\overline{Y}$.

Comparison of Monod–Wyman–Changeux and Koshland–Nemethy–Filmer Models

The concept of symmetry (the concerted transition) in the MWC model is seen by some as a simplifying assumption tantamount to a physical law; others view it as an unnecessary restriction on the behavior of proteins. Some view the KNF model as conceptually difficult, whereas others readily grasp the implicit thermodynamic interactions and linkages. Regardless of the subjective responses to the two models, certain objective evaluations may be made. Putative differences in the two models based on a preexisting equilibrium compared to a ligand-induced fit are no longer considered to be relevant.[1,6] The same thermodynamic model applies for the two concepts with simply different occupation of various molecular populations.[1]

Initially, quite different relationships of the state, $\overline{R}$, and binding, $\overline{Y}$, functions were predicted for the MWC and KNF models. In principle, a plot of $\overline{Y}$ versus $\overline{R}$ should give a straight line with a slope of unity for the restricted KNF model and a curvilinear plot with an initial slope larger than unity for the MWC model (cf. Figs. 2 and 3). Nonexclusive binding in both models and unliganded hybrids in the KNF model, however, diminish the effectiveness of using this distinction to differentiate the models. In fact (as pointed out by several authors[28–30]), the two models are simply extremes of the most general model, depicted here by all forms in Fig. 3, with the restricted KNF being the diagonal and the MWC being the first and fifth columns without any hybrid intermediates.

Three molecular fitting parameters arise in both the nonexclusive MWC model (c, L, and K_R) and the KNF model (K_tK_S, K_{BB}, and K_{AB}). Although for the latter model the number of Adair constants used to initially fit the primary data is equal to n, the relation between constants restricts values

[28] M. Eigen, *Q. Rev. Biophys.* **1,** 3 (1968).

[29] E. Whitehead, *Prog. Biophys. Mol. Biol.* **21,** 321 (1970).

[30] G. G. Hammes and C.-W. Wu, *Science* **172,** 1205 (1971).

that Ψ_i can assume and still be consistent with the model. For example, the sequential tetrameric (square or tetrahedral) KNF model requires $\Psi_1^2\Psi_4 = \Psi_3^2$, whereas the concerted MWC model has the relation among Ψ values shown in Eq. (5) that demonstrates the lack of negative cooperativity since Ψ'_n/Ψ'_{n-1} must always be greater than unity. The relationship of these coefficients in the binding equation can, in principle, be used to discriminate between models[23,31] since the two models predict slightly different relative values (Table I). This experimental comparison requires extremely precise and extensive data and is tantamount to distinguishing between models based on the shape of very similar saturation curves.[31] For this reason, comparison of Ψ ratios, or saturation curves, is not an efficient nor widely used procedure.

$$\frac{\Psi'_n}{\Psi'_{n-1}} = \frac{(1 + Lc^n)}{(1 + Lc^{n-1})K_R} \tag{5}$$

General equations for the effect of activators and inhibitors, nonexclusive binding, the relationship of the KNF or MWC parameters to the intrinsic constants for a dimer, differential catalytic rates in the R and T or A and B forms, differential catalytic rates with two substrates, and the influence of products on differential catalytic rates have previously been presented and discussed for both the MWC and KNF models.[1] Most attempts to determine whether the MWC or KNF model best applies to a specific enzyme have been based on various detailed physical assessments of structural or energetic changes on partial ligand binding, rather than on binding curves or simple state functions. Hemoglobin and ATCase are discussed in detail in the next section to demonstrate how these methods may be used.

Oligomer Association–Dissociation Models

If, on binding ligand, the change in subunit affinity (K_{BB} and K_{AB} in the KNF model) is sufficiently great or if the initial equilibrium involves a dissociated species (R or T in the MWC model), then cooperativity can be directly due to the association or dissociation reaction itself.[32–34] Simply, the ligand binds preferentially either to the associated or the dissociated form. Systems that favor binding to the associated species or ones that favor the dissociated species can equally well occur, but they imply

[31] A. Cornish-Bowden and D. E. Koshland, Jr., *Biochemistry* **9,** 3325 (1970).

[32] L. W. Nichol, W. J. H. Jackson, and D. J. Winzor, *Biochemistry* **6,** 2449 (1967).

[33] C. Frieden, *J. Biol. Chem.* **242,** 4045 (1967).

[34] B. I. Kurganov, *Khim. Tekhnol. Polim.* **11,** 140 (1967).

different molecular mechanisms of protein–ligand interaction.[1] Either exclusive or nonexclusive binding to the two association states can occur. If the substrate favors dissociation of an oligomer, lower enzyme concentrations will enhance activity but will depress cooperativity. The converse relationship also holds; maximum cooperativity is generated at intermediate enzyme concentrations. The equations (presented elsewhere[1,32–35] and extensively discussed[35–37]) are similar in form to those of the MWC or KNF models but with the addition of a term dependent on the protein concentration.

Interestingly, both positive and negative cooperativity can occur with this mechanism either because of the possibility of differing numbers of ligand bound per subunit in the monomer and oligomer state or because of the maintenance of two active forms, monomer and oligomer, at certain protein concentrations. The critical test for this mechanism is the dependence of the homotropic cooperativity (e.g., the n_H) on the concentration; the specific activity of the enzyme will also vary with its concentration. The cooperativity, in most cases, should also occur in equilibrium binding studies, since no kinetic terms are directly involved. Estimates of the K_D for the association reaction can be calculated by fitting the appropriate equations. Similarly, the effects of heterotropic effectors can occur as a shift of the monomer–oligomer equilibrium toward a more active or a less active association state. This type of allosterism has been discussed adequately in other places[1,35] and will not be belabored herein. Numerous enzymes show changes in oligomerization state on substrate binding,[35,38] and several appear to generate cooperativity by this mechanism; muscle 6-phosphofructo-1-kinase is a good example of an enzyme in which the subunit interactions are weakened in the tetrameric state and dissociation to dimers and monomers occurs (see review by Lee *et al.*[39]). Assessing the applicability of this mechanism in physiological situations depends on accurately determining the true concentration of the particular enzyme in its native environment.

Kinetic Considerations of Site–Site Cooperativity Models

Both the MWC and the KNF models of cooperativity are based on thermodynamic considerations and strictly apply only to equilibrium bind-

[35] B. I. Kurganov, "Allosteric Enzymes: Kinetic Behaviour." Wiley, New York, 1982.

[36] L. W. Nichol and D. J. Winzor, *Biochemistry* **15,** 3015 (1976).

[37] A. Levitzki and J. Schlessinger, *Biochemistry* **13,** 5214 (1974).

[38] K. E. Neet, *Bull. Mol. Biol. Med.* **4,** 101 (1979).

[39] J. C. Lee, L. K. Hesterberg, M. A. Luther, and G.-Z. Cai, *in* "Allosteric Enzymes" (G. Herve, ed.), p. 231. CRC Press, Boca Raton, Florida, 1988.

ing measurements. The most rigorous utilization of these models, therefore, has been with such systems as O_2 binding to hemoglobin[40] or with ATCase where substrate binding measurements have been made in parallel with kinetics.[41] Nevertheless, application to kinetic assays is common because of the simplicity of the experiments and the fundamentalism of the models. The assumption is frequently made that rapid equilibrium binding of substrates would reduce the kinetic measurements to the equivalent of thermodynamic equilibrium situations, that is, a simple proportionality exists between substrate binding and reaction rate. However, this simple relation is not generally adequate, nor supported by theoretical considerations.[6] Alternatively, attempts have been made to extend the equilibrium models to incorporate rate constants for each of the E · S complexes in either cooperativity model.[1,42–44] The problems of such simplified considerations have been pointed out by Ricard and co-workers,[6,23] who have developed a general theory of "structural kinetics."

Theory of Structural Kinetics. Structural kinetics[45–48] attempts to apply principles derived from the study of enzyme mechanisms to the effects on subunits in a cooperative, oligomeric enzyme. In other words, the kinetic properties of allosteric enzymes are derived from those of an ideal, isolated subunit utilizing the thermodynamics of structural changes. The catalytic principles[49,50] include transition state complementarity, induced fit, strain, release, and evolution. The fundamental principles of structural kinetics partition the free energy of conversion of substrate to product into three terms. (i) The "intrinsic energy component" is equivalent to the ΔG^* that would occur if an idealized free subunit, that is, with no subunit interactions, were to catalyze the same reaction. (ii) The "protomer arrangement energy" (α) describes the contribution to the free energy of catalysis of the energy of interaction of the subunits on association together with no conformational change of each subunit. (iii) The "quaternary constraint contribution" (σ) accounts for the contribution of intersubunit strain on catalysis owing to a conformational change affecting

[40] G. K. Ackers and J. H. Hazzard, *Trends Biochem. Sci.* **18,** 385 (1993).

[41] J. O. Newell, D. W. Markby, and H. K. Schachman, *J. Biol. Chem.* **264,** 2476 (1989).

[42] H. Paulus and J. K. DeRiel, *J. Mol. Biol.* **97,** 667 (1975).

[43] A. Goldbeter, *Biophys. Chem.* **4,** 159 (1976).

[44] J. Ricard, C. Mouttet, and J. Nari, *Eur. J. Biochem.* **41,** 479 (1974).

[45] J. Ricard and G. Noat, *J. Theor. Biol.* **96,** 347 (1982).

[46] J. Ricard and G. Noat, *J. Theor. Biol.* **111,** 737 (1984).

[47] J. Ricard and G. Noat, *J. Theor. Biol.* **117,** 633 (1985).

[48] J. Ricard and G. Noat, *J. Theor. Biol.* **123,** 431 (1986).

[49] A. R. Fersht, R. J. Leatherbarrow, and T. N. C. Wells, *Trends Biochem. Sci.* **11,** 321 (1986).

[50] W. P. Jencks, *Adv. Enzymol.* **43,** 219 (1975).

"strain" in the catalytic site. The overall free energy of catalysis by the oligomer is simply the sum of these three energy terms, or, correspondingly, the phenomenological rate constant is the product of the intrinsic rate constant of the subunit and the equilibrium constants for subunit association and subunit conformational change. These three energy terms are essentially the same as those of the KNF model[17] to describe quaternary effects on substrate binding, but here applied to catalysis.

Three additional postulates are introduced into structural kinetics that simplify the rate equation. (i) In each catalytic transition state the quaternary constraints are relieved, similar to the relief of intrasubunit strain in theories of catalysis. (ii) The minimum number of conformations is assumed (as in the MWC and KNF binding models), that is, each free subunit can exist in only two conformations, the unliganded or the liganded (with substrate or product). (iii) The conformation of the liganded transition state complex is the same, or energetically indistinguishable, for all transition states. The three postulates effectively take into account the generally accepted belief that maximum catalytic efficiency occurs when the enzyme conformation is complementary to the transition state of the reaction. These reasonable assumptions allow straightforward derivation of rate expressions containing conceptual terms. The full structural rate equation for the dimer is given in Eq. (6):

$$\frac{v}{[\mathrm{E}]_0} = \frac{2\bar{k}^*\sigma_{\mathrm{AA}}\bar{K}^*[\mathrm{S}] + 2\bar{k}^*\sigma_{\mathrm{AA}}\dfrac{\alpha_{\mathrm{AA}}}{\alpha_{\mathrm{AB}}}\bar{K}^{*2}[\mathrm{S}]^2}{1 + 2\dfrac{\sigma_{\mathrm{AA}}}{\sigma_{\mathrm{AB}}}\dfrac{\alpha_{\mathrm{AA}}}{\alpha_{\mathrm{AB}}}\bar{K}^*[\mathrm{S}] + \dfrac{\sigma_{\mathrm{AA}}}{\sigma_{\mathrm{BB}}}\dfrac{\alpha_{\mathrm{AA}}}{\alpha_{\mathrm{BB}}}\bar{K}^{*2}[\mathrm{S}]^2} \tag{6}$$

where α can be considered a dissociation constant for AA or AB; σ is an isomerization constant, and σ^*, k^*, and K^* are the apparent intrinsic isomerization, catalytic, and reciprocal Michaelis constant, respectively, of the isolated subunit. The observed catalytic rate constant for a subunit, k, is equal to $k^*\sigma_{\mathrm{AB}}(\alpha_{\mathrm{AB}}/\alpha_{\mathrm{AA}})$. The constants $1/\alpha_{\mathrm{AA}}$, $\alpha_{\mathrm{AA}}/\alpha_{\mathrm{AB}}$, $\alpha_{\mathrm{AA}}/\alpha_{\mathrm{BB}}$, and σ^* of the Ricard structural kinetic model are analogous to K_{AA}, K_{AB}, K_{BB}, and K_t, respectively, of the KNF model. When Eq. (6) is compared to the KNF formulation [the dimer equivalent of Eq. (1) and Table I], both have the same form but with a different arrangement and relationship between terms.

Two limiting cases of structural kinetics are revealing. If $\sigma_i = 1$, then the system is "loosely" coupled with no intersubunit interactions other than association and no propagation of substrate-induced conformational changes. Cooperativity is determined for a dimer by the term $\alpha^2_{\mathrm{AB}}/\alpha_{\mathrm{AA}}\alpha_{\mathrm{BB}}$ with different dependencies on this term for binding and kinetics (compare the analogous term for the KNF model given above). However, both binding cooperativity and kinetic cooperativity have the same sign (i.e., catalysis enhances any preexisting binding cooperativity). On the other hand, if $\sigma_i \neq 1$ then the dimer is "tightly" coupled. If quaternary constraints are relieved in the unliganded and fully liganded states ($\sigma_{AA} = \sigma_{BB} = 1$), then the conformation of the idealized monomer is the same as those of the oligomer; if the constraints are not relieved, then distinct conformations can occur in the oligomer. Binding cooperativity for the tightly coupled enzyme depends on the term $\alpha_{AB}^2\sigma_{AB}^2/\alpha_{AA}\alpha_{BB}\sigma_{AA}\sigma_{BB}$, but the kinetic cooperativity also depends on the ratio σ_{AA}/σ_{AB}. Therefore, kinetic cooperativity does not have to have the same sign as binding cooperativity, and "inversion" of observed cooperativity may occur. If subunits of the oligomer are so tightly coupled that conformational changes are propagated to the entire molecule, then a concerted molecular change occurs, analogous to the MWC binding model. In the absence of binding cooperativity, kinetic cooperativity for this model can only be positive[6] as was originally reported for kinetic extensions to the MWC model.[42,43] The intermediate cases of $\sigma \neq 1$ are more difficult to generalize. Tight and loose coupling situations for a dimer and for a tetramer with "square" or "tetrahedral" geometry have been analyzed[51,52] and reviewed.[6,23]

Application of the theory to tetrameric chloroplast fructose bisphosphatase has utilized precise rate data to discriminate between models[53]; the tetrahedral structural kinetic model fits the data well, whereas Adair, MWC, or KNF binding models do not. Quaternary constraints are not relieved in the unliganded and fully liganded states (only in the transition states) of fructose bisphosphatase, and free energies associated with individual subunit interaction steps have been calculated.[53]

The reader is referred to two reviews[6,23] for further discussion of the analysis by structural kinetics. The frequent finding that catalytic sites have shared contributions from neighboring subunits (see below) may influence application of the theory of structural kinetics. At a minimum, the concept of the idealized, isolated subunit that can catalyze the reaction with maximum efficiency would not hold. Future detailed interpretations of cooperative, allosteric enzymes will require the extension, refinement, and application of the concepts of structural kinetics to enzymes with known catalytic mechanisms and crystallographic structures.

[51] J. Ricard, M.-T. Giudici-Orticoni, and J. Buc, *Eur. J. Biochem.* **194,** 463 (1990).

[52] M.-T. Giudici-Orticoni, J. Buc, and J. Ricard, *Eur. J. Biochem.* **194,** 475 (1990).

[53] M.-T. Giudici-Orticoni, J. Buc, M. Bidaud, and J. Ricard, *Eur. J. Biochem.* **194,** 483 (1990).

Kinetic Cooperativity and Cooperativity in Monomeric Enzymes

Several models have been proposed for the generation of cooperativity by kinetic mechanisms, rather than by equilibrium effects of site–site interactions discussed in the previous paragraphs. With some enzymes, such as AMP deaminase, a thorough analysis of the kinetic mechanism has shown that the allosteric and cooperative kinetics can be attributed to direct effects on catalytic turnover of individual active sites in the tetramer without invoking binding interactions.[54] Other models based on the type of kinetic mechanism[9,55–57] and on a slow conformational change, or hysteresis,[58] of the enzyme[7,8,59] are more prevelant in the literature and are discussed here. The kinetic cooperativity observed in these cases will have an appearance similar to that arising from other mechanisms and cannot be distinguished simply from appearance of the steady-state velocity plots; however, the curves from kinetic mechanisms do tend to be less symmetrical around the $S_{0.5}$ than binding mechanisms. Kinetic models of these types are dependent on catalytic turnover; therefore, cooperativity will not be observed for the same substrate in equilibrium binding measurements. This marked distinction is the most diagnostic feature of these mechanisms. Single-site mechanisms which purport to generate cooperativity of equilibrium binding invariably neglect thermodynamic constraints, microscopic reversibility, or detailed balance and are, therefore, invalid. Kinetic mechanisms are maintained at nonequilibrium by the conversion of substrate to product. Because interactions between multiple active sites are not required in kinetic mechanisms, cooperativity can be produced in monomeric enzymes[3] or in oligomers with independent sites.[1,2,3] These kinetic mechanisms that generate cooperativity should not be confused with the effect of kinetic constants on cooperativity generated by site binding mechanisms (see above and Refs. 42–44).

Steady-State Random Kinetic Mechanism. When an enzyme catalyzes a reaction with two (or more) substrates (A and B) by a steady-state (nonequilibrium) random mechanism, the potential exists for nonMichaelian concentration dependence.[9,29,55,56] Second-order terms for each substrate occur in the general rate equation.[24] No cooperativity will occur if the kinetic mechanism is rapid equilibrium, which reduces to an equilibrium

[54] D. J. Merkler and V. L. Schramm, *J. Biol. Chem.* **265,** 4420 (1990).
[55] B. D. Wells, T. A. Stewart, and J. R. Fisher, *J. Theor. Biol.* **60,** 209 (1976).
[56] G. Pettersson, *Biochem. J.* **233,** 347 (1986).
[57] G. Pettersson, *Eur. J. Biochem.* **154,** 167 (1986).
[58] C. Frieden, *J. Biol. Chem.* **245,** 5788 (1970).
[59] J. Ricard, J. C. Meunier, and J. Buc, *Eur. J. Biochem.* **80,** 581 (1977).

binding situation, or if the reaction is ordered with one substrate binding exclusively before the other(s). Kinetic cooperativity may occur in the true steady-state random mechanism because the relationship among the individual rate constants determines the flux through each alternate pathway. At low substrate concentrations the reaction may proceed through one limb (e.g. A binds before B), and at higher substrate concentrations the other limb (i.e., B binds before A) may predominate. This shift in flux can produce nonhyperbolic kinetics of either the positively or negatively cooperative type. The degree of cooperativity of each substrate will depend on the concentration of the other substrate and will diminish either at very low or at very high concentrations of the fixed substrate, since the reaction flux will then proceed through a single pathway. Simulated analysis has indicated[55] that "positive cooperativity" of one substrate will be associated only with noncooperative or substrate inhibition kinetics of the other substrate, whereas "negative cooperativity" (sometimes called substrate activation) with both substrates can occur. Not all steady-state random mechanisms will generate this type of cooperativity because a special relationship must exist among the various rate constants. The requirements and details of this mechanism have been discussed.[1,3,55,56]

The steady-state random mechanism probably generates deviations from Michealian behavior with many enzymes.[60] The extent of the deviation will frequently be low,[61,62] that is, producing Hill coefficients that are not far from unity. No clear examples have been reported of significant cooperativity from this mechanism that contributes to physiological behavior of an enzyme, although several possibilities have been suggested[56,57,63–65] and in some cases contested.[66–68] Nevertheless, if cooperative enzyme kinetics are observed with a multisubstrate reaction, the first consideration should be that of a steady-state random mechanism until the exact kinetic mechanism is worked out. In a closely related situation, relevant substrate inhibition of *Escherichia coli* phosphofructokinase has been demonstrated to be due to a steady-state random mechanism by

[60] C. M. Hill, R. D. Waight, and G. Bardsley, *Mol. Cell. Biochem.* **15,** 173 (1977).

[61] J. S. Gulbinsky and W. W. Cleland, *Biochemistry* **7,** 566 (1968).

[62] W. W. Cleland, *in* "The Enzymes" (P. D. Boyer, ed.), Vol. 2, 3rd ed., p. 1. Academic Press, New York, 1970.

[63] A. Cornish-Bowden and J. T. Wong, *Biochem. J.* **175,** 969 (1978).

[64] K. M. Ivanetich, R. D. Goold, and C. N. Sikakana, *Biochem. Pharmacol.* **39,** 1999 (1990).

[65] P. Trost and P. Pupillo, *Arch. Biochem. Biophys.* **306,** 76 (1993).

[66] A. Cornish-Bowden and A. C. Storer, *Biochem. J.* **240,** 293 (1986).

[67] J. Ricard, J.-M. Soulie, and M. Bidaud, *Eur. J. Biochem.* **159,** 247 (1986).

[68] J. Ricard and G. Noat, *Eur. J. Biochem.* **152,** 557 (1985).

construction of a mutant that converts the enzyme to rapid equilibrium kinetics and eliminates MgATP inhibition.[69]

Ligand-Induced Slow Transition (Hysteretic, Mnemonic) Mechanism. Enzyme hysteresis,[58] in the form of a substrate-induced slow conformational change, slow association–dissociation, or slow binding, has been observed on a standard assay time scale of minutes with many enzymes.[1,7,35,58,70] These slow transitions can have the form of either bursts or lags, depending on whether the activity decreases or increases with time, respectively. The assay transient can be characterized by the initial velocity, v_i, the steady-state velocity, v_{ss}, and the relaxation time for the transient, τ [Eq. (7)]:

$$P = v_{ss}t - (v_{ss} - v_i)(1 - e^{-t/\tau})\tau \tag{7}$$

where P is the product concentration at time t. The hysteretic or slow transition can be readily observable in standard assays if τ is on the order of seconds to minutes; rapid mixing techniques may be required if τ is between 10 msec and 1 sec.

Of vital importance when working with enzymes that show a hysteretic response is to eliminate possible artifacts. All enzymes will show nonlinear progress curves if substrate depletion or production inhibition occurs. Calculation of final substrate and product concentrations, measurement of product inhibition constants, readdition of substrate to the steady-state portion of the assay, and addition of product in the initial assay are effective means of checking for such possibilities. Other simple considerations such as temperature change, dilution, or inactivation of the enzyme can also lead to nonlinearity and should be evaluated. Last, the utilization of different assay systems, different preincubation conditions, and alternate substrates can help validate the presence of true enzymatic hysteresis.

If the slow transition is on the same order of magnitude (milliseconds) as the substrate binding and the catalytic cycle, then cooperativity of v_{ss} can be generated,[7,8,29,71] provided that the appropriate relationship among all rate constants exists.[2,7,8] Either bursts or lags can be associated with either positive or negative cooperativity in v_{ss}. This type of cooperativity mechanism has been termed a ligand-induced slow transition[7] (LIST), a mnemonic mechanism,[8,59] or, more recently, rate-dependent recycling of free enzyme conformational states.[72] These concepts emphasize the relatively slow isomerization between two enzyme forms under nonequilib-

[69] R.-L. Zheng and R. G. Kemp, *J. Biol. Chem.* **267,** 23640 (1992).

[70] C. Frieden, *Annu. Rev. Biochem.* **48,** 471 (1979).

[71] B. R. Rabin, *Biochem. J.* **102,** 22c (1967).

[72] I. A. Rose, J. V. B. Warms, and R. G. Yuan, *Biochemistry* **32,** 8504 (1993).

rium conditions. These mechanisms are sometimes considered to involve "temporal" cooperativity in contrast to the "spatial" cooperativity of site–site mechanisms. Thus, a hysteretic response in an enzyme can give rise to kinetic cooperativity in a single-site, monomeric enzyme.[3,7] Alternatively, the inherent cooperativity of a site–site mechanism of an oligomeric enzyme can be altered because of the additional power of substrate added to the rate equation.

In the LIST mechanism, the initial steady-state velocity, v_i, can be measured after the pre-steady-state adjustments of substrate binding (this volume, [1]) occur but before the slow enzyme isomerization. Multiple turnovers occur during v_i (i.e., it is not a stoichiometric burst). The v_i, extrapolated to zero time, should either be hyperbolic, if the enzyme starts in a single form, or give the appearance of negative cooperativity (owing to heterogeneity), if the enzyme initially exists in two kinetically significant forms; in other words, no true temporal cooperative interactions exist in that "frozen" initial form. The rate of the transient, τ^{-1}, will be a second-order term in substrate concentration with complex combinations of rate constants[7,58] and, in principle, can be used to help evaluate parameters of the mechanism. However, unrealistically good data are required to determine the concentration dependency of τ with enough accuracy to define the parameters well. Graphical methods can be used to assess the v_i, v_{ss}, and τ,[2,58] but a more rigorous procedure is to utilize a nonlinear fitting algorithm to Eq. (7) and obtain unbiased estimates of the parameters with associated standard errors.[73]

The cooperativity from the LIST mechanism occurs in v_{ss} after a linear steady state is reached. A second-order rate equation for the steady-state velocity is generated because of substrate addition steps to two distinct enzyme forms (Fig. 4). Conceptually, cooperativity occurs because the E conformer (Fig. 4, front upper left) can partition between formation of EA ($v = k_1$[A]) or isomerization to E′ ($v = k_{t1}$). At low [A], the isomerization will dominate, and v_{ss} will be mainly due to the cycle favored by the $E \rightleftharpoons E'$ distribution. At high [A], E or E′ will be trapped in the form from which product is released and that same cycle will predominate. Either positive or negative cooperativity can occur with either substrate. The cooperativity of v_{ss} for substrate A will depend on the concentration of substrate B, since the kinetic partitioning of EA (Fig. 4, front upper right) between EAB ($v = k_2$[B]) and E′A ($v = k_{t2}$) depends on the second substrate. At low [B], $EA \rightleftharpoons E'A$ should be nearly able to equilibrate and thus no cooperativity will be found; at high [B], EA or E′A will tend to form EAB or E′AB rapidly, and cooperativity of A in v_{ss} can occur. Thus,

[73] K. E. Neet, R. P. Keenan, and P. S. Tippett, *Biochemistry* **29,** 770 (1990).

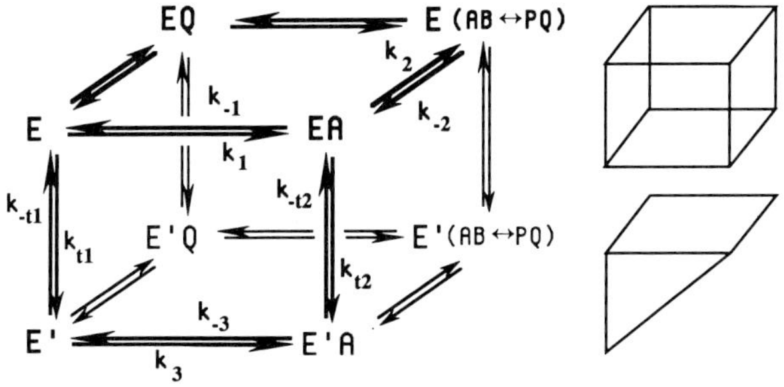

FIG. 4. Ligand-induced slow transition (LIST) model for a two-substrate monomeric enzyme with ordered addition of substrates. E and E′ represent two conformations of the enzyme with the vertical steps (k_{t1}, k_{t2}) being the isomerization. The heavy arrows and bold forms represent the mnemonic mechanism contained within the LIST model with one (upper) catalytic cycle and $k_{t2} \gg k_3[A]$ so that E′ + A → A on the diagonal and the E′A intermediate is not kinetically significant. The small diagrams at right distinguish the full LIST (upper) and the mnemonic (lower) pathways. For glucokinase: A, glucose; B, MgATP; P, MgADP; and Q, glucose 6-phosphate.

the three conditions for the generation of v_{ss} cooperativity are as follows: (a) the rate constants for isomerization must be of the same magnitude as k_{cat}, (b) substrate binding cannot be at rapid equilibrium (i.e., k_{cat} cannot be slow), and (c) the flux through the two catalytic cycles must change with changing substrate concentration. Procedures to identify the key catalytic pathways in the mechanism and the requirements among the rate constants that contribute to the latter condition have been discussed.[2,74] Similar considerations for a single-substrate enzyme have been made.[2,7]

The mnemonic mechanism[8,59] emphasizes the "memory" that the enzyme conformation retains of the substrate after product conversion and release. This mechanism can be considered a special case of the LIST mechanism[2,3,23]; an even more general mechanism with six explicit conformations has been presented.[75] The mnemonic simplification allows relatively straightforward predictions in many cases. The mnemonic mechanism for a two-substrate enzyme can be described by the heavy lines in Fig. 4 and the condition that A binding to E′ leads directly to EA (no E′A). Cooperativity is due to the lack of relaxation of E (less stable) back to E′ (more stable) at higher concentrations of A (i.e., a slow isomerization transition). At low [A] the pathway proceeds primarily through E′ + A → EA → E (AB ⇌ PQ), whereas at high concentration the first substrate

[74] E. P. Whitehead, *Biochem. J.* **159,** 449 (1976).
[75] J. Ricard, *Biochem. J.* **175,** 779 (1978).

binds before the isomerization can occur and the pathway is predominantly through $E + A \rightarrow EA \rightarrow E(AB \rightleftharpoons PQ)$. Either negative or positive cooperativity of A (as measured by the Hill coefficient) is enhanced by increasing concentrations of B.

A major difference between the two kinetic mechanisms is that only one catalytic cycle occurs in the mnemonic mechanism whereas the LIST mechanism has two catalytic cycles (i.e., both E and E′ are catalytic active). For a single-substrate enzyme the two mechanisms are virtually indistinguishable. With two substrates, ordered binding is implied in the mnemonic mechanism, and, therefore, cooperativity can only be generated with the first substrate. The LIST mechanism can produce cooperativity with either substrate and can accommodate random binding of two (or more) substrates. Product inhibition can help discriminate between the mechanisms. The second product released in the mnemonic (ordered, one catalytic cycle) mechanism is predicted to increase negative cooperativity or decrease positive cooperativity of the first substrate while showing linear slope inhibition.[76] In the LIST mechanism predictions of product inhibition effects on cooperativity cannot be made, but product slope inhibition can itself be cooperative (or nonlinear) as in a simple Ordered Bi Bi mechanism.[2] A complete set of rate constants has not been experimentally obtained for either the LIST or the mnemonic mechanisms for any enzyme, primarily owing to the inability to make adequate kinetic measurements of each individual step.

The mnemonic mechanism has been extensively applied to the interpretation of the negatively cooperative kinetics of glucose phosphorylation found with wheat germ hexokinase L_I, a monomeric enzyme. The predictions of initial velocity kinetics, product inhibition, and effects of substrates on conformation of the enzyme are consistent with the mnemonic mechanism.[59,76–78] Other mechanisms for kinetic cooperativity of wheat germ hexokinase have been proposed[57] and rebutted.[23,68]

Relatively few monomeric enzymes have been shown to have cooperativity,[3] and few oligomeric enzymes have convincingly been proved to utilize a LIST or mnemonic mechanism as part of their cooperativity. Newer approaches analyzing reaction flux by the isotope counterflow method may prove useful in establishing the presence of alternate conformational pathways.[72,79] Cooperativity and slow transitions in enzymes that

[76] J. C. Meunier, J. Buc, A. Navarro, and J. Ricard, *Eur. J. Biochem.* **49,** 209 (1974).
[77] J. Buc, J. Ricard, and J. C. Meunier, *Eur. J. Biochem.* **80,** 593 (1977).
[78] J. C. Meunier, J. Buc, and J. Ricard, *Eur. J. Biochem.* **97,** 573 (1979).
[79] M. Gregoriou, I. P. Trayer, and A. Cornish-Bowden, *Biochemistry* **20,** 499 (1981).

have been reasonably well characterized include rat liver glucokinase[80] (discussed in detail below), wheat germ hexokinase,[76] octopine dehydrogenase,[81] and fumarase (fumarate hydratase).[72] Furthermore, although many enzymes have hysteretic properties,[2,7,35,58,70] the contribution of the hysteresis, itself, to a physiological function (e.g., see glucose-6-phosphatase[82]) has yet to be proved in most cases. Thus, the concepts of hysteresis and kinetic cooperativity are currently an intriguing aspect of enzyme kinetics and mechanism, but their importance, frequency of occurrence, and contribution to biological function remain speculative, for the most part. Emerging techniques for the future (see below) may help in such evaluations.

Structural Basis of Slow Conformational Changes. Since the proposal in 1980 of four structural types of conformational changes that could be slow enough to account for hysteretic mechanisms,[1] little progress has been made in proving or refining these suggestions. Possible three-dimensional changes proposed were cis–trans isomerization of proline residues, cleft closure by domain movement, segmental movement of polypeptide chains, and sliding of α helices past or around are another. Each has the potential for a high activation energy. The crystallographic description of more complex and intricate conformational changes provide additional possibilities, but little can be said at the moment about the kinetics of such slow changes.

A slow cleft closure with glucokinase is highly probable, based on the glucose-induced domain movement of the homologous yeast hexokinase,[83] but no direct data have yet been presented. The hysteresis of *E. coli* dihydrofolate reductase[84] may also be due to slow domain movement, but attempts to modify the $NADP^+$/dihydrofolate-dependent slow isomerization by mutagenic substitution of several amino acid residues in the hinge region have not yet been successful.[85] Proline isomerization certainly is involved in slow steps in protein folding,[86] but no ligand-induced isomerization of proline has been documented. Standard X-ray crystallographic methods used to define the structural basis of allosteric transitions are not capable of direct extension to time-dependent changes. Laue diffraction

[80] D. Pollard-Knight and A. Cornish-Bowden, *Mol. Cell. Biochem.* **44,** 71 (1982).

[81] M. O. Monneuse-Doublet, A. Olomucki, and J. Buc, *Eur. J. Biochem.* **84,** 441 (1978).

[82] K. L. Nelson-Rossow, K. A. Sukalski, and R. C. Nordlie, *Biochim. Biophys. Acta* **1163,** 297 (1993).

[83] W. S. Bennett and T. A. Steitz, *Proc. Natl. Acad. Sci. U.S.A.* **75,** 4848 (1978).

[84] M. H. Penner and C. Frieden, *J. Biol. Chem.* **260,** 5366 (1985).

[85] P. M. Ahrweiler and C. Frieden, *Biochemistry* **30,** 7801 (1991).

[86] L. C. Wood, T. B. White, L. Ramdas, and B. T. Nall, *Biochemistry* **27,** 8562 (1988).

methods, now being used to study time-dependent changes during catalysis,[87,88] could be utilized in future studies to describe slow (minutes) conformational changes of allosteric, hysteretic enzymes at high resolution. Solution nuclear magnetic resonance (NMR) methods, because of the time required for data collection subsequent to ligand addition, will probably remain too slow for the study of ligand-induced slow conformational changes.

Experimental Evaluation of Models

Concerted versus Sequential Models: Classic Case of Oxygen Binding to Hemoglobin

Hemoglobin is the prototypical cooperative allosteric protein that has been studied with numerous techniques since the R- and T-states were first described by crystallography.[89,90] A simple concerted transition between the two states can be explained in detail by the triggering of the Fe^{2+} movement in the heme ring being transmitted over 25 to 40 Å to the other hemes by shifts in the tertiary and quaternary structure.[90] The $\alpha_1\beta_1$ dimer rotates about 15° relative to the other dimer with changes in the ionic and hydrogen bonds between the FG bend (residues 94–102) and the C helix (residues 40–45) at the $\alpha_1\beta_2$ and $\alpha_2\beta_1$ contacts.

More recent data, however, indicate that the simple concerted (MWC) model is not sufficient to explain completely the energetics of the reaction and that intermediate microstates occur. Precise O_2 binding data at protein concentrations in which the hemoglobin heterotetramer dissociates to $\alpha\beta$ dimers and measurement of subunit affinities by gel filtration from Ackers laboratory[40,91] have allowed calculation of the free energy for individual oxygen-binding steps. Other studies with nonlabile ligands, allosteric modifiers, and hemoglobin mutants have provided further details of the subunit interactions during ligand binding. The clearest scenario, at least with cyanide binding, is that a symmetry switch occurs in each $\alpha\beta$ dimer when one of the subunits is ligated.[40] The cooperative free energy (i.e., a measure of the free energy altering dimer–dimer interactions on ligation) is propagated to the other subunit of the $\alpha\beta$ dimeric half-tetramer as about 3

[87] B. L. Stoddard, P. Koenigs, N. Porter, K. Petratos, G. A. Petsko, and D. Ringe, *Proc. Natl. Acad. Sci. U.S.A.* **88,** 5503 (1991).

[88] L. N. Johnson, *Protein. Sci.* **1,** 1237 (1992).

[89] M. F. Perutz, *Nature (London)* **228,** 726 (1970).

[90] M. F. Perutz, *Q. Rev. Biophys.* **22,** 139 (1989).

[91] G. K. Ackers, M. L. Doyle, D. Myers, and M. A. Daugherty, *Science* **255,** 54 (1992).

kcal/mol in two steps per tetramer (Fig. 5). The full T–R transition of the quaternary structure of the tetramer only occurs when at least one subunit of each half-tetramer is ligated.

A careful study of the energetics of 60 mutant or modified hemoglobins

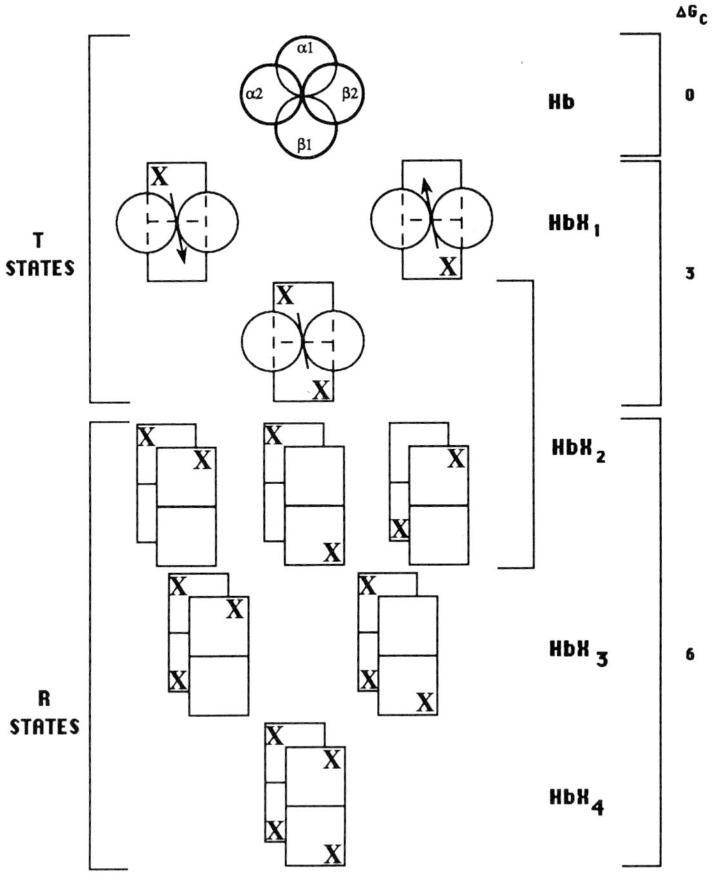

FIG. 5. Model for ligand binding to the substates of cyanomethemoglobin. The 10 possible ligation substates of hemoglobin are shown with the $(\alpha_1\beta_1)(\alpha_2\beta_2)$ composition of the dimer of dimers indicated in the top, unligated line. Oxygenation is shown increasing vertically downward, with the stoichiometry indicated. When ligand binds to either α or β subunit, the effect is transmitted across the dimer interface, indicated by the arrow, and changes the conformation of the dimer from circle (T-state) to square (R-state). The T (four substates) and R (six substates) macroquaternary states are bracketed. The cooperative free energy, ΔG_c, at each stage is indicated at right. Quarternary switching from the T- to the R-state, indicated by rotation of the $\alpha_2\beta_2$ dimer relative to the $\alpha_1\beta_1$ dimer, occurs when at least one subunit of each $\alpha\beta$ dimer is ligated. (Adapted from Ackers *et al.*[91] and Ackers and Hazzard.[40])

confirmed that the $\alpha_1\beta_2$ interface was uniquely coupled to the cooperativity.[92] Detailed proton NMR studies support the conclusion that intermediate states exist between R and T (see review by Ho[93]). The resulting mechanism is neither purely MWC nor sequential (KNF) but encompasses aspects of each within the dimer of dimer context. This conclusion is reminiscent of early pseudosymmetrical dimer models for hemoglobin and enzymes.[94,95] The description of this mechanism for hemoglobin is facilitated by utilization of the definition of molecular constants proposed for the KNF model. Furthermore, the microstate model may also explain macroscopic deviations from the MWC model (e.g., a 10-fold difference in K_T in the presence of the effector 2,3-bisphosphoglycerate[90,96]). However, a similar detailed analysis of the heterotropic effects on hemoglobin is needed to test the model further.

Concerted versus Sequential Models: Case of Aspartate Transcarbamylase

Aspartate transcarbamylase (ATCase; EC 2.1.3.2, aspartate carbamolytransferase) from *E. coli* is the most extensively studied allosteric enzyme. Many different experimental approaches have been utilized over the years to test the theoretical models of cooperativity with ATCase: How much concerted symmetry is retained in the molecular allosteric mechanism? This section briefly reviews the molecular structure of the enzyme and the allosteric transitions involved, but it mainly focuses on methods that have been utilized to understand the conformational changes and provides a critical view of the proposed models as applied to ATCase. The enzyme plays a central role in the regulation of the pyrimidine pathway in bacteria; its structure and allosteric regulation are described in many introductory biochemistry texts. More extensive reviews and interpretations in terms of allosteric models and structure have appeared.[96–100] Determinations of the crystal structure of the T-state, the T-state with CTP bound, the R-state with *N*-phosphonacetyl-L-aspartate (PALA) bound,

[92] G. J. Turner, F. Galacteros, M. L. Doyle, B. Hedlund, D. W. Pettigrew, B. W. Turner, F. R. Smith, W. Moo-Penn, D. L. Rucknagel, and G. K. Ackers, *Proteins* **14,** 333 (1992).
[93] C. Ho, *Adv. Protein Chem.* **43,** 153 (1992).
[94] G. Guidotti, *J. Biol. Chem.* **242,** 3704 (1967).
[95] O. M. Viratelle and F. J. Seydoux, *J. Mol. Biol.* **92,** 193 (1975).
[96] L. N. Johnson, *in* "Receptor Subunits and Complexes." (A. Burgen and E. A. Barnard, eds.), p. 39. Cambridge Univ. Press, Cambridge, 1992.
[97] H. K. Schachman, *J. Biol. Chem.* **263,** 18583 (1988).
[98] E. R. Kantrowitz and W. N. Lipscomb, *Science* **241,** 669 (1988).
[99] E. R. Kantrowitz and W. N. Lipscomb, *Trends Biochem. Sci.* **15,** 53 (1990).
[100] N. M. Allewell, *Annu. Rev. Biophys. Biophys. Chem.* **18,** 71 (1989).

and the R-state with phosphonoacetamide plus malonate bound[101–104] have been extremely useful in interpreting kinetic and mutational studies.

When ATCase is fit to the MWC model,[102] the equilibrium constant (L_0) between the R- and T-states is 250 (or an energetic difference of 3.3 kcal/mol); this value is about 2.5 kcal/mol in the presence of ATP and 4.2 kcal/mol with CTP. The ratio for substrate dissociation constants in the two states, *c*, for PALA is 0.03 and that for Asp is 0.05. The Hill coefficient is 1.8–2.0 in the absence of effectors, 2.3 in the presence of CTP, and 1.4 in the presence of ATP. No significant kinetic contribution occurs, since similar values have been measured both by equilibrium binding studies with PALA in the absence of carbamyl phosphate[41] (i.e., a purely thermodynamic situation) and by the kinetics of catalysis of the carbamoylation of aspartate.[105]

Structural Model. The enzyme ATCase (310 kDa) exists as a dimer of catalytic trimers (3 × 33 kDa) that are held together by three dimeric (2 × 17 kDa) regulatory subunits. The dihedral molecular symmetry axes of the heterododecamer allow description of interactions among the catalytic polypeptides (C1–C2–C3, C4–C5–C6), between the regulatory peptides (R1–R2, etc.), and between the regulatory and catalytic chains (C1–R1, etc.).

Significant changes in the tertiary structure of the subunits occur on binding the bisubstrate analog PALA and the consequent transition from the T- to the R-state. The aspartate moiety of the substrate binds to Arg-167, Arg-229, and Gln-231 in the aspartate domain and to Lys-84′ (from the neighboring C-polypeptide chain in the trimer), changing several intrachain ionic interactions and triggering a 2 Å closure (residues 50–55) between the Asp domain and the carbamoyl phosphate domain with a 5 Å movement of the 70–75 sequence culminating in a 10 Å movement of Lys-84′. The movement of Arg-229 displaces the sequences around 230 to 245 (the 240s loop) by 8 Å and disrupts interaction with the other catalytic trimer (the C1–C4 interface) by eliminating the E239(C1)–Y165,K164(C4) bonding that only occurs in the T-state. Major changes in quaternary structure result. A screw movement occurs between the two catalytic trimers as a result of a "differential gear" effect[90] between R- and C-subunit interactions (Fig. 6). This movement results in a 12 Å separation

[101] H. M. Ke, R. B. Honzatko, and W. N. Lipscomb, *Proc. Natl. Acad. Sci. U.S.A.* **81,** 4037 (1984).

[102] K. L. Krause, K. W. Volz, and W. N. Lipscomb, *J. Mol. Biol.* **193,** 527 (1987).

[103] R. C. Stevens, J. E. Gouaux, and W. N. Lipscomb *Biochemistry* **29,** 7691 (1990).

[104] J. E. Gouaux and W. N. Lipscomb, *Biochemistry* **29,** 389 (1990).

[105] G. J. Howlett, M. N. Blackburn, J. G. Compton, and H. K. Schachman, *Biochemistry* **16,** 5091 (1977).

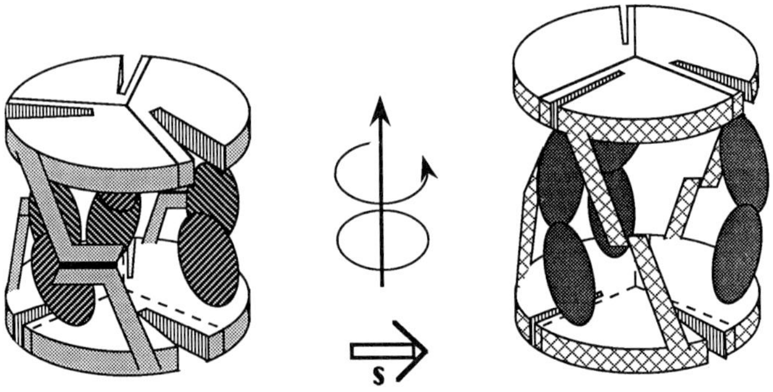

FIG. 6. Model of the allosteric cooperative transition of ATCase as determined from crystal structures. The two disks represent the two catalytic subunits containing three polypeptide chains each. The darker ellipsoids represent the regulatory dimers. The interaction of the 240s loop between the two catalytic subunits is shown by the vertical extensions of the cross-hatched or shaded volume with a more extensive interaction in the T-state (heavy bar). The concerted transition from the T- to the R-state causes an expansion of 15 Å along the vertical axis and a rotation of 15° of the upper catalytic subunit with respect to the lower; the resultant conformational change closes the active site cleft to promote interactions with the substrate in the R-state. (Adapted from Kantrowitz and Lipscomb.[99])

along the 3-fold axis and a net rotation of the two C-trimers of about 10° relative to one another. The C1–R1–R6–C6 (and symmetry related) contacts are maintained as the R1–R6 dimer rotates about 15° around the local 2-fold symmetry axis. The T-state has C1–R1, C1–R4, and C1–C4 (and corresponding symmetry related) interactions intact, whereas the R-state has only C1–R1 interactions. Thus, the cleft closure around the substrate displaces an adjacent (240s) loop, weakens the C1–C4 interaction, releases the "inhibition" by the regulatory subunits (C1–R4), and allows the other C-chains to relax into the higher affinity (Asp or PALA), closed domain conformation.

Hydrodynamic and structural studies in solution have proved sensitive and useful in support of the gross structural changes seen in the crystallographic studies. The R-state has been found to be more swollen or larger than the T-state by sedimentation velocity,[105] low-angle X-ray scattering,[106,107] and analytical gel-exclusion chromatography.[108] Changes on the order of 3–6% in molecular size have been observed subsequent to sub-

[106] M. F. Moody, P. Vachette, and A. M. Foote, *J. Mol. Biol.* **133,** 517 (1979).

[107] G. Hervë, M. F. Moody, P. Tauc, P. Vachette, and P. T. Jones, *J. Mol. Biol.* **185,** 189 (1985).

[108] S. Bromberg, D. S. Burz, and N. M. Allewell, *J. Biochem. Biophys. Methods* **20,** 143 (1990).

strate binding. Such methods should be useful with other enzymes that undergo relatively large quaternary structural changes. Direct comparisons of $\bar{R}$ to $\bar{Y}$ have not been completely satisfactory for distinguishing allosteric mechanisms for ATCase.[1,109,110]

Homotropic Cooperativity. Not only is the MWC model useful for interpreting kinetic data for ATCase, but several lines of evidence strongly support the details of the concerted mechanism for the positive cooperativity seen with aspartate or PALA. The binding of one molecule of PALA can produce about five new R-state catalytic sites in the hexamer under the correct conditions, as measured by arsenolysis of carbamoyl aspartate.[111] This observation is an elegant and straightforward demonstration of one of the tenets of the MWC model. Spectroscopic studies of hybrid molecules containing nitrated catalytic polypeptides showed that unliganded chains change conformation,[112] consistent with the symmetry requisite of the MWC model. Boundary spreading of sedimentation velocity patterns detected no intermediate forms between the different hydrodynamic sizes of the R- and T-states in the presence of a 1.5 molar ratio of PALA to ATCase.[113,114] The changes in sedimentation velocity of the wild-type[105] and mutant[115] enzymes parallel the activity changes as judged by the correspondence of the calculated [R] to [T] ratio. Linkage between heterotropic and homotropic ligands have been observed in many instances, for example, with mutations* such as cE239Q,[116] rN111A,[117] rK143A,[115] cD236A,[118] or cD162A[119] that effect both types of cooperativity (but see exceptions below), suggesting that the same equilibria are fundamental to both processes. These results are consistent in that cD236 and rK143 interact across the R1–C4 interface in the T-state.[104] The double mutant cK164E/E239K (and cK164E only) that lost both heterotropic and homotropic effects was shown to have sedimentation properties consistent with the R-state.[120]

Finally, the molecular model discussed above is so dependent on qua-

[109] M. N. Blackburn and H. K. Schachman, *J. Biol. Chem.* **16,** 5084 (1977).
[110] D. K. McClintock and G. Markus, *J. Biol. Chem.* **244,** 36 (1969).
[111] J. Foote and H. K. Schachman, *J. Mol. Biol.* **186,** 175 (1985).
[112] Y. R. Yang and H. K. Schachman, *Proc. Natl. Acad. Sci. U.S.A.* **77,** 6187 (1980).
[113] W. E. Werner, J. R. Cann, and H. K. Schachman, *J. Mol. Biol.* **206,** 231 (1989).
[114] W. E. Werner and H. K. Schachman, *J. Mol. Biol.* **206,** 221 (1989).
[115] E. Eisenstein, D. W. Markby, and H. K. Schachman, *Biochemistry* **29,** 3724 (1990).
[116] M. M. Ladjimi and E. R. Kantrowitz, *Biochemistry* **27,** 276 (1988).
[117] E. Eisenstein, D. W. Markby, and H. K. Schachman, *Proc. Natl. Acad. Sci. U.S.A.* **86,** 3094 (1989).
[118] C. J. Newton and E. R. Kantrowitz, *Proc. Natl. Acad. Sci. U.S.A.* **87,** 2309 (1990).
[119] C. J. Newton, R. C. Stevens, and E. R. Kantrowitz, *Biochemistry* **31,** 3026 (1992).
[120] J. O. Newell and H. K. Schachman, *Biophys. Chem.* **37,** 183 (1990).
* The letter c or r denotes a mutant in the catalytic or regulatory chains, respectively.

ternary structural changes via the screw shift of the catalytic trimers through the three analogous C1–R1–R6–C6 connections that sequential conformational changes with part T-like and part R-like hybrid states is almost inconceivable; the reorientation of subunits with respect to one another precludes nonsymmetrical intermediates. A sequential mechanism using the ATCase R and T structures would be like trying to sit on a tripod stool with one leg 20% longer than the other two. Thus, structural and kinetic data are in good agreement with respect to the most appropriate allosteric model for homotropic cooperativity in ATCase.

Heterotropic Cooperativity. The situation is less clear with the heterotropic effectors, which bind at a distance of some 60 Å from the active site, partly because the structural changes have only relatively recently been described at high resolution. The uncoupling of heterotropic effects from homotropic effects (i.e., nucleotide activation or inhibition with no aspartate homotropic cooperativity) has been taken as evidence that the simplest MWC model does not strictly hold. Several situations have been found in which this linkage is lost, namely, with the alternate substrates L-alanosine[121] and cysteine sulfinate,[122] the mutants rY77F,[123] cK83Q,[124] cE50Q,[125] and C-terminal deletion of two residues of the regulatory chain (pAR5-related mutations).[126,127] However, these uncoupling observations must be interpreted cautiously because of the different ways in which homotropic cooperativity can be lost. If sigmoid binding with a mutant is lost because $L_0 \approx 1$, the potential still exists for an effector-induced transition to a full R- or T-state with activation or inhibition. For the alternate substrates, Schachman[97] has used an extended MWC model with $c = 1$ and $V_R > V_T$ to explain the apparent uncoupling with alternate substrates.

Evidence for a global conformational change produced by ATP or CTP was obtained by sedimentation velocity analysis of the rK143A mutant that had nearly equal proportions of R- and T-state enzyme ($L_0 = 2.7$),[115] consistent with the effectors simply shifting the R–T equilibrium. Detailed catalytic mechanism studies, including ^{13}C kinetic isotope effects, have

[121] J. Baillon, P. Tauc, and G. Hervë, *Biochemistry* **24,** 7182 (1985).

[122] J. Foote, A. M. Lauritzen, and W. N. Lipscomb, *J. Biol. Chem.* **260,** 9624 (1985).

[123] F. Van Vliet, X. G. Xi, C. de Staercke, B. de Wannemaeker, A. Jacobs, J. Cherfils, M. M. Ladjimi, G. Hervë, and R. Cunin, *Proc. Natl. Acad. Sci. U.S.A.* **88,** 9180 (1991).

[124] E. A. Robey, S. R. Wente, D. W. Markby, A. Flint, Y. R. Yang, and H. K. Schachman, *Proc. Natl. Acad. Sci. U.S.A.* **83,** 5934 (1986).

[125] M. M. Ladjimi, S. A. Middleton, K. S. Kelleher, and E. R. Kantrowitz, *Biochemistry* **27,** 268 (1988).

[126] X. G. Xi, F. Van Vliet, M. M. Ladjimi, B. de Wannemaeker, C. de Staercke, N. Glansdorff, A. Piërard, R. Cunin, and G. Hervë, *J. Mol. Biol.* **220,** 789 (1991).

[127] X. G. Xi, F. Van Vliet, M. M. Ladjimi, B. de Wannemaeker, C. de Staercke, A. Piërard, N. Glansdorff, G. Hervë, and R. Cunin, *J. Mol. Biol.* **216,** 375 (1990).

supported the conclusion that only one active form—the R-state—is present in the presence of nucleotides.[128] On the other hand, crystallographic studies of wild-type ATCase suggest that the C-trimers move by only 0.4 Å (CTP) or 0.5 Å (ATP) along the 3-fold axis in the presence of nucleotide,[103] at least in the crystal environment. An intriguing suggestion has been made,[129] based on electrostatic calculations, that chains of ionizable groups link subunit interfaces and binding sites, thus providing a means of transmission of allosteric information independent of conformational changes per se. If this were the case, the circle and squares of Fig. 2 would represent "states" of the protein and not actual conformations.

The inhibitor CTP (as well as the substrate carbamoyl phosphate[5]) has been shown to have "negative cooperativity" of binding,[130,131] and asymmetry is present in CTP-ligated ATCase crystals.[132] Although solution studies cannot distinguish (see above) between preexisting heterogeneity of the structure (inherent) and negative cooperativity (induced), a high-resolution (2.5 Å) structure that enabled visualization of the N terminus of the R-chains has demonstrated, within the limits of the atomic structural analysis, that part if not all of the CTP binding heterogeneity is due to negatively cooperative effects propagated through the R1–R6 interface to the Lys-60 that directly interacts with CTP.[133] The nucleotide perturbation model for heterotropic interactions proposed by Stevens and Lipscomb[134] on the basis of the small effector-induced conformational changes in the crystal structure[103] provides a detailed explanation for transmittal of the effector signal across the C1–R1–R6–C6 interfaces. This rationale for heterotropic interactions has some elements of induced fit by ATP and CTP with states attained that are neither R nor T, but asymmetrical hybrids are not implicit in the mechanism. The presence of negative cooperativity with CTP and the implications of the nucleotide perturbation model suggest that the simplest MWC model is not sufficient to explain entirely the heterotropic effects.

Summary. The concerted model of Monod–Wyman–Changeux satisfies most of the data relevant to the homotropic interactions of ATCase and eliminates the necessity of invoking a sequential model, that is, symmetry

[128] L. E. Parmentier, M. H. O'Leary, H. K. Schachman, and W. W. Cleland, *Biochemistry* **31,** 6570 (1992).

[129] M. P. Glackin, M. P. McCarthy, D. Mallikarachchi, J. B. Matthew, and N. M. Allewell, *Proteins,* **5,** 66 (1989).

[130] P. Suter and J. P. Rosenbusch, *J. Biol. Chem.* **252,** 8136 (1977).

[131] S. Matsumoto and G. G. Hammes, *Biochemistry* **12,** 1388 (1973).

[132] K. H. Kim, Z. Pan, R. B. Honzatko, H. M. Ke, and W. N. Lipscomb, *J. Mol. Biol.* **196,** 853 (1987).

[133] R. P. Kosman, J. E. Gouaux, and W. N. Lipscomb, *Proteins* **15,** 147 (1993).

[134] R. C. Stevens and W. N. Lipscomb, *Proc. Natl. Acad. Sci. U.S.A.* **89,** 5281 (1992).

appears to be conserved as originally proposed on theoretical grounds. However, the implications of the heterotropic effector interactions are that more than two states must exist to account for the structural changes that occur on ATP or CTP binding, perhaps multiple T-like substates. Whether these states retain molecular symmetry has not yet been conclusively proved. Sequential changes of subunit conformation, or significant concerted conformational changes,[4] along the reaction coordinate for effector binding could account both for the "global" effects on nucleotide binding[115,133] and for the negative cooperativity of the CTP binding.[130,131]

Use of Mutants to Distinguish Models

Mutational analysis (this volume, [4]) has been very useful with ATCase, hemoglobin, and other proteins in testing the importance or the role of individual amino acid residues, particularly in contact regions that transmit effects between subunits or domains. However, assignment of interaction surfaces from mutational data without a three-dimensional structure is clearly dangerous because of the transmission of substitution effects through the flexibility of a protein molecule. Mutations have provided a range of molecules with varying allosteric properties (e.g., altered L_0, n_H, c, heterotropic linkage) that probe the MWC versus the KNF models when used in conjunction with other methods. Study of the energetics of hemoglobin mutants has been extremely enlightening.[92] Hybrid ATCase[135] and hemoglobin[91] molecules containing varying mixtures of wild-type polypeptide chains with mutant or modified chains have elucidated other details; such methods will probably see more extensive usage in the future. Nevertheless, it is clear that mutants do not provide a simple, definitive conclusion regarding models by themselves. Spectroscopic, binding, catalytic, thermodynamic, crystallographic, and hydrodynamic studies as well as cleverly designed experiments have been needed to provide the kind of understanding that currently exists with ATCase and hemoglobin.

Allosteric Structures and Model Testing

The study of crystal structures leads to a detailed description of a limited number of conformations, usually termed R and T as in the MWC model, but may impose certain symmetry constraints. A comparison of the structural changes in the R- and T-states of five well-characterized proteins, namely, hemoglobin, ATCase, phosphofructokinase, glycogen

[135] E. Eisenstein, M. S. Han, T. S. Woo, J. M. Ritchey, I. Gibbons, Y. R. Yang, and H. K. Schachman, *J. Biol. Chem.* **267,** 22148 (1992).

phosphorylase, and fructose-bisphosphatase, has been made.[90,96,136–138] The transmission of homotropic and heterotropic regulatory information over molecular distances has been described in each case by the movement of polypeptide backbone and side chains. As expected, the interactions across the subunit interfaces are very important. Common features are as follows: a minimal, but critical, movement at the active site (as little as 0.5 Å in hemoglobin); the rotation of subunits with respect to one another by 7 to 17 Å around a cyclic axis; the alteration of salt bonds and hydrogen bonds at the subunit interface in the T–R transition; only two apparent modes of docking of the subunits, implying a concerted mechanism; and a more highly constrained and extensive subunit interface in the T-state. Variable features include the following: translation along the subunit rotational axis may occur (ATCase); the involvement of each type of tertiary structure at the subunit interface, that is, α helices (hemoglobin, fructose-bisphosphatase), β strands (phosphofructokinase), unordered loops (ATCase), and both α and β structures (phosphorylase); and the occurrence of the substrate binding directly at the interface to facilitate communication (only in phosphofructokinase and fructose-bisphosphatase).

The study of the kinetics and structural properties in solution add to the critical tests of cooperative models that can be applied. Hemoglobin and ATCase come close to fitting the MWC model but vary from it when sufficient solution data are obtained (discussed above). The same may be true for phosphofructokinase, which fits the MWC model at one level but also has elements of association–dissociation in the mammalian enzyme[39] and kinetic effects in the bacterial enzyme.[69,139] Other well-studied enzymes, such as glycogen phosphorylase,[96,140] pyruvate kinase,[141] and fructose-1,6-bisphosphatase,[138] have for the moment been fit to the MWC model but are still being tested. When enough types of data are obtained, each individual enzyme appears to have its own peculiarity, and the general models serve as a starting framework, rather than a final solution.

Kinetic Cooperativity in Monomeric Enzymes: Case of Glucokinase

Vertebrate glucokinase (hexokinase D or IV, EC2.7.1.2) is essential for the uptake of glucose by the liver and is involved in the regulation of

[136] L. N. Johnson and D. Barford, *J. Biol. Chem.* **265,** 2409 (1990).

[137] W. N. Lipscomb, *Chemtracts* **2,** 1 (1991).

[138] J.-Y. Liang, Y. Zhang, S. Huang, and W. N. Lipscomb, *Proc. Natl. Acad. Sci. U.S.A.* **90,** 2132 (1993).

[139] D. Deville-Bonne, F. Fourgain, and J.-R. Garel, *Biochemistry* **30,** 5750 (1991).

[140] L. N. Johnson, J. Hajdu, K. R. Acharya, D. I. Stuart, P. J. McLaughlin, N. G. Oikonomakos, and D. Barford, *in* "Allosteric Proteins" (G. Herve, ed.), p. 81. CRC Press, Boca Raton, Florida, 1988.

[141] T. G. Consler, M. J. Jennewein, G.-Z. Cai, and J. C. Lee, *Biochemistry* **31,** 7870 (1992).

blood glucose levels by the liver[142] and pancreas.[143] Positive cooperativity with glucose[144,145] (n_H = 1.5) is thought to be important for the enhanced responsiveness of the enzyme to changes in blood glucose levels.[142,143] A portion of non-insulin-dependent diabetes mellitus of the young, NIDDM type II, has been shown to be due to a genetic defect in glucokinase.[146,147]

Because glucokinase is a monomer (52 kDa) under all conditions with no evidence of multiple sites or cooperativity of binding,[148,149] the cooperativity cannot be generated by site–site interactions. Evidence has been presented that the kinetic mechanism is essentially ordered[79,80] and that a random mechanism[56] or abortive complex formation[57] is unlikely to explain the cooperativity.[66] A glucose-induced slow conformational change, in the form of a mnemonic mechanism[80,150] and subsequently a LIST mechanism,[151–153] has been proposed as the cause of the kinetic cooperativity. The predicted relationship with the second substrate occurs with glucokinase since the cooperativity decreases from an n_H of 1.5 to 1.0 as the MgATP concentration is decreased.[144,154] Glycerol added to assays depresses cooperativity[152,155] and decreases the rate of the molecular transition sufficiently so that a slow transient has been observed in reaction assays.[73] Consistent with effects on an enzyme isomerization, a burst was observed under these conditions if the glucokinase was preincubated in high glucose and a lag occurred if the enzyme was initially in low glucose. The steady-state velocities after the transient showed the

[142] L. Hue, *Adv. Enzymol.* **52,** 247 (1981).

[143] F. M. Matschinsky, *Diabetes* **39,** 647 (1990).

[144] A. C. Storer and A. Cornish-Bowden, *Biochem. J.* **159,** (1976).

[145] H. Niemeyer, M. L. Cardenas, E. Rabajille, T. Ureta, L. Clark-Turri, and J. Penaranda, *Enzyme* **20,** 321 (1975).

[146] N. Vionnet, M. Stoffel, J. Takeda, K. Yasuda, G. I. Bell, H. Zouali, S. Lesage, G. Velho, F. Iris, and P. Passa, *Nature (London)* **356,** 721 (1992).

[147] M. Stoffel, Ph. Froguel, J. Takeda, H. Zouali, N. Vionnet, S. Nishi, I. T. Weber, R. W. Harrison, S. J. Pilkis, S. Lesage, M. Vaxillaire, G. Velho, F. Sun, F. Iris, Ph. Passa, D. Cohen, and G. I. Bell, *Proc. Natl. Acad. Sci. U.S.A.* **89,** 7698 (1992).

[148] M. J. Holroyde, M. B. Allen, A. C. Storer, A. S. Warsy, J. M. E. Chesher, I. P. Trayer, A. Cornish-Bowden, and D. G. Walker, *Biochem. J.* **153,** 363 (1976).

[149] M. L. Cardenas, E. Rabajille, and H. Niemeyer, *Arch. Biochem. Biophys.* **190,** 142 (1978).

[150] A. C. Storer and A. Cornish-Bowden, *Biochem. J.* **165,** 61 (1977).

[151] M. L. Cardenas, E. Rabajille, and H. Niemeyer, *Eur. J. Biochem.* **145,** 163 (1984).

[152] M. L. Cardenas, E. Rabajille, I. P. Trayer, and H. Niemeyer, *Arch. Biol. Med. Exp.* **18,** 273 (1985).

[153] K. E. Neet, P. S. Tippett, and R. P. Keenan, *in* "Enzyme Dynamics and Regulation", (P. B. Chock, C. Y. Huang, C. L. Tsou, and J. H. Wang, eds.), p. 28. Springer-Verlag, New York, 1988.

[154] M. L. Cardenas, E. Rabajille, and H. Niemeyer, *Arch. Biol. Med. Exp.* **12,** 571 (1979).

[155] D. Pollard-Knight, B. A. Connolly, A. Cornish-Bowden, and I. P. Trayer, *Arch. Biochem. Biophys.* **237,** 328 (1985).

cooperativity and other features of the linear glucokinase assay without glycerol.[73] The LIST mechanism is also consistent with the effects of the inhibitors *N*-acetylglucosamine[73,151,154] and palmitoyl-CoA[156] on cooperativity,[153] the requirement for two catalytic cycles,[73,153] and the observed negative cooperativity with MgATP[73] or glucose[73,154] under certain conditions. Furthermore, a glucose-induced conformational change in the glucokinase molecule has been observed by protein intrinsic fluorescence measurements[152,157] and shows a slow transition in glycerol similar to that in the kinetic assay under similar conditions.[157] The half-times for the kinetic assay transient and the fluorescent conformational change are both between 0.5 and 5 min, have similar dependencies on glucose concentration, and provide strong evidence that the basis for the glucokinase cooperativity is a ligand-induced slow transition.[157] The current model is shown in Fig. 4 with ordered addition of glucose (A) before MgATP (B); the vertical lines represent the enzyme isomerizations.

Thermodynamic Analysis of Cooperativity, Linkage, and Multiple Ligand Binding

Thermodynamic Analysis of Cooperativity from Graphs

Hill plots of positive cooperativity can also be used to assess the energy of interaction between the first and the last ligand to bind to the protein, if the data are extensive enough. Typical Hill plots have a linear segment with a slope of one at each extreme (i.e., very low and very high ligand saturation). These regions represent the noninteractive binding of the first molecule to the unliganded protein and the last molecule to the last empty site. The perpendicular distance between the asymptotes is equal to the total free energy of interaction and is, therefore, a measure of the difference in free energy of binding of the first and last molecules.[1,6,29,158] This difference in free energy is the energy invested in cooperative interactions within the oligomer.

Concept and Utility of Linkage Relationships

Heterotropic effects of binding are related to homotropic effects by the thermodynamic linkage principles.[158] For example, if an inhibitor increases the K_d^S for a substrate, then the reciprocal effect must occur and the substrate would increase the K_d^I for the inhibitor. This concept is a direct consequence of the principle of detailed balance of reactions (or thermodynamic "boxes"). Consider a protein that binds n_1 molecules

[156] P. S. Tippett and K. E. Neet, *J. Biol. Chem.* **257,** 12846 (1982).
[157] S.-X. Lin and K. E. Neet, *J. Biol. Chem.* **265,** 9670 (1990).
[158] J. J. Wyman, *Adv. Protein Chem.* **19,** 223 (1964).

of X_1 and n_2 molecules of X_2. The simplest expression to describe the relationship of the binding isotherms for each ligand, Y_1 and Y_2, is by the differential equation that relates the variation in binding of X_1 and X_2. It can be shown[158] that

$$n_1 \left(\frac{\partial \overline{Y}_1}{\partial \ln[X_2]} \right)_{[X_1]} = n_2 \left(\frac{\partial \overline{Y}_2}{\partial \ln[X_1]} \right)_{[X_2]} \tag{8}$$

If there is no effect of X_1 on X_2, then the partial differentials are equal to zero, no interaction occurs, and there would be no effect of X_2 on X_1 either.

For the Bohr effect of hemoglobin the difference between the free energy of oxygenating a mole of hemoglobin at acidic or alkaline pH is exactly equal to the difference between the free energy of protonating a mole of hemoglobin and a mole of oxyhemoglobin.[158] Because this principle is thermodynamic in nature, it is independent of a particular model and is quite useful in designing experiments and testing the consistency of data. However, the linkage principle applies rigorously only to equilibrium situations, that is, kinetic parameters for a reaction may not follow linkage considerations owing to the unique contribution of kinetic constants. Furthermore, reciprocal effects arising from linkage on the cooperativity, n_H, of binding are predictable, but they are usually model dependent (see cooperativity models above).

In a practical sense, it is often useful to know if two different ligands affect an enzyme by binding to the same enzyme form (mutually exclusive binding) or whether they bind to different sites and have additive or synergistic effects. Exclusive binding is usually interpreted that both ligands bind to the same site, although this structural constraint is not required to explain the experimental result. The most rigorous procedure is to measure binding of one ligand in the presence of the other and directly look for displacement. The interaction of multiple inhibitors (or activators), however, and whether they are inhibiting (or activating) nonexclusively or exclusively can also be assessed by appropriate design of experiments. If v_0/v_i is plotted against ($[I_1]$ plus $[I_2]$), a straight line is obtained for mutually exclusive binding; a curve or deviation from linearity supports concurrent binding to the same enzyme form.[159] An alternate, more rigorous graphical method[160] is to measure v_0/v_i as a function of $[I_1]$ at constant $[I_2]$ and plot the result as a function of $[I_1]$. In this case a series of parallel lines is obtained for exclusive binding and straight lines with different slopes occur for nonexclusive, interacting binding. A complete discussion of these methods has been provided.[161] A more general, but

[159] K. Yagi and T. Ozawa, *Biochim. Biophys. Acta* **42,** 381 (1960).
[160] T. Yonetani and H. Theorell, *Arch. Biochem. Biophys.* **106,** 243 (1964).
[161] T. Yonetani, this series, Vol. **87,** p. 500.

rigorous, method that is not limited to Michaelis–Menten enzymes or to specific mechanisms has been introduced[162,163] and is based on the linearity of simple graphs of $\log[(f_i)^{-1} - 1]$ versus $[I_1 + I_2]$, where f_i is the fractional inhibition. This procedure is straightforward but appears to have been underutilized.

Ligand Binding Sites and Intramolecular Communication Distance: Kinetic Consequences

Two major findings that were not entirely anticipated in early concepts of regulatory enzymes have evolved from structural studies of allosteric proteins. Binding sites for ligands, either substrates or effectors, have frequently been found to lie on interfaces between two subunits of both allosteric and other oligomeric enzymes, rather than being contained within a single subunit. Distances between binding sites can vary from large distances (e.g., 65 Å), to vanishingly small (i.e., overlapping or isosteric sites) in the case of some covalent modifications. Neither of these observations could have been, nor can be, readily inferred from kinetic or other solution studies.

Intersubunit Ligand Binding Sites

The placing of ligand binding sites on intersubunit interfaces provides for an exquisite means of communicating cooperative or allosteric effects between subunits in an oligomeric enzyme. Examples include 2,3-bisphosphoglycerate with hemoglobin,[89,90] glutathione with glutathione reductase,[164] the glutamine analog methionine sulfoximine with glutamine synthetase (glutamate–ammonia ligase),[165,166] fructose 6-phosphate at the active site and ADP at the effector site of phosphofructokinase,[167,168] fructose 1,6-bisphosphate and fructose 2,6-bisphosphate with fructose-1,6-bisphosphatase,[169,170] and the bisubstrate analog PALA with aspartate

[162] T.-C. Chou and P. Talalay, *J. Biol. Chem.* **252,** 6438 (1977).

[163] T.-C. Chou and P. Talalay, *Eur. J. Biochem.* **115,** 207 (1981).

[164] G. E. Schulz, R. H. Schirmer, W. Sachsenheimer, and E. F. Pai, *Nature (London)* **273,** 120 (1978).

[165] M. M. Yamashita, R. J. Almassy, C. A. Janson, D. Cascio, and D. Eisenberg, *J. Biol. Chem.* **264,** 17681 (1989).

[166] R. J. Almassy, C. A. Janson, R. Hamlin, N.-H. Xuong, and D. Eisenberg, *Nature (London)* **323,** 304 (1986).

[167] W. R. Rypniewski and P. R. Evans, *J. Mol. Biol.* **207,** 805 (1989).

[168] Y. Shirakihara and P. R. Evans, *J. Mol. Biol.* **204,** 973 (1988).

[169] H. Ke, C. M. Thorpe, B. A. Seaton, F. Marcus, and W. N. Lipscomb, *Proc. Natl. Acad. Sci. U.S.A.* **86,** 1475 (1989).

[170] J.-Y. Liang, S. Huang, Y. Zhang, H. Ke, and W. N. Lipscomb, *Proc. Natl. Acad. Sci. U.S.A.* **89,** 2404 (1992).

transcarbamylase.[101,102] Such shared sites almost by necessity have to be responsive to changes in subunit interactions. Whether the high incidence of interfacial binding would effect how one views the validity of the general allosteric models will be left to the reader to consider—pros and cons for either MWC or KNF models can be debated vigorously. The lack of activity or loss of cooperativity of a dissociated enzyme can certainly be more easily understood if the active site is shared by neighboring subunits. In this case, of course, kinetics must be done on the intact oligomeric enzyme. Nevertheless, the formalism of the analysis in terms of kinetic mechanism or allosteric model does not change.

Comparison of kinetics and binding constants between oligomer and subunits can give insight into the makeup of binding sites.[41,171] However, the subunits must be dissociated gently enough to maintain a folded structure, and elaborate proof is needed for "nativeness" of any inactive subunit that results. Proteolytic fragments and kinetic analysis of refolding of the bifunctional aspartokinase–homoserine dehydrogenase I have shown that the kinase activity occurs in the monomeric species but that the dehydrogenase activity is generated only in the dimeric intermediate.[171,172] Cooperativity due to subunit association mechanisms (see previous section) could also be readily understood with intersubunit substrate binding sites. In the final analysis, crystal or NMR structures are required for definitive answers regarding the subunit contributions of binding sites.

The concept that a distinct subunit would likely contain the allosteric effector binding sites, suggested from the early example of the separable regulatory and catalytic subunits of ATCase, has not held up well as a general finding; relatively few other similar examples currently exist, such as the cAMP-dependent protein kinase[173] and calmodulin in phosphorylase kinase.[174] In most cases a single polypeptide chain contains both the allosteric and catalytic site, as in phosphofructokinase,[168] pyruvate kinase,[175] and homoserine dehydrogenase.[171]

In many other instances binding sites lie on the interface of two domains of the same polypeptide, that is, a crevice or cleft, as foretold by early crystal studies of lysozyme and yeast hexokinase. Thus, evolution has apparently found it easy to form a functional binding site by juxtaposition of two rather separate subsites from distant domains of a protein. Ligand-

[171] A. Dautry-Varsat and J.-R. Garel, *Biochemistry* **20,** 1396 (1981).

[172] H. Vaucheret, L. Signon, G. LeBras, and J.-R. Garel, *Biochemistry* **26,** 2785 (1987).

[173] S. S. Taylor, *J. Biol. Chem.* **264,** 8443 (1989).

[174] C. A. Pickett-Giles and D. A. Walsh, *in* "The Enzymes" (P. D. Boyer, ed.), 3rd Ed., Vol. 17, p. 396. Academic Press, New York, 1987.

[175] L. Engstrom, P. Ekman, E. Humble, and O. Zetterqvist, *in* "The Enzymes" (P. D. Boyer, ed.), 3rd Ed., Vol. 18, p. 47. Academic Press, New York, 1987.

TABLE II
ATOMIC DISTANCES IN SELECTED ALLOSTERIC PROTEINS

Enzyme (protein)	Effect	Site(s)	Distance (Å)	Ref.
Hemoglobin	Homotropic	O_2	25–40	89
Aspartate transcarbamylase	Homotropic	Aspartate	40–50	101, 102
Aspartate transcarbamylase	Heterotropic	ATP/CTP to aspartate	60	
Phosphofructokinase	Homotropic	Fru-6-P	45–55	168
Phosphofructokinase	Heterotropic	ADP to Fru-6-P	20–25	
Glutamine synthetase[a]	Homotropic	Gln to Gln	45–52	165, 166
Glutamine synthetase	Heterotropic	Covalent adenylyl to Gln	22	
Glycogen phosphorylase	Homotropic	Glucoside to glucoside	66	96, 140, *b*
Glycogen phosphorylase	Heterotropic	AMP to glucoside	32	
Glycogen phosphorylase	Heterotropic	Covalent P to AMP	12	
Fructose-1,6-bisphosphatase	Heterotropic	AMP to Fru-1,6-P_2	28	138, 169, 170
Fructose-1,6-bisphosphatase	Heterotropic	Fru-2,6-P_2 to Fru-1,6-P_2	0	
Isocitrate dehydrogenase	Heterotropic	Covalent P to isocitrate	0	*c*

[a] Glutamate–ammonia ligase.
[b] S. R. Sprang, K. R. Acharya, E. J. Goldsmith, D. I. Stuart, K. Varvill, R. J. Fletterick, N. B. Madsen, and L. N. Johnson, *Nature (London)* **336,** 215 (1988).
[c] J. H. Hurley, A. M. Dean, and D. E. Koshland, Jr., *Science* **249,** 1012 (1990).

induced movement of two domains or alteration of the spatial relationship of two subunits also appears to be a relatively common way by which a protein can achieve significant effects on catalysis or regulation through conformational changes.

Interaction Distances

Distances between binding sites can be measured, in principle, in solution by such techniques as NMR and fluorescence. Many examples exist in which these distances are in reasonable agreement with those subsequently determined from crystal structures. Transmittal of regulatory information over long distances is exemplified by ATCase and glycogen phosphorylase with distances of 60 Å or more (Table II). The movement of polypeptide backbone and amino acid residues has been described in detail in several cases. Communication of regulatory information over shorter distances also occurs in fructose-1,6-bisphosphatase and isocitrate dehydrogenase (Table II).

A recent demonstration has been made that isosteric regulatory effects can occur, that is, regulation at the same site rather than another (allosteric) site.[176] Phosphorylation of isocitrate dehydrogenase introduces a nega-

[176] A. M. Dean and D. E. Koshland, Jr., *Science* **249,** 1044 (1990).

tive charge from the phosphate at the residue, Ser-113, that would normally hydrogen bond to the negatively charged γ-carboxyl of isocitrate; substrate repulsion occurs directly without a conformational change.[177] Identical types of kinetic responses may be observed with isosteric regulation as with allosteric effects in which the phosphorylation site is more distant (e.g., 12 Å away in glycogen synthase), and significant conformational changes are required to transmit the regulatory information. Although these examples concern regulation by covalent modification, similar effects could occur with regulation by noncovalent modifiers (i.e., by apparent "allosteric" effectors). Thus, more fuel is provided for long-standing kinetic controversies. Crystallographic data indicate that fructose 2,6-bisphosphate, a negative effector of fructose-1,6-bisphosphatase, binds to the fructose 1,6-bisphosphate (active) site[138,169,170]; in this case, of course, the effector and substrate have similar structures. Whether product inhibition of mammalian brain hexokinase I by glucose 6-phosphate occurs at an allosteric site or at the active site has been discussed for 30 years.[178] Inhibition at the active site (an isosteric effect), as proposed by Solheim and Fromm[179] from kinetic arguments, is probably correct (but see Refs. 180 and 181) since site-specific mutagenesis[182] and a truncated recombinant hexokinase[178] show that the glucose 6-phosphate binding site is in the same C-terminal domain as the active site, rather than being on a distinct domain. Furthermore, NMR data indicate the glucose 6-phosphate binds to hexokinase I within 3 to 19 Å of the glucose site.[183] Interpretation of similar types of purely kinetic data in terms of the underlying structure must be made with caution.

Cooperativity in Other Macromolecular Systems

Ligand Binding to Membrane Receptors

Numerous receptor systems have demonstrable cooperativity, either positive or negative. Because the macromolecular receptor is binding a ligand (frequently a hormone or growth factor that is itself a peptide or protein), the mathematical analysis of binding to the receptor is similar to that described above for enzymes.[14] Analysis of cellular response data

[177] J. H. Hurley, A. M. Dean, and D. E. Koshland, Jr., *Science* **249,** 1012 (1990).
[178] M. Magnani, M. Bianchi, A. Casabianca, V. Stocchi, A. Daniele, F. Altruda, M. Ferrone, and L. Silengo, *Biochem. J.* **285,** 193 (1992).
[179] L. P. Solheim and H. J. Fromm, *Arch. Biochem. Biophys.* **211,** 92 (1981).
[180] A. D. Smith and J. E. Wilson, *Arch. Biochem. Biophys.* **292,** 165 (1992).
[181] T. K. White and J. E. Wilson, *Arch. Biochem. Biophys.* **259,** 402 (1987).
[182] M. Baijal and J. E. Wilson, *Arch. Biochem. Biophys.* **298,** 271 (1992).
[183] G. K. Jarori, S. B. Iyer, S. R. Kasturi, and U. W. Kenkare, *Eur. J. Biochem.* **188,** 9 (1990).

is not analogous to that of enzyme kinetic data, however, since the signal transduction by coupling and amplification complicates any rigorous interpretation of cooperativity of the responses. The experimental methods for obtaining binding data are different since the receptor will be either in the membrane of a whole cell, in a membrane fragment, or solubilized in detergent. The reader is referred to Vols. 109, 146, and 147 in this series for a detailed discussion of receptor techniques. A significant difference when dealing with receptors on cells in culture is that dynamic internalization of the ligand–receptor complex occurs and usually a steady-state binding situation, rather than a true equilibrium, must be assumed.[184] Furthermore, the fluidity of the membrane allows dynamic alterations in receptor–receptor interactions (e.g., dimerization) that can influence cooperativity,[185–187] much as in the enzyme oligomerization model discussed above.

Phenomenological (Adair-type) models are initially applied to describe the binding of proteins to membrane receptors. From graphs of the binding data (analogous to those in Fig. 1) the possible presence of cooperativity can be deduced. The MWC model is frequently utilized to explain both homotropic and heterotropic effects, and R- and T-states of receptors are commonly invoked.[188] A good example is the nicotinic acetylcholine receptor,[189] which is an ion channel regulated by the effect of agonists, competitive antagonists, and noncompetitive inhibitors at several allosteric sites. Multiple, presumably pseudosymmetric, states of the heterologous pentameric acetylcholine receptor are postulated to exist.[190] Whether this model will withstand further testing, when conformational changes and molecular interactions are described at high resolution, remains to be seen.

An additional complexity of membrane receptors is the frequent heterogeneity of receptors in a membrane that gives rise to biphasic binding curves. This heterogeneity may be due to pharmacological subtypes (isoreceptors) (e.g., neuronal dopamine receptors), cross-talk with related receptors (e.g., the nerve growth factor family of neurotrophin receptors),

[184] H. S. Wiley and D. D. Cunningham, *Cell (Cambridge, Mass.)* **25,** 433 (1981).

[185] C. DeLisi and R. Chabay, *Cell Biophys.* **1,** 117 (1979).

[186] A. DeLean and D. Rodbard, *in* "The Receptors" (R. D. O'Brien, ed.), p. 143. Plenum, New York, 1979.

[187] S. Jacobs and P. Cuatrecasas, *Biochim. Biophys. Acta* **433,** 482 (1976).

[188] J.-P. Changeux, J. P. Thiery, Y. Tung, and C. Kittel, *Proc. Natl. Acad. Sci. U.S.A.* **57,** 335 (1967).

[189] J. L. Galzi, F. Revah, A. Bessis, and J.-P. Changeux, *Annu. Rev. Pharmacol. Toxicol.* **31,** 37 (1991).

[190] C. Lena and J.-P. Changeux, *Trends Neurosci.* **16,** 181 (1993).

functional diversity (e.g., interleukin 2 receptors), or environmental variation (e.g., insulin receptors). Because the biphasic curves have the same appearance as "negative cooperativity" (see earlier discussion), much controversy has arisen about the cause of the biphasic curves. One interesting test with membrane receptors has been to compare simple dissociation rates from dilution of the ligand–receptor complex to the apparent dissociation rate of labeled ligand in the presence of an excess of unlabeled ligand.[187,191] If negative cooperativity exists, the increased receptor occupancy in the presence of excess ligand should increase the rate of dissociation of the bound radioactive ligand (assuming the association rate is constant), whereas the rate for independent heterogeneous receptors should be unaffected by the cold chase. Although some success was reported with this method, others have found it not to be a useful discriminator.[185,192] Further use of this interesting approach is warranted.

Proteins Binding to DNA

An active area of research is the study of nucleic-binding proteins, many of which show positive cooperativity of binding. Well-studied examples include Gene-32 protein of T4 bacteriophage, the single-stranded DNA-binding protein of *E. coli,* and numerous transcriptional factors. Methodology often utilizes labeled oligonucleotides and gel-shift assays, so the analysis of data can be somewhat different than that with enzymes and low molecular weight ligands. Because the protein–protein cooperativity can be influenced by the presence of multiple binding sites within the DNA template, a different model[193] has been applied to this system. The equilibrium binding of a protein to an infinite lattice, such as DNA, can be described by three thermodynamic parameters: n, the binding site size of the protein for the DNA (number of bases "covered"); K, the intrinsic binding constant for a nucleotide sequence; and ω, the cooperativity term. The cooperativity parameter, ω, is the relative affinity for a protein to bind adjacent to an already bound protein ligand as opposed to an isolated DNA site and can be greater than 1 (positive cooperativity) or less than 1 (negative cooperativity). The binding equations, methods to measure the parameters with either varying ligand or lattice concentration, and a rigorous interpretation of the results have been presented.[193]

The steroid receptor complexes also bind specific DNA sequences with positive cooperativity. The molecular structures of the glucocorti-

[191] P. De Meyts, A. R. Bianco, and J. R. Roth, *J. Biol. Chem.* **251,** 1877 (1976).

[192] C. P. de Vries, T. W. van Haeften, and E. A. van der Veen, *Endocr. Res.* **17,** 331 (1991).

[193] S. C. Kowalczykowski, L. S. Paul, N. Lonberg, J. W. Newport, J. A. McSwiggen, and P. H. von Hippel, *Biochemistry* **25,** 1226 (1986).

coid[194] and the estrogen[195] receptor oligonucleotide complexes have been determined by X-ray crystallography and provide a molecular model to explain the cooperativity. The monomeric glucocorticoid or estrogen DNA-binding domain dimerizes on binding to DNA. The apparent specificity (and affinity) of the glucocorticoid DNA-binding domain depends on the correct spacing (3 base pairs) between the two hexameric half-sites. The cooperativity (ω or dimerization) on DNA binding is due to the ordering of a region of the binding domain that is disordered in solution and makes positive protein–protein contacts on the correctly spaced palindromic response element.[194,195] The steroid receptors have the most molecular detail and the most specific proposal of any cooperative system other than soluble enzyme systems.

Macromolecular Assembly

Self-assembling, homomeric proteins frequently show (positive) cooperativity, that is, no (or few) intermediates exist between the monomer state and the high polymeric state. Examples include the classic case of hemoglobin S (which leads to sickling), tubulin, and other cytoskeletal proteins. Analysis of these equilibria differs from that of ligand binding systems because the change in the protein–protein association constant is the criterion for cooperativity. Methods employed usually involve a measure of the size distribution of the system by sedimentation equilibrium, viscosity, gel filtration, etc. The exchange of labeled monomers into the polymer can also be utilized to estimate relative affinities.

Future Directions

Considering the fact that the field of allosteric enzymes is three decades old, there is surprisingly poor understanding of the details of most regulatory enzymes. The two examples discussed, hemoglobin and ATCase, have a nearly complete molecular description but still lack complete agreement on the ultimate mechanism. Others (e.g., phosphofructokinase, glycogen phosphorylase, isocitrate dehydrogenase) have atomic coordinates and a molecular hypothesis that needs further testing. Finally, less well-characterized proteins (e.g., glucokinase, wheat germ hexokinase, chloroplast fructose-bisphosphatase) have kinetic models but lack information at the molecular structure level. The examples discussed in this chapter

[194] B. F. Luisi, W. X. Xu, Z. Otwinowski, L. P. Freedman, K. R. Yamamoto, and P. B. Sigler, *Nature (London)* **352,** 497 (1991).

[195] J. W. R. Schwabe, L. Chapman, J. T. Finch, and D. Rhodes, *Cell (Cambridge, Mass.)* **75,** 567 (1993).

quite clearly show the necessity of having one or more three-dimensional coordinate structures from X-ray crystallography in order to interpret allosteric enzymes. However, it is equally clear that such a structure is not in itself sufficient and that extensive solution studies of conformation, structural kinetics, and energetics are required for a thorough understanding of the enzyme and to substantiate an appropriate model.

Physical techniques that can characterize protein conformations and interactions in solution are making tremendous progress in application to allosteric mechanisms. Such techniques not only are able to reveal hybrid or asymmetrical intermediates that crystallography may not, but can also provide information on the dynamics of the allosteric transition. The most visible technique of NMR (and the one with the most information content next to X-ray crystallography) is already making significant contributions to smaller proteins and will undoubtedly continue to be important to allosteric systems as the study of larger proteins becomes possible. Sedimentation and gel filtration techniques that have made such important contributions to ATCase and hemoglobin may not be broadly applicable to other systems. However, microcalorimetry [both differential scanning (DSC) and isothermal titration] and quasi-elastic light scattering (QEL) may provide similar types of detailed information on the energetics of protein–protein interactions between subunits. Raman spectroscopy and fluorescence relaxation may complement steady-state fluorescence as a means of studying the kinetics of (undefined) conformational changes. Fourier transform infrared spectroscopy (FTIR) may supplant circular dichroism as a means for determining secondary structure in solution and provide more dynamic details of allosteric transitions. A limiting amount of protein for such demanding studies has become much less of a problem with the advent of recombinant DNA methodology. Indeed, the molecular biology tools of protein overexpression and site-specific mutagenesis will continue to be of tremendous benefit to the practicing enzymologist. Application of these physical methods to membrane receptor systems will remain more difficult.

Important questions for the enzymologist extend beyond the mechanistic approaches posed above into the classic arena of the physiological importance of the regulatory aspects of enzymes. What is really the importance of the allosteric site of a key enzyme for the regulation of a metabolic pathway? What would be the biological result of losing the positive cooperativity of an enzyme such as phosphofructokinase or glucokinase? Is the slow isomerization of an hysteretic enzyme essential for life? What is the role of negative cooperativity in physiological function? What would happen to the organism if an enzyme could not form tetramers?

Fortunately, the means to answer these questions appears to be at hand with a combination of molecular biology, cell biology, genetics, and biology to assist the enzymologist. Most enzyme mutants have only been studied mechanistically *in vitro*, but these same mutants can readily be studied in a physiological setting in mammalian cell culture or in transgenic animals. Homologous recombination of a mutant regulatory enzyme, deficient only in its allosteric binding site or in its substrate cooperativity, into a metabolically important cell (e.g., hepatocyte) or into a mouse embryo is technically possible. The former system would allow study of metabolic flux in a living cell with altered regulation; the latter system would determine the consequences of altered regulation of pathways for a living organism. Such regulation-deficient transgenic mice should be much more informative than "knock-out" experiments, although somewhat more difficult to construct. The answer may well be that the animal adapts to the lack of regulation, much to the chagrin of the enzymologist. In any event, models of various metabolic disorders (including rather mild effects) may be provided with the production of such genetically altered animals. The next few decades may have as many advances and increase our knowledge as much as the past 15 to 30 years; it should be an equally exciting era for the regulatory enzymologist.

Acknowledgments

The author gratefully acknowledges support by grants from the National Institutes of Health and from the Juvenile Diabetes Foundation. I also thank the following colleagues who critically read portions of the manuscript: Drs. Marilu Cardenas, Athel Cornish-Bowden, Antony M. Dean, Robert G. Kemp, Jacque Ricard, and Howard K. Schachman.

[21] Kinetic Basis for Interfacial Catalysis by Phospholipase A_2

By MAHENDRA KUMAR JAIN, MICHAEL H. GELB, JOSEPH ROGERS, and OTTO G. BERG

First you must have the images, then come the words.
Robert James Waller
In *The Bridges of Madison County*

A. Introduction

Interfacial catalysis is an obligatory mode of action of enzymes whose natural substrates show a propensity to form organized interfaces. Accordingly, catalytic turnover by virtually all lipolytic enzymes occurs at the interface of organized aggregates formed in aqueous dispersions of amphipathic and "lipidic" substrates. This chapter focuses on the kinetics of interfacial catalysis by a water-soluble lipolytic enzyme that must attach to the substrate aggregate as a prerequisite for catalytic turnover (Fig. 1, left). Under these conditions both the enzyme and the substrate are present at the interface, and the turnover occurs at the interface. Similar considerations also apply to the behavior of integral membrane enzymes that act on substrates at the interface (Fig. 1, middle left). Membrane proteins that act on substrates in the aqueous phase (Fig. 1, middle right) or water-soluble enzymes that operate on water-soluble substrates that partition between the aqueous and membrane phases (Fig. 1, right) can be analyzed by standard kinetic methods and are not discussed further in this review.

The reason that the kinetic behavior of interfacial enzymes is challenging to study is that for an analytical description not only must one keep track of the concentration of the interacting species at the interface where catalysis occurs, but one must also take into account the kinetics of transfer of the enzyme, substrate, and product between the aggregates that

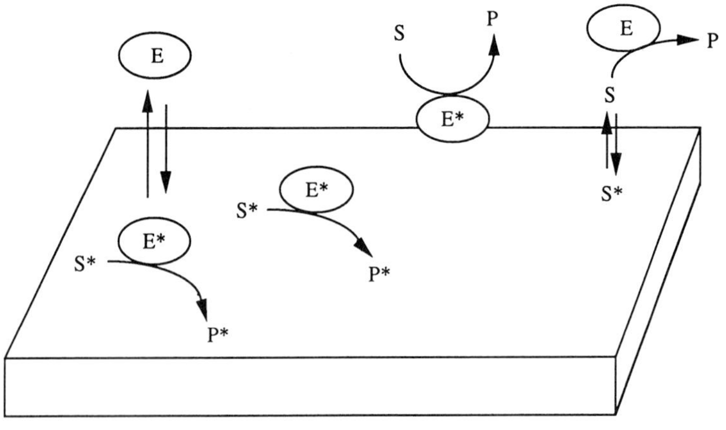

FIG. 1. Possible effects of an interface between two phases on the activity of an enzyme. From left: The enzyme in the aqueous phase (E) binds to the interface (E*) where it carries out catalysis on the substrate at the interface (S*) to release products at the interface (P*). Two of the possible special cases of this type of interfacial catalysis are shown where the enzyme is present only at the interface and acts on the substrate at the interface (middle left) or in the aqueous phase (S) (middle right). If an enzyme present in the aqueous phase acts on substrate partitioned between the interface and the aqueous phase (right), the interface may effectively lower the substrate concentration available to the enzyme in the aqueous phase. Only the first case (left) is considered in this review.

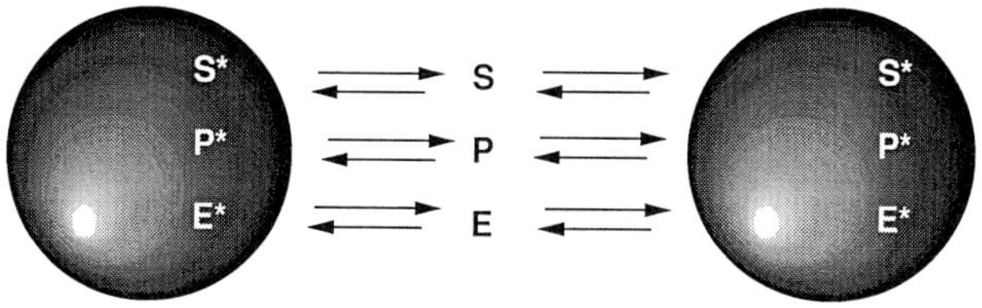

FIG. 2. The exchange of enzyme, substrate, and product between aggregates controls the local concentrations of substrate and product that the enzyme bound to the aggregate "sees." These local concentrations, which are not necessarily the same as the global concentrations for the entire reaction mixture, control the microscopic steady-state kinetics of the bound enzyme.

provide the interface (Fig. 2). The bulk concentration of the components changes the overall number of aggregates and therefore the fraction of the enzyme at the interface. On the other hand, the "local" concentration of the interacting species at the interface of an aggregate that the bound enzyme (E*) "sees" during the reaction progress depends on the size of the aggregates, the miscibility of the components at the interface, and the kinetics of exchange of components between aggregates. Although such issues that occur in "real-life" situations complicate the overall observed kinetic behavior, our emphasis is on the processes that occur at the interface, and we focus on the concepts that provide a general analytical framework for elaboration of the primary rate and equilibrium constants that quantitatively describe interfacial catalysis.

This review focuses on a class of 14-kDa secreted phospholipases A_2 (PLA2) which catalyze the hydrolysis of the *sn*-2-acyl chain of naturally occurring phospholipids to form lysophospholipids and fatty acids (Fig. 3). In describing the kinetic behavior of PLA2 our emphasis is on the formulation of experimentally testable questions and on the design of strategies and protocols necessary to resolve the primary rate and equilibrium parameters. Because only a few interfacial enzymes have been adequately characterized, the formalism developed here should be considered

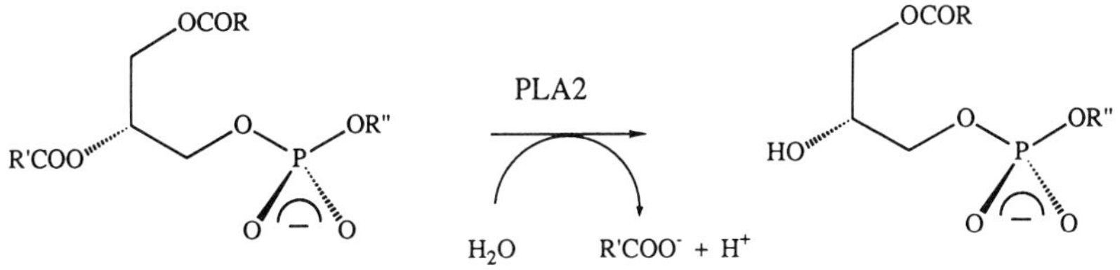

FIG. 3. Hydrolytic reaction catalyzed by PLA2.

as a prototype to build a general conceptual understanding of the kinetic and molecular aspects of interfacial catalysis by membrane-bound proteins.

The Michaelis–Menten formalism is intrinsic in most discussions of interfacial catalysis.[1-5] The formalism, originally articulated by de Haas and co-workers,[2] has guided the experimental work on interfacial catalysis for over two decades. Only relatively recently has a complete analytical solution to the problem of interfacial catalysis emerged which provides unequivocal experimental evidence for the interfacial nature of the catalytic process and the underlying assumptions necessary to understand such processes while taking into account the exchange of reaction components between substrate aggregates.[5] No attempt is made here to give a historical account of the developments in the field of interfacial catalysis; such information has been adequately reviewed from time to time.[1-5] Detailed elaboration of key concepts and experimental procedures can be found in the references that are given throughout the text.

B. Properties of Components

Interfacial catalysis by PLA2 is a kinetic system involving two complex components, the enzyme and the substrate-containing aggregate. Before embarking on a detailed kinetic analysis of interfacial catalysis, we summarize the relevant properties of PLA2 and phospholipids in aqueous dispersions.

Organization and Dynamics of Phospholipids in Aqueous Dispersions

Owing to the hydrophobic effect, amphipathic molecules in aqueous dispersions form organized aggregated structures in which the apolar chains are effectively separated from bulk water.[6] Because amphiphiles in aggregates are held together by weak interactions, their aqueous dispersions exhibit extensive polymorphism. The effective shape and size of the polar and nonpolar regions of a phospholipid molecule and the presence

[1] M. K. Jain, *Curr. Top. Membr. Transp.* **4,** 175 (1973).

[2] R. Verger and G. H. de Haas, *Annu. Rev. Biophys. Bioeng.* **5,** 77 (1976).

[3] H. M. Verheij, A. J. Slotboom and G. H. de Haas, *Rev. Physiol. Biochem. Pharmacol.* **91,** 91 (1981).

[4] E. A. Dennis, *in* "The Enzymes" (P. D. Boyer, ed.), 3rd Ed., Vol. 16, p. 307. Academic Press, New York, 1983.

[5] M. K. Jain and O. G. Berg, *Biochim. Biophys. Acta* **1002,** 127 (1989).

[6] M. K. Jain, "Introduction to Biological Membranes." Wiley, New York, 1988; G. Cevc and D. Marsh, "Phospholipid Bilayers: Physical Principles and Methods." Wiley, New York, 1987.

of other additives determine the overall morphology of the aggregate (Fig. 4). Naturally occurring phospholipids with long acyl chains have a cylindrical shape with a polar end. When dispersed in water they organize into closed structures, such as bilayer-enclosed vesicles, to minimize the contact of chains with bulk water. For the same reason the concentration of solitary phospholipid molecules in the aqueous phase (cmc, critical micelle concentration) is estimated to be less than 1 pM, and it remains

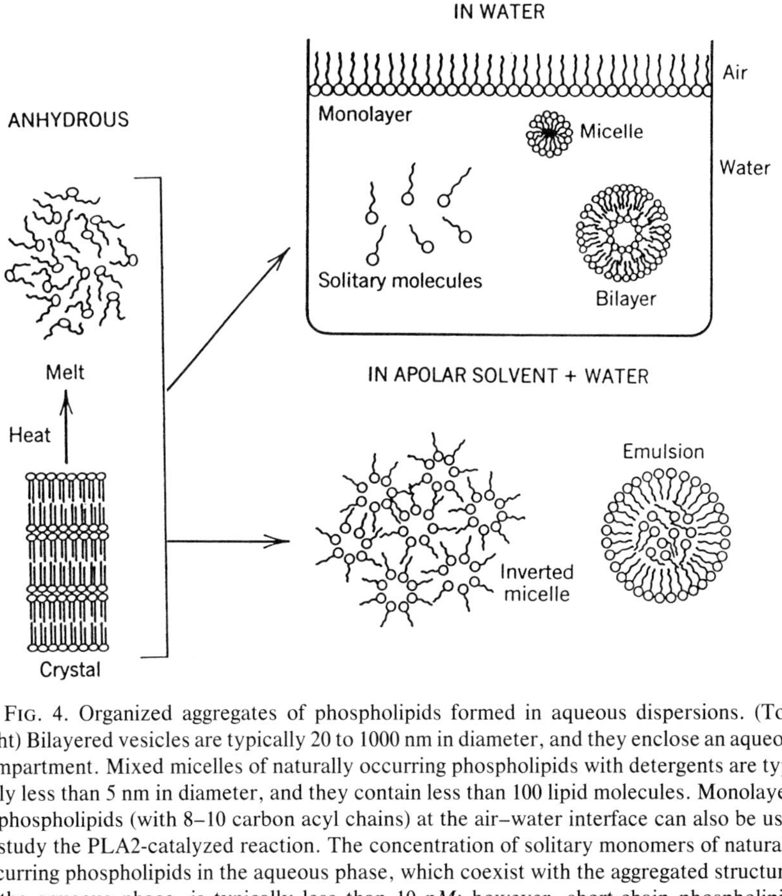

FIG. 4. Organized aggregates of phospholipids formed in aqueous dispersions. (Top, right) Bilayered vesicles are typically 20 to 1000 nm in diameter, and they enclose an aqueous compartment. Mixed micelles of naturally occurring phospholipids with detergents are typically less than 5 nm in diameter, and they contain less than 100 lipid molecules. Monolayers of phospholipids (with 8–10 carbon acyl chains) at the air–water interface can also be used to study the PLA2-catalyzed reaction. The concentration of solitary monomers of naturally occurring phospholipids in the aqueous phase, which coexist with the aggregated structures in the aqueous phase, is typically less than 10 pM; however, short-chain phospholipids dispersed as monomers have also been used for monitoring the PLA2-catalyzed hydrolysis (middle, right). Phospholipids dispersed in apolar solvents such as hexane containing small amounts of water form inverted micelles or emulsions. Such structures are useful for preparative reactions involving the action of lipolytic enzymes, but their hydrolysis is difficult to characterize kinetically.

constant at higher bulk concentrations of phospholipids. Phospholipid molecules in a bilayer rotate about the long axis (once every 10^{-10} sec), and they move rapidly in the plane of the bilayer (net displacement of about 300 nm per second); however, the half-time for transbilayer movement or intervesicle exchange is of the order of several hours.

The products of the PLA2-catalyzed hydrolysis of phospholipids, lysophospholipids and fatty acids (Fig. 3), form micelles when dispersed alone in aqueous solutions; however, the products form bilayer-enclosed vesicles when codispersed in equimolar proportion.[7] During the course of hydrolysis of phospholipid vesicles by PLA2, only the phospholipid molecules present in the outer monolayer are hydrolyzed,[8] and the bilayer organization is retained.[9] The rate of intervesicle exchange of the substrate and products during the progress of the reaction is much slower than the rate of hydrolysis of the substrate molecules present on the surface of the vesicle. Such properties make bilayer-enclosed vesicles a nearly ideal substrate interface for the study of the kinetics of interfacial catalysis.

Besides the characteristic morphological features shown in Fig. 4, key differences between bilayers and other aggregated structures of phospholipids may also be noted. Transfer of phospholipids between micelles occurs on the time scale of seconds compared to several hours for bilayer vesicles. Transfer of solutes across the unstirred layer of water (about 1000 nm thick) below the monolayer of phospholipids at the air–water interface occurs on the time scale of a few minutes; although the underlying rate for the transfer from the aqueous phase to the monolayer is probably diffusion limited, it is also constrained by the geometry of the interface.[5] The size of micelles is typically less than 50 amphiphiles per particle and is governed largely by the composition. Yet vesicles of defined size and narrow monomodal dispersity can be prepared from a wide variety of phospholipids.[10] The relevance of such differences in the size of phospholipid aggregates and the interaggregate exchange rate to interfacial catalysis is discussed later in the appropriate contexts.

Structural Features of Phospholipase A_2

The water-soluble, compact globular proteins have about 120 amino acid residues and isoelectric points between 3 and 11.[3] Various forms of

[7] M. K. Jain, C. J. A. Van Echteld, F. Ramirez, J. Gier, G. H. de Haas, and L. L. M. Van Deenen, *Nature (London)* **284,** 486 (1980); M. K. Jain and G. H. de Haas, *Biochim. Biophys. Acta* **642,** 203 (1981).

[8] M. K. Jain, J. Rogers, D. V. Jahagirdar, J. F. Marecek, and F. Ramirez, *Biochim. Biophys Acta* **860,** 435 (1986).

[9] J. C. Wilschut, J. Regts, H. Westenberg, and G. Scherphof, *Biochim. Biophys. Acta* **508,** 185 (1978); J. C. Wilshut, J. Regts, and G. Schephof, *FEBS Lett.* **98,** 181 (1979).

[10] F. Szoka and D. Papahadjopoulos, *Annu. Rev. Biophys. Bioeng.* **9,** 467 (1980).

PLA2 are widely distributed, and they are abundant in pancreatic digestive fluid, venoms, and inflammatory exudates. They are divided into three evolutionary classes[11] which differ not only in amino acid sequences but also in the folding patterns where five of the seven disulfide bridges are conserved.[3,12] The chain is folded with several helices and at least one sheet, and the exposed surface is predominantly hydrophilic.[13,14] The crystallographic structural features of PLA2 without[13,15] (E) or with a short-chain phospholipid analog in the active site[16,17] (EL) are essentially identical.[18]

The substrate binding site of PLA2 is a slot that starts at the surface of the enzyme which contacts the interface (i-face, interfacial recognition surface of PLA2) and runs all the way through the enzyme (Fig. 5).[3,15–18] The end of the polar head group of a bound phospholipid protrudes through the face of the enzyme that is opposite from the i-face (Fig. 5). The key catalytic residues are His-48 and Asp-99.[19] As shown in Fig. 6, calcium bound in the active site is required for binding of the substrate[16,19] and also for electrophilic catalysis in the chemical step.[20]

[11] F. F. Davidson and E. A. Dennis, *J. Mol. Evol.* **31,** 228 (1990).

[12] C. J. Van den Bergh, A. J. Slotboom, H. M. Verheij, and G. H. de Haas, *J. Cell. Biochem.* **39,** 379 (1989).

[13] M. J. Dufton, D. Eaker, and R. C. Hider, *Eur. J. Biochem.* **127,** 537 (1983).

[14] R. Renetseder, S. Brunie, B. W. Dijkstra, J. Drenth, and P. B. Sigler, *J. Biol. Chem.* **260,** 11627 (1986).

[15] B. W. Dijkstra, K. H. Kalk, W. G. J. Hol, and J. Drenth, *J. Mol. Biol.* **147,** 97 (1981).

[16] D. L. Scott, S. P. White, Z. Otwinowski, W. Yuan, M. H. Gelb, and P. B. Sigler, *Science* **250,** 1541 (1990).

[17] M. M. G. M. Thunnissen, E. Ab, K. H. Kalk, J. Drenth, B. W. Dijkstra, O. P. Kuipers, R. Dijkman, G. H. de Haas, and H. M. Verheij, *Nature (London)* **347,** 689 (1990).

[18] Other crystallographic results are available: complex of bee venom PLA2 with MG14, D. L. Scott, Z. Otwinowski, M. H. Gelb, and P. B. Sigler, *Science* **250,** 1563 (1990); complex of cobra venom PLA2 with MG14, S. P. White, D. L. Scott, Z. Otwinowski, M. H. Gelb, and P. B. Sigler, *Science* **250,** 1560 (1990); E form of synovial fluid PLA2, J. P. Wery, R. W. Schevitz, D. K. Clawson, J. L. Bobbitt, E. R. Dow, G. Gamboa, T. Goodson, R. B. Hermann, R. M. Kramer, D. B. McClure, E. D. Mihelich, J. E. Putnam, J. D. Sharp, D. H. Stark, C. Teater, M. W. Warrick, and N. D. Jones, *Nature (London)* **352,** 79 (1991); complex of synovial fluid PLA2 with MG14 form, D. L. Scott, S. P. White, J. L. Browning, J. J. Rosa, M. H. Gelb, and P. B. Sigler, *Science* **254,** 1007 (1991); complex of bovine PLA2 with alkylphosphocholine, K. Tomoo, H. Ohishi, M. Doi, T. Ishida, M. Inoue, K. Ikeda, and H. Mizuno, *Biochem. Biophys. Res. Commun.* **187,** 821 (1992); Notexin in E form, B. Westerlund, P. Nordlund, U. Uhlin, D. Eaker, and H. Eklund, *FEBS Lett.* **301,** 159 (1992). Structures of these ligands are given in Fig. 8.

[19] H. M. Verheij, J. J. Volwerk, E. H. J. M. Volwerk, W. C. Puijk, B. W. Dijkstra, J. Drenth, and G. H. de Haas, *Biochemistry* **19,** 743 (1980). This article also describes a "pseudo-triad" mechanism for catalysis where the catalytic water is activated by hydrogen bonding to His-48. An alternative mechanism involving a "calcium-coordinated oxyanion" has also been suggested by K. Seshadri, S. Vishveshwara, and M. K. Jain, *Proc. Indian Acad. Sci.* **106,** (1994).

[20] B.-Z. Yu, O. G. Berg, and M. K. Jain, *Biochemistry* **32,** 6485 (1993).

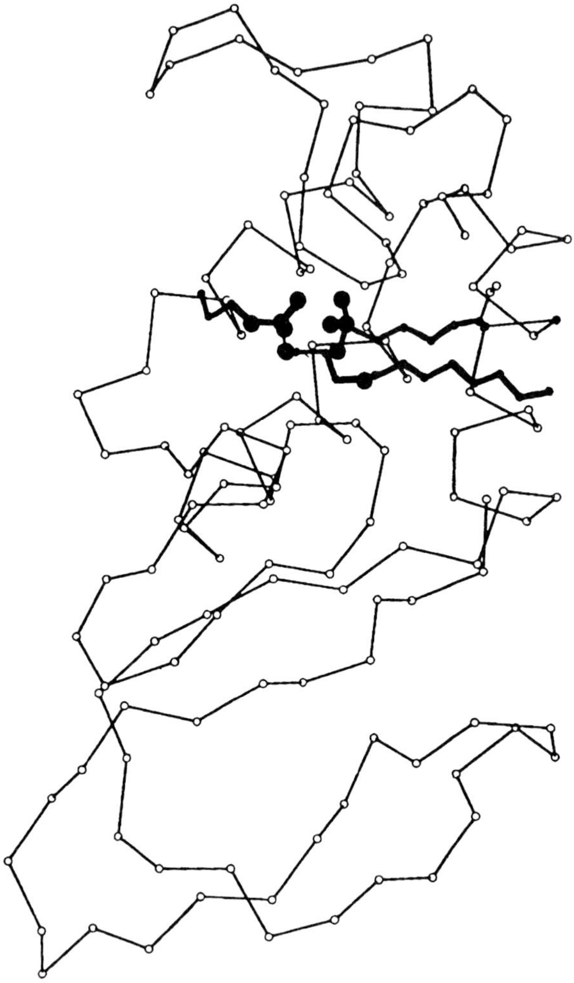

FIG. 5. Topology of a transition state analog inhibitor of PLA2, MG14 (see Fig. 8), in the catalytic site of PLA2 from honey bee venom. The C_α-backbone of the enzyme is shown, and the inhibitor is depicted by bold lines. The i-face of the enzyme is on the right and is perpendicular to the page. The inhibitor fits into a slot in the enzyme that runs from the i-face to the opposite side of the molecule, underscoring the fact that the i-face and the catalytic site are formed by different protein structural elements. Similar structural patterns are seen in the complexes of inhibitors with the enzymes from pig pancreas, cobra venom, and human inflammatory exudates.

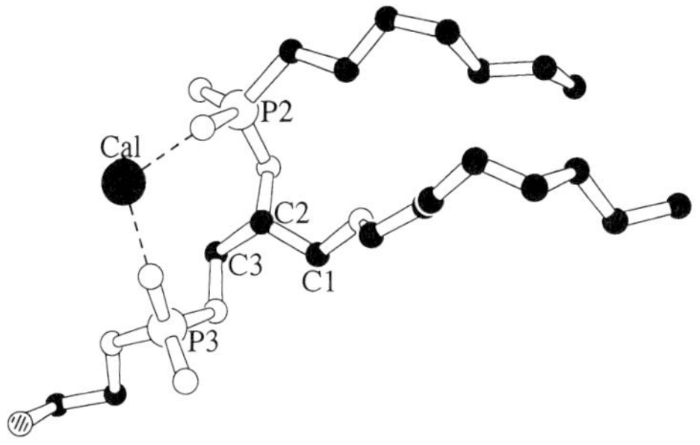

FIG. 6. Direct interaction of MG14 with the catalytic site calcium ion. P2 designates the phosphonate (see Fig. 8) that replaces the enzyme-susceptible *sn*-2 ester, and P3 is the phosphate diester of the polar head group. The hatched atom at the left end of the molecule is the ammonium group of the ethanolamine chain. In addition to the coordination of the calcium to two oxygen atoms of the inhibitor, the protein donates five oxygen atoms: two from D49, and three from backbone carbonyls of Y28, G30, and G32.

The binding of the enzyme to the interface is believed to occur via the i-face. According to the interpretation shown in Fig. 7, the i-face of the bee venom and pig pancreatic PLA2s consists of a ring of hydrophobic residues which is surrounded by a collar of six to eight cationic residues.[21] The entrance to the catalytic site is through the i-face which is exposed to bulk water when the enzyme is in the aqueous phase. The i-face is shielded from the bulk water when it is in contact with an interface.[22] Not only are the catalytic site and the i-face topologically distinct, but the binding of the enzyme to the interface (E to E*) is a process separate from the binding of a single phospholipid substrate molecule into the catalytic site of the enzyme at the interface (E* to E*S). Direct evidence for this assertion comes from the results, to be discussed below, which show that the enzyme can bind to the interface even if the catalytic site is devoid of substrate and presumably filled with water molecules. This point is critical for a proper understanding of interfacial catalysis by PLA2.

Neutral Diluents, Substrates, and Transition State Mimics

The optimum structural requirements of glycerides as substrate for PLA2 are an anionic group at the *sn*-3 position, an *sn*-2 ester linkage, and an acyl or ether chain at the *sn*-1 position (Fig. 3). Available evidence suggests that the conformation of the glycerol backbone of phospholipids is essentially the same in aqueous dispersions, in organic solvents, in

[21] F. Ramirez and M. K. Jain, *Proteins* **9,** 229 (1991).

[22] M. K. Jain and B. P. Maliwal, *Biochemistry* **32,** 11838 (1993).

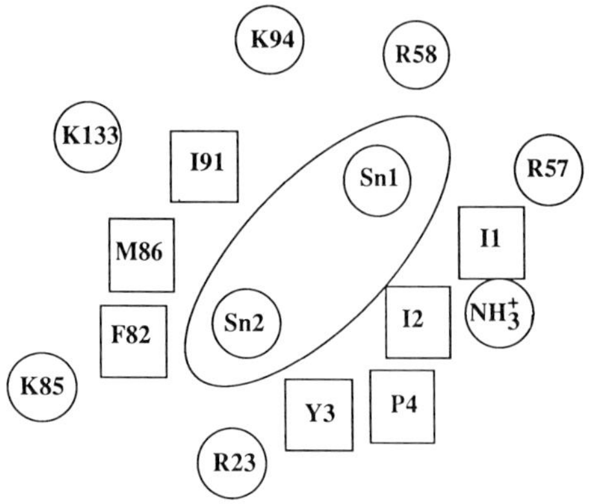

FIG. 7. Schematic representation of the i-face of the PLA2 from bee venom. Sn1 and Sn2 designate the methyl groups at the tips of the alkyl chains of MG14 that lie in the catalytic slot, and thus the view of the i-face is from the plane of the membrane facing toward the aqueous phase. A number of lysines and arginines form a collar of positive charge that surrounds the catalytic slot, and there is an inner ring of hydrophobic amino acids. The N terminus of the protein, designated as NH_3^+, is also part of the i-face.

crystals,[23] and at the catalytic site of PLA2.[16-18] Thus, suggestions about a conformational change in the substrate under the kinetic conditions of interfacial catalysis are discounted.

Based on the separate identities of the catalytic site and the i-face, it is apparent that structural features of ligands that dictate catalytic active site interactions are different from those that influence the binding of the enzyme to the interface. For example, the *sn*-2 phosphonate-containing phospholipid MG14, shown in Figs. 6 and 8, forms micelles only at high concentrations, and it binds tightly to the catalytic site of the enzyme at the interface. As discussed later such transition state mimics function as inhibitors by competing with the substrate for binding to the catalytic site of the enzyme.

Neutral diluents shown in Fig. 9 exemplify the other extreme. These amphiphilic compounds form aggregates to which the enzyme can bind, but a neutral diluent molecule does not have significant affinity for the catalytic site of the bound enzyme.[24] Known neutral diluents are phospholipid analogs and may be viewed as extremely poor inhibitors. Thus, subtle changes in the structure of phospholipids can have a profound influence

[23] H. Hauser, I. Pascher, R. H. Pearson, and S. Sundell, *Biochim. Biophys. Acta* **650,** 21 (1981).

[24] M. K. Jain, B.-Z. Yu, J. Rogers, G. N. Ranadive, and O. G. Berg, *Biochemistry* **30,** 7306 (1991).

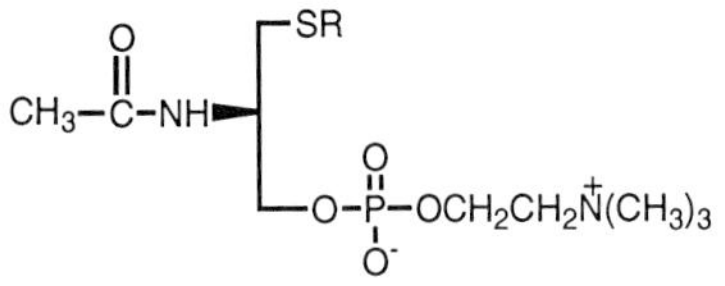

RM-2 R=$C_{18}H_{37}$

RM-3 R=C_8H_{17}

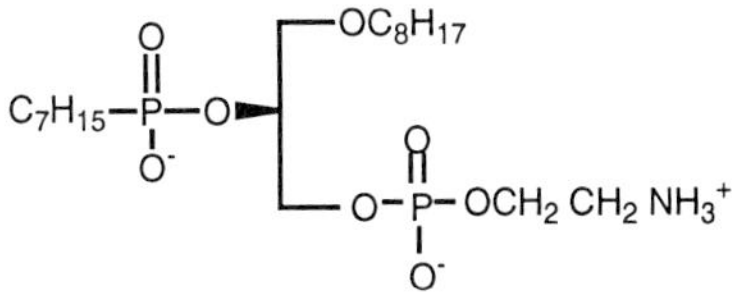

MG-14

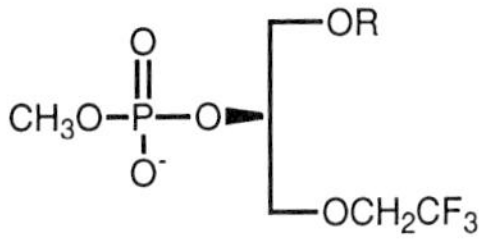

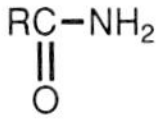

Acylamide

MJ33 R=$C_{16}H_{33}$

MJ72 R=C_8H_{17}

FIG. 8. Structures of catalytic site-directed inhibitors of PLA2.

on the affinities for the catalytic site of the enzyme at the interface. Although most phospholipid dispersions may bind PLA2, a neutral diluent for one PLA2 may be an inhibitor for another, and vice versa. Protocols for distinguishing a neutral diluent from an inhibitor are described later in this review. It will also be shown that neutral diluents are extremely useful as tools to alter the surface mole fraction of phospholipid substrates and also the bulk concentration of the interface as in aqueous dispersions.

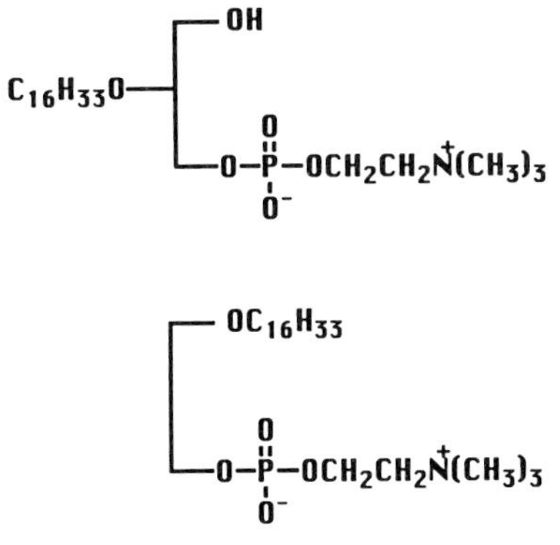

FIG. 9. Structures of neutral diluents for pig pancreatic PLA2: (top) 2-hexadecyl-*sn*-glycero-3-phosphocholine; (bottom) 1-hexadecylpropanediol-3-phosphocholine.

This provides a matrix for determining equilibrium constants for the binding of ligands to the catalytic site of PLA2.

Detergents have been used as "surface diluents[25] to change the surface concentration of phospholipid substrates at the interface. Mixed micelles prepared by dispersion of phospholipids in detergents are used to study the kinetics of interfacial catalysis by PLA2 where the characteristics of the reaction progress depend on the nature and mole fraction of the detergent.[3-5] Nevertheless, commonly used detergents, such as Triton X-100 and bile salts, are not suitable neutral diluents.[24] For example, changing the mole fraction of detergent in codispersions with phospholipids changes the phase behavior, size, and polymorphism of the aggregates. With mixed micelles it is difficult to determine the concentration of the detergent in the particles. Because the cmc's of detergents utilized are high, their partitioning behavior between the aqueous phase and the aggregate will also change as a function of the bulk concentration. Furthermore, as described in Section H, interfacial catalysis by PLA2 on detergent–phospholipid mixed micelles is difficult to interpret in terms of standard kinetic parameters.

C. Key Considerations for Interpretation of Interfacial Catalysis

The Michaelis–Menten formalism for describing the steady-state kinetic behavior of water-soluble enzymes acting on water-soluble substrates is based on two key considerations. First, at any given time in a homogeneous reaction environment, all substrate and product molecules are equally accessible to a given enzyme molecule. Second, all enzyme molecules in the reaction mixture experience an identical environment at any given point in time. When these conditions hold at all time points along the reaction progress curve, the ensemble averaging of the individual events necessary to describe the observed steady-state kinetic behavior of the mixture is accomplished in a straightforward manner. These conditions must also be satisfied for the interpretation of the kinetics of interfacial catalysis. This is a nontrivial problem because additional kinetic possibilities come into play because of the local environment created by the clustering of components into aggregates and by the kinetics of the exchange processes intrinsic in a two-phase system (cf. Fig. 2).

Such complexities are illustrated in the following example. Consider the action of PLA2 on a population of phospholipid vesicles. Suppose

[25] A. Pluckthun and E. A. Dennis, *J. Biol. Chem.* **260,** 11099 (1985); A. Pluckthun, R. Rohlfs, F. F. Davidson, and E. A. Dennis, *Biochemistry* **24,** 4201 (1985); H. S. Hendrickson and E. A. Dennis, *J. Biol. Chem.* **259,** 5735 and 5740 (1984).

that the enzyme can bind to a vesicle, hydrolyze some of the substrates in that vesicle, and then leave the vesicle and bind to a new one (Fig. 10, bottom). As time evolves, the enzyme may bind to a vesicle that has not yet had a bound enzyme. This enzyme is effectively at zero time since it encounters a vesicle with only substrates and no products. At the same time another enzyme may bind to a vesicle that has already been substantially hydrolyzed. This enzyme will experience substrate depletion and product inhibition. Thus, exchange of enzyme between the ensemble of vesicles leads to a scrambling of time such that the environments seen by different enzyme molecules present in the reaction mixture are different at any point in time.

The same can be said for a population of aggregates where intervesicle exchange of substrates and/or products is significant. The situation would be analytically straightforward if all of the interacting species exchange very rapidly such that, at any given point in time, all of the enzyme

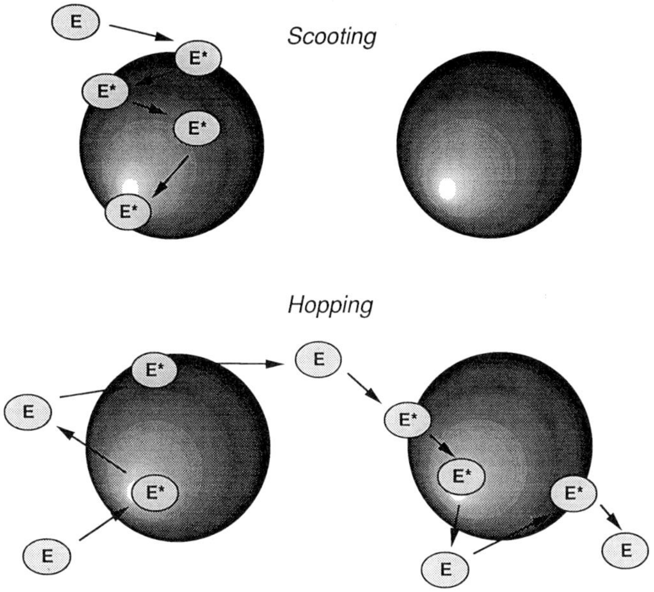

FIG. 10. Schematic illustration of the two extreme modes of interfacial catalysis on vesicles. In the scooting mode (top) the enzyme bound to the interface does not dissociate during several thousand turnover cycles. Under these conditions the bound enzyme "sees" only the substrate and inhibitor at the interface to which it is bound, and the excess vesicles which do not contain enzyme are not hydrolyzed. In the hopping mode (bottom) the enzyme desorbs from the interface after each or a few turnover cycles. Thus excess vesicles are ultimately accessible for hydrolysis.

molecules "see" the same time-averaged environment.[26] However, the intrinsic rates of intervesicle exchange of PLA2[27] and phospholipids[28] are slow on the time scale required for the enzyme to hydrolyze a significant fraction of substrate present in a vesicle. Thus, what one would monitor under these conditions is the intervesicle exchange rate rather than the rate of the lipolysis reaction. As discussed in Section H, the problem is not solved by using mixed micelles since the intermicelle exchange processes are also too slow to keep pace with the enzymatic reaction. If the exchange rates are not rapid on the time scale of the hydrolysis of the individual aggregates, the equations that describe the reaction progress under limiting-exchange conditions become very complex since they must contain not only the parameters that describe the interfacial catalysis by the vesicle- or micelle-bound enzyme, but also the rate parameters that describe all of the relevant exchange processes. Finally, to make matters more complex, the intrinsic rate parameters become apparently time-dependent if the affinity of the enzyme for the interface changes with the changing ratio of the substrate to products, as is indeed the case with zwitterionic vesicles of phosphatidylcholine[29] (Section H).

The conceptual basis for difficulties encountered in the interpretation of catalysis on interfaces under the conditions of time scrambling goes to the very core of enzymology, that is, the steady-state condition. The environment experienced by the enzyme determines the true steady state that controls the observed turnover rate by each enzyme. With exchange of enzyme, substrate, and products among the ensemble of aggregates there are two steady-state conditions which control the observed reaction progress: a local, or microscopic, condition that describes the action of the enzyme bound to a single aggregate, and a global, or macroscopic condition, determined by the total (bulk) concentration of the components. The microscopic environment seen by the enzyme on the substrate aggregate need not be the same as the experimentally observable global environment defined in terms of the total substrate and products present in the entire reaction mixture.

In short, a fundamental challenge in the interpretation of the behavior of interfacial enzymes arises from the fact that they work on aggregates

[26] O. G. Berg, B.-Z. Yu, J. Rogers, and M. K. Jain, *Biochemistry* **30,** 7283 (1991).

[27] M. K. Jain, J. Rogers, and G. H. de Haas, *Biochim. Biophys. Acta* **940,** 51 (1988).

[28] D. A. Fullington, D. G. Shoemaker, and J. W. Nichols, *Biochemistry* **29,** 879 (1990); J. W. Nichols, *Biochemistry* **27,** 3925 (1988).

[29] R. J. Apitz-Castro, M. K. Jain, and G. H. de Haas, *Biochim. Biophys. Acta* **688,** 349 (1982); M. K. Jain, M. R. Egmond, H. M. Verheij, R. J. Apitz-Castro, R. Dijkman, and G. H. de Haas, *Biochim. Biophys. Acta* **688,** 341 (1982); M. K. Kain, B.-Z. Yu, and A. Kozubek, *Biochim. Biophys. Acta* **980,** 23 (1989).

of substrates rather than on a bulk collection of solitary substrate monomers, as is the case in homogeneous solutions. As a rapid exchange of the components between the aggregates appears to be experimentally unachievable, our approach, as elaborated next, has been to eliminate the kinetic contribution of the exchange processes to the observed kinetics of PLA2-catalyzed reactions.

Kinetic Variables at Interface

As will become apparent in this review, analysis of interfacial catalysis can be accommodated in terms of the kinetic scheme shown in Fig. 11. According to this minimal adaptation of Michaelis–Menten formalism for interfacial catalysis, the binding of the enzyme in the aqueous phase to the interface (E to E*) precedes the catalytic turnover in the interface. In addition, the E to E* step is distinct from the binding of a single phospholipid substrate to the catalytic site of the interface-bound enzyme, the E* + S to E*S step. This distinction has broad and significant consequences. Two key aspects of this deceptively simple but remarkably versatile scheme may be emphasized at the outset. First, the residence time of E* at the interface determines the catalytic processivity of the enzyme at

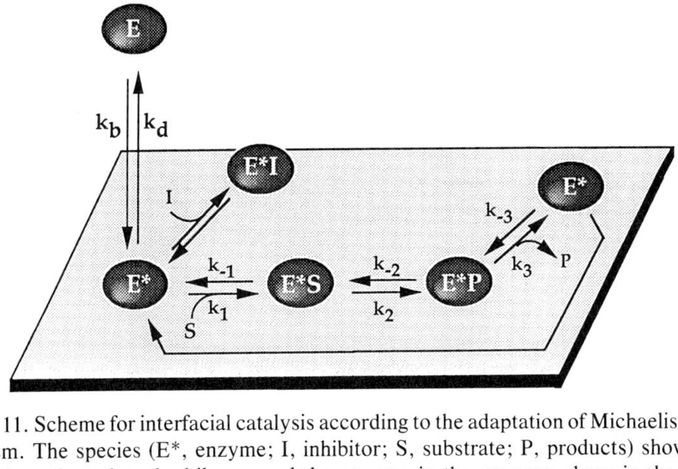

FIG. 11. Scheme for interfacial catalysis according to the adaptation of Michaelis–Menten formalism. The species (E*, enzyme; I, inhibitor; S, substrate; P, products) shown in the box are in or bound to the bilayer, and the enzyme in the aqueous phase is shown as E. During steady-state turnover, E is in equilibrium with E*. E* brings about catalytic turnover by the steps shown in the box. Factors that regulate catalysis will modulate the steps in the box, whereas a shift in the E to E* equilibrium will modulate the apparent rate by increasing or decreasing the fraction of the total enzyme available for interfacial catalysis. The values of the various rate and equilibrium constants and parameters characteristic for this scheme are summarized in Table I.

the interface, that is, the E to E* step controls the number of catalytic turnover cycles that the enzyme undergoes during each visit from the aqueous phase to the interface. Second, the organization and dynamics of the aggregate control not only the E to E* equilibrium, but also the local concentration of reactants that the bound enzyme "sees" during the catalytic turnover at each time point during the progress of the reaction. Indeed, depending on the rate of exchange of the enzyme, substrate, and products between aggregates, it is possible to generate enzymatic reaction progress curves of numerous shapes[5] which can be rationalized in terms of the scheme in Fig. 11. However, the analytical interpretation of interfacial catalysis in terms of the primary rate and equilibrium parameters of the scheme (Fig. 11) is possible only under certain conditions where the parallel exchange processes implicit in the scheme are constrained so as to become kinetically insignificant.

Constraining Variables for High Processivity

Complexities of the reaction progress resulting from the type of time scrambling described above can be avoided by the proper choice of experimental conditions. Consider the action of a lipolytic enzyme on phospholipids in a bilayer vesicle along the lines illustrated in Fig. 10. In the extreme case of fast hopping, the enzyme remains on a vesicle only for a brief instance during each visit, for example, for a single turnover cycle, after which it hops to a new vesicle. In this way, at any particular point in time, the substrate present in all vesicles, on the average, are equally accessible and become enzymatically acted on to a similar extent. In some ways this fast-exchange situation is similar to solution-phase enzymology. The same fast-exchange condition may be satisfied by rapid intervesicle exchange of substrates and products. Unfortunately, the rate of hopping of PLA2[26,27] or the exchange of substrate[28,30] is not rapid enough under most experimental conditions to support the intrinsic interfacial catalytic turnover number of a few hundred per second.

The most feasible approach is to study the action of lipolytic enzymes on vesicles in the absence of intervesicle hopping of the enzyme and of the exchange of substrates and products. With the help of Fig. 10 (top), imagine the reaction mixture in which there are many more vesicles than enzyme molecules such that according to Poisson distribution statistics at most one enzyme is present on each enzyme-containing vesicle. Enzyme remains bound to the surface of a vesicle longer than the time needed for all of the substrate in the outer layer of the vesicle to become hydrolyzed.

[30] M. K. Jain, J. Rogers, H. S. Hendrickson, and O. G. Berg, *Biochemistry* **32,** 8360 (1993).

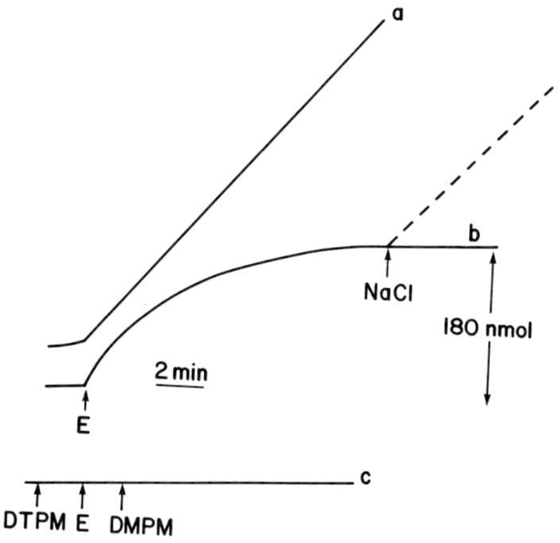

FIG. 12. Reaction progress curves for the hydrolysis of sonicated DMPM vesicles by pig pancreatic PLA2. (Bottom) PLA2 (30 pmol) is added to 4 ml of 0.3 m*M* $CaCl_2$, 1 m*M* NaCl, pH 8.0, containing 1 μmol of DTPM as vesicles. This is followed by the addition of substrate (DMPM) vesicles. (Middle) Hydrolysis of DMPM vesicles is initiated immediately after the addition of PLA2; however, the reaction ceases after only a small fraction of the substrate has been hydrolyzed. If 0.2 *M* NaCl is added after the cessation of the reaction, hydrolysis of excess vesicles begins immediately (dashed line). (Top) If PLA2 is added to DMPM vesicles in the presence of 0.2 *M* NaCl, hydrolysis is initiated immediately, and all of the accessible substrate is hydrolyzed at the end of the reaction.

This virtually infinite processive behavior is called interfacial catalysis in the scooting mode.[31] Under these conditions, all enzyme-containing vesicles will behave, on the average, identically in time. Thus experimental observations, such as the total product formed, will simply be the sum of products formed in each of the enzyme-containing vesicles. In other words, measuring the total product formed as a function of time is equivalent to measuring the total product formed in any of the enzyme-containing vesicles as a function of time. The kinetics under the scooting condition thus provide a simple way to link the microscopic events of the reaction progress by a single enzyme on a single substrate vesicle to the macroscopic steady-state kinetics that describe the behavior of the ensemble of enzyme-containing vesicles.

The reaction progress curve for the action of PLA2 in the scooting mode on vesicles of the negatively charged phospholipid 1,2-dimyristoyl-*sn*-glycero-3-phosphomethanol (DMPM) is shown in Fig. 12 (unless stated otherwise, the PLA2 utilized is purified from porcine pancreas). Similar

[31] M. K. Jain, J. Rogers, D. V. Jahagirdar, J. F. Marecek, and F. Ramirez, *Biochim. Biophys. Acta* **860,** 435 (1986); G. C. Upreti and M. K. Jain, *J. Membr. Biol.* **55,** 113 (1980).

behavior is observed with zwitterionic phosphatidylcholine vesicles containing anionic phospholipids.[32] These progress curves are generated by measuring the appearance of protons released from the fatty acid product with a pH-stat titrator; identical results are obtained using radiometric measurements. An interesting feature of such progress curves is that in the presence of more vesicles than enzymes the reaction ceases when only a fraction of the total substrate present in the reaction mixture is hydrolyzed, despite the fact that the equilibrium for the reaction is virtually completely in favor of products. Cessation of the hydrolysis is due to the irreversible binding of the enzyme to vesicles such that the enzyme, once bound, does not hop to the excess vesicles present in the reaction mixture.

This conclusion is based on the following observations. (a) Addition of more substrate does not initiate further hydrolysis although the enzyme added initially is still catalytically active as shown by reinitiation of the hydrolysis by promoting intervesicle exchange of the enzyme or the substrate (Section D).[31] (b) In the presence of vesicles and an excess of enzymes such that the enzyme-containing vesicles contain at most one enzyme per vesicle (Poisson distribution), the total amount of product formed (extent of reaction) increases linearly with increasing amounts of enzyme.[31] Addition of an aliquot of enzyme during a reaction already in progress will not necessarily cause a proportional increase in the observed rate because some of the enzyme molecules added later will bind to vesicles that are already partially hydrolyzed. (c) Addition of sufficient enzyme such that every DMPM vesicle has at least one enzyme (>5 enzymes per vesicle) results in the hydrolysis of about 60% of the total DMPM; this is the amount present in the outer monolayer of small sonicated vesicles. This indicates that the cessation of the reaction seen when there are more vesicles than enzymes is not due to inhibition by reaction products. The effect of products on the reaction progress is explicitly considered in Section E. (d) The extent of hydrolysis increases with increasing size of the vesicles.[31] Under conditions of at most one enzyme per vesicle, the extent of the reaction per enzyme is experimentally found to equal the number of substrate phospholipids present in the outer monolayer of the vesicles (N_S). This has been established by measuring the size of DMPM vesicles by a number of independent methods.[33] (e) The enzyme-containing vesicles retain their integrity, and the solute trapped in the aqueous compartment of the vesicles is not released even when all of the outer layer substrate is hydrolyzed.[26] (f) PLA2 added initially to nonhydrolyzable

[32] F. Ghomashchi, B.-Z. Yu, O. G. Berg, M. K. Jain, and M. H. Gelb, *Biochemistry* **30,** 7318 (1991).

[33] M. K. Jain, G. Ranadive, B.-Z. Yu, and H. M. Verheij, *Biochemistry* **30,** 7330 (1991).

vesicles of the diether phospholipid 1,2-ditetradecyl-*sn*-glycero-3-phosphomethanol (DTPM) is not available for the hydrolysis of DMPM vesicles added subsequently.[31] Similarly, DTPM added during the progress of the reaction on DMPM has little effect on the course of hydrolysis.

Phospholipases A_2 from various sources give the same extent of hydrolysis per enzyme, and in all cases this number is equal to N_S.[33] This and other observations show that monomeric PLA2 at the interface is fully catalytically active. Enzyme dimers or trimers observed in crystallographic unit cells[34] seem to have little relevance for the events of interfacial catalysis. The formation of enzyme aggregates in the absence of a phospholipid interface is probably due to the exposure of the hydrophobic region of the i-face to the aqueous environment, and these interactions are probably disrupted on binding of the enzyme to the interface.

Having established that PLA2, once bound to anionic vesicles, is not available for the hydrolysis of excess vesicles, the steady-state time course of the reaction progress (Fig. 12) in the scooting mode is described only by the steps shown in the box (see Fig. 11), that is, the enzyme does not desorb from the vesicle interface during several thousand catalytic turnover cycles. Under these conditions the intervesicle exchange of the substrate and products is also negligible on the time scale of the reaction progress.

Lipolysis in the scooting mode provides absolute proof for catalysis at the interface. As elaborated next, under such conditions, interfacial catalytic turnover can be described by the interfacial rate and equilibrium constants. With most, if not all, other forms of phospholipid aggregates which have been used to study interfacial catalysis, the observed time course of the enzymatic reaction is convoluted with exchange processes. The effect of these on the observed reaction progress is discussed later in this review.

It is important to point out here that constraining the reaction progress to occur in the scooting mode is an "experimental trick" used to avoid the problem of time scrambling so that the kinetic and equilibrium constants which describe the action of the enzyme at the interface can be measured. In other situations, PLA2 may operate in a less processive mode, but the catalytic turnover of naturally occurring phospholipids in aqueous dispersions will still occur only by the enzyme at the interface

[34] J. P. Demaret, S. Chevetzoff, and S. Brunei, *Protein Eng.* **4,** 171 (1990); M. M. G. M. Thunnissen, P. A. Franken, G. H. de Haas, J. Drenth, K. H. Kalk, H. M. Verheij, and B. W. Dijkstra, *J. Mol. Biol.* **232,** 839 (1993); M. M. G. M. Thunnisen, P. A. Franken, G. H. de Haas, J. Drenth, K. H. Kalk, H. M. Verheij, and B. W. Dijkstra, *J. Mol. Biol.* **232,** 839 (1993).

and will be controlled by the same interfacial rate and equilibrium constants that are determined from the scooting mode analysis.

D. Kinetic Formalism for Interfacial Catalysis in Scooting Mode

The rationale, assumptions, and experimental evidence leading to the formalism for the interpretation of interfacial catalysis are developed below. In terms of the scheme given in Fig. 11, this requires conceptualization of the traditional Michaelis–Menten formalism to accommodate the constraints of the interface, the evaluation of the contributions of the exchange processes which ultimately determine the residence time of the enzyme at the interface in a given environment, and the consideration of the relationship of this environment to the bulk-averaged environment. For example, the fraction of enzyme bound to the interface (E*) versus in the solution (E) is controlled by the bulk concentration of the interface, expressed in units of molarity. On the other hand, the surface concentration of ligands (substrate, products, inhibitors) at the interface determines the fraction of interface-bound enzyme that contains bound ligands. Surface concentration is most conveniently expressed in terms of the mole fraction of a ligand at the interface. This quantity is proportional to the concentration or number density in the surface as long as all of the components in the interface have similar cross-sectional molecular areas. This distinction of bulk and surface units follows from the conceptual separation of the i-face of PLA2 from its catalytic site.

Additional Features of Hydrolysis of Vesicles in Scooting Mode

It has already been mentioned in Section C that the cessation of the reaction progress for the action of PLA2 on DMPM vesicles is due to depletion of phospholipid substrate in the outer layer of enzyme-containing vesicles. It is important to demonstrate that other factors do not contribute to the reaction cessation. As shown in Fig. 12, addition of NaCl after the reaction cessation leads to a reinitiation of the reaction progress. Depending on the concentration, salt promotes the exchange of the enzyme from one DMPM vesicle to another,[35,36] and this brings the enzyme in contact with all of the substrate vesicles. Similar results are seen following the addition of calcium at concentrations exceeding 1 m*M*, which causes the fusion of DMPM vesicles. Fusion brings more substrate into substrate-depleted enzyme-containing vesicles. These results clearly show that the

[35] M. K. Jain, B. P. Maliwal, G. H. de Haas, and A. J. Slotboom, *Biochim. Biophys. Acta* **860,** 448 (1986).
[36] M. K. Jain, J. A. Rogers, O. G. Berg, and M. H. Gelb, *Biochemistry* **30,** 7340 (1991).

enzyme does not become inactivated during the course of the reaction. Additional consideration of these effects is discussed further later in this review (Section H).

A complete kinetic analysis of the reaction progress requires that the DMPM vesicles are of the same size (i.e., a narrow monomodal dispersity) or that the size dispersity is known *a priori*. If the vesicles are not of uniform size, the observed reaction progress curve will be a summation of the different curves for each of the different size vesicles.[26] Thus, large vesicles will give a larger extent of reaction (N_S) since the bound enzyme is exposed to a larger amount of substrate in the outer monolayer. Determining the size dispersity is difficult, and it is best to use carefully sonicated small unilamellar vesicles of DMPM of uniform size. The size of the vesicle depends on the fatty acyl chain purity of the lipid preparation; however, vesicles containing about 6000 DMPM molecules in the outer monolayer can be routinely prepared.[37] It is best to calibrate the size of the vesicles by measuring the extent of reaction catalyzed by a preparation of fully native PLA2. Large unilamellar vesicles (typically 0.1 μm diameter) prepared by high-pressure extrusion through polycarbonate filters of the corresponding pore size can also be used.

In summary, kinetic analysis of PLA2 acting in the scooting mode on an ensemble of monodisperse vesicles provides a method for obtaining the steady-state kinetic parameters that describe the interfacial catalysis according to the steps within the box of the kinetic scheme shown in Fig. 11. During the interfacial catalytic turnover the enzyme bound to the vesicle responds not to the bulk concentration of the substrate, but to the surface concentration because this is what the enzyme "sees" during the reaction cycle.

E. Determination of Interfacial Equilibrium Parameters

Interpretation of the reaction progress in terms of the scheme in Fig. 11 requires knowledge of several equilibrium dissociation constants (K_L* where L is substrate, products, cofactor, or inhibitor) for ligands bound to PLA2 at the interface. Asterisks are used to emphasize the fact that the equilibrium is for the species at the interface. The usual methods for the determination of equilibrium dissociation constants for ligands bound to enzymes in solution is not readily adopted for interfacial equilibria. If the i-face and the catalytic site on PLA2 are structurally and functionally

[37] M. K. Jain and M. H. Gelb, this series, Vol. 197, p. 112. It may be noted here that the size of the smallest vesicles prepared by sonication depends not only on the time of sonication and energy input, but also on the structure and composition of the phospholipids.

separate, it follows that the enzyme in the interface can exist in discrete states (E* or E*L). As described next, on aqueous dispersions of a neutral diluent E* and E*L can be distinguished by the protection of the catalytic site of PLA2 from alkylation of His-48 by an alkylating agent.

Interfacial Equilibrium Constants by Protection Method

Studies with PLA2 at the interface of neutral diluents prove that the enzyme at the interface can exist in the E* and E*L forms. The importance of identifying suitable neutral diluents can hardly be overemphasized. It is like the characterization of a nonreactive solvent. Not only do neutral diluents permit the spectroscopic study of E* and E*L forms of an interfacial enzyme, but the equilibrium constants between these forms of the enzyme can be determined in the presence of catalytic site-directed ligands. The set of equilibria for PLA2 bound to the interface of a neutral diluent in the presence of a catalytic site-directed ligand are illustrated in Fig. 13. Because neutral diluents do not bind to the catalytic site of PLA2, the enzyme bound to an aggregate of a neutral diluent will be in E* form. The His-48 in the catalytic site of PLA2 is selectively alkylated by phenacyl

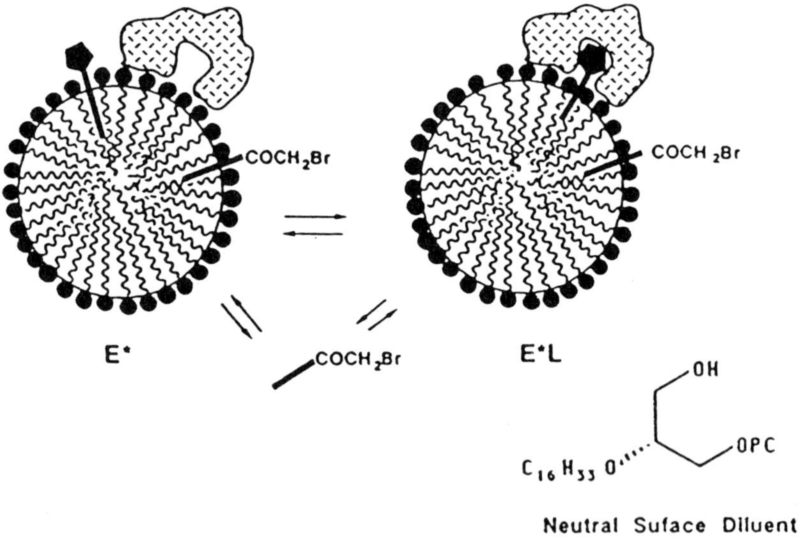

FIG. 13. Equilibrium binding of a catalytic site-directed ligand to E* to form E*L. An alkylating agent, such as a bromomethyl ketone, partitioned between the micelle of a neutral diluent and the aqueous phase reacts only with the catalytic residue of E*.

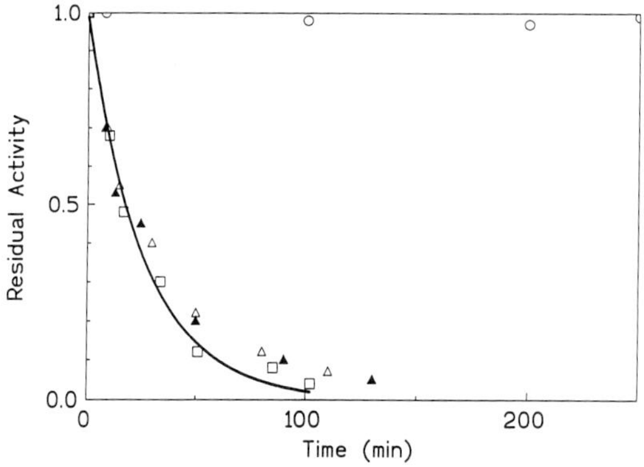

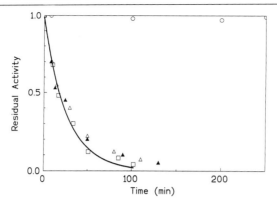

FIG. 14. Time course of inactivation of pig pancreatic PLA2 (30 μM) with *p*-nitrophenacyl bromide (1 m*M*). Activity was measured for (□) enzyme in the aqueous phase, (△) enzyme bound to 3.3 m*M* 1-hexadecylpropanediol-3-phosphocholine, (▲) enzyme bound to 3.3 m*M* 2-hexadecyl-*sn*-glycero-3-phosphocholine, and (○) in 0.5 m*M* products of hydrolysis of DMPM. Reaction mixtures also contained 0.5 m*M* $CaCl_2$ and 50 m*M* NaCl in 50 m*M* cacodylate buffer at pH 7.3 and 23°.

bromides, which renders the enzyme inactive.[38] As shown in Fig. 14, binding of a ligand to the catalytic site of E* bound to a neutral diluent protects the enzyme from alkylation, whereas in the absence of ligand the rate of enzyme inactivation is the same as for the enzyme in the aqueous solution (E). Moreover, as discussed later, by spectroscopic methods it can be demonstrated that the enzyme binds to the aqueous dispersions of a neutral diluent. By these criteria 2-hexadecyl-*sn*-glycero-3-phosphocholine and 1-hexadecylpropanediol-3-phosphocholine were found to be effective neutral diluents for porcine and bovine[39] pancreatic PLA2, and hexadecyl-1-phosphocholine is an effective neutral diluent for the PLA2 from human synovial fluid.[40] The cmc's of these neutral diluents are about 10 μM, whereas the dissociation constants for the enzyme bound to micelles of these neutral diluents (K_d) is between 1 and 3 m*M*. By the method described below, the estimated K_L^* for the neutral diluent molecule bound to pig pancreatic PLA2 at the interface is over 2 mole fraction. In other words, the enzyme bound to a micelle of neutral diluent at mole fraction 1 will be mainly in the E* form.

[38] J. J. Volwerk, W. A. Pieterson, and G. H. de Haas, *Biochemistry* **13,** 1446 (1974).

[39] C. M. Dupureur, B.-Z. Yu, M. K. Jain, J. P. Noel, T. Deng, Y. Li, I. L. Bycon, and M. D. Tsai, *Biochemistry* **31,** 6402 (1992); C. M. Dupureur, B.-Z. Yu, J. A. Mamone, M. K. Jain, and M. D. Tsai, *Biochemistry* **31,** 10576 (1992).

[40] T. Bayburt, B.-Z. Yu, H. K. Lin, J. Browning, M. K. Jain, and M. H. Gelb, *Biochemistry* **32,** 573 (1993).

The half-time for alkylation of PLA2 bound to dispersions of neutral diluents increases in the presence of a catalytic site-directed ligand in the reaction mixture. The equilibrium dissociation constant for the E*L complex is calculated from the following relationship:

$$\frac{t_L}{t_L - t_0} = \frac{1 + K_L^*/X_L}{1 - k_L/k_0} \tag{1}$$

where t_0 is the half-time for the inactivation by an alkylating agent of E* bound to an aggregate of the neutral diluent in the absence of a ligand and t_L is the half-time in the presence of a ligand at mole fraction, X_L, in the codispersions of the neutral diluent and the ligand. Dissociation constants for water-soluble ligands are expressed in bulk concentration units (m*M*), whereas the K_L^* values of ligands dispersed in the interface of a neutral diluent are expressed in mole fraction units (which are in effect unitless).

Values of K_L^* are calculated from the slope of the plot of $t_L/(t_L - t_0)$ versus $1/X_L$. This relationship is considerably simplified if the ratio of the inactivation rate constants for E*L and E* with the alkylating agent, k_L/k_0, is much less than 1, as is the case with many ligands that bind to the catalytic site of pig pancreatic PLA2. The K_L^* values for several ligands are summarized in Table I. For example, K_{Ca}^*, the dissociation constant for calcium bound to E* at the interface of 1-hexadecylpropanediol-3-phosphocholine, was obtained from the half-times for alkylation of PLA2 in the absence (t_0) or presence (t_L) of known concentrations of calcium.

As shown schematically in Fig. 15, binding of the catalytic site-directed ligands is observed only in the presence of calcium or certain other divalent cations, that is, the binding of the cation to the enzyme is obligatory before the ligand can bind. The effective dissociation constant for the cation at a given mole fraction of the ligand, $K_C^*(L)$, is obtained from the half-time for alkylation in the absence of calcium (t_0) and the half-time for alkylation in the presence of calcium and ligand (t_{CL}) at mole fraction X_L in the neutral diluent. $K_C^*(L)$ is related to the dissociation constant of the cation in the absence of the ligand, K_C^*:

$$\frac{t_0[C]}{t_{CL} - t_0} = K_C^*(L) = \frac{K_C^*}{1 + X_L/K_L^*} \tag{2}$$

where [C] is the bulk cation concentration. The first equality is based on the fact that, in Eq. (1), $k_L/k_0 \ll 1$. The constant K_L^* is the dissociation constant for the ligand at saturating concentration of divalent cations which acts as cofactor for the binding of the ligand. It is related to the

TABLE I
DEFINITION AND VALUES OF EQUILIBRIUM CONSTANTS, KINETIC RATE CONSTANTS, AND PARAMETERS[a]

Parameter	Value	Definition
k_1	1350 sec^{-1} (MF^b)$^{-1}$	Rate constant for E* + S to E*S
k_{-1}	35 sec^{-1}	Rate constant for E*S to E* + S
$k_2 = k_{cat}$	400 sec^{-1}	Rate constant for decomposition of E*S to products
k_b	4 sec^{-1}	Diffusion-independent on-rate constant for the E to E* step
k_d	<0.00002 sec^{-1}	Off-rate constant for the E* to E step
K_{Ca}	0.35 mM	Dissociation constant for ECa in the aqueous phase
K_{Ca}*	0.24 mM	Dissociation constant for E*Ca
K_{Ca}*(S)	0.06 mM	Michaelis constant for Ca dependence of v_0
K_d	0.9 mM	Dissociation constant for E* bound to 1-hexadecylpropanediol-3-phosphocholine to give E
K_d^I	0.03 mM	Dissociation constant for E*I bound to 1-hexadecylpropanediol-3-phosphocholine to give EI
K_I	0.008 mM	Dissociation constant for E · MJ33
K_I*	0.0008 MF	Dissociation constant for E* · MJ33
	0.0011 MF	Dissociation constant for E* · MG14
K_L*		Interfacial dissociation constant for E* · L complex
K_s*	0.025 MF	k_{-1}/k_1 (for DTPM)
K_P*	0.03 MF	k_3/k_{-3} (for products of hydrolysis of DMPM)
K_m*	0.35 MF	Michaelis–Menten constant for DMPM
K'	<0.01 μM (MF)$^{-1}$	Dissociation constant for MJ33 from bilayer
$N_s k_i$	35 sec^{-1}	Defined in Eq. (7)
v_0	320 sec^{-1}	Initial rate per enzyme for hydrolysis of DMPM at $X_s = 1$

[a] Described in this review for the hydrolysis of DMPM vesicles by PLA2 from pig pancreas at pH 8.0 and 22°.
[b] MF, Mole fraction in the interface.

effective dissociation constant of the ligand in the presence of subsaturating amounts of cation, $K_L^*(C)$:

$$\frac{t_C X_L}{t_{CL} - t_C} = K_L^*(C) = K_L^*(1 + K_C^*/[C]) \tag{3}$$

The value of $K_L^*(C)$ is obtained from the half-times for alkylation in the presence of cation (t_C) and in the presence of cation plus ligand (t_{CL}). The constant K_C^* can be obtained directly by monitoring the inactivation half-times as a function of [C] in the absence of the catalytic site-directed ligand. Also, based on Eq. (2) or (3), K_L^* and K_C^* can be obtained from

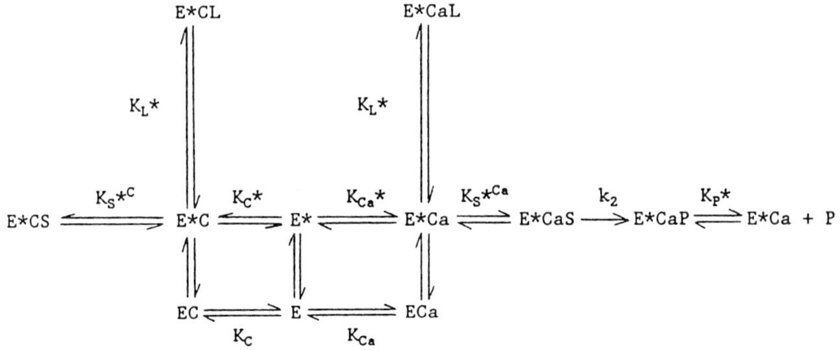

FIG. 15. Scheme for interfacial catalysis which elaborates the cation-dependent equilibria for the binding of catalytic site-directed ligand (L) and substrate (S) to the enzyme at the interface. The experimentally demonstrated key feature of this scheme is that several divalent ions (designated by C) bind to the enzyme and support the binding; however, only certain ions such as calcium (Ca) promote the lipolysis reaction.[16] All possible equilibria in the aqueous phase are not developed in this scheme.

the linearized plots of the effective dissociation constants $K_C^*(L)$ or $K_L^*(C)$ as a function of [C] or X_L. Results obtained by both protocols are virtually the same and are consistent with the value obtained from the kinetic studies.[20]

Binding of Phospholipase A_2 to Interface

Binding of pig pancreatic PLA2 to the interface can be demonstrated by taking advantage of the spectral properties of Trp-3 localized on the i-face of the enzyme. As shown in Fig. 16, the change in the fluorescence emission from Trp-3 depends on the nature of the phospholipid as well as the concentration of the dispersions. Such plots are also useful for determining the equilibrium dissociation constant K_d for E* to E and K_d^I for E*I to EI.[20,22,29] These results (Table I) show that the binding of the enzyme to a zwitterionic interface is quite weak and that binding is promoted by anionic additives as well as by catalytic site-directed ligands.

A significant increase in the resonance energy transfer from Trp-3 emission to an acceptor dansyl group localized at the interface is also observed when the enzyme binds to the interface.[41] The time course of the E to E* step has been monitored by rapid kinetic techniques.[27] Such studies with the pig pancreatic PLA2 show that the interfacial binding is described by a second-order rate constant that depends on the bulk

[41] M. K. Jain and W. L. C. Vaz, *Biochim. Biophys. Acta* **905,** 1 (1987).

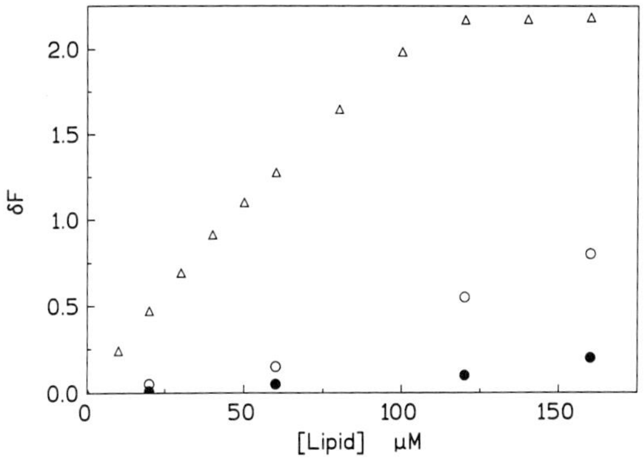

FIG. 16. Change in fluorescence emission intensity of pig pancreatic PLA2 (2 μM) as a function of the concentration of (△) DMPM, (●) DTPC, and (○) DTPC plus 15% products of hydrolysis of DMPC. The lipids were added as vesicles.

concentration of the lipid and is most probably the diffusion-limited binding of the enzyme to the interface. This is followed by a first-order process (4 sec^{-1}) which is probably related to the desolvation of the i-face of the enzyme and the portion of the interface that contacts the i-face.

Spectroscopic Signatures of E and E*L*

The protection method effectively distinguishes the E*L form from the E and E* forms. Although tedious, this method is generally useful for the determination of interfacial dissociation constants because His-48 is a key catalytic site residue in virtually all secreted PLA2s. Under suitable conditions it is also possible to monitor spectral perturbations of Trp-3 in pig pancreatic PLA2 which are characteristic of the E to E* or E* to E*L transitions. For example, the binding of the enzyme to the interface is not accompanied by a change in the near-UV absorbance, but such a change is observed when the bound enzyme interacts with a catalytic site-directed ligand. These results show that the binding of the enzyme to the interface is independent of the events of the catalytic cycle, although, as expected, the overall position of the binding equilibrium is determined by two sequential steps: E to E* to E*L. To interpret changes in the UV-difference spectra one relies on the changes that are observed on transfer of a chromophore to a solvent of differing dielectric,[42] which can be rationalized further in terms of the changes characteristic of the local

[42] J. W. Donovan, *in* "Physical Principles and Techniques of Protein Chemistry" (S. J. Leach, ed.), Part A, p. 101. Academic Press, New York, 1969.

molecular environment.[22] For example, changes in the absorption spectrum are observed during the E* to E*L transition ($\delta\varepsilon$ 2500 M^{-1} cm^{-1} at 292 nm with shoulders at 288 and 298 nm). A qualitatively different difference spectrum and smaller changes in the intensity ($\delta\varepsilon$ 300 M^{-1} cm^{-1} at 284 nm) are observed on the binding of the enzyme to the dispersions of a neutral diluent (E to E* transition). On the other hand, a major change in the fluorescence emission spectrum is observed during the E to E* transition, and a smaller change is observed on the formation of E*L. Such observations show that the spectral signatures of E, E*, and E*L forms are distinct. Detailed studies show that the spectral changes during the E to E* transition are due to changes in the ionic environment of Trp-3 resulting from dampened segmental motions of the interfacial binding region, whereas the changes in the E* to E*L step are primarily due to the dehydration of the enzyme–lipid complex.

A major limitation of many biophysical studies of PLA2 reported so far is that a distinction between the E* and E*L forms of the enzyme at the interface has not been possible. For example, spectroscopic and NMR studies on pancreatic PLA2 have been carried out with micelles of hexadecylphosphocholine.[43] In such dispersions all of the enzyme may be bound to the interface ($K_d \approx 0.5$ mM); however, the bound PLA2 is present as a mixture of the E* and E*L forms because the K_L^* for hexadecylphosphocholine is approximately 0.65 mol fraction.[44]

F. Interfacial Rate Parameters

Having established the scooting mode analysis of interfacial catalysis as a way to monitor the steady-state kinetics of the action of PLA2 on DMPM and other anionic vesicles and having defined and discovered satisfactory neutral diluents for measuring equilibrium constants for the interaction of ligands with PLA2 in the interface, it is now possible to obtain exact values or limit-estimates of all of the rate constants given in the scheme in Fig. 11. As elaborated next, a set of complementary protocols were used to deconvolute the rate constants.

[43] N. Dekker, A. R. Peters, A. J. Slotboom, R. Boelens, R. Kaptein, and G. H. de Haas, *Biochemistry* **30,** 3135 (1991); A. R. Peters, N. Dekker, L. Van den Berg, R. Boelens, R. Kaptein, A. J. Slotboom, and G. H. de Haas, *Biochemistry* **31,** 10024 (1992); M. C. E. Van Dam-Mieras, A. J. Slotboom, W. A. Pieterson, and G. H. de Haas, *Biochemistry* **14,** 5387 (1975).

[44] M. K. Jain, W. Tao, J. Rogers, C. Arenson, H. Eibl, and B.-Z. Yu, *Biochemistry* **30,** 10256 (1991).

Cleavage of sn-2 Ester Bond Is Irreversible

The chemical step of the lipolysis reaction catalyzed by PLA2 is effectively irreversible ($k_{-2} \ll k_2$) as shown by studies in which the PLA2-catalyzed hydrolysis of DMPM is carried out in ^{18}O-labeled water, with the ^{18}O content of the fatty acid product being monitored by mass spectrometry.[45] The hydrolysis of DMPM leads to the incorporation of a single ^{18}O into the fatty acid product. If the chemical step is reversible, ^{18}O-labeled fatty acid would be converted back to DMPM and some of the DMPM would retain ^{18}O in the carbonyl group of the *sn*-2 ester. Subsequent hydrolysis of this material would yield doubly ^{18}O-labeled fatty acid. Only myristic acid containing a single ^{18}O was recovered over the entire time course of hydrolysis of DMPM vesicles, which indicates that the chemical step is irreversible. Note that this experiment would detect chemical reversibility even if the resynthesized ester is not released from the enzyme before being converted to products.

Integrated Michaelis–Menten Equation

If it is assumed that the PLA2 species at the interface of DMPM vesicles are at steady-state during the reaction progress, the Michaelis–Menten equation can be applied to the reaction at the interface (steps within the box in Fig. 11). The integrated rate expression for the irreversible reaction is given by Eq. (4):

$$k_i t = -\ln\left(1 - \frac{P_t}{P_{max}}\right) + \left(\frac{N_S k_i}{v_0} - 1\right)\left(\frac{P_t}{P_{max}}\right) \tag{4}$$

and the fit to the observed reaction progress curve is shown in Fig. 17.[26] The amount of product formed at time t, P_t, is expressed as a function of the maximum amount of product formed, P_{max}, when all of the substrate in the outer monolayer of enzyme-containing DMPM vesicles has been hydrolyzed. The validity of Eq. (4) also requires that there is no intervesicle exchange of E, S, and P, that the vesicles have a uniform size [narrow monomodal dispersity ($\sigma^2 < 0.2$)],[26] and that there are sufficiently more vesicles than enzymes such that the vesicles contain at most one enzyme (Poisson distribution).

According to Eq. (4), three parameters are obtained from the reaction progress curve. The constant N_S is defined as

$$P_{max} = C_E N_S \tag{5}$$

[45] F. Ghomashchi, T. O'Hare, D. Clary, and M. H. Gelb, *Biochemistry* **30,** 7298 (1991).

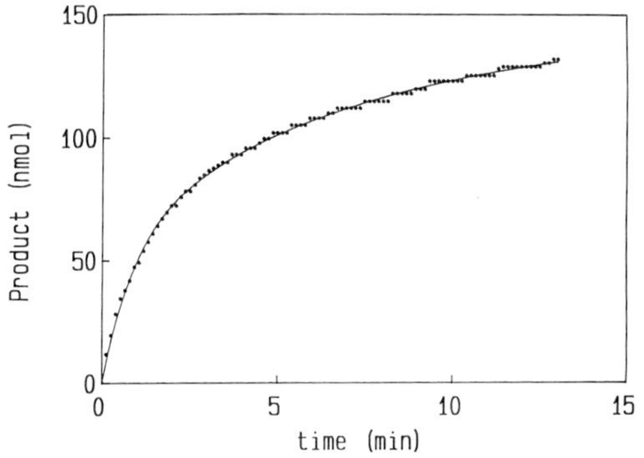

FIG. 17. Reaction progress curve for the hydrolysis of extruded DMPM (0.35 m*M*) vesicles by 1.5 pmol of pig pancreatic PLA2 in 0.6 m*M* $CaCl_2$, 1 m*M* NaCl. A theoretical fit to Eq. (4) gives $v_0 = 320\ \text{sec}^{-1}$, $N_s k_i = 35\ \text{sec}^{-1}$, and $N_s = 95{,}000$.

where C_E is the total amount of enzyme. Thus N_S represents the number of substrate molecules in the outer monolayer of the target vesicles as also described in Section C.

The initial velocity per enzyme, v_0, and the first-order relaxation constant, k_i, are obtained by curve-fitting the progress curve by the nonlinear regression method. More precisely, v_0 is the reaction rate at the initial concentration of DMPM substrate in pure DMPM vesicles (mole fraction 1). The parameters $N_S k_i$ and v_0 are related to k_{cat} through $K_m{}^*$ and $K_P{}^*$ as follows:

$$v_0 = \frac{k_{cat}}{1 + K_m{}^*} \quad \text{or} \quad \frac{k_{cat} X_S}{X_S + K_m{}^*} \quad \text{at} \quad X_s < 1 \tag{6}$$

$$N_S k_i = \frac{k_{cat}}{K_m{}^*(1 + 1/K_P{}^*)} \tag{7}$$

Here $K_m{}^*$ is the interfacial Michaelis constant expressed as the mole fraction of DMPM substrate in the interface at which one-half of the total amount of enzyme is in the E* form and the other one-half is in the E*S and E*P forms. In Eq. (7) $K_P{}^*$ is the equilibrium constant for the dissociation of the reaction products from the E*P complex. The constant k_{cat} is the velocity of the reaction per enzyme at saturating substrate mole fraction (i.e., when the fraction of the enzyme in the E* form is near zero). Note that if $K_m{}^* > 1$ mole fraction, the value of k_{cat} can only be estimated by extrapolation of enzymatic rates measured under conditions in which E* is not saturated with substrate. In the presence of a neutral diluent at the substrate interface the mole fraction of substrate can be varied between 1 and 0.

Equation (4) contains both a zero-order term (linear in P_t/P_{max}) that contributes mainly at early times and a first-order term (logarithmic in P_t/P_{max}) that dominates near the end of the reaction. The shape of the reaction progress curve is sensitive to the zero-order and the first-order terms in Eq. (4). When N_S is small, the first-order term dominates except for the very early time period, and thus the shape of the curve has a first-order appearance as shown in Fig. 12. When N_S is large, the initial portion of the progress curve is linear, and a first-order portion is seen near the end of the reaction (Fig. 17). Thus, curvature in the progress curve is dominated by the depletion of substrate below K_m^* and by the accumulation of products which leads to product inhibition. The shape of the progress curve in the scooting mode thus depends on the value of k_i [Eq. (7)], which is not only a function of k_{cat}/K_m^* but is also influenced by the probability of binding products to the enzyme. Small K_P^* causes the early appearance of the first-order course of the reaction progress. The products bind about 10-fold more tightly to the pig pancreatic PLA2 compared to K_m^* for the substrate, and this contributes to the rapid appearance of a first-order progress curve for the action of the enzyme on small DMPM vesicles (Fig. 12). These patterns are analogous to progress curves for solution-phase enzymes in which the shape of the curves depends on the total amount of substrate present in solution.

Determination of K_m^*

There are several independent ways to determine K_m^*. The mole fraction of DMPM substrate in the vesicle can be changed by the addition of a neutral diluent such as 2-hexadecyl-*sn*-glycero-3-phosphocholine to the vesicles up to 0.3 mole fraction; higher mole fractions disrupt vesicles to form mixed micelles. From the decrease in v_0 owing to surface dilution of substrate, a K_m^* value of 0.3 mole fraction is obtained according to Eq. (6).[24] The K_m^* can also be obtained by monitoring the effect of competitive inhibitors on the kinetics in the scooting mode. This is because the degree of inhibition in the presence of an inhibitor at the interface depends on the mole fraction of inhibitor and substrate which are known *a priori*, the equilibrium constant for the dissociation of inhibitor from the E*I complex (K_I^*), which can be independently measured by the protection method (Section E), and K_m^*.[24] Competitive inhibitors are discussed in detail in Section G. By using these methods, K_m^* for the hydrolysis of DMPM in vesicles by pig pancreatic PLA2 is in the range 0.25–0.5 mole fraction. These results attest to the internal consistency of Eqs. (2)–(4), (6), and (7).

Chemical Step Is Rate-Limiting for Maximal Turnover

The observed value of v_0 is 300 sec^{-1} for pig pancreatic PLA2 acting on DMPM vesicles, and it does not depend on the isothermal or thermotropic phase transition properties of the vesicles. According to Eq. (6) the calculated value of k_{cat}, 400 sec^{-1}, is considerably less than the value of about 20,000 sec^{-1} expected if the rate of hydrolysis is limited by the rate of lateral diffusion of phospholipids in the plane of the bilayer.[5,26] The rate constant k_{cat} is a function of the rate constants for the chemical step (k_2) and the release of products from E*P (k_3).

To probe whether one or both of these steps is rate limiting, the kinetics of the PLA2-catalyzed hydrolysis of DMPM and the analog of DMPM in which the *sn*-2 ester is replaced by a thioester were studied. The reasoning behind the use of such an element effect is that the sulfur substitution may change the rate constant for the chemical step more than the rate constant for the physical dissociation of the products. This was found to be the case[46]: the v_0 for the thiol analog was about 40 sec^{-1} compared to a value of 300 sec^{-1} for DMPM. Because the K_m* for the thiol analog is about 5-fold smaller than the K_m* for the oxy-ester, the effect of sulfur substitution is to reduce k_{cat}. There is no measured element effect on K_P*, which suggests that k_3 is also not effected, as long as it is assumed that sulfur substitution does not change k_3 and k_{-3} by the same factor.

The overall conclusion from these results is that k_{cat} is limited by the chemical step and not by the release of products or the diffusion of substrate to the catalytic active site, or by any other physical process. Strictly speaking, this result holds for the hydrolysis of the thiol ester; however, the fact that the element effect on k_{cat} remains constant when measured with catalytic site mutants of pig pancreatic PLA2 that have greatly reduced k_{cat} values suggests that the chemical step is rate limiting for the hydrolysis of the oxygen ester as well.

Forward Commitment to Catalysis Is Large

Isotope effects on enzymatic reactions are useful in determining what fraction of E*S goes through the chemical step to form reaction products versus releasing S to give E*. The relative rates for the hydrolysis of DMPM with either ^{12}C or ^{14}C at the carbonyl carbon of the *sn*-2 ester have been measured.[45] The heavy-atom isotope ratio approaches one for the hydrolysis of DMPM vesicles, which establishes that $k_2 \gg k_{-1}$. In other words, once the substrate is dislodged from the interface into the

[46] M. K. Jain, B.-Z. Yu, J. Rogers, M. H. Gelb, M. D. Tsai, E. K. Hendrickson, and S. Hendrickson, *Biochemistry* **31,** 7841 (1992).

catalytic site of E*, essentially all of it is hydrolyzed rather than returned intact to the interface. This result together with the fact that $k_3 > k_2$ indicates that K_m^* is approximately equal to k_2/k_1, and thus K_m^* is much greater than the dissociation constant for the E*S complex, $K_S^* = k_{-1}/k_1$. This result indicates that the PLA2 turnover cycle operates at steady state rather than at rapid equilibrium. Consistent with these observations is the fact that the binding of DTPM, the DMPM analog in which the *sn*-1 and *sn*-2 ester is replaced by an ether linkage, displays a dissociation constant of 0.03 mole fraction (Table I), which is 10-fold less than the K_m^* for DMPM.

Role of Divalent Cations in Interfacial Catalysis

The effect of substrate on the apparent Michaelis constant for calcium during the hydrolysis of DMPM, defined here as apparent $K_{Ca}^*(S)$, was determined from the dependence of v_0 on the calcium concentration.[20] It is related to K_m^* by

$$K_{Ca}^*(S) = \frac{K_{Ca}^*}{1 + 1/K_m^*} \tag{8}$$

Because K_{Ca}^* is obtained independently by the protection method (Section E), this relationship provides an independent estimate of K_m^* of 0.3 mole fraction. Equation (8) emphasizes the ordered binding of calcium and substrate. It is the same as Eq. (2) except that in Eq. (8) K_m^* for the substrate controls the Michaelis–Menten constant for the cofactor.

Enhanced binding of a catalytic site-directed ligand to PLA2 is observed only in the presence of certain divalent cations.[20] For example, Ba^{2+} and Sr^{2+} do not support the hydrolysis of DMPM nor the binding of catalytic site-directed ligands. Cadmium supports synergistic binding of the catalytic site-directed transition state mimics; however, the rate of catalytic turnover is less than 1% of the value observed with calcium. Such observations suggest that certain divalent cations like calcium play a role in the chemical step, presumably by stabilizing the coordination geometry of the transition state (Section B).

G. Uses of Kinetic Parameters

The scheme in Fig. 11 provides a useful basis for many aspects of the structure and function relationships of interfacial enzymes. With protocols for monitoring interfacial catalysis in the scooting mode and those for the determination of interfacial equilibria, it is now possible to study detailed features of interfacial catalysis. Thus, not only do kinetics in the scooting mode offer absolute proof for processive interfacial catalysis, but the steps of the catalytic turnover cycle in the scheme in Fig. 11 have virtually the same mechanistic significance as those in traditional enzyme kinetics. Some of the uses of the protocols outlined in the preceding sections are summarized below.

Phospholipase A_2 from Different Sources Can Be Assayed at Same Interface

The DMPM vesicles are ideal as substrate because many, and probably all, forms of PLA2 bind to the vesicles with high affinity, and the observed rate depends only on the interfacial catalytic turnover. As K_d is quite small (<1 p*M* for pig pancreatic PLA2), assays with micromolar quantities of fluorescent substrate have been developed.[20] Not only can PLA2s from different sources be assayed on the same substrate, but the assay is not influenced by amphiphilic solutes when present up to 0.3 mole fraction in DMPM vesicles. For example, the scooting mode assay is reliable for PLA2 present in complex mixtures such as serum. In most of the commonly used protocols, the enzyme binds weakly to the interface, and thus the E to E* equilibrium is susceptible to thermotropic or solute-induced perturbations in the interface. In short, a reliable PLA2 assay must yield the amount of active enzyme, and this is only achieved reliably if all of the enzyme is tightly bound to the interface.

Monomeric Enzyme Is Fully Catalytically Active

There has been a lively controversy for several years about whether PLA2 operates at an interface as monomer or as a functional aggregate. Many PLA2s tend to aggregate in aqueous solution[47] or in the presence of amphiphiles.[48] Enzyme dimers and trimers present in a crystal lattice[34] have also fueled this controversy, even though on the basis of the contact surfaces it appeared unlikely that such aggregates could bind to interfaces.

The scooting mode analysis provides unequivocal evidence that PLA2s are fully active as monomers.[33] This is because dimeric enzyme, on a weight basis, would hydrolyze half the number of vesicles that monomeric enzyme would hydrolyze. Based on Eq. (5), the number of catalytic units, C_E, can be obtained from the experimentally determined values of P_{max} and N_S. An independent measure of N_S, the number of DMPM molecules

[47] S. Brunie, J. Bohim, D. Gewirth, and P. B. Sigler, *J. Biol. Chem.* **260,** 9742 (1985).

[48] J. H. Van Eijk, H. M. Verheij, R. Dijkman, and G. H. de Haas, *Eur. J. Biochem.* **132,** 183 (1983); J. D. R. Hille, M. R. Egmond, R. Dijkman, M. van Oort, B. Jirginsons, and G. H. de Haas, *Biochemistry* **22,** 5347 (1983); T. L. Hazlett and E. A. Dennis, *Biochemistry* **24,** 6152 (1985).

in the outer layer of the vesicles, comes from a determination of the size of DMPM vesicles. The vesicle size, determined by a multitude of methods, agrees well with the value of N_S calculated from P_{max} and C_E as long as the value of C_E is for the monomeric enzyme. This analysis has been applied to 23 different PLA2s, and all give the same value of N_S irrespective of the aggregation tendency of the enzymes. The conclusion that the monomeric PLA2 is fully catalytically active is reinforced by resonance energy transfer experiments with suitably labeled PLA2s showing that there is no enzyme aggregation at the interface even when there are more than 20 enzyme molecules bound to a single vesicle.

Competitive Substrate Specificity

In any substrate specificity study, the goal is to determine the relative rates of hydrolysis of the various substrates present in the mixture. This is accomplished in a competitive mode in which the enzyme acts on a mixture of substrates and "chooses" the ones that it prefers. In the scooting mode this is accomplished by using, for example, two competing substrates, S1 and S2, present in DMPM vesicles. The relative amounts of the different products are measured. The following equation applies to this competitive substrate analysis:

$$\frac{(k_{cat}/k_m{}^*)_1}{(k_{cat}/K_m{}^*)_2} = \frac{\ln (X^0_{S1}/X^t_{S1})}{\ln (X^0_{S2}/X^t_{S2})} \tag{9}$$

where $X_S{}^0$ and $X_S{}^t$ represent the mole fractions of the two substrates at times 0 and t, respectively. Equation (9) is a steady-state equation, and thus the analysis must be carried out with the enzyme in the scooting mode for the reasons stressed throughout this review. In addition, there should be an excess of vesicles over enzymes so that the enzyme-containing vesicles will have at most one enzyme per vesicle. The ratio $k_{cat}/K_m{}^*$ is often called the specificity constant since it relates the ratio of products formed to the ratio of substrates present.[49]

Often substrate specificity studies on lipolytic enzymes are performed by determining the relative velocities in separate experiments for the action of the enzyme on lipid aggregates composed of a single lipid species. Such studies have little meaning because the difference in observed velocities will often have a contribution from the different fractions of enzyme bound to different aggregates (E → E*). The latter problem can be circumvented by examining the action of the enzyme on mixed lipid aggregates; however, one should not lose sight of the fact that the enzyme must be

[49] A. Fersht, "Enzyme Structure and Mechanism." Freeman, San Francisco, 1977.

operating in the scooting mode for Eq. (9) to be valid. In the hopping mode, the various vesicles will take on different relative compositions as a function of reaction time, and the summation of the products formed in each enzyme-containing vesicle to give the total product formed will be difficult to determine.

Nevertheless, when any of the following three conditions are met, the need to be in the scooting mode is obviated. (a) If the $k_{cat}/K_m{}^*$ values for the two substrates are very similar, then the ratio of products will be very similar to the ratio of substrates in all vesicles. As it is not known *a priori* whether this condition is valid, one should use caution in using this approximation. (b) If the enzyme is hopping rapidly among the vesicles such that the residence time of the enzyme on a vesicle is small compared to the time needed to change significantly the composition of the vesicle, then the composition of all vesicles will be on the average similar at any point in time. It has already been mentioned that the condition of fast hopping is difficult to achieve. (c) If the enzymatic reaction is allowed to proceed for short times such that only a small amount of substrate in the vesicles is hydrolyzed, then the composition of each vesicle will not change much unless the $k_{cat}/K_m{}^*$ values for the two substrates are vastly different. Caution is required in applying this condition. For example, if the amount of product made at an early time point is 5% of the total, it is not clear whether each vesicle has become hydrolyzed by 5% or whether 5% of the vesicles have been completely hydrolyzed.

Substrate specificity studies consistent with the assumptions in Eq. (9) are easily conducted experimentally. One approach is to prepare two different substrate species, one with tritium-labeled fatty acid and the other with carbon-14-labeled fatty acid, each at the *sn*-2 position. Nonradiolabeled DMPM vesicles are then doped with small amounts of the radiolabeled substrates. Enzyme is added, and the reaction is allowed to proceed until sufficient amounts of the radiolabeled substrates have been hydrolyzed. The amounts of radiolabeled substrates and products are used to calculate the relative $k_{cat}/K_m{}^*$ values according to Eq. (9). If the reaction is carried out for a short duration so that the values of v_0 are being measured, then the relative $k_{cat}/K_m{}^*$ values can be calculated from the ratio of isotopes according to the following equation:

$$\frac{^{3}v_0}{^{14}v_0} = \frac{^{3}\mathrm{P}}{^{14}\mathrm{P}} = \frac{^{3}(k_{cat}/K_m{}^*)}{^{14}(k_{cat}/K_m{}^*)}\left(\frac{^{3}S}{^{14}S}\right) \tag{10}$$

where the subscripts 3 and 14 designate tritium- and carbon-14-labeled species, respectively. Although DMPM is being hydrolyzed in these experiments, it is not radiolabeled, and thus it does not effect the analysis of the radiolabeled species.

Equations (9) and (10) are valid for any pair of competing substrates even if there are several different substrate species present in the vesicles as long as all enzyme species are in the steady state. The final note of caution is that one should work under conditions in which there is no significant asymmetric distribution of the radiolabeled phospholipid substrates between the inner and outer monolayer of the vesicles. If asymmetry exists, the appropriate correction is needed such that only the ratio of radiolabeled substrates in the enzyme-accessible outer layer is used to evaluate the substrate specificity. Asymmetry can be spotted by letting the reaction proceed until all of the outer layer substrate is hydrolyzed and comparing the ratio of isotopes in the products to that in the substrates. In cases where different *sn*-2 fatty acids are compared, gas chromatography can be used to determine the ratio of the fatty acids released. This approach has been used for determining the substrate preference of the cytosolic 87-kDa PLA2.[50]

Relative substrate specificities for a variety of PLA2s have been determined.[32,40] During the catalytic turnover cycle these secreted enzymes do not appear to show significant discrimination on the basis of the head group or the acyl chain unsaturation, although the difference between the $E \rightarrow E^*$ equilibria at interfaces of the phospholipids is several orders of magnitude.[51]

Covalent Modifiers of Phospholipase A_2

The effect of alkylating agents, such as phenacyl bromides, that irreversibly modify the catalytic site of PLA2 are not a problem to analyze since the modified enzyme will be devoid of activity when tested in any assay. Covalent modifiers that react with residues on the i-face of the enzyme may reduce the enzymatic activity either by rendering the enzyme incapable of binding to the interface or by changing the catalytic properties of E*. Manoalogue reacts covalently with lysine residues of PLA2.[52] Analysis of the modified enzyme on DMPM vesicles shows that the enzyme is still tightly bound to the vesicle but the turnover rate of the enzyme at the interface is reduced.[53] This is apparent from the fact that the amplitude of the reaction progress curve per enzyme (N_S) is the same as that for

[50] A. M. Hanel, S. Schuttel, and M. H. Gelb, *Biochemistry* **32,** 5949 (1993).

[51] M. K. Jain, J. Rogers, J. F. Marecek, R. Ramirez, and H. Eibl, *Biochim. Biophys. Acta* **860,** 462 (1986); M. K. Jain and J. Rogers, *Biochim. Biophys. Acta* **1003,** 91 (1989).

[52] R. A. Deems, D. Lombardo, B. P. Morgan, E. D. Mihelich, and E. Dennis, *Biochim. Biophys. Acta* **917,** 258 (1987); L. J. Reynolds, E. D. Mihelich, and E. A. Dennis, *J. Biol. Chem.* **266,** 16512 (1991).

[53] F. Ghomashchi, B.-Z. Yu, E. D. Mihelich, M. K. Jain, and M. H. Gelb, *Biochemistry* **30,** 9559 (1991).

the unmodified enzyme, but the time needed for the reaction cessation is longer for the modified enzyme. These results show that even though several different molecular species of the modified enzyme are produced, most are catalytically active. For efficient interfacial catalysis, the vesicle-bound enzyme forms a tight interaction with the interface in which the region of the interface that is covered by enzyme is desolvated.[21,41] One of the possible interpretations is that modification of PLA2 with manoalogue reduces the ability of the enzyme to "settle snug" against the vesicle interface.

Kinetic Characterization of Competitive Inhibitors of Interfacial Catalysis

The scooting mode analysis provides a useful way to analyze catalytic site-directed inhibitors by kinetic[54–56] and equilibrium binding methods.[24] As shown in the scheme in Fig. 11 such inhibitors compete with substrate for binding to the catalytic site of E*. This method has been useful for characterization of several synthetic[56] and naturally occurring[57] specific inhibitors of PLA2. By these criteria and the analysis outlined below, the transition state mimics shown in Fig. 8 are competitive inhibitors of interfacial catalysis. As discussed later, most commonly used agents that reduce the rate of PLA2-catalyzed hydrolysis are not specific competitive inhibitors, but they promote the desorption of the weakly bound enzyme from an interface.[58]

The presence of a competitive inhibitor in DMPM vesicles will reduce the initial velocity from v_0 to a value v_I that relates to the mole fraction of the inhibitor, X_I:

$$\frac{v_0}{v_\mathrm{I}} = 1 + \left(\frac{1 + 1/K_\mathrm{I}^*}{1 + 1/K_\mathrm{m}^*}\right)\left(\frac{X_\mathrm{I}}{1 - X_\mathrm{I}}\right) \tag{11}$$

$$\frac{(N_\mathrm{S}k_\mathrm{i})_0}{(N_\mathrm{S}k_\mathrm{i})_\mathrm{I}} = 1 + \left(\frac{1 + 1/K_\mathrm{I}^*}{1 + 1/K_\mathrm{P}^*}\right)\left(\frac{X_\mathrm{I}}{1 - X_\mathrm{I}}\right) \tag{12}$$

[54] M. K. Jain, G. H. de Haas, J. F. Marecek, and F. Ramirez, *Biochim. Biophys. Acta* **860,** 475 (1986).

[55] M. K. Jain, W. Yuan, and M. H. Gelb, *Biochemistry* **28,** 4135 (1989).

[56] M. K. Jain, W. Tao, J. Rogers, C. Arenson, H. Eibl, and B.-Z. Yu, *Biochemistry* **30,** 10256 (1991).

[57] M. K. Jain, F. Ghomashchi, B.-Z. Yu, T. Bayburt, D. Murphy, D. Houck, J. Brownell, J. C. Reid, J. E. Solowiej, S. Wong, U. Mocek, R. Jarrell, M. Sasser, and M. H. Gelb, *J. Med. Chem.* **35,** 3584 (1992).

[58] M. K. Jain, M. Streb, J. Rogers, and G. H. de Haas, *Biochem. Pharmacol.* **33,** 2541 (1984); M. K. Jain and D. V. Jahagirdar, *Biochim. Biophys. Acta* **814,** 319 (1985).

$$\frac{v_0}{N_S k_i} = \frac{[1/X_I(50) - 1]}{[1/n_I(50) - 1]} = \frac{(1 + 1/K_P^*)}{(1 + 1/K_m^*)} \tag{13}$$

Equation (12) is for the reduction in the parameter $N_S k_i$ by the presence of a competitive inhibitor. In Eq. (13), $X_I(50)$ and $n_I(50)$ are the mole fractions of inhibitor that reduce v_0 and $N_S k_i$, respectively, by 50%.

Equations (11) and (12) are standard steady-state equations for competitive inhibition adapted for interfacial kinetics in the scooting mode. It may be noted from these equations that for a reduction of v_0 the inhibitor competes with substrate, whereas under the substrate-limiting conditions the inhibitor competes with products for a reduction in $N_S k_i$. For the action of the pig pancreatic PLA2 on DMPM vesicles $K_P < K_m^*$ (Table I) and thus $n_I(50) > X_I(50)$. For a kinetic proof of competitive inhibition in the interface, Eq. (13) must be satisfied.

Site-Directed Mutagenesis

Having established the conditions for the kinetic characterization of interfacial catalysis by PLA2, structure–function relationships can be established using site-directed mutagenesis for the role of certain amino acid residues in promoting the lipolysis reaction. For example, several mutants of the conserved residues Tyr-52,[59] Phe-22, and Phe-106[60] have been constructed, and their global structure has been shown to be the same as that of the wild-type enzyme. Even though these conserved residues are replaced, the activity of the enzyme is retained. Such studies also rule out a catalytic role for the hydrogen bond network involving Tyr-52 or for the aromatic sandwich involving Phe-22 and Phe-106.

H. Kinetic Complexities

Hydrolysis of Zwitterionic Vesicles

It is often stated that PLA2s are sensitive to the "quality of the interface."[2] This is based on the observations that under certain conditions the physical state of the phospholipid interface has a marked effect on the enzymatic activity. It is turning out that most, and perhaps all, of these effects are due to the modulation of the E to E* equilibrium. For example, in the hydrolysis of phosphatidylcholine vesicles by PLA2, an anomalous activa-

[59] C. M. Dupureur, B.-Z. Yu, M. K. Jain, J. P. Noel, T. Deng, Y. Li, I. L. Byeon, and M. D. Tsai, *Biochemistry* **31,** 6402 (1992).

[60] C. M. Dupureur, B.-Z. Yu, J. A. Mamone, M. K. Jain, and M. D. Tsai, *Biochemistry* **31,** 10576 (1992).

tion occurs at temperatures near the gel-to-fluid phase transition, and enzymatic activity decreases at temperatures both below and above the phase transition temperature. These effects are observed as a change in the shape of the reaction progress curve,[61] and their origin can be hypothesized on the basis of the following observations. The pig pancreatic PLA2 binds very weakly to vesicles of 1,2-dimyristoyl-*sn*-glycero-3-phosphocholine (DMPC) (cf. Fig. 16), whether at, below, or above the phase transition temperature; the K_d is in the millimolar range compared to the picometer range for DMPM vesicles. The binding to phosphatidylcholine vesicles is enhanced in the presence of greater than 0.06 mole fraction of the reaction products in a 1:1 mole ratio, and this binding is accompanied by an increase in the rate of hydrolysis.[29]

Phase separation of vesicle components is induced by reaction products,[62] and it seems to occur to the maximal extent when the temperature is near the phase transition temperature of the bilayer. Patches of high negative charge density are formed from fatty acids in bilayers[29] and monolayers.[63] Epifluorescence microscopy has shown that PLA2 binds to these anionic patches. Thus, it is reasonable to propose that at zwitterionic interfaces most of the PLA2 is in the aqueous phase and that the E to E* equilibrium is shifted in favor of E* when the phase separation of anionic amphiphiles occurs. The hydrolysis of zwitterionic vesicles also displays a lag in the reaction progress curve, and the addition of reaction products eliminates the lag.[29] The simplest explanation for the lag is that most of the enzyme is in the E form at the beginning of the reaction, but as the reaction products accumulate in the vesicles, more and more E* is formed with acceleration of the rate. These effects are not seen with DMPM vesicles because all of the enzyme is tightly bound to the vesicle under all conditions. In short, these studies underscore the importance of keeping track of where the enzyme is during the progress of reaction.

Nonspecific Inhibitors

Based on the possibility that the E to E* equilibrium is readily perturbed under most commonly used assay conditions, one should be careful in suggesting whether a compound is an activator or inhibitor of PLA2. Numerous inhibitors of interfacial catalysis by PLA2 have been described, but only a small subset of them are true specific inhibitors. There is a

[61] G. C. Upreti and M. K. Jain, *J. Membr. Biol.* **55,** 113 (1980).

[62] M. K. Jain and D. V. Jahagirdar, *Biochim. Biophys. Acta* **814,** 313 (1985).

[63] A. Reichert, H. Ringsdorf, and A. Wogenknecht, *Biochim. Biophys. Acta* **1106,** 178 (1992); D. W. Grainger, A. Reichert, H. Ringsdorf, C. Salesse, D. E. Davies, and J. B. Lloyd, *Biochim. Biophys. Acta* **1022,** 146 (1992).

major problem in analyzing inhibitors under the conditions in which the binding of the enzyme to the interface is weak since most apolar and amphiphilic compounds, when present in zwitterionic vesicles, do influence the E to E* equilibrium possibly by preventing the segregation of reaction products in zwitterionic vesicles or by disrupting the bilayer organization to form mixed micelles. Besides detergents and organic solvents, such nonspecific inhibitors include mepacrine, aristolochic acid, quinacrine, vitamin A analogs, and possibly most reversible inhibitors used for pharmacological studies and those claimed in the patent literature. Such bilayer perturbing agents do not alter the kinetics of hydrolysis of DMPM vesicles even when present at the interface at a mole fraction of 0.1.[24] On the other hand, the inhibitors shown in Fig. 8 inhibit the action of PLA2 on DMPM vesicles when present at mole fractions of one inhibitor per several thousand DMPM substrates.

Nonspecific Activators

Just as there are nonspecific inhibitors of PLA2, there are also nonspecific activators that increase the fraction of interface-bound enzyme under conditions in which the binding to the inferface is weak. Also, agents that increase the rate of substrate replenishment in cases where the availability of substrate limits the reaction velocity are nonspecific activators. The cationic peptide polymyxin B activates in this fashion. As mentioned earlier, the action of PLA2 on small DMPM vesicles shows signs of substrate depletion such that the initial time period in which the mole fraction of substrate is 1 is very short (Fig. 12). Polymyxin B promotes rapid intervesicle exchange of DMPM, and this keeps the mole fraction of substrate in the enzyme-containing vesicles near 1; thus, the apparent rate remains near v_0 for a significantly longer period of time.[36] The cationic peptide melittin produces a similar effect. Cations such as Ba^{2+} or Sr^{2+} do not substitute for Ca^{2+} as a catalytic cofactor, but they do cause apparent activation by promoting vesicle fusion.[31] Agents that promote intervesicle transfer of phospholipids in mixed micelles will also lead to apparent activation.

Phospholipid–Detergent Mixed Micelles

Codispersions of phospholipids and detergents (mixed micelles) have been used as substrates of PLA2.[3–5,25] With a suitable choice of detergent, substrate, and enzyme the reaction progress curves show an initial linear region (apparent v_0). Several considerations are relevant for the interpretation of such apparently simple results. (i) Mixed micelles contain typically less than 50 phospholipid molecules per aggregate, compared to more than 5000 phospholipid molecules present in the smallest vesicles. Therefore, for PLA2 with an intrinsic turnover number (v_0) of about 300 sec^{-1}, all of the substrate present in an enzyme-containing micelle will be hydrolyzed in less than 0.1 sec unless the substrate is replenished in much less than 0.1 sec. (ii) The rate of exchange of detergent between micelles is rapid (half-time of about 1 msec); however, the rate of intermicellar exchange of phospholipids is considerably slower (half-time of several seconds), and it occurs only by fusion–fission of micelles.[28] Thus, enzymatic turnover in mixed micelles may be limited by the rate of substrate replenishment rather than by the catalytic power of PLA2. (iii) Detergents do not ideally mix with phospholipids, and therefore the composition, size, and dispersity of the resulting particles change with the bulk composition. As mentioned earlier, this peculiarity of polymorphism also creates problems with the use of detergents as "surface diluents." (iv) The properties of micelles containing the enzyme are not generally known; therefore, it is often assumed that the size, organization, and dynamics are the same as those of micelles without enzyme. Based on preliminary evidence this assumption does not appear to be valid.[48]

Direct evidence for the contribution of substrate replenishment to the overall kinetics is shown by the fact that in micelles the rate-limiting step of the catalytic turnover cycle is altered; although the chemical step is rate limiting for the hydrolysis of DMPM vesicles by PLA2 (Section F), this is not the case for the hydrolysis of mixed micelles.[30] As shown in Fig. 18, addition of small amounts of the negatively charged detergent deoxycholate to DMPC vesicles causes the rate of the reaction to increase owing to a higher fraction of enzyme bound to the interface. At higher mole fractions of detergent, the bilayers begin to disrupt into smaller disks and micelles, and the apparent rate of hydrolysis decreases significantly, for example, from about 200 to about 10 sec^{-1} for pig pancreatic PLA2. This decrease in the rate parallels a decrease in the O/S element effect (Section F) which decreases from seven to less than two at high detergent concentrations. This change shows that the chemical step is no longer rate determining for the overall reaction, but rather a physical process, perhaps substrate replenishment, dominates the overall observed rate. This would be the case if, under such conditions, the observed reaction kinetics are not determined by global steady state assumed from global concentrations of substrate and enzymes. Thus, in effect, the steady-state mole fraction of substrate in enzyme-containing mixed micelles will be lower than that in mixed micelles without enzyme.

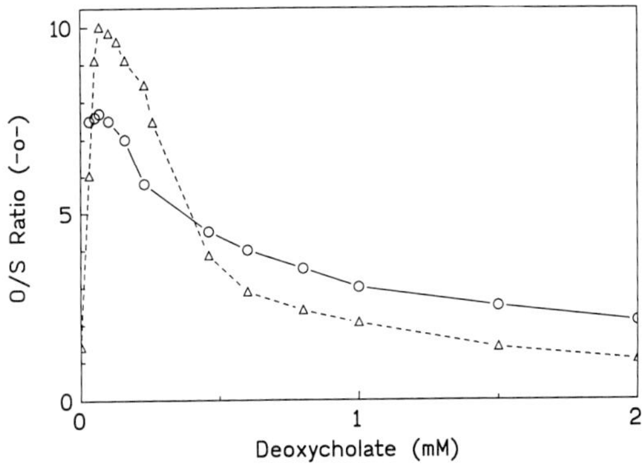

FIG. 18. Dependence of the rate of hydrolysis of 1 m*M* DMPC vesicles by pig pancreatic PLA2 (with an arbitrary ordinate; however, the maximum turnover number is 210 sec^{-1}), shown as triangles, as a function of the deoxycholate concentration. The O/S ratio shown with circles (with the values on the ordinate) represents the relative rate of hydrolysis of the 1-thiol ester analog versus the 1,2-dithiol ester analog of DMPC as a function of the deoxycholate concentration.

The effective steady-state rate of hydrolysis under the substrate-limiting conditions is given by

$$\text{Rate} = k_{\text{exch}} N_{\text{T}}(1 - X_{\text{S}}) = \frac{k_{\text{cat}} X_{\text{S}}}{K_{\text{m}}^{*} + X_{\text{S}}} \tag{14}$$

where k_{exch} is the rate of replenishment and N_{T} is the total number of exchangeable molecules in the micelle. The product inhibition term is not included; however, in effect, the X_{S}-dependent term will adjust to the level where the rate of catalysis (on the right-hand side) exactly matches that of replenishment (on the left-hand side).

The half-time for the spontaneous transfer of long-chain diacylphospholipids between micelles as monomers through the aqueous phase is several hours.[28,64] The effective replenishment must involve either the exchange of the enzyme between the micelles or fusion and fission of the enzyme-containing micelle with excess substrate micelles. In either case the event brings a fresh load of substrate to the enzyme-containing micelle. If N substrate molecules are replenished after an average time T and if T is large on the time scale during which the enzyme can hydrolyze the substrates, then the effective turnover would be N/T. For $N = 50$ and the observed rate of hydrolysis of 10 sec^{-1}, T would be around 5 sec. The

[64] T. Arvinte and K. Hildenbrand, *Biochim. Biophys. Acta* **775,** 86 (1984).

half-time for the fusion and fission of mixed micelles of bile salts is also of the order of seconds.[28] This is consistent with the estimate that it would not take more than a second to hydrolyze a large fraction of the substrate molecules initially present in the micelle.

Hydrolysis of Short-Chain Phospholipids Dispersed as Monomers

Phospholipids with six to eight carbons in the acyl chains remain dispersed as solitary monomers in aqueous solutions even up to concentrations of several millimolar. The observed rate of PLA2-catalyzed hydrolysis of these substrates dispersed as monomers is quite low, and the rate does not increase appreciably even in the presence of micelles of a neutral diluent. Kinetic and spectroscopic evidence shows that, even under these conditions, premicellar aggregates of the substrate with the enzyme are formed.[65]

Kinetic Problems at Monolayer Interfaces

Monolayers of medium-chain phospholipids (8 to 12 carbons) at the air–water interface have been used for the characterization of lipolytic enzymes.[1,2] In spite of the elegant simplicity of such an analysis, many of the intrinsic limitations have not been explicitly considered.[5] The unstirred layer of water at the monolayer interface is significant; therefore, it could contribute toward the pre-steady-state diffusion of the enzyme from the bulk aqueous phase to the monolayer. Because the rate of hydrolysis of monolayers is measured as the rate of dissolution of the products from the monolayer to the aqueous phase, the unstirred layer could also contribute to the lag phase that is typically observed. Other considerations based on adsorption of the enzyme to the hydrophobic surface of the monolayer trough are also known to create uncertainties. Yet another significant problem with monolayers as substrate is that the ratio of bulk water to interface is extremely large, and therefore even under optimal conditions only a small fraction (estimated as <5%) of the enzyme is bound to the monolayer interface.

I. Kinetic Basis of Interfacial Activation

As mentioned in the preceding section, PLA2 hydrolyzes short-chain phospholipids dispersed as monomers below the cmc's much more slowly than it hydrolyzes the same substrates present above the cmc's as micelles. An interesting question is whether the intrinsic catalytic properties of

[65] J. Rogers, B.-Z. Yu, and M. K. Jain, *Biochemistry* **31**, 5056 (1992).

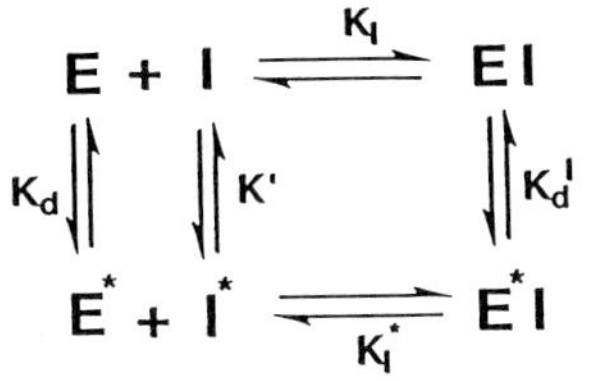

FIG. 19. Minimal scheme to show the relationship between the various equilibria in a mixture of enzyme (E), inhibitor (I), and a neutral diluent that forms the interface. See Section I for definitions of the equilibrium constants.

PLA2 bound to the interface are different from those of the enzyme in the aqueous phase. As described throughout this review, it has been possible to determine the intrinsic rate and equilibrium constants for PLA2 at the interface (steps within the box of the scheme in Fig. 11) that are deconvoluted from the effects associated with the intervesicle exchange of enzyme, substrate, and products. Thus, it is now possible to address the regulatory role of the interface on the intrinsic activity of PLA2. Some of the possibilities are considered below; they have been discussed in detail[66] in terms of the equilibria at the interface shown in Fig. 19.

Data on the binding of catalytic site-directed inhibitors to E versus E* can be used to shed some light on the question of whether the interface has an effect on the intrinsic affinity of the enzyme for ligands. The equilibrium constants for the dissociation of an inhibitor from EI or E*I, K_I or K_I*, respectively, cannot be directly compared because they have different units. Instead, it is useful to consider the effect of the inhibitor on the binding of the enzyme to the interface. This is expressed by the ratio of the equilibrium constants for the dissociation of E* or E*I from the interface (K_d/K_d^I); if this ratio is greater than 1, the binding of the inhibitor to the enzyme increases the affinity of the enzyme for the interface. According to Eq. (15), which is a thermodynamic relation (detailed-balance condition derived from Fig. 19), K_d/K_d^I is related to K_I and K_I*, respectively, and the dissociation constant of free inhibitor from the interface into the aqueous phase, K', defined as the ratio of the bulk aqueous concentration of inhibitor to the mole fraction of inhibitor in the interface:

$$\frac{K_d}{K_d^I} = \frac{K_I}{(K_I{}^*K')} = \frac{[E][E^*I]}{[E^*][EI]} \tag{15}$$

Note that if $K_d/K_d^I > 1$, $[E^*I]/[E^*] > [EI]/[E]$, and thus, regardless of the total amount of inhibitor in the system, the fraction of enzyme in the

[66] M. K. Jain, B.-Z. Yu, and O. G. Berg, *Biochemistry* **32,** 11319 (1993).

interface that is bound to ligand is higher than that for the enzyme in the aqueous phase. Using the thermodynamic cycle (Fig. 19) in this way provides a direct evaluation of the intrinsic inhibitor affinities and automatically accounts for differences in local concentrations of inhibitor in the aqueous phase and at the interface.

For all catalytic site-directed inhibitors studied,[65] it is found that K_d/K_d^I is approximately 50. The strengthening of the enzyme–interface binding could come about in three different ways: (i) it could be due to a residual inhibitor–interface interaction where part of the inhibitor remains attached to the interface even when the inhibitor has entered the catalytic site ("inhibitor-anchoring"); (ii) the interface could somehow be influenced by the inhibitor moving into the catalytic site such that it makes better contact with the enzyme ("interface modulation"); or (iii) the enzyme undergoes a conformational change when the inhibitor binds such that it makes better contact with the interface ("enzyme modulation"). The first two possibilities would suggest that the nature of the inhibitor (e.g., the length of the hydrocarbon chains) would influence the strength of the effect; this is not found in the experiments.[65] Thus, the most likely effect is as allosteric modulation of the enzyme when it binds the inhibitor.

According to the formalism of allosteric activation for a single-subunit enzyme, one may distinguish activation in the substrate binding and activation of the chemical step. The equilibrium scheme discussed here does not address the effect on the chemical step, but it does suggest that there is allosteric intefacial activation of the ligand affinity of pig pancreatic PLA2. Although the basis for this enhanced affinity can be interpreted in several different ways, biophysical studies[22] can provide some guidance. It appears that E exists as an ensemble of isoenergetic conformations, whereas E* and E*I forms have a much more restricted range of conformations.[22] Thus, the interfacial binding of enzyme may restrict the enzyme conformations to a subset that has higher affinity for the ligand. In effect, the interfacial binding would take up a large part of the entropy loss associated with the tightening of the conformation that would otherwise be accompanying the ligand binding.

A possible direct effect of the interface on the chemical step has not yet been directly dealt with. Because the enzyme is fully active in the interface as a monomer (Section G), multimer formation cannot be the cause of interfacial activation. It has already been mentioned that, under certain conditions, the rate of substrate replenishment in enzyme-containing aggregates can influence the observed reaction rate. If the enzyme is present in a premicellar aggregate with short-chain phospholipid substrate and if the rate of exchange of premicellar substrate with bulk soluble substrate is rate limiting, the anomalous increase in rate above the cmc of the substrate may be due to the formation of micelles that contain a relatively large amount of substrate. It is also conceivable that the release of products from the enzyme is rate limiting for the enzyme in solution but not for the enzyme at the interface. Such aspects are being investigated by addressing the question of whether the ES form undergoes the chemical change.

J. Other Interfacial Enzymes

The scooting mode analysis has been applied to other interfacial enzymes, but all of the rate constants in the scheme in Fig. 11 have been determined only for pig pancreatic PLA2. Although virtually all PLA2s undergo catalysis in the scooting mode, enzymes from synovial fluid[40] and enzyme from bovine pancreas and mutants[39] have been studied to characterize the kinetic basis of some of the key functional differences.

Studies have shown that an 87-kDa cytosolic phospholipase A_2 from mammalian cells operates in the scooting mode on phosphatidylcholine vesicles containing 10 mol % reaction products,[67] and such an analysis provided a description of the substrate preferences of the enzyme.[50] A bacterial lipase/transacylase functions in the scooting mode on DMPM vesicles and hydrolyzes embedded phospholipids and diglycerides.[68] The phosphatidylinositol-specific phospholipase C from *Bacillus cereus* appears to hydrolyze phosphatidylinositol present in vesicles in a moderately processive manner in which the enzyme undergoes approximately 50 turnover cycles during each visit to the inferface.[69] Although the enzyme is inhibited by DTPM, dioleoyl phosphatidylcholine is a neutral diluent. Interestingly, the k_{cat} for the turnover but not the K_m of a water-soluble substrate displayed by this phospholipase C is enhanced severalfold when the enzyme binds to the interface, which suggests a k_{cat} type of interfacial activation.

K. Epilog

The ability to monitor interfacial kinetics and the establishment of a kinetic model provides a basis for understanding the cellular and biochemical functions of PLA2 (e.g., in ischemia–reperfusion injury[70]). The kinetic

[67] F. Ghomashchi, S. Schuttel, M. K. Jain, and M. H. Gelb, *Biochemistry* **31,** 3814 (1992).

[68] M. K. Jain, C. D. Krause, J. T. Buckley, T. Bayburt, and M. H. Gelb, *Biochemistry* **33,** 5011 (1994).

[69] J. J. Volwerk, E. Filthuth, O. H. Griffith, and M. K. Jain, *Biochemistry* **33,** 3464 (1994).

[70] A. B. Fisher, C. Dodia, A. Chander, and M. K. Jain, *Biochem. J.* **288,** 407 (1992); A. B. Al-Mehdi, C. Dodia, M. K. Jain, and A. B. Fisher, *Biochim. Biophys. Acta* **1167,** 56 (1993).

parameters also provide a functional basis for the quantitative interpretation of site-directed mutagenesis and X-ray crystallographic results. In short, the protocols reviewed here can be used to characterize virtually all aspects of the behavior of interfacial enzymes, including those related to the primary rate and equilibrium parameters, activation and inhibition, substrate specificity, and structure–function correlations. Resolution of the kinetic steps has also provided a method to study interfacial activation. For pig pancreatic PLA2, there is allosteric modulation by the interface of the affinity of the enzyme for catalytic site-directed ligands.

Interfacial catalysis can be adequately described in terms of the Michaelis–Menten formalism as long as the enzyme is confined to the interface and the parallel processes of intervesicle exchange of substrates and products are eliminated. Most of the kinetic complexities reported in the literature arise when the enzyme partitions between the interface and the aqueous phase. Modification of the Michaelis–Menten formalism by the addition of only a single step, the E to E* equilibrium, has marked consequences on the observed behavior of PLA2, and the elimination of this single step greatly simplifies the kinetic problems. In retrospect, it is humbling to recall that in earlier studies, the residence time and processivity of the enzyme at the interface were not always appreciated and explicitly considered. The moral of the story of interfacial catalysis is that one must worry about what the enzyme "sees" and not just what is present in the bulk reaction mixture.

Acknowledgments

The authors acknowledge support for their work by grants from Sterling Inc. and National Institutes of Health.

Author Index

Numbers in parentheses are footnote reference numbers and indicate that an author's work is referred to although the name is not cited in the text.

A

B

C

D

E

F

H

I

J

K

L

M

N

O

P

Q

R

S

X

Y

Z

Subject Index

B

E

F

G

H

J

K

L

M

N

O

Q

R

S

T

U

X

ISBN 0-12-182150-1

90038

9 780121 821500